Non-Conventional Materials and Technologies:
NOCMAT for the XXI Century

17th NOCMAT 2017 - Non-Conventional Materials and Technologies, 26-29 November 2017, Mérida, Yucatán, México

Khosrow Ghavami[1] and Pedro Jesús Herrera Franco[2]

[1] PUC-Rio, Brasil

[2] CICY Merida, Yucatan, Mexico

Peer review statement

All papers published in this volume of "Materials Research Proceedings" have been peer reviewed. The process of peer review was initiated and overseen by the above proceedings editors. All reviews were conducted by expert referees in accordance to Materials Research Forum LLC high standards.

Published under License by **Materials Research Forum LLC**
Millersville, PA 17551, USA

Published as part of the proceedings series
Materials Research Proceedings
Volume 7 (2018)

ISSN 2474-3941 (Print)
ISSN 2474-395X (Online)

ISBN 978-1-945291-82-1 (Print)
ISBN 978-1-945291-83-8 (eBook)

Distributed worldwide by

Materials Research Forum LLC
105 Springdale Lane
Millersville, PA 17551
USA
http://www.mrforum.com

Manufactured in the United State of America
10 9 8 7 6 5 4 3 2 1

Table of Contents

Preface

Materials science has witnessed amazing developments over the last 20 years. With the development of materials at the nano-scale, such as carbon nanotubes, graphene, and other forms such as nanofibers, research in several fields of science have reported results in electronic devices of smaller size but with undreamed capacity. Health and biological sciences have reported advances in the development of new methods to treat illnesses such as several forms of cancer using such nanostructured materials. The exceptionally high specific stiffness and strengths reported for carbon nanotubes, combined with their fiber-like structure, has stimulated research in the development of polymer hierarchical composites reinforced with carbon nanotubes for, both, structural and smart auto-sensing capabilities. Even today, a considerable amount of research is being performed to use these extraordinary properties observed at the nano-scale in a macroscopic composite. However, in other areas, because of "modern man" activities, the environment is resenting the results of pollution. The average consumer comes into daily contact with all kinds of man-made plastic materials that have been developed specifically to defeat natural decay processes. These are lightweight, single-use plastic products and packaging materials, derived mainly from petroleum, which account for approximately 50 percent of all plastics produced, which are not deposited in containers for subsequent removal to landfills, recycling centers. Instead, they are improperly disposed of at or near the location where they end their usefulness to the consumer.

The depletion of petroleum resources together with an increasing awareness and better environmental regulations are acting in a synergistic way to provide impetus for the development of new materials and products that are more amenable and independent of petroleum resources. The so called "green composites" fit well into this new paradigm shift. Biobased materials include industrial products for durable goods applications, made from renewable agricultural and forestry feed stocks, including wood, agricultural waste, grasses and natural plant fibers. These lignocellulosic materials composed mainly of carbohydrates such as sugars and lignin, lignin and cellulose, as well as vegetable oils and proteins. Even when most of the chemical products and materials came from renewable resources until the early part of the 20th century, the use of petroleum hampered the growth of biobased products. This has posed a new challenge to scientists and engineers to develop the technology needed to make biobased materials revolution a reality. Now, sustainability goes together with moderation, meeting basic needs and a general well-being of the population with justice and human common social activities. Sustainability makes possible that a greater part of the population can follow up a life's dream or at least can have a decent meaningful life. Sustainability means as well that remaining natural resources should be preserved for future generations. Thinking about this and knowing that modern sciences have been studied and taught in schools and universities, considering only industrial non-renewable materials such as cement and steel, a new research line was started at PUC-Rio in 1979 by the present chairman of IC-NOCMAT, to investigate local renewable materials such as bamboo, vegetable fibers, soil composite and recycled materials such as rice husk ash, sugar cane ash among others, besides studying the application of local technologies such as Taipa, rammed earth, Kah-Gel (soil reinforced with wheat straw). This research line was denoted as "Non-Conventional Materials and Technologies- NOCMAT". The NOCMATs are referring to preindustrial Materials and Technologies which have been the normal practice during the centuries before the establishment of centralized industries in 20'"century. With the implementation of centralized industries and the new concept of building individual houses at a large distance from each other, with shopping centers far from their homes, common social life of the old civilization was transferred. Now the inhabitants of new towns depend strongly on cars and living in isolation.

The 17th IC-NOCMAT Mérida/Yucatán-México follows previous events in Winnipeg/Canada (2015). The 14"' International Conference on Non-Conventional Materials and Technology, after 30 years activities at UFPB, returned to Joao Pessoa continuing the series of events which started in Rio de Janeiro in 1984, passed through Changsha, China, (2011), Cairo, Egypt (2010), Bath, England (2009), Hanoi, Vietnam (2002), Bhubaneswar, India (1997). Other conferences were organized in Joao Pessoa (2003), Pirassununga (2004), Rio de Janeiro (1999, 2005), Salvador (2006), Maceio (2007) and Santiago de Cali-Colombia (2008. The main concern of the conference is to disseminate research results of different groups from all five continents in the field of non-conventional materials and technologies (NOCMAT), presenting alternatives to the existing non-renewable polluting conventional materials. It addresses scientists, researchers and especially the young engineers, architects, designers and planners to get acquainted with new materials and technologies which certainly will influence their future choices, to make use of traditional sustainable materials such as soil, natural fibers, bamboo and agricultural, mineral residues besides discussing about the climatic constructions which was the basic tradition in India, Iran, Peru etc.

During the last 35 years of research and development into NOCMAT, we have shown many viable materials and technologies which could contribute towards the sustainability and ecological development. Unfortunately big industries are not interested in solving the problems in a humane way without big and immediate interest and wealth for those in power.

The application and investigation into NOCMAT have not been as we have expected. We do not claim that we have the answers to solve all the problems of our time immediately but we show the direction and invite those in political power to invest in the application of NOCMAT which will contribute to the solution of the existing ecological and environmental problems. Large investments are still needed in R&D of those materials to extend the life cycles of the low-cost energy saving materials which in general produces oxygen and absorbs CO_2 generated by centralized industries.

The organizers of the 17th NOCMAT would like to thank all the people who contributed to the success of the event, especially the students of G-NOCMAT group at PUC-Rio. Our thanks are also to Ursula Schuler Ghavami, who has contributed whenever it was necessary.

Special thanks to Pedro J. Herrera-Franco local Chairman of the conference, and coworkers at CICY, specially, Emmanuel Flores-Johnson, Alex Valadez-González, Carlos Rolando Rios, Francis Avilés Cetina, Gonzalo Canché Carrillo, Gonzalo Carrillo Baeza, who contributed in logistics, design of the folders and the book of abstracts and administrative activities. We hope that all participants have returned home with some new information about NOCMAT which they will pass on to their colleagues and students so that we could propagate new knowledge.

Best wishes and we hope to see you during the next NOCMAT
Khosrow Ghavami,
Chairman 17 NOCMAT 2017

Scientific Committee

Organizing Committee

Dr. Pedro Jesús Herrera Franco – pherrera@cicy.mx
Dr. Gonzalo Canché Escamilla – gcanche@cicy.mx
Dr. Pedro Iván González Chi – ivan@cicy.mx
Dr. Alex Valadez González – avaladez@cicy.mx
Dr. Francis Avilés Cetina – faviles@cicy.mx
Dr. Carlos Rolando Rios Soberanis – Rolando@cicy.mx
Dr. Javier Guillén Mallette – jguillen@cicy.mx
Dr. Ricardo Herbé Cruz Estrada – rhcruze@cicy.mx
Dr. Jorge Alonso Uribe Calderón – Jorge.uribe@cicy.mx
Dr. Emmanuel Alejandro Flores Johnson – emmanuel.flores@cicy.mx

Non-Conventional Materials and Technologies – NOCMAT for XXI Century   Materials Research Forum LLC
Materials Research Proceedings 7 (2018) 1-6   doi: http://dx.doi.org/10.21741/9781945291838-1

Mechanical Characterization of Foamed Concrete Reinforced with Natural Fibre

E.A. Flores-Johnson [a], Y.Z. Yan [b], J.G. Carrillo [c], P.I. González-Chi [c], P.J. Herrera-Franco [c], Q.M. Li [b,*]

[a] CONACYT – Unidad de Materiales, Centro de Investigación Científica de Yucatán, Calle 43, No. 130 Col. Chuburná de Hidalgo, Mérida, Yucatán 97205, México, emmanuel.flores@cicy.mx

[b] School of Mechanical, Aerospace and Civil Engineering, The University of Manchester, Pariser Building, Manchester M13 9PL, UK; yinzhong.yan@manchester.ac.uk, qingming.li@manchester.ac.uk

[c] Unidad de Materiales, Centro de Investigación Científica de Yucatán, Calle 43, No. 130 Col. Chuburná de Hidalgo, Mérida, Yucatán 97205, México; jgcb@cicy.mx, ivan@cicy.mx, pherrera@cicy.mx

Abstract. In this work, an experimental investigation of the mechanical properties of foamed concrete with and without fibre reinforcement is presented. To reinforce the foamed concrete, natural fibres, i.e., henequen and coir fibres were used. Target dry density of 800 kg/m^3 was investigated. Plain foamed concrete without fibre reinforcement was also studied. Quasi-static uniaxial compressive tests were performed on foamed concrete samples. It is found that while the henequen fibre-reinforcement enhances slightly the mechanical behaviour of the foamed concrete, the coir fibre reinforcement results in a reduction of mechanical performance when compared with plain foam concrete. This observation is explained by transfer of load from the matrix to the fibres, which indicates that the fibre/matrix bonding plays an important role in the structural performance of reinforced foamed concrete. Differences of density in the foamed concrete samples can also explain the difference in mechanical performance. The results show that this light material has the potential to be used as a core for the design of sandwich panel for construction structural applications; however, the results in this study are limited and further research has to be carried out to fully understand this type of material.

Keywords: Foamed Concrete, Natural Fibre, Mechanical Property

Introduction

Foamed concrete is a cellular material made of cement mortar and air voids. This material is low cost, easy-to-fabricate and lightweight. However, this material is currently used mainly for filling purposes and thermal insulation and its application as a construction material is largely restricted by its mechanical properties, e.g. poor strength, stiffness and ductility [1]. It has been reported that adding fibre reinforcement to foam concrete can improve its mechanical properties. These studies have been primarily performed on foam concrete employing synthetic fibres as reinforcement such as polypropylene [2] and polyvinyl alcohol [3]. However, environmental awareness is forcing the construction industry to use environmentally friendly materials derived from natural resources.

The demand for natural fibres composites made with easily available renewable sources and good mechanical properties combined with the volatility in oil prices and environmental advantages with government supports [4] makes foam concrete reinforced with natural fibres an interesting material for the construction industry even though this imply a serious challenge for the industry to maintain low-cost production and performance.

The performance of foamed concrete in structural applications has not been fully investigated. Foamed concrete could be an ideal core material for composite sandwich structures due to its low density. Othuman Mydin and Wang [5] and Flores-Johnson and Li [3] studied sandwich panels made with profiled thin steel face sheets and foamed concrete core under uniaxial compression and four-point bending. They found that the flexural strength of the panels is increased considerably when compared to the strength of plain foamed concrete. These results demonstrate that foamed concrete may be used as a structural material when employed as a sandwich panel.

The use of natural fibre as reinforcement of foamed concrete has shown encouraging results [6]; however, there are not sufficient studies to fully understand foamed concrete materials reinforced with natural fibres. To address this issue, in this work the compressive properties of plain foamed concrete and fibre-reinforced foamed concrete with natural henequen and coir fibres are presented and discussed.

Materials and methods

Materials and sample preparation

The foamed concrete mix design was prepared using the following to target a dry density of 800 kg/m³: Portland cement (1 kg), Fly ash (0.25 kg), water (0.5 kg), foam (44.32 g), superplasticizer (2.3 g) and fibre (2.98 g). Two different natural fibre reinforcements were used supplied by local producers from Mexico, i.e., henequen fibre (supplied by Desfibradora La Lupita) and coir fibre (supplied by Coirtech). A fibre length of 40 mm was used. Additionally plain foam concrete without fibre-reinforcement was prepared.

The foam was prepared separately from the slurry mix using foaming agent from EAB Associates (UK) and a foam generator JFG200 from Propump Engineering (UK). The slurry and the foam were then mixed in a mixer until a targeted wet density of 910 kg/m³ was achieved. The mix was then poured into moulds to obtain cylinders with a diameter of 100 mm and a height of 200 mm. Subsequently, the cylinder was cut into four shorter cylinder samples with a height of 50 mm. Uniaxial compression test was performed using an Instron universal testing machine at a fixed crosshead displacement of 2.4 mm/min (Fig. 1).

FIGURE 1 –UNIAXIAL COMPRESSION TEST SET-UP.

Results and discussion

Non-Conventional Materials and Technologies – NOCMAT for XXI Century   Materials Research Forum LLC
Materials Research Proceedings 7 (2018) 1-6   doi: http://dx.doi.org/10.21741/9781945291838-1

Experimental results

Figures 2a and 2b show henequen and coir fibre, respectively. It can be seen that henequen fibres are straighter than coir fibres, which could influence the adhesion to the foamed concrete. Figures 2c and 2d show optical microscopy images of henequen and coir fibres, respectively. For both fibres, a rough surface with imperfections can be observed, which is typical of natural fibres.

A diameter of 0.2 mm and 0.25 mm was estimated for the henequen and coir fibres, respectively. The main difficulty to utilize the henequen fibres is the lack of a good adhesion to matrices. The hydrophilic nature of natural fibres adversely affects adhesion to a hydrophobic matrix and as a result, it may cause loss of strength [7-8].

Fibre surface treatments are recommended to improve fibre-matrix adhesion. There are several methods to modify the natural fibre surface such as graft copolymerization of monomers onto the fibres surface, the use of maleic anhydride copolymers and the use of alkyl succinic anhydride. Coupling agents such as silanes, titanates, zirconates can also be used to promote fibre-matrix adhesion [7-8]. We will employ some of these fibre surface treatments in future research to improve fibre-matrix with the foamed concrete.

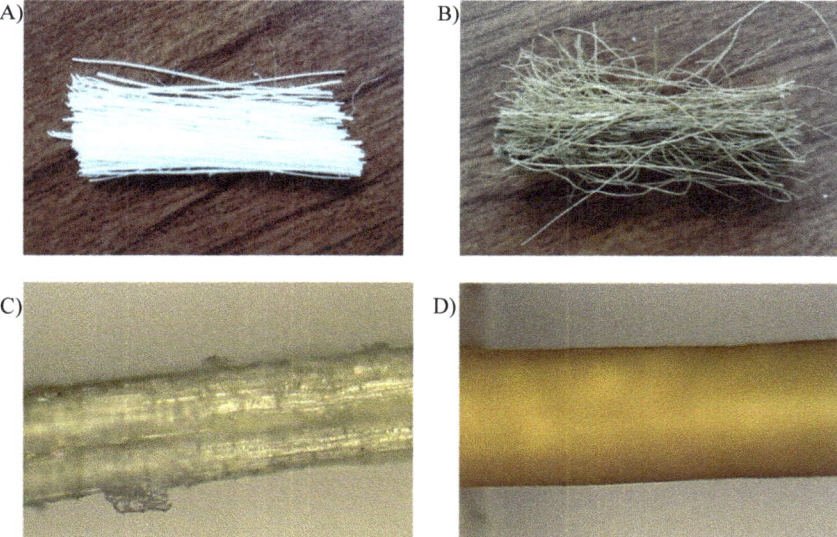

***FIGURE 2** –A) HENEQUEN FIBRE. B) COIR FIBRE. C) OPTICAL MICROSCOPY IMAGE OF HENEQUEN FIBRE. D) OPTICAL MICROSCOPY IMAGE OF COIR FIBRE.*

Figures 3a, 3b and 3c show typical force-displacement curves under uniaxial compression for plain foamed concrete, foamed concrete with henequen reinforcement and foamed concrete with coir reinforcement, respectively. In all cases, an initial elastic response at very low strains is observed until a peak force is reached followed by a sudden drop in force, which corresponds to the failure of the specimen. It can be seen that plain foamed concrete and foamed concrete with henequen reinforcement exhibit similar behaviour with a slight increase of performance when the henequen

3

fibre is used. For coir reinforcement a decrease of performance is observed. This observation is explained by transfer of load from the matrix to the fibres, which indicates that the fibre/matrix bonding plays an important role in the structural performance of reinforced foamed concrete. Also, density differences were recorded between plain foamed concrete and fibre-reinforced foamed concrete, which could explained the differences in mechanical performance.

Figures 4a and 4b show the henequen fibre reinforced foamed concrete during testing and after testing, respectively. It can be seen that the fibres tent to maintain the integrity of the foamed concrete; however, further research including fibre treatments should be performed to fully understand the role of the fibre-reinforcement.

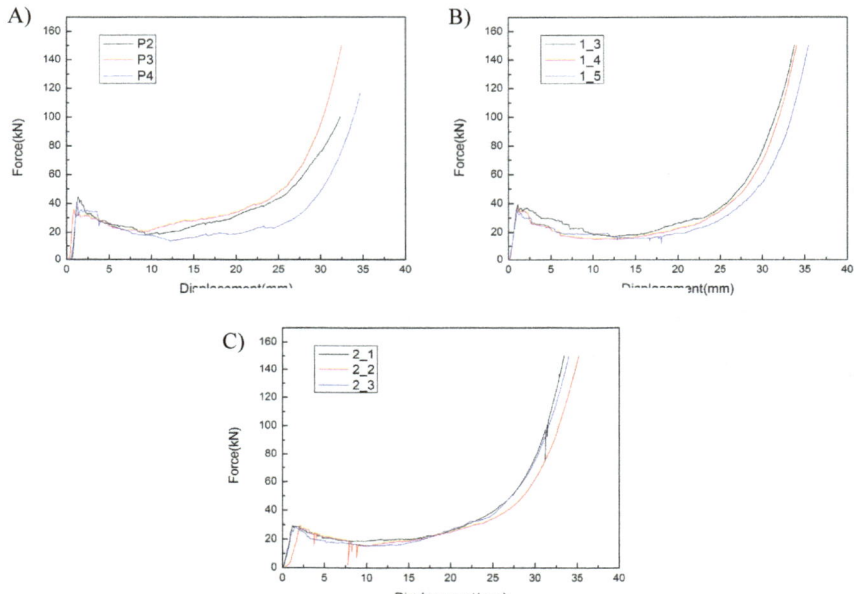

FIGURE 3 – *TYPICAL FORCE-DISPLACEMENT CURVES OF FOAM CONCRETE SPECIMENS: A) PLAIN FOAM CONCRETE. B) FOAM CONCRETE WITH HENEQUEN FIBRE REINFORCEMENT. C) FOAM CONCRETE WITH COIR FIBRE REINFORCEMENT.*

A)  B)

FIGURE 4 –HENEQUEN FIBRE REINFORCED FOAMED CONCRETE: A) DURING TESTING. B) AFTER TESTING.

Summary and conclusions

The uniaxial compressive response of plain foamed concrete and foamed concrete reinforced with natural henequen and coir fibres was investigated. The results show that plain foamed concrete and foamed concrete with henequen reinforcement exhibit similar behaviour with a slight increase of performance when the henequen fibre is used. It was observed that when coir reinforcement is used there is a decrease of performance. This observation is explained by transfer of load from the matrix to the fibres, which indicates that the fibre/matrix bonding plays an important role in the structural performance of reinforced foamed concrete, and also by differences recorded between the plain foamed concrete and the fibre-reinforced foamed concrete. The results show that this light material has the potential to be used as a core for the design of sandwich panel for construction structural applications; however, the results in this study are limited and further research has to be carried out to fully understand this type of material since there are few studies addressing this research topic.

Acknowledgements

The authors acknowledge the support from British Council Newton Fund Institutional Links (grant No. 275902862) and CONACYT-British Council México Institutional Links 2016 (grant No. 277730).

References

[1] Narayanan N, Ramamurthy K. Structure and properties of aerated concrete: a review. Cem Concr Compos 2000;22(5):321-329. https://doi.org/10.1016/S0958-9465(00)00016-0

[2] Jones MR, McCarthy A. Preliminary views on the potential of foamed concrete as a structural material. Mag Concr Res 2005;57(1):21-31. https://doi.org/10.1680/macr.2005.57.1.21

[3] Flores-Johnson EA, Li QM. Structural behaviour of composite sandwich panels with plain and fibre-reinforced foamed concrete cores and corrugated steel faces. Compos Struct 2012; 94(5):1555-1563. https://doi.org/10.1016/j.compstruct.2011.12.017

[4] Faruk O, Bledzki AK, Fink H-P, Sain M, Progress Report on Natural Fiber Reinforced Composites. Macromol Mater Eng 2014; 299: 9–26. https://doi.org/10.1002/mame.201300008

Non-Conventional Materials and Technologies – NOCMAT for XXI Century   Materials Research Forum LLC
Materials Research Proceedings 7 (2018) 1-6   doi: http://dx.doi.org/10.21741/9781945291838-1

[5] Othuman Mydin MA, Wang YC. Structural performance of lightweight steel-foamed concrete–steel composite walling system under compression. Thin Walled Struct 2011;49(1):66-76. https://doi.org/10.1016/j.tws.2010.08.007

[6] Othuman Mydin MA, Rozlan NA, Ganesan S. Experimental Study on the Mechanical Properties of Coconut Fibre Reinforced Lightweight Foamed Concrete, J Mater Environ Sci 2015;6(2):407-411.

[7] Herrera-Franco PJ, Valadez-González A. Mechanical properties of continuous natural fibre-reinforced polymer composites. Compos Part A 2004; 35(3):339-345. https://doi.org/10.1016/j.compositesa.2003.09.012

[8] Herrera-Franco PJ, Valadez-González A. A study of the mechanical properties of short natural-fiber reinforced composites. Compos Part B 2005; 36(8):597-608. https://doi.org/10.1016/j.compositesb.2005.04.001

Non-Conventional Materials and Technologies – NOCMAT for XXI Century   Materials Research Forum LLC
Materials Research Proceedings 7 (2018) 7-19   doi: http://dx.doi.org/10.21741/9781945291838-2

Reinterpretation of Vernacular Constructive Process with the use of Soil-Cement- Sisal Fiber Mortar and a Bamboo Framework

Kathelyn S.G. Souza[a], Anne B. C. Rocha[a], Adriana P. S. Martins[b*], Romildo D. T. Filho[b]

[a] Faculdade de Arquitetura e Urbanismo – FAU/UFRJ, Brazil

[b] Núcleo de Materiais e Técnicas Sustentáveis – NUMATS/UFRJ, Brazil

kathelyngandra@gmail.com, anne.beatris@gmail.com

adripsmartins@gmail.com*, toledo@coc.ufrj.br

Abstract. Constructive techniques based on low environmental impact materials are increasingly gaining importance in developed and developing countries as a way to mitigate the effects of intensive use of industrialized materials such as steel, cement, concrete, among others, responsible for high energy consumption and emission of greenhouse gases. The aim of this work is to promote a re-reading of a vernacular constructive technique widely used in the colonial period, the wattle and daub, aiming at its application in social interest habitatio. Innovations were incorporated in relation to the materials and the traditional production process, in order to allow the pre-fabrication of modular units that fit the pillars of the building. A prototype of the modular unit was produced and mechanically tested using a self-compacting soil–cement-sisal fiber mortar and a bamboo framework (*Bambusa Tuldoides*) as reinforcing material. Bamboo is a native plant in Brazil with accelerated growth, without any need of replanting, low cost, good mechanical properties and high capacity of carbon sequestration in its biomass. The bamboo framework was assembled with sticks rods of dimensions 1.55 x 0.03 x 0.005 m (length, width, thickness, respectively), obtained from the subdivision of the culm into four parts. A three-point bending test was performed to characterize the mechanical behavior of the modular unit (prototype), designed with dimensions of 1.5 x 0.12 x 0.30 m (length, thickness, height, respectively). The purpose of this research is to revive and valorize a constructive technique that is widespread in some Latin American countries, with great economic, technological and environmental advantages.

Keywords: Bamboo Reinforcement, Self-Compacting Soil-Cement Mortar, Soil-Cement Composites

Introduction

According to a report by the United Nations Human Settlements Program (UN-HABITAT) [1], Latin America's urban population will reach 89% by 2050. This trend of population concentration in cities makes it mandatory to mobilize all sectors of society around crucial issues such as economic development, housing, basic urban services, environment and governance. The inclusive and sustainable city model we want should promote the provision of quality and low-cost housing, which is more accessible to the vast majority of the population living in urban centers. In recent years, researchers from all over the world have been working to increase the use of local materials in buildings, as a means to lower costs, reduce the environmental impacts associated with conventional materials and add value to products that were not used in view of their full potential. This work has as main objective the development of a prototype based on bamboo, sisal fibers and raw earth, which will be used as a prefabricated modular unit for the production of masonry in social housing. It has as its conceptual framework an ancient technique used in some Latin American countries called "quincha". Bamboo is a native plant in Brazil that does not require replanting, has accelerated growth, low cost, good mechanical properties, high carbon sequestration capacity and resilience to climatic changes [2, 3]. The Brazilian government has been carrying out actions in recent years aimed at intensifying research, technological development, production and application of bamboo on a large scale. In 2011, a bilateral Brazil-China agreement was signed to transfer technology in bamboo [4]. Also in 2011 the Law 12484 was enacted, which instituted the National Policy of Incentive to Sustainable Management and Bamboo Cultivation (PNMCB) [5]. In 2016 Brazil joined an international network (INBAR) created by the UN to implement a global sustainable development agenda based on bamboo [4]. Chithambaram and Kumar [6] developed bamboo reinforced panels for low-cost habitation in India and bamboo insertion significantly increased the serviceability limit of these panels, filled with cement mortar, fly ash and sand. Jain et al. [7] used tissue-shaped bamboo in an epoxy resin matrix and obtained a composite with optimized and isotropic strength properties. Ghavami [2] produced structural elements (beams, slabs and pillars) using the reinforcement of bamboo (stem) instead of steel, obtaining excellent results of strength and durability. However, the author stresses the absence of international technical standards to support the structural project. The bamboo culms (Bambusa vulgaris) used in the construction of ecological houses in Manaus were subjected to bending tests by Ribeiro et al. [8], with final compressive load values of 7.9 kN and modulus of elasticity of 9.6 MPa. The behavior of bamboo-reinforced masonry panels was similar to that of panels reinforced with steel under seismic loads, evidencing the contribution of bamboo in increasing shear strength and ductility for earthquakes and extreme winds [9].These authors proposed the use of bamboo as an alternative to steel, to reinforce masonry in low cost solutions. The weak matrix-bamboo adhesion and low modulus of elasticity of the bamboo generated cracking in composites under tensile force [10].The bamboo absorbed water from the fresh mixture and expanded, then lost water during drying and retracted, generating voids at the interface, consequently favoring cracking. Chand et al. [11] aimed at proposing alternatives for the deficit of low cost dwellings in India, produced panels based on a stabilized soil matrix reinforced with straw, bamboo weft and Ipomea carnea. These authors considered the results satisfactory and promising for the intended application.

The present work investigates the behaviour of a modular unit of masonry (prototype) under bending loads, as well as the constitutive laws of the reinforcement material of the prototype (bamboo) under loads of compression, tension and bending. In this way, the traditional technique of wattle and daub is rescued, contributing to create modern and attractive systems of masonry from the economical, technological and environmental point of view.

Materials and methods

Construction system
The project presented in this paper was proposed as part of the Technology of Construction with Earth discipline, in the Faculty of Architecture and Urbanism-UFRJ. The main idea of the project is the prefabrication of bamboo-earth panels with all parts included. This type of frame is similar to the quincha which would improve the flexural strength and shear strength, which are constant preoccupation for transporting precast parts and also during service life [12].

For the filling and covering of the bamboo frame it was used a self-compacting mixture of soil-cement-sisal fiber developed by Martins [13]. The proposal of the work is to foresee an industrial precast construction system which parts can be easily transportated by trucks and mounted with cranes. The panel was designed as a type of tight fitting module in a supporting structure. The pannel had the following dimensions: 1,50m x 0,3m x 0,12m. The pillars would be constructed of wood with a groove to facilitate the passage of wires and ducts of the building systems. The concept of the pillar, modular units and of the structural system are presented in Figures 1 and 2.

FIGURE 1- *Detail of pillar.*

FIGURE 2 - *SYSTEM FORMED BY THE PANELS AND PILLARS.*

Bamboo-reinforced soil prototype production

The mould used in the production of the panel was prismatic with external dimensions of 155 x 14.5 x 35 cm (length, width, height, respectively) and was made of a 2.5 cm thick pine wood (Figure 3a). The prototype was made of a self-compacting soil-cement-sisal fibers composite and a frame of bamboo.

The bamboo frame

The species of bamboo selected for the production of the frame was the *Bambusa Tuldoides*. The bamboo samples were donated by the Horto City Hall of the Federal University of Rio de Janeiro, and was extracted from the medial region of the culm of a 5 year old bamboo. The culms were cut to a length of 3 m, air dried and used without any preservative treatment. After drying the culm was cut to the length of 1.55 m and then split into four parts producing bamboo sticks of 32,5 mm: 6,5 mm (width and tickness). Two segments of bamboo culms were inserted in the frame, 37.5 cm apart from the side faces, and the sticks were then interlaced as can be seen in Figure 3.

Non-Conventional Materials and Technologies – NOCMAT for XXI Century   Materials Research Forum LLC
Materials Research Proceedings 7 (2018) 7-19   doi: http://dx.doi.org/10.21741/9781945291838-2

(a)   (b)

*FIGURE 3 - (A) MOULD USED TO PRODUCE THE PROTOTYPE AND BAMBOO CULMS
PLACED IN POSITION (B) BAMBOO FRAME FULLY MOUNTED.*

Self-compacting soil-cement-sisal fiber composite

The used composite was developed by Martins [13] and was composed of cement, fly ash, metakaolin, superplasticizer Glenium 51 and water. The matrix mass proportion was 1:6.97:0.10:0.10 (cement: soil: flay ash: metakaolin) and the sisal fiber content used was 1% (in relation to the mass of dry soil) (20mm long). The soil used was classified as "SC" (clay sand) following the Unified System [14], with a liquidity limit of 34% [15] and a plasticity limit of 16.4% [16]. Figure 4 shows the grain size curves of the materials used. The cement, metakaolin, fly ash and superplasticizer were comercial products available in the market. Chemical and physical properties of these materials can be find elsewhere [13].

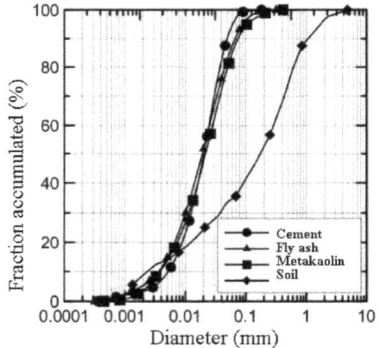

*FIGURE 4 - GRANULOMETRIC CURVES OF THE MATERIALS USED IN THE
PRODUCTION OF SELF-COMPACTING MORTAR (CEMENT, FLY-ASH, METAKAOLIN
AND SOIL) [13].*

Mechanical characterization of the bambusa tuldóides

Aiming to obtain the constitutive relationships of the bamboo used in this research, mechanical tests were performed on air dried samples, with equilibrium moisture content ranging from 8.5 to 9.5%. The tests were conducted on a Shimadzu AGX electromechanical equipment fitted with a load cell of 5 kN, operating in a temperature controlled environment (21 ° C). These tests were performed with displacement control at a rate of 1mm / min, until the complete fracture of the samples (compression and tensile) or displacements of 20 mm (bending). In the tensile tests a clip-gage was coupled in the central portion of the samples. In the bending tests, 3 roller supports were used, one being positioned in the center of the span and the other two in a distance of 20 cm to each other,. The deflection in the center of the span and the displacements of the beam were recorded continuously by LVDT. The equivalent bending stress was calculated by Equation 1. The samples for compression were culm segments with a diameter of 32 mm and a height of 64 mm. The samples for the tensile and bending tests were (250 x 10 x 5 mm, length, width, thickness, respectively) extracted from the internodal region and in the tangential direction, discarding the outermost (shell) and innermost regions (rich in parenchyma). At least 3 samples were tested for each type of mechanical test. Figures 5 (a), 5 (b) and 5 (c) illustrate the setup of tests parallel to fibers compression, parallel to fibers tensile and three point bending, respectively.

 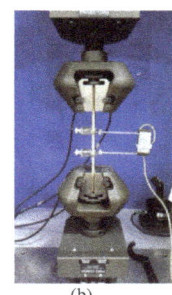 
            (a)   (b)   (c)

FIGURE 5 - *SETUP OF THE TESTS OF (A) COMPRESSION PARALLEL TO THE FIBERS, (B) TENSILE TESTS PARALLEL TO THE FIBERS AND (C) THREE-POINT BENDING.*

Flexural test of the prototype

The prototype was subjected to three-point bending test in a Shimadzu servo-controlled equipment with 1000 kN load cell and displacement control at a rate of 1 mm / min in a temperature controlled environment (21 ° C). The distance between the end supports was 62 cm. Two LVDT's were coupled: one to obtain the displacements in the center of the span and another to obtain the displacements of the stroke, according to the setup shown in figure 6. Only one prototype was tested.

FIGURE 6 - *SETUP OF THE THREE POINT FLEXURAL TEST OF THE PROTOTYPE.*

Non-Conventional Materials and Technologies – NOCMAT for XXI Century   Materials Research Forum LLC
Materials Research Proceedings 7 (2018) 7-19   doi: http://dx.doi.org/10.21741/9781945291838-2

The equivalent bending stress was obtained by the expression:

$$\sigma = \frac{3Pl}{2bh^2} \tag{1}$$

where:

σ - equivalent bending stress;

P - applied load; l – distance between supports; b - sample width; h - height of the sample.

Results and discussion

Rheological and mechanical characterization of the self-compacting soil:cement:fiber composite

According to Martins [13], the mixture used in the production of the prototype showed the slump test and slump test flow results to be 280 and 695 mm, respectively, and these results are compatible with the desired self-compacting condition. The mechanical behavior of the fibrous matrix under uniaxial compression, under direct tension and under bending at four points is shown in Figures 7 (a), 7 (b) and 7 (c), respectively.

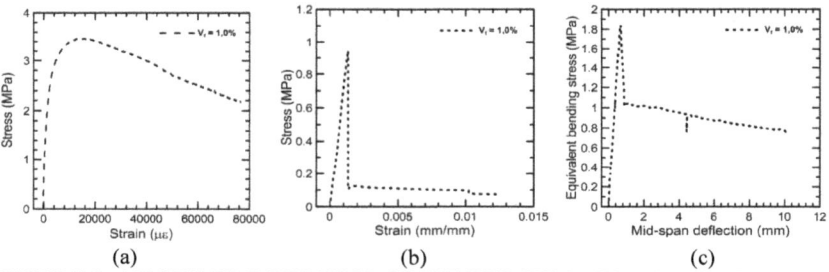

(a)   (b)   (c)

FIGURE 7 - MECHANICAL BEHAVIOR OF THE FIBROUS MATRIX (A) UNDER UNIAXIAL COMPRESSION, (B) UNDER DIRECT TENSION AND (C) UNDER FLEXION AT FOUR POINTS [13].

Compression test results parallel to the bamboo fibers

The stress-strain behavior of the samples under compression loads parallel to the fibers, the mode of sample rupture, and the parameters obtained are shown in Figures 8 (a), 8 (b) and Table 1, respectively.

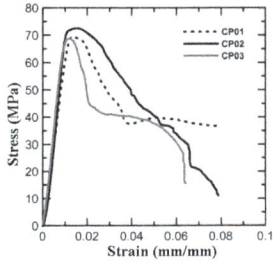
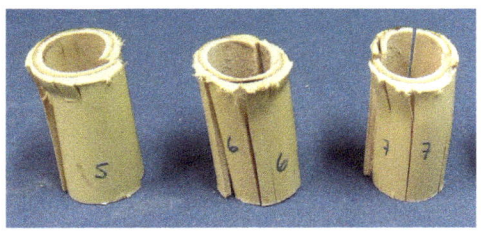

(a)   (b)

FIGURE 8 - *COMPRESSION STRESS-STRAIN CURVES PARALLEL TO FIBERS FOR
SPECIMENS OF BAMBUSA TULDÓIDES (A) AND MODE OF RUPTURE (B).*

The average values of compressive strength and modulus of elasticity were 70.58 MPa and 7.55 GPa, respectively (Table 1). Ghavami and Marinho [17] obtained compressive strength results and modulus of elasticity of 31.77 MPa and 12.25 GPa, respectively, for samples of *Guadua angustifolia* extracted from the central part of the culm. Xu *et al.* [18] found that their samples (*Phyllostachys Pubescens*) presented strength and modulus of elasticity of 45.4 MPa and 6.1 GPa, respectively. Values of compressive strength and modulus of elasticity of 69 MPa and 9.3 GPa, respectively, were found by Chung and Yu [19].

Janssen [20] explains that under compression, the fracture will be governed by the rupture of the lignin matrix (weak component) rather than the rupture of the cellulose fibers (strong component). Compressive strength is influenced by the stiffness and volumetric fraction of these components (matrix and fibers), which vary along the wall thickness [21]. Due to the lateral restraining effect at the loaded ends, the samples under compression developed multiple longitudinal cracks (Figure 8 (b)), and splitted as the individual sections buckled. This fracture behavior was also observed by Xu *et al.* [18].

TABLE 1 – *RESULTS OBTAINED IN THE COMPRESSION TESTS - SAMPLES OF BAMBUSA
TULDÓIDES.*

Sample	σ_{rupt} (MPa)	ε_{rupt} (mm/mm)	E (GPa)
CP01	69,54	0,013	6,37
CP02	73,48	0,015	8,37
CP03	68,72	0,012	7,90
Average	70,58	0,013	7,55
Standard Deviation	2,54	0,002	1,05

Tensile test results parallel to fibers
Figure 9 shows the tensile stress-strain curves obtained for the bamboo samples and the fracture patterns of the specimens.

Non-Conventional Materials and Technologies – NOCMAT for XXI Century   Materials Research Forum LLC
Materials Research Proceedings 7 (2018) 7-19   doi: http://dx.doi.org/10.21741/9781945291838-2

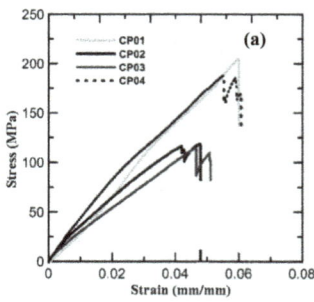
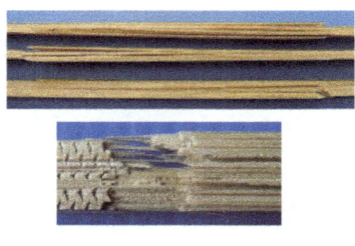

FIGURE 9 - *(A) TENSILE STRESS-STRAIN CURVES PARALLEL TO FIBERS FOR SAMPLES OF BAMBUSA TULDÓIDES, (B) MODE OF RUPTURE OF THE SPECIMENS, (C) DETAIL OF THE FRACTURED REGION.*

The tensile strength (mean value) obtained was 165.30 MPa, close to the 150.09 MPa obtained by Krause [22], for samples of the *Dendrocalamus giganteus* species extracted from the medial region of the culms. Tensile strength values of 210 and 193 MPa were obtained by Jakovljevic *et al.* [23] for the species *Pseudosasa amabilis* and *Pleioblastus amarus,* respectively. Tan *et al.* [24] have shown, through nanoidentation studies, that for functionally graded materials such as bamboo, the tensile strength depends on the fiber density, being higher in the region closest to the shell, lower in the region furthest from the shell, and intermediate in the regions between the bark and the inner surface.

Krause *et al.* [25] have pointed out that, specifically for tension loading, the resistance gradient from the periphery to the interior becomes smaller when considering the specific resistance. At the microstructural scale, the high tensile strength of bamboo is attributed to the polylamellate structure of the fiber wall, high cellulose content and small angle of microfibrils [7]. The mode of failure of the samples under tension occurred with rupture of fiber bundles associated to longitudinal delaminations, according to Figure 9 (b). In this Figure one can see the thicknesses of the samples, to facilitate the visualization of the longitudinal delaminations. Figure 9 (c) shows the sample in the direction of its length and width, giving a detail of the fractured region.

TABLE 2 - *PARAMETERS OBTAINED TENSILE TRACTION TESTS - SAMPLES OF BAMBUSA TULDÓIDES.*

Sample	σ_{rupt} (MPa)	ε_{rupt} (mm/mm)	E (GPa)
CP01	205,75	0,06	11,67
CP02	118,48	0,048	7,65
CP03	114,73	0,046	6,37
CP04	187,42	0,055	15,63
Average	156,60	0,052	10,33
Standard Deviation	46,80	0,0064	0,54

Tree-point bending test result
The flexural stress *versus* displacement curves in the center of the span are shown in Figure 10 (a). In Figure 10 (b) it's shown the configuration of the specimen for a displacement in the order of 20 mm. The obtained parameters are shown in Table 3.

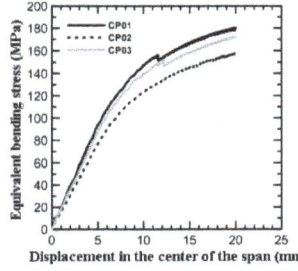

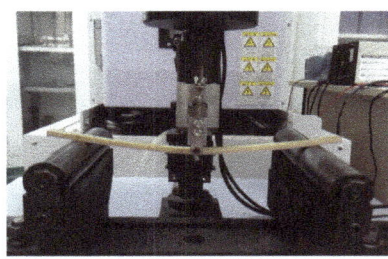

(a)   (b)

FIGURE 10 - *(A) EQUIVALENT BENDING STRESS VERSUS DISPLACEMENT CURVES IN THE CENTER OF THE GAP AND (B) TEST CONFIGURATION FOR 20 MM DISPLACEMENT.*

TABLE 3 - *PARAMETERS OBTAINED FROM THE BENDING TESTS. LOAD AND STRESS VALUES FOR DISPLACEMENTS OF 20 MM.*

Sample	$P_{d=20mm}$ (kN)	$\sigma_{d=20mm}$ (MPa)	E (GPa)
CP01	0,15	181,76	16,68
CP02	0,13	157,48	14,18
CP03	0,14	172,16	15,97
Average	0,14	170,47	15,61
Standard Deviation	0,01	12,23	1,29

The mean values obtained for bending strength (Table 3) were close to those of the bamboo tested by Sharma [26]. Krause [22] obtained average values of bending strength and modulus of elasticity of 28.6 and 12.76 MPa, respectively, for the species *Dendrocalamus giganteus*, in the medium height of the stem. The values of last load, rupture modulus and elasticity modulus of 7.9 kN; 88MPa and 9.6 GPa, respectively, were obtained by Ribeiro *et al.* [8] testing culms of *Bambusa vulgaris*.

Three-point flexural test results of the prototype

Figure 11 (a) shows the load *versus* displacement curve of the prototype tested after 163 days of casting. Figure 11 (b) shows the configuration of the test after large deflections (of the order of 33 mm). The test parameters can be seen in Table 4.

The force *versus* displacement behavior was linear until the first crack, which occurred at a load of 10.34 kN. After the cracking of the matrix, a load transfer from the matrix to the reinforcement (sisal fibers and bamboo frame) occurred, followed by an increase in the bearing capacity of the material (deflection hardening) up to the ultimate tension (4.26 MPa).

After large displacements, the presence of the hybrid reinforcement, constituted by sisal fibers and the bamboo weft, allowed a gradual decrease of the stress, instead of abrupt rupture, with both reinforcements acting in order to control the opening of the fissures.

The failure mode included a large fissure that was subdivided into two from half the width of the sample (Figure 11b). Even after the complete fracture of the material and the opening of large cracks in the tension region, sisal fibers were quite effective in forming bridges, interconnecting the cracked edges and maintaining a certain integrity of the material. The energy associated with the fracture process was 516 J.

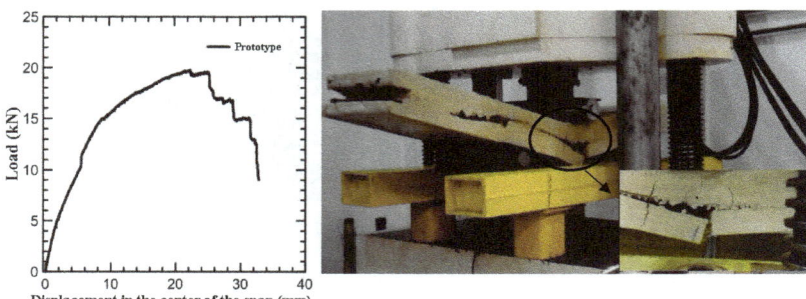

(a)   (b)

FIGURE 11 - (A) FORCE VERSUS DISPLACEMENT CURVE IN THE CENTER OF THE SPAN FOR THE PROTOTYPE (B) CONFIGURATION OF THE PROTOTYPE AFTER DISPLACEMENTS OF THE ORDER OF 33 MM.

TABLE 4 - PARAMETERS OBTAINED IN THE THREE-POINT FLEXURAL TEST OF THE PROTOTYPE.

$P_{1^a\ fissura}$ (kN)	$\sigma_{1^a\ fissura}$ (kN)	P_{rupt} (kN)	σ_{rupt} (MPa)	d_{rupt} (mm)	E (MPa)	$T_{d=33mm}$** (J)
10,34	2,23	19,77	4,26	22,24	1,81	516
Obs.: $T_{d=33mm}$** _ tenacity for the displacement of 33 mm						

(a)   (b)

FIGURE 12 - (A) MODE OF FRACTURE OF THE PROTOTYPE UNDER FLEXURAL LOADS AND (B) ENLARGEMENT OF THE FISSURED REGION.

It is observed that the bamboo frame provided a considerable increase of strenght, deformability and capacity of releasing energy to the material, increasing its serviceability limit and showing promising results in relation to the idealized constructive system.

The prototype thus produced can be engineered for situations of exposure to seismic loads, extreme wind loads, impact loads, handling stresses, among others. Prefabricated panels produced by

Non-Conventional Materials and Technologies – NOCMAT for XXI Century   Materials Research Forum LLC
Materials Research Proceedings 7 (2018) 7-19   doi: http://dx.doi.org/10.21741/9781945291838-2

Puri *et al.* [27] and reinforced with bamboo frame showed flexural strength and rupture deflections similar to the prototype tested in this research. These authors observed detachment at the matrix-reinforcement interface during material rupture, a fact also shown in Figure 11 (b). Similar to the panels developed in this research, Chithambaram and Kumar [6] obtained flexural behavior with great ductility and triplicate load capacity in relation to panels without bamboo. Moroz *et al.* [9] emphasized the potential of bamboo to maintain the structural integrity of masonry when subjected to lateral loads from earthquakes or strong winds. The panels produced by Chand *et al.* [11] showed bending strength of 0.003 MPa, which was much lower than the one produced in this research and was considered not acceptable by some soil construction standards [28].

Conclusion

In the present research a prototype was developed and characterized with regards to its mechanical behavior in order to investigate its potentiality to construct low cost and low environmental impact masonry systems as alternative to aliviate the existing habitation deficit, mainly in developing countries. The system is similar to the vernacular technique called "quincha", but incorporates modern concepts of pre-fabrication of the elements and hybridization of fibrous reinforcement. The results of the flexural test with the prototype showed a great potential of the material to support loads in service situations and in extreme situations (earthquakes, strong winds, among others). The flexion strength of the prototype was 4.26 MPa, and the bamboo incorporation promoted high tensile strength, ductility and toughness gains, increasing the serviceability of the material. The mechanical properties of bamboo (*Bambusa tuldóides*) were also investigated, in order to understand the constitutive laws of the reinforcement material. The values of 42.53; 165.30 and 170.47 MPa, for compression strength parallel to the fibers, parallel tension to the fibers and bending, respectively, were obtained. The investigations should be complemented in future by durability studies, adhesion (bamboo-matrix) and microstructural characterization.

References

[1] United Nations, 2017, UN – HABITAT – United Nations Human Settlements Programme. Available on https://nacoesunidas.org/cidades-al-caribe-2012/. Accessed August /2017.

[2] Ghavami, K., 2005, "Bamboo as reinforcement in structural concrete elements", *Cement&Concrete Composites*, v.27, pp. 637-649. https://doi.org/10.1016/j.cemconcomp.2004.06.002

[3] Inbar, 2017, International Network For Bamboo and Rattan – Policy Synthesis Report N°1. Available on http://resource.inbar.int/download/showdownload.php?lang=cn&id=167785. Accessed August /2017.

[4] Embrapa – Empresa Brasileira de Pesquisa Agropecuária, 2017, Available on https://www.embrapa.br/busca-de-noticias/-/noticia/16896451/embrapa-investe-no-conhecimento-e-conservacao-de-especies-nativas-de-bambu. Accessed August /2017.

[5] Planalto, 2017, Presidência da República/Lei 12484/2011, It deals with the National Policy for Incentives for Sustainable Management and Bamboo Farming. Available on http://www.planalto.gov.br/ccivil_03/_ato2011-2014/2011/lei/l12484.htm. Accessed August /2017.

[6] Chithambaram,S.J., Kumar, S., 2017, "Flexural behaviour of bamboo based ferrocement slab panels with flyash", *Construction and Building Materials*, v.134, pp. 641-648. https://doi.org/10.1016/j.conbuildmat.2016.12.205

Non-Conventional Materials and Technologies – NOCMAT for XXI Century   Materials Research Forum LLC
Materials Research Proceedings 7 (2018) 7-19   doi: http://dx.doi.org/10.21741/9781945291838-2

[7] Jain, S., Kumar, R., Jindal, U.C., 1992, "Mechanical behavior of bamboo and bamboo composite", *Journal of Materials Science*, v.27, pp. 4598-4604. https://doi.org/10.1007/BF01165993

[8] Ribeiro, R.A.S., Ribeiro, M.G.S., Miranda, I.P.A., 2017, "Bending strength and nondestructive evaluation of structural bamboo", *Construction and Building Materials*, v.146, pp. 38-42. https://doi.org/10.1016/j.conbuildmat.2017.04.074

[9] Moroz, J.G., Lissel, S.L., Hagel, M.D., 2014, "Performance of bamboo reinforced concrete masonry shear walls", *Construction and Building Materials*, v.6, pp.125-137. https://doi.org/10.1016/j.conbuildmat.2014.02.006

[10] Mansur, M.A., Aziz, M.A., 1983, Study of bamboo-mesh reinforced cement composites, *The International Journal of Cement Composites and Lightweight Concrete*, v.5, n.3, pp.165-171. https://doi.org/10.1016/0262-5075(83)90003-9

[11] Chand, N., Khazanchi, A.C., Rohatgi, P.K., 1986, Structure and properties of Ipomea Carnea: Its performance in polymer, clay and cement based composites, *The International Journal of Cement and Lightweight Concrete*, v.8, n.1, pp. 11-20. https://doi.org/10.1016/0262-5075(86)90020-5

[12] Gutièrrez, A. D., Sistema Construtivo "Quinchas Prefabricadas", *Informes de la Construcción*. Vol. 36. Spain 1984.

[13] Martins, A.P.S., 2014, Development, Mechanical Characterization and Durability of Self-Compacting Soil-Cement Composites Reinforced with Sisal Fibers. Doctoral Thesis, COPPE/UFRJ, Rio de Janeiro, Brazil.

[14] ASTM – American Society for Testing and Materials, 2008, Standard Classification of Soils for Engineering Purposes: D2487-08. West Conshohocken Philadelphia, USA.

[15] Brazilian Association of Technical Standards, 1984, Soil – Determination of the Liquidity Limit: NBR 6459. Rio de Janeiro.

[16] Brazilian Association of Technical Standards, 1984, Soil – Determination of the Plasticity Limit: NBR 7180, Corrected Version 1988. Rio de Janeiro.

[17] Ghavami, K., Marinho, A. B., 2005, "Physical and mechanical properties of the whole culm of bamboo of the Guadua angustifolia species" *Rev. Bras. Eng. Agríc. Ambient.* [online], v.9, n.1, pp.107-114. ISSN 1415-4366. http://dx.doi.org/10.1590/S1415-43662005000100016 .

[18] Xu, Q., Harries, K., Li,X., Liu, Q., Gottron, J., 2014, "Mechanical properties of structural bamboo following immersion", *Engineering Structures*, n.81, pp. 230-239. https://doi.org/10.1016/j.engstruct.2014.09.044

[19] Chu Chung, K.F., Yu, W.K., 2002, "Mechanical properties of structural bamboo for bamboo scaffoldings, *Engineering Structures*, n.24, pp. 429-442. https://doi.org/10.1016/S0141-0296(01)00110-9

[20] Janssen, J.J.A., 2000, Designing and Building with bamboo, Inbar Repository. Available on http://resource.inbar.int/download/showdownload.php?lang=cn&id=167649. Accessed August /2017.

Non-Conventional Materials and Technologies – NOCMAT for XXI Century   Materials Research Forum LLC
Materials Research Proceedings **7** (2018) 7-19   doi: http://dx.doi.org/10.21741/9781945291838-2

[21] Lo, T.Y., Cui, H.Z., Tang, P.W.C., Leung, H.C., 2008, "Strength analysis of bamboo by microscopic investigation of bamboo fibre", *Construction and Building Materials*, n. 22, pp. 1532-1535.

[22] Krause, J.Q., 2015, Micro e macromecânica de lâminas de bambu *Dendrocalamus Giganteus* para aplicações estruturais. Doctoral Thesis, PUC-Rio, Rio de Janeiro, Brazil.

[23] Jakovljevic, S., Lisjak, D., Alar, Z., Penava, F., 2017, "The influence of humidity on mechanical properties of bamboo for bicycles", *Construction and Building Materials*, v.150, pp. 35-48. https://doi.org/10.1016/j.conbuildmat.2017.05.189

[24] Tan, T., Rahbar, N., Allameh, S.M., Kwofie, S., Dissmore, D., Ghavami, K., Soboyejo, W.O., 2011, "Mechanical properties of functionally hierarchical bamboo structures", *Acta Biomaterialia*, v.7, pp. 3796-3803. https://doi.org/10.1016/j.actbio.2011.06.008

[25] Krause, J.Q., Silva, F.A., Ghavami, K., Gomes, O.F.M., Toledo Filho, R.D., 2016, "On the influence of Dendrocalamus giganteus bamboo microstructure on its mechanical behavior", *Construction and Building Materials*, v.127, pp. 199-209. https://doi.org/10.1016/j.conbuildmat.2016.09.104

[26] Sharma, B., Gatoo, A., Bock, M., Ramage, M., 2015, "Engineered bamboo for structural applications", *Construction and Building Materials*, n.81, pp. 66-73. https://doi.org/10.1016/j.conbuildmat.2015.01.077

[27] Puri, V., Chakrabortty, P., anand, S., Majumdar, S., 2017, "Bamboo Reinforced Prefabricated Wall Panels for Low Cost Housing", *Journal of Building Engineering*, v.9, pp. 52-59. https://doi.org/10.1016/j.jobe.2016.11.010

[28] NZS, 1998, Standards New Zealand NZS 4298, Materials and Workmanship for Earth Buildings, Wellington, New Zealand.

Non-Conventional Materials and Technologies – NOCMAT for XXI Century   Materials Research Forum LLC
Materials Research Proceedings 7 (2018) 20-25   doi: http://dx.doi.org/10.21741/9781945291838-3

Bamboo-Piles Analysis for Slope Stability

Angel Mauricio Jaime Davila[a*], Khrosrow Ghavamia[b], Celso Romanel[c]

[a]Pontificia Universidade Católica do Rio de PUC-Rio, Brazil; angel_jaime@live.com.mx

[b]Pontificia Universidade Católica do Rio de Janeiro – PUC-Rio ghavami@puc-rio.br

[c]Pontifica Universidade Católica do Rio de Janeiro – PUC-Rio romanel@puc-rio.br

Abstract. Severe world damages due extreme natural phenomena consequence of abrupt climate change, increasing rainfall index had not stopped since the start of the present century and landslides incidents are more frequent day by day. These scenarios presents big economic losses and in worst human losses to the most vulnerable human settlements, which must cases implies irregular urban settlements. Different piles made of less process materials as trunk trees trunks to more conventional materials as steel and cement piles, are located and configured to support soil masses in order to prevent landslides accidents. The present paper studies the use of non–**conventional material**, in this case the bamboo as a pile element (**Bamboo-pile**), of the species *Dendrocalamus Gignteus*, (DG) taking advantage of its natural properties, its geographical avaibility, GEE absorbing capabilities and its mechanical properties that full fit for this purpose. Analyses was in **Finite Element** Software PLAXIS three-dimensional tests. Unstable slope condition reinforced with bamboo-piles to forward observe depth surface rupture on the slope, flexural and shear stresses on the pile developed by the soil-pile interaction. The safety factor evaluates the slope stability, and results, conclusion and further consideration are discussed. The results show that the capabilities of the bamboo-piles are an effective alternative as technique for slope stability.

Keywords: Soil Stabilization, Bamboo-Pile, Non-Conventional Materials, Finite Elements

Introduction

Over time, the world has suffered many environmental disasters caused by natural phenomena, such as floods, earthquakes, tsunamis, landslides, volcanic eruptions, cyclones, hurricanes, among others. Climate change tends to increase the frequency and the magnitude of these phenomena. Unfortunately, the cities where these phenomena presents more frequently suffer human life losses and economic losses caused by damages on infrastructure.

Action plan development has been the objective of different researchers and public institutions altogether. These plans aim to mitigate impacts of climate change. However, there are still difficulties that weaken these plans as the disordered occupation without a safety criterion takes place.

In the current transition and global technological improvement era, sectors, protagonists of civil development, have shown interest in technologies inspired by more natural elements, with cleaner production processes and low energy consumption, taking advantage of the most recent studies coming out from universities and other entities all over the world. The main objective of the international agreement COP21 is the reduction of greenhouse gas emissions through the practice of processes that involve fewer emissions of greenhouse gases (COP21, 2015).

Objectives

The objective of this work is to elaborate a finite element analysis of the capacities of bamboo-pile as non-conventional material that can be used in slope stabilization to replace those conventionally used, such as concrete and steel. Where landslides are probable to occur, bamboo-pile technique

Non-Conventional Materials and Technologies – NOCMAT for XXI Century   Materials Research Forum LLC
Materials Research Proceedings 7 (2018) 20-25   doi: http://dx.doi.org/10.21741/9781945291838-3

might be suitable to applied in places with little available area, difficult access and even need no specialized labor, in order to provide a way to prevent and minimize the incidence of these phenomena. Thus, in order to evaluate this technique, the PLAXIS 3D software is used as a software to simulate the situations. It models the soil-pile system that forms the slope structure, the safety factors, as well as the iteration efforts, which are calculated by the program, and the results are analyzed. Finally, conclusions and considerations based on the results of the tests aiming to improve the efficiency in the future use of this promising technique are presented.

FIGURE 1 -MACIÇO DA PEDRA BRANCA LANDSLIDE OVER GUANABARA TURISTIC ROAD RIO DE JANEIR, BRASIL 2011

History of environmental catastropehs in latinamerican regions
Authors such as Maskrey 1993 describe a natural disaster as a product of the correlation between dangerous natural phenomena and certain vulnerable socioeconomic and physical conditions. According to documented data, the mortality rate for disasters is ten times higher in poor countries than in rich countries. Latin American countries are certainly vulnerable to these events; for example, Mexico has 71% of GDP, 15% of the territory and 68.2% of the population at risk due to climate change events (Olivera, 2013).

Slope

Slope Force Development Factors
The soil force development is recognized as the difference between the total forces and effective forces; the first refers to the force of the structure or body of the soil and the pressure of the pore together, whereas the second only refers to the force of the structure or body of the soil. While a force is being applied to the soil body, a normal force and the diverting force develop over time in the soil body. In this interval of time, the forces that are transferred to the soil body draw a path through the soil body. Thus these forces start behaving in different directions, developing efforts in the soil body, such as: vertical compression, horizontal compression and shear forces (Kourkoulis, R.; Gelagoti, F.; Anastasopoulos, I., and Gazetas, G., 2011). The different states of the forces are shown in Figure 3.1.

Non-Conventional Materials and Technologies – NOCMAT for XXI Century   Materials Research Forum LLC
Materials Research Proceedings **7** (2018) 20-25   doi: http://dx.doi.org/10.21741/9781945291838-3

FIGURE 2 - PERU RAINFALLS 2017

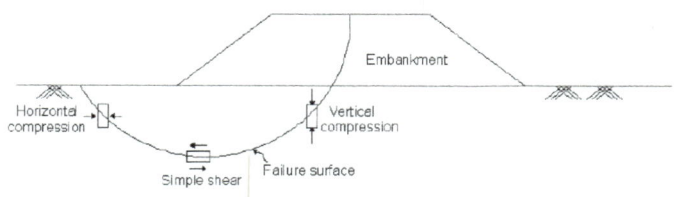

FIGURE 3 - STRESS CONDITION UNDER EMBANKMENT (Zdravkovic et al., 2002)

Slope Instability

Some reasons for slope instability are cutting or filling of slopes, pore pressure excess caused by high phreatic levels or sewage trajectory interruption, scouring provoked by water erosion superficially, resistance degradation by creeping process and increasing rainfall index.Some relevant conditionals of instability material as residual soils that progressively evolution provoked by the intemperism and the mineral variation, foundation and void index are also implied in the residuals soils behavior affecting its resistance stiffness compressibility an permeability. In soil mass movements somehow all elements, are subjected to initial dention and depend on the load or unload action for strain to happen to forward produce a soil mass movement. Other authors describes this movements because of a gravity effect. In the meantime it is known that the soil movements are influenced by factors such as climatic and geomorphological, geotechnics, anisotropic, pedagogical, geomorphologic and hydraulic acting simultaneously.

A classification of soil mass movements could be the subsidence or collapse, slip and creeping. Slips commonly happen in natural slopes, in the cut or creation of land fields, some of these characteristics are fast movements of short time lapse. The geometry of slope or land field creeping condition and the drainages as well as the soil properties gives the shape of the slip surface. There are normal four groups of slips, rotational, translational, block and debris falls. The first two accounts for are mainly defined by weak planes and disconnectivity or discontinuity between different types of soils or soil properties and its geometry, rotational are smaller when compare to translational. For Block falls erosion is the main cause of origin with the loss of support along the joints. In the case for debris falls dry and saturation most cases provoke it.

Non-Conventional Materials and Technologies – NOCMAT for XXI Century   Materials Research Forum LLC
Materials Research Proceedings **7** (2018) 20-25   doi: http://dx.doi.org/10.21741/9781945291838-3

(a)   (b)

Historic and chronologic documentation of collapse provoked by de reduction of resistance is also affected by hydraulic superficial ways some time difficult to see at first sight, for this by being a variation of the extreme mechanism, they are frequently reason monitoring is imprescindible if the slope control is an objective to be achieved. Other intermediary mechanisms are characterized by being a variation of extern mechanism, they are frequently related to sear strength lost due to water trip carrying fine soil particles. Water parameters accounts for a very important slope knowledge such as rain index, retaining capacity and hydraulic conductivity (Jose, 2014).

Analysis for Slope Slip Surface Development
When seeking for a slip surface, variational calculus, dynamic program, genetic algorithms, besides different developed techniques technique had been employed. In general, slip surface is divided in 2 segments by number of nodal points and straight line or a lightly curved line connects each two points and then forward is evaluated the Safety factor, one of most popular methods is Spencer´s 1967. Jianping Sun et al 2008 points out that there is advantage analyzing complex slope that needs more free degrees, fewer nodal points are needed to reach the same accuracy compare to other methods for defining slip surface by spline curves and this may be considered in order to save time and memory storage for CPU process.

Piles are consider the oldest traditional method used by man to overcome for foundations where is needed to reach a soil with sufficient stability capacities to absorb the vertical and lateral loads. Modern literature about piles it is said to star at the end of IXX century. As time passed theories and practices had been developed, thus these helps to evaluate the possible mistakes that would occur when designing pile use regarding quantity, size and economic issues it represent (H.G. Poulos, E. H. Davis , 1980).

For this work, the horizontal forces are the main issue to overcome, as piles will need to add stability to the slope by absorbing lateral forces. Some examples besides natural unstable slopes and the creation landfills are the creation of piers and offshore structures and even for structures, which are constructed in earthquakes areas.

The pile purpose is to increase the safety factor by adding the reactive portion to the resistant forces that are acting contrary to the ground. For these reason the acceptable deflection of the pile is essential for pile design. Slope stabilization by piles technique requires soil displacements to activate pile forces as mentioned before, as it is a passive technique. Pile stiffness is mainly the property that responds and mostly defines pile behavior. When estimating ultimate lateral resistance of a single pile base resistance is negligible and effects of socketing and pile-batter are analyzed. Several authors have classified pile types according to this factor in rigid, flexible and intermediary (Bello, 1997).

Non-Conventional Materials and Technologies – NOCMAT for XXI Century   Materials Research Forum LLC
Materials Research Proceedings 7 (2018) 20-25   doi: http://dx.doi.org/10.21741/9781945291838-3

Pile modelling by finite elements fe-method

Currently the finite element method is one of the most recurrent analysis methods, since it facilitates the labor and offers precision. Kourkolis et al. (2010) proposes in one of his publications a hybrid method based on two steps for the modeling of the stabilizing piles by means of the finite element technique, which holds wide analytical acceptance. The first step is to develop a slope in which the safety factor is increased by means of a study of strength provided by one single pile. In the second case, with the configuration of the pile or set of piles, it is sought to provide the study of strength prescribed in step one.

In order to model slope, some authors have considered some factors based on the analytical studies; these factors are relevant when modeling individual or group of piles. The most frequently mentioned factors are the pile material, the bedding, pile diameter, soil homogeneity, and spacing. Plaxis (Finite Element Code for Soil and Rock Analysis) is a finite element package developed for applications to geotechnical problems by the Technical University of Delft, The Netherlands since 1987, and succeeded in 1993 by the commercial company Plaxis. It was developed with the purpose of being a practical numerical tool for the use by geotechnical engineers who are not necessarily specialists in numerical procedures. This software development philosophy has resulted in a quite simple user-engineer interaction (pre- and post-processing routines are very easy to manipulate).

The software currently implements the following constitutive laws:

- Linear elastic
- The Mohr-Coulomb model (elasto-perfectly plastic behavior),
- Elasto-plastic model with isotropic hardening (hyperbolic dependence on soil stiffness on the stress state),
- The Soft-Soil Model
- The constitutive law for creep (time dependent behavior).

In this work, the materials that constitute the pile-bamboo and its protective layer were considered homogeneous, isotropic and linearly elastic, requiring the definition of only 2 parameters (E, v). The soil of the mass was represented by the model of Mohr Coulomb, which requires knowledge of the following 4 basic parameters: modulus of elasticity E, Poisson coefficientv, soil cohesion (c), shear strength angle (ϕ).Plaxis generates the finite element mesh automatically. The geometry, different materials, the water level etc. impose respective restrictions. The finite element type selected for the numerical models given below was the quadratic triangular element of 15 and 6 nodes respectively (Dao, 2011).

Critic slope definition

This chapter presents the performed test to evaluate the bamboo-pile capacities by finite element method using the numerical software PLAXIS 3D. The determination of the tests were based on the information discussed in the previous chapters. Two types of tests; the first evaluates the influence of the bamboo-pile position and the second evaluates the spacing for a bamboo-pile group. Some parameters of test data was abstracted from previous work Lobato 1997 who also abstracted from Zou et al 1995 used to perform equilibrium limit state that develops a critical rupture surface with a safety factor close to failure Figure 5.1. Soil properties are 200 Mpa for elastic module, Poisson ratio 0.25, 20 degrees for friction angle and cohesion of 20 KN/m^3.

The analyses executed were based on two stages divided in two phases each one. The first two phases corresponds to the linear elastic stage, which contains the gravitational activity at initial phase and phase 1 as helps to verify the linear elastic behavior looking forward to obtain the initial stress state or *in situ*, imagining that this configures a certain slope accommodation or/and densification as a

24

Non-Conventional Materials and Technologies – NOCMAT for XXI Century   Materials Research Forum LLC
Materials Research Proceedings 7 (2018) 20-25   doi: http://dx.doi.org/10.21741/9781945291838-3

whole body. In second stage which correspond to the Mohr Coulomb Model with an associated plasticization. In figures 5.2a-d it is noticed that there is a tendency where a natural more or less circular shaped surface rupture.

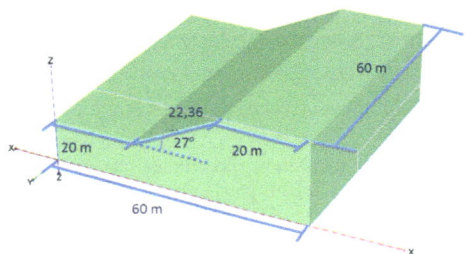

FIGURE 5 - SLOPE DIMENSION

Conclusion

Bamboo-pile finite element to analyses were made for a single pile bamboo which did not have any great effect for the critic slope dimension. For row pile analyses more than one position where tested over the slope at top, top middle, middle, low middle and low position. Lower position did the greater effect over de safety factor as it raised from 1.036 to 1. 078.

For these star research it is taken a relative short approach for a future combination and other implications that help to raise the safety factor value to more considerable influence for the stability of a critical slope to achieve the improvement for these promising technique to help for climate and growing population awareness.

References

[1] Bello L. A., Estudo Numerico sobre o ouso de estacas de bambu-cal na estabilizañ{ao de taludes. Tio de Janeiro. PUC-Rio.

[2] Dao T.; Validation of Plaxis Embedded Piles For Lateral Loading.Delft University of Technology 2011.

[3] Ferronato Pretto, Jose Henrique; Analise de tensão x deformação de uma encosta natural estudo de caso: Morro do boi – Balneario Camboriú, Curitiba 2014.

[4] H. G. Poulos, E.H. Davis; Pile Analises foundatio and Desing. Canda: The university of Sydney.

[5] Kourkolis, R.; Gelagoti, F.; Anastasopoulos, I.; Gazetas, G.; M.ASCE; Hybrid Metod for analysis and desing of selo stabilising piles. Journal Of Geotechnical and Geoenvironmental Engineering v. 663-677, 2011.

[6] Olivera, B. 2013; El Ártico y los efectos del cambio climático en México. México (CDMX); Green Peace.

[7] Sun, J., Li, J., & Liu, A. Q.;Search for Critical Slip Surface in Slope Stability Analysis J. Geotech. Geonviron, Eng,, pp 252-256, 2008.

Non-Conventional Materials and Technologies – NOCMAT for XXI Century   Materials Research Forum LLC
Materials Research Proceedings 7 (2018) 26-34   doi: http://dx.doi.org/10.21741/9781945291838-4

Speeding up Post-Disaster Reconstruction: Material Choice or Roof Design?

Giulia Celentano*[a], Edwin Zea Escamilla[b], Verena Göswein[c], Vincent Hischier[a], Guillaume Habert[a]

[a] Chair of Sustainable Construction, Swiss Federal Institute of Technology, ETH Zurich, CH, Zürich, Switzerland, celentano@ibi.baug.ethz.ch

[b] Center for Corporate Responsibility and Sustainability, University of Zurich, CH

[c] Instituto Superior Técnico, Universidade de Lisboa, PT

Abstract. The consequences of urbanization and climate change are dangerously converging. The most affected populations are the urban poor, settled in informal settlements, vulnerable to increasingly frequent disasters. This severely contributes to the existing housing gap of the affected regions, already struggling with housing demand. The speed of shelter delivery becomes key for an efficient response. The present study aims to understand the impact of material choice on post-disaster shelters delivery through a multiscale analysis of their construction speed. The scales considered for the study are: Constructive technology, Shelter Unit and Post-disaster settlement. At the scale of the Constructive technology, nine different solutions suitable for the Nepal earthquake reconstruction are compared, covering a range from local to industrialized. Successively, twelve different shelter designs delivered worldwide by the International Federation of the Red Cross have been studied under the same lens, at the Shelter unit scale as well as for the case of the Post-disaster camp. The study shows that a clear correlation between material procurement and speed can be identified at the element scale. This correlation becomes secondary at the shelter scale, where it is visible that materials play a limited role in affecting the construction time, that is mainly driven by the complexity of the roof design. Moving to the settlement scale, the procurement choice of materials seems to be impacting the speed again. The study indicates how no univocal solution fits for the three different scales of the study, providing efficient guidelines for post-disaster reconstruction. Beyond that, it highlights that effective construction can be developed with a variety of materials, but its emergency responsiveness can seriously be compromised by a non-appropriate design.

Keywords: Post-Disaster Reconstruction, Shelters, Material Selection, Speed, Large Scale Reconstruction

Introduction

Since 2009, an estimated one person per second has been displaced by a disaster (1). Ongoing urbanization, combined with the escalating consequences of climate change (2,3), is leading to an increase in the impact of disasters on the world's habitat.

Currently, eighty percent of cities, home to 1.9 billion people, are located in areas that are highly exposed to the occurrence of natural disasters, and it is projected that low-income countries will be the most severely affected (2). Thus, the issue of how to speed up large-scale post-disaster reconstruction has become a key challenge.

Post-disaster reconstruction projects are prone to deliver inadequate building solutions due to the complexity of the situation. The main bottlenecks in reconstruction programs have been clearly identified: (i) supply chain dysfunction, (ii) resources shortage, (iii) corruption, (iv) lack of coordination among agencies, (v) poor construction skills and (vi) infrastructure breakdown.

Non-Conventional Materials and Technologies – NOCMAT for XXI Century   Materials Research Forum LLC
Materials Research Proceedings 7 (2018) 26-34   doi: http://dx.doi.org/10.21741/9781945291838-4

Solutions to reduce these obstacles for delivering large-scale affordable reconstruction projects are still under discussion (4–8).

It is then important to investigate approaches capable of increasing the speed of shelter delivery at a large scale, as well as the role of material selection in these approaches. To facilitate understanding of the state of the art for this matter, current positions have been reviewed in this paper according to the scales of the study: (i) constructive technology, (ii) shelter unit and (iii) post-disaster settlement. The results from this review are presented in the following sections

Scale 1: Constructive Technology

Every post-disaster reconstruction project is faced with the challenge of quickly responding to the crisis at hand using available resources, resulting in either a global or local material choice (9). Material selection carries consequences extending past the field of materials science. Local materials such as bamboo, earth, or stone have been identified by many as the most effective choice for post-disaster reconstruction. Used in vernacular construction techniques, they are in fact strictly related to their territory and culture and thus must not only be available in the direct aftermath of a disaster but also be climatically appropriate (10). Furthermore, they allow for social acceptance and local labor adoption, maintaining the use of traditional techniques and facilitating maintenance in the future due to the availability of materials and skills (11,12). The main reason the main international organizations back this choice is its social, cultural and participatory value. In contrast to these benefits, it is important to mention that the application of local building technologies may be constrained by a number of factors, (10), possibly leading to poor construction quality and higher risk of failure in the case of further disaster.

For these reasons, it is considered by some to be a viable option to adopt globally available solutions consisting of industrialized systems, based, for instance, on concrete elements or steel structures. The adoption of industrialized solutions can lead to very different results, ranging from optimized projects to more resource-consuming results from an environmental and economic perspective (9). At this scale, the discussion regarding material selection is mainly focused on the socio-cultural consequences of its implementation rather than capacity to boost the construction. The issue of speed, in fact, seems not to be considered a relevant point of this discussion.

Scale 2: Shelter Unit

Material selection is one of the first steps in decision-making for shelter delivery, together with design choice definition. Moving to the scale of the shelter unit, diverse architectural solutions have been adopted for large-scale reconstruction programs. Those can be grouped into two main manufacturing strategies: in situ construction or prefabricated technologies. Both approaches, have proven to be viable. The first is mainly used for the formerly mentioned local materials, embedding direct involvement of community members in both design and construction phases, leading to a higher sense of ownership and community involvement (10). The latter, aiming to achieve large-scale production of easy-to-assemble building products, derives from decades of research conducted by industries and aims to reduce the building time to assembly time while overcoming the incapacity of the local production system to cope with the emergency, shifting the production in areas untouched by the disaster.

The here-mentioned prefabricated and modular systems include steel or pre-cast concrete frames with a variety of infill walls, as well as pre-cast panel systems or fast-to-assemble wall packages. The key issue with prefabrication is often identified as cultural acceptability. Despite multiple examples of prototypical successful practices, many have been arguing against the adoption of prefabricated solutions. The International Federation of the Red Cross recently officially backed the adoption of more local solutions, still without rejecting prefabrication-based designs for some reconstruction programs (as seen in Haiti, 2008).

Non-Conventional Materials and Technologies – NOCMAT for XXI Century   Materials Research Forum LLC
Materials Research Proceedings 7 (2018) 26-34   doi: http://dx.doi.org/10.21741/9781945291838-4

30 years after the beginning of this debate, a common agreement on the efficiency of prefabrication is still under discussion. Social inclusion and acceptance, environmental footprint and economic cost are investigated in relationship to the advantages of prefabrication, while no main effort is invested in the implementation of the construction speed itself.

At this scale again, it is then possible to say that, despite the acknowledged relevance of construction speed, the discussion is instead shifted to other aspects of the process.

Scale 3: Post-disaster settlement

Housing, as one of the basic human needs, is always placed as the top priority in a country's recovery agenda. However, the complexity of a post-disaster environment determines that the delivery process of a housing rebuilding project is often more complex than it is for conventional projects.

The inadequacy of construction resources along with the disaster's impact on the recovery process has been largely overlooked. Despite consistent research on the topic, there is no agreement on how material choice affects the large-scale construction in terms of speed. Numerous International Organizations, such as IFRC and UN agencies, also highlighted the key role of resource availability (4,13,13–16).

Disadvantages for the local purchase are also highlighted by UNHCR, such as the possibility of higher prices, poor quality, and inability to meet specifications and to supply the volume of goods demanded.

The review of this scale shows the multifaceted characteristics of resource procurement.. Their implementation, backed by solid studies, should be integrated into a holistic resource planning framework for disaster reconstruction (4).

Data and methods

For the three different scales of the analysis, construction techniques were assessed regarding their building time and cost related to their material procurement. In more detail, this has been identified by the terms *local* and *imported* for their literal translations related to the procurement. The ratios of imported materials of different constructive technologies have been analyzed in correlation with the construction speed and economic cost at the three scales of constructive technology (referring to one square meter of wall build), shelter unit (referring to a complete shelter) and post-disaster reconstruction camp (considering the complete camp development).

Scale 1: Construction Technology

The assessment of different technologies at this scale was conducted according to different constructive options suitable for the Thame Valley post-earthquake reconstruction in Nepal (2015). This specific context represents a relevant case study due to the isolated location of the reconstruction, and accessible only by foot via a two-day hike from the closest city served by proper road access, i.e., Lukla. This case study presents an interesting occasion for reflecting on the consequences of material selection on construction speed.

Different technologies have been chosen to cover a range from local to imported solutions, suitable for the site. These are shown in Figure 1 and ranked from those totally available on site to the fully imported ones, are Rammed Earth, Stone Masonry, Compressed Earth Blocks (CEB), Bahareque, Chicken wire with stone, Earthbags, OSB cladded timber frame, Iron sheet cladded timber frame, and Iron sheet cladded steel frame. Information on the construction time are obtained through a literature review and consultations with experts. The speed of construction was measured in working days per person over square meter of wall (person*day/m_2). Cost estimation is based on the Indian market prices at the time of the reconstruction, following the indications of local experts. The unit adopted for the economic cost assessment is USD per square meters of wall built ($/$m_2$) (17).

Non-Conventional Materials and Technologies – NOCMAT for XXI Century   Materials Research Forum LLC
Materials Research Proceedings **7** (2018) 26-34   doi: http://dx.doi.org/10.21741/9781945291838-4

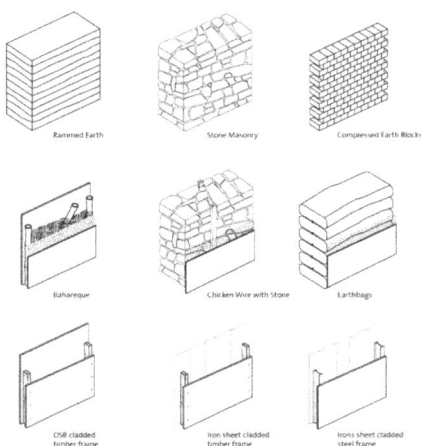

FIGURE 1 - REPRESENTATION OF THE ADOPTED TECHNOLOGIES

Scale 2: Shelter Unit

The assessments of the scale of the building are based on the IFRC data collected in the two documents (18,19) related to shelters delivered worldwide between 2004 and 2011. The shelters considered for the assessment have been implemented in the following countries: Afghanistan (1), Bangladesh (2), Burkina Faso (3), Haiti (4), Indonesia (5), Pakistan (6), Peru (7), Philippines (8), Sri Lanka (9) and Vietnam (10).

The single-family-use shelter designs are different, as are their building technologies, mainly based on bamboo, concrete, steel or wood structures.

For each of the shelters, the detailed material inventory was given, together with the material origin. The ratio of imported materials was calculated over the total amount (kg/kg), and expressed as a percentage. Details on the construction time were also available from the IFRC reports and described the number of working days per team, as well as the number of team members. This information was adapted into the unit of person*day/m2 as a way to compare the different shelters. Information on the cost was given in Swiss Francs (ChF).

Correlations between the ratio of imported materials, construction speed and cost were then investigated.

Scale 3: Post-disaster settlement

Where data were available, the same shelters were the subject of further study at the settlement scale. Details on the settlement projects are obtained by consultations of diverse reports from the Shelter Cluster. Due to the extensive research that occurred in the affected areas of the 2004 Indian Ocean tsunami, the reconstruction program in Aceh became a relevant benchmark for post-disaster shelter delivery on a large scale (11,20–22). For this reason, the case study of Aceh (Indonesia) has been added to the trend.

The assessment at the settlement scale considered project cost and construction time over the elapsed project delivery among the overall settlement completion. Due to the different scales of the settlements considered, the elapsed time has been expressed as a percentage over the total project conclusion. The influence of material procurement has been studied by dividing the shelters into two

Non-Conventional Materials and Technologies – NOCMAT for XXI Century   Materials Research Forum LLC
Materials Research Proceedings **7** (2018) 26-34   doi: http://dx.doi.org/10.21741/9781945291838-4

groups: locally based and imported. The correlation between project cost, construction time and material procurement was then plotted.

Results

The results are presented in the following paragraphs according to the three different scales of the assessment (Constructive technology, shelter unit and settlement scale).

Scale 1: Constructive Technology

The results for material procurement (X-axis), construction speed (Y-axis) and cost (secondary Y-axis) are shown in Figure 2. From this figure, it is possible to see that local technologies such as Rammed earth have a low construction speed and some of the lowest values for cost. The opposite trend can be observed for the global materials, ranked as the fastest and most expensive. Due to the logarithmic trend of the construction speed, it is possible to observe that a minimum input of imported materials, as in the case of Compressed Earth Blocks, Chicken wire with stone, Masonry or Bahareque, allows for an important reduction in the construction time while still maintaining its affordability. In contrast, the economic cost of the construction increases linearly when moving from local to imported technologies. At the material scale, we can then conclude that a limited amount of imported material allows for a drastic improve in the construction speed without significantly impacting its economic cost.

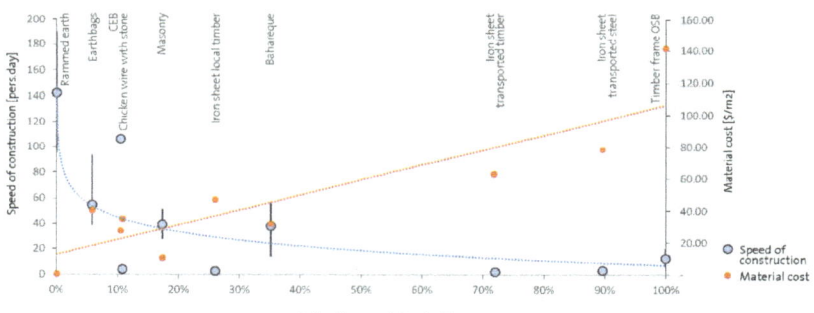

FIGURE 2 - *CONSTRUCTION TIME AND PROJECT COST VS RATIO OF IMPORTED MATERIAL*

Scale 2: Shelter Unit

At shelter unit scale, the correlation between material origin, construction time and project cost does not show a clear correlation. Shelters mainly built with local materials (in blue in the figure) here show both a low project cost and a low constructive time. Local materials seem to be more effective from a cost perspective, as well as for their speed efficiency. Imported solutions instead cover a broad range from a cost perspective but a more homogeneous result in terms of time delivery. Due to the lack of clear correlation between material procurement and speed at the building scale, further analysis has been carried out to identify the drivers of construction speed at the current scale.

Shelters have been investigated according to their type of footing and number of total constructive elements, representing the ease of assembly. In none of the cases any relevant correlation with construction time has been identified,

Non-Conventional Materials and Technologies – NOCMAT for XXI Century   Materials Research Forum LLC
Materials Research Proceedings 7 (2018) 26-34   doi: http://dx.doi.org/10.21741/9781945291838-4

Even though the shelters differ in many aspects, all of them consist in a one-floor single-family shelter based on a single room, aiming for a fast replicable solution, and easy to implement by non-trained labors. The element of the roof is the one that most shows variations in terms of design, ranging from a low slope flat roof to gable, hip and mansard types, making it an interesting object to be considered in the study due to its possible impact on the construction time. The shelters have thus been ranked according to the number of structural elements composing the roof and studied in correlation with the construction time, as shown in Figure 3.

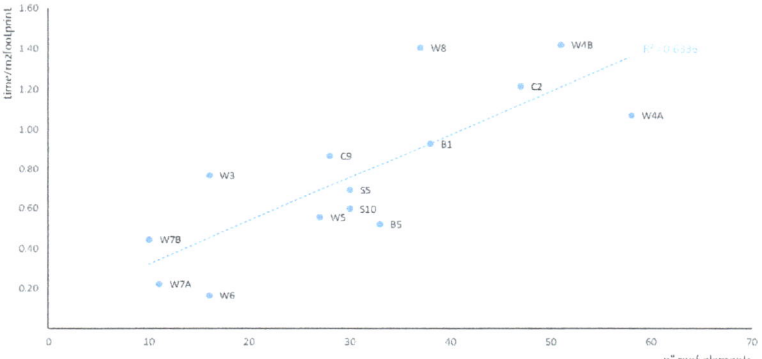

FIGURE 3 - NUMBER OF ROOFING ELEMENTS VS. CONSTRUCTION TIME (SHELTER SCALE)

Figure 3 clearly shows that the roof design has a strong impact on the time of delivery of the full construction unit.

To conclude, it is possible to say that, at the shelter scale, the ease of assembly of the roof, based on the complexity of its design, has a major impact on the construction speed of the shelter. Beyond that, it is also shown that the main speed driver identified at the building scale, consisting of material origin, has no particular relevance at the current scale, where it becomes secondary compared to the roof design.

Scale 3: Post-disaster settlement
The assessment then considers the speed and cost of delivery of the post-disaster shelter at the settlement scale in relationship to the material procurement. The settlement completions follow an S-Curve trend, divided into the three phases of Build Up, Steady State and Run Down.

Shelters have been grouped according to their material origin in Figure 4, where yellow indicates the shelters based on transported materials and blue the ones mainly built with locally available resources.

Two observations can be made according to the material procurement:

(i) The Build Up phase is shorter for local shelters
(ii) The Steady State curve is steeper for local shelters.

Non-Conventional Materials and Technologies – NOCMAT for XXI Century   Materials Research Forum LLC
Materials Research Proceedings 7 (2018) 26-34   doi: http://dx.doi.org/10.21741/9781945291838-4

This results in a faster take-off for local shelters, followed by a boosted speed during their project running time.

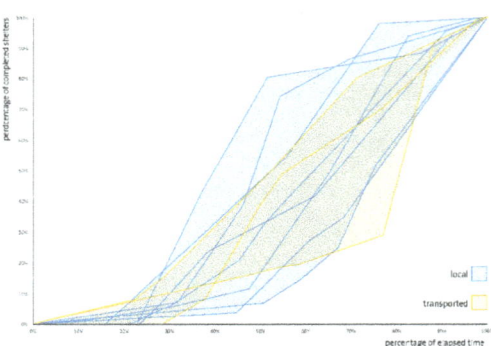

FIGURE 4 - SHELTER SETTLEMENTS CONSTRUCTION DEVELOPMENT OVER TIME

It is then possible to state that, at the settlement scale, the material origin becomes relevant again, as shown at the material scale. The impact of the roof design at this stage is, instead, secondary.

Discussion and conclusion

Post-disaster reconstruction necessarily confronts the issue of speed of delivery due to the urge posed by the emergency. Different approaches to material selection according to its procurement (*local* or *global*) have been extensively explored to cope with the shelter demand in the shortest possible timeframe. This paper, based on data provided by IFRC, together with consultations from experts and literature, looked at how resource procurement affects the speed of shelter delivery at three different scales: building technology, shelter unit and post-disaster settlement. The results show that different drivers of the speed of delivery can be identified at each scale. At the settlement scale, the adoption of mainly local materials over imported ones appears to be responsible for improving the construction time. At the shelter scale, the speed of construction is instead impacted by the complexity of the design and the resulting ease of shelter assembly, here identified as the number of roofing elements. Local materials can provide faster construction, while the influence of the material procurement is negligible over the influence of the shelter design. At the building technology scale, material procurement emerges once again as the main driver for construction speed, where fast and affordable construction can be achieved thanks to mainly locally procured technology, implemented in a minimal percentage by the adoption of imported materials.

Furthermore, to achieve results at a large scale, when selecting materials, it is fundamental to take into account supply chain management and logistic planning, which impact the speed far more than assembly time or material characteristics.

To conclude, the assessment shows that, when dealing with the construction of one single wall, there is an interest in an appropriate input of industrialized materials. A limited amount of industrialization (approximately 10-20%) boosts the construction speed significantly, without heavily impacting the construction cost. Further industrialization does not provide additional time savings, but it does severely damage the cost. This driver, however, becomes secondary when dealing with the higher scale of the shelter unit, where the complexity of the roof assembly emerges as the main factor affecting the construction speed. The wall constructive technique seems to be negligible for

Non-Conventional Materials and Technologies – NOCMAT for XXI Century   Materials Research Forum LLC
Materials Research Proceedings 7 (2018) 26-34   doi: http://dx.doi.org/10.21741/9781945291838-4

this scale. Finally, when dealing with the construction of multiple shelters, the material supply for the entire building (walls and roof elements) again becomes a priority.

According to the scale of the project, different decisions regarding design and material selection should then be considered in a manner to achieve fast construction.

Knowing the different drivers' impacts on the construction speed at the three scales of constructive technique, shelter unit and post-disaster settlement sets a solid basis for efficient guidelines towards rapid reconstruction.

Acknowledgment

The authors would like to thank Scott Llyod from TEN for giving us the opportunity to work on the preliminary study of the reconstruction project in the Thame Valley in Nepal. Particular acknowledgement is extended to Corinne Trehern from IFRC for the fruitful discussion on the topic. Finally, HILTI foundation and ETH Global are thanked for their partial financial support.

References

[1] UNHCR. Emergency Handbook: Shelter Solutions. UNHCR; 2017.

[2] World Bank. Atlas of Sustainable Development Goals 2017: From World Development Indicators. The World Bank; 2017.

[3] UNHCR. Climate Change and Disasters [Internet]. UNHCR. [cited 2017 Jun 12]. Available from: http://www.unhcr.org/climate-change-and-disasters.html

[4] Chang Y, Wilkinson S, Potangaroa R, Seville E. Interpreting resourcing bottlenecks of post-Wenc huan earthquake reconstruction in China. Int J Strateg Prop Manag. 2010 Dec;14(4):314–31. https://doi.org/10.3846/ijspm.2010.24

[5] UNEP, SKAT. After the Tsunami. Sustainable building guidelines for South-East Asia. 2007.

[6] Lyons M, Schilderman T, Boano C. Building Back Better. 2010.

[7] Bilau AA, Witt E, Lill I. A Framework for Managing Post-disaster Housing Reconstruction. Procedia Econ Finance. 2015;21:313–20. https://doi.org/10.1016/S2212-5671(15)00182-3

[8] UNISDR. Sendai Framework for Disaster Risk Reduction 2015 - 2030. Geneva, Switzerland: United Nations; 2015.

[9] Zea Escamilla H. Global or local construction materials for post-disaster reconstruction? Sustainability assessment of twenty post-disaster shelter designs. 2015.

[10] Lyons M. Building Back Better: The Large-Scale Impact of Small-Scale Approaches to Reconstruction. World Dev. 2009 Feb;37(2):385–98. https://doi.org/10.1016/j.worlddev.2008.01.006

[11] Jo Da Silva. Lessons from Aceh: Key Considerations in Post-Disaster Reconstruction. Arup/The Disasters Emergency Committee; 2010.

[12] Sphere Project. The Sphere Handbook | Shelter and settlement standard 4: Construction. 2011.

[13] IFRC. Appeals and Reports - IFRC. Geneva, Switzerland; 2017.

[14] UNDP. Survivors of the Tsunami. One Year Later. Washington, DC; 2005.

[15] UNDP. Construction boom analysis: pilot analysis for bricks. Banda Aceh, Indonesia. Geneva, Switzerland: United Nations; 2006.

[16] UNEP. Environment and reconstruction in Aceh: two years after the tsunami. 2007.

[17] Celentano G, Goeswein V, Zea Escamilla E, Habert G. Sustainable post-disaster reconstruction in remote areas. A selection method for appropriate constructive techniques. In: 5th Eco-Materials International Conference. Riobamba, Ecuador; 2016.

[18] IFRC. Transitional Shelters: Eight designs. Geneva, Switzerland; 2013.

[19] IFRC. Post-disaster shelter: Ten designs. 2011.

[20] Zuo K, Wilkinson S, Potangaroa R. Supply chain and material procurement for post disaster construction: the Boxing Day Tsunami reconstruction experience in Aceh, Indonesia. In: Building resilience. Heritance Kandalama, Sri Lanka; 2008.

[21] Nazara S, Resosudarmo BP. Aceh-Nias reconstruction and rehabilitation: Progress and challenges at the end of 2006. ADB Institute Discussion Papers; 2007.

[22] Steinberg F. Housing reconstruction and rehabilitation in Aceh and Nias, Indonesia— Rebuilding lives. Habitat Int. 2007 Mar;31(1):150–66. https://doi.org/10.1016/j.habitatint.2006.11.002

Non-Conventional Materials and Technologies – NOCMAT for XXI Century   Materials Research Forum LLC
Materials Research Proceedings 7 (2018) 35-44   doi: http://dx.doi.org/10.21741/9781945291838-5

A Bamboo Beam-Colum Connection Capable to Transmit Moment

Richard Moran [a], Jesús Muñoz [b, *], Hector F. Silva[c], José J. García[d]

[a]Escuela de Ingeniería Civil y Geomática, Universidad del Valle.
richard.moran@correounivalle.edu.co

[b*] Escuela de Ingeniería Mecánica, Universidad del Valle. jesus.munoz@correounivale.edu.co

[c] Departamento de Artes Integradas. hector.silva@correounivalle.edu.co

[d] Escuela de Ingeniería Civil y Geomática, Universidad del Valle. josejgar@gmail.com

Abstract. Guadua *angustifolia* Kunth (GAK) is the most common bamboo species in Colombia. As a material, it is well recognized for its high axial strength, lightness, low cost and tubular cross section. Therefore, it has a great structural potential. Additionally, it is an alternative to reduce the high pressure to the forest exploitation and the use of traditional materials. Despite its mechanical attributes, GAK connections are difficult to construct due to hollow cylindrical shape of the culms, the variations of shape and dimensions, and the low mechanical properties associated with the transverse directions. Hence, typical GAK connections are custom based constructed, usually by drilling the culms, which tends to induce the formation of longitudinal cracks. These connections are considered unable to transmit moment, which preclude using walls without diagonals for several types of applications. To overcome this problem, a new beam-column connection is presented that uses three steel angles and five pairs of thin light steel semi-rings, which can accommodate a range of culm sizes. Tests and finite element simulations of this connection have shown a consistent and improved performance when compared to traditional fish-mouth and grouted GAK connections. Strength and ductility of the proposed connection were 373% and 595% higher than those reported in other study for a connection composed of screw bars, plates, fish-mouth cuts, and injection with mortar.

Keywords: Guadua *Angustifolia* Kunth, Moment connections, Finite Element Method

Introduction

Guadua *angustifolia* Kunth (GAK) is the most common bamboo species in Colombia. This plant is a giant grass characteristic from tropical zones [1] and is considered as one the fastest growing plant in the world [2], with a maturation cycle at least three times lower than that of lumber trees. Therefore, it is a great alternative to substitute wood and reduce the high pressure over the native forest. GAK crops can be easily handled with little care [3] and they bring many benefits to the environment to regulate water cycles, control the soil erosion and capture CO_2 [4, 5, 6].

From the mechanical point-of-view, GAK is a hollow cylinder reinforced in the axial direction with strong cellulose fibers immersed in a weak and flexible lignin matrix [7, 3]. Therefore, the GAK mechanical behavior is anisotropic, being strong and stiff in the axial direction and weak in the fiber planes [8, 9]. Additionally, fiber distribution through the wall thickness is not uniform but it increases from the inner side to the external surface [10, 11]. As a material, GAK is featured by its lightness, high axial strength, tubular shape, low-cost and high sustainability [3, 4]. Therefore, it has an enormous potential as structural material.

In Colombia, this material has historically played a key role for the development of the Antioquia region during the eighteenth and nineteenth centuries, which was mainly stimulated by the availability of GAK [12]. Many houses, particularly in rural areas, were built using ancestral

Non-Conventional Materials and Technologies – NOCMAT for XXI Century   Materials Research Forum LLC
Materials Research Proceedings 7 (2018) 35-44   doi: http://dx.doi.org/10.21741/9781945291838-5

technologies based on GAK and many of them stay today in good conditions. Some of them have even resisted natural disasters like the Armenia (Colombia) earthquake in 1999 [13].

Even though, GAK has been included as a construction material in the latest version of the Colombian construction norm [14], it is not still extensively used to construct housing, due in part to the wide variations in mechanical and geometrical properties, the transverse weakness, the slow and tedious construction processes, and durability concerns. These issues combined with the misuses of the material have created the idea that GAK is a low category material. Therefore, one of the challenges to promote its use is to develop practical and efficient construction processes keeping a satisfactory level of quality.

One of the main difficulties to build GAK structures are the connections, which are custom-made, which increases costs and construction times [15, 1, 9, 16, 4]. Typically, to construct these connections, it is necessary to drill and carve curved cuts in the culms, leaving the material prone to develop longitudinal cracks [17, 18, 9, 15]. In many cases, culms must be injected with mortar to improve the transverse capacity, which adds weight to the structure. Aditionally, GAK and bamboo connections are considered unable to transmit moment [14, 19, 20]. This restriction limits the versatility of the constructive processes with GAK, especially when it is desired to use frames and panels with openings for doors and windows [21].

In an effort to improve the joints for GAK, our group developed a thin and light steel semi-ring connections system [22], which has allowed to connect culms with great versatility and preserving the integrity of the material [23, 24]. The semi-rings can accommodate a range of culm sizes, solving the problem of the joints customization. In this paper we present a new beam-colum connection that is able to transmit moment, this connection is composed of thin steel semi-rings and steel angles. Static tests and simulations of this connection showed a promising mechanical beahavior compared to typical GAK connections composed of screw bars, curve cuts and mortar injection.

Methods

The developed connection is based on steel semi-rings (Figure 1), which are made with 25.4 mm (1") wide and 1/8" (3.175 mm) thick platens. Semi-rings are secured in pairs with 9.5 mm (3/8 ") bolts through the lugs. Bolts are tightened to close the lugs and generate a high radial compression on the culm, in this way, it is possible to transfer load from semi-rings to the GAK culm to build diverse types of joints. Benitez [22] performed static tests of this system (Figure 1) and found average axial and transverse maximum forces of 14729 N (COV = 0.22) and 21600 N (COV = 0.19) respectively, with a ductile failure mode.

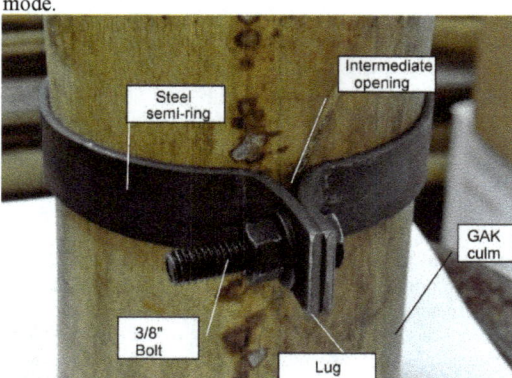

FIGURE 1- STEEL SEMI-RINGS ASECURED TO A GAK CULM [24]

Non-Conventional Materials and Technologies – NOCMAT for XXI Century   Materials Research Forum LLC
Materials Research Proceedings **7** (2018) 35-44   doi: http://dx.doi.org/10.21741/9781945291838-5

This system works based on mechanical interference, like shrinkage fits widely used to set shafts to wheel hubs in rotatory machinery [25]. Hence, in order to get a positive interference, the inner perimeter of a closed pair of semi-rings must be lower than the external perimeter of the culm. Benitez [21] determined that the circumferential interference should be at least 10 mm to assure a proper performance of the system. In the tested connections, interference was calculated as the internal perimeter of the closed semi-ring pair minus the external perimeter of the culm minus the sum of the two intermediate openings [24].

The proposed beam- column connection shown in Figure 2 was constructed using five pairs of semi-rings, two short angles of 31.75 mm (1 1/4 inch) wide and one diagonal brace constructed with angle of 31.75 mm (1 1/4 inch) wide. In some cases, semi-rings with welded nuts were used to allow connecting elements in various planes. The weight of all steel components of this connection is around 2.3 Kg.

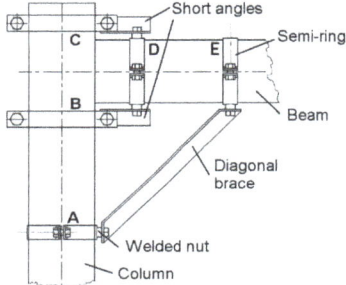

FIGURE 2 - PROPOSED BEAM- COLUMN CONNECTION

The set up for the static tests of this connection is shown in Figure 3, which has been used to test timber moment connections [26, 27] in conventional testing machines. In this assembly, the machine applies a compression load P that produces a moment PL / 2 and a vertical load P at the center of the connection. This force can be decomposed into axial and shear components.

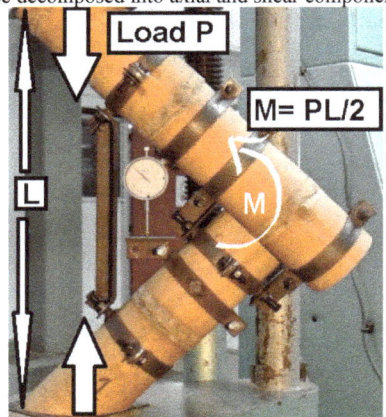

FIGURE 3 - SET UP USED TO THE STATIC TESTS [24]

Three connections specimens were tested. Average diameters and thicknesses of the culms were 112 mm (COV = 0.008) and 15.7 mm (COV = 0.0015), respectively. All GAK culms were borax treated and oven dried with practices well stablished [28]. Moisture content was measured at four locations of each specimen using a test probe (EXTECH Moisture Meter MO50), which yielded an average moisture of 13.6% (COV = 0.07).

All specimens were loaded at a rate of about 50 N/s, while the displacement was recorded with a dial gauge with a resolution of 0.001 inch between divisions and a range of 1 inch (25.4 mm). The tests were performed on a universal machine ZD (formerly Veb Werkstoffprüfmaschinen, Leipzig, Germany) with a force resolution of 39 N. The rotation angle of the connection was estimated using the dial gauge registers. With reference to the semi-ring labels of the Figure 2, the interference in the semi-rings are given in the Table 1.

TABLE 1 - CIRCUMFERENTIAL SEMI-RINGS INTERFERENCE FOR EACH TESTED CONNECTION

Specimen Number	Semi-ring pair interference (mm)				
	A	B	C	D	E
I	31	23	28	32	35
II	26	22	34	28	27
III	26	22	22	33	24

Ductility was calculated as a ratio of u_f and u_y, where u_y is the angle at yielding and u_f is the angle at failure. The procedure prescribed in the standard EN12512 was followed [29] to estimate u_y.

The minimum stiffness value Sj needed to consider that the connection can transmit moment was calculated as [30],

$$S_j = \frac{EI}{2l} \qquad (1)$$

where E is the axial elastic modulus of the base material (E = 9500 MPa for GAK [14]), I is the inertia moment of the culm and l is the length of the beam where the connection is going to be used. A stiffness value lower than Sj would indicate that the connection is articulated.

A finite element model (FEM) was developed to analyze the mechanical performance of this connection. The FE software ANSYS 16.1 (SAS IP, Inc, Cheyenne, United States) was used to develop the FEM that considers the anisotropy of GAK, nonlinear displacements and frictional contacts among elements.

In the model, culms were assumed to have an external diameter of 110 mm and a thickness of 10 mm, which are representative of the tested connections. The steel semi-rings, bolts and angles were assumed to have a Young's modulus of 200 GPa and a Poisson's ratio of 0.3. Orthotropic elastic properties of GAK were taken from literature [31, 8, 28] as shown in the Table 2, where the axes 1, 2 and 3 were respectively aligned along radial, circumferential, and axial directions of the culm. Letters E and G refer to Young's and shear moduli, while the Greek letter v describes the Poisson's ratios.

TABLE 2 - ORTHOTROPIC ELASTIC CONSTANTS OF GAK USED IN THE FEM

E_{11} (MPa)	E_{22} (MPa)	E_{33} (MPa)	v_{12}	v_{13}	v_{23}	G_{12} (MPa)	G_{13} (MPa)	G_{23} (MPa)
860	860	13450	0.22	0.01	0.01	352	581	581

Non-Conventional Materials and Technologies – NOCMAT for XXI Century   Materials Research Forum LLC
Materials Research Proceedings 7 (2018) 35-44   doi: http://dx.doi.org/10.21741/9781945291838-5

The FEM connection was composed of 54927 elements following a stress-based convergence analysis. The mesh consisted mostly of the eight-node hexahedral elements except in some places where tetrahedrons were automatically defined by the software. The interference applied on all semi-ring pairs was 7 mm. A static non-linear analysis was performed with two loading steps. The first step simulated the tightening of the bolts that closed the semi-rings completely around the culms. The second step simulated joint loading. Contacts among culms and semi-rings were defined with a tangential behavior using the penalty method and with a coefficient of friction of 0.35, other contacts were assumed bonded.

All steel components were modeled with a plastic hardening model. Properties were assumed similar to those that are reported (Table 3) for A-36 steel (Payne, 2000).

TABLE 3 - PLASTIC HARDENING MODEL PARAMETERS INCLUDED FOR ALL STEEL COMPONENTS

Reference stress σ_0 (MPa)	Kinematic hardening parameter C_1 (MPa)	γ
351.6	3447	50

To compare the performance of the proposed connection with traditional GAK joints, results of static tests of the ECMF_3 connection (Figure 4) reported by Camacho et al. [32] were used to calculate the average stiffness, strength and ductility.

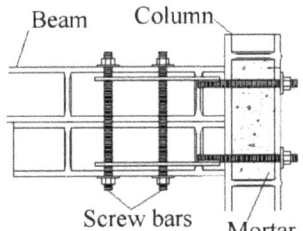

FIGURE 4 - MOMENT CONNECTION FOR GAK, NAMED ECMF_3 BY CAMACHO ET AL.
[32]

Results

The initial stiffness of the connection was determined following a procedure similar to that used in timber connections [29], where the secant slope of the moment curve was determined at 10% and 40 % of the maximum moment. The average initial stiffness of the connection was 89980 N.m/rad (COV = 0.35). Curves of moment versus rotation angle (Figure 5) exhibited an initial high stiffness until the semi-ring began to slip causing a reduction of the slope.

Non-Conventional Materials and Technologies – NOCMAT for XXI Century   Materials Research Forum LLC
Materials Research Proceedings 7 (2018) 35-44   doi: http://dx.doi.org/10.21741/9781945291838-5

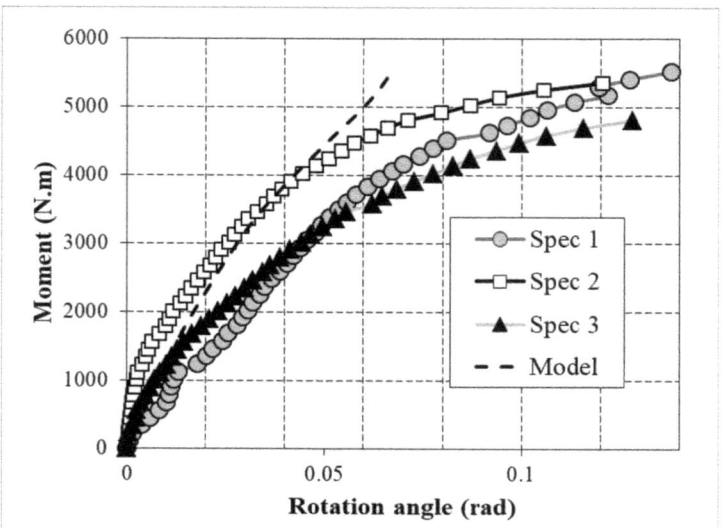

***FIGURE 5** - CURVES OF MOMENT VERSUS ANGLE OF ROTATION OF THE CONNECTION*

Specimen II showed the higher stiffness (125981 N.m/rad), while Specimen I exhibited the lower value (64940 N.m/rad). The high dispersion in initial stiffness is thought to be due to wide differences in clearances in the assembly holes. Finite element curve shows a stiffness 14.7% higher than the average experimental. Based in Eqn. 1, the threshold stiffness to consider the connection as articulated is 13886 N.m/rad (considering a beam span of 2 m), therefore this joint can be considered rigid or partially restrained depending of the beam span.

The average connection ductility was 2.8 (COV =0.34), which means that this connection has a low ductility based on the classification presented by Brühl [28] for timber connections. Average connection strength was 5234 N.m (COV = 0.07). Steel components of the connection did not fail while GAK members failed in all tests out of the connection zone (Figure 6). This is not a surprising result, since bending failures in GAK culms are expected at a lower moment, e.g., for a culm with a diameter and thickness of 110 mm and 10 mm respectively, the admissible bending moment is 1157 N.m (as prescribed by the Colombian Seismic- Resistant Construction Norm [14] where the admissible bending strength of the GAK is 15 MPa).

Compared with the connection ECMF_3 reported by Camacho et al. [32], the stiffness, strength and ductility of the proposed connections were 6%, 373% and 595% higher respectively.

Non-Conventional Materials and Technologies – NOCMAT for XXI Century   Materials Research Forum LLC
Materials Research Proceedings **7** (2018) 35-44   doi: http://dx.doi.org/10.21741/9781945291838-5

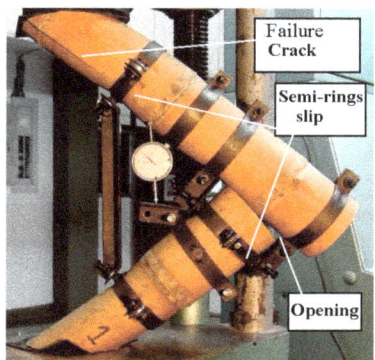

FIGURE 6 - FAILURE OF SPECIMEN NUMBER ONE

FEM curve was near to the stiffer moment-angle experimental curve (Figure 5). High stress concentrations in GAK were shown to be near the contact zones and areas of load transmission (Figure 7).

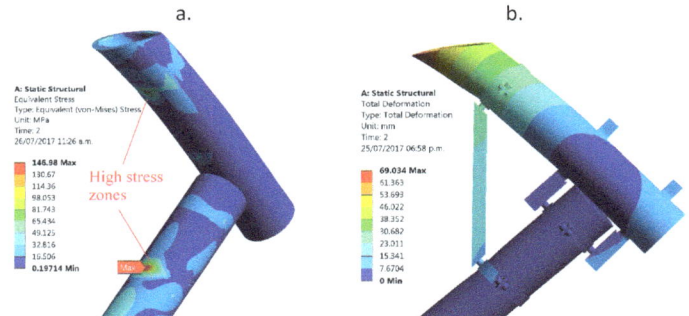

FIGURE 7- FEM RESULTS, A. VON MISES STRESS CONTOURS ON GAK, B.DISPLACEMENT CONTONURS

Discussion and conclusions

A new beam-column moment resisting connection for structural GAK members was presented. This connection is based on thin and light steel semi-rings, which submit the GAK culms to a high radial compression. Semi-rings are not harmful to GAK integrity, on the contrary, they avoid longitudinal splitting and premature failure of the culms as verified in simulations and experiments [22, 23, 24], and as shown in this study where all failures occurred out of the connection.

In general, the mechanical behavior of this connection is suitable compared with that of other bamboo moment connections analyzed in other studies [19, 32, 24]. This joint is not articulated. It can be considered as partially restrained or rigid depending on the beam span. Results of strength and ductility of the present connection were substantially higher to those of the ECMF_3 connection reported by Camacho et al. [31], which can be considered representative of the typical GAK connections that use curved cuts, screw bars and mortar injection. In addition, the semi-rings are light and do not add considerable weight to the structure.

Although the stiffness of the proposed connection is only 6% higher than that of the ECMF_3 joint, the difference is larger (221%) when the secant stiffness of the two connections is compared in a moment range of 0 to 1400 N.m which is the range allowed by the Colombian Norm [14] for a typical GAK culm.

Connections based on semi-rings are amenable to build prefabricated GAK structures since semi-rings can be constructed in varied sizes to cover the wide dimensional variation of the culms and they can be easily installed in the culms even without typical fish-mouth cuts.

Our current efforts are oriented at improving the performance of these connections and complete quasi-static tests. Since GAK is beautiful, low-cost and sustainable material, our hypothesis is that by improving its construction process will appreciate the material. In that way, farmers will find economic incentives to increase the cultivated areas.

Acknowledgments

The first author is grateful to Administrative Department of Science, Technology and Innovation COLCIENCIAS for financial support through national doctoral grant No. 617. The other authors greatly appreciate the support given by Universidad del Valle to undertake this study.

References

[1] P. Laroque, Design of a low cost bamboo footbridge. Master Thesis, Massachusets: Massachusets Institute of Technology, 2007.

[2] G. Singh, Richa y M. Sharma, «Bamboo - A Miracle Plant,» *International Journal of Current Research in Biosciences and Plant Biology,* vol. 4, n° 1, pp. 110-112, 2017. https://doi.org/10.20546/ijcrbp.2017.401.013

[3] J. Janssen, «Designing and building with bamboo. Inbar, Technical report No. 20,» INBAR, Eindhoven, 2000.

[4] P. Van der Lugt, A. Van den Dobbelsteen y J. Janssen, «An environmental, economics and practial assessment of bamboo as a building material for supporting structures,» *Construction and Building Materials,* pp. 648-656, 2006. https://doi.org/10.1016/j.conbuildmat.2005.02.023

[5] R. Dhillon y G. Wuehlisch, «Mitigation of global warming through renewable biomass,» *Biomass and bioenergy,* pp. 75-89, 2013. https://doi.org/10.1016/j.biombioe.2012.11.005

[6] H. Archila, M. Ansell y P. Walker, «Low carbon construction using Guadua bamboo in Colombia,» *Key Engineering Materials,* pp. 127-134, 2012. https://doi.org/10.4028/www.scientific.net/KEM.517.127

[7] J. Janssen, Bamboo in building structures. Doctoral dissertation, Eindhoven: Eindhoven University of Technology, 1981.

[8] R. Moran, K. Weeb, K. Harries y J. García, «Edge bearing tests to assess the influence of radial gradation on the transverse behavior of bamboo,» *Construction and Building Materials,* pp. 574-584, 2017. https://doi.org/10.1016/j.conbuildmat.2016.11.106

[9] L. Villegas, R. Moran y J. García, «A new joint to assemble light structures of bamboo slats,» *Construction and Building Materials,* pp. 61-68, 2015. https://doi.org/10.1016/j.conbuildmat.2015.08.113

[10] K. Ghavami, C. Rodrigues y S. Paciomik, «Bamboo: funtionally graded composite material,» *Asian journal of civil engineering (Building and housing)*, pp. 1-10, 2003.

[11] K. Ghavami y A. Marinho, «Propiedades físicas e mecánicas do colmo inteiro do bambu da espécie Guadua angustifólia,» *Revista Brasileira de Engenharia Agrícola e Ambiental*, pp. 107-114, 2005. https://doi.org/10.1590/S1415-43662005000100016

[12] J. Robledo, Un siglo de bahareque en el antiguo Caldas, Texas: Ancora Editores, 1993.

[13] D. Salas, Actualidad y futuro de la arquitectura de bambú en Colombia. Tésis de doctorado, Barcelona: Universidad Politénica de Cataluña, 2006.

[14] AIS, Reglamento colombiano de Construcción Sismo Resistente NSR-10, Bogotá: AIS, 2010.

[15] O. Arce-Villalobos, Fundamentals of the design of bamboo structures. Doctoral dissertation, Eindhoven: Eindhoven University of Technology, 1993.

[16] M. Vahanvati, «The Challenge of connecting bamboo,» de *10th World Bamboo Congress*, Damyang, 2015.

[17] L. Moreira y K. Ghavami, «Limits states analysis for bamboo pin connections,» *Key Engineering Materials*, pp. 3-12, 2012. https://doi.org/10.4028/www.scientific.net/KEM.517.3

[18] K. Ghavami y L. Moreira, «Development of a new joint for bamboo space structures,» *Transactions on the Built Environment*, pp. 3-12, 1996.

[19] C. Davies, Bamboo connections, London: Bath University, 2008.

[20] International Standard Organization, «ISO 22156:2004 (E) Bamboo - Structural design,» ISO, Geneva, 2004.

[21] S. Andrade, F. Lamus y N. Torres, «Connections between a column and its foundation for Guadua angustifolia structures under lateral loads,» *Key Engineering Materials*, pp. 227-237, 2015. https://doi.org/10.4028/www.scientific.net/KEM.668.227

[22] C. Benitez, Comportamiento mecánico de culmos de Guadua angustifolia ajsutados con abrazaderas metálicas, Cali: Universidad del Valle, 2017, p. 75.

[23] R. Moran, C. Benitez, H. Silva y J. García, «Desing of steel connector for structural bamboo members,» de *AMDM 2016 Terce Congreso Internacional Sobre Tecnologías Avanzadas de Mecatrónica, Diseño y Manufactura*, Cali, 2016.

[24] R. Morán, J. Muñoz, H. Silva y J. García, «Conexiones resistentes a momento para culmos de Guadua angustifolia,» de *VI Simposio Internacional del Bambú y la Guadua*, Bogotá, 2016.

[25] E. Oberg, F. Jones, H. Horton y H. Ryfel, Machinery´s Handbook, New York: Industrial Press Inc., 1997.

[26] S. Maleki, A. Rostampsour, D. Mosayeb, M. Faezipour y M. Tajvidi, «Bending moment resistance of corner joints constructed with spline under diagonal tension and compression,» *Journal of Forestry Research*, vol. 23, n° 3, pp. 481-490, 2012. https://doi.org/10.1007/s11676-012-0288-7

[27] A. Awaludin, Y. Sasaki, A. Oikawa y T. Hirai, «MOMENT RESISTING TIMBER JOINTS WITH HIGH STRENGTH STEEL DOWELS: NATURAL FIBER REINFORCEMENTS,» de *WCTE World Conference on Timber Engineering*, Riva del Garda, 2010.

[28] J. García, C. Rangel y K. Ghavami, «Experiments with rings to determine the anisotropoc elastic constants of bamboo,» *Construction and building of materials,* pp. 52-57, 2012. https://doi.org/10.1016/j.conbuildmat.2011.12.089

[29] F. Bruhl y U. Kuhlmann, «CONNECTION DUCTILITY IN TIMBER STRUCTURES,» de *World Conference on Timber Engineering,* Auckland, 2012.

[30] Eurocode 3, ENV 1993-1-1 Design of steel structures, United Kingdon: ENV, 1993.

[31] M. Richard y K. Harries, «On inherent bending in tension tests of bamboo,» *Wood Science and Technology,* vol. 49, n° 1, pp. 99-119, 2015. https://doi.org/10.1007/s00226-014-0681-9

[32] V. Camacho y I. Páez, Estudio de conexiones en guadua solicitadas a momento flector, Bogotá: Universidad Nacional de Colombia, 2002.

Non-Conventional Materials and Technologies – NOCMAT for XXI Century   Materials Research Forum LLC
Materials Research Proceedings 7 (2018) 45-53   doi: http://dx.doi.org/10.21741/9781945291838-6

Concretes Prepared with High-Density Limestone: Influence on the Transition Zone

Eduardo da Cruz Teixeira[a,*], Camila Macêdo Medeiros[a], Emerson Renildo da Silva Santos[a], Ulisses Targino Barbosa[b], Normando Perazzo Barbosa[c]

[a]IF Sertão PE – campus Salgueiro, Brazil; educrtx@hotmail.com,
camilamedeirosm@gmail.com, emerson_renildo@hotmail.com

[b]IFPB, Brazil; dartarios@yahoo.com.br

[c]UFPB, Brazil; nperazzob@yahoo.com.br

Abstract. The comprehension of the microstructure of concretes provides subsidies and improvement tools of its properties, mainly of durability and mechanical strength. Therefore, this project aims its research to a macro and microstructural analysis in concretes prepared with high-density limestone from the Cariri region of Ceará - Brazil. A comparative analysis with concretes prepared with granite gravel and low-density limestone was performed as a secondary objective. The concretes were prepared using the mix ratio 1:2:2:0.5. For the conventional concrete, granite was used as coarse aggregate and the alternatives concrete were prepared with high and low-density limestone. The test specimens were submitted to the following evaluations: XRD, in order to identify the phases of the constituent materials of the used aggregates; XRF, for the identification of the chemical elements of the aggregates; test of compressive strength of the concretes, the real specific density and SEM of the concrete samples for the analysis of the transition zone. The two types of high density limestones found in the Cariri deposits located in Brazil, presented CaO in their chemical composition predominating in 73% for white limestone and 65% for gray limestone, but the gray type presented in its composition 4.5% silica whereas the white type presented only 1.84%. Concretes prepared with high-density limestone presented an average compressive strength of 16% higher than the conventional concretes. The micrographs obtained in the SEM showed a transition zone with a small number of cracks or fissures and low porosity in the concretes prepared with limestone, in the concretes prepared with granite gravel were found cracks along its interface and accentuated porosity.

Keywords: Concrete, Limestone, Microestructure

Introduction

The concrete has been used by humanity for a long time, since the ancient times, from that time, the idea of making a paste that, as it hardens, becomes an artificial stone, has been developing and improving itself through new research that happens throughout the cycle.

This work addresses a research of analysis in the most important material in the Civil Construction Industry, the concrete, due to its large volume used in comparison to the other materials, regarding the analysis of the macro and microstructural properties.

According to (Mehta e Monteiro [1]), the concrete is the most used element on civil constructions.

As mentioned by (Neville [2]), The Portland cement concrete is a porous, heterogeneous and complex material, there are two main constituents in its microstructure: the hardened cement paste and the aggregates, basically the simple concrete is formed by mixing cement, sand, gravel and water. However, (Farran [3]), through his studies proved that exists another phase, the transition zone.

Non-Conventional Materials and Technologies – NOCMAT for XXI Century   Materials Research Forum LLC
Materials Research Proceedings 7 (2018) 45-53   doi: http://dx.doi.org/10.21741/9781945291838-6

According to (Monteiro [4]), the scientific studies that address the improvement of the concrete must take into account the relevance of its microstructure, which has a direct connection with the improvement of the general properties of the material.

Researches has shown that the relationship between the concrete microstructure and its properties, especially in the transition zone, interface among the cement paste and the aggregate, considered the most fragile of concrete, according to (Mehta and Monteiro [1]). These authors affirms that the identification and behavior of the constituent phases of the concrete in its microstructure generally offers possibilities of mechanisms capable of improving the general properties of the concrete.

Studies developed by (Kaefer [5]) showed that the porosity and heterogeneity in relation to the rest of the paste are the characteristics of the transition zone, as well as being poor in the C-S-H compound (hydrated calcium silicate), a partially crystalline structure responsible for the concrete mechanical strength.

This work was carried out through a macro and microstructural analysis of concrete specifically in the transition zone, for verification of the morphological formations.

The aggregate used in this research, to replace the conventional aggregate of our region, was the limestone of high strength and density from the state of Ceará, where some tests have already been carried out with the same material, but without the investigation by electronic microscopy of the transition zone in concrete prepared with this aggregate.

The specific objectives of this work were the characterization of the raw materials; comparison of the microstructure of concretes prepared with distinct aggregates, specifically in the transition zone; and analysis of the compressive strength of the prepared concretes.

Methodology

Concrete

(Mehta and Monteiro [1]) explained that concrete is defined as a composite material consisting of a binder medium in which particles or fragments of the aggregate are bonded, where the binder is the mixture of cement and water, for hydraulic cement concrete.

According to (Kaefer [5]), a very important component of the concrete is the aggregate, which are, the coarse and fine, acquired directly from the nature (sand and gravel) or even by the reuse of industrial and urban waste such as blast furnace slag and recycled concrete from demolitions.

The paste and the aggregate are distinct parts and have different characteristics, considering the coarse aggregate as chemically inert, the durability of the concrete depends on the paste and the transition zone (paste / coarse aggregate) and the cohesion between them.

According to data from the Departamento Nacional de Produção Mineral – DNPM, the region where the deposits of high-strength limestone are found has approximately 97 million of cubic meters, equivalent to 241 million tons, encompassing 13 municipalities, being Nova Olinda and Santana do Cariri the most important.

(Ribeiro et al. [6]) the limestone of the Cariri region of Ceará has a high mechanical strength, around 120 MPa, especially due to its crystalline formation, which gives it high hardness.

The research of (Ribeiro et al. [6]) used high-density limestone as a new mineral input in the asphalt composition, resulting in good physical results of its use in paving.

Material and methods

Materials

In this research, limestone rocks from the state of Ceará were used as aggregate as shown in Figure 1, where, according to (Ribeiro et al. [6]), exists a typical limestone formation of high mechanical strength.

FIGURE 1- HIGH-DENSITY DEPOSIT OF LIMESTONE, FARIAS BRITO – CE

During the sample collections were found two differentiated formations of the high-density limestone, one of light coloration, known in the region as "pedra branca", and the other of rocky limestone formation with gray coloration, known as "pedra cinza". For reasons of detail of its composition and the influence in the transition zone of the prepared concretes with these two types of high density limestone, we used these two types found in the Cariri region of Ceará and used it independently for analysis. Figures 2 and 3 show the different aspects, for easy in loco identification, of the aggregates used in this research.

From the local deposits of the city of João Pessoa, the conventional low-density limestone was used in the research in order to compare the results obtained in the analysis.

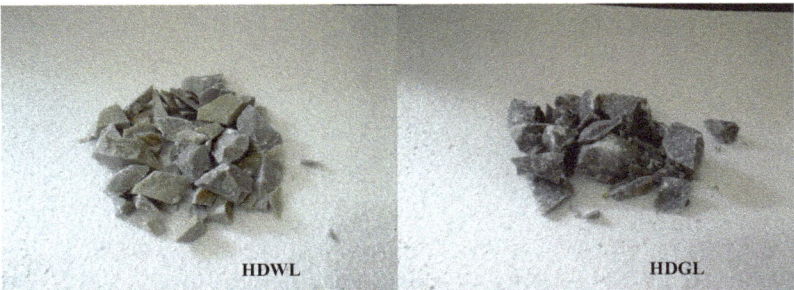

FIGURE 2 - HDWL (HIGH-DENSITY WHITE LIMESTONE); GRAY HIGH-DENSITY
LIMESTONE).

For the concrete preparation were used the Portland cement CPII Z - 32 MPa, natural quartz sand, granite gravel with maximum characteristic dimension of 9.5 mm and the unconventional aggregates studied: HDWL, HDGL, LDL (low-density limestone).

For each concrete type, six test specimens were produced: they were prepared with the four coarse aggregate types for each one and with water/cement factors equal to 0.5, to analyze the behavior of the transition zone with the different aggregates. The specimens suggested by (Rossignolo [7]) were of size 35 mm in diameter X 70 mm in height, as shown in Figure 4.

Non-Conventional Materials and Technologies – NOCMAT for XXI Century   Materials Research Forum LLC
Materials Research Proceedings **7** (2018) 45-53   doi: http://dx.doi.org/10.21741/9781945291838-6

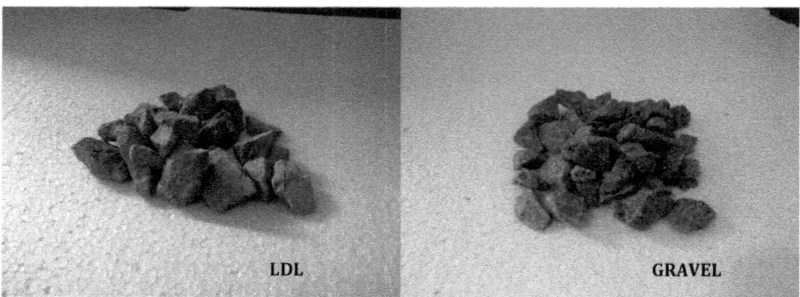

FIGURE 3 – LDL (LOW-DENSITY LIMESTONE); GRAVEL (GRANITE GRAVEL).

FIGURE 4 – TEST SPECIMENS USED IN THE RESEARCH.

As shown in Table 1 for each test specimens, were used the following quantities of material;

TABLE 1 – MATERIAIS UTILIZADOS NOS CORPOS DE PROVA.

Materials	Weight (g)
Cement	30.70
FA[1]	61.41
CA[2]	61.41
Water	15

[1] Fine Aggregate [2] Coarse Aggregate

48

Methods
The preparation of the concrete samples followed the methodology suggested by (Rossignolo [7]) with adaptations. In his thesis this author carried out researches in the transition zone of concretes prepared with expanded clay, with the methodology that will be used in this research. This is used as standard procedure at the Núcleo de Química (NQ) of Departamento de Materiais de Construção (DMC) of the Laboratório Nacional de Engenharia Civil in Lisbon, Portugal, where were carried out the characterization essays of the samples of Rossignolo.

For the best comparative analysis all the concretes were prepared with the same mix ratio 1:2:2: 0.5. Firstly, physical, chemical and mineralogical characterization of the raw materials were performed, such as: X-ray fluorescence, specific density, X-ray diffraction and pH.

For the preparation for XRD, XRF, pH and specific density, the raw samples of the aggregates used in the research were washed in running water with the aid of a brush. The samples were placed in a ventilated stove at 35 °C for 5 days.

After drying the samples, they were submitted to a manual grinding with the use of steel hammer, then the crushed material went to the Mill, with alumina grinding balls. Was used a ratio of 1:4, the ratio of sample density/ball density, approximately 60 g of the sample with 240g of alumina balls.

The HDWL, HDGL, and the granite gravel were submitted to gridding for 45 minutes, and the LDL was submitted to gridding for 35, due to its less hard and dense characteristics. The powders produced by the milling of the aggregates (samples) passed through a mechanical sieving, all samples were placed in ceramic bowls of high resistance with the alumina balls in a volume of 500 ml.

The powder, sieved through the ABNT nº 200 (0,074 mm) sieve, was stored to perform the XRF, XRD and pH tests, and were performed at the Laboratório de Caracterização Engenharia de Materiais da UFCG.

In the preparation for the analysis in the SEM (secondary electrons mode), the concrete samples were cut in dimensions of 3.0 cm x 3.0 cm, with 1.0 cm of thickness.

After the cutting, the samples were immersed in isopropyl alcohol for 24 hours in order to interrupt the hydration of the cement matrix. The samples were then submitted to the drying process, where remained for 12 hours in a ventilated stove, at temperature of 35°C and 24 hours in a desiccator at a temperature of 23°C ± 2°C.

Results and discussion

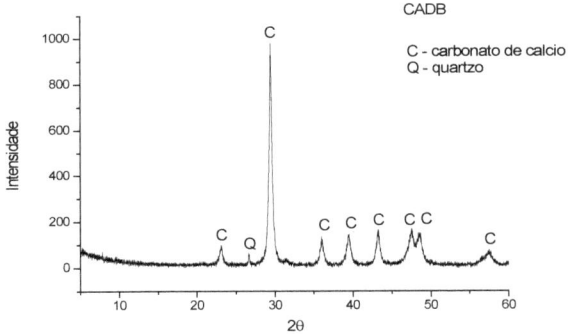

FIGURE 5 – XRD OF THE HIGH-DENSITY WHITE LIMESTONE. (HDWL, intensidade = intensity, calcium carbonate- C, Q-quatz)

Non-Conventional Materials and Technologies – NOCMAT for XXI Century   Materials Research Forum LLC
Materials Research Proceedings 7 (2018) 45-53   doi: http://dx.doi.org/10.21741/9781945291838-6

Analyzing the diffractogram of Figure 5, it is verified that the high-density white limestone is formed by the following mineralogical phases: calcium carbonate with basal interplanar distance of 3.03 Å and quartz with basal interplanar distance of 3.34 Å.

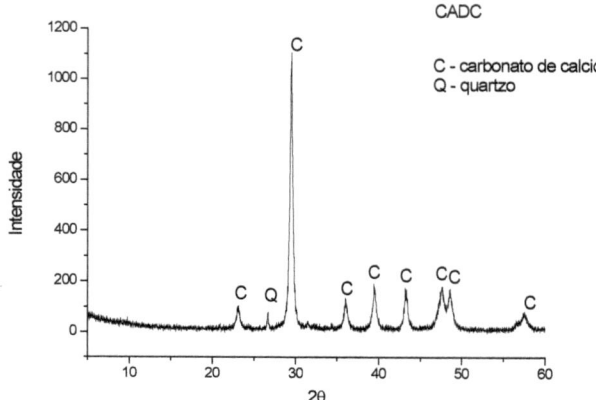

FIGURE 6 – XRD OF HIGH-DENSITY GRAY LIMESTONE (HDGL)

Analyzing the XRD curve of Figure 6, it is verified that the HDGL is formed by the following mineralogical phases: calcium carbonate with basal interplanar distance of 3.03 Å and quartz with basal interplanar distance of 3.34 Å.

TABLE 2 – CHEMICAL COMPOSITION OF THE SAMPLES

Compost	LDL1	HDGL	HDWL	GRANITE GRAVEL
CaO	58.67	65.27	73.34	1.65
CO_2	29.51	25.86	23.24	0.00
SiO_3	6.07	4.58	1.84	77.92
MgO	2.08	1.30	0.00	0.48
Al_2O_3	1.73	1.22	0.66	11.61
Fe_2O_3	0.64	0.47	0.07	1.61
K_2O	0.49	0.23	0.03	2.66
Na_2O	0.00	0.16	0.00	3.54

[1] Percentage values (%)

The chemical composition of the aggregates studied in this research is in agreement with the results of the literature, as well as with the expectations.

Non-Conventional Materials and Technologies – NOCMAT for XXI Century   Materials Research Forum LLC
Materials Research Proceedings **7** (2018) 45-53   doi: http://dx.doi.org/10.21741/9781945291838-6

The concretes prepared with the different types of aggregates were submitted to the compressive strength test on specimens with a size of 35 mm diameter X 70 mm high. Were prepared 6 test specimens of each concrete type and obtained an arithmetic mean of the results.

Table 3 shows the results of mechanical strength of the concretes prepared with the different aggregates.

TABLE 3 – MECHANICAL STRENGTH OF THE CONCRETES

Concrete type	Strength (MPa)
HDWL	16.89
HDGL	22.95
LDL	15.91
GRAVEL GRANITE	19.24

The concrete prepared with the conventional aggregate, the gravel granite, had an average mechanical strength of 19.24 MPa, as predicted, the concrete prepared with the low-density limestone, presented mechanical strength of 15.91 MPa, approximately 17. 30% less resistant, the concretes prepared with the high-density limestone presented positive results, the use of high-density gray limestone stood out with a 16.16% increase in the mechanical strength in comparison to the conventional, the use of high-density white limestone presented a small increase in the mechanical strength compared to the conventional concrete.

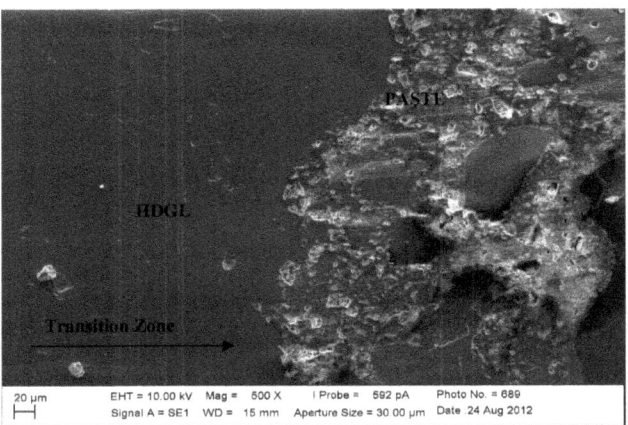

FIGURE 7 – MICROGRAPHY BY SECONDARY ELECTRONS OF THE TRANSITION ZONE OF CONCRETE PREPARED WITH HDGL (MAGNITUDE: X500)

The micrograph by secondary electrons of the transition zone of concrete prepared with HDGL (High-Density Gray Limestone) presented, in general, a transition zone with relative uniformity, which is visualized in Figure 7, porous, but still without fissures or cracks. The uniformity of the

Non-Conventional Materials and Technologies – NOCMAT for XXI Century   Materials Research Forum LLC
Materials Research Proceedings **7** (2018) 45-53   doi: http://dx.doi.org/10.21741/9781945291838-6

transition zone of this concrete was good, however, porosity were found along with the transition zone at the interface.

The micrograph by secondary electrons of the transition zone of concrete prepared with HDWL (High-Density White Limestone) presented, in general, a transition zone very similar to concrete prepared with HDGL: relatively uniform, as can be seen in Figure 8, without cracks or fissures, however, the transition zone of this concrete had less regions with porosity.

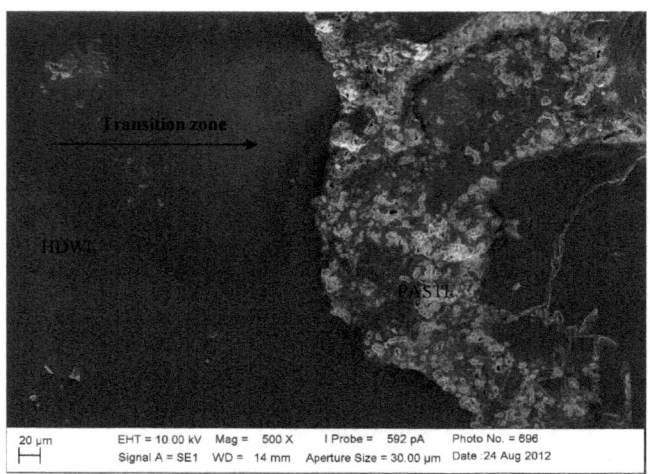

FIGURE 8 – *MICROGRAPHY BY SECONDARY ELECTRONS OF THE TRANSITION ZONE OF CONCRETE PREPARED WITH HDWL (MAGNITUDE: X500).*

Conclusions

In relation to the concretes prepared with conventional aggregate, the use of limestone of high resistance as coarse aggregate in concretes alters the formation and characteristics of the transition zone, reducing porosity, presenting a small number of cracks or fissures and better uniformity.

The use of low-density limestone as a coarse aggregate in concretes also alters the formation and the characteristics of the transition zone in comparison to the concretes prepared with conventional aggregate: much lower porosity, excellent interaction between paste and aggregate, absence of cracks or fissures, however, it has an inferior mechanical strength than the conventional concretes.

In agreement to the characteristics of the transition zone, the mechanical strength of the concretes prepared with the two types of high-density limestone were superior in comparison to the concrete prepared with the conventional aggregate, the granite gravel, the HDGL presented better resistance, 16 % higher than the concrete with granite gravel.

References

[1] MEHTA, P. K.; MONTEIRO, P. J. M. Concreto: microestrutura, propriedades e materiais. São Paulo: PINI, 2008.

[2] NEVILLE, A. M. Propriedades do concreto. São Paulo: PINI, 1997.

Non-Conventional Materials and Technologies – NOCMAT for XXI Century   Materials Research Forum LLC
Materials Research Proceedings **7** (2018) 45-53   doi: http://dx.doi.org/10.21741/9781945291838-6

[3] FARRAN, J. Contribution minérlogique á l'adhérence entre constituants hidrates dês cimens et lês materiaux enrobés Revue dês materiaux des constrctions, 1956.

[4] MONTEIRO, P. J. M. Caracterização da microestrutura do concreto: Fases e interfaces; aspectos de durabilidade e de microestrutura. Tese livre docência, Escola Politécnica da Universidade de São Paulo, São Paulo, 1993.

[5] KAEFER, L. F. BRASIL. Escola politécnica da USP. Considerações sobre a microestrutura do concreto. São Paulo, 2002 (Boletim Técnico).

[6] RIBEIRO, R. C. C. Utilização do calcário do Cariri cearense como agregado mineral em pavimentação asfáltica. Rio de Janeiro: CETEM, 2006.

[7] ROSSIGNOLO, J. A. Concreto leve de alto desempenho modificado com SB para pré-fabricados esbeltos: dosagem, produção, propriedades e microestrutura. 2003. Tese (doutorado em Ciências e Engenharia de Materiais), Escola de Engenharia de São Carlos, Universidade de São Paulo, São Carlos.

Non-Conventional Materials and Technologies – NOCMAT for XXI Century   Materials Research Forum LLC
Materials Research Proceedings 7 (2018) 54-64   doi: http://dx.doi.org/10.21741/9781945291838-7

Behaviour under Bending Loads of Workable Bamboo Bio-Concrete

Andreola. V. M.[1, a], Da Gloria. M. Y. R.[1, b], Toledo Filho. R. D.[1, c]

[1] Federal University of Rio de Janeiro, Brasil

[a]vanessaandreola@gmail.com

[b]ydagloria@yahoo.fr

[c]toledo@coc.ufrj.br

Abstract. This work presents the results obtained in an experimental program that evaluated the flexural behavior of workable bio-concretes of bamboo. Leftovers and fragments were reduced to particles using an industrial grinder and a knife mill. The resulting material was used in its entirety, including the fines passed through the 1.18 mm sieve. The cement: bamboo ratios used were 1: 3.0 and 1: 2.5 corresponding to a bio-aggregates volume of 45 and 50%, respectively. The water to cement (W/C) ratios studied were 0.40 and 0.50. The use of fines bio-aggregates and viscosity modifying agent (VMA) helped to control the exudation and segregation of the fresh bio-concretes which presented spreading of 290 mm ± 20mm. The modulus of rupture (MOR) at 28 days varied from 1.26 to 1.85 MPa while the modulus of elasticity (MOE) varied from 1.42 to 2.76 GPa. There was significant influence of particle volume and W/C ratio on the hardened properties of the bio-concretes. There was a post-fissuration behavior with gradual reduction of load and increase of deformation, demonstrating the benefit of the particles as reinforcement element and control of cracking.

Keywords: Bio-Concrete Bamboo Particles, Workability, Deflection Load

Introduction

In the last decade, there was a growing concern of the national and international scientific community about the development of bio-concretes using agroindustrial residues. These materials can be obtained both by replacing the cementitious materials and the aggregates. In the specific case of the replacement of the aggregates, the use of residues of vegetal biomass stands out.

Due to the diversity and characteristics of his soil and climate, Brazil is a country with a tendency and potential for agricultural production. This results in a significant occurrence of agroindustrial residues [Manzatto *et al.*, 2002]. The cumulative volumes of residues stimulated the researches on alternatives for the use and application of these materials in the production of composites for various sectors of civil construction [Cassilha *et al.*, 2004].

The research for the construction system optimization is increasing due to the need of cost and waste reduction. The cement composites reinforced with bio-aggregates are part of products   able to improve the industrialization of the civil construction, since they can be used in the execution of several constructive components [Amziane *et al.*, 2017].

Among the available plant materials, bamboo is a raw material that is being exploited by several industries. Featuring good mechanical characteristics, the processing of the material can be used in the construction and furniture making [Nogueira 2008]. However, the bamboo lamination process produces a considerable amount of waste that corresponds to 40% of the stem mass. These residues can give rise to new products, such as agglomerated, particulate, structural or insulation panels which have characteristics and properties similar to those normally marketed [Kravchenko 2013].

Non-Conventional Materials and Technologies – NOCMAT for XXI Century   Materials Research Forum LLC
Materials Research Proceedings 7 (2018) 54-64   doi: http://dx.doi.org/10.21741/9781945291838-7

Biomass particles are usually characterized by good insulating properties due to the high porosity of their structure. They are highly hygroscopic and have low density. When combined with cement, they allow the production of lightweight concrete without overloading the structure, but allowing good insulation and thermal inertia [Nozahic *et al.*, 2012].

The extractives present in the composition of the biomass are the main responsible for the inhibition of cement solidification. [Pomárico 2013] observed, for example that the inhibition of the cement paste was due to the use of this natural biomass. Due to the crystallinity of the cellulose, it may not react with the cement, while lignin has no inhibitory effect and simple sugars (glucose, galactose and mannose) may inhibit hydration by different mechanisms [Pimienta, 1994; Mendes 2015].

To minimize or even remove all the inhibitory substances present in biomasses, [Sarmiento 1996] proposed treatments of different natures to reduce chemical incompatibility. Among the technical options, the author emphasized washing, roasting and impregnation. [Beraldo 2011], [Da Gloria 2015] and [Andreola 2017] adopted as treatment, the hot washing of the biomass to remove the water-soluble extractives. The washing was carried out at 80 °C for 1 hour and the procedures followed were adequate, with no delay in the handle of the bio-concretes.

[Wolfe *et al.*, 1999] produced cementitious panels with wood particles and observed that residues combined with cement showed good mechanical strength and were indicated for use in external construction panels.

Usually, bio-concretes are composed of cement, biomass in the particles form, water and chemical additives. The cement acts as binder, providing strength and durability, while the biomass improve the lightness, the energy absorption capacity, and the thermal insulation.. The additives act as set facilitators [Matoski 2013].

An important property of bio-concretes is the production method. Both nacional and international manufacturing procedures are based on manual or industrial pressing. These procedures require specific systems and equipment that demand high energy at each stage.

In the manual pressing process, the fresh bio-concrete is compacted manually in the molds using sockets. In the process of industrial pressing, biomasses are mixed with cement and water. According to [Moslemi 1999], the whole forms a dry mattress and is cold pressed with the aid of presses that reach certain temperatures and loads. After pressing the mattresses are stapled and go through the healing process.

In these pressing processes, there are continuous energy expenditures. In the case of manual pressing, the energy used cannot be quantified and reproduced with precision and control. Another production alternative is to use conventional techniques of concrete production. In this way, the energy consumption will be lower and consequently a high workability and moldability will be obtained.

In this context, this study proposed a new method to design bio-concretes with high workability. This method turns the pressing of fresh material unnecessary, and offers possibilities of application of bio-concretes that use different sources of vegetal biomass. The validation of the technique at fresh state of the mixture were done through the spreading test. The influence of the particles volume and the W/C ratios were accessed by the analysis of the mechanical behavior under 3 points flexural stress.

Materials and methods

Bio-aggregate: raw material
In this study, bamboo of the specie *Dendrocalamus asper* coming from Rio Grande do Sul (Brazil) was used. The aggregates of bamboo were obtained from the waste of the stems cutting process. Once in the laboratory, the stems wastes (length around 45 cm) were reduced first in particles with

Non-Conventional Materials and Technologies – NOCMAT for XXI Century   Materials Research Forum LLC
Materials Research Proceedings 7 (2018) 54-64   doi: http://dx.doi.org/10.21741/9781945291838-7

length between 5 and 19 mm using an industrial crusher. Then, in a knife mill, these particles were refined to obtain aggregates with a maximum diameter of 4.75 mm. (Figure 1).

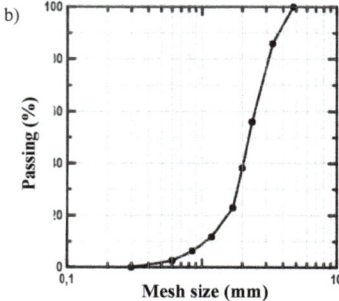

FIGURE 1 - A) AGGREGATES OF BAMBU; B) GRANULOMETRIC CURVE THE AGGREGATES.

Specific mass and apparent density (NM 52/2009), water absorption (NM 53/2003) and moisture content (NM 9939/2011) were used to characterize the bio-aggregates . The values found are shown in Table 1.

TABLE 1 - SPECIFIC OVERALL AND APPARENT MASS, WATER ABSORPTION AND MOISTURE CONTENT OF BAMBOO BIO-AGGREGATES.

Specific mass (g/cm^3)	Apparent Density (g/cm^3)	Water absorption (%)	Moisture content (%)
0.58	1.43	110.57	11.01

Bio-aggregates treatment

Considering that cement setting may be inhibited when combined with plant biomass, the bio-aggregates were washed in hot water to remove the water-soluble extractives from the material. The washing was carried out at 80 °C for 1 hour with a ratio biomass: water of 1: 10, as suggested by [Beraldo 1994]. In order to access the amount of the water-soluble extractives and also the most efficient washing number, 10 washings were done according to the methodoly proposed by [da Gloria & Toledo 2016].

After each washing, the waste water was separated, filtered through the 150 μm mesh, refreshed and stored in plastic bottle of 350 ml. From the waste water, 3 samples of 35 g were weight to the nearest 0.1 mg into a recipient. Then the samples were dried at 40° ± 2°C during 48 h in order to eliminate the water by evaporation and determine the residual extractives mass. The extractives contents were calculated as expressed in the Equation 1.

$$Ce = \frac{m_1}{m_2} x100$$ (1)

Where:
Ce = concentration of extractives (%)
m_1 = mass of the extractives after drying (g)

m_2 = mass of waste water (g)

Design of the bamboo bio-concretes (BCB)
As binder, the Portland cement labelled CP V - ARI with high initial strength and density of 3.17 g/cm^3, was used. In order to improve the workability and control the segregation, a viscosity modifying agent (VMA) named Rheomac UW 410 was used.

For the production of the bio-concretes, the cement: bamboo ratios of 1: 3.0 and 1: 2.50, corresponding to bio-aggregates volumes of 45 and 50%, respectively, and water to cement ratio (W/C) 0.4 and 0.50 were studied.

To produce the bio-concretes, it is fundamental to use rational methods to compensate the absorption water of the bio-aggregates. In order to improve the production of bamboo bio-concretes with high workability (290 mm ± 20 mm), the water compensation of the mixture was established based on the concepts suggested by [Wolfe 1999], [Souza 2006] and applied in researches of [Santos et al., 2017] and [Andreola 2017].

The water in mixtures with biomasses should be enough to keep the biomass saturated, to allow the cement hydration and also to guarantee the consistency of the mixture. Then, the total amount of water was determined considering the water needed to hydrate the cement and an additional amount (called compensating water) that varied depending on the water absorption capacity of the biomass used.

Regarding the segregation of the fresh mixtures, it was added 15% of fines bio-aggregates (maximum diameter of 1.18 mm), combined with 0.125% of VMA and absorption water corresponding to the mass gain of 24 hours were adequate to control exudation and segregation, resulting in mixtures of workable bio-concretes. The mass relations between the materials used are shown in Table 2.

TABLE 2 - *MASS RATIO OF THE MATERIALS USED IN EACH MIXTURE.*

Mixture	Cement	Bamboo	W/C	Compensating water	VMA
BCB 2.50	1	0.40	0.50	0.515	0.00125
BCB 3.00	1	0.33	0.50	0.427	0.00125
BCB 2.50	1	0.40	0.40	0.515	0.00125
BCB 3.00	1	0.33	0.40	0.427	0.00125

Production of bamboo bio-concretes (BCB)
The bio-concretes were produced in a mixer of 80 liters capacity. First, the dried bamboo particles were mixed with the cement 2 minutes, followed by the gradual addition of the total water for 2 minutes. After 5 minutes, the VMA was progressively added and the total time of mixing was 8 min. This time was enough to obtain a homogeneous mixture.

The fresh bio-concretes were molded in 90 x 25 x 10 cm (length, width, thickness) prismatic molds in three layers. Each layer was vibrated mechanically on a vibrating table (68 Hz, T = 30 seconds). The bio-concretes were protected against moisture loss and demolded after 24 hours. After demolding the specimens were cured in a humid chamber (UR = 100% ± 2%, T = 21 °C ± 2 °C) where they remained until reaching 28 days of age.

From the specimens of 90 x 25 x 10 cm, were extracted smaller many prismatic specimens of 35 x 5 x 1.5 cm (length, width, thickness) according to the European standard EN 310/1993 (Figure 2 a).

Non-Conventional Materials and Technologies – NOCMAT for XXI Century   Materials Research Forum LLC
Materials Research Proceedings 7 (2018) 54-64   doi: http://dx.doi.org/10.21741/9781945291838-7

The bending test was performed using the Shimadzu AGX-100 kN universal test machine with displacement velocity of 0.1 mm/min. The central deflection were measured through a Linear Variable Differential Transformer (LVDT) coupled to a device, positioned at the middle of the specimen. The span was 25 cm and the test assembly illustrated in the Figure 2 (b).

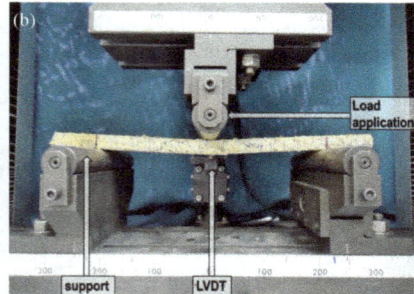

FIGURE 2 *- (A) PRISMATIC SPECIMENS. (B) STATIC BENDING TEST ASSEMBLY.*

The rupture modulus (MOR) and the modulus of elasticity (MOE) were determined according to the European standard EN 310/1993 while the toughness indices were calculated according to the method proposed by [Rilem 1984].

$$MOR = \frac{3 \times F_{max} \times l_1}{2 \times bt^2} \tag{2}$$

$$MOE = \frac{l_1{}^3 (F_2 - F_1)}{4 \times bt^3 (a_2 - a_1)} \div 1000 \tag{3}$$

$$T = \frac{A}{b \times t} \tag{4}$$

Where:
MOR = flexural strength (MPa)
MOE = modulus of elasticity (GPa)
F_{max} = maximum load (N)
l_1 = distance between supports (mm)
b = width of the body of evidence (mm)
t = thickness of the body of evidence (mm)
F_2 = 40% of the maximum load (N)
F_1 = 10% of the maximum load (N)
a_2 = deflection corresponding to 40% of the maximum load (mm)
a_1 = deflection corresponding to 10% of the maximum load (mm)

Results and discussions

Bio-aggregate: quantification of extractives
After the washing cycles, a gradual variation of the waste water color was observed. The first waste water was very dark and as the washings were done, the water got clearer (Figure 3).

Non-Conventional Materials and Technologies – NOCMAT for XXI Century   Materials Research Forum LLC
Materials Research Proceedings 7 (2018) 54-64   doi: http://dx.doi.org/10.21741/9781945291838-7

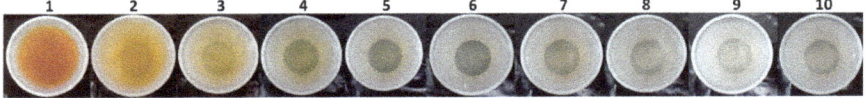

FIGURE 3 - WASTE WATER AFTER THE 10 WASHING CYCLES.

As observed by [da Gloria 2016] and [Andreola 2017], the change of color can be justified by reduction of biomass extractives. According to the Figure 5, the reduction of the extractives is proportional to the number of washings, with significant reduction of the extractives after the first cycles. After the third washing, about 88.9% of the extractives was removed, while after the sixth cycle, 97% of the extractives was removed. Based on this result, it was decided to wash the bio-aggregates three times before using them for the production of bio-concretes. After the washings, the bio-aggregates were placed in a room at 40 °C (± 2 °C) for 48 hours, time required to return to natural moisture.

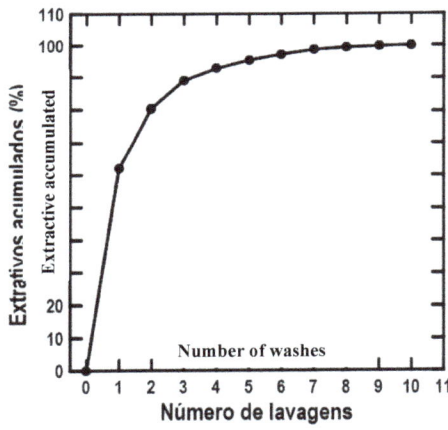

FIGURE 4 - EXTRACTIVE ACCUMULATED VERSUS WASH CYCLES.

Workability of bamboo bio-concretes (BCB)

The spreading of the fresh bio-concretes are shown in the Figure 6 and the consistence indexes in the Table 4. According to the results, the bio-concretes showed spreading around 290 mm ± 20 mm as desired. A good workability was observed without exudation and segregation. Therefore, it can be affirmed that the production of bio-concretes of easy moldability and rationality using the compensating water and the dry bio-aggregate is feasible.

Non-Conventional Materials and Technologies – NOCMAT for XXI Century   Materials Research Forum LLC
Materials Research Proceedings 7 (2018) 54-64   doi: http://dx.doi.org/10.21741/9781945291838-7

FIGURE 5 - *FLOW TABLE TEST: (A) BEGINNING OF THE TEST (B) SPREADING OF THE BIO-CONCRETES.*

TABLE 3 - *CONSISTENCE INDEXES OF THE BIO-CONCRETES.*

Mixture	W/C	Consistence (mm)
BCB 2.50	0,50	292
BCB 3.00	0,50	310
BCB 2.50	0,40	280
BCB 3.00	0,40	285

Density and homogeneity of bamboo bio-concretes (BCB)
The average values of the densities of the bio-concretes produced are presented in Table 5. The densities were analyzed in the condition of the saturated bio-concrete, removed from the humid chamber (U = 100% and T = 21°C).

TABLE 4 - *DENSITIES OF THE BIO-CONCRETES.*

Mixture	W/C	ρ (kg/m³)
BCB 2.50	0,50	1058.65
BCB 3.00	0,50	1093.60
BCB 2.50	0,40	1081.94
BCB 3.00	0.40	1190.15

In the saturated condition, the bio-concretes presented variations from 1058.65 to 1190.15 kg/m³. [Da Gloria 2015] studying wood bio-concretes of cement to wood ratio 2.5 and 3.0, obtained density of 1100 kg/m³ and 1250 kg/m³, respectively. [Macêdo 2012] produced bio-concretes of density 1330 kg/m³ and 1557 kg/m³ for the especies Cedar and Brazilian Cherry, respectively. It should be mentioned that the densities found by the cited researchers are higher than those of the present study, due to the molding method used, where the compaction occurred by manual or industrial pressing. The pressing reduce the voids and improve consequently the densities.

Regarding the densities obtained (lower than 1800 kg/m³), the bio-concretes can be classified as lightweight, according to the [Rilem 1978] recommendations. The low densities found can be explained by the use of high volume of lightweight bio-aggregates, a high total water to cement ratio, and the molding process, which did not reduce the voids. The pores at the hardened state can be observed in the Figure 7.

Non-Conventional Materials and Technologies – NOCMAT for XXI Century   Materials Research Forum LLC
Materials Research Proceedings **7** (2018) 54-64   doi: http://dx.doi.org/10.21741/9781945291838-7

FIGURE 6 - (A) SMOOTH SURFACE (B) FRACTURED SURFACE

Three-point flexural strength of bamboo bio-concretes (BCB)

The average values of the modulus of rupture (MOR) and modulus of elasticity (MOE) are presented in the Table 5. The load versus the central deflection curves are shown in the Figure 8. From the curves, it can be seen that after the peak load there was not a sudden break, but a post-cracking behavior with gradual reduction of load and increased deformation. The bamboo particles acted as reinforcement and controlled the crack opening under bending.

TABLE 5 - MOR, MOE AND TOUGHNESS OF THE BIO-CONCRETES, WITH THE RESPECTIVE COEFFICIENTS OF VARIATION (% IN PARENTHESES).

Mixtures	**W/C**	**MOR (MPa)**	**MOE (GPa)**	T_f (kJ/m²)
BCB 2.50	0.50	1.26 (8.70)	1.42 (6.91)	0.04 (8.71)
BCB 3.00	0.50	1.78 (6,59)	2.41 (8.24)	0.09 (4.43)
BCB 2.50	0.40	1.46 (3.40)	1.88 (3.19)	0.06 (6.11)
BCB 3.00	0.40	1.85 (5.06)	2.76 (9.24)	0.11 (5.74)

The bio-concrete BCB 2.50/0.40 presented MOR 13.6% and MOE 24.4% higher than 2.50 W/C 0.50. The BCB 3.00 W/C 0.40 presented MOR 3.7% and MOE 12.6% higher than 3.00 W/C 0.50. BCB 3.00 W/C 0.50 presented MOR 29.2% and MOE 41% higher than 2.50 W/C 0.50. The BCB 3.00 W/C 0.40 presented MOR 21% and MOE 31.8% higher than the 2.50 W/C 0.40. The BCBs 3.00 W/C 0.40 and W/C 0.50 presented higher MOR and MOE due to the higher cement volume in the set.

Regarding the toughness, the results indicated that its increase was higher for the mixture with lower particle volume and higher W/C ratio (BCB 3.00/0.40). For the BCB 2.50 and 3.00/0.40 mixtures there were a toughness increase of 0.89%.

For illustrative comparison purposes, the researches using biomass and cement with industrial and manual pressing methods were selected. For example, [Souza 2006] obtained MOR of 4.72 MPa and 7.56 MPa and MOR of 5.24 and 5.65 GPa for BCB 3.00 with particles of wood of the species Jatobá and Quaruba, respectively. [Marzuki 2011] obtained MOR and MOE of 8.95 MPa and 4.0 GPa when using wood particles and BCB 2.50. In these surveys, both the MOR and the MOE found are higher than the values of this research. This can be explained by the method of molding adopted, where the pressing reduced the voids and increased the strength of the bio-concretes.

Non-Conventional Materials and Technologies – NOCMAT for XXI Century   Materials Research Forum LLC
Materials Research Proceedings 7 (2018) 54-64   doi: http://dx.doi.org/10.21741/9781945291838-7

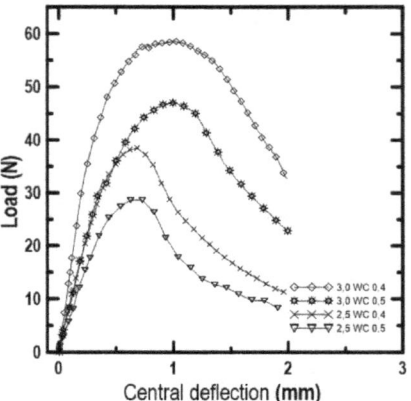

FIGURE 7 - LOAD-DEFLEXION CURVES

Conclusion

This study presented the results of an experimental campaign in which workable bio-concrete of bamboo were produced. The results indicated that it is possible to produce bio-concretes of high workability through the rational dosage, adopting necessary quantities of water in the mixture. It was also concluded that the addition of fines (diameter low than 1.18 mm) combined with the viscosity modifying agent were important to control both exudation and segregation.

For the removal of the extractives from the bamboo biomass, the study showed that three washing in hot water (T = 80°C for 1h) guaranteed a significant reduction of 88.9% of the extractives. In this way, it was not observed delay in the cement setting, allowing the demolding of the bio-concretes after 24 hours.

Based on the density of the bio-concretes, all the mixtures (analyzed in the saturated condition) can be classified as light materials, according to the classification recommended by the association [Rilem 1978]. The tensile modulus showed a modulus of modulus of elasticity (MOR) ranging from 1.26 to 1.85 MPa, while modulus of elasticity (MOE) ranged from 1.42 to 2.76 GPa, higher in the bio-concretes that presented lower volumes of bio-aggregate and W/C, which is the case of BCB 3.00 W/C 0.40, where the value obtained was 0.11 kJ/m².

References

[Amziane 2017], Amziane, S..; Collet, F.; Bio-aggregates Based Building Materials. RILEM State-of-the-Art Reports (2017).

[Andreola 2016], Andreola, V. M.; Da Gloria, M. Y. R.; Toledo Filho, R. D. II Congresso Luso-Brasileiro de Materiais de Construção Sustentáveis, Universidade Federal da Paraíba João Pessoa, PB (2016).

[Andreola 2017], Andreola, V. M. Caracterização física, mecânica e ambiental de bio-concretos de bambo. Tese de Mestrado, Programa de Pós Graduação em Engenharia Civil, Universidade Federal do Rio de Janeiro, RJ (2017).

Non-Conventional Materials and Technologies – NOCMAT for XXI Century   Materials Research Forum LLC
Materials Research Proceedings 7 (2018) 54-64   doi: http://dx.doi.org/10.21741/9781945291838-7

[Beraldo 1994], Beraldo, A. L. Généralisation et optimisation de la fabrication d'un composite biomasse végétale-ciment à variations dimensionnelles limitées vis-à-vis des variations de l'humidité. Tese de Doutorado, Université Henri Poincaré, Nancy 1, France (1994).

[Beraldo 2011], Beraldo, A. L. Compuestos de biomasa vegetal y cemento. Aprovechamiento de residuos agro-industriales como fuente sostenible de materiales de construcción, v. 1, pp. 301-326 (2011).

[Cassilha 2004], Cassilha, A. C.; Podlasek, C. L.; Junior, E. F. C.; Silva, M. da;. C.; Mengatto. S. N. F.. Indústria moveleira e resíduos sólidos: considerações para o equilíbrio ambiental. Revista Educação & Tecnologia, n. 8 (2004).

[Da Gloria 2015], Da Gloria, M. Y. R. Desenvolvimento e caracterização de painéis sanduíches de concreto com núcleo leve e faces em laminados reforçados com fibras longas de sisal. Tese de Mestrado, Programa de Pós Graduação em Engenharia Civil, Universidade Federal do Rio de Janeiro, RJ (2015).

[Da Gloria & Toledo 2016], Da Gloria, M. Y. R.; Toledo Filho, R. D Influence of the wood shavings/cement ratio on the thermo-mechanical properties of lightweight wood shavings-cement based composites. 6th Amazon & Pacific Green Materials Congress and Sustainable Construction Materials LAT-RILEM Conference (2016).

[Kravchenko 2013], Kravchenko, G. A.; Ferreira, E. M.; Pasqualetto, A. Utilização de Resíduos do Processamento de Chapas Laminadas de Bambu para Produção de Chapas Recompostas. 4th International Workshop Advances in Cleaner Production (2013).

[Macêdo 2012], Macêdo, A. N. et al. Chapas de cimento-madeira com resíduos da indústria madeireira da Região Amazônica (2012).

[Manzatto 2002] Manzatto, C., V.; Freitas Junior, E. de; Peres, J. R. R. Uso agrícola dos solos brasileiros. Embrapa Solos. Infoteca-E, (2002) 174 p.

[Matoski 2013], Matoski, A. Utilização de pó de madeira com granulometria controlada na produção de painéis de cimento-madeira (2013).

[Mendes, 2015], Mendes, L. M., Loschi, F. A. P. L., De Ramos, L. E., et al. Potencial de utilização da madeira de clones de Eucalyptus urophylla na produção de painéis cimento-madeira, Cerne, v. 17, n. 1, pp. 69-75 (2015).

[Moslemi 1999], Moslemi, A. A. Emerging technologies in mineral-bonded wood and fiber composites. Advanced Performance Materials, v. 6, n. 2, p. 161-179 (1999).

[Nogueira 2008], Nogueira C., de L. Painel de bambu laminado colado estrutural. Tese de Doutorado. Universidade de São Paulo (2008).

[Nozahic 2012], Nozahic, V., Amziane, S., Torrent, G., Saïdi, K., & De Baynast, H. Design of green concrete made of plant-derived aggregates and a pumice–lime binder. Cement and Concrete Composites, 231-241 (2012).

[Pimienta 1994], Pimienta, P. Étude de faisabilité des procédés de construction à base de béton de bois. CSTB (1994).

Non-Conventional Materials and Technologies – NOCMAT for XXI Century   Materials Research Forum LLC
Materials Research Proceedings 7 (2018) 54-64   doi: http://dx.doi.org/10.21741/9781945291838-7

[Rilem 1978], Rilem, L. C. Functional classification of lightweight concrete. Mater. Struct, v. 11, p. 281-283 (1978).

[Rilem 1984], Rilem, T. C. 49 TFR. Testing methods for fibre reinforced cement-based composites. v.17, n.102, 15p. (1984).

[Santos 2017], Santos, D.O.J; da Gloria M. Y. R.; Andreola, V. M.; Pepe, M.; Toledo Filho R. D. (2017) Compressive stress strain behavior of workable bio-concretes produced using bamboo, rice husk and wood shavings particles. 2nd International Conference on Bio-based Building Materials & 1st Conference on ECOlogical valorisation of GRAnular and FIbrous materials, Clermont-Ferrand, France (2017).

[Sarmiento 1996], Sarmiento, C. Argamassa de Cimento Reforçada com Fibras de Bagaço de Cana de Açúcar e sua Utilização como Material de Construção. Tese de Mestrado, Faculdade de Engenharia Agrícola, Universidade Estadual de Campinas, São Paulo (1996).

[Souza 2006], Souza, A. A. C. Utilização de resíduos da indústria madeireira para fabricação de chapas cimento-madeira. Dissertação de Mestrado. Universidade Federal do Pará (2006).

[Wolfe 1999], Wolfe, R. W.; Gjinolli, A. Durability and strength of cement-bonded wood particle composites made from construction waste. Forest Products Journal, v. 49, n. 2, p. 24 (1999).

Non-Conventional Materials and Technologies – NOCMAT for XXI Century   Materials Research Forum LLC
Materials Research Proceedings 7 (2018) 65-84   doi: http://dx.doi.org/10.21741/9781945291838-8

Deferred Deflexions with Time of Bamboo *Guadua Angustifolia* Kunth under Permanent Loads

Sergio Chavarro [a], Caori Takeuchi [b*]

[a] Universidad Nacional de Colombia, Colombia; sachavarrom@unal.edu.co

[b] Universidad Nacional de Colombia, Colombia: cptakeuchit@unal.edu.co

Abstract. In the research "Deflections of *Guadua angustifolia* Kunth culms under bending stresses with permanent loads in the environment of Bogota. Burger's and Findley's models", it was found, that compared with the Burgers' models, the Findleys' models were more appropriate to describe de curve of deferred deflections in time for bamboo *Guadua angustifolia* Kunth under permanent loads with different stress levels. To determine this, bending tests in guadua culms with short and long duration loads were performed. In the short-duration tests, the failure stress was determined. On the other hand, long-duration tests were performed on samples with different stress levels ranging between 95% and 30% of the average failure stress obtained from a representative short-duration tests with a standard speed loading. Deferred deflections and failure times of the long-duration samples were also registered. In some cases, for culms with low stress level, the load was retired after 280 days without that the failure of the culms had presented. The Burger and Findley's models were applied to each one of the time-deflection curves of the long duration tests. The Findley's and Burger's models estimated with records of deflections of the total duration of the tests, had correlation coefficients of 0.99 and 0.88 respectively.

Keywords: Duration of Load, Permanent Loads, Deferred Deflections, Bamboo *Guadua angustifolia* Kunth

Introduction

Renewable materials attract attention for be an option to diminish the negative impact of construction industries that exploits natural resources. A new alternative is the bamboo *Guadua angustifolia* Kunth (GAK) that helps to preserve bodies of water and decreases the levels of carbon dioxide of the environment. In addition, it is a material that has engineering potential due to its resistance, flexibility and low cost of obtaining and being a renewable material, as long as there is a responsible exploitation within the framework of sustainable forestry.

Currently the GAK is a material that has shown to be useful in the regional development of Colombia and other countries of the Andean region. It has been employed in constructions although in the past it has had little technical support. Today there are design specifications for structures that use this material. However, it is important to study the behaviour of this material under special circumstances.

One of these circumstances is the effect of the permanent loads on the flexural elements since, even its failure stress values have been found [1], [2] and the behaviour of the long-duration deformations in time has not been studied. Another aspect is that, currently, the Reglamento Colombiano de Construccion Sismo Resistente NSR-10 [3], considers that the allowable stress of the GAK is affected by the same factor of duration of the load of the wood. Also, the NSR-10 considers for GAK, combinations of increased loads for the evaluation of the deferred deflections. Both design criteria are not supported by researches with long duration tests with failure and time deformation records for GAK culms.

Non-Conventional Materials and Technologies – NOCMAT for XXI Century   Materials Research Forum LLC
Materials Research Proceedings 7 (2018) 65-84   doi: http://dx.doi.org/10.21741/9781945291838-8

In the research work "Deflections of *Guadua angustifolia* Kunth culms under bending stresses with permanent loads in the environment of Bogota. Burger's and Findley's models" [4] short and long duration bending tests in bamboo guadua culms were performed to study the behaviour of their deflections and failure times, under permanent loads. With the average failure stress found in a representative short duration tests at a standard loading speed, the permanent loads values to which the long duration specimens would be subjected, were established according to an expected level of stress. With the recording of the deflections of the middle of the beams, the deferred deflections in time of the GAK were modelled using the mathematical models of Burger and Findley.

Regarding to the effect of the load duration on the resistance, the failure times and stress levels of each test were recorded and contrasted with the curve that describes the relationship between the duration of the load and the level of stress in the woods.

EXPERIMENTAL METHODOLOGY

Material

The samples of the short and long duration bending tests were made using bamboo guaduas with ages ranging from 2.5 to 3.0 years and without any chemical treatment or industrialized drying. The guaduas were cleaned and air-dried during 60 days.in vertical position under covered condition in the Bogota City environment

Short duration bending test

The short duration bending test was a standard four-point bending test defined in the NTC-5525 [5].

The samples had a length ranges between 2.7 and 4.6 meters, external diameters between 5.5 and 10.9 centimetres and wall thicknesses between 0.8 and 2.8 centimetres. The length between the supports was greater than 30 times the external diameter according to NTC 5525.

The application of the load was made on the nodes closest to the thirds of the length of the culm. The Figure 1 shows the typical assembly used in the short term tests.

FIGURE 1 - *ASSEMBLY USED IN THE SHORT DURATION TEST. SORCE [4]*

A set of 23 samples with a standard loading speed (0.5mm /s) and a set of 5 samples with a load application speed of 2.0 in/min (0.846mm/s) were tested. Deflections were recorded at the points of application of the load and in the centre of the beam. The Figure 2 shows the nomenclature of the short duration tests.

Non-Conventional Materials and Technologies – NOCMAT for XXI Century   Materials Research Forum LLC
Materials Research Proceedings 7 (2018) 65-84   doi: http://dx.doi.org/10.21741/9781945291838-8

FIGURE 2 - NOMECLATURE OF THE SHORT DURATION TEST. SORCE [4]

After testing and having the deflection and load registers, two samples near the failure section were obtained to determine the moisture content of the bamboo when the test was carried out.

Long duration bending test

In order to calculate the load to be applied, some expected levels of stress were established. The stress level was the percentage of the stress exerted on the long-duration sample and the average failure stress (MOR) of the representative short duration samples tested with a speed of 0.5 mm/s. The lowest stress levels were to record the deflections over long periods of time, and the highest stress levels were to take readings of the failure time of the samples. The estimated stress levels in the test samples are shown in Table 1.

TABLE 1 - ESTIMATED STRESS LEVEL FOR LONG-DURATION TESTE SPECIMENS. SOURCE [4]

ESTIMATED STRESS LEVEL	NUMBER OF SAMPLES
30% - 40%	15
40% - 60%	6
>= 60%	5

The test consisted in leaving the loads applied permanently on the bamboo guadua elements and recording the deflections in the time and the failure time of the sample.

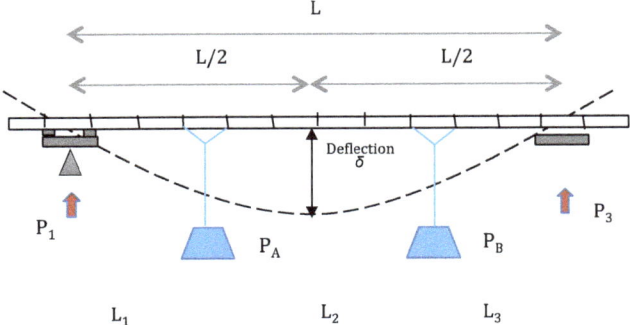

FIGURE 3 - SCHEME FOR THE ASSEMBLIES OF LONG DURATION TEST. SORCE [4]

Non-Conventional Materials and Technologies – NOCMAT for XXI Century   Materials Research Forum LLC
Materials Research Proceedings 7 (2018) 65-84   doi: http://dx.doi.org/10.21741/9781945291838-8

The Figure 3 shows the loading scheme, where two P_A and P_B loads were applied on the nodes closed to the thirds of the element, resulting in reactions at the ends, P_1 and P_3. The distances L_1, L_2 and L_3 correspond to the distances between supports and loads. The deflection δ was measured in the centre of the beam.

The Figure 4 shows the assemblies used in long duration test.

FIGURE 4 - ASSEMBLIES OF LONG DURATION TEST. SORCE [4]

A total of 26 samples with permanent loads were tested. The length of the different culms ranged between 5.0 and 6.4 meters and the external diameters from 5.0 to 9.5 centimetres. The diameters were relatively uniform throughout the element. All elements had the length to diameter ratio required for bending tests. Wall thicknesses varied from 0.4 to 2.4 centimetres. The Figure 5 shows the nomenclature of the long duration tests.

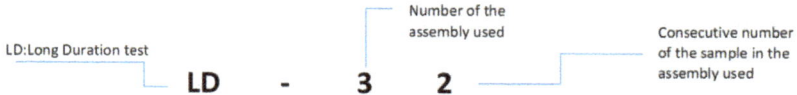

FIGURE 5 - NOMECLATURE OF THE LONG DURATION TEST. SORCE [4]

After 280 days, those specimens that had not previously failed were unloaded. Deflection records were taken for all specimens.

At the end of the test, the moisture content and the section dimensions were determined at the point of failure or at the centre of the element (if there was no failure).

Calculus

Failure stress (MOR) of the short duration test

It was determined the average failure stress (MOR_{prom}) of the short duration test of the samples loaded at a standard speed and with a moisture content less or equal than 20%.

The failure stress (MOR) was calculated with the Ecuation (1).

$$MOR = \frac{M_{failure}}{\frac{\pi}{32}\left(D^3 - \frac{(D-2t)^4}{D}\right)} \qquad (1)$$

Where $M_{failure}$ is the value of the failure bending moment, D the external diameter of the failure section and t the wall thickness.

Deflection in long duration test
With the deflection records the parameter k was calculated. The parameter k is the ratio between the deferred deflections and the initial deflection. For those samples that failed instantly there were not record of the deflections and for that reason the parameter k was not obtained.

The equation (2) relates the parameter k with the deferred deflections, $\delta_{deferred}$, and the initial deflection, $\delta_{inmediate}$.

$$\delta_{deferred} = k \cdot \delta_{inmediate} \tag{2}$$

Stress level for long duration test
For each sample it was determined the work stress level N.E during the test, as the ratio between the stress in the middle section σ_T, and the average failure stress, MOR_{prom}

The variation of the deflection speed
The decrement of the deflection speed of each specimen was calculated and verified. The total time of the test was considered and divided into 4 almost equal time intervals in order to analyse the behaviour of the specimen in the total duration of the test. With these time intervals and the deflections recorded in the limits of these intervals, the speed of the deflections was calculated.

Application of the Burger's and Findley's models to time-deflection curves for long-duration tests

Burger's model
In order to define the Burger's model it was necessary to select only the specimens that could be unloaded. The information collected from the deflections during the whole test time was considered, to know the elastic and viscous deformations.

Among the numerous viscoelastic creep models, the Burger model or four-element model, is widely used to analyse the viscous-elasticity of materials.

This model considers the elastic deformations, the viscous (time dependent) deformations that are not recovered and the deferred elastic deformations that consider recovery after discharge. The Figure 6 shows the physical model of the long-duration deflections where the deflections were attributed to an elastic behaviour (δ_e), viscous (δ_v) and deferred elastic (δ_{de}).

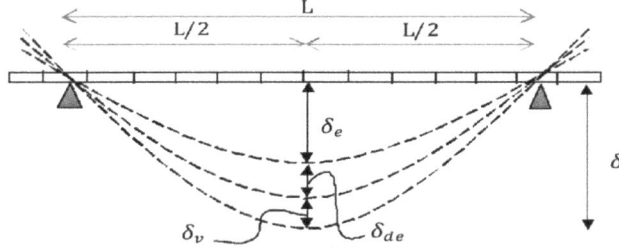

FIGURE 6 - *DEFLECTION IN LONG DURATION TEST. SORCE [4]*

Non-Conventional Materials and Technologies – NOCMAT for XXI Century   Materials Research Forum LLC
Materials Research Proceedings 7 (2018) 65-84   doi: http://dx.doi.org/10.21741/9781945291838-8

Taking into account that $L = L_1 + L_2 + L_3$ and $P = P_A + P_B$, the deflection δ and the load P were the variables to define the test model in the research.

The constitutive equation of the Burger's model are (equation (3)):

$$\delta = \delta_e + \delta_v + \delta_{de} = \frac{P}{k_e} + \frac{P}{r_v}t + \frac{P}{k_{de}}\left(1-e^{-t/\tau}\right) \tag{3}$$

Where δ is the total deformation, δ_e is the instantaneous elastic deformation, δ_v is the viscous deformation, δ_{de} is the deferred elastic deformation, k_e is the Maxwell's spring constant, r_v is the Maxwell's damping constant, k_{de} is the Kelvin's spring constant, τ is the relaxation-time constant and t is the time (Figure 7).

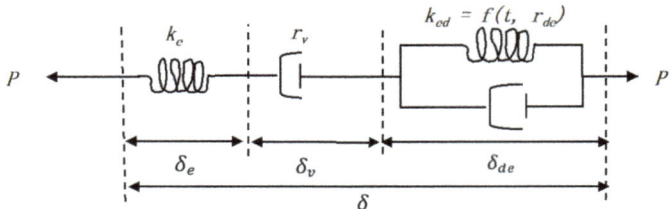

FIGURE 7 - *BURGER'S MODEL. SOURCE [4]*

The Figure 8 shows the Load deflection typical curve according with the Burger's model.

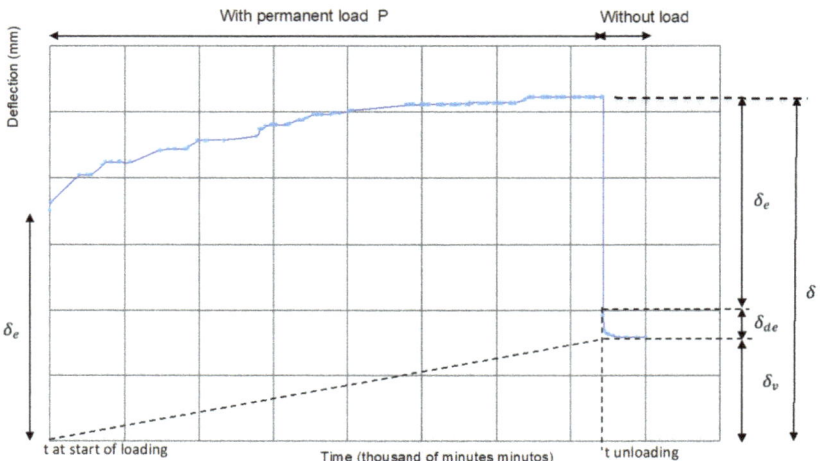

FIGURE 8 - *LOAD VS DEFLECTION TYPICAL CURVE (P VS.Δ). SOURSE [4]*

Non-Conventional Materials and Technologies – NOCMAT for XXI Century   Materials Research Forum LLC
Materials Research Proceedings 7 (2018) 65-84   doi: http://dx.doi.org/10.21741/9781945291838-8

Findley's model
The Findley model is based on the behaviour of materials such as polymers or metals where it has been used for describing their creep. It is a mathematical model that has empirically been associated with an empirical formula shown in the equation (4).

$$\delta = \delta_e + \delta_t \cdot t^n \qquad (4)$$

Where δ is the total deformation, δ_e is the instantaneous elastic deformation, δ_t the time dependent deformation, n is the potential constant (usually less than 1.0) and t is the time.

This mathematical equation fits with the materials that have a variable deflection rate and do not develop a constant rate of deformation over time

Relationship between stress levels and failure time of long-duration test samples
Investigations related the behaviour of the resistance of the woods with respect to the duration of the load and the speed of application of the load [6]. The literature references that short duration and impact tests, and long duration tests were independently associated with trend lines Wood [7] calculated all the data in a way that related the impact data, short and long duration with a unique empirical curve, which in this case was hyperbolic, obtaining the following expression, equation (5).

$$N.E. = \frac{108.4}{\left(t_{falla}\right)^{0.04635}} + 18.3 \qquad (5)$$

The t_{falla} is the failure time in seconds that is expected for a given stress level. This curve mainly was defined from three pairs of data: the first pair is for a duration of the load of 0.015 seconds with a level of stress of 150%; the second is a 100% stress level for a load duration of 7.5 minutes; and the third is an stress level of 69% for a load duration of 3750 hours (156 days). Due to the absence of similar information, this hyperbola was taken as the general expression of the relationship between resistance and duration of loading for the wood species most used in construction.

RESULTS

Failure stress of the short duration test
The failure stress and the moisture content of the short duration test are shown in the Table 2.

TABLE 2 - *FAILURE STRESS (MOR) AND MOISTURE CONTENT OF THE SAMPLE OF SHORT DURATION TESTS AT A STANDARD SPEED LOADING. SOURCE [4]*

SAMPLE	MOR (MPa)	MOISTURE CONTENT (%)
CD- I - 1	111.2	15.7%
CD- I - 2	67.5	27.0%
CD- I - 3	62.2	29.0%
CD- I - 4	62.2	37.2%
CD- I - 5	70.4	21.7%
CD- I - 6	69.5	13.8%
CD- I - 7	85.3	14.7%
CD- I - 8	74.9	25.4%
CD- I - 9	120.3	20.6%
CD- I - 10	79.4	16.8%

CD- I - 11	71.8	27.9%
CD- I - 12	66.0	16.0%
CD- I - 13	54.1	26.4%
CD- I - 14	85.4	24.4%
CD- I - 15	69.5	14.6%
CD- I - 16	55.8	39.7%
CD- I - 17	113.0	15.7%
CD- I - 18	115.7	16.8%
CD- I - 19	106.2	15.6%
CD- I - 20	75.7	15.1%
CD- I - 21	97.9	14.0%
CD- I - 22	64.6	16.1%
CD- I - 23	51.6	21.5%

The Figure 9 shows the records of a total of 224 couples of environment temperature and relative humidity data accompanied by an exponential trend curve. Based on the methodology proposed for the particular case of the woods [8], mean equilibrium moisture content (EMC) of 12.1% was determined with a standard deviation of 2.9% and a variance of 0.1% for the environment from Bogota. The Reglamento Colombiano de Construccion Sismo Resistente NSR-10, defines for the city of Bogotá an EMC value equal to 16%. With these values it was assumed that short duration tests with moisture contents less than or equal to 20% would be the most representative.

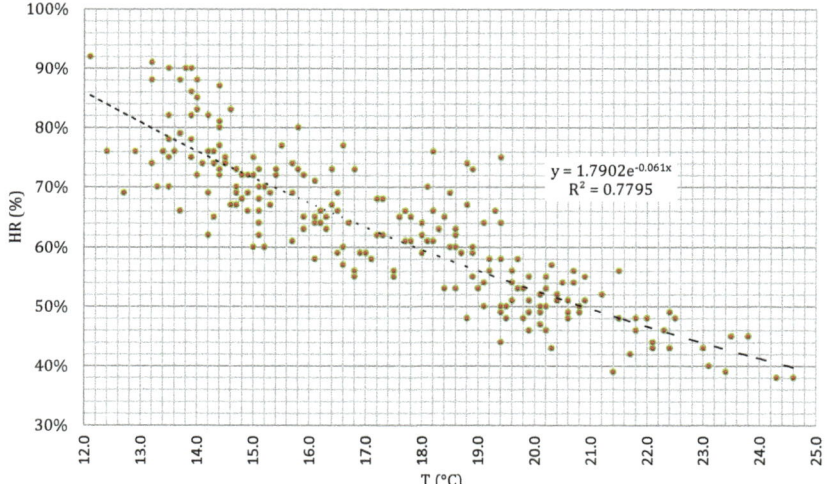

FIGURE 9 - ENVIROMENT TEMPERATURE VS. RELATIVE HUMIDITY SOURCE [4]

Although the incidence of moisture content in the bending tests has not been studied, there are other studies carried out by the Universidad Nacional de Colombia (Gutierrez, 2011) to find the moisture content modification factor for GAK for tensile strength. In this study, with a greater amount of experimental data (478), it was found that the tensile strength parallel to the fibre presents an important variation when the moisture content is below 10% or above 21% and that there is not a considerable variation between the humidity between these two points. Gutierrez also concludes that within the range of 11% to 21%, the material normally works as a structural element and a decrease of the resistance does not exceed 4%, which means that in this range of moisture content the tension strength parallel to the fibre is indifferent to the change in moisture content.

The Table 3 shows a summary of the results of the short duration tests at a standard loading speed for two groups: the first group with the 23 samples and the second with the samples with moisture content less that 20%.

TABLA 3 - STATISTICAL VALUES FOR MOR FOR THE 2 GROUPS OF SHORT DURATION TESTS LOADED AT A STANDARD SPEED. SOURCE [4]

Source: Chavarro (2016)

	ALL TEST	TEST WITH MOISTURE CONTENT < 20%
Maximun	120.3 MPa	115.7 MPa
Minimum	51.6 MPa	64.6 MPa
Q1	65.3 MPa	69.5 MPa
Q2	71.8 MPa	82.4 MPa
Q3	91.7 MPa	107.4 MPa
Number of data, n	23	12
Average, m	79.6 MPa	87.8 MPa
Standard Deviation, s	20.6 MPa	18.9 MPa

The average value of MOR_{prom} = 87.8 MPa was considered to calculate the stress level of the long duration samples.

Deferred deflection in the long duration test

The Figure 10 shows the variation of the factor k (the ratio between the deferred deflections and the initial deflection) over time for the unloaded bamboo guaduas. The curves of LD-11, LD-42, LD-62 LD-72 and LD-151 samples tended to be horizontal in the final third of the time of the test; these tests had a last value of $k_{ultimate}$ between 1.46 and 1.54; as these samples were not failed it is possible that the real last k values could be higher than the ones recorded. The samples LD-33, LD-91, LD-121 y LD-142 had a steeper initial slope that remained for longer time than in the other tests.

The Figure 11 shows the variation between the factor k over time for the failed bamboo guaduas, so, for these samples were possible to have the complete record of the deflections from the beginning to the moment of the failure. These failed samples are distinguished by having the steeper slopes in the first part of the test.

Among the three tests with a stress level N.E (Table 4) between 45% and 60%, the LD - 22 and LD - 51 samples (Figure 11) failed before develop the horizontal tendency; in contrast, the LD - 102 sample (Figure 10), was the only one that was unloaded and reached a nearly uniform stress with a last k of 1.72. The three curves had a deflection rate greater than the curves shown in the tests with stress levels less than or equal to 35%.

The test samples with stress levels greater than 80% N.E are characterized by a strongly steep slope and with deferred deflection speed greater than the other long duration tests. Additionally, it is observed that they do not develop a change in this velocity until failure. The last $k_{ultimate}$ ranged between 1.58 and 1.62.

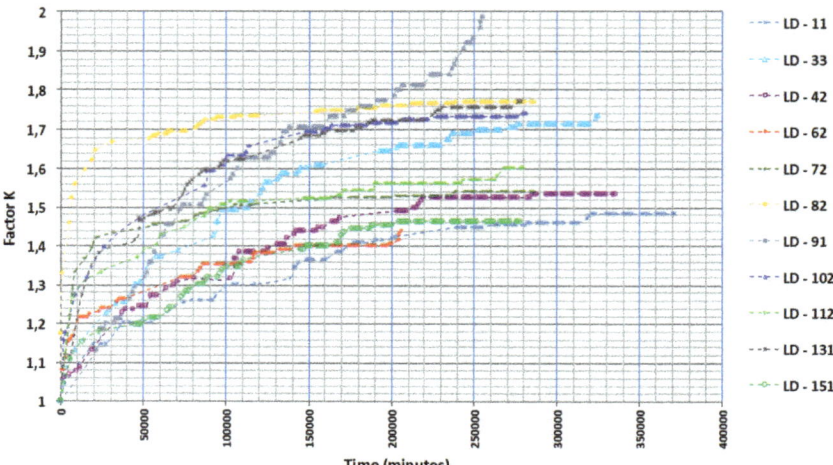

FIGURE 10 - KULTIMATE VS TIME FOR DIFFERENT STRESS LEVEL FOR THE UNLOADED SAMPLES SOURCE [4]

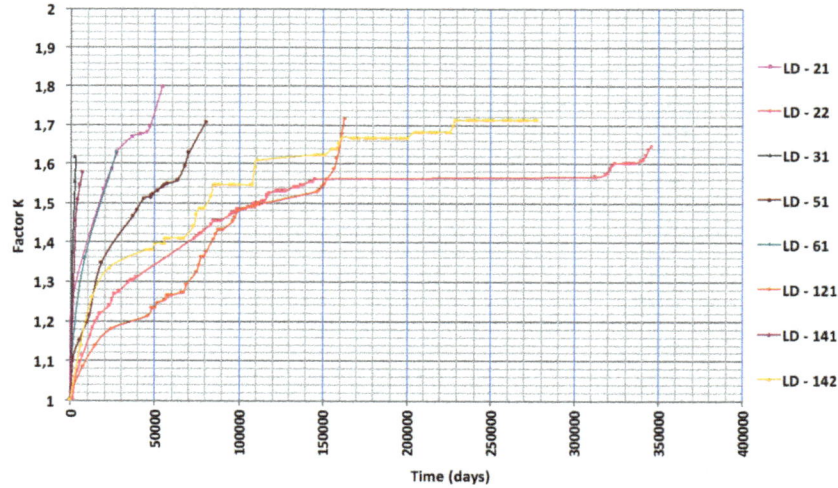

FIGURA 11 - KULTIMATE VS TIME FOR DIFFERENT STRESS LEVEL FOR THE FAILED SAMPLES SOURCE [4]

In general, it is observed that a stress level affects the change in deflections deferred over time.

Non-Conventional Materials and Technologies – NOCMAT for XXI Century   Materials Research Forum LLC
Materials Research Proceedings 7 (2018) 65-84   doi: http://dx.doi.org/10.21741/9781945291838-8

Stress level, failure time and moisture content of the long duration test
The Table 5 shows for each sample the bending moment, the stress and the stress level (where *N.E.* $_{real} = \sigma_{T\,real} / MOR_{prom}$). The duration of the test until failure or the test termination are also shown.

The average moisture content, CH, of the tests reported in the Table 5 was 15.8%. The only samples that did not have a moisture content less than 20% were the LD-71 that failure in 26 minutes (the moisture content had not been regulated with the environment after cutting) and the LD-121 (considered as atypical among all the data).

It is observed that the longest time of failure was 126 days at a stress level of 52.2%. Twelve specimens were unloaded so the time of failure was not recorded

TABLA 4 - STRESS LEVEL, FAILURE TIME AND MOISTURE CONTENT OF THE LONG DURATION SAMPLES. SOURCE [4]

Sample	C.H. (%)	Real values				Failure time		
		M	(KN-m)	σ$_T$ (MPa)	N.E. (%)	(minutes)	(hours)	(days)
LD-11	14.7	1.21	39.1	44.5	Do not fail		
LD-21	15.1	0.63	34.0	38.7	73091	1218.2	50.8
LD-22	12.3	1.72	45.9	52.2	346890	5781.5	240.9
LD-31	14.9	1.63	84.8	96.5	3883	64.7	2.7
LD-32	15.2	1.48	90.2	102.7	153	2.6	0.1
LD-33	14.3	1.52	29.3	33.3	Do not fail		
LD-41	15.9	1.11	57.0	64.9	7332	122.2	5.1
LD-42	14.8	1.19	30.1	34.3	Do not fail		
LD-51	14.5	1.76	46.6	53.1	93287	1554.8	64.8
LD-52	14.9	2.02	80.1	91.2	10	0.2	-
LD-61	15.0	0.49	39.5	45.0	27259	454.3	18.9
LD-62	14.0	1.01	31.0	35.3	Do not fail		
LD-71	21.7	2.47	74.7	85.1	26	0.4	-
LD-72	14.7	1.45	19.3	22.0	Do not fail		
LD-81	15.8	1.42	46.4	52.8	21933	365.6	15.2
LD-82	14.3	2.12	38.3	43.6	Do not fail		
LD-91	15.4	1.33	22.7	25.9	Do not fail		
LD-101	15.8	1.46	63.3	72.1	1312	21.9	0.9
LD-102	14.2	2.18	44.1	50.2	Do not fail		
LD-111	16.2	2.10	62.5	71.2	1	-	-
LD-112	14.7	1.91	54.2	61.7	Do not fail		
LD-121	34.8	0.99	26.8	30.6	164114	2735.2	114.0
LD-131	14.6	0.95	32.3	36.8	Do not fail		
LD-141	14.3	2.39	79.6	90.7	8266	137.8	5.7
LD-142	14.3	0.89	24.0	27.3	Do not fail		
LD-151	14.6	0.94	27.4	31.2	Do not fail		

Deflection rate

The Table 6 shows the behaviour in the deflection rate of each sample. In general in all the samples the rate of deflection decreased except in LD - 21, LD - 31, LD - 91 and LD - 121 samples. Analysis of this parameter in the LD - 31 sample can generate errors of interpretation since the test had only a duration of 5 hours. In the case of the LD-21 sample, the few data obtained do not accurately assure the definition of the deflection velocity. The samples LD - 91 and LD - 121 have a particular slope change at the end of the deflection curve. In the particular case of the LD - 91 sample it was observed a crack, so it is possible that a failure occurred prior to the collapse. In the LD-121 sample, although no cracking failure occurred, the material was already in a process of creep where the deformations increase prior to failure.

TABLE 6 - DEFLECTION RATE FOR LONG DURATION TEST SOURCE [4]

Sample	$\Delta\delta/\Delta t$ (mm/day)			
	$0 - 1/4\ t_{failure}$	$1/4\ t_{failure} - 1/2\ t_{failure}$	$1/2\ t_{failure} - 3/4\ t_{failure}$	$3/4\ t_{failure} - t_{failure}$
LD-11	0.72	0.41	0.12	0.08
LD-21	6.21	1.47	2.06	-
LD-22	2.27	0.86	0.65	0.15
LD-31	162.28	189.26	59.51	87.40
LD-33	0.91	0.43	0.17	0.11
LD-42	0.64	0.24	0.12	0.02
LD-51	6.34	1.94	1.63	1.19
LD-61	26.37	7.49	3.46	-
LD-62	0.79	0.22	0.15	0.14
LD-72	2.18	0.25	0.05	0.06
LD-82	2.66	0.18	0.10	0.02
LD-91	0.77	0.34	0.22	0.36
LD-102	1.22	0.42	0.09	0.04
LD-112	0.90	0.16	0.09	0.08
LD-121	1.06	1.00	0.86	0.94
LD-131	1.62	0.50	0.16	0.15
LD-141	29.44	8.77	4.82	4.22
LD-142	0.58	0.26	0.09	0.04
LD-151	0.58	0.31	0.18	0.00

Application of Burger's and Findlay's models to the time-deflections curves

Burger's model

In order to define the Burger's model it was necessary to select the 12 unloaded samples to know the elastic (δ_{de}) and viscous deflections (δ_v) with the information gathered during the whole test time. The Table 7 presents the values of the parameters of the Burger model obtained.

TABLE 7 - *BURGER'S MODEL PARAMETERS. SOURCE [4]*

Test	P (KN)	Instantaneous elastic deflection		Viscous deflection		deferred elastic deflection			δ_{final} (mm)	R^2
		δ_e (mm)	k_e (KN/mm)	δ_v (mm)	r_v (KN / mm)	t (min)	K_{de} (KN/mm)	δ_{de} (mm)		
LD - 11	1.45	176	0.008	79	6806	402.3	0.241	6	261	0.94
LD - 33	1.96	136	0.014	70	9118	4808.8	0.065	30	236	0.96
LD - 42	1.35	114	0.012	54	8407	455.6	0.193	7	175	0.94
LD - 62	1.22	125	0.010	52	4811	0.5	0.124	3	180	0.93
LD - 72	1.80	240	0.008	102	5055	10.0	0.064	28	370	0.77
LD - 82	2.70	195	0.014	105	7353	12.1	0.060	45	345	0.77
LD - 91	1.68	75	0.022	39	10967	119707.6	0.042	35	149	0.99
LD - 102	2.70	131	0.021	63	12046	3018.5	0.079	34	228	0.92
LD - 112	2.59	105	0.025	54	13398	550.0	0.098	9	168	0.91
LD - 131	1.25	152	0.008	82	4222	9616.8	0.036	35	269	0.95
LD - 142	1.11	66	0.017	37	8356	3325.9	0.111	10	113	0.95
LD - 151	1.11	112	0.010	51	6062	550.0	0.102	1	164	0.96

The Figure 12 shows the percentages corresponding to the instantaneous elastic, viscous and elastic deferred deflections, with respect to the total deflection of each one of the tests.

The elastic deflections vary between 69.4% and 50.3% with an average of 61.2% of total deflection. The deflections that were recorded as viscous are those registered after discharge and after a time in which the guadua was not recovered its initial position. Viscous deflections ranged from 32.7% to 26.2% with an average of 29.8%. The elastic deferred deflections vary between 23.5% and 0.6% with an average of 9.0%. There was no direct relationship between stress levels and percentages of the immediate, deferred elastic and viscous deflections respect the total deflection.

Both types of deflections vary in a reduced or defined range of total deflections, so that parameters such as the Maxwell spring constant (K_e) and the Maxwell damping constant (r_v) have a linear proportion. In addition, the parameter K_e is proportional to the load P and the parameter r_v is proportional to the product between the load P and the duration of the load condition t.

Non-Conventional Materials and Technologies – NOCMAT for XXI Century   Materials Research Forum LLC
Materials Research Proceedings 7 (2018) 65-84   doi: http://dx.doi.org/10.21741/9781945291838-8

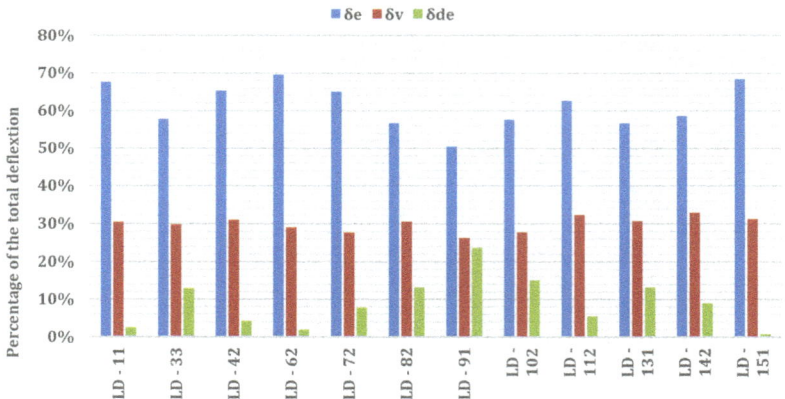

FIGURE 12 - *PERCENTAGE OF ELASTIC (ΔE), VISCOUS (ΔV) AND ELASTIC DEFERRED (ΔDE) DEFLECTIONS RESPECT THE TOTAL DEFLECTION. SOURCE [4]*

The Figure 14 shows the parameters of the Burger's model. The parameter K_e and r_y show a proportional relation. A linear regression results in a coefficient of determination $R^2 = 0.88$ (correlation coefficient R of 0.94).

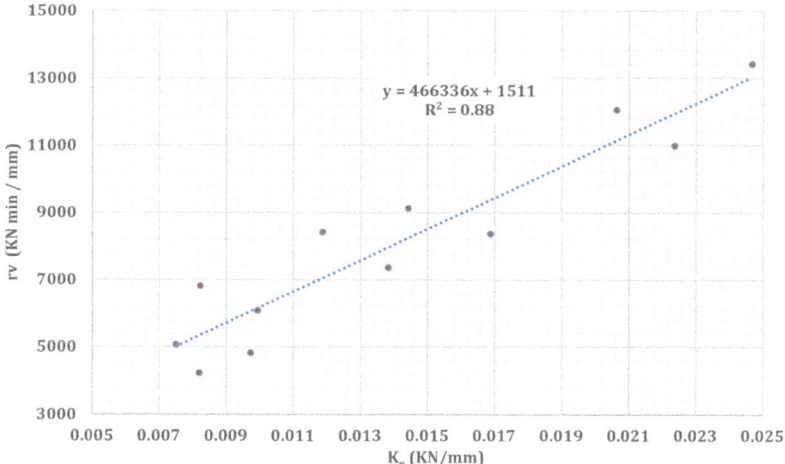

FIGURE 13 - *MAXWELL SPRING CONSTANT KE VS. MAXWELL'S DAMPING CONSTANT RV SOURCE [4]*

Another relationship between parameters is shown in the Figure 14. A direct relationship between the Kelvin's spring constant K_{de} and the relaxation time, t, it was not found. However, under a logarithmic scale a potential trend curve with a correlation coefficient R = 0.72 it was determined.

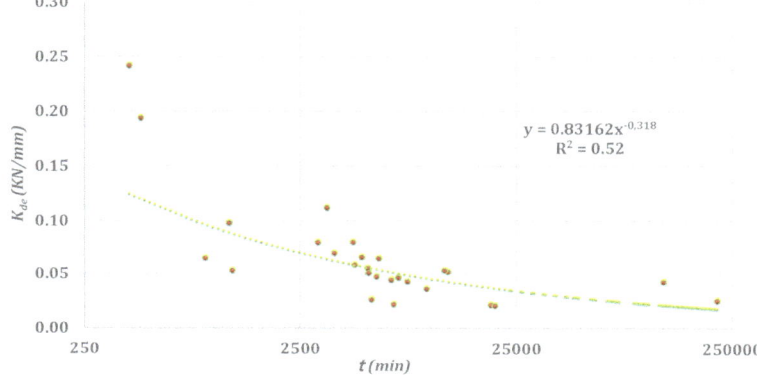

FIGURE 14 - KELVIN'S SPRING CONSTANT (KDE) VS RELAXATION TIME (T) SOURCE [4]

Findley's model

For the same tests, the mathematical model of Findley was applied. The parameters of the Findley model considering the entire test time are presented in Table 8

TABLE 8 - FINDLEY'S MODEL PARAMETERS SOURCE [4]

Test	Elastic deflection δ_{de} (mm)	Time depended deflection δ_t (mm)	Exponenital Constant (n)	R^2
LD - 11	176	0.285	0.450	0.99
LD - 33	136	0.286	0.467	0.99
LD - 42	114	0.185	0.464	0.99
LD - 62	125	1.663	0.284	0.99
LD - 72	240	12.710	0.190	0.98
LD - 82	195	29.258	0.134	0.99
LD - 91	75	0.016	0.680	0.99
LD - 102	131	1.691	0.329	0.98
LD - 112	105	4.235	0.215	0.99
LD - 131	152	1.554	0.350	0.98
LD - 142	66	0.442	0.378	0.99
LD - 151	112	0.201	0.450	0.98

Non-Conventional Materials and Technologies – NOCMAT for XXI Century   Materials Research Forum LLC
Materials Research Proceedings 7 (2018) 65-84   doi: http://dx.doi.org/10.21741/9781945291838-8

In the Figure 15 it is observed that between the time-dependent deflection and the exponential constant there is a logarithmic relation with a coefficient of determination of the curve is $R^2 = 0.97$ (correlation coefficient R of 0.98).

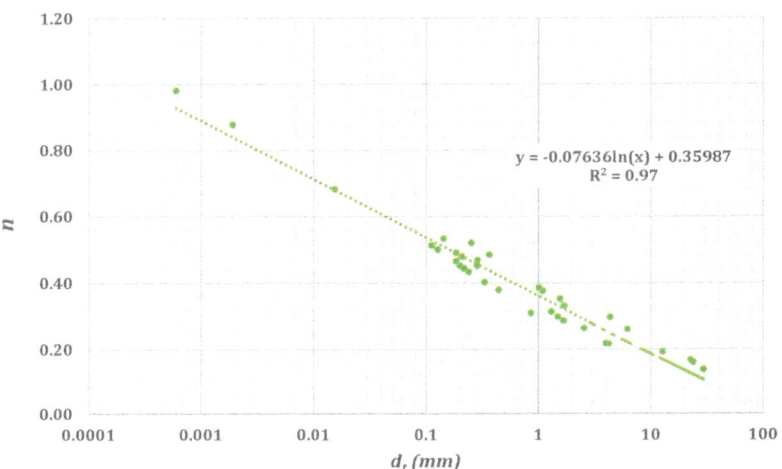

$$y = -0.07636\ln(x) + 0.35987$$
$$R^2 = 0.97$$

FIGURE 15 - *TIME-DEPENDENT DEFLECTION (DT) VS. EXPONENTIAL CONSTANT (N) SOURCE [4]*

Estimation of deferred deflections from short time ranges using the Burger and Findley's model

The parameters of the Burger and Findley's models were obtained with the total recorded data of the deflections and the total time of the test. The models have been developed, adjusting as closely as possible to the data recorded and were corroborated with the coefficient of determination.

Four Burger's models and 4 Findley's models, described below, were made for each of the long-duration unloaded samples.

The Burger 1 and Findley 1 Models consider the total time of the test duration
The Burger 2 and Findley 2 Models consider the data from $t = 0$ to $t = t_{final} / 2$
The Burger 3 and Findley 3 Models consider the data from $t = 0$ to $t = t_{final} / 4$
The Burger 4 and Findley 4 Models consider the data from $t = 0$ to $t = t_{final} / 8$

For these samples, graphs were made for each model and test, to know how effective the parameters obtained allows the prediction of the behaviour of the remaining deflections of the test. As example the LD-11 test models are shown in Figure 16.

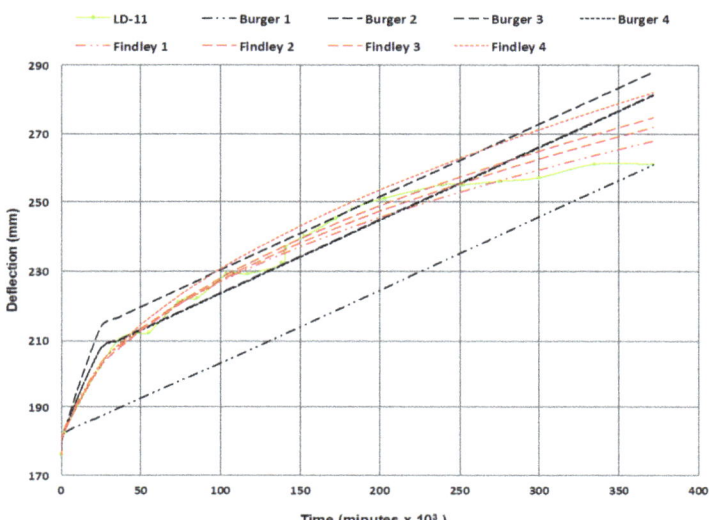

FIGURE 16 - BURGER'S AND FINDLEY'S FOR LD-11 TEST SOURCE [4]

The Figure 17 shows the coefficient of determination of the Burger's models for the unloaded test samples.

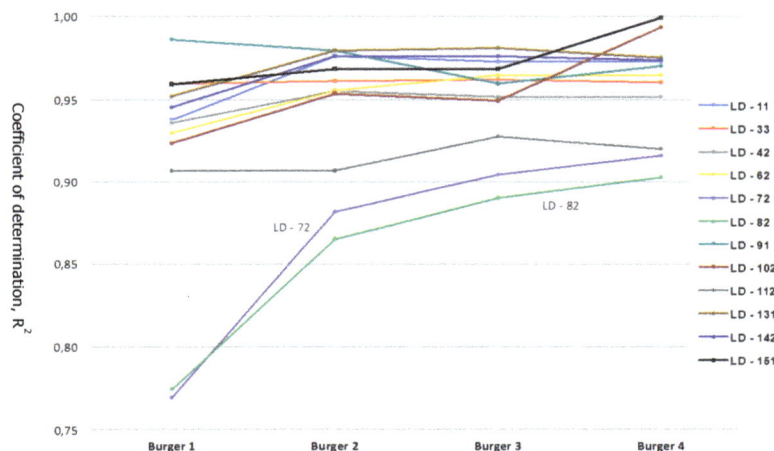

FIGURE 17 - DETERMINATION COEFFICIENT FOR 4 BURGER'S MODELS SOURCE [4]

In the Figure 18 it can be seen that the Burger 1 models of the tests LD-72 and LD-82 have a coefficient of determination R^2 of 0.7689 and 0.7738 respectively. These tests are also those that have the most significant change in the deflection speed with respect to the other tests since they go from 2.18 to 0.25 mm/day (reduction to 8.7 times) and from 2.66 to 0.18 mm/day (reduction to 14.77 times) in only 80640 minutes (56 days).

The Burger 4 models had at least a coefficient of determination of 0.9019 showing good results in the general description of the deflection curve. These models correctly describe the curve from 25% of the time to the total time (as is possible to see in the Figure 16 for LD-11 test). The Burger 4 models do not correctly describe the beginning of immediate deflections.

The Burger 1 models, which are the ones that consider the totality of the recorded data, are not adequate to describe deflections during the time. It was observed in all these models that at the beginning represent the real deflection but later the viscous deflection is developed with a linear behaviour that is out of the real deflections curve.

The Figure 19 shows the coefficient of determination of the Findley's models for the unloaded test samples. In the Findley's models the ratio of the estimated deflections and the actual deflections may vary depending on the velocity of deflection of the sample. In many cases, the Findley's models are more faithful to the description of the curve. For example, in the Figure 19, for all Findley 1 models (where the total deflection register was used), the lowest determination coefficient was 0.98 and in contrast, in the Burger 1 models the lowest coefficient of determination of 0.77.

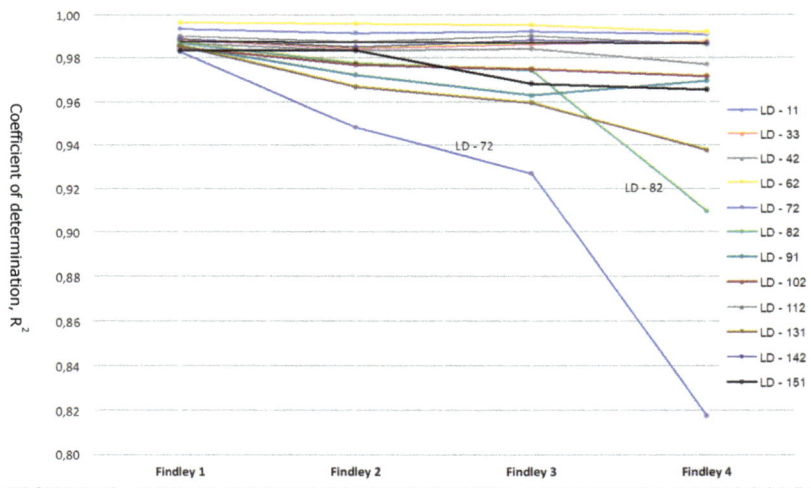

FIGURE 19 - *DETERMINATION COEFFICIENT FOR 4 FINDLEY'S MODELS SOURCE [4]*

Relationship between stress levels and failure time of long-duration test samples
The Figure 20 shows the stress level of short duration tests with standard application of speed load (CD standard), and load application speed of 2.0 inches / minute (CD 2.0"/min) and long duration tests recorded until failure time (LD failed) and those unloaded (LD unloaded). The Figure 20 also includes the empirical curve that describes the time of failure and level of strength for the wood.

Non-Conventional Materials and Technologies – NOCMAT for XXI Century   Materials Research Forum LLC
Materials Research Proceedings **7** (2018) 65-84   doi: http://dx.doi.org/10.21741/9781945291838-8

It is appreciated that the data of the long duration tests of the guadua can be well approximated for durations of the load of 7 days. From this point forward, the stress level and failure time recorded in the tests are predominant below to the empirical curve, which means that for a given stress level, more time is required to fail a piece of wood than In a sample of guadua.

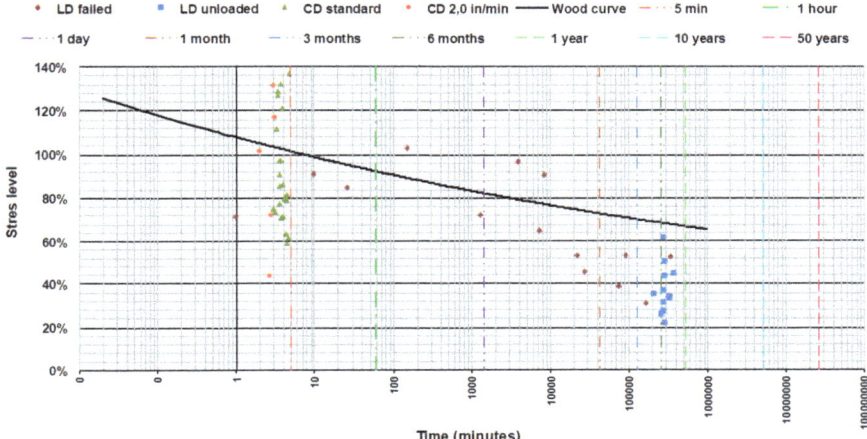

FIGURE 20 - STRESS LEVEL VS. FAILURE TIME FOR SHORT DURATION SAMPLES (CD) AND LONG DURATION SAMPLES (LD) SOURCE [4]

Conclusions

The Burger's model could be not as accurate to describe the whole deferred deflection curve for an element of guadua as the Findley's model.

The Burger's model presents an area where elastic deformations are identified and differentiated in a given range (instantaneous and deferred elastic deflections) and later, typical deflections of a viscous behaviour of the material. In contrast, the Findley's model with only the exponential and the constant value that multiplies the power determines how soft or strong the deflection curve should be.

In order to carry out a descriptive investigation that explains the precise relationship between the stress level and the failure time, more data are required to complement different ranges, such as stress levels between 60% and 100%. The data that are recorded as unloaded specimens question the necessity to know the behaviour of *Guadua angustifolia* under stress levels less than 50%.

The empirical curve crosses the group of the results of the short duration tests in the middle, which shows that the average stress (87.8 MPa) used for the calculation of the stress levels, is a good reference point.

References

[1] González, E., Osorio, J. A., & García, E. A. (2002). Resistance to flexion of *Guadua angustifolia* Kunth on a natural scale. Revista Facultad Nacional de Agronomía, 55(2), 1555–1572.

[2] Sánchez, J. y Prieto, E. (2002) Bending behavior of *Guadua angustifolia*. Graduate Thesis. Bogotá. Universidad Nacional de Colombia.

[3] AIS Asociación Colombiana de Ingeniería Sísmica (2010). Reglamento Colombiano de Construcciones Sismo Resistente NSR-10. Bogotá. Colombia.

[4] Chavarro, S. (2016) Deflections of *Guadua angustifolia* Kunth culms under bending stresses with permanent loads in the environment of Bogota. Burger's and Findley's models. Master Thesis. Bogotá. Universidad Nacional de Colombia.

[5] ICONTEC Instituto Colombiano de Normas Tecnicas (2007). NTC 5525. Métodos de ensayo para determinar las propiedades físicas y mecánicas de la *Guadua angustifolia* Kunth. Bogotá. Colombia

[6]. Gerhards Charles Effect of duration and rate of loading on strength of wood and wood-base materials. [Informe]. - Madison : USDA Forest Service, 1977

[7]. Wood Lyman W. Relation of strength of wood to duration of load [Informe]. - Madison : U.S. Department of Agriculture, 1951.

[8] Simpson, W. (1998) Equilibrium moisture content of Wood in outdoor locations in the United States and worldwide Madison : USDA Forest Service. https://doi.org/10.2737/FPL-RN-268

[9] Gutiérrez, M. (2011) Correction factor for moisture content for tensile strength parallel to the fibre of *Guadua angustifolia* Kunth. Master final work. Bogotá. Universidad Nacional de Colombia.

Non-Conventional Materials and Technologies – NOCMAT for XXI Century   Materials Research Forum LLC
Materials Research Proceedings 7 (2018) 85-94   doi: http://dx.doi.org/10.21741/9781945291838-9

Analysis of the Mechanical Properties of Composite Boards with Natural Resin and Guadua Unidirectional Fibers

J.P. Patiño [a], M.L. Sánchez [b*], W.A. Patiño [c]

[a] Universidad Militar Nueva Granada, Bogotá, Colombia, Programa de Ingeniería Civil,
Universidad Militar Nueva Granada. Carrera 11, No.101 80, Bloque D.
u1102014@unimilitar.edu.co

[b*] Universidad Militar Nueva Granada, Bogotá, Colombia, Programa de Ingeniería Civil,
Universidad Militar Nueva Granada. Carrera 11, No.101 80, Bloque D.
martha.sanchez@unimilitar.edu.co

[c] Universidad Militar Nueva Granada, Bogotá, Colombia, Programa de Ingeniería Civil,
Universidad Militar Nueva Granada. Carrera 11, No.101 80, Bloque D.
u1101753@unimilitar.edu.co

Abstract. During the last years the study of vegetal fibers for its use as reinforcement of composites of polymeric matrix has gained importance, mainly in the area of civil engineering. The high mechanical strength, together with its biodegradability, makes this type of fibers a competitive option, compared with the synthetic fibers usually used in the design of polymeric composites. In this work the mechanical behavior of composite boards, made with natural materials, is analyzed. For the preparation of the composite there were used fibers that were 30 cm long, unidirectional, extracted from the top of bamboo culms of the species Guadua Angustifolia Kunth, aged between 4 and 6 years. The culms used to obtain the fibers were immunized with a boric acid solution. As the matrix of the composite, was used a vegetal polyurethane. The experimental characterization was based on the performance of static bending, tensile and compression tests according to the specifications of ASTM D1037-12 "Standard Test Methods for Evaluating Properties of Wood-Base Fiber and Particle Panel Materials". The results obtained were compared with results reported in the literature for composites made from vegetable fibers.

KEYWORDS: Bamboo, Vegetal Fibers, Mechanical Properties, Static Bending, Tensile, Compression

Introduction

In the last decades, there has been a great increase in the use of natural materials for the elaboration of compounds in the construction field, this is due to the efficiency in its physical and mechanical properties, in addition to its low cost. Bamboo, henequen, banana and coconut fibers stand out. These fibers can be combined with polymeric resins (vegetal or synthetic) in order to make a natural compound.

Not only at a national level, but at the global level, a greater use of eco-friendly materials is sought, in order to reduce emissions of polluting gases to the environment and the reduction in the use of non-renewable resources. In addition, the use of such eco-friendly materials is expected to generate a series of favorable impacts on health, economy and the environment. (Bledzki & Gassan, 1999 [1])

Recent studies show the possibility of using plant fibers in the field of civil engineering. These studies show the great advantage that the material can offer, highlighting the ease of disposition of the fibers, their fast renewal, economy and high specific properties. As important characteristics of these fibers are that they are non-toxic and biodegradable, which makes them less abrasive than other synthetic fibers. (Da Silva, 2003 [2])

Non-Conventional Materials and Technologies – NOCMAT for XXI Century   Materials Research Forum LLC
Materials Research Proceedings 7 (2018) 85-94   doi: http://dx.doi.org/10.21741/9781945291838-9

One of the most recent applications of vegetable fibers is the reinforcement in the development of agglomerates. The agglomerates are composed materials that are formed from the union of two or more materials, in order to obtain their distinct physical and mechanical characteristics. They are characterized by not being soluble with each other, which is why they originate two distinct phases in the material, a continuous phase called matrix and a dispersed phase responsible for the mechanical properties of the material. (Fiorelli, 2011 [3])

Recent research has focused on the use of vegetal resins as a substitute for resins commonly used for the manufacture of wood pellets. These natural adhesives have been studied, however, to date, their application is not yet widespread enough in the area of civil construction. (Moran & Alvarez, 2008 [4])

Methodology

There were used natural fibers from the bamboo extracts of the species Guadua Angustifolia Kunth. The Guadua Angustifolia Kunth used in the project has a harvest age between four and six years, adequately preserved following the specifications of the Colombian Technical Standard NTC 225-04 "Standard for Preservation and Drying, Guadua Angustifolia Kunth".

Extraction of fibers

The extraction of the fiber was done by a chemical-mechanical method, because the removal of the lignin and the hemicellusa is done from an alkaline chemical method with a solution of sodium hydroxide. This process is done to modify the surface of the fiber (Cuéllar & Muñoz, 2010 [6]) and to improve the interaction between fiber-matrix.

An alkaline treatment was done, this consists in breaking the bonds of the molecules of the lignin using a chemical solution of sodium hydroxide with a concentration of 5%. The material should be submerged for a period of 48 hours at room temperature. After the time is up, the material should be washed with distilled water for the purpose of removing sodium hydroxide.

Once the material is dry at room temperature, a mechanical method is used for the extraction, which consists in the separation of the fibers using a material crushing system. The system is based on toothed rollers coupled by an engine as shown in FIGURE 1 this allows the obtaining of the fibers of the guadua.

FIGURE 1 - CRUSHER

Non-Conventional Materials and Technologies – NOCMAT for XXI Century   Materials Research Forum LLC
Materials Research Proceedings 7 (2018) 85-94   doi: http://dx.doi.org/10.21741/9781945291838-9

After the crushing process, the material is obtained in the presentation set forth in FIGURE 2. The presentation obtained is adecuate to have an individual extraction of the necessary fibers for the preparation of the compound material, taking into account that the necessary length of each fiber is of 300 mm as shown in FIGURE 3.

FIGURE 2 – PRESENTATION AFTER THE CRUSHING

FIGURE 3 - FIBERS WITH 300 MILLIMETERS

Elaboration of the compound

The compound was made with a ratio of 70% to 30%, where 70% corresponds to the percentage of fiber, which expressed in terms of mass corresponds to 362.80 grams and 30% corresponds to the vegetal resin used as a binder.

Once the mass corresponding to fiber was obtained, small groups of fiber were assembled in order to arrange them with an specific order in the mold previously elaborated with the dimensions of 300mm by 300mm, as shown in FIGURE 4

FIGURE 4 - FIBERS WELL ACCOMPLISHED IN THE MOLD

The resin used has the properties set out in TABLE 1, which shows the color, density, state, boiling point and shape of each of the two components that make it up.

TABLE 1- RESIN PROPERTIES (DATA PROVIDED BY THE SUPPLIER)

PROPERTY	COMPONENT A	COMPONENT B
Fisic state	Liquid	liquid
color	Brown	Yellow - amber
form	Viscose	viscose
Boiling point	190°C	313°C
Relative density	1.25gr/cm3	0.98gr/cm3

To guarantee the volume of the resin, the volume ratio provided by the supplier was used as explained in Eqn. 1 and Eqn 2. The mass distribution of the two components is shown in TABLE 2 in order to be 30% of the compound.

$$V_R = V_A + V_B \tag{1}$$

$$V_A = 1.5V_B \tag{2}$$

Taking:
V_R= Total volume of the resin
V_A= Volume of component A
V_B= Volume of component B

TABLE 2 - Distribution of the resin

Material	Density (gr/cm3)	Volume (cm3)	Mass (gr)
"A"	1.25	86.4	108
"B"	0.98	129.6	127

For the preparation of the compound the manual molding method was used. This method consists of coupling the fibers in the mold, making sure that they are all unidirectional. First, 20% of resin is added to the mold, then the first layer of fibers is introduced and 40% of additional resin is subsequently added ensuring that all of these are impregnated; as the last step, the second layer of fibers was placed and the remaining resin was added. As shown in FIGURE 5.

Non-Conventional Materials and Technologies – NOCMAT for XXI Century   Materials Research Forum LLC
Materials Research Proceedings **7** (2018) 85-94   doi: http://dx.doi.org/10.21741/9781945291838-9

FIGURE 5 - COMPOUND ELABORATION

Once the mold is ready with the material, it is pre-compacted with a uniformly distributed load. The mold is then brought to the hydraulic press shown in FIGURE 6 by applying a load of 18 tons for a time of 18 hours.

FIGURE 6- HYDRAULIC PRESS

After 18 hours the mold was removed from the press, the board was demolded and allowed to cure for 24 hours at room temperature, so that the resin reached its maximum strength. Obtaining the board shown in FIGURE 7.

Non-Conventional Materials and Technologies – NOCMAT for XXI Century   Materials Research Forum LLC
Materials Research Proceedings 7 (2018) 85-94   doi: http://dx.doi.org/10.21741/9781945291838-9

FIGURE 7 - CURED BOARD

Physical and mechanical properties
Physical properties
Moisture content

The moisture content is performed according to the recommendations of ASTM D 4442-07. This standard shows a method of oven drying. Here are some requirements for performing the test:

1- It is required 10 samples taken from the material for the test
2- Use an oven with temperature controlled at $105 \pm 2 \,°$ C
3- A precision balance of at least 0.01 gr resolution is required
4- the relative humidity of the samples does not exceed 70%.
5- In order to guarantee a good quality of the data, check that the oven does not contain any moisture accumulation in the interior. It is recommended to check it before starting the test and to check its ventilation.

The sample to be evaluated must be moisture-balance of the environment; once this is done, the initial mass of the sample is determined. The sample is then baked for 24 hours at a constant temperature of 105 ° C. After 24 hours, the sample is removed and its dry weight is determined. Having these two weights, and with the help of Eqn. 3, the value of the moisture content

$$\%CH = \frac{A-B}{B} * 100 \tag{3}$$

Where:
   %CH is the percentage of moisture content
   A is the mass of the fiber bundle in natural moisture condition
   B is the mass of the fiber bundle in kiln-dried condition

Absorption capacity

The absorption capacity was performed following the recommendations of ASTM D570. The standard indicates that the mass of the material must be found in its natural state, and then immersed in distilled water for 24 hours. The sample is then withdrawn from the water and dried with a surface absorbent towel; at that time, the mass should be recorded in a dry saturated condition. Finally, proceed to take the sample to the oven until it has a constant mass, this will be the dry mass. Taking

Non-Conventional Materials and Technologies – NOCMAT for XXI Century   Materials Research Forum LLC
Materials Research Proceedings 7 (2018) 85-94   doi: http://dx.doi.org/10.21741/9781945291838-9

these weights, we continue the use of the Eqn. 4. Finally, the material is brought to the furnace for 24 hours and its mass is found in dry condition to be able to have the value of absorption capacity with the help of Eqn. 5.

$$\%AE = \frac{P_i - P_{sss}}{P_i} * 100 \tag{4}$$

$$\%A = \frac{P_s - P_{sss}}{P_s} * 100 \tag{5}$$

Where:
- •% A: absorption capacity
- •% AE: Effective absorption capacity
- • Pi: fiber bundle mass in natural moisture condition
- • Ps: dough in dry oven condition
- • Psss: mass of the fiber bundle in saturated dry surface condition

Density
Establishing the volume and measurement of the mass of the material, the determination of the density of the material was made. First, the "long, wide and thick" measurements of each sample were taken. With this data the value of the volume of each sample can be found by multiplying the length, width and thickness of the sample. It is found the mass of each of the samples. With the mentioned data, the value of the density can be calculated with the aid of Eqn.

$$6D = \frac{M}{V} \tag{6}$$

Mechanical properties
Flexural Bending
To perform the sample flexure test, the recommendations of ASTM D 1037 are used, in which they determine the modulus of rupture and the modulus of elasticity. In the standard measures that must fulfill the material for the test, it must have a thickness greater than or equal to 6 mm, a width of 76 mm, and the length between the supports should be 24 times its thickness, clarify that, if the thickness is less than 6 mm, the width of the test must be 51mm. Once the probe measurements are taken, the probe is placed in the loading frame, as shown in FIGURE 8.

FIGURE 8 - FLEXIBLE TEST ASSEMBLY

The supports are at a distance of 24 times the thickness of the specimen, the speed of the test is also given by the thickness, the standard suggests a speed of 3 mm/min. To find the value of the deflection, a linear voltage differential transducer (LVDT) is used. Once the data are acquired, the following equations are applied to find the failure modulus Eqn. 7 and modulus of elasticity Eqn. 8.

$$Rb = \frac{3Pmax\ L}{2bd^2} \tag{7}$$

$$E = \frac{L^3}{4bd^3} * \frac{\Delta P}{\Delta y} \tag{8}$$

Analysis of results

Physical properties
TABLE 3 shows the results obtained in relation to the physical properties together with their respective standard deviation.

TABLE 3 – PHYSICAL PROPERTIES

PHYSICAL PROPERTIES OF THE PANELS		
	2 hours	24 hours
Absorption capacity (%) ASTM D 570.	13,51 ± 1,46	25,34 ± 2,61
Effective absorption (%)	4,98 ± 1,41	15,68 ± 2,20
Percentage of swelling (%)	5,36 ± 0,01	9,83 ± 0,02
Density (gr/cm³)	1,11 ± 0,04	
Moisture content(%) ASTM D 4442-07.	8,32 ± 0,51	

Mechanical properties
Flexion
The sample was preloaded at 50Newtons and the test speed was 3mm / min. The test had duration of approximately 900 seconds (15 minutes), the values of the test show that the maximum load of the probe was of 1789.82 Newton with a deflection of 13.77mm. **TABLE 4** shows the values of the modulus of rupture and the apparent modulus of elasticity of the material, it is also possible to show the maximum load of the material with **FIGURE 9**

TABLE 4 - MOR AND MOE VALUES

FLEXION RESULTS		
Rupture module	155.17	MPa
ΔY/Δx	0.2954	kN/mm
Apparent elasticity module	13784.12	MPa

Non-Conventional Materials and Technologies – NOCMAT for XXI Century   Materials Research Forum LLC
Materials Research Proceedings **7** (2018) 85-94   doi: http://dx.doi.org/10.21741/9781945291838-9

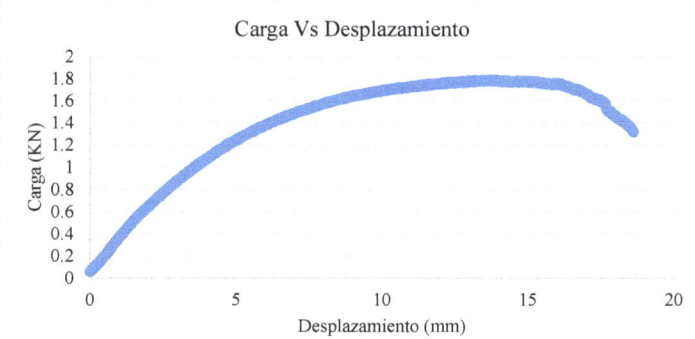

FIGURE 9 - *GRAFICS LOAD VS DISLACEMENT*

FIGURE 10 shows the type of failure presented in the corresponding test specimen of the bending test.

FIGURE 10 - *TIPE OF FAILURE IN THE SPECIMEN ON BENDING TEST*

As part of the analysis of the results we wanted to compare the results obtained with bending versus other materials with similar thicknesses and densities, shown in TABLE 5.

TABLE 5 - *BOARD VERSUS OTHER MATERIALS*

MATERIALS	DENSITY	MOR	MOE
	gr/cm^3	MPa	MPa
Composed with guidua unidirectional fibers	1.1	155.17	13874.12
Composed with randomly distributed fibers of Guadua (Sanchez, Martha 2017 [7])	1.06	25.6	2000
HDF boards (Poblete, Hernan 2006 [8])	1.04	51.9	4406

Conclusions

The use of non-conventional materials developed from natural raw materials, allows reducing the emission of polluting gases to the environment, contributing substantially to the improvement of the quality of life of living beings. The implementation of the proposed methodology contributes to the development of new ecological materials, which allows reducing the impact that the processes of manufacture of traditional materials generate to the environment.

Acknowlegment

This paper is a derivative product of the project (INV-ING-2392) financed by the Vice-rectory of Research of Universidad Militar Nueva Granada-validity (2017).

References

[1] A.K. Bledzki A.K, J. Gassan. Composites reinforced with cellulose based fibres. Progress in Polymer Science. 24(2), (1999), 221–274. https://doi.org/10.1016/S0079-6700(98)00018-5

[2] R. V. Da Silva. Compósito de resina poliuretano derivada de óleo de mamona e fibras vegatais. Tese Doutorado, 2003, 157 p.

[3] Fiorelli J. Et. Al.-Painéis de partículas à base de bagaço de cana e resina de mamona – produção e propriedades. Acta Scientiarum. Technology. Maringá, 33(4), 2011, 401-406.

[4] Moran J.I., Alvarez V.A., Cyras V.P., and Vasquez A.. Extraction of cellulose and preparation of nanocellulose from sisal fibers. Cellulose, 15, 2008, 149–159. https://doi.org/10.1007/s10570-007-9145-9

[5] Ortiz, M. L. (2009). Aproximaciones a la comprensión de la degradación de la lignina. Orinoquia, 13(2), 137–144.

[6] Cuéllar, A., & Muñoz, I. (2010). Fibra De Guadua Como Refuerzo De Matrices Poliméricas. Dyna, 162, 137–142. Retrieved from http://www.scielo.org.co/pdf/dyna/v77n162/a15v77n162.pdf

[7] Martha L. Sánchez, Luz Y. Morales, Juan D. Caicedo. Physical and mechanical properties of agglomerated panels made from bamboo fiber and vegetable resin. Construction and Building Materials 156 (2017) 330–339. https://doi.org/10.1016/j.conbuildmat.2017.09.003

[8] Hernán Poblete W., Roque Vargas C. RELATIONSHIP BETWEEN DENSITY AND PROPERTIES OF HDF MANUFACTURED BY A DRY PROCESS. Maderas. Ciencia y tecnología 8(3): 169-182, 2006.

Non-Conventional Materials and Technologies – NOCMAT for XXI Century   Materials Research Forum LLC
Materials Research Proceedings 7 (2018) 95-107   doi: http://dx.doi.org/10.21741/9781945291838-10

Nonconventional Curing for Fiber-Cement Material

Valdemir dos Santos[a*,] Gustavo Henrique Tonoli[b] and Holmer Savastano Jr[a]

[a*] Research Nucleus on Materials for Biosystems, University of São Paulo, Brazil,
valdemir@usp.br

[a] Research Nucleus on Materials for Biosystems, University of São Paulo, Brazil

[b] Federal University of Lavras, Lavras-MG, Brazil

Abstract. The proposal presented in this paper is to use accelerated carbonation via carbonated water to modify the microstructure of the cementitious matrix. This particular process generates stabilization of the pore volume in the early age of the composite and avoiding the degradation of the cellulosic fiber, which is susceptible to the alkaline attack of the cementitious matrix. Chemical and morphological characterizations were performed, such as X-ray diffraction (XRD), scanning electron microscopy (SEM) and thermogravimetric analysis (TG). The compounds were also physically (water absorption and density) and mechanically analyzed (four-point bending test: modulus of rupture, modulus of elasticity, limit of proportionality and specific energy) after 28 days of wet curing and after 200 soak & dry cycles. The results show that there is a need for a brief period of hydration and then it is possible to carbonate the materials. Carbonation yields substantial mechanical and physical gains, as well as greater protection of the fiber that does not mineralize.

Keywords: Carbonated Water, Carbonation, Carbonated, Cellulose Fiber, Vegetable Fiber

Introduction

The fiber-cement industry produces various building elements, such as pipes, tiles and flat plates for walls [1-2]. However, a major challenge in the fiber-cement industry is the control of hygroscopic migration, porosity and pore size distribution generated during the production and post-production process [3]. Different water migration mechanisms occur during the curing process, which generates flaws and microcrackings among other problems [4].

In the last years, researchers and manufacturers have explored alternatives to avoid uncontrolled hygroscopic migration, such as the use of ashes [5], use of different size and scale distributions of fibers -nano, micro and/or macro- [6-7], and chemical treatments of fibers to improve the hydrophobic capacity such as hornification [8] among others. However, three main factors must be observed in the preparation of composites based on Portland cement and vegetable fibers to avoid the hygroscopic migration: (i) the physical and chemical incompatibility between the fibers and the matrix; (ii) dimensional variation of the fibers and (iii) the alkaline attack on the fibers.

The vegetable fiber degradation in cementitious materials occurs mainly because of the high alkalinity of the water present in the pores of the cementitious matrix [9]. The high porosity of matriz allows the accumulation of water with the presence of $Ca(OH)_2$ released in the hydratation, which leads to the formation of a region with high alkalinity [10]. In this context, the high alkalinity causes degradation and loss of mechanical strength of the composite [11] and the uncontrolled hygroscopic migration also occurs, which generates tension in the pores and can initiate the cracking process [12].

The route proposed here to mitigate these process is through the use of carbonated water which meets concomitant hydration and carbonatation of the fiber-cement, minimizing alkalinity of the cement by consuming some of the hydroxides and turning the Ca ions into carbonates that will fill up the pores, step by step.

The carbonation process stabilizes the mechanical structure of the material due to the densification of the cementitious matrix [13]. In denser structures, capillary permeability and porosity are reduced due to deposition of non-soluble materials in its pores [14] and the hygroscopic migration is decreased.

Another important factor is the shrinkage produced by densification of the matrix, which can be responsible for cracks generation in the microstructure of the fiber-cement, in special during the curing process [15]. Shrinkage can be decreased through carbonated water, due control of the nucleation and growth of $CaCO_3$ during the hydration. However, carbonation process depends on the presence of enough water in the pores for effective occurrence [16].

Therefore, the objective of this work was to analyze the effects of curing with carbonated water on the physical-mechanical properties at early ages and after accelerated ageing cycles of fiber-cement composites reinforced with cellulosic fibers.

Materials and methods

Materials
Curauá (*Ananas erectifolius*) fibers were obtained from Pematec Triangel Industry, Pará/PA, Brazil. Some properties of the curauá fibers used in the present work were determined by Soltana (2017), such as: cellulose content ~ 68%; hemicelluloses content ~ 10%; lignin content ~ 14%; cross section ~ 0.114 mm^2; thickness ~ 75 μm; real density ~ 1.42 g/cm^3; ultimate tensile strength ~ 550 MPa; Young's modulus ~ 64 GPa. Curauá bundles were cut manually into 10 mm length, according to the length of the commercial polyvinyl alcohol fiber. Unbleached kraft pulp of Pinus sp. was produced by Arauco do Brasil S/A and refined to 70° Schopper Riegler (SR), Method for the determination of the drainability of a pulp suspension in water. Commercial polyvinyl alcohol (PVA), cut into 10 cm length was used to contribute as a reinforcement. CPV-ARI (high early strength cement, according to NBR 5733 equivalent to OPC Type I ASTM C150) [19] Portland cement was used, with an average particle size of 8 Portland cement was used, with an average particle size of 8 μm and a specific density of 3.1 g.cm^{-3}. This type of cement is produced without additions and has a compressive strength of approximately 26 MPa. Limestone with an average particle size of 8 μm (inert) and specific density of 2.8 g.cm^{-3} was used for dimensional stability of the composite and reducing production costs.

The concentration of $CO_{2(g)}$ in water was in the range (4.2-3.8) mg x L^{-1} to carbonated water.

X-ray diffraction (XRD)
A Rigaku MiniFlex 600 X-ray diffractometer was used with CuKα radiation and wavelength of 0.15456 nm (λ Kα1), in the range between 2θ = 5-75°, spaced every 0.02°, at 5°/min speed and 1 s of exposure. Diffractogram peaks of the crystalline phases were compared with the JCPDS and ICSD libraries.

Thermogravimetric analysis (TGA)
Thermogravimetric analysis (TGA) were carried out in a NETZCH Geratbau GnbH equipment, using alumina crucibles, under air flow of 10 cm^3/min, heating rate of 10°C/min, and temperature range of 25 to 900°C.

Formulation of the fiber-cement composites.
The mix design (Table 1) used in the production of the composites were: water, ordinary Portland cement CPV - ARI [equivalent ASTM category. Type III (ASTM C150)][19], limestone, commercial refined pine pulp, PVA and curauá fibers.

Non-Conventional Materials and Technologies – NOCMAT for XXI Century   Materials Research Forum LLC
Materials Research Proceedings 7 (2018) 95-107   doi: http://dx.doi.org/10.21741/9781945291838-10

TABLE 1 – MIX DESIGN USED IN THE MANUFACTURE OF FIBER-CEMENT COMPOSITES.

Materials	Content (% by mass)
CPV – ARI	72.0
Limestone filler	21.2
Cellulose pulp (pine)	3.0
Curauá fiber	2.0
PVA fiber	1.8
Total	**100.0**

The composites were produced by a vacuum-pressing method, a laboratory scale adaptation of the Hatschek industrial process. 200 mm x 200 mm plates with a thickness between 5 and 6 mm were molded. The cellulosic pulp was the disintegrated by mechanical stirring at 1750 rpm for 30 min. Subsequently, the curauá fiber was added and the suspension was stirred in a high energy mixer (3,000 rpm) for 5 min. When the carbonated water was used during molding, it was used after the agitation stage (3,000 rpm). Then, cement and limestone were added and stirred at 1500 rpm for 1 min. The slurry mixture was placed in a suction chamber (vacuum pressure of around 550 mmHg) for 5 min, until excess water was drained. Finally, the drained plate was pressed in a hydraulic press at 5 MPa for 5 min [21] and placed in sealed plastic bags for setting and initial curing in a saturated environment at 100% relative humidity (RH) for 48 h.

After this point, two conditions of hydration/curing were tested, varying the use of tap and carbonated water. The samples were introduced into the water tank using the specific solvent (water or carbonated water) to hydration and cure of fiber-cement. The curing process lasted 762 h, or 28 days.

Accelerated ageing cycles
The objective of using accelerated aging cycles was to simulate the natural exposure of the composites to weathering conditions. Thus, specimens were successively immersed in water at 20 ± 5°C for 170 min and, after a 10 min interval, temperature was heated up to 60 ± 5°C for 170 min in an air-ventilated oven. Further 10 min interval, at room temperature, was applied before the subsequent cycle has elapsed, as recommended by Standard EN 494 Standard (1994). A total of 200 cycles were performed in order to better identify the modification in the physical-mechanical behavior of the composites.

Physical characterization of the composites
Physical characterization was performed for determination of water absorption (WA) and bulk density (BD) and according to ASTM C1185-08 [17].

Mechanical characterization of the composites
4-points bending test was performed using a lower span of 135 mm and an upper span of 45 mm, and a displacement rate of 5 mm/min. Modulus of rupture (MOR) and specific energy (SE) were determined following the recommendations of Rilem [20].

Scanning electron microscopy (SEM).
Samples were morphologically characterized by scanning electron microscopy (SEM) on backscattering (BSE) mode, using a Hitachi Tabletop Microscope, TM3000 model. The preparation of samples for SEM were performed using epoxy resin to vacuum-impregnate the sample pores

Non-Conventional Materials and Technologies – NOCMAT for XXI Century   Materials Research Forum LLC
Materials Research Proceedings 7 (2018) 95-107   doi: http://dx.doi.org/10.21741/9781945291838-10

before cutting. After the resin was cured, samples were manually cut and polished with silicon carbide based abrasive paper using pure alcohol as a lubricant.

Results and discussion

X-ray diffraction (XRD)

Figures 1 and 2 depict the XRD of the fiber-cement using tap water and carbonated water for curing, respectively. The cement hydration process begins with the solvation of the calcium and sulfates in water that react in the first minutes with part of the (C_3A) of the cement grain to form ettringite (AFt) [27], however, it is consumed at the beginning of the hydration process, followed by precipitation of CH [24].

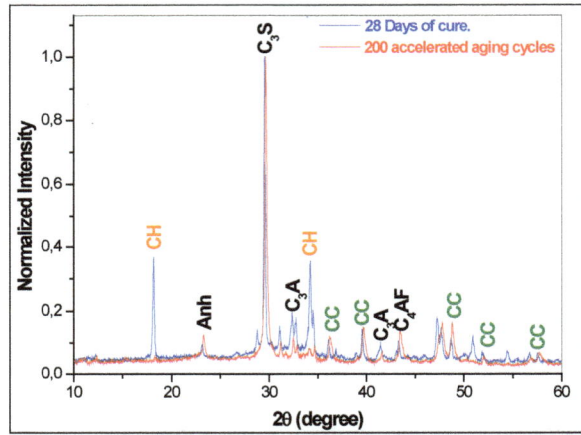

FIGURE 1 - *TYPICAL X-RAY DIFFRACTION (XRD) PATTERNS OF THE FIBER-CEMENT SAMPLES HYDRATED WITH WATER. TESTED AFTER 28 DAYS OF CURE AND AFTER 200 ACCELERATED AGING CYCLES. THE MAIN DIFFRACTION PEAKS OF THE PHASE ARE ASSIGNED AS FOLLOWS: ALITE (C_3S), ALUMINATE (C_3A), ANHYDRITE (ANH), FERRITE (C_4AF) AND CALCITE (CC).*

The increased presence of CH in solution implies in high pH in the pore solution. According to Taylor [25], high pH favors the presence of Anh (anhydrite) instead of ettringite (AFt), Figures 1 and 2. Furthermore, the use of carbonated water consumes part of the CH (Figure 2) and decrease the pH values. Consequently, it leads to an increase of $CaCO_3$ (calcite) concentration. The reactions are shown in Eqs. 1, 2 and 3.

The reduction of alkalinity (pH) by carbonation (increased presence of CC in Figure 2) increases the dimensional stability and densification of the cement matrix [26], which causes the reduction of its permeability, capillarity and porosity [27].

Non-Conventional Materials and Technologies – NOCMAT for XXI Century   Materials Research Forum LLC
Materials Research Proceedings **7** (2018) 95-107   doi: http://dx.doi.org/10.21741/9781945291838-10

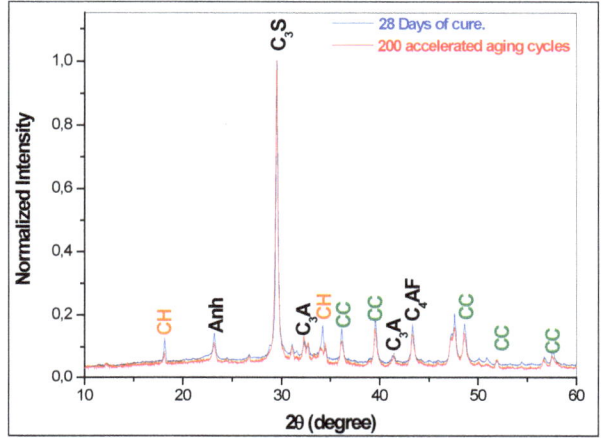

***FIGURE 2** - TYPICAL X-RAY DIFFRACTION (XRD) PATTERNS OF THE FIBER-CEMENT
SAMPLES HYDRATED WITH CARBONATED WATER. TESTED AT AFTER 28 DAYS OF
CURE AND 200 ACCELERATED AGING CYCLES. THE MAIN DIFFRACTION PEAKS OF
THE PHASE ARE ASSIGNED AS FOLLOWS: ALITE (C_3S), ALUMINATE (C_3A), ANHYDRITE
(ANH), FERRITE (C_4AF) AND CALCITE (CC).*

Thermogravimetry

The weight loss (%) of the fiber-cement cured with tap water and carbonated water, are presented in
the Figures 3 and 4 respectively. According to Skinner [28], during the hydration, just part of the
SiO_4 tetramer (in twist) would be required for the formation of C-H from C-S-H phases, by local
rearrangement [29].

The portlandite (CH) content in the sample are solvent's dependents (water or carbonated water)
because carbonated water releases CO_2 that reacts with CH to form $CaCO_3$ as showed in Eq. 1-3.

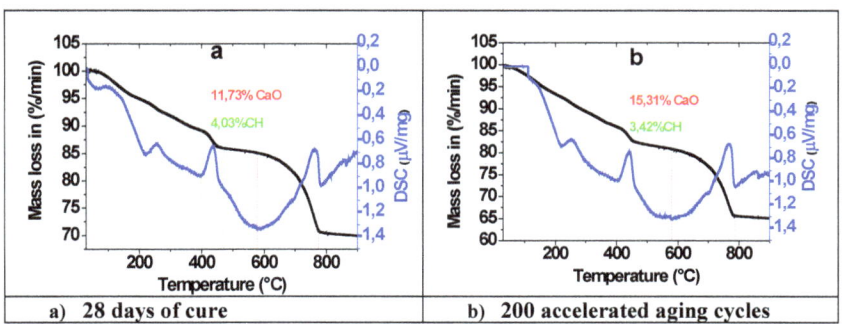

***FIGURE 3** - WEIGHT LOSS AND DIFFERENTIAL SCANNING CALORIMETRY (DSC) OF
THE FIBER-CEMENT HYDRATED WITH WATER. TESTED AFTER 28 DAYS OF CURE AND
200 ACCELERATED AGING CYCLES.*

The presence of CO_2 during the hydration of cement (with CH present), generates insoluble carbonates and their polymorphs [30], decreasing the porosity of the fiber-cement matrix. The reduction of CH by reaction with CO_2 in the fiber-cement matrix can be observed when comparing Figures 3 and 4.

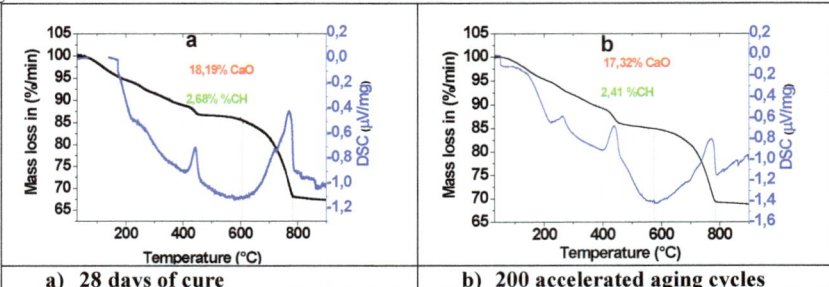

| a) 28 days of cure | b) 200 accelerated aging cycles |

FIGURE 4 - WEIGHT LOSS AND DIFFERENTIAL SCANNING CALORIMETRY (DSC) OF THE FIBER-CEMENT SAMPLES HYDRATED WITH CARBONATED WATER. TESTED AT AFTER 28 DAYS OF CURE AND 200 ACCELERATED AGING CYCLES.

$CaCO_3$ starts to degrade at around 600°C while CH at 400°C, furthermore, the pathway of oxidation is different [30] as showed in Figures 3 and 4. Fiber-cement samples using carbonated water produces 50% more CaO (formed by the oxidation of CH and $CaCO_3$) when compared to reference water, indicating more efficiency in carbonation.

Physical properties of the fiber-cement composites
The average and standard deviation values of water absorption (WA) and bulk density (BD) of the composites in the different conditions of curing with tap water and carbonated water are presented in Figure 5. The carbonation process led to densification of the fiber-cement and contributed to the reduction of water absorption (WA), in special when using carbonated water (Figure 5).

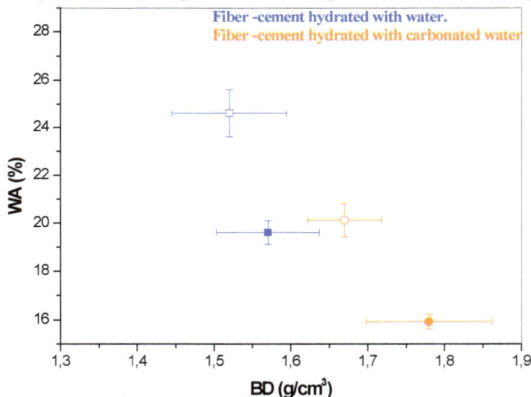

FIGURE 5 - AVERAGE VALUES OF WATER ABSORPTION (WA) AND BULK DENSITY (BD) OF THE FIBER-CEMENT COMPOSITES WITH WATER. TESTED AT 28 DAYS (EMPTY SYMBOL) AND AFTER 200 CYCLES OF ACCELERATED AGING (FULL SYMBOL).

Non-Conventional Materials and Technologies – NOCMAT for XXI Century   Materials Research Forum LLC
Materials Research Proceedings **7** (2018) 95-107   doi: http://dx.doi.org/10.21741/9781945291838-10

The densification of the composites minimized the pathways of the hygroscopic migration due to the filling of pores by the growth of $CaCO_3$ microstructure [32] that turns denser the structure and causes reduction of WA. These denser materials have a stabilized volume due to the refinement of pore sizes, which generates less shrinkage by drying.

The bounce observed in the density when compared 28 days to 200 aging cycles (Figure 5) can be explained by the released of CO_2 gas during the process of hydration by the use of carbonated water. These gas concentration have been trapped in to the fiber-cement, like in a chamber, impaired the microstructure and generate higher porosity in the composite. However, during the accelerated ageing cycles occurred the partial fill up of pores by $CaCO_3$ and consequently the densification.

After 200 accelerated ageing cycles, a densification of the composites was observed due to the continued hydration of the anhydrous components and their re-precipitation in the matrix pores and in the cavities of the cellulosic fibers [9]. Additionally, more $CaCO_3$ were formed due to natural carbonation by the absorption of CO_2 from the environment.

Mechanical properties

The results of MOR *vs.* SE (Figure 6) confirm that the high density imply on the increase of the mechanical strength [29], which made the composite stiffer and stronger, as also observed by Almeida [32]. This happened due to the increase in bulk density and decrease in porosity, leading to decrease of specific energy (SE).

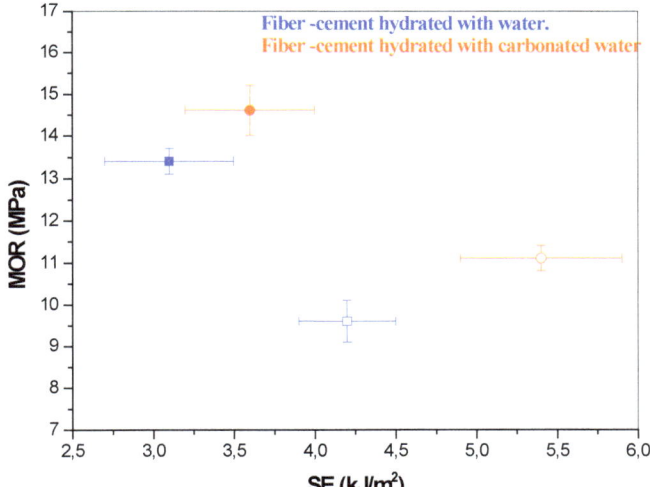

FIGURE 6 - AVERAGE VALUES OF MODULUS OF RUPTURE (MOR) AND SPECIFIC ENERGY (SE) OF FIBER-CEMENT HYDRATED WITH WATER. TESTED AT 28 DAYS (EMPTY SYMBOL) AND AFTER 200 CYCLES OF ACCELERATED AGING (FULL SYMBOL).

The carbonate formation contributes to protect the cellulosic fiber due to the low pH reduction of the fiber-cement matrix, by the consumption of hydroxyl anion (^-OH) as shown in Eqs. 1, 2 and 3.

Non-Conventional Materials and Technologies – NOCMAT for XXI Century   Materials Research Forum LLC
Materials Research Proceedings 7 (2018) 95-107   doi: http://dx.doi.org/10.21741/9781945291838-10

These trends were also reported in literature [9] and [33] and observed in the fiber-matrix interface (Figures 7 and 8).

$$CO_{2(gás)} + 2OH^-_{(aquoso)} \Leftrightarrow CO_3^{2-}{}_{(aquoso)} + H_2O_{(liquido)} \qquad \text{(Eq. 1)}$$

$$Ca(OH)_{2(solido)}\downarrow \Leftrightarrow Ca^{2+}{}_{(aquoso)} + 2OH^-_{(aquoso)} \qquad \text{(Eq. 2)}$$

$$Ca^{2+}{}_{(aquoso)} + CO_3^{2-}{}_{(aquoso)} + H_2O_{(liquido)} \Leftrightarrow CaCO_{3(solido)}\downarrow + H_2O_{(liquido)} \qquad \text{(Eq. 3)}$$

The higher SE values for carbonated composites in relation to control (cured with tap water) after ageing cycles (Figure 6) corroborate to this hypothesis of preservation of the cellulosic fibers, which were less affected by the alkaline attack of the carbonated cementitious matrix.

After 200 immersion-drying cycles, the composites cured with tap water presented lower SE values. This decrease is caused by the antagonism of the fiber/cement adhesion after cure process. Additionally, the embrittlement of the Curauá fibers is accelerated due to their mineralization, caused by the re-precipitation of cement hydration products in the cell wall of the cellulosic fibers and in its interior [34].

Microstructure

The SEM/BSE micrographs of composites cured with tap water and/or carbonated water are showed in Figures 7 to 9. The vegetable fibers are free from hydration products in their interior, in those composites cured with carbonated water (Figure 7). It occurs because part of $^-$OH ions was consumed and caused pH reduction. This reduction helps to preserve natural fibers reducing the hydrolysis process. In addition, part of the Ca^{++} released during the hydrolysis is transformed in insoluble $CaCO_3$. Thus, carbonated water minimized the cellulosic chain mineralization and hydrolysis process, since it consumed part of the $^-$OH and Ca^{++} ions (Eqs. 1, 2 and 3).

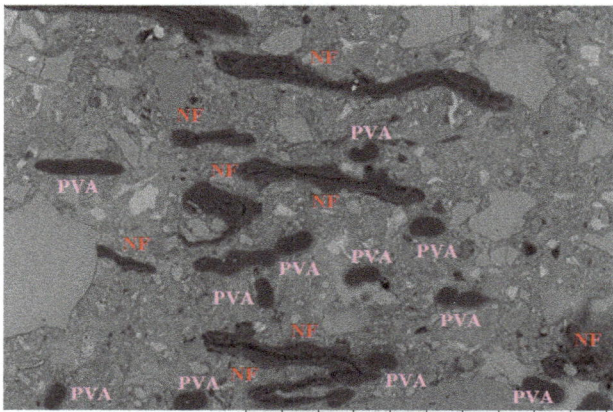

NL   D7.9   x500   200 um

FIGURE 7 - SEM IMAGES ON BACKSCATTERING (BSE) MODE OF POLISHED SURFACE OF COMPOSITE WITH CARBONATED WATER CURED FOR 28 DAYS. LEGENDS: NATURAL FIBER (PULP) - NF AND PVA.

Non-Conventional Materials and Technologies – NOCMAT for XXI Century   Materials Research Forum LLC
Materials Research Proceedings **7** (2018) 95-107   doi: http://dx.doi.org/10.21741/9781945291838-10

In contrast, the cure with tap water presents an excess of Ca^{++} ions in the fiber-cement matrix, which led to their re-precipitation into the fibers [35-36]; as observed within the fibers (the white dots) in Figure 8. This re-precipitation of solids into the fiber voids (both fiber cell wall and fiber lumen) induces the embrittlement of the cellulose fibers. Furthermore, there is an increase of the pH of reaction, allowing the hydrolysis of the fiber chain and resulting on a loss of mechanical strength.

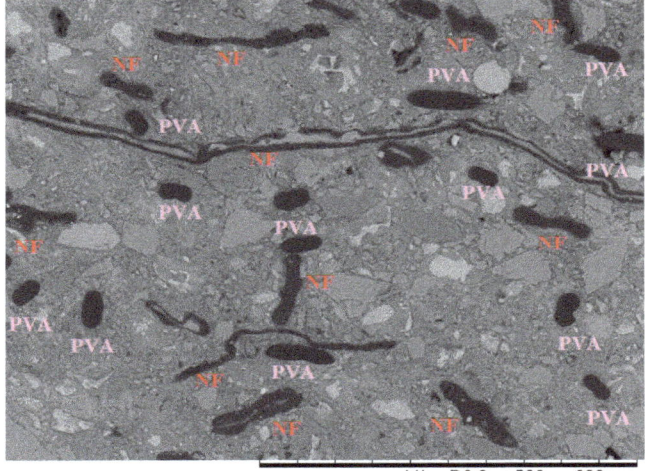

NL   D8.3   ×500   200 um

***FIGURE 8** - SEM IMAGES ON BACKSCATTERING (BSE) MODE OF POLISHED SURFACE OF COMPOSITE WITH WATER CURED FOR 28 DAYS. LEGENDS: NATURAL FIBER (NF) AND PVA.*

Figure 8 reveals that for composites made with tap water, the ions migration of the alkaline solution goes to the inner of the fibers. So the performance of the fiber-cement composites is impaired in the long term [9]. However, no attack was observed in the PVA fibers, since they are industrially fabricated for a good synergy with cement matrix.

The accelerated ageing cycles did not cause a significant difference on the microstructure of the composites when compared to 28 days of cure (Figures 7 and 9). There was only a slight re-precipitation of hydration products within the fibers after 200 ageing cycles (SM), when carbonated water were used, and there were no substantial difference using tap water for 28 days or 200 cycles. Additionally, no differences on the microstructure of the composites were detected for PVA fibers under the different conditions.

Non-Conventional Materials and Technologies – NOCMAT for XXI Century   Materials Research Forum LLC
Materials Research Proceedings 7 (2018) 95-107   doi: http://dx.doi.org/10.21741/9781945291838-10

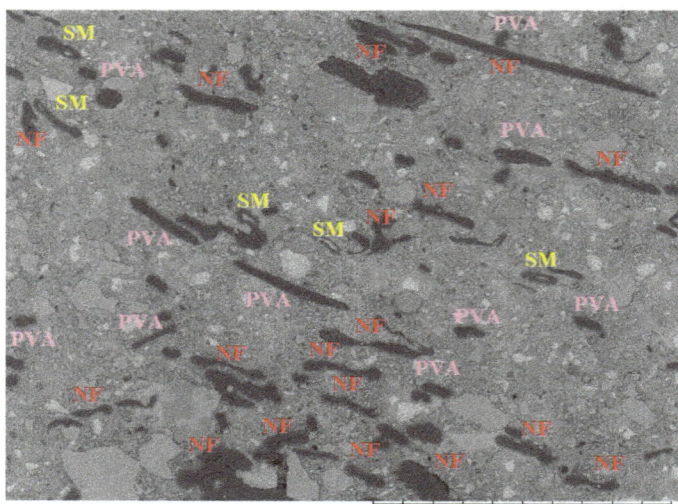

FIGURE 9 - *SEM IMAGES ON BACKSCATTERING (BSE) MODE OF POLISHED SURFACE OF COMPOSITE WITH CARBONATED WATER CURED FOR 28 DAYS, FOLLOWED BY 200 SOAK & DRY CYCLES CYCLES. LEGENDS: NATURAL FIBER (NF), START OF MINERALIZATION (SM) AND PVA.*

The accelerated ageing cycles did not cause a significant difference on the microstructure of the composites when compared to 28 days of cure (Figures 7 and 9). There was only a slight re-precipitation of hydration products within the fibers after 200 ageing cycles (SM), when carbonated water were used, and there were no substantial difference using tap water for 28 days or 200 cycles. Additionally, no differences on the microstructure of the composites were detected for PVA fibers under the different conditions.

Conclusion

The carbonated water performed as a great vehicle for hydration and cure of fiber-cement composites.

The use of carbonated water promoted the decrease of pH and the increase of the calcium consume altering the amount of $CaCO_3$ produced.

The carbonation process contributed to improve the preservation of the vegetable fibers, as well as densified the matrix and the transition zone between fiber and cement. The densification of the matrix contributed to the reduction of water absorption.

The use of carbonated water improved mechanical performance under flexural test in 16%, as showed in MOR measure, it occurred by the higher presence of $CaCO_3$ during the hydration due to the new curing system proposed in this work.

The toughness of the composites measured thru the higher specific energy (SE) values for carbonated composites in relation to cured with tap water after ageing cycles indicates preservation of the cellulosic fibers

The cure proposed too increased the density in 14% by the partial fill up the cement's pores with $CaCO_3$.

Acknowledgments

The Brazilian National Council for Scientific and Technological Development – CNPq (project #150384/2016-5), also grateful to Coordination of Improvement of Higher Education Personnel – CAPES and São Paulo Research Foundation – FAPESP (process #2017/10123-3, #2012/51467-3) and Research Nucleus on Materials for Biosystems.

References

[1] ARDANUY, M., CLARAMUNT, J., TOLEDO Filho, R. D. Cellulosic fiber reinforced cement-based composites: A review of recent. Construction and Building Materials. 2015, v. 79, p. 115-128. https://doi.org/10.1016/j.conbuildmat.2015.01.035

[2] XIE, X., ZHOU, Z. MAN J, XIAOLING, X, ZEYONG W., HUI, D. Cellulosic fibers from rice straw and bamboo used as reinforcement of cement-based composites for remarkably improving mechanical properties. Composites Part B. 2015, v. 78, p. 53.161.

[3] AKERS, S. A. S., PARTL, M. "Hygral & Thermal Expansion/Shrinkage Properties of Asbestos Free Fibre Cement". Cement & Concrete Composites. 1990, vol. 12, p.23. https://doi.org/10.1016/0958-9465(90)90032-S

[4] IDIART, A. E. Coupled analysis of degradation processes in concrete specimens at the meso-level. Barcelona: Universitat Politécnica de Catalunya, 2009.

[5] FRÍAS, M., VILLAR, E., SAVASTANO JR, H. Brazilian sugar cane bagasse ashes from the cogeneration industry as active pozzolans for cement manufacture. Cement e Concrete Composites. 2011, v. 33, p. 490-496. https://doi.org/10.1016/j.cemconcomp.2011.02.003

[6] CORREIA, V. C.; SANTOS, V.; RODIER, L. B.; GHAVAMI, K.; SAVASTANO JR, H. Bamboo fiber at macro-, micro- and nanoscale for application as reinforcement. Green Materials. 2016, v. 4, p. 26. https://doi.org/10.1680/jgrma.15.00026

[7] CORREIA, V.; C.; SANTOS, V.; SAIN, M.; SANTOS, S. F.; LEÃO, A. L.; SAVASTANO Jr, H. Grinding process for the production of nanofibrillated cellulose based on unbleached and bleached bamboo organosolv pulp. Cellulose. 2016, v. 23, p. 2971-2987. https://doi.org/10.1007/s10570-016-0996-9

[8] BALLESTEROS, J. E. M.; SANTOS, V.; MÁRMOL, G.; FRÍAS, M.; FIORELLI, J. Potential of the hornification treatment on eucalyptus and pine fibers for fiber-cement applications. Cellulose. 2017, v. 24, p. 2275-2286. https://doi.org/10.1007/s10570-017-1253-6

[9] TONOLI, G. H. D.; PIZZOL, V. D.; URREA, G.; SANTOS, S. F.; MENDES, L. M.; SANTOS, V.; JOHN, V. M.; FRÍAS, M.; SAVASTANO, H. Rationalizing the impact of aging on fiber-matrix interface and stability of cement-based composites submitted to carbonation at early ages. Journal of Materials Science. 2016, v. 51, p. 7929-7943. https://doi.org/10.1007/s10853-016-0060-z

[10] LIN, X. Effect of Early Age Carbonation on Strength and PH of Concrete. M.Eng Thesis, McGill University, Canada, 2007.

[11] ACHOUR, A. CHOMARI, F. BELAYACHI, N. Properties of cementitious mortars reinforced with natural fibers. Journal of Adhesion Science and Technology. 2017, v.1, p. 1-25.

[12] MOUNANGA, P. KHELIDI, A.; LOUKILI, A.; BAROGHEL-BOUNY, V. Predicting Ca(OH)$_2$ content and chemical shrinkage of hydrating cement paste using analytical approach. Cement and Concrete Research. 2014, v. 34, p. 255-265. https://doi.org/10.1016/j.cemconres.2003.07.006

[13] SOROUSHIAN, P.; WONB, JP.; HASSAN, M. Durability characteristics of CO$_2$-cured cellulose fiber reinforced cement composites. Construction and Building Materials. 2012, v. 34, p.44-53. https://doi.org/10.1016/j.conbuildmat.2012.02.016

[14] LESTI, M.; TIEMEYER, C. and PLANK, J. CO$_2$ stability of Portland cement based well cementing systems for use on carbon capture & storage (CCS) wells. Cement and Concrete Research. 2013, v 45, p. 45-54. https://doi.org/10.1016/j.cemconres.2012.12.001

[15] ABABNEH, A.; AL-ROUSAN, R.; ALHASSAN, M. A.; SHEBAN, M. A. Assessment of shrinkage-induced cracks in restrained and unrestrained cement-based slabs. Construction and Building Materials. 2017, v.131, p. 371–380. https://doi.org/10.1016/j.conbuildmat.2016.11.036

[16] GOMES, C. E. M. & SAVASTNO H. Jr. Study of hygral behavior of non-asbestos fiber cement made by similar hatschek process. Materials Research. 2014, v 17, p.121-129. https://doi.org/10.1590/1516-1439.224713

[17] AMERICAN SOCIETY FOR TESTING AND MATERIALS ASTM C150: Standard Specification for Portland Cement. West Conshohocken, PA, 2007.AMERICAN SOCIETY FOR TESTING AND MATERIALS. ASTM C1185 – 08: Standard Test Methods for Sampling and Testing Non-Asbestos Fiber-Cement Flat Sheet, Roofing and Siding Shingles, and Clapboards, West Conshohocken, PA, 2016.

[18] AMERICAN SOCIETY FOR TESTING AND MATERIALS. ASTM C 948-81: test method for dry and wet bulk density, water absorption, and apparent porosity of thin sections of glass-fiber reinforced concrete. West Conshohocken, PA, 1981.

[19] ASSOCIAÇÃO BRASILEIRA DE NORMAS TÉCNICAS. NBR 5733: Cimento Portland de alta resistência. Rio de Janeiro, 1991. p. 1453.

[20] RILEM, T. C. Testing Methods for Reinforced cement based composite. Materials and Structures, v.,17, (1984), p. 441-456.

[21] DIAS, C. M. R.; SAVASTANO-Jr., H. and JOHN, V. M. Exploring the potential of functionally graded materials concept for the development of fiber cement. Construction and Building Materials. 1010, v.24, p. 140-146.

[22] RILEM, T. C. Testing Methods for Reinforced cement based composite. Materials and Structures, v.,17, (1994), p. 441-456.

[23] BAQUERIZO, L. G.; MATSCHEI, T.; SCRIVENER, K. L.; SAEDIPOURS, M.; THORELL, A.; WADSO, L. Methods to determine hydration states of minerals and cement hydrates Cem. Concr. Res. 2014, v. 65, p. 85–95. https://doi.org/10.1016/j.cemconres.2014.07.009

[24] REAUNUDIN, G.; FILINCHUK, Y.; NEUBAUER, F.; NUENHOEFFER-GOETZ, A. A comparative structural study of wet and dried ettringite, Cem. Concr. Res. 2010, v.40, p. 370–375. https://doi.org/10.1016/j.cemconres.2009.11.002

[25] TAYLOR, H. F. W.; FAMY, C.; SCRIVENER, K. L. Delayed ettringite formation. Cem. Concr. Res. 2001, v.31, pp. 683–693. https://doi.org/10.1016/S0008-8846(01)00466-5

[26] NASSAR, R. U. D., & SOROUSHIAN, P. (2012). Strength and durability of recycled aggregate concrete containing milled glass as partial replacement for cement. *Construction and Building Materials*. 2012, *v.29*, p. 368-377. https://doi.org/10.1016/j.conbuildmat.2011.10.061

[27] TONOLI, G, H, D.; SANTOS, S, F.; JOAQUIM, B, A, P.; SAVASTANO JR, H. Effect of accelerated carbonation on cementitious roofing tiles reinforced with lignocellulosic fibre. Construction and Building Materials. 2010, v. 24, p.193-201. https://doi.org/10.1016/j.conbuildmat.2007.11.018

[28] SKINNER, B. L.; CHAE, S. R.; BENMORE, C. J.; WENK, H. R.; MONTEIRO, P. J. M. Nanostructure of Calcium Silicate Hydrates in Cements. Physical review letters. 2010, v.104, p. 195502. https://doi.org/10.1103/PhysRevLett.104.195502

[29] GRANGEON, S.; CLARET, F.; LEROUGE, C.; WARMONT, F.; SATO, T.; ANRAKU, S.; NUMAKO, C.; LINARD, Y.; LANSON, B. On the nature of structural disorder in calcium silicate hydrates with a calcium/silicon ratio similar to tobermorite. Cement and Concrete Research. 2013, v. 52, p. 31-37. https://doi.org/10.1016/j.cemconres.2013.05.007

[30] GOPI, S.; SUBRAMANIAN, V. K.; PALANISAMY, K. Aragonite–calcite–vaterite: A temperature influenced sequential polymorphic transformation of $CaCO_3$ in the presence of DTPA. Materials Research Bulletin. 2013, v 48, p. 1906-1912. https://doi.org/10.1016/j.materresbull.2013.01.048

[31] SCRIVENER, K.; SNELLINGS, R.; LOTHENBACH, B. A Practical Guide to Microstructural Analysis of Cementitious Materials. CRC Press is an imprint of Taylor & Francis Group, an Informa business, 2016.

[32] ALMEIDA, A. E.F.S, TONOLI, G. H.D., SANTOS, S.F., SAVASTANO-Jr H. Improved durability of vegetable fiber reinforced cement composite subject to accelerated carbonation at early age. Cem Concr Compos. 2013, v.22, p.49-58. https://doi.org/10.1016/j.cemconcomp.2013.05.001

[33] PIZZOL, V. D., MENDES, L M, FREZZATTI, L, SAVASTANO Jr H, TONOLI, G. H. D. Effect of accelerated carbonation on the microstructure and physical properties of hybrid fiber-cement composites. Minerals Eng. 2014, v. 59, p.101-106. https://doi.org/10.1016/j.mineng.2013.11.007

[34] MOHR B. J.; NANKO, H.; KURTIS, K.E. Durability of kraft pulp fiber-cement composites to wet/dry cycling. Cement and Concrete Composites. 2005, v. 27, p.435-448. https://doi.org/10.1016/j.cemconcomp.2004.07.006

[35] BRAQUERIZO, L.; MATSCHEIA, T.; SCRIVENE, K. L. Impact of water activity on the stability of ettringite. Cement and Concrete Research. 2016, v. 79, p. 31-34. https://doi.org/10.1016/j.cemconres.2015.07.008

[36] PIZZOL V. D., MENDES, L. M, SAVASTANO-Jr H, FRIAS, M, DAVILA, F.J, CINCOTTO, M. A, JOHN, V. M, TONOLI. G. H. D. Mineralogical and microstructural changes promoted by accelerated carbonation and ageing cycles.

Non-Conventional Materials and Technologies – NOCMAT for XXI Century   Materials Research Forum LLC
Materials Research Proceedings 11 (2018) 108-118   doi: http://dx.doi.org/10.21741/9781945291838-11

Stacked-MFC into a Typical Septic Tank used in Public Housing

Jorge Arturo Dominguez Maldonado [a], Gerardo Cámara-Chalé [b],
Rodrigo Moreno Cervera [c], Liliana Alzate-Gaviria [d*]

[a] Joe2@cicy.mx, [b] gerardo_camara@hotmail.es, [c] ingermc26@gmail.com, [d] Lag@cicy.mx

Yucatan Center for Scientific Research (CICY), Renewable Energy Unit, Calle 40 No. 130,
Colonia Chuburná de Hidalgo, 97200 Mérida, Yucatán, México

Abstract. We assessed the viability of incorporating a Microbial Fuel Cell (MFC) Stack in a typical septic tank, eight MFCs with proton exchange membrane without catalysts were installed in a real system. Both chemical oxygen demand (COD) removal and electricity generation using super capacitors for electricity storage were investigated under continuous flow mode. Three MFCs A1, B4 and C2 with 109.40±34.25 mW/m3, 131.58±27.75 mW/m3 and 124.01±27.57 mW/m3, respectively, were chosen for testing. The organic loading rate was 0.24, 0.52 and 1.05 Kg DQO/m3-d corresponding to the 200, 500 and 1000 ppm. Total COD removal and total columbic efficiency were 89.67±5.19% and 48.07±2.33%.The results of this study suggest that MFCs may be suitable for deployment in a septic tank. This research has demonstrated the great challenges in applying a stack of MFC in scale-up; however, the configuration seems to be indicated for this kind of depuration system. It is necessary to develop more research in real scale to test the feasibility of this implementation.

Keywords: Microbial Fuel Cell, Scale-up, Septic Tank, Wastewater Treatment

Introduction

Water is one of the basic elements for the existence of life because it is impossible for all biological processes to be performed without it. During centuries, water was considered an infinite resource but is becoming scarce in areas where previously it was not. Mexico treats about 60% of wastewater; however, in the state of Yucatan this percentage is less than 10% [1]. Water has now been recognized as a strategic issue involving national security and has become a central element to the current environmental and economic policies, as well as a key factor of social development. Surface water must be kept free from wastewater discharges in order to avoid affecting their natural capacities of assimilation and dilution and to ensure that all water resources in the country regain their health.

Mérida city has soils of permeable calcareous sedimentary rocks that hinder the installation of sewerage systems, due in part to the impact of a meteorite of the Baptistine family 65 million years ago which formed a crater of 180 kilometers in diameter forming one of the biggest impact areas of the world (Chicxulub Crater).

Wastewater generated in Yucatan homes is arranged in septic tanks but is a potential source of water pollution. However, their contribution to freshwater eutrophication and impacts on human health are uncertain and difficult to quantify [2].

The only source of drinking water is the karstic aquifer beneath the city. This aquifer is vulnerable to contamination by these effluents [3]. One solution to this problem is to implement technology of microbial fuel cells that are basically bioreactors that use bacteria as electro catalysts to convert residual biomass into bioenergy [4].

One additional challenge in the use of MFCs for domestic wastewater treatment that has not received sufficient attention is the performance of reactors containing many electrodes [5-9].

Likewise, the Nafion 117 membrane is still undeniably the most commonly used membrane in MFC applications because of its good conductivity and should be modified to improve its properties, such as preventing biofouling formation on its surface easily and reducing oxygen and substrate crossover. Until now, none of the membrane separators have been able to avoid the formation of biofilm on their surfaces. Hence, the membranes have to be replenished once they are severely fouled; indirectly increasing the MFC's cost [10].

The objective of this study was to design, assemble, install and check performance (in terms of removal of organic matter and energy generation) of a stack of microbial fuel cells in a conventional septic tank design used in Yucatán, Mexico.

Experimental

MFC-Stack construction

MFC construction type PEM, passive air cathode MFCs (2 L anode and 0.5 L cathode), was constructed as described. Each MFC contained granular carbon and stainless steel mesh (size 400£400, alloy 316) using metal mesh current collectors. Each MFC contained two cylindrical chambers: anode chamber with a diameter of 10 cm by 40 cm with SW as affluent and the cathode with a diameter of 8 cm by 20 cm and synthetic greywater (SGW) as influent. The stack of 15 MFC's was separated by a pro- ton exchange membrane (PEM) Nafion 117 with a diameter of 7.7 cm. Assembling was done on a flat support with fixing holes which prevented deforming and leaking. The cathode is contained in the anode chamber so that the membrane is in contact with both anolyte and catholyte. The anode's bottom was reduced in size because it was necessary to adapt it to a second chamber of the septic tank. The influent of anode was at the bot- tom (upstream) and effluent at the top. The top (downstream) fed the cathode

MFC-Stack within typical septic tank

The material used for civil works was of conventional construction such as cement, sand, gravel and smooth finished walls. The total volume of the septic tank was 2.93 m^3 with a useful volume of 2.44 m^3.

SMFC consisted of 15 individual cells distributed in the second chamber, which initially correspond to the filter holes.

The septic tank was covered with polycarbonate lids to prevent entry of rain water, insects and trash, while allowing for the achievement of anaerobic conditions and quick access for physical chemical and electrochemical tests.

MFC-Stack operation

Leak testing of the septic tank was performed according to Mexican law (NOM-006-CNA-1997) [11].

The oxygen used in cathodes comes from SGW with a concentration of approximately 5 ppm, according to literature [12, 13].

SW and mixture inoculum [14, 15] were added in the first chamber of the septic tank.

The system was initially operated in batch mode for 15 days and 22 days in continuous flow (Start-up). A carbon source (glucose) was provided every 24 hours and thereafter continuously fed for 90 days with SW concentration as follows: 200, 500 and 1000 ppm [16].

The design was made for a house inhabited by 5 people and the hydraulic system had a gravity discharge flow from its tributary until effluent.

Energy storage

Due to the fact that stack-MFC shares the same anolyte, it is not possible to add the voltages of the 15 MFC's. Therefore, the energy generated by each MFC was harvested externally through a card (supercapacitors connected in series) that allows storage of energy and voltage rise at the same time

Non-Conventional Materials and Technologies – NOCMAT for XXI Century   Materials Research Forum LLC
Materials Research Proceedings 11 (2018) 108-118   doi: http://dx.doi.org/10.21741/9781945291838-11

[17]. The electronic component used was a charge pump circuit topology consists of analog switches to control the connections of voltages to several capacitors. The basic 2x boost charge pump is shown in Figure 1. For this simple boost application, analog switches SW1, SW2 and SW3 are closed when the control is asserted, charging C1 and C2 to the voltage present at VIN. This is typically called the charge phase. During the second phase switches SW1, SW2 and SW3 are opened when the switch control is de-asserted, connecting C1 and C2 in series resulting in VOUT which is effectively double the voltage of VIN. The pulsing or switching noise is filtered by the capacitor COUT [18].

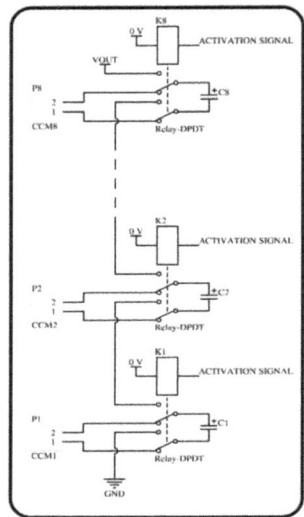

Figure 1. Electronic Circuit with capacitor arrangement.

Calculations and measurements Polarization curves and impedance
To measure the internal resistance, the SMFC was connected to a Biologic VSP potentiostat in a three-electrode mode: one electrode serving as the working electrode (anode or cathode) and the others acting as reference (saturated calomel) and counter electrodes (platinum mesh). The electrochemical impedance spectra (EIS) measurements were performed by using an Interface instrument Biologic VSP. After this, each MFC was operated under open circuit conditions for 2 h. The impedance spectra were recorded in the frequency range from 0.01 to 100,000 Hz by applying a sinusoidal excitation signal of 10 mV. The data were fitted to an equivalent electrical circuit by using the Ec-Lab® (version 10.37) impedance analysis software.

Chemical analyses
The pH and temperature were measured with a Thermo Scientific Orion® multiparameter meter. Chemical Oxygen Demand (COD) was measured with the potassium dichromate in digestion solution technique (high range COD reagent from 0 to 15,000 ppm).

Non-Conventional Materials and Technologies – NOCMAT for XXI Century   Materials Research Forum LLC
Materials Research Proceedings 11 (2018) 108-118   doi: http://dx.doi.org/10.21741/9781945291838-11

Results and discussion

The start-up was done with 200, 500 and 1000 ppm respectively and finished when three periods or cycles of electricity generation were carried out and SMFC were acclimated with the same external resistance (1000 Ω) [19]. The nomenclature used for naming the each MFC is from 1 to 5 for rows and A, B and C for the three columns. Each fed-batch cycle was typically around 4-5 days [8]. The organic loading rates were 0.24, 0.52 and 1.05 Kg DQO/m^3 d, corresponding to the affluents 200, 500 and 1000 ppm. Likewise, average total COD removal and total columbic efficiency were 89.67±5.19% and 48.07±2.33%, respectively.

Energy storage and the maximum power harvesting

Fifteen MFCs share the same electrolyte in continuous flow conditions. Although the resistors make it straightforward to measure the MFC power generation, this scheme cannot be used for practical purpose [20]. For efficient harvesting and using of the MFC energy, a power conversion circuitry is indispensable for capturing the electrical energy from MFCs and shaping it into a usable form [21]. Furthermore, the MFC units in the stack connected either serial or parallel connections; this increased the total available electrode surface area but did not change the reactor volume that involves the voltage reversal [22]. Therefore, in this study an external connection was made in series of 15 supercapacitors initially connected to each MFC, using a charge pump power supply that is a DC to DC converter that uses supercapacitors as energy storage elements to efficiently create either a higher or lower output voltage from each MFC [18]. Likewise, the charge pumps may not affect parameters as microbial activity and community much because these devices just receive whatever amount of power provided by the MFC [23].

Figure 2 A, B and C show the polarization and power density curves obtained in the steady-state. The maximum power density produced by the MFC with 0.24, 0.52 and 1.05 Kg DQO/m^3 d were 146.67, 151.2 and 152.7 mW/m^3, with an average of 302, 323 and 351 mV respectively. The corresponding external resistors at the peak power density were 73 Ω, 53 Ω and 39 Ω.

The charge pump was able to harvest the stack-MFC energy. During the initial load conditions the voltage of the super-capacitors was 0 V causing a short circuit on the CCM [17, 24], when increase R-value the capacitor stores more energy and so increase voltage of the capacitor and the cell.

Charge mode; around the time (t) = 18.12 hours the capacitor voltage equals the open circuit voltage of the MFC that is when R is much greater than the internal resistance of the MFC and no electron flow to the capacitor [24].Once loaded 15 capacitors were connected in series with each other to amplify the voltage at 3.11 V, and in the discharge process of capacitors transfer energy to a battery with a stable voltage of 2.2 V.

The initial voltage transfer from capacitors to the battery was 2.2 V up to the theoretical value of battery charge of 2.4 V (t = 378 s). The maximum current transferred by connecting the capacitors to the battery was 1.98 mA, reaching a minimum value of 1.14 mA and maximum power transferred of 4.51 mW.

Likewise, the integration of the power transferred over time, it allowed to obtain the total energy transferred from the capacitors to the battery (1.117 J).

Performance of MFC units

Three of 15 MFCs corresponding to the highest (A1), intermediate (B4) and lower (C2) power density were chosen for each organic loading rate evaluated.

As shown in Figure 2A, to the substrate concentration of 200 ppm, the MFC A1 had the highest power density of 146.67±2.12 mW/m^3 and lower internal resistance 99.68±5.44 Ω; MFC B4 was an intermediate power density of 102.24±6.03 mW/m^3 and an internal resistance of 122.24±12.78 Ω and the low power density corresponded to the MFC C2 with 79.30±3.11 mW/m^3 and an internal resistance of 139.66±11.69 Ω.

At a substrate concentration of 500 ppm (Figure 2B), the MFC A1 had the highest power density of 151.20 ± 9.77 mW/m^3 and lower internal resistance 87.93 ± 8.83 Ω. Meanwhile, MFC B4 had a power density intermediate of 111.96 ± 4.32 mW/m^3 and an internal resistance of 95.09 ± 10.32 Ω and the lower power density corresponded to the MFC C2 of 91.12 ± 7.12 mW/m^3 with an internal resistance of 96.78 ± 12.64 Ω.

When it used 1000 ppm, as shown in Figure 2C, MFC A1 had the highest power density 152.70 ± 8.34 mW/m^3 with an internal resistance of 76.68 ± 6.85 Ω. MFC B4 had an intermediate power density of 121.63 ± 9.42 mW/m^3 and an internal resistance of 77.48 ± 6.54 Ω and the lower power density corresponded to MFC C2 with 97.71 ± 2.69 mW/m^3 and an internal resistance of 77.74 ± 8.52 Ω, respectively.

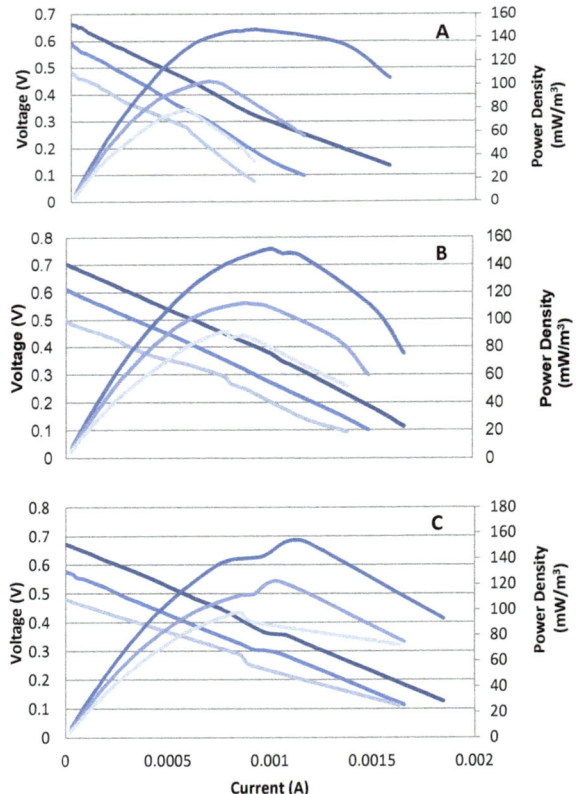

Figure 2. Polarization and power density curves in three Microbial fuel cells (A1, B4, and C2) using different COD. A) 200 ppm, B) 500 and C) 1000 ppm

Non-Conventional Materials and Technologies – NOCMAT for XXI Century   Materials Research Forum LLC
Materials Research Proceedings **11** (2018) 108-118   doi: http://dx.doi.org/10.21741/9781945291838-11

These values are almost similar to those obtained by Zhang et al. [25] who used MFC-C-AC (180 mW/m^3) in tubular MFCs that contained only activated carbon powder (5 mg/cm^2) as a cathode catalyst for more than 400 days in situ investigation in an aeration tank. Likewise, Zhang et al. [25] obtained values slightly higher in an unstable electric current constructed with ion exchange membrane and cathode catalysts, MFC-C-Pt and MFC-A-Pt, that had 0.1 mg/cm^2 of Pt (10%Pt on carbon black) and 4mg/cm^2 of activated carbon as cathode catalysts for oxygen reduction, obtaining power densities of 370 mW/m^3 and 270 mW/m^3, respectively. Rabaey et al. [26] and Zhang et al. [27] obtained from laboratory tubular shape MFCs ranges from 2 to 60 W/m.

The volumetric loading rate was increased by increasing the concentration of sucrose and buffer in the feed while maintaining constant nutrient levels. The increase in buffer concentration increased the concentration of ions in the anolyte and, therefore, decreased R_{int} from 120.53±20.05 to 77.30±2.05 Ω over the operating period. He et al. [28] found that the internal resistance of their MFC decreased from 5.44 to 2.41 Ω when the organic load was increased from 0.57 to 3.40 kg COD/m^3-day. However, while they used an organic load rate of 4.29 kg COD/m3-day, its resistance increased to 2.89 Ω, likely due to the fact that substrate concentration was higher than capacity of the MFC for conversion to electricity which caused lower power density and methane production.

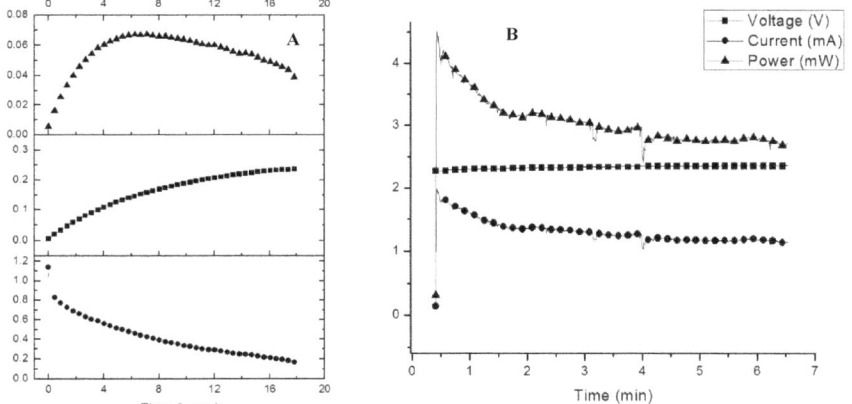

Figure 3. Voltage, current and power density curves for Microbial fuel cell as a function of time.

The stack MFC increased the power density when the substrate concentration changed from 200 ppm to 500 ppm. However, in the concentration of 1000 ppm a difference in the power densities was not observed. Likewise, COD removal was similar at 500 and 1000 ppm (about 85%). This may explain why the power density does not increase significantly with increasing concentration of the substrate because the electrons obtained from degradation of the substrate are the same. Cheng and Logan [29], obtained the maximum power density when they increased the substrate concentration at 0.15 g/L, the power density was 27 W/m^3. When the substrate concentration increased, power increased by 33% to 36 W/m^3 (0.5 g/L), and by 56% to 42 W/m^3 at 1 g/L; however, further increases in the acetate concentration to 2 g/L did not appreciably affect power density; therefore they concluded that at low substrate concentrations power can be hindered by the anode.

Non-Conventional Materials and Technologies – NOCMAT for XXI Century   Materials Research Forum LLC
Materials Research Proceedings **11** (2018) 108-118   doi: http://dx.doi.org/10.21741/9781945291838-11

Electrochemical Impedance Spectroscopy (EIS)

To obtain quantitative data on the resistances, overall internal resistance (R_{int}) and its sources were analyzed by fitting experimental data into an equivalent circuit without a diffusion resistance [28], because the diffusion resistance showed that transport limitation played a minor role, compared to ohmic and kinetic limitations (Figure 4 A, B, C). According Yuan et al. [30], at low over potentials the main contributor to the impedance is the interfacial charge transfer resistance, which is potential dependent. Dependence on electrode potential is given by the Tafel equation ($V \sim \log Rp^{-1}$).

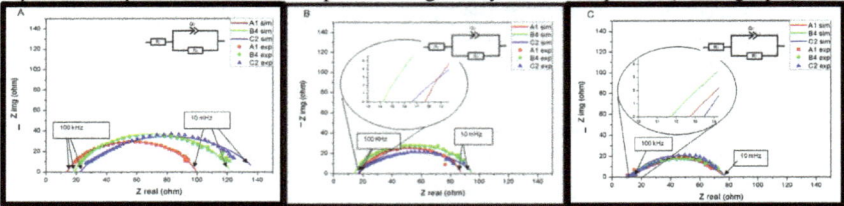

Figure 4. Nyquist plots and equivalent circuit A) 200 ppm, B) 500 and C) 1000 ppm.

The simulated data in the Nyquist plot were generated from the experimental data at the frequency range between 0.01 to 100,000 Hz. A Nyquist plot anode showed that R_{int} of SMFC at the volumetric loading rate of 0.24 kg COD/m³ day was 120.53±20.05 Ω and significantly decreased R_{int} in 23% and 36% by 93.27±7.68 Ω (0.52 kg COD/m³) and 77.30±2.05 Ω (1.05 kg COD/m³), respectively, likely due to the fact that the biofilm facilitates charge accumulation at the electrode interface. Although the values may vary based on the experimental conditions (electrode material, inoculum type, etc.), the behavior is similar to that observed in previous works [31, 32].

Rabaey et al. [26] with an Upflow microbial fuel cell obtained a lower overall R_{int} of 4 Ω compared to this study. The lower electrolyte resistance was likely caused by a higher anolyte conductivity with sodium acetate (780 mg/L) as compared to sucrose as an artificial wastewater and a lower volume to PEM surface area ratio that was compared.

The trend of ohmic resistance is decreased with increasing substrate concentration from 200 to 1000 mg/LCOD. This is likely due to the amount of ions that increase with the concentration of synthetic wastewater (SW). This favors the conductivity of the anolyte resulting in a reduced ohmic resistance (from 18.27±4.43 Ω to 12.63±0.87 Ω). The relationship between conductivity and ohmic resistance is consistent with that obtained in other studies [33, 34], using higher buffer concentrations [35] enhances MFC performance through reduction of ohmic resistance. The buffer affects the MFC performance in several ways due to its chemical composition and interaction with the electrodes, bacteria, and membrane (if present). In addition, the buffer helps to reduce changes in pH in the bulk solution and in the biofilm, and therefore it maintains the pH in the range suitable for the growth of microorganisms [36], while the effect of solution conductivity on maximum power density is consistent with the literature [34, 36, 37, 38].

As shown in Table 1, when the volumetric loading rate was increased from 0.24 to 0.52 kg COD/m³ d, R_{ct} decreased from 102.26±15.62 to 77.13±5.93 Ω. Likewise, a similar correlation between anode charge transfer resistance and power output was observed between 500 and 1000 mg/L, 77.13±5.93 Ω for 131.58±27.75 mW/m³ and 64.67±1.17 Ω corresponding to 124.01±27.57 mW/m³, respectively, thereby indicating the increased electron transfer ability through the selection of anodophilic microbes [28].

Also, it was found that there is only one semicircle over the low frequency range in the cathode EIS results, which indicates that there was no obvious diffusion effect. The diameter of the semicircle still represents the resistance. Thus, the cathode internal resistance is divided into ohmic resistance and charge transfer resistance in series [39]. The R_{int} represented the cathode with a slightly varying resistance $66.78\pm0.89\Omega$ over the operating period of the three different organic loading rates 0.24, 0.52 and 1.05 Kg DQO/m^3 d (catholyte composition was constant).

Comparison of the anode and cathode impedance reveals that anode internal resistances are higher. The larger semicircle relates approximately to the larger reaction resistance due to the slow kinetics [40]. However, anode and cathode transfer resistances were similarly limited and performance can be further improved by optimizing both the anode and the cathode reaction rates.

Table 1. *Summary of the power density and fit of experimental results from EIS for A1, B4 and C2.*

MFC	$P_{s, max}$ (mW/m^3)	Romh (Ω)	Rct (Ω)	Rint (Ω)*
200 mg/L (0.24 Kg COD/m^3 d)				
A1	146.67±2.12	13.71±0.45	85.97±4.99	99.68±5.44
B4	102.24±6.03	18.54±1.16	103.7±11.62	122.24±12.78
C2	79.3±3.11	22.56±2.01	117.1±9.68	139.66±11.69
Average	109.40±34.25	18.27±4.43	102.26±15.62	120.53±20.05
500 mg/L (0.52 Kg COD/m^3 d)				
A1	151.2±9.77	17.64±0.69	70.29±8.14	87.93±8.83
B4	111.96±4.32	14.21±2.65	80.88±7.67	95.09±10.32
C2	91.12±7.12	16.57±4.57	80.21±8.07	96.78±12.64
Average	131.58±27.75	16.14±1.75	77.13±5.93	93.27±7.68
1000 mg/L (1.05 Kg COD/m^3 d)				
A1	152.7±8.34	12.72±1.44	63.96±5.41	76.68±6.85
B4	121.63±9.42	11.72±0.98	66.02±7.54	77.48±6.54
C2	97.71±2.69	13.46±2.33	64.02±4.21	77.74±8.52
Average	124.01±27.57	12.63±0.87	64.67±1.17	77.30±2.05

P_{S-max}: maximum power density

±: standard deviation

*Internal resistance calculated from impedance analysis

Conclusions

This research has demonstrated the great challenges in applying a stack of MFC in scale-up. However, it is very important to identify the potential application into a septic tank as polishing treatment. R_{int} showed that anode resistances are higher than cathode; however, the configuration seems to be indicated for this kind of system depuration in developing countries. The difference in electrochemical performance between laboratory and scale up exposed that factors as conductivity of the anode and cathode solutions, biofilm in the anode, and the high activation energy play an important role. An electrochemical model must be developed for scale up which explains performance and electrochemical data. It is necessary to develop experiments in scale up and to test the feasibility of real implementation.

Acknowledgements
Support for this work was provided by the Mexican Council for Science and Technology
(CONACYT) grant to carry out this study through the projects CONAVI-101284, Basic Science-
106416, FINNOVA-217189 and by the scholarship 41777 awarded for master's studies.

References

[1] CONAGUA: *National Water Commission*, 2015. [Online]. Available:
http://www.conagua.gob.mx/.

[2] Withers P, May L, Jarvie H, Jordan P, Doody D, Foy R, Bechmann M, Cooksley S, Dils R,
Deal N. Nutrient emissions to water from septic tank systems in rural catchments: Uncertainties
and implications for policy. *Environ Sci Policy.* 71-82 (2012).
https://doi.org/10.1016/j.envsci.2012.07.023

[3] Metcalfe C, Beddows P, Gold Bouchot G, Metcalfe T, Li H, Van Lavieren H. Contaminants
in the coastal karst aquifer system along the Caribbean coast of the Yucatan Peninsula, Mexico.
Environ Pollution. 991-7 (2011). https://doi.org/10.1016/j.envpol.2010.11.031

[4] Shah V. *Emerging Environmental Technologies* Volume II. Springer Netherlands, pp. 174.
(ISBN 978-90-481-3352-9), 2010.

[5] Cusick R, Bryan B, Parker D, Merrill M, Mehanna M, Kiely P, Liu G, Logan B. Performance
of a pilot-scale continuous flow microbial electrolysis cell fed winery waste water. *Appl
Microbiol Biotechnol.* 2053-2063 (2011). https://doi.org/10.1007/s00253-011-3130-9

[6] Jiang D, Curtis M, Troop E, Scheible K, McGrath J, Hu B, Suib S, Raymond D, Li B. A
pilot-scale study on utilizing multi-anode/cathode microbial fuel cells MAC MFCs to enhance
the power production in wastewater treatment. *Int J hydrogen energy.* 876-884 (2011).
https://doi.org/10.1016/j.ijhydene.2010.08.074

[7] Kim J., Premier G., Hawkes F., Rodríguez J., Dinsdale R., Guwy A., Modular tubular
microbial fuel cells for energy recovery during sucrose wastewater treatment at low organic
loading rate. *Bioresour Technol.* 1190-1198 (2010).
https://doi.org/10.1016/j.biortech.2009.09.023

[8] Ahn Y, Logan B. Domestic wastewater treatment using multi-electrode continuous flow
MFCs with a separator electrode assembly design. *Appl Microbiol Biotechnol.* 409-416 (2013).
https://doi.org/10.1007/s00253-012-4455-8

[9] Yazdi H., Alzate-Gaviria L., Ren Z. Pluggable microbial fuel cell stacks for septic
wastewater treatment and electricity production. *Biores Technol.* 258-263 (2015).
https://doi.org/10.1016/j.biortech.2014.12.100

[10] Leong J, Daud W, Ghasemi M, Liew K, Ismail M. Ion exchange membranes as separators in
microbial fuel cells for bioenergy conversion: A comprehensive review. *Renewable and
Sustainable Energy Reviews.* 575-587 (2013). https://doi.org/10.1016/j.rser.2013.08.052

[11] Official Mexican Standard NOM-006-CNA-1997. Septic tanks - Specifications and test
methods. Available from http://www.profepa.gob.mx/innovaportal/file/3302/1/nom-006-
conagua-1997.pdf

[12] Gross A, Kaplan D, Baker K. Removal of chemical and microbiological contaminants from domestic greywater using a recycled vertical flow bioreactor (RVFB). *Ecol Engine.* 107-114 (2007).

[13] Eriksson E, Auffarth K, Henze M, Ledin A. Characteristics of grey wastewater. *Urban Water.* 85-104 (2002). https://doi.org/10.1016/S1462-0758(01)00064-4

[14] Canto-Canché B, Tzec-Simá M, Vázquez-Loría J, Espadas-Álvarez H, Chí-Manzanero B, Rojas-Herrera R, Valdez-Ojeda R, Alzate-Gaviria L. Simple and inexpensive DNA extraction protocol for studying the bacterial composition of sludges used in microbial fuel cells. *Genetics and Molecular Res.* 282-292 (2013). https://doi.org/10.4238/2013.February.4.2

[15] Sanchez-Herrera D, Pacheco-Catalan D, Valdez-Ojeda R, Canto-Canche B, Dominguez-Benetton X, Domínguez-Maldonado J, Alzate-Gaviria L. Characterization of anode and anolyte community growth and the impact of impedance in a microbial fuel cell. *BMC Biotechnol.* 102 (2014). https://doi.org/10.1186/s12896-014-0102-z

[16] G. Tchobanoglous, F. Burton, D. Stensel Wastewater Engineering: Treatment, Disposal and Reuse, 3rd ed., Metcalf & Eddy, McGraw-Hill, New York, 1991.

[17] Dewan A., Donovan C., Heo D., Beyenal H., Evaluating the performance of microbial fuel cells powering electronic devices. *J Power Sources.* 90-96 (2010). https://doi.org/10.1016/j.jpowsour.2009.07.001

[18] J. Fleming, (Apr. 2010); doi:10.1117/12.850316; http://dx.doi.org/10.1117/12.850316.

[19] Ahn Y, Logan B. A multi-electrode continuous flow microbial fuel cell with separator electrode assembly design. *Appl Microbiol Biotechnol.* 2241-2248 (2012). https://doi.org/10.1007/s00253-012-3916-4

[20] Ieropoulos I, Greenman J, Melhuish C. Microbial fuel cells based on carbon veil electrodes: Stack configuration and scalability. *Int J Energy Res.* 1228-1240 (2008). https://doi.org/10.1002/er.1419

[21] Park J, Ren R. High efficiency energy harvesting from microbial fuel cells using a synchronous boost converter. *J Power Sources.* 322-327 (2012). https://doi.org/10.1016/j.jpowsour.2012.02.035

[22] Wang B, Han J. A single chamber stackable microbial fuel cell with air cathode. *Biotechnol Lett.* 387-393 (2009). https://doi.org/10.1007/s10529-008-9877-0

[23] Wang H, Park J, Zhiyong J. Practical Energy Harvesting for Microbial Fuel Cells: A Review. *Environ Sci Technol.* 3267–3277 (2015). https://doi.org/10.1021/es5047765

[24] Shantaram A, Beyenal H, Raajan R, Veluchamy R, Lewandowski Z. Wireless sensors powered by microbial fuel cells. *Environ Sci Technol.* 5037-5042 (2005). https://doi.org/10.1021/es0480668

[25] Zhang F, Ge Z, Grimaud J, Hurst J, He Z. In situ investigation of tubular microbial fuel cells deployed in an aeration tank at a municipal wastewater treatment plant. *Biores Technol.* 316-321 (2013). https://doi.org/10.1016/j.biortech.2013.02.107

Non-Conventional Materials and Technologies – NOCMAT for XXI Century   Materials Research Forum LLC
Materials Research Proceedings 11 (2018) 108-118   doi: http://dx.doi.org/10.21741/9781945291838-11

[26] Rabaey K, Clauwaert P, Aelterman P, Verstraete W. Environ. Tubular Microbial Fuel Cells for Efficient Electricity Generation. *Sci Technol.* 8077-8082 (2005).

[27] Zhan F. Novel cathode materials for microbial fuel cells. A Thesis in Environ. Engin. The Pennsylvania State University The Graduate School College of Engineering, U.S, 2010.

[28] He Z, Wagner N, Minteer S, Angenent L. An Upflow Microbial Fuel Cell with an Interior Cathode: Assessment of the Internal Resistance by Impedance Spectroscopy. *Environ Sci Technol.* 5212-5217 (2006). https://doi.org/10.1021/es060394f

[29] Cheng S, Logan B. Increasing power generation for scaling up single-chamber air cathode microbial fuel cells. *Bioresource Technol.* 4468–4473 (2011). https://doi.org/10.1016/j.biortech.2010.12.104

[30] Yuan X, Wang H, Sun J, Zhang J. AC impedance technique in PEM fuel cell diagnosis: A review. *Int J Hydrogen Energy.* 4365-4380 (2007). https://doi.org/10.1016/j.ijhydene.2007.05.036

[31] Marcus A, Torres C, Rittmann B., Conduction-Based Modeling of the Biofilm Anode of a Microbial Fuel Cell. *Biotech Bioeng.* 1171-1182 (2007). https://doi.org/10.1002/bit.21533

[32] Ha P, Moon H, Kim B, Ng H, Chang I. Determination of charge transfer resistance and capacitance of microbial fuel cell through a transient response analysis of cell voltage. *Bios Bioelec.* 1629-1634 (2010).

[33] Feng Y, Wang X, Logan B, Lee H. Brewery wastewater treatment using air-cathode microbial fuel cells. *Appl. Microbiol. Biot.* 873–880 (2008).

[34] Huang L, Logan B. Electricity generation and treatment of paper recycling wastewater using a microbial fuel cell. *Appl. Microbiol. Biot.* 349–355 (2008).

[35] Min B, Roman O, Angelidaki I. 'Importance of temperature and anodic medium composition on microbial fuel cell (MFC) performance. *Biotechnol. Lett.* 1213–1218 (2008). https://doi.org/10.1007/s10529-008-9687-4

[36] Gil G, Chang I, Kim B, Kim M, Jang J, Park H, Kim H. *Biosens. Bioelectron.* 327–334 (2003). https://doi.org/10.1016/S0956-5663(02)00110-0

[37] Nam J, Kim H, Lim K, Shin H, Logan B. Variation of power generation at different buffer types and conductivities in single chamber microbial fuel cells. *Biosensors and Bioelectronics.* 1155–1159 (2010). https://doi.org/10.1016/j.bios.2009.10.005

[38] Liu H, Cheng S, Logan B. Power generation in fed-batch microbial fuel cells as a function of ionic strength, temperature, and reactor configuration. *Environ. Sci. Technol.* 5488–5493 (2005). https://doi.org/10.1021/es050316c

[39] Yin Y, Huang G, Tong Y, Liu Y, Zhang L. Electricity production and electrochemical impedance modeling of microbial fuel cells under static magnetic field. *J Power Sources.* 58-63 (2013). https://doi.org/10.1016/j.jpowsour.2013.02.080

[40] Reshetenko T, Kim H, Lee H, Jang M, Kweon H. Performance of a direct methanol fuel cell (DMFC) at low temperature: Cathode optimization. *J Power Sources.* 925-932 (2006). https://doi.org/10.1016/j.jpowsour.2006.02.058

Non-Conventional Materials and Technologies – NOCMAT for XXI Century   Materials Research Forum LLC
Materials Research Proceedings 7 (2018) 119-127   doi: http://dx.doi.org/10.21741/9781945291838-12

Cement-Based Composites Reinforced with Nanofibrillated Cellulose from Bamboo Organossolv Pulp

Viviane da Costa Correia*, Holmer Savastano Jr.

Research Nucleus on Materials for Biosystems, Department of Biosystems Engineering - Faculty of Animal Science and Food Engineering, University of São Paulo, Brazil

*vivianecostcor@usp.br, holmersj@usp.br

Abstract. Nanofibrillated cellulose has good mechanical performance and high specific surface, which contributes to improve the adhesion between fiber and matrix. The use of cellulose nanofibres as reinforcement of cementitious materials may contribute for improving particle packing, as consequence it helps to decrease porosity and crack growth rate at nanoscale, with corresponding strengthening of the composite. Thus, the aim of this work was to study the effect of 1%, 2% and 3% (by weight) of bamboo nanofibrillated cellulose as reinforcement of cementitious composites in comparison to composites reinforced with 8% of bamboo pulp. The cementitious composites were produced by slurry-dewatering method and they were subjected to physical, mechanical e microstructural analyses at 8 days of age. The results indicated that on average 3% of nanofibrillated cellulose showed higher strength than composites reinforced with 1% and 2% of fibers. Modulus of rupture of composites reinforced with 8% of bamboo pulp was lower than composites reinforced with nanofibrillated cellulose. In relation to toughness, composites reinforced with nanofibrillated cellulose were more fragile than composites reinforced with pulp. This is due to the lower amount of nanofibrillated cellulose used as reinforcement, and due to the lower length of these fibers in comparison to pulp fibers, which reduces the pull-out energy involved in the fracture. Considering the intrinsic characteristics of the nanofibrillated cellulose, this vegetable fiber showed to be a promising material for use as nanoreinforcement of the cement-based composites, however, more studies are necessary about the optimum amount of nanofibers for improving the toughness of the cementitious composites.

Keywords: Nanofibrillated Cellulose, Bamboo, Cement-Based Composites

Introduction

A cementitious material originally exhibits brittle performance, and when it is cracked, has its durability impaired. The aggressive agents from the environment, such as water, acids, chloride and sulfate ions migrate into the cracks, which accelerate degradation of the material. Cracking starts at the nanoscale and propagates along the micro and macroscale until the material ruptures. One possible way to reduce the problem of cracking and degradation of cement matrices is the incorporation of nanofibers as a reinforcement of these cementitious materials. In this application, nanofibers may act as stress transfer bridges containing and retarding the propagation of cracks.

Some studies showed the best mechanical performance of cementitious materials through the incorporation of nanofibers and carbon nanotubes [1, 2, 3, 5, 6]. Li et al. [2] produced cement pastes reinforced by 0.5% of carbon nanotubes. The authors showed that the flexural strength of these pastes were 25% higher than control pastes without nanotubes at 28 days. Konsta-Gdoutos et al. [4] produced cement pastes incorporating 0.08% of carbon nanotubes. They showed that the flexural strength of the material is increased by 34% in relation to the pastes without nano reinforcement. Additionally, the work developed by Makar et al. [1] confirmed that carbon nanotubes act as stress transfer bridges between the nanoreinforcement and the cement matrix.

Non-Conventional Materials and Technologies – NOCMAT for XXI Century   Materials Research Forum LLC
Materials Research Proceedings 7 (2018) 119-127   doi: http://dx.doi.org/10.21741/9781945291838-12

Fiber-cement reinforced with vegetable nanofibers

The use of vegetable fibers, in the macro and micro scales, as reinforcement of cementitious materials is established and its effectiveness has already been proven. Cement-based materials reinforced with cellulosic pulp (microfibers) have been commercially available in the form of building components since the 1980s [7]. These materials can be applied as hollow load-bearing walls, roofing tiles, and ceiling plates [8]. However, the use of vegetable fibers at the nanoscale is still in an early stage, but it is considered promising.

Vegetable nanofibers can be obtained in the nanofibrillated cellulose form. Nanofibrillated cellulose is a natural material produced from the isolation of the cell wall of lignocellulosic materials using shear force, without the need for chemicals as in acid hydrolysis. The production of nanofibrillated cellulose requires intensive mechanical treatment of fibrillation, where much of the amorphous phase is maintained. Through nanofibrillation morphology of fibers are modified, in which the width of fibers are decreased and their bonding potential are increased due it high specific surface area [9, 10]. As a consequence of nanodimensions, high specific surface area and high aspect ratio, nanofibrillated cellulose is a potential reinforcement material.

Ardanuy et al. [11] used 3.3% (by mass) of sisal nanofibers as reinforcement of mortar in comparison to mortar reinforced with same content of cellulosic pulp. The flexural strength of mortar reinforced with nanofibers increased of 26.4% and the modulus of elasticity increased of 41.5%, compared to the mortar reinforced with cellulosic pulp. Onuaguluchi et al. [12] produced cement pastes reinforced with nanofibers from bleached pine pulp with amount varying between 0.05 and 0.4%. The results showed that the increase of flexural strength and energy absorption of pastes reinforced with 0.1% of nanofiber, was 106% and 184%, respectively, in comparison to pastes without nanofibers.

According to the authors the improvement of these properties is attributed to the characteristics of high surface area of nanofibers, which increases the interface between fiber and matrix, and the high hydrophilicity of the cellulose, which promotes better adhesion with the cement. Additionally, the authors reported that pastes with contents above 0.1% did not present a good mechanical behavior due to the heterogeneity of the dispersion of the nanofibres when incorporated in greater quantity.

To determine the optimum content of nano fibers reinforcement, Thomson et al. [13] produced hybrid composites, by the slurry-dewatering method followed by pressing process, reinforced by 8% of cellulosic pulp and nanocrystalline cellulose with amount varying between of 0.5 and 2.0%.. The authors showed that the modulus of rupture of composites increased up to 0.5% of nanofibers and the modulus of elasticity increased with the nanofibers content up to 2.0%.

The results of these studies confirm that the use of vegetable fibers at nanoscale is effective to increase the mechanical properties of cementitious materials. However, there is no generally accepted opinion regarding the optimum nanoreinforcement content used in order to improve the properties of the material. Hence, the aim of this work was to study different levels of nanofibrillated cellulose, defining the optimum content of nanoreinforcement, and to measure its effect on the physical, mechanical, microstructural performance of the cement-based composites.

Materials and methods

Production and characterization of pulp and nanofibrillated cellulose

Unbleached cellulosic pulp was obtained from bamboo culms by the organosolv pulping method, according to Correia et al. [10]. The nanofibrillated cellulose was produced from unbleached bamboo organosolv pulp by the grinding method with 15 cycles of nanofibrillation, according to Correia et al. [10] using a commercial grinder, Supermasscolloider Mini, model MKCA 6-2, produced by Masuko Sangyo Co., Ltd., Japan.

Non-Conventional Materials and Technologies – NOCMAT for XXI Century   Materials Research Forum LLC
Materials Research Proceedings 7 (2018) 119-127   doi: http://dx.doi.org/10.21741/9781945291838-12

Chemical, physical and morphological characterization of fibers was carried out according to Correia et al. [10]. Chemical composition, physical and morphological characteristics of pulp and nanofibrillated cellulose are presented in Table 1.

TABLE 1 – *CHEMICAL COMPOSITION, PHYSICAL AND MORPHOLOGICAL CHARACTERISTICS OF PULP AND NANOFIBRILLATED CELLULOSE*

	Chemical composition	
Components (%)	Pulp	Nanofibrillated cellulose
Extractives	6.92 ± 1.47	9.07 ± 1.25
Lignin	9.85 ± 3.86	9.67 ± 2.37
Holocellulose	85.39 ± 1.42	85.28 ± 1.96
Cellulose	82.75 ± 0.42	77.82 ± 1.02
Hemicellulose	2.64 ± 0.42	7.46 ± 1.02
Physical and morphological characteristics		
	Pulp	Nanofibrillated cellulose
Density (g/cm^3)	1.51	1.42
Specific surface area (m^2/kg)	77.37	188.20
Width (nm)	2,720.0	13.88

Production of fiber-cement

The specific density of the matrix components and the fibers was measured by helium gas pycnometer (Quantachrome, Multipycnometer 1000). The values of specific density of Ordinary Portland Cement and limestone are 3.07 and 2.76 g.cm^{-3} respectively.

The cementitious composites were produced from the modification of the methodology adopted by Savastano Jr. et al. [14], using the suction method with negative pressure (~ 600 mmHg), at laboratory scale, and subsequent pressing. In the methodology described by Savastano Jr. et al. [14] the pressing used is 3.2 MPa for 5 min, however, because of the high surface area of the nanofibrillated cellulose and its high water absorption capacity, the pressing was increased to 5 MPa for 10 min [13], to reduce the water/cement ratio of the composites.

The composition of the cement matrix was based on Correia et al. [15], containing 75% of Portland Cement type CP V-ARI and 25% of crushed limestone. The limestone was used based on the industrial production of fiber cement, which is used to reduce the cost of the cement matrix and also to improve the packing between the cement and limestone particles, contributing to the reduction of porosity.

Plates with nominal dimensions of 20 cm x 20 cm were produced from different formulations, with contents of 1%, 2% and 3% of nanofibrillated cellulose, and 8% of pulp. After production, plates were kept in hermetically sealed plastic containers for 2 days at temperature of 30 ° C ± 5 ° C and then it was subjected to accelerated carbonation for 5 days at temperature of 45 °C, relative humidity of 70% and 15% of CO_2, as adopted by Almeida et al. [16].

Physical and mechanical characterization of fiber-cement

The composites were subjected to physical and mechanical tests, which were based on studies developed by [15, 16, 17, 18]. Physical tests were performed based on the principle of Archimedes for determination of water absorption (AA), apparent porosity (PA) and apparent density (AD), according to the American Society for Testing and Materials - ASTM C-948 [19].

Mechanical tests were performed in wet conditions at 8 days using a universal testing machine Emic DL-30000 equipped with a 1-kN load cell. A four-point bending configuration was employed to evaluate the modulus of rupture (MOR), limit of proportionality (LOP), modulus of elasticity (MOE) and specific energy (SE).

The microstructure of the composites was evaluated by Scanning Electron Microscopy (SEM) technique of polished surfaces, where the distribution and dispersion of the nanofibers along the matrix and the nanofiber-matrix interface were analyzed. The adhesion of nanofiber with matrix and the occurrence of stress transfer bridges between the nanofibers and the matrix were also analyzed.

Results and discussion

This section presents a study of effect the amount of nanofibrilllated cellulose, as reinforcement, on physical and mechanical performances of the cement matrix. In addition, the effect of nanofibrillated cellulose on cement matrix was compared to a composite produced with cellulosic pulp, which is a type of reinforcement that has already been used in the commercial cement-based composites.

Figure 1 presents the physical results of water absorption, apparent void volume and bulk density of composites reinforced by 1%, 2% and 3% of nanofibrillated cellulose and 8% of pulp at 8 days. The water absorption and porosity of composites are directly associated with water/cement ratio of composites as presented in Table 2.

TABLE 2 - WATER/CEMENT RATIO OF COMPOSITES REINFORCED WITH 1%, 2%, 3% OF NANOFIBRILLATED CELLULOSE AND 8% OF PULP

Water/cement ratio			
1% nano	2% nano	3% nano	8% pulp
0.22	0.23	0.24	0.28

Figure 1 (A) and 1 (B) shows an increase of water absorption and porosity with increase fibers content. The decrease of bulk density (Figure 1(C)) of composites with 8% of pulp is due the higher amount of fibers used, as density of fibers is lower than cement.

Despite the higher water retention capacity of nanofibrillated cellulose in comparison to cellulosic pulp, the higher fiber content used as reinforcement promoted the increase of water retention during the production of composites due to the hydrophilicity of the vegetable fibers. The retention of water during the elaboration of composites is reflected directly in the higher water absorption and, consequently, a greater porosity of the composites, due to the increase of fiber content used.

Figure 2 presents the modulus of rupture, the limit of proportionality, the modulus of elasticity and the specific energy of composites reinforced by 1%, 2% and 3% of nanofibrillated cellulose and 8% of pulp at 8 days.

The results indicated that on average, the composites containing 3% of nanofibrillated cellulose showed a higher strength than composites reinforced by 1% and 2% of fibers. This performance was due to the increase of fiber content, since higher concentration of reinforcement promotes the stress transfer between the matrix and fibers. According to Yoo et al. [20] the fiber bridging strength clearly increased with increasing fiber volume fraction, because there are more fibers at the crack surfaces, leading to a higher bonding area between the fiber and matrix.

However, modulus of rupture of composites reinforced by 8% of bamboo pulp was lower than composites reinforced by nanofibrillated cellulose. Additionally, it is proposed that 8% of pulp decreased the strength of composites due to the increase of composites porosity. The effect of the increase of the porosity was preponderant in comparison to the effect of the reinforcement. The nanofibers contributed to increase the physical and chemical adhesion; friction; mechanical anchorage, which was induced by high surface area. Figure 3 (A) shows a micrograph of nanofibrillated cellulose, confirming the high specific surface area while Figure 3 (B) shows the adhesion of nanofibrillated cellulose with cement matrix.

Non-Conventional Materials and Technologies – NOCMAT for XXI Century   Materials Research Forum LLC
Materials Research Proceedings **7** (2018) 119-127   doi: http://dx.doi.org/10.21741/9781945291838-12

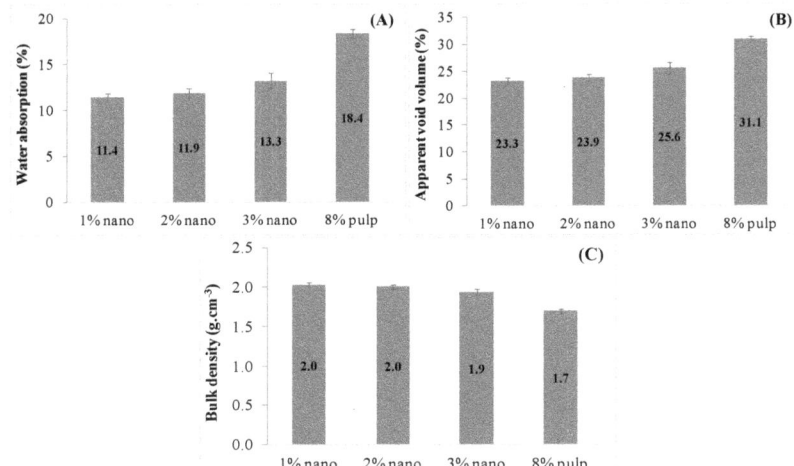

FIGURE 1 - *PHYSICAL PERFORMANCE OF CEMENTITIOUS COMPOSITES*
REINFORCED WITH 1%, 2%, 3% OF NANOFIBRILLATED CELLULOSE AND 8% OF PULP

The results of limit of proportionality indicate the performance of composites in pre-cracking step and the ability of the matrix to support the load until the first crack. Modulus of elasticity is also related to matrix stiffness. The amount of fibers added as reinforcement has a direct influence on the modulus of elasticity and on the limit of proportionality, since the composites with higher fiber contents presented higher porosity and, consequently, a less stiff matrix.

In relation to toughness, composites reinforced by nanofibrillated cellulose were more fragile than composites reinforced by pulp, which is clearly shown in Figure 4. This is due to the lower amount of nanofibrillated cellulose used and also to the lower length of nano fibers in comparison to pulp fibers, which reduces the pull-out energy involved in the fracture [21].

Figure 4 confirms results presented in Figure 2, where composites reinforced by 3% of nanofibrillated showed a higher mechanical performance in relation to the modulus of rupture, however, such composites presented a fragile rupture. Composites with 8% of pulp had a higher deformation and consequently, higher toughness, but, lower modulus of rupture. According to Bentur and Mindess [21], modulus of rupture and specific energy of fiber-cement depends strongly of properties and amount of fibers and of the fiber / matrix bonding.

Non-Conventional Materials and Technologies – NOCMAT for XXI Century   Materials Research Forum LLC
Materials Research Proceedings 7 (2018) 119-127   doi: http://dx.doi.org/10.21741/9781945291838-12

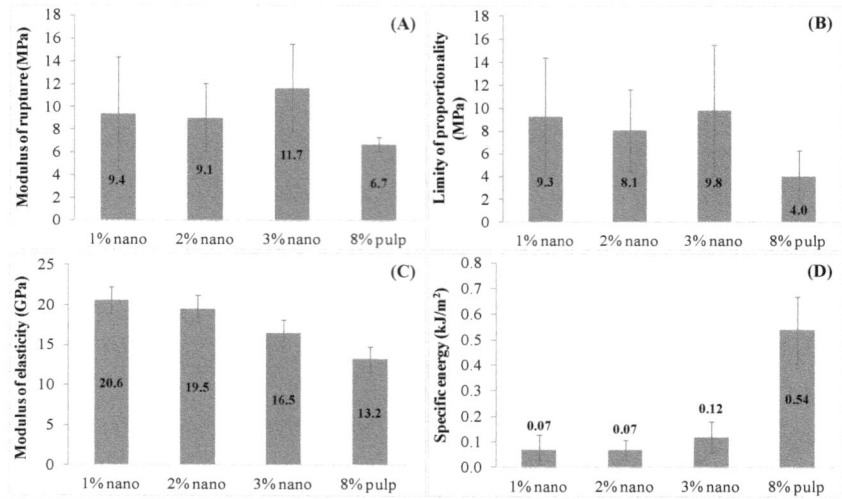

FIGURE 2 - *MECHANICAL PERFORMANCE OF COMPOSITES REINFORCED WITH 1%, 2%, 3% OF NANOFIBRILLATED CELLULOSE AND 8% OF PULP*

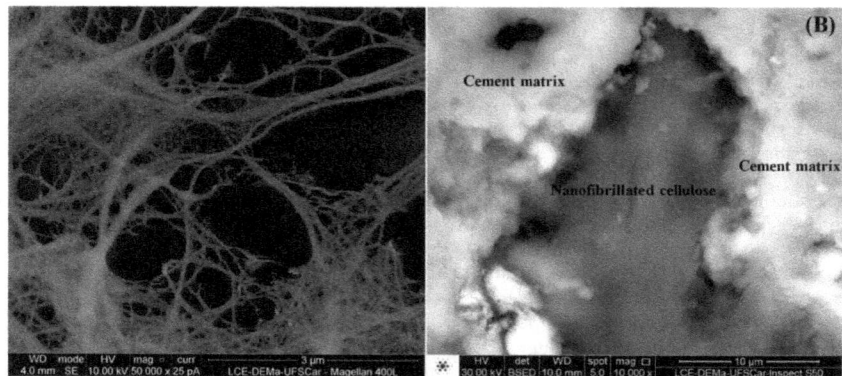

FIGURE 3 – *HIGH SPECIFIC SURFACE AREA OF NANOFIBRILLATED CELLULOSE (A) AND GOOD ADHESION OF NANOFIBRILLATED CELLULOSE WITH CEMENT MATRIX (B)*

Non-Conventional Materials and Technologies – NOCMAT for XXI Century   Materials Research Forum LLC
Materials Research Proceedings 7 (2018) 119-127   doi: http://dx.doi.org/10.21741/9781945291838-12

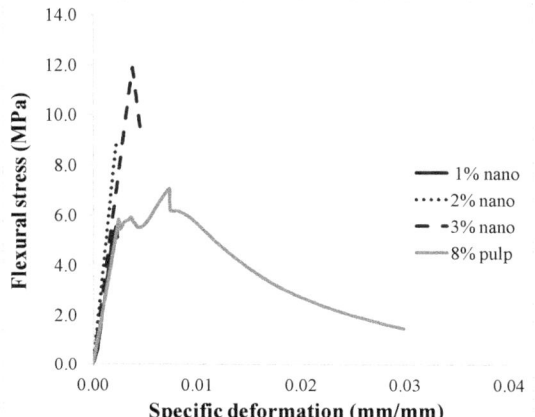

FIGURE 4 - TYPICAL STRESS X STRAIN CURVES AT UNDER FLEXURE TESTS OF COMPOSITES REINFORCED WITH 1%, 2% AND 3% OF NANOFIBRILLATED CELLULOSE AND 8% OF PULP

Conclusion

Composites reinforced by nanofibrillated cellulose presented a higher mechanical strength than composites reinforced by pulp. The high specific surface of nanofibrillated cellulose promotes a better adhesion with the matrix, which improves the mechanical anchorage. However, the amount of nanofibers used was not sufficient to increase the toughness of the composites. Thus, the use of hybrid reinforcement, as combination between the mechanical anchoring of the nanofibrillated cellulose with the matrix and the greater length of pulp fibers, can be the solution for a cementitious material reinforced at the nano and micrometric scales and with high mechanical strength and toughness.

Acknowledgments

The authors would like to thank Research Foundation of the Sao Paulo State - FAPESP, Brazil, for Post-doctoral grant n° 2015/21079-0, and by Financial support (Thematic Project 2012/51467-3).

References

[1] Makar JM, Margeson JC, Luh J. Carbon nanotube/cement composite—early results and potential applications. Proceedings of International Conference on Construction Materials: Performance, Innovation and Structural Implications. Ottawa: Institute for Research in Construction, National Research Council Canada, 2005. p. 1–10.

[2] Li GY, Wang PM, Zhao X. Mechanical behavior and microstructure of cement composites incorporating surface-treated multi-walled carbon nanotubes. Carbon 2005; 43: 1239-1245. https://doi.org/10.1016/j.carbon.2004.12.017

[3] Yakovlev G, Keriene J, Gailius A, Girniene I. Cement based foam concrete reinforced by carbon nanotubes. Mater Sci+ 2006; 12 (2): 147-151.

Non-Conventional Materials and Technologies – NOCMAT for XXI Century   Materials Research Forum LLC
Materials Research Proceedings 7 (2018) 119-127   doi: http://dx.doi.org/10.21741/9781945291838-12

[4] Konsta-Gdoutos MS, Metaxa ZS, Shah SP. Multi-scale mechanical and fracture characteristics and early-age strain capacity of high performance carbon nanotube/cement nanocomposites. Cement Concrete Comp 2010; 32: 110-115. https://doi.org/10.1016/j.cemconcomp.2009.10.007

[5] Metaxa ZS, Konsta-Gdoutos MS, Shah SP. Carbon nanofiber cementitious composites: Effect of debulking procedure on dispersion and reinforcing efficiency. Cement Concrete Comp 2013; 36: 25-32. https://doi.org/10.1016/j.cemconcomp.2012.10.009

[6] Galao O, Baeza FJ, Zornoza E, Garcés P. Strain and damage sensing properties on multifunctional cement composites with CNF admixture. Cement Concrete Comp 2014; 46: 90-98. https://doi.org/10.1016/j.cemconcomp.2013.11.009

[7] Coutts RSP. Fibre–matrix interface in air-cured wood-pulp fibre– cement composites. J Mater Sci Lett 1987; 6: 140-142. https://doi.org/10.1007/BF01728964

[8] Correia VC, Santos SF, Tonoli GHD, Savastano Jr H. Characterization of Vegetable Fibres and their Application in Cementitious Composites. In: Harries KA, Sharma B, editors. Nonconventional and Vernacular Construction Materials: Characterisation, Properties and Applications. Woodhead Publishing, 2016. p. 83-110. https://doi.org/10.1016/B978-0-08-100038-0.00004-4

[9] Kamel S. Nanotechnology and its applications in lignocellulosic composites, a mini review. Express Polym Lett 2007; 1(9): 546-575. https://doi.org/10.3144/expresspolymlett.2007.78

[10] Correia VC, Santos V, Sain M, Santos SF, Leão AL, Savastano Jr H. Grinding process for the production of nanofibrillated cellulose based on unbleached and bleached bamboo organosolv pulp. Cellulose 2016; 23: 2971 - 2987. https://doi.org/10.1007/s10570-016-0996-9

[11] Ardanuy M, Claramunt J, Arévalo R, Parés F. Nanofibrillated cellulose (NFC) as a potential reinforcement for high performance cement mortar composites. BioResources 2012; 73: 3883-3894.

[12] Onuaguluchi O, Panesar DK, Sain M. Properties of nanofibre reinforced cement composites. Constr Build Mater 2014; 63: 119-124. https://doi.org/10.1016/j.conbuildmat.2014.04.072

[13] Thomson SL, O'Callaghan DJ, Westland JA, Su B. Method of making a fiber cement board with improved properties of the product. US Patent 2010/0162926 AL, 01 July 2010.

[14] Savastano Jr H, Warden PG, Coutts RSP. Brazilian waste fibres as reinforcement for cement-based composites. Cement Concrete Comp 2000; 22 (5): 379-384. https://doi.org/10.1016/S0958-9465(00)00034-2

[15] Correia VC, Santos SF, Mármol G, Curvelo AAS, Savastano Jr H. Potential of bamboo organosolv pulp as reinforcement element in fiber-cement. Constr Build Mater 2014; 72: 65-71. https://doi.org/10.1016/j.conbuildmat.2014.09.005

[16] Almeida AEFS, Tonoli GHD, Santos SF, Savastano Jr H. Improved durability of vegetable fiber reinforced cement composite subject to accelerated carbonation at early age. Cement Concrete Comp 2013; 42: 49-58. https://doi.org/10.1016/j.cemconcomp.2013.05.001

Non-Conventional Materials and Technologies – NOCMAT for XXI Century   Materials Research Forum LLC
Materials Research Proceedings 7 (2018) 119-127   doi: http://dx.doi.org/10.21741/9781945291838-12

[17] Tonoli GHD, Fuente E, Monte C, Savastano Jr H, Rocco Lahr FA, Blanco A. Effect of fibre morphology on flocculation of fibre-cement suspensions. Cement Concrete Res 2009; 39: 1017-1022. https://doi.org/10.1016/j.cemconres.2009.07.010

[18] Tonoli GHD, Joaquim AP, Arsène M-A, Bilba K, Savastano Jr H. Performance and Durability of Cement Based Composites Reinforced with Refined Sisal Pulp. Mater Manuf Process 2007; 22: 149-156. https://doi.org/10.1080/10426910601062065

[19] ASTM C 948 standard test method for dry and wet bulk density, water absorption, and apparent porosity of thin sections of glass-fibre reinforced concrete, 1982.

[20] Yoo D-Y, Kim S, Park G-J, Park J-J, Kim S-W. Effects of fiber shape, aspect ratio, ad volume fraction on flexural behavior of ultra-high-performance fiber-reinforced cement composites. Compos Struct 2017; 174: 375-388. https://doi.org/10.1016/j.compstruct.2017.04.069

[21] Bentur A, Mindess S. Introduction. In: Bentur A, Mindess S, editor. Fibre Reinforced Cementitious Composites, 2nd ed. New York: Taylor &Francis, 2007. p. 1-10.

Non-Conventional Materials and Technologies – NOCMAT for XXI Century   Materials Research Forum LLC
Materials Research Proceedings 7 (2018) 128-138   doi: http://dx.doi.org/10.21741/9781945291838-13

Study of Axial Compression Resistance of Cardboard Tubes to Elaboration an Innovative Structural System

SALADO, Gerusa de Cássia [a]*; DIAS, Nathália Shimidt [b]

[a] University of Campinas - UNICAMP, Brazil; gerusa@ft.unicamp.br

[b] University of Campinas - UNICAMP, Brazil; nathalia.schimidt@hotmail.com

Abstract. Problem statement: The society has been worried about the environment, especially due to the intense exploitation of natural resources, the large quantity of residues and their effects on the planet. In this context, the study and development of new construction systems and materials have become fundamental to make use of residues or products discarded by the population. An alternative is to recycle paper and use it to manufacture cardboard tubes to serve as structural and sealing elements in the Architecture and Civil Construction areas, as Japanese architect Shigeru Ban has made in his projects worldwide for more than thirty years. **Approach:** This paper addresses the compression resistance of cardboard tubes, including 190 mm and 950 mm lengths, to elaboration a new structural system. These cardboard tubes were tested in laboratory. They underwent tests of compression resistance. **Results:** The test pieces made of cardboard tubes presented good results in all the tests, resisting to a maximum of 1700 kgf compression load or 6.14 MPa compression stress for specimens of 190 mm length and 2000 kgf or 5.69 MPa, respectively, for specimens of 950 mm length. **Conclusions:** The cardboard tubes tested in this research demonstrated good resistance to be used in the elaboration a new structural system. Constructions made of paper tubes may offer several advantages, such as lightness, cleanliness, speed of execution, recyclability and the reuse of discarded paper. This material is of big importance because of large production in Brazil and also of enough resistance to be used in constructions. It can be a good alternative for some kinds of constructions, like temporary, popular, emergency and movable structures, while also contributing to environmental sustainability.

Keywords: Compression Resistance Test, Compressive Strength Test, Cardboard Tubes, Paper Tubes, Sustainable Constructions

Introduction

The growth of the world population and consumption has led to an intense industrial activity, natural resources exploitation and enormous quantity of residues, resulting in environmental problems and degradation of the planet.

In this context, civil construction industries are the biggest consumers of natural and energy resources [1], using approximately 75% of natural resources extracted from the planet and producing 40% to 60% of all solid urban residues [2]. Based on this information, the civil construction sector should be responsible for recompensing the planet, using residues and decreasing the natural resources extraction.

In Brazil, only in 2015, about 45 million tons of construction rubbish were produced [3]. Paper is also a numerous residue in Brazil, which produced, in 2015, 10.4 million tons of paper and consumed 9.2 million tons [4]. Also, in 2015 approximately 63.4% of papers were recycled in Brazil, of its recyclable papers total consumption [3]. The figure below presents paper growing manufacturing in Brazil since 2000.

Non-Conventional Materials and Technologies – NOCMAT for XXI Century   Materials Research Forum LLC
Materials Research Proceedings **7** (2018) 128-138   doi: http://dx.doi.org/10.21741/9781945291838-13

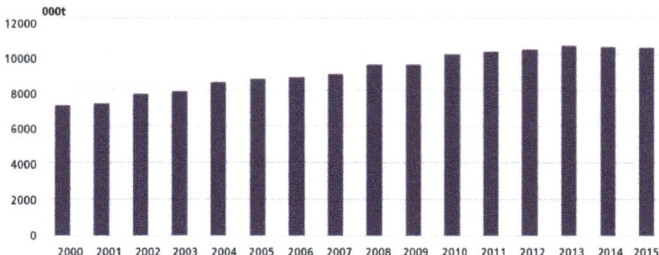

GRAPH 1 – *EVOLUTION OF PAPER MANUFACTURING IN BRAZIL. ADAPTED FROM [4].*

Considering that paper is a numerous residue in Brazil and worldwide, it can be recycled and used in civil construction, contributing to the reduction of natural resources extraction and landfills. A suggestion is to incorporate residues into construction elements, made of paper tubes, which are very resistant elements due to their cylindrical shape [5].

The main points for using cardboard tubes in civil construction are low cost, due to the low technology necessary in their manufacture, and no generation of construction rubbish. Besides, after a building demolition the cardboard tubes can be either re-used, if they are still in good conditions, or recycled.

Construction elements made of paper tubes offer several advantages to the sector of civil construction, such as lightness, cleanliness and speed of execution and are a good alternative for some types of constructions, such as temporary, popular, emergency and movable. These construction elements are of large importance due to their recyclability, contributing towards environmental sustainability.

Japanese architect Shigeru Ban has already adopted such techniques in his projects worldwide for over thirty years [6], as shown in the figures below.

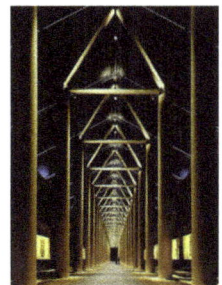  

FIGURE 1, 2, 3 – *NOMADIC MUSEUM, NOMADIC PAPER DOME, PAPER PAVILION. [7], P.193, 90, 103.*

Non-Conventional Materials and Technologies – NOCMAT for XXI Century   Materials Research Forum LLC
Materials Research Proceedings **7** (2018) 128-138   doi: http://dx.doi.org/10.21741/9781945291838-13

FIGURE 4, 5 – ODAWARA PAVILION, PAPER DOME. [8], P.104, 82.

Based on this idea, researches have been developed in Brazil for the elaboration of new construction systems made of paper tubes – Fig. 6 and 7 [9] [10]. However, the study of compression resistance of cardboard tubes used in these construction systems is of large importance.

FIGURE 6, 7 - STRUCTURAL PANELS CONSTRUCTION SYSTEM DEVELOPED IN BRAZIL.
[9], P.109, 115.

The objective of this paper is to study the axial compression resistance of cardboard tubes of 95 mm nominal external diameter, 75 mm internal diameter and 10 mm thickness [10]. The specimens have undergone analyses of compression strength test, including lengths of 190 mm and 950 mm.

The test procedures were based on technical norms to other materials, as wood, concrete, and tests by Shigeru Ban for his projects [6] because there is not a technical norm to the study of axial compression resistance on cardboard tubes and this test aims to application of paper tubes into civil construction projects.

Compression strength test

Objective
The objective was to simulate centered vertical actions to identify the cardboard tube axial compressive resistance and observe its behavior and deformations for load increased.

Specimens
Based on [11], which also addresses the wood compression resistance, 12 specimens were tested made of hollow paper tubes of 95 mm nominal external diameter, 75 mm internal diameter and 10

mm thickness. The specimen's length was based on [12], which determines double of external diameter, in this case 190 mm.

Besides, five specimens of 950 mm length, or ten times of external diameter, were tested. The objective was identifying some relationship between specimens' results of both lengths.

The specimens' moisture content was 9.1%.

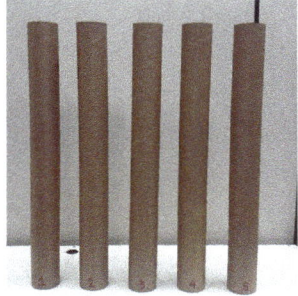

FIGURE 8, 9 – SPECIMENS OF 190 MM AND 950 MM LENGTH BEFORE TEST.

Materials and equipments

The equipments used to test specimens of 190 mm length were an Universal Machine of Tests, by INSTRON/EMIC, model EMIC 23-600, electro-mechanical, with capacity of 600 kN (60,000 kgf) and a computer with an indicator system to read the loads.

The equipments used to test specimens of 950 mm length were a press MTS SINTECH 5/G with capacity of 3,000 kgf, two dial indicators to gauge the possible horizontal displacements with capacity of 50 mm extension each other and a computer with an indicator system to read the loads.

FIGURE 10, 11 – EQUIPMENTS USED TO TEST SPECIMENS OF 190 MM AND 950 MM LENGTH, RESPECTIVELY.

Other materials and equipments used in the test preparation were an analogical pachymeter with precision of 0.05 mm, an aluminum rule with precision of 0.50 mm, a digital balance with precision of 0.01 g and an electrical oven with capacity to 105°C.

FIGURE 12, 13, 14 – ELECTRICAL OVEN, BALANCE AND OTHER EQUIPMENTS USED IN THE TEST PREPARATION.

Preparation

The test pieces were cut with a circular saw at 90° ± 1° to the axis of the core. The burrs on the tops were removed with fine sandpaper, the tops were not capped.

After, test pieces weight was gauge with balance and real length, thickness, external and internal diameters were measured with analogical pachymeter. These data made possible to calculate real section area and apparently density.

The specimens were conditioning under room temperature, shadow and humidity free. A cardboard tube piece had its weight gauged and was left inside oven at 105°C up until weight stabilization, according to [13] requirements. The moisture content of pieces was calculated.

Procedure

Testing carried out in a standard atmosphere identical to that used for conditioning the specimens.

The first specimen was placed centrally between the pressure platens, so that its longitudinal axis was at 90° or perpendicular to the platens. Only for specimens of 950 mm two dial indicators were placed in the middle of the length of the pieces at 90° or perpendicular each other and to the specimen to gauge the possible horizontal displacements at the moment of the maximum strength.

Compression was accomplished by evenly moving one platen towards the other, at a constant relative rate 1.0 mm/min for specimens of 950 mm length. Specimens of 190 mm length were subject at a 0.01 MPa/s constant load.

The test pieces were subject to a load until the first maximum of the compression resistance is markedly exceeded followed by 10% reduction.

This procedure was repeated for the remaining test pieces.

Non-Conventional Materials and Technologies – NOCMAT for XXI Century   Materials Research Forum LLC
Materials Research Proceedings **7** (2018) 128-138   doi: http://dx.doi.org/10.21741/9781945291838-13

FIGURE 15, 16, 17 – SPECIMEN OF 190 MM LENGTH AT THE BEGINNING, DURING AND AFTER EXCEEDS THE MAXIMUM RESISTANCE.

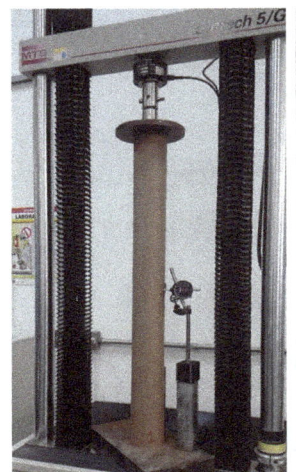  

FIGURE 18, 19, 20 – SPECIMEN OF 950 MM LENGTH AT THE BEGINNING AND AFTER EXCEEDS THE MAXIMUM RESISTANCE, AND THE DIAL INDICATORS USED TO GAUGE THE HORIZONTAL DISPLACEMENTS.

Non-Conventional Materials and Technologies – NOCMAT for XXI Century   Materials Research Forum LLC
Materials Research Proceedings 7 (2018) 128-138   doi: http://dx.doi.org/10.21741/9781945291838-13

Results

The results and behavior of specimens of 190 mm length for compression strength test can be analyzed from table 1 and graph 2 below:

TABLE 1 – RESULTS FOR COMPRESSION STRENGTH TEST - SPECIMENS OF 190 MM LENGTH.

S	Φ_e (mm)	Φ_i (mm)	e (mm)	l (mm)	S_L (cm^2)	V (cm^3)	m (g)	ρ (g/cm^3)	P_{max} (kgf)	f_c (MPa)	E (GPa)
01	97.7	77.5	10.3	190.0	27.80	528.12	380.74	0.721	1735	6.24	3.23
02	97.7	77.2	10.5	190.0	28.16	535.04	381.60	0.713	1639	5.82	2.81
03	97.3	77.4	10.1	190.0	27.31	518.79	376.28	0.725	1723	6.31	2.92
04	97.1	77.4	10.1	190.0	27.00	512.99	377.93	0.737	1695	6.28	2.99
05	97.5	77.3	10.1	190.0	27.73	526.91	379.40	0.720	1690	6.09	3.25
06	97.6	77.6	10.1	190.0	27.52	522.89	383.54	0.734	1596	5.80	2.81
07	97.4	77.2	10.2	190.0	27.70	526.31	379.82	0.722	1697	6.13	3.06
08	97.5	77.0	10.1	190.0	28.10	533.82	382.84	0.717	1662	5.91	3.04
09	97.8	77.1	10.1	190.0	28.43	540.26	380.65	0.705	1580	5.56	2.59
10	97.5	77.5	10.1	190.0	27.49	522.29	377.13	0.722	1747	6.36	3.23
11	97.7	77.2	10.1	190.0	28.16	535.04	379.99	0.710	1756	6.24	2.99
12	97.3	77.4	10.1	190.0	27.31	518.79	380.74	0.734	1896	6.94	3.36
Media	97.5	77.3	10.1	190.0	27.73	526.77	380.01	0.722	1701.3	6.14	3.02

Where,

S = specimens;

Φ_e = external diameter;

Φ_i = internal diameter;

e = thickness;

l = length;

S_L = net area perpendicular to P_{max};

V = volume;

m = weight;

ρ = apparently density;

P_{max} = maximum load resisted in the elastic region;

f_c = compressive stress at P_{max};

E = modulus of elasticity or Young's modulus.

134

Non-Conventional Materials and Technologies – NOCMAT for XXI Century   Materials Research Forum LLC
Materials Research Proceedings 7 (2018) 128-138   doi: http://dx.doi.org/10.21741/9781945291838-13

GRAPH 2 – RESULTS FOR COMPRESSION STRENGTH TEST - SPECIMENS OF 190 MM LENGTH.

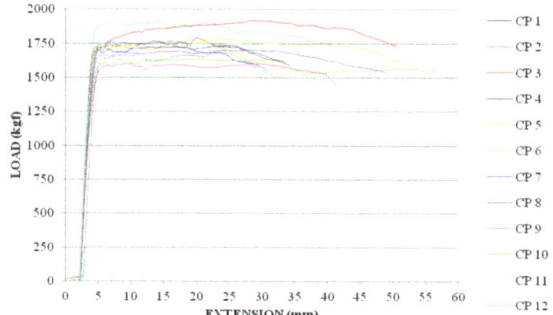

The specimens of 190 mm length condition after test can be observed from the following illustrations.

FIGURE 21 – SPECIMENS OF 190 MM LENGTH AFTER TEST.

FIGURE 22 – SPECIMENS 4 AND 6 OF 190 MM LENGTH AFTER TEST.

Non-Conventional Materials and Technologies – NOCMAT for XXI Century   Materials Research Forum LLC
Materials Research Proceedings 7 (2018) 128-138   doi: http://dx.doi.org/10.21741/9781945291838-13

The results and behavior of specimens of 950 mm length for compression strength test can be analyzed from table 2 and graph 3 below:

TABLE 2 – *RESULTS FOR COMPRESSION STRENGTH TEST - SPECIMENS OF 950 MM LENGTH.*

S	Φ_e (mm)	Φ_i (mm)	e (mm)	l (mm)	S_L (cm²)	V (cm³)	M (g)	ρ (g/cm³)	P_{max} (kgf)	f_c (MPa)	E (GPa)	D (mm)	
01	100.6	74.7	12.3	950.1	37.25	3542.48	2550.80	0.720	2004	5.38	1.01	D1	0.67
												D2	-3.94
02	101.4	76.2	12.9	950.1	35.15	3342.83	2553.90	0.764	1925	5.48	1.15	D1	3.72
												D2	-2.12
03	100.4	76.0	12.5	950.0	33.81	3211.46	2435.80	0.758	2006	5.93	1.38	D1	1.33
												D2	-2.88
04	100.3	76.3	12.2	950.0	33.29	3162.35	2424.50	0.767	1992	5.98	1.34	D1	0.41
												D2	-1.26
05	101.7	76.2	12.7	950.0	35.63	3384.77	2553.10	0.754	2016	5.66	0.84	D1	-0.15
												D2	1.97
Media	100.9	75.9	12.5	950.0	35.03	3328.78	2503.62	0.753	1988.6	5.69	1.14	D1	1.20
												D2	1.65

Where,
D = horizontal displacement in the middle of the length of the specimen at P_{max};
$D1$ = value in the dial indicator 1 at P_{max};
$D2$ = value in the dial indicator 2 at P_{max}.

GRAPH 3 – *RESULTS FOR COMPRESSION STRENGTH TEST - SPECIMENS OF 950 MM LENGTH.*

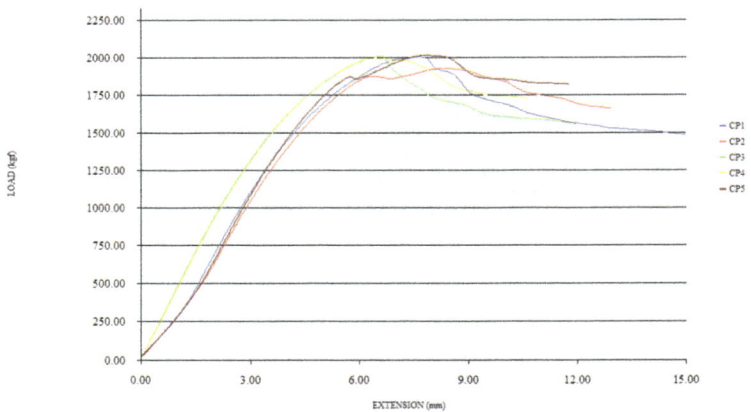

Non-Conventional Materials and Technologies – NOCMAT for XXI Century   Materials Research Forum LLC
Materials Research Proceedings 7 (2018) 128-138   doi: http://dx.doi.org/10.21741/9781945291838-13

The specimens of 950 mm length condition after test can be observed from illustration below.

FIGURE 23 – SPECIMENS OF 950 MM LENGTH AFTER TEST.

Conclusions

The cardboard tubes tested in this research demonstrated good resistance to be used in the elaboration a new structural system. The test pieces made of cardboard tubes presented good results in all the tests, resisting to a maximum of 1700 kgf compression load or 6.14 MPa compression stress for specimens of 190 mm length and 2000 kgf or 5.69 MPa, respectively, for specimens of 950 mm length.

About specimens' behavior, the test pieces of 190 mm length showed a little deformation in the elastic region and after exceed the first maximum load presented plastic deformation. The test pieces of 950 mm length showed bigger deformation in the elastic region than and also plastic deformation after to reach the first maximum load.

Both lengths are short and have low slenderness ratio ($\lambda < 40$), don't suffering buckling, however the test pieces of 950 mm length presented bigger deformation in the elastic region than because they are longer than. Such deformations were small (elastic region) and gauged by dial indicators.

The results of this study prove that the paper tubes tested are enough resistance to be used into constructions as structural and sealing elements.

Constructions made of paper tubes may offer several advantages, such as lightness, cleanliness, speed of execution, recyclability and the reuse of discarded paper. This material is of big importance because of large production in Brazil and also of enough resistance to be used in constructions. It can be a good alternative for some kinds of constructions, like temporary, popular, emergency and movable structures, while also contributing to environmental sustainability.

Acknowledgements

The authors would like to acknowledge FAPESP for the financial support provided to this research.

References

[1] CIB. Agenda 21 para a construção sustentável. Publicação 237. São Paulo: CIB, 2000.

[2] Sartor, C.; Lamberts, R. Habitare: resultados de impacto 1995/2007. Florianópolis: Coan Impressão Gráfica, 2008.

[3] ABRELPE. Panorama dos Resíduos Sólidos no Brasil 2015. Information on http://www.abrelpe.org.br/Panorama/panorama2015.pdf. Acessed in 15/08/2017.

[4] Farinha e Silva, C. A.; Bueno, J. M.; Neves, M. R. A indústria de celulose e papel no Brasil. Guia ABTCP Fornecedores & Fabricantes Celulose e Papel 2016-2017. Information on http://www.poyry.com.br/sites/www.poyry.com.br/files/media/related_material/16out27a-abtcp.pdf. Acessed in 15/08/2017.

[5] Salado, G. Construindo com tubos de papelão: um estudo da tecnologia desenvolvida por Shigeru Ban. MS dissertation. University of São Paulo, Brazil (2006).

[6] McQuaid, M. Shigeru Ban. Nova York: Phaidon Press, 2003.

[7] Miyake, R. Shigeru Ban: Paper in Architecture. Nova York: Rizzoli International Publications, 2009.

[8] The JA (The Japan Architect). Shigeru Ban. Edição especial n30. Tóquio: A+U Publishing CO, summer 1998, 184p.

[9] Salado, G. Painel de vedação vertical de tubos de papelão: estudo, proposta e análise de desempenho. PhD thesis. University of São Paulo, Brazil (2011).

[10] Dias, N. S. Estudo e proposta de sistema estrutural com tubos de papelão. Scientific Initiation. University of Campinas, Brazil (2017).

[11] ABNT (Associação Brasileira de Normas Técnicas/Brasil). NBR 7190 - Design of wooden structures. Rio de Janeiro: ABNT (1997).

[12] ABNT (Associação Brasileira de Normas Técnicas/Brasil). NBR 5739 - Concrete - Compression test of cylindric specimens - method of test. Rio de Janeiro: ABNT (2007).

[13] ABNT (Associação Brasileira de Normas Técnicas/Brasil). NBR 14257 - Paper and board - Cores - Determination of moisture content - Oven drying method. Rio de Janeiro: ABNT (1998).

Non-Conventional Materials and Technologies – NOCMAT for XXI Century   Materials Research Forum LLC
Materials Research Proceedings 7 (2018) 139-147   doi: http://dx.doi.org/10.21741/9781945291838-14

Low-Energy Walling for Low-Income Housing in East Africa

Terry Thomas

Engineering Dept., Warwick University UK, t.h.thomas@warwick.ac.uk

Abstract. Walling comprises much of the mass of Africa low-rise housing). An energy-sustainable walling technology needs to minimise both the volume and the energy-intensity of the materials used. The first requires that walls are thin or hollow. The second requires minimal use of energy-intensive cement or brick-firing firewood. This paper explores the scope for both sorts of minimisation while yet accepting a say 80-year durability requirement and enhancing seismic performance for possible Asian applications. Minimum wall thickness in housing is only weakly determined by load-bearing performance. It is to achieve adequate stiffness and resistance to lateral forces that requires walls to be commonly more than 100mm thick. Thin walls may be stiffened in various ways – most obviously by buttresses and cross walls. Relatively neglected is the use of wavy or stepped (crenelated) wall plans despite their saving up to 40% of the material needed for free-standing walling. The paper therefore explores the material saving potential of non-straight walling.For various reasons 'stabilised' soil (SS) – for example formed from an earthen mix containing about 6% cement – is regarded as more attractive than plain soil. Moreover the need to use densification, by application of high pressure, points away from rammed earth' construction and towards the use of masonry. Such units are assembled using cement-intensive mortar but there is scope for mortarless construction as is already in use in several African countries. The paper assesses the viability of mortarless walling.Finally the paper considers the option of combining pressed, but very lean, SS blocks with mortar and renders that have been reinforced with natural fibres to enhance their shear strength and crack resistance.

Keywords: Walling, Masonry, Africa, Low-cost, Housing

Introduction

Low-income tropical housing has for many years been in transition from subsistence construction using mainly local materials to buildings employing some factory-made materials or components. In Indian terminology the transition is from 'kutcha' to 'pukka'. It is generally associated with enhanced building durability, improved performance, and sometimes greater resilience to meteorological and seismic extremes. However it has incurred an increase in embodied energy and usually therefore greater GHG emissions. The progression has however had little immediate impact on energy use over a building's lifetime, but that will change as mechanical cooling becomes more common with urbanisation, electrification and global urban warming.

Walling comprises much of the mass of low-rise housing in East Africa ('low-rise' being still the norm there). A sustainable walling technology needs to perform well and last long yet minimise both the volume and the energy-intensity of the materials used. For walls of a given length and height, reducing volume requires that walling is thin or hollow. Reducing embodied energy also requires minimal use of energy-intensive cement or brick-firing firewood. This paper explores the scope for both sorts of minimisation while yet accepting a say 80-year durability requirement and enhancing certain aspects of performance.

Wall performance criteria

Walls have many functions, including aesthetic, acoustic, thermal, privacy and structural ones. Of course cost-minimisation is of central importance in low-income tropical housing; by contrast thermal performance (e.g. insulation) is as yet of little importance. The structural functions include

Non-Conventional Materials and Technologies – NOCMAT for XXI Century   Materials Research Forum LLC
Materials Research Proceedings 7 (2018) 139-147   doi: http://dx.doi.org/10.21741/9781945291838-14

support of upper floors/roofs, resistance to impact and penetration, and resistance to lateral forces. Walls rarely fail by crushing or buckling and deliberate penetration is more easily made via openings than via masonry, so it is this last requirement of lateral strength and stiffness that largely determines choice of wall thickness and materials. It is mainly achieved by the limiting slenderness ratio S (height/thickness) of masonry to about 16 and, since vertical spans in housing are normally around 2.5m, by thus ensuring wall thickness exceeds 150mm. Greater masonry thicknesses are needed for walls over 10m high and indeed beyond this height framed buildings with infill 'curtain' walling have largely replaced load-bearing masonry. In East Africa, rural housing heights never exceed 10m and as yet urban housing rarely does so.

Lateral stiffness and strength

The lateral stiffness of a long wall depends in its height, thickness, material and hollowness. Such a wall cantilevered up from its foundation and subject to a force perpendicular to its face has a *stiffness* of

$$F/\delta = K = \lambda \, E \, I \, / \, H^3$$

δ is displacement, H is height, E is the material's stiffness (Young's modulus). Force F and 2nd moment I are per unit length of wall. $\lambda = 3$ If the force is applied to the top edge of the wall (as e.g. a roof out-thrust) and $\lambda = 8$ if the force is spread evenly over the wall face as in wind-loading or seismic loading.

As we usually have little designer control over the material's stiffness E, we obtain sufficient wall stiffness by ensuring an adequate value for I. For a plain wall I. is proportional to B^3, where B is wall thickness. Doubling thickness thus gives an 8-fold increase in lateral stiffness or a 4-fold increase in stiffness per unit of material used.

However we are normally more interested in lateral strength than in stiffness and in maximising the ratio of strength to wall mass, since mass translates approximately into cost. Masonry walls may fail by shear failure (sliding at the courses), especially if loaded on their top edge, so mortared bricks are sometimes made with depressions ('frogs') to increase the brick-to-mortar sliding friction. Blocks for *mortarless* assembly require an interlock so that sliding cannot progress far until the interlock is broken off. More commonly however masonry fails not by sliding but by toppling. First cracking occurs on the face to which some normal 'yield' force F_y is applied. Later, those cracks grow into hinges, as the force is increased to a 'failure' or 'toppling' value F_f. In the absence of any adhesion (ability to transmit tensile stress) between mortar and block, the failure force F_f is typically 3 times the yield force F_y. In such absence of adhesion, it is the wall's weight which gives it lateral resistance and the failure force per unit length of wall is:

$$F_f = \gamma \, I \, / B$$

where $\gamma = 2\rho g \approx$ typically 40,000 for top-edge loading; $\gamma = 80,000$ for loading over the whole wall face

For a solid simple wall, doubling its thickness B will give an 4-fold increase in section modulus $Z = I/y_{maxand}$ thereby in F_y and a 2-fold increase in normalised performance (lateral strength per unit of walling material).

A full analysis of the yield force for a practical wall with some bonding at the top, with some mortar adhesion and with cross-walls or buttresses, is complex [cite CP4]. However in general we will improve normalised performance by increasing the wall's modulus/mass (Z/M). Well-established ways of doing this include

✓ Using hollow blocks which reduces material mass (& cost), more than it reduces Z (& so strength).

Non-Conventional Materials and Technologies – NOCMAT for XXI Century   Materials Research Forum LLC
Materials Research Proceedings 7 (2018) 139-147   doi: http://dx.doi.org/10.21741/9781945291838-14

- ✓ Providing periodic buttresses where land-access allows – very common with boundary walls
- ✓ Sharing walls between adjacent houses ('terracing', row housing)
- ✓ Making rooms small and hence cross-walls closer. This also simplifies floor spans in 2-storey housing.

Rarely practiced though of considerable promise is to make walls wavy or crenelated in plan. Wavy ('crinkle-crankle') estate-boundary walls were employed in the USA, UK and Netherlands in the 18th century and some still stand

Wavy and crenelated walling

For many good reasons, especially economy of design effort and simplicity of assembly, masonry walling is generally straight or rectangular. Curvature in plan is not uncommon but high radii are generally chosen, to keep the brick-to-brick rotation angle below about 6^0. Both intentional curvature in elevation and double curvature are very uncommon in masonry, although the latter was skilfully practiced by the architect Eladio Dieste in Uruguay, Spain etc [Remo, 2000] using crude clamp-fired bricks. There have been some attempts [Russel Gentry, 2009] to extend computer structural modelling to cover such shapes.

(above) Eastern England: 18th C estate boundary wall, still standing (to right) 90m x 1.9m wall in seismic Fort Portal, Uganda.
GPS = latitude 0.65279N; longitude 30.25442E
Under construction in 2015, capping followed; expansion joints were wrongly incorporated,

FIGURE 1 - WAVY WALLS ALONG PLOT BOUNDARIES

We need to explore the stiffness and strength of various simple wall geometries, comparing them with some 'datum'. A convenient datum is a 'standard' 20 cm thick wall (bricks laid as headers or in say Flemish bond). As well as wavy walls we consider crenelated walls (i.e. stepped forwards and backwards after every few bricks).

TABLE 1 - STRAIGHT, WAVY & CRENELATED WALLS: WIND-PRESSURE P_F, OR SEISMIC ACCELERATION A, TO INITIATE CRACKING

Row	Brickwork Thickness (B)	Shape	Wall-length (L)	Wall Depth $(2\,y_{max})$	Stiffness and I NTD	Modulus $Z=I/y_{max}$ NTD	Pressure P_f NTD	Acc'n a NTD	Bricks per m NTD
	Mm		m	m					
A	'Datum' 200	straight	2	0.2	1	1	1	1	1
B	100	straight	4	0.1	0.25	0.5	0.125	0.25	0.5
C	**100**	wavy	**3.96**	**0.3**	**1.95**	**1.29**	**2.60**	**2.58**	0.51
D	100	wavy	3.85	0.5	6.91	2.77	5.75	5.53	0.52
E	100	wavy	3.64	0.7	15.12	4.33	9.49	8.66	0.54
F	100	cren'td	3.60	0.2	1.00	1.00	0.56	1.00	0.56
G	**100**	cren'td	**3.20**	**0.3**	**2.95**	**1.97**	**1.23**	**1.97**	0.62
H	100	cren'td	2.80	0.4	5.80	2.90	2.07	2.90	0.71

Unrendered wall, height 1.8m, made of 480 solid bricks each 200mm (including mortar) x 100mm in plan. Datum (Row A) is a 200m deep straight wall. 'Wall depth' is the maximum distance from front face to rear face. The other parameters are averaged over one sinusoidal cycle. Brick density is taken as 2000kg m^{-3}. 'NTD' indicates 'normalised to datum's value'. The values of L, I and Z are *per wall* of 480 bricks, not per unit length of wall; "cren'td" denotes a crenelated wall plan

The table shows the predicted pressure/acceleration (P_f and a) sufficient to initiate cracking at joints in an unplastered wall in the absence of (reliable) mortar adhesion. It also shows the 2nd moment of area of each wall's footprint, which for a given wall height is proportional to lateral stiffness.

Table 1 portrays expectation – i.e. calculations based on theory. One small experiment (Uganda 2014) compared crenelated and straight stabilised-soil walling where theory (Row F compared with Row B) suggested a 4:1 stiffness ratio. The measured ratio was 2.7:1. A separate experiment (Uganda 2015) comparing wavy and straight country-brick walls similar to those of table rows B and E for which theory predicts a stiffness ratio of 60:1. Measured relative stiffness was about 30:1. More thorough confirmation of theory by experiment is needed.

Both strength (e.g. withstand wind pressure) and stiffness increase with wall depth. In the case of seismic acceleration the entries also reflect the variation in wall mass of walls A to H, lower mass being desirable. To attain performance comparable with a standard *200mm straight wall* ('Datum' = row A), in terms of both stiffness and seismic resistance (compare bold table entries), we need a wavy or crenelated wall approximately 300mm deep and 100mm thick – as shown in rows C and G. These designs have a much lower brick count than the Datum per unit length of wall – see last column. If we increase the depth of wavy or crenelated walls further (rows D, E & H) we obtain a massive increase in stiffness (which is proportional to 2nd moment I) and a substantial increase in strength, while only modestly increasing the brick count per meter run. So to minimise the ratio of material use (bricks) to performance (strength and stiffness), wavy or crenelated wall plans look very attractive. Yet they are little used. Instead buttressing of walling is popular whose performance-to-materials ratio is not so good.

Wavy walls have of course some disadvantages. Building curves with straight bricks has its limitations – a 'quilted' appearance may be unwanted and for elegance the radius of curvature should exceed about 10 times the brick length (which would exclude row E above). Rooms with curved walls are stylish but difficult to furnish, fitting openings like doorways and tee joints takes more skill, so first attention should be given to boundary walls. (Such 'site works' typically account for a substantial fraction of housing costs and associated GHG emissions.) A wavy boundary wall on a rectangular plot may sacrifice the equivalent of a 100 to 200mm strip of land to a neighbour. None of these costs seem large enough to miss the substantial saving of about 40% of materials when 'wavy' is substituted for straight. And there can be other benefits – for example removing the need for expansion joints and reducing the impact of long-term sinkage/creep of wall foundations.

Crenelation is perhaps less radical but also slightly less beneficial. It fits well with use of interlocking blocks but fits badly where quoins need brick-cutting.

Hollow walling

Some fired bricks are about 15% hollow and many cement blocks are up to 45% hollow. However the 'cavity-wall' construction dominant in 20th century Europe had other objectives (damp control) and has now been replaced by solid walling that employs different materials for inside (insulation) and outside (durable) faces. Vertical passages through bricks/blocks have advantages over horizontal ones since they can be used for reinforcement in seismic zones or for threading electrical/plumbing services. Unlike horizontal passages/holes they do not incur extra rendering costs at corners and openings. Cavities of up to 50% placed close to a brick's centreline have rather little negative effect on its 2nd moment of area or modulus (and thus a wall's lateral stiffness and resistance to overturning), they directly reduce crushing strength and they may result in loss of mortar down the voids during assembly of masonry. It is probably for this latter reason that the 300mm x 150mm cement blocks so widely used in the tropics usually have a continuous 10mm 'plate' across their top face. Moreover the thin walls of hollow blocks can be easily punctured unless the cement mix is expensively strong (sand:cement ratio <5). The traditional Chinese 'rat trap' brick bond (also promoted In S India by Laurie Baker) is hollow but looks rather precarious - like a tower of playing cards. It achieves material savings because bricks (placed both horizontally and vertically) are thin like tiles.

It seems there is good justification for employing blocks or brick bonds that are more-hollow than are common at present and also for replacing horizontal passages in extruded bricks by vertical passages. For a given mass of clay or concrete, a 50% hollow brick will give a wall that is 100% thicker, about 6 times laterally stiffer and about 3 times laterally stronger than a wall of solid bricks. As the mortaring of very hollow blocks is difficult, they may better match the use of a 'glue' jointing-mastic rather than cement mortar.

Mortarless masonry

To be workable, mortar needs to be much richer in cement content than the 'cement' or stabilised soil (SS) blocks it connects. Moreover in some countries mortar is applied thickly (typically 20mm thick to erect 200 x 100 x 75mm 'country-brick' walling in Uganda, 15mm thick for erecting walling of 300 x 150 x 150 cement blocks). As mortar in that country costs 2 times more *per litre* than solid blocks of SS or very lean hollow 'cement' blocks, mortaring adds about 13% to the material cost of each square meter of block wall. For the much cheaper fired-brick walling it adds about 90%. So substantial savings would follow if mortaring could be omitted.

There are thus two routes to reducing the material costs and embodied energy of walling of a specified thickness, namely to reduce *per litre* block costs or reduce mortar thickness. The former has been long explored and includes pursuing the hollowness mentioned above. Leanly-stabilised soils (SS) can be strengthened by pressure prior to curing – i.e. for a given final strength, vigorous pressing reduces the cement content required to attain some specified compressive strength. For example to attain a 7-day wet strength of 1.4 MPa the cement fraction required for a SS block falls from typically 7.8% to 4.3% as the moulding pressure is increased from 1 MPa to 10MPa[Gooding 1994]. Careful architecture may even remove the need for 'wet strength' (which is the main justification for stabilisation).

Non-Conventional Materials and Technologies – NOCMAT for XXI Century   Materials Research Forum LLC
Materials Research Proceedings 7 (2018) 139-147   doi: http://dx.doi.org/10.21741/9781945291838-14

The second approach of avoiding mortar entirely is quite widely practiced but requires skill in both bricklaying and in brick production. Mortar has many functions to justify its use over several millennia. It:

i.   compensates for brick imperfections in achieving plumb walling and straight courses
ii.   cushions the bricks so that wall's weight is spread over the full brick surface
iii. seals the joints between bricks to exclude wind rain, light, vermin, sound
iv.   provides shear strength that prevents individual bricks being 'pushed through' and reduces wall-failure by racking, for example during impact or shaking or criminal entry.

This is a daunting list of attributes for mortarless brick-laying to achieve, yet presses for forming interlocking blocks of stabilised soil for mortarless laying are on the market in many countries. Interlocks are thought essential for satisfying functions (iii) and (iv) above. However interlocks themselves introduce several problems. They may prevent brick-reversing - which is a useful technique for achieving function (i) above – and they often prevent the production of neat corners or tee joints. Forming quoins, joints and openings then requires either that bricks be roughly cut on site or that a set of several different preformed units be made (full, ½ and ¾ blocks, closers etc as shown in Fig 2(b)). Top and bottom interlocks allow block reversing, end interlocks do not. Figure 2 shows a set of interlocking stabilised-soil blocks used in East Africa for over a decade and a Tanzanian house so produced. Unfortunately only the crudest form of interlock is compatible with commercial clay-brick extrusion and clamp-fired hand-moulded bricks are generally too geometrically irregular for mortarless assembly. So mortarless assembly of *fired* bricks is apparently nowhere practised. Mud mortaring may suffice in a dry climate.

(a)   *house built ca 2012 with mortarless SS walls*

Non-Conventional Materials and Technologies – NOCMAT for XXI Century   Materials Research Forum LLC
Materials Research Proceedings **7** (2018) 139-147   doi: http://dx.doi.org/10.21741/9781945291838-14

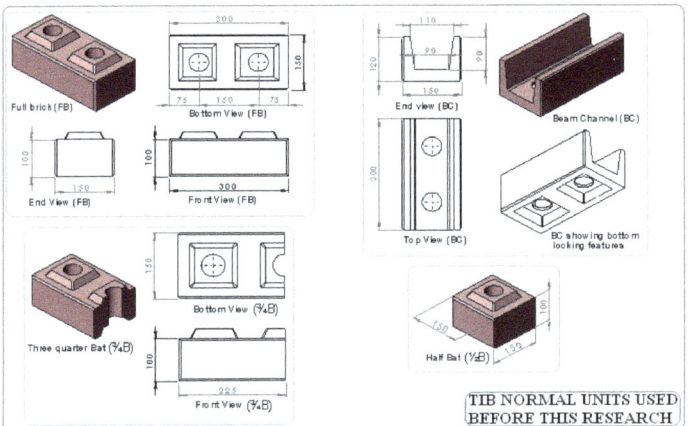

(b)   associated interlocking block set developed by Nat Housing & Building Research Agency in Dar es Salaam. Note that interlock is on two axes and that blocks are reversible. Other block sets exist in India, S Africa, Thailand

FIGURE 2 - *TANZANIAN MORTARLESS HOUSE-WALLING*

Mortarless columns or walls rarely fail by crushing because most walling materials have a large safety factor in compression. Absence of mortar slightly reduces crushing strength, e.g. 42% reduction was observed in experiments with SS blocks where 2-block prisms were compared with single blocks. Fortunately as vertical pressures approach failure values the contact area, between an unmortared block and that below it, increases. With 300mm x 150mm lean, SS, manually-pressed blocks the area was found to increase from 8% to 25% to 40% of plan area as pressure was increased from 2 to 20 to 200 kPa

In the absence of mortar, functions (i) and (ii) above require blocks be very accurate, with tight tolerances – e.g. < 0.3mm - on height, taper (especially lateral taper), twist or hogging. This accuracy depends mainly on the press used to form the blocks rather than on subsequent handling and curing. (Curing does however affect strength.) Presses commonly come in two forms – manual and motorised. The former (usually cinva-ram type) cost around $\$_{US}1000$; the latter around $\$_{US}15000$. These presses impart pressures of around 1MPa and 10MPa respectively, substituting the latter for the former may save about 50% of cement for a given block compressive strength. Both purport to produce uniformly-sized blocks, though the motorised blocks are generally better looking and more free from flaws. There is some evidence that very high pressures result in the interior of blocks being largely cement-free and therefore easily eroded if the more cementitious surface layer is penetrated. The unfortunately high elasticity of *manual* presses means that inadvertently varying block pressure (by poor control of batching) causes variation in block height of up to 3mm.

Because the lower cost of manual presses matches the resources of artisanal builders, research has been addressed at minimising the consequences of the irregular blocks such presses produce. As the standard deviation of the overhang ('out of plumb') of an *n*-course wall is of the order of $C\ n^{1.5}$, attention has focussed on reducing C and putting limits on n. These measures include:

i. Improvements to press design to reduce variation in height, taper etc and hence C.
ii. Reversing alternate blocks prior to construction to offset any bias taper; also measuring and rejecting some obviously poorly-dimensioned blocks.
iii. Selectively reversing blocks *during construction*, while laying to a level or plumb bob. This reduces C by a factor of about 33%.[Kintingu 2009 p.174] or by over 50% if the mason may select a replacement block from the pile
iv. Grooving the blocks to prevent any contact between them in a central zone (approx. 35%wide) to prevent rocking – this reduces C by a further 50% [Kintingu 2009 p.182]. It also improves lateral strength and stiffness.
v. Limiting the number of courses (to e.g. 10) before a mortared joint is inserted to 'straighten' the wall. Such joints are typically placed at windowsill or lintel height. Generally ring beams are required for walling extending over 1 storey high, especially in high gable-end walls.
vi. Using the 'cross-walls' to control the verticality of wall ends. (Walls are already much straighter than isolated brick columns, so columns are best avoided.
vii. Laying blocks only part-cured to take advantage of their greater plasticity when green However this can lead to an unacceptably high breakage rate and is not recommended: experiments by the author (Uganda 2014)indicated little benefit from this procedure.
viii. Machining blocks, or at least vigorously cleaning them, before or after curing. Vertical pressure on blocks also improves their contact area and hence their straightness and lateral stiffness. With good-quality 300 x 150mm SS blocks, it was found that increasing the vertical pressure on a block-to-block interface from 2 kPa to 20 kPa to 200 kPa resulted in contact areas of respectively 8%, 25% and 40% of the blocks' plan area. 50kPa is the pressure on the bottom course of a single-storey wall. There has even been proposals to vertically clamp unmortared walls using rebars passed through vertical cavities in order to increase lateral stiffness, strength and straightness.
ix. Employing a pseudo mortar – e.g. 1mm of fine dry sand, - to better seat each block on those below it. Very thin and fluid mortar might be injected *after* wall assembly.

Because of the occurrence of 'rocking' contact between blocks, mortarless walls and columns are much (e.g. ten-fold) less stiff and strong against lateral forces than mortared ones. They are vulnerable to being blown over if storms occur during construction before a ring beam can be added.

Mesh-and-infill walling
The opposite strategy to minimising mortar is to use it more effectively, by forming a honeycomb matrix or exoskeleton. Within such a matrix the infill blocks can be quite primitive – for example can be hollow and made of unstabilised pressed soil. (This partly matches the construction of many large medieval buildings where columns surfaced with ashlar (stone) masonry contained a core of cheaper rubble masonry.) The contribution of such a matrix is much enhanced if it possesses reliable tensile strength and if its horizontal and vertical elements (mortar, perpends and surface render) are bonded together. There has long been interested in adding fibre to mortar mixes – metal, glass, plastics, carbon, vegetable or animal fibres - to increase the cured tensile strength. Unfortunately many fibres are not cheap, some may decay, and plastering/mortaring with fibrous mixes is not easy. So the fibre content needs to be kept low – say under 1% by mass - and the fibres need selection for durability and ability to bond to their mix. The world has a great supply of agricultural waste fibres, such as wheat straw, but these are poorly suited to reinforcing building materials. Classic fibres such as sisal, jute, cotton (and in former years horsehair) command a good price, have good properties and yet are not sufficiently durable to be used in external renders. For interior use they last over 100 years.

Non-Conventional Materials and Technologies – NOCMAT for XXI Century   Materials Research Forum LLC
Materials Research Proceedings 7 (2018) 139-147   doi: http://dx.doi.org/10.21741/9781945291838-14

The benefit of enhancing the (reliable) tensile strength of render is illustrate by Table 2, where the moment-to-crack M, applied to the lowest course in a single-storey mortarless wall, is related to the tensile strength of the inside and outside renders. A tensile strength of 2000 kPa, giving a 27-fold improvement in resistance to wind or 'quakes, should be realisable with fibrous renders. The improvement in lateral stiffness is however much less.

TABLE 2 - *EFFECT OF TENSILE YIELD STRENGTH Σ_Y OF A PLASTER ON THE MOMENT M TO INITIATE CRACKING*

Rendered?		No	Yes (Datum)	Yes	Yes	Yes
Grooved?		No	Yes	Yes	Yes	Yes
σ_y of render	kPa	0	0	500	1000	2000
Moment M	kPa-m per m of wall	1.65	2.43	19.55	35.2	66.5
M / M_{DATUM}	(improvement)	0.68	1	8.0	14.5	27.4

Wall is 2.3m high, 0.1 m thick plus 2 x 1cm renders, blocks are 35% grooved

Research is underway into how fibrous mortars/renders can be optimised and used in conjunction with very cheap blocks. In general the former should provide lateral strength and seismic damping while the later should carry the basic vertical (compressive) loads on a wall.

Conclusions

Low-rise masonry house-walling, although a technology practised for millennia, offers some scope for further improvement – lower cost, lower embodied energy, lower GHG emissions and greater durability. Of considerable promise are omission of the cement-intensive mortar between blocks, replacement of straight walls by thinner curved or crenelated ones, use of higher pressures in the formation of cementitious blocks, location of superior (but more costly) materials at the wall surfaces and of inferior materials and voids in the interior of walls, and enhancing the tensile strength of masonry by the incorporation of fibres in mixes.

References

[1] Gooding D E M, Improved processes for the production of soil-cement building blocks. PhD thesis, Warwick University, 1994

[2] Kintingu SH, Design of interlocking bricks for enhanced wall construction flexibility, alignment accuracy and load bearing, PhD thesis, Warwick University, 2009

[3] Russell Gentry T *et al*, 'Parametric design, detailing and structural analysis of doubly-curved load-bearing block walls', 11th Canadian Masonry Symp, Toronto, Canada, 2009 (contact lead author via College of Architecture, Georgia Inst Tech, Atlanta)

[4] Remo P, *Eladio Dieste - The Engineer's Contribution to Contemporary Architecture,* Thomas Telford Pubs.(ICE, London) , 2000. (see also the Dieste Wikipedia entry: https://en.wikipedia.org/wiki/Eladio_Dieste)

Non-Conventional Materials and Technologies – NOCMAT for XXI Century   Materials Research Forum LLC
Materials Research Proceedings 7 (2018) 148-166   doi: http://dx.doi.org/10.21741/9781945291838-15

Application of Small Diameter Bamboos in Architecture

Sven Mouton[a*], Karen Allacker[a], Khosrow Ghavami[b], Han Verschure[a]

[a] Faculty of Engineering Science, KU Leuven, Heverlee 3001, Belgium

[b] Faculdade de Enghenaria, Pontifícia Universidade Católica do Rio de Janeiro (PUC), Rio de Janeiro 38097, Brazil

Abstract. Having built several constructions with widely used bamboo species in Ubatuba-Brazil during the years 2004-2006, it was found that one of the main problems in the construction of bamboo structures was the local accessibility of these species. The unavailability of these species leads to higher costs for transportation and to higher environmental impacts. These widely used bamboo species typically have diameters ranging from 12 to 20 cm when used in constructions. Locally available bamboo species have smaller diameters, ranging from 4 to 8 cm, such as *Bambusa Tuldoides (Taquaral)* or *Phyllostachys Aurea (Mirim)*. These are however rarely used in construction. The main objective of this paper is to present several experimental case studies using small diameter bamboos in order to share the insights gained to construct with such small diameter bamboo species.

Keywords: Bamboo Architecture, Bamboo Construction, Small Diameter Bamboo, Jointing Techniques

Introduction

In bamboo construction, mostly thicker bamboo culms (12 to 20 cm diameter) are employed for load-bearing elements such as trusses, girders, purlins, posts and rafters whilst walls are often made of bamboo strips covered with cement or plaster to improve sound insulation or heat-protection. The majority of such bamboo constructions are related to rural community needs in South China, Eastern-South Asia and Latin America and are built in accordance with local traditions being simple in design and construction. [1] Even though bamboo is gaining more popularity amongst architects as a construction material, merely three to four bamboo species, all with diameters over 100 mm are applied, for instance Guadua and Dendrocalamus, while alone in Brazil there exist 135 bamboo species and 17 genera. [2] Only few architects built with small diameter bamboos that are more commonly used in furniture or for decoration. Vo Trong Nghia, a Vietnamese architect, is one of the most famous adaptors. He bundles several small bamboos forming a thicker column. Fig 1 shows one of his works in Vietnam using 50 mm diameter culms of the native species Bambusa Oldhamii. Especially in tropical countries, small diameter bamboos are easily found near roads and in smaller forests while large diameter bamboos are often located on private domains, commercially exploited bamboo forests or on remote locations. Accordingly, when acquiring bamboo one thicker bamboo culm (> 10 cm diameter) has the same cost as 6 smaller ones (< 8 cm diameter). In the coastal region of São Paulo and Rio de Janeiro states in Brazil for example, an average of 32 R$ is paid for one 10-20 cm culm compared to 28 R$ for six smaller ones. Besides the lower material cost and a wider availability also the lightweight of the culms is advantageous. All these factors facilitate the application of small diameter bamboos in construction, also in the West (where on the contrary to large diameter bamboos, they can also be found and grown on local plantations).

Non-Conventional Materials and Technologies – NOCMAT for XXI Century   Materials Research Forum LLC
Materials Research Proceedings **7** (2018) 148-166   doi: http://dx.doi.org/10.21741/9781945291838-15

FIGURE 1 - NAMAN BEACH BAR ©VO TRONG NGHIA

The objective of this paper is to demonstrate that the use of small bamboos in architecture is plausible and in some cases preferable. The influence of the diameter of the bamboos on the design is moreover discussed. Designing with bamboo boils down to three main choices: choosing a structural design method, designing the joints and finally selecting the bamboo type and diameters. In this article, the third choice has been fixed on small diameter bamboos, thereby determining if and how the first two parameters will differ from building with large diameter bamboos.

This paper moreover aims at transcending the scope of laboratory settings by discussing actual field constructions with a variety of building techniques.

Methodology

The use of small bamboos in architecture is demonstrated by empirical research consisting of three case studies. The first case is a veranda with a bitumen flat roof. The second case is an art-gallery with a flat roof consisting of aluminum. The third case study is a guesthouse with a green-roof. These case-studies are all built in the southeast of Brazil and have been executed during 2016-2017 by the first author of this paper. Phyllostachys Aurea 50mm diameter was applied for the structural frames. In the art-gallery also Bambusa Tuldoides from the proper farm was imbedded in the project. A different structural design method and a different roof-load (from bitumen roofing to a green roof) was used to emphasize the wide range of possibilities. The analysis consists of a comparison of the cases in terms of modular frames used, jointing and connection techniques and of linking the results with learnings from literature and the first author´s former field experiences.

The empirical method emerges from the following <u>hypothesis</u>: ´When is the use of small diameter bamboo viable in bamboo construction and how does the technique differ from building with large diameter bamboo?' which is subsequently tested using <u>observation</u> and experiments (actual building of examples). Case studies, in their true essence, explore and investigate contemporary real-life cases through detailed contextual analysis of a limited number of events or conditions, and their relationship. A descriptive methodology as formulated by Yin (1984) was used to research these three case-studies where the goal set by the researcher is to describe the data as they occurred [3].

The scope of this research is limited to the structural bamboo elements of the case studies, as bamboo was not applied as filler nor as finishing elements in any of the case studies. Laboratory tests to determine the mechanical properties of the actual bamboo culms are beyond the scope of this paper and therefore existing data from literature will be referred to.

Non-Conventional Materials and Technologies – NOCMAT for XXI Century   Materials Research Forum LLC
Materials Research Proceedings 7 (2018) 148-166   doi: http://dx.doi.org/10.21741/9781945291838-15

Material

The bamboos applied in the three case studies have a diameter ranging from 40 to 60 mm. The species used is Phyllostachys Aurea, commonly known in Brazil as Mirim. In the art-gallery also Bambusa Tuldoides Munro was used for the arcs and to cover the aluminum roof panels providing a more natural look to the building. The bamboo was bought at a local wood supplier at lengths of at least 295 cm, their origin is a plantation in the countryside of Rio de Janeiro. The culms have been treated/caramelized by diesel oil and a blowtorch. The Bambusa Tuldoides Munro was cut at the proper farm where the art gallery was erected and received a likewise curing method.

In order to protect the bamboo against all forms of degradation, it is advisable to use a protective layer such as varnish. As recommended by M. Seixas, a protective layer that allows the material to breathe should be used as this avoids cracking or the formation of internal humidity due to the ability of air exchange by differential pressure that allows the bamboo to continue to dry. [4] In the three case-studies discussed Osmocolor© varnish was applied and has shown good results.

Regarding its mechanical properties, tests executed at the laboratory of PUC-Rio on Phyllostachys Aurea joints, corresponding to the subject of this paper, showed that columns demonstrated a highest load of 13.57kN, with very little displacement due to the shear failure of the bamboo. Tension tests showed an average load of 2,84kN with an average displacement of 9mm [5]. In 2002 Cruz also tested specimens of Phyllostachys Aurea in tension, and the average maximum tensile load of these specimens tested was slightly more than tested via the Mungpoo frame [6]. Although other bamboo species demonstrate a higher resistance in relation to the culm and wall thickness, this species is widely spread and therefore more available. Architecture needs to design the trusses and columns in such a way that the small diameter becomes a virtue.

No distribution maps of Phyllostachys or Tuldoides for South America were found. The map in Fig 2 however indicates that Guadua (most used bamboo species for construction) is less available in Central and Southern Brazil.

FIGURE 2 - THE DISTRIBUTION MAP OF GUADUA IN SOUTH AMERICA
©GUADUABAMBOO.COM

Non-Conventional Materials and Technologies – NOCMAT for XXI Century   Materials Research Forum LLC
Materials Research Proceedings 7 (2018) 148-166   doi: http://dx.doi.org/10.21741/9781945291838-15

It is found that Dendrocalamus is locally available, but in few amounts and usually on private territory. Phyllostachys Aurea is Angiosperm from the Poeceae-Graminae family, and as the Tuldoides, they are leptomorphic and monopodial or more popularly known as 'runners', which could pose a problem towards aggressive overtake of other species. It grows in semi-tropical climates and is widely cultivated around the world.

Bamboo is not very well codified in construction codes except for the well-known Columbian (seismic) building-code NSR-10, chapter G12 that mainly holds a guideline to building with Guadua, for joints in NTC 5407 and the ISO 22156:2004. The latter is a code based on limit-state design and on the performance of the structure. It only includes requirements for mechanical resistance, serviceability and durability of structures, and has been discussed within the timber structures technical committee ISO TC165. Codes specifically for smaller diameter bamboos are not yet available. This however does not mean that experiments should not be undertaken (with sufficient care), referring to existing codes. The Columbian building code NSR-10 is one of the most practical and adequate codes and hence could be used as a guideline for smaller species. Clearly more research is needed to validate the analogy for these smaller diameter species.

Case study 1 – bamboo veranda

Location
The bamboo veranda is built in the coastal town Picinguaba in the South-East of Brazil, 43 km from Ubatuba and 36 km from Paraty. The town is situated in the Atlantic Rainforest, more specifically in the National Park ´Serra do Mar – Nucleo Picinguaba` in the state of Sao Paulo. The Atlantic Rainforest is a South American forest which extends along the Atlantic coast of Brazil largely between the megacities São Paulo and Rio de Janeiro. Over 85% of this principal biome has been deforested, threatening many plants and animals with extinction Therefore, the Brazilian government implemented several National Parks intending to preserve the original rainforest. Picinguaba is a fisherman´s village inhabited with Caiçaras, descendants from the indigenous people mixed with Europeans and Africans, living their traditional way of life based on subsistence agriculture, hunting but mainly fishing [7]. The latitudes of Picinguaba are 23°23´37.8"S, 44°50´17.3"W.

Climate
The climate of Brazil varies considerably from the tropical north (the equator traverses the mouth of the Amazon) to more temperate zones south of the Tropic of Capricorn (23°26' S latitude). The climate in Ubatuba, due to its geographical location at the sea surrounded by the mountainous Atlantic rainforest, is warm and humid. According to Köppen and Geiger, Ubatuba´s climate is classified as Af, meaning a tropical rainforest climate, also known as an equatorial climate. A tropical climate is usually found along the equator featuring tropical rainforests however, several exceptions exist such as Ubatuba which is not only far removed from the equator, but is actually located just outside the tropics as the Tropic of Capricorn passes the city center. [8]

A tropical climate is influenced by the low-pressure area of the calm zones with little air movement resulting in cloudy and humid conditions throughout the year. [9] The relative humidity in Ubatuba is between 84-88%, which makes the air moist and damp. The average annual temperature remains high throughout the year with slight variations being 25,2°C in the warmest month February and 19°C in the coldest month July, Fig 3. The rain season extends over several months with high levels of precipitation, even in the driest month. In a year, the average rainfall is 2552 mm. The driest month is June, with 84 mm of rain. The greatest amount of precipitation occurs in January, with an average of 343 mm (Fig. 3).

Non-Conventional Materials and Technologies – NOCMAT for XXI Century   Materials Research Forum LLC
Materials Research Proceedings 7 (2018) 148-166   doi: http://dx.doi.org/10.21741/9781945291838-15

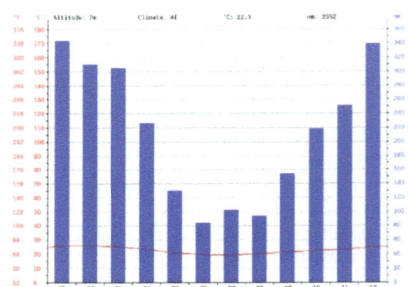

FIGURE 3 - CLIMATE GRAPHIC UBATUBA   ©https://en.climate-data.org/

Building in the Tropics means a confrontation in terms of construction and function with extreme climate conditions. Here the architect needs to deal with heat, strong solar radiation, high levels of air humidity and torrential rainfall. The challenge is to design buildings that offer users comfortable spaces without requiring mechanical cooling systems. [11] The following section describes how the particular design of the bamboo veranda addresses these difficulties

Architectural Design
Organization
The client´s main requirements were to extend the indoor with an outdoor living space offering protection from weathering conditions and to implement a laundry area separated from the rest. As Picinguaba is located in a tropical climate, as mentioned above, no walls or windows were needed in the design. A large enough roof structure to withstand sun and rain was sufficient. Fig 4 shows how the veranda offers space for a large dining table, a cosy sitting area and a laundry area. A light bamboo structure was designed to maintain a spacious area and merely the laundry area was divided from the rest with a small bamboo wall.

FIGURE 4 – VERANDA ©Sven Mouton   *FIGURE 5 - ISOMETRY BAMBOO MODULE*
                                            ©Sven Mouto

Weathering protection
The orientation of the veranda was logically positioned at the back of the house accessed by two large sliding doors with the additional advantage that this location was oriented away from the main sun direction. Providing sufficient ventilation in the veranda was not an issue as the entire structure

Non-Conventional Materials and Technologies – NOCMAT for XXI Century   Materials Research Forum LLC
Materials Research Proceedings 7 (2018) 148-166   doi: http://dx.doi.org/10.21741/9781945291838-15

was maintained open. Because the veranda is enclosed between several other buildings no high wind velocities can affect the structure.

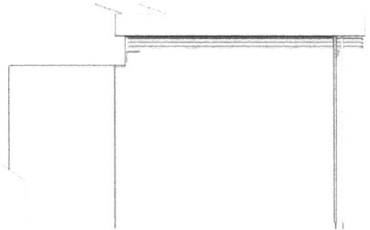

FIGURE 6 - SECTION ©Sven Mouton.

Structural design
General
The veranda measures 10m x 3m. To maintain a light structure the bamboo species Phyllostachys Aurea, diameter 50mm was applied. There are four columns consisting each of four bamboos, and are all put on the outermost line of the veranda. The other side of the roof is connected to the outer wall of the house. Thus, without any supporting columns here. There are eight bamboo trusses with 1.5m distance in between while the column´s distance is 3m. Hence, a supporting column is provided every two trusses. The other truss supports on the lateral bamboos serving as wind brace between the columns. Fig 5 illustrates the isometric view of one module in detail. Smaller whitewashed bamboos, diameter 30mm, are each fixed with a 30/4.5 screw onto the upper bamboo of the trusses. The receiving bamboos are predrilled to avoid cracking. Subsequently OSB plates are put on top of these smaller bamboos where after they are covered with a bitumen layer with an aluminium coating. This coating provides a cooling effect through reflection of solar rays, and thus reduces heating up below.

Columns
The fact that all bamboos need lifting from the ground for protection is basic bamboo building knowledge. The challenge is however to do this in an aesthetical and elegant manner. Already many solutions have been tried and tested. [12] In this case wire rods 5/8 serve as a connector piece between the floor and culm giving the impression that the bamboos are 'floating'. As a tile flooring with underneath a 10 cm thick concrete subfloor was already existing, only 10 cm deep holes for the wire rods (no extra foundations) were drilled into the flooring which afterwards were filled with high cement grout before inserting the rod, stretching out for about 20 cm. Galvanized steel wire rods were used, because of its high resistance to salinity and its lower price (half the price of inox wire rods).

After drying, a bolt and 60 mm diameter washer were screwed onto the rod, about 15 cm above the ground. Then a bamboo was put on top, resting on this large washer and bolt (Fig 7). Normally grout would be injected to stabilize and secure the wire rod inside the bamboo. Nevertheless, in this case none of the bamboos were filled with cement. This method depends on the already available compression, where the weight of the roof pushes the bamboos down onto the washer and dislocation of the bamboo culms is no longer possible. It has to be noted that given the specific location, no high wind velocities could pull the wire rods from the ground. Therefore, this system cannot just be implemented at any location and local wind velocities have to be carefully investigated.

Non-Conventional Materials and Technologies – NOCMAT for XXI Century   Materials Research Forum LLC
Materials Research Proceedings 7 (2018) 148-166   doi: http://dx.doi.org/10.21741/9781945291838-15

Before finalizing the roof, movement of the bamboo structure is still possible but the further in the building process the firmer the structure gets. The advantage of using a bolt and washer to support the bamboo culm is that it can be screwed up or down to align all culm heights and afterwards arrange the inclination of the roof. Screwing up can also increase the pressure and therefore the rigidity of the structure. It is also possible to replace the washer by a two cm thick hardwood circle of similar diameter. Using this kind of connection holds some advantages over the classical grout/cement-filled connections. For example, tests on the Chinese Pavilion built by M. Heinsdorff, where the 'classical concrete infill joint' was used, proved that a longitudinal crack was formed pushing the steel connector and concrete infill into the bulge in the node when high compression load was applied. [13] The problem is that in this type of connection the loads are not placed on the outside wall where fibers are to be found, but on the internal internode that holds the concrete. It is assumed that concrete forms a 'unity' with the entire bamboo by connecting to the wall. But when opening an internode that has been filled with concrete, it can be observed that the cement filling can be taken out as a whole without it sticking greatly to the internal walls.

High compression loads should therefore not be inflected on the internal diaphragm, even though it has a bulge in the node where the concrete can 'attach' to preventing it from further entering the culm or being pulled out from it. Applying loads on the wall of the bamboo culm by the above described principle of bolts and washers also diminishes the need for concrete.

FIGURE 7 - *COLUMN BASE ©Sven Mouton*   **FIGURE 8** - *DEFLECTION TRUSSES ©Sven Mouton*

Trusses

For trusses, normally a large diameter bamboo would be required but by connecting three smaller bamboos to each other, drilling two small wire rods 3/16 through these bamboos, they form one large diameter truss. The trusses are made on the ground and then lifted onto their place. One side of the truss rests on an iron angle, bolted onto the outside wall of the house. The other side is alternately supported by a column or a wind brace. There is however a slight deflection observable which became visible after being several months in use (Fig 8). Recommendable would be to connect four instead of three bamboos for each truss. Nevertheless, this poses no danger to the construction as it is merely a small deflection.

A traction cable is diagonally stretched between the trusses supporting on the columns. Only by stretching this cable the construction became rigid. It is necessary to put sufficient tension on the traction cable (5mm) before functioning properly. This can only be done by a device that screws tension on the cable. In a later phase, experiments could also be made using lianas or sisal, which could be tensioned by using a tourniquet.

In order to avoid deformation or bending of the horizontal member between two columns (where the trusses in-between support upon), a system of pre-tensioning was tried. Similar as in pre-

Non-Conventional Materials and Technologies – NOCMAT for XXI Century   Materials Research Forum LLC
Materials Research Proceedings **7** (2018) 148-166   doi: http://dx.doi.org/10.21741/9781945291838-15

tensioned steel, both small bamboos were bended upwards before being connected in order to form an upward bulge. When placed between the columns, the truss and its inherent weight were placed in the middle of these bamboos, at the highest point of the bulge. At full weight (with the roof- and ultimate roof load), the two small bamboos appeared to bend 'back' into a horizontal position, equalling out the bending they would have had anyway when there would have been no pre-tensioning. In one of the compartments the structure proved insufficient as the two bamboos bended even further down from the horizontal position. It can be concluded that for this building two bamboos were sufficient to avoid bending, but for some it was insufficient. More rigid calculations, with sufficient 'safety-coefficient' should exclude this error in future constructions. The structural integrity was however never compromised, the bending has stopped and is only a (small) aesthetical disadvantage.

In fig 9 a detail of the applied connection method is shown. The use of four culms forming one column, with cross-bracing of both lateral trusses appeared sufficient to act as wind-bracing. As studied by Demets in his doctoral thesis [14], placing lateral structures in between four culms provides a form of wind bracing in itself, much like the fixed connections of the spatial trusses by Prof. Eng. Arthur Vierendeel (Fig. 10). Such trusses do not have the usual triangular voids as seen in typical wind-bracing, but rather employ rectangular openings and rigid connections in the elements, which (unlike a conventional truss) must also resist substantial bending forces. This rigid connection is also formed by the four bamboo columns that are cross-connected by structures arriving at it from two different sides. Once more no cement filling was applied in these connections and 3/16 steel wire rods were used for the connections.

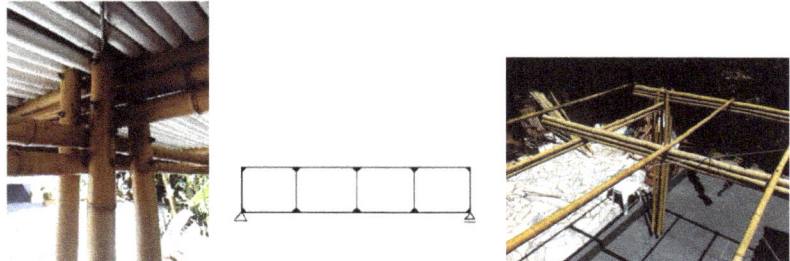

FIGURE 9 - *DETAIL OF JOINT ©Sven Mouton;* **FIGURE 10** - *SPATIAL TRUSS VIERENDEEL;* **FIGURE 11** - *ROOF STRUCTURE*

Roof
The roof height of the veranda is 2.6 m, and was determined by the height of the roof of the house. The roof weight with its bitumen layer, OSB plates, small bamboos and wooden edge is in total about 28kg/m². Later on, for aesthetic and water draining reasons, another load of 6kg/m² expanded clay grains were applied. The inclination of the roof towards the drain is sufficiently so that a layer of no more than 1 cm rainwater is formed during rain-storms. Wind-load as stated before is negligible because of the specific enclosed location (in a non-hurricane zone). The total load is therefore 34kg/m² x 30m²=1020kg, carried by seven trusses connected to the wall and four columns (consisting of four small bamboos each). Figures 11 shows the roofstructure in more detail.

Non-Conventional Materials and Technologies – NOCMAT for XXI Century   Materials Research Forum LLC
Materials Research Proceedings 7 (2018) 148-166   doi: http://dx.doi.org/10.21741/9781945291838-15

Case study 2 – art gallery

Location

The art gallery is built at Fazenda Catuçaba, a 450-hectare coffee farm turned into a luxurious hotel in the mountain range of the Atlantic Rainforest and just three hours from São Paulo. The small village of Catuçaba, only five km from the farm, was once a stop on the way to the Minas Gerais region in the times of the gold rush. Catuçaba is part of the municipality São Luis de Paraitinga, at 20km distance. In the 19th century, the Paraiba valley where the farm is located was responsible for over 90% of the coffee production of the world, until in 1888 the prohibition of slavery ended this. The coffee mono culture and the influx of European immigrants from the beginning of the 20th century have been the basis of the prosperity of São Paulo. The hotel is surrounded by the Serra do Mar State Park, which stretches all the way out to Ubatuba 50 km further. The latitudes of Catuçaba are 23°24′88.4"S, 45°19′99.3"W.

Climate

No climate records of Catuçaba were found. Thus, the climate data of its municipality are used to indicate the climate zone. São Luis de Paraitinga is located in a Cfb climate zone (temperate oceanic climate) in the Köppen classification. This is a typical climate of west-coasts in higher middle latitudes, and generally features cool summers and cool but not cold winters, with a relatively narrow annual temperature range and few extremes of temperature. Oceanic temperatures are defined as having a monthly mean temperature below 22° C in the warmest month, and above 0°C in the coldest month. [15] The average temperature in São Luis de Paraitinga is 14.7°C in the coldest month July and 21.3°C in the warmest month January. The altitude of the town is 795m (2,608ft) above sea level. It typically lacks a dry season, as precipitation is more evenly dispersed throughout the year. The highest average rainfall is measured 236 mm in January and the fewest in July with an average of 26 mm. It is a predominant climate type comparable to Western Europe. [16] Contrary to the equatorial climate of Picinguaba (case study 1 and 3), this area is much dryer and more moderate and therefore easier to build in. The bamboo specie Bambusa Tuldoides Munro vegetates well in this milder climate and therefore can be found on various locations at the farm.

Architectural Design

Organization

The clients' idea of an art gallery was that art and nature are intertwined and art should be presented in a natural environment. Throughout the hotel property various art installations can be found. By being in nature one should be more receptive to see and feel art, according to the hotel owner. The art gallery has to host changing exhibitions offering clients a variation of art. The location was set between the main farmhouse (colonial Portuguese style) and the Occa, a communal space built by an Amazonian Indian tribe where originally several families could live together. Aside from this no further requirements were made. Fig 12 to 15 show 3D designs and AutoCAD drawings of the design. The art-gallery, arising between the colonial farmhouse and the indigenous communal space, had to unite these two different styles. On that account, the design from the outside outlines the colonial Portuguese style with its white walls and blue doors similar to the farmhouse, and shows an Indian heart/core on the inside (being the bamboo structure). The narrow passage, the arcs and courtyard in the middle refer to ancient monasteries and hereby tries to invoke a divine sensation. A small fountain is situated in the centre of the patio from where the water runs back into the river. By providing 13 doors one can enter the gallery from each point augmenting the transparency but maintaining a particular curiosity to see what's inside. The floor level on one side extends the building and partly floats over the river giving a visitor the ability to look at the art gallery from a distance.

Non-Conventional Materials and Technologies – NOCMAT for XXI Century   Materials Research Forum LLC
Materials Research Proceedings 7 (2018) 148-166   doi: http://dx.doi.org/10.21741/9781945291838-15

Weathering protection
A flat roof of 2.9 m high was foreseen to avoid unnecessarily elevating the structure and maintaining a low profile, offering the farmhouse and occa a primary role. Three existing six meter high palm trees are integrated in the design and even one of these palm trees goes through the roof. The passageway is covered, with a 0.5 m eave (after the columns) to the patio, but generally maintained open. As no extreme climate conditions are present, this doesn´t affect the art exposed in the gallery. If in future it appears necessary the centre could be closed with large glass sliding doors (the required foundations have already been placed). The floor level is elevated 30 cm from the ground for protection.

FIGURE 12+13 - 3D DESIGN ART GALLERY   © *Sven Mouton*

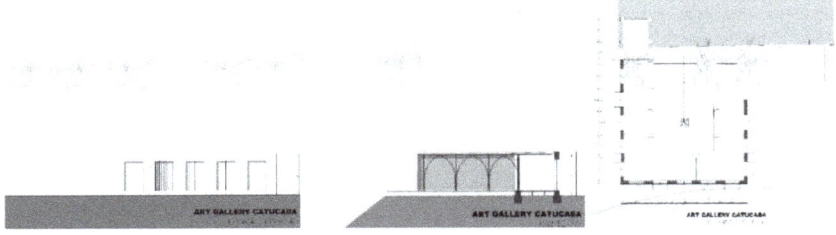

FIGURE 14-16 - AUTOCAD DRAWINGS ART GALLERY © *Sven Mouton*

Structural Design

General

The art-gallery is nearly quadratic (15 x 11 m) with a passage way of 2.10m wide. All measures are designed according the golden proportions of Da Vinci. [17] The floor plan is divided into 15 squares of 2.10m x 2.10m and every square, except the latter two, has a proper door opening centred in the middle. The columns are placed outside these squares on an 80cm broad border in blue stone and weathering steel. The walls consist of bricks finished with a cement layer on the outside. On the inside the bricks remained in sight but both sides were painted white. For the structure, identical bamboo culms as in the veranda were used, being Phyllostachys Aurea 50mm diameter. Only for the arcs the locally found bamboo species Bambusa Tuldoides Munro was used with a diameter of 40mm. These were cut on site and curved, anointed with gasoil and simultaneous heating the cane with a blowtorch. The arches support the wind braces and trusses between two columns and therefore are part of the structural system. The roof is made with aluminum prefab plates that afterwards were

Non-Conventional Materials and Technologies – NOCMAT for XXI Century   Materials Research Forum LLC
Materials Research Proceedings 7 (2018) 148-166   doi: http://dx.doi.org/10.21741/9781945291838-15

covered with Bambusa Tuldoides Munro culms for aesthetical reasons. The roof structure is finished with a fine aluminium boarding on the outside and a timber sheathing on the inside.

Columns

As mentioned above the columns are lifted 30cm above the ground, on a blue stone/weathering steel basis. Four galvanized steel wire rods 5/8 are put 20cm into the cement basis while still wet, extending about 30cm from the ground (for later putting the bamboo culm 15cm above the ground and 15cm inside the bamboo). A two cm thick hardwood plate with four holes drilled into was placed onto this cement basis to cover the concrete and give a more natural look to the basis. Again four bamboo culms were used to form one column using the similar bolt and washer system as applied in the previous case-study, lifting the bamboos culms another 15cm from floor level. Fig 17 illustrates the column in detail. In total there are ten columns, all placed on one side of the building (the outer).

The arches are connected to a connector piece on top and onto the centre of the column (about 1m above the ground). Bending the dried poles of Phyllo. Aurea appeared to be impossible. The required curving couldn´t be attained to the full extent before cracking it. Whilst some sources advice to insert fine sand into the culm to avoid wall-cracking, this also did not suffice. Only with fresh cut bamboos ('green') the curving was achieved for these fibres are still flexible. All of the internodes needed to be drilled with a small hole, as well as all the diaphragms punctured in order to release the hot inside-air (to avoid cracking when trying to escape - because hot air expands). A formwork with rebars was placed on the ground in order to bend the pole piece by piece securing it in between. A blowtorch, while anointing with gasoil, was moved up and down, only continuing to the next node when it felt the node 'gave way'. Cutting the pole to the right size was only done when a satisfactory curve was obtained enabling to choose the best piece. As one arch is formed by three pieces, each bend culm was left inside the rebar formwork to later add the next one and only take out the formed arches when all three were connected by wire rods 3/16. By connecting the arches together, it was found that a rigid shape was obtained, whereas separately they still wanted to return to their natural shape. A small wooden connector-piece for the top of the three arches was used to subsequently connect them in one connection to the ´violin` connector piece, as can be seen in Fig. 18.

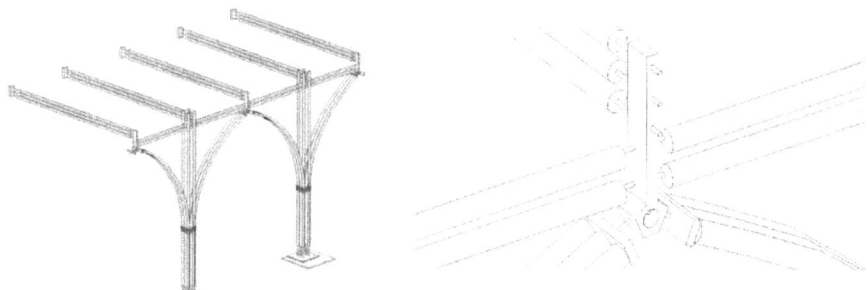

FIGURE 17 - *ISOMETRIC VIEW OF THE COLUMNS AND TRUSSES ©sven mouton,*
FIGURE 18 - *'VIOLIN' CONNECTOR PIECE ©Sven Mouton*

Non-Conventional Materials and Technologies – NOCMAT for XXI Century   Materials Research Forum LLC
Materials Research Proceedings **7** (2018) 148-166   doi: http://dx.doi.org/10.21741/9781945291838-15

Trusses
The trusses are formed by three bamboo culms and a wooden batten (50mmx20mm) placed on top, identically connected to each other as in the first case study; drilling twice a hole simultaneously through the bamboos and batten and inserting a 5/16 galvanized steel wire rod to connect them. Between the upper bamboo culm and wooden batten two bolts and washers are foreseen in order to facilitate aligning of the wooden batten. No grout/cement is used. This wooden batten enables a convenient placement of plasterboard underneath and the aluminium panels above. The trusses are all prefabricated on the ground before lifting onto their place. Similar to the veranda, the bamboo trusses are on one side connected to the wall and on the other side supported by a column, with the difference that in this case a wooden piece of 20x10x5cm was encapsulated into the wall onto which the culms were bolted. The advantage of this system is that all three culms are separately connected to the wall and don´t transfer weight onto each other as is the case in the veranda where the two upper culms support onto the latter, transferring all the roof weight. The majority of the 27 trusses have a length of 2.7m. Just two diagonally placed trusses in both corners have a wider free span which is 3.2m long.

Connections
As can be seen in Fig. 19 there are two different types of connections to be found in the passageway. First the connection principle of Demets was used where four columns are interconnected to a laterally placed truss supported by contrary adhered wind braces, as explained in the previous case study. Advantageous is that such joints disperse the foreseeable connecting culms over several bamboos averting too much connections arriving at one bamboo thereby weakening it. Moreover, it is a firm connection that is able to withstand wind forces already after two columns. The second joint doesn´t connect the column to a truss but merely the arches in the middle (Fig 18). Designing a bamboo structure also means designing connector pieces by drawing lines from the directions of the columns, trusses and arches to 'sculpt' the point where they connect to each other. In this case a 'violin'-shaped hard wood connector piece was designed where all wire rods from the end-points of the bamboo culms join, making a safe passage of all transferring forces. Attention needs to be given to the possibility of screwing the bolts tight so that all wire rods are kept in place. In this case, a round form (the 'hole' of the 'violin'-shaped joint) was cut out for the bolts to be tightened (Fig 18). The strength of this joint is essential, no softwood should ever be used, but only hardwood with first durability-class [18]. In this case the Brazilian hardwood Cumaru was used. Connector pieces in joints are of high importance to deduct tension and compression forces onto the bamboo members and generally they are made of steel, metal, bamboo or even rubber but very few times with hardwood, although the environmental and cost factors obviously favour wood above steel or metal. [19] Perhaps in future constructions more wooden connector pieces could be developed to explore this possibility more. Although strength tests are needed for the individual designs of connectors, it is perceived that hardwood connectors are in no way the weakest part of the structure (on the contrary).

Roof
The roof cladding consists of insulated aluminium panels of 300x100x6cm supported by wooden rafters bolted onto the bamboo trusses, as described above. The construction cannot deform because these wooden rafters and aluminium panels provide the construction with enough rigidity that a traction cable is no longer necessary. From the inside the roof was finished with plasterboard. As the art-gallery is a freestanding construction, a higher wind velocity can occur and add horizontal loads to the building; even lifting the building. The roof weight hence needed to be calculated precisely. The weight of the roof is 15kg/m² for the aluminum panels, along with the own weight of the trusses and roof-covering reaching 23kg/m². For the total area of 68.9m² this leads to a total weight of 2618kg on the load-bearing structure.

FIGURE 19+20 - *THE ART GALLERY AFTER BEING CONCLUDED ©Sven Mouton*

Case study 3 – guesthouse

Location

The location of the third case study is identical to the first, being the coastal town of Picinguaba in Southeast Brazil. Both constructions are built in the same town and hence are faced with identical climatic conditions, being tropical warm and humid. Compared to the veranda, the guesthouse is less protected by surrounding structures. Wind loads are hence higher, enlarging the need for extra roof weight. This is achieved by a green roof.

Architectural Design

Organization

The clients required a multifunctional guesthouse where family and friends could stay over for a short vacation and the owners could also use it as a separate room to work or for the children to play. One bedroom with double and single bed and a sleeping couch sufficed, the bathroom needed to be comfortable but not too large. As the guesthouse is close to the residents' house they wanted a soundproof barrier and enough privacy for both residents and guests. A large stone that was present on site was preferably integrated into the house.

As the project location was remote from the town centre and everything had to be carried to the site by carriers, the principal idea was to use as little construction material as needed by re-using materials and applying natural materials extracted from the site. The 6.3m long rammed earth wall serves as noise barrier and is made with locally excavated red earth. As the terrain lies on a slope, levelling of the terrain was required thereby delivering base material without extra energy needed. The formwork used for the rammed earth was later on applied in the roof structure. The beams holding the framework together became the ring beam. The green roof was finished with locally found black earth and plants. The large granite rock that was present on the left-side of the terrain was integrated into the bedroom forming half of the wall. Beyond an existing palm tree was cut out of the roof to maintain it. Bamboo culms were used to form the roof structure. To ensure privacy large windows were positioned opening to the back and right side of the building, the front was maintained relatively closed. Also, the large rammed earth wall extending for another 1.70m outside the guesthouse offers a small private and covered patio. Fig. 21 and 22 show an isometric view and floor plan of the guesthouse.

Weathering protection
Because of its thermic inertia, the green roof marks a difference in low and high-pressure areas in and around the construction encouraging ventilation. The small windows in the front provide cross-ventilation. In every room windows can be opened up to ensure ventilation but can also be closed whenever it is colder outside. Screens are foreseen so mosquitos cannot enter when sleeping with the windows open. The roof has a 1.10m eave to the front and a 2m 'free-hanging' eave to the right side to protect the walls and bamboo structure. As stated, the terrain being situated on a slope needed to be excavated in order to obtain an equal ground level. Running rainwater from the mountain needed to be deviated through a 50cm broad canal behind the house, which later was finished off with a wooden 'walkway/deck'. This was absolutely necessary to avoid water from entering during heavy rainfall.

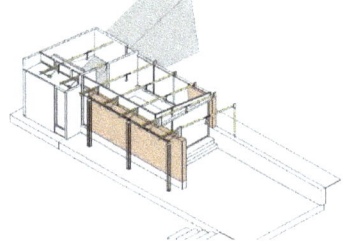

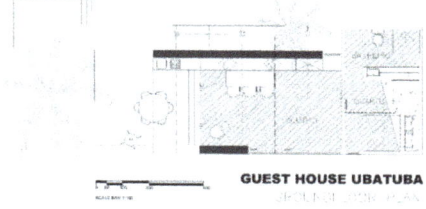

***FIGURE 21** - ISOMETRIC VIEW OF THE GUESTHOUSE*

***FIGURE 22** - GROUND FLOOR PLAN THE GUESTHOUSE ©Sven Mouton*

Structural Design
General
The guesthouse has a total surface of 36.8m² with a living room of 8.5m by 3.5m enlarging to 5m width for the bathroom and bedroom. The walls are made of cemented bricks painted white or rammed earth. The largest rammed earth wall has a length of 6.3m and is 2.45 m high (2.9m with brick foundation). The height of the roof runs from 4m to 3.3m which is an inclination of 6° on the total length. Preferably a green roof has an inclination between 0 to 15% [20]. The bamboo structure can be divided into two parts; inside the guesthouse an inverted truss is used, outside another bamboo structure supports the 1.10m roof eave at the front.

***FIGURE 23** - SECTION OF THE GUESTHOUSE **FIGURE 24** - CONNECTOR PIECE*
***FIGURE 25** - COLUMN ©Sven Mouton*

Non-Conventional Materials and Technologies – NOCMAT for XXI Century   Materials Research Forum LLC
Materials Research Proceedings 7 (2018) 148-166   doi: http://dx.doi.org/10.21741/9781945291838-15

Columns

The roof structure is mainly supported by a large ring beam formed by the rammed earthen walls with a concrete end layer that continues into a wooden beam (the former reinforcement of the rammed-earth formwork) to form a closed ring beam. There are merely four columns in the construction. Three columns are put at the front of the building to support the roof eave at the front. Although the columns exist again of four bamboo culms the principle applied is not similar to the one from Demets as discussed in the previous case studies. The tops of the bamboo culms are connected to a steel plate of 150x150x8mm and merely one steel bar goes to the joint above making the whole work as one column. The wind-bracing and upper bamboo structure culms are connected to that one steel piece and aren´t interconnected (Fig 24). The basis of these three columns is identical to the other two case studies as it appeared easy to make and work with. The fourth column is formed by only one bamboo culm, Phyllostachys Pubescens with diameter 120mm instead of 50mm (Fig. 25). This last column is connected to the basis and top by a wooden plug with an outside diameter equal to the diameter of the bamboo and the inner part less than the internal diameter, wherein a galvanized wire rod 5/8 is drilled and secured with a bolt and small washer. No grout was used.

Trusses

The trusses can be divided into internal and external trusses. Both were separated from each other because it is complicated and undesirable to have a circular tube continuing from inside a building to the outside for the difficulty in filling the interspace. This way both trusses can also be shorter and both use the walls and columns on either side with a shorter distance than if the trusses were continuous. Outside the building, a similar truss as in the previous two cases was utilised; two small diameter bamboos interconnected in three places that also hold up the roof panels in this interconnection. These small diameter bamboos divide about 1/4 of the entire green roof-weight to the three columns (each existing of four bamboos) beneath.

Inside the guesthouse a new technique of an inverted truss was tried (Fig 26). The diameter of the bamboo used was 12-14cm (also from Phyllostachys Pubescens) which is larger than for the previous cases because the weight of the green roof demanded such diameter. Although the diameter is larger than intended for the scope of this paper, it is included in this paper to highlight the importance of using the right diameter at the right place. The inverted truss is an innovative solution to answer to the problem of buckling/elasticity in the middle of a horizontal bamboo member. It was noticed that bamboo, like any beam, is deforming mainly in the center point, which can also be observed in a typical bending moment and shear force diagram. Where the obvious solution to this problem is making triangulated typical trusses or augmenting the diameter/section of the beam, it could be done in a more subtle way as well, by providing the 'inverted truss' as seen in Fig. 27. Referring to tensile steel structures (such as 'tensigrety'), a wire rod with tensioned cables attached is foreseen in the middle point of the truss, taking the load of the central roofpanels. Because the tensioned cables are brought to the same point where the bamboo-truss is connected to the wall, these cannot change position in the orthogonal field, and thus also preventing deformation/bending of the bamboo in the central point (as in the typical bending moment diagram). Using this technique, a smaller diameter/section of the entire truss can be foreseen, because the bending doesn't have to be taken into account anymore when designing the total section. Clearly, the total 'strength' of the truss needs to be calculated more carefully because failure would be more sudden as there is no possibilty anymore of bending before failure. The result however is a 'lighter' look of the truss where otherwise two or three larger size diameters would have had to be connected underneath each other to form a truss as was done in the previous two case studies.

Similar to the columns discussed before, the endpoints where the truss had to be connected to the wall were attached with wooden 'plugs' so that the wire rod could remain centralized inside the bamboo, reducing the need for grout. Because of the large weight of the green roof, it was decided to

Non-Conventional Materials and Technologies – NOCMAT for XXI Century   Materials Research Forum LLC
Materials Research Proceedings **7** (2018) 148-166   doi: http://dx.doi.org/10.21741/9781945291838-15

fill the internode below the wooden plug with a small amount of concrete. This way the wire rod could extend beyond the wooden plug and be fixed inside the grout as well, connecting it more to the entire culm. Tension bands however are necessary in these kind of trusses; the moment-forces at the endpoints of the truss are very high, making cracking nearly inevitable if the cane is not hold together with a tensioning band. With this band, the structral integretity of the circular tube-form is maintained and all forces can be divided correctly. The wire rod at the end of the truss is connected to the wooden ringbeam by a small steel plate; a hole is drilled into the wooden ringbeam that can hold the bolt used to secure the truss to the small steel plate, and this steel plate is screwed tight to the wooden beam. In Fig 28 the wooden plugs at both ends can be observed in detail.

FIGURE 27 - 'INVERTED TRUSS' FIGURE 28 - WOODEN PLUG DETAIL
FIGURE 29 - CONNECTIONS IN THE FRONT ©Sven Mouton

Connections
For the second time, hardwood is used as a connector piece. Different in this case study is that this connector is three-dimensional whereas the previous was two-dimensional (Fig 18 and Fig 24). The upper bamboo culms going to the front of the roof are connected to a small steel plate on top of the wooden sculptured 'bird'. The upper culms that go to the back are connected to a small steel plate at the back of this 'bird'. The lateral placed wind braces are connected side-ways. A wire rod ½ goes entirely through the connector piece, bolted on both sides to secure its proper location (Fig 29). The wire rods applied in this project were galvanized in fire (artisanal). The wind-braces are connected by the same system as described above for the fourth column. This method of connecting is also applied to connect the bamboo ends of the inverted trusses to the wooden ring beam. A wooden plug is inserted in the bamboo for about 5 to 10 cm and has a thicker outside diameter equal to the outer bamboo diameter. As discussed in Sassu (2012) [19], timber cylinders inside bamboos avoids the fragility resulting from longitudinal fracture of bamboo fibers. Filling the open end-internode of a bamboo cane also avoids water or insects entering in the bamboo, similar as a concrete infill would do. Wooden plugs have been used before, but rarely with wire rods, bolts and washers for tensioning them against another material. As for these 'new' kinds of joints, it can be observed that it is not safe to repeat them without proper testing. In this case, as empirical study/experiment at own risk of researcher, it can be allowed. For further use of these types of joints it can be noticed that, following NSR-10-code from Columbia, it is sufficient to perform 20 times the required test on a joint to have it accepted as a 'conform' joint. [21]

Roof
In the last case study, a larger roof load was tested, see Fig 30. The green roof has following layers above the truss; (1-) wooden battens that hold up the (2-) multiplex boards (formerly used for the casting of the rammed earth), and also serve to attach the (3-) dry wall panels for finishing off the ceiling. Above these multiplex boards (4-) a waterproof layer of PDA was placed with sufficient overhang to form upstanding boards at the borders. Above this waterproof layer, (5-) expanded clay granulates were placed to form a water retaining buffer, divided by a (6-) root proof geo-membrane

Non-Conventional Materials and Technologies – NOCMAT for XXI Century   Materials Research Forum LLC
Materials Research Proceedings 7 (2018) 148-166   doi: http://dx.doi.org/10.21741/9781945291838-15

from the (7-) earth layer with sufficient humus for (8-) plants and vegetables to grow on it. This total package equals around 82 kg/m², making the total load of the roof 4206 kg, substantially higher than previous two cases.

FIGURE 30 - GREEN ROOF

Discussion
The three case-studies discussed in this paper are about bamboos of a smaller size (4-6cm) than is commonly used in bamboo architecture (few exceptions aside). The question can be raised if small size bamboo can only be applied in small scale constructions. However, larger scale constructions such as Vo Trong Nghia´s buildings with small bamboo indicate that it is possible to apply them not only to larger loads but also to greater spans. Further testing about the validity of the applied techniques in these case-studies is necessary before transferring them to larger constructions. The larger spans such as in the projects of Vo Trong Nghia are mostly composed of bundles of small diameter bamboos tied together (with ropes or tensioning band), whereas the examples in the case-studies of this paper rather assume similar techniques as would be used when designing with larger diameters (doubling the amounts needed).

Various 'new' techniques were introduced in the case studies discussed such as hardwood connector pieces, wooden plugs, the wire-rod basis where the bamboos support on a large washer and bolt, the inverted truss system. These techniques cannot merely be adhered without any proof that they are save to use. For example, when considering washers or wooden cylinders to support the base of a bamboo culm, it is important that the wall is completely supported in order for forces to be transferred evenly to the ground. A non-perfectly horizontal cut could induce cracking because some parts would be more supported than others or could transfer the loads askew to the foundations. Even though in none of the three cases this problem arose, this could be due to the light loads. Further tests should confirm the validity of this statement. The inverted-truss contravenes buckling of the bamboo culm and hence a smaller diameter can be worked with. Again, further research needs to investigate exactly how much forces this inverted truss can transfer. The wooden connector pieces seemed a cheap, light and proper solution. However, each building is different and therefore the connector pieces would differ as well. Testing in laboratory settings would test one type of connector piece, but another might not be viable, endangering the construction. Here maybe several guidelines regarding the use of hard wooden connector pieces should be drawn up, serving as a guideline for architects and bamboo builders. This might be important because if the steel and metal connector pieces could all be executed in timber it would drastically lower the building (eco-) costs. Also, more research needs to be conducted on the actual mechanical properties of these small diameter bamboos to improve the accuracy of the calculations. Moreover, a durability analysis after five to ten years would present useful information about the weathering of these small bamboos. For now, few cracking is noticed, either small or perpendicular running over several internodes. But no other signs of mould, fungi are observed. It could be that smaller bamboos are less sensitive to these factors, or that this will slowly appear in time.

Non-Conventional Materials and Technologies – NOCMAT for XXI Century   Materials Research Forum LLC
Materials Research Proceedings 7 (2018) 148-166   doi: http://dx.doi.org/10.21741/9781945291838-15

Although the exact availability and presence of particular bamboo species and which one to use could be discussed, it is believed that each case is specific and one should work with what one has available (geographically and financially). Moreover, also personal preferences play a role. These descriptive case-studies indicate that applying small diameter bamboos is possible and show how it is done in these particular cases, hopefully inspiring others to do the same and incentive more research into these apparently more abundant species and diameter-sizes. Perhaps codes and norms, such as the NSR-10, could be extended to smaller diameter bamboo.

Another element of discussion is the elimination of the use of mortar in the connections. Considering the end of life of used components, it is recommendable to perform non-mortared connections such as presented in this paper. Care should however be taken when transferring the proposed coneection to larger (heavier) constructions, that are subdued to large upward wind-loads, for it is mainly by (the correct amount of) compression that the proposed jointing/footing methods have proven valid. The constructions only gained complete rigidness after applying the full weight of the roof, making the wall transferring the full load only through the wall and not the internal diaphragm as would be the case with connections made by mortar or grouted connections. Further research needs to be conducted to confirm this statement.

In order to achieve a lower building cost, prefabrication of bamboo construction elements should always be considered, even if it is only at the construction site itself. Working on heights to connect bamboos should always be avoided, both towards building cost as safety. This is design-dependent, but in specific projects it could be considered advantageous to not have all pieces prefabricated for this heightens the hand-sculpted/craft effect of specific projects. In mass-housing this for example should be avoided at all costs, also because control of a specific solution cannot be repeated without knowing who will be executing the work. In the cases discussed in this paper, an in between was found, where the trusses themselves were assembled on the floor, but all other connections were made on height.

Conclusion

The specificity of working with small diameter bamboos differs only in a few aspects with large diameter bamboos. Where normally one large diameter bamboo culm is sufficient for a truss or a beam, now three or even four culms have to be provided, implying a creativity in connecting these together. The Demets` principle proved to function well for large (12-14cm used in the community centre in Cambury [22]), as well as small diameters (4-6cm) and is a recommended system for joints between columns, wind braces and trusses. When replacing the large diameter culms by culms with a diameter of three to four cm, the main aspects of bamboo building techniques remain similar, even for larger constructions. In general, it can be concluded that one should work with what is available and is required for the specific structural design. Three factors play a role in designing with bamboo; defining the structural design method, choosing the bamboo type and diameters to work with and designing the joints. When a specific structural method requires a large diameter bamboo, as was the case with the inverted truss in the guesthouse, one should not be limiting oneself. Finally, the two types of diameters can easily be combined.

References

[1] ZHENG, W., WENJING, G., `Search Report: Current status and prospects of new house construction materials from bamboo`, *Research Institute of Wood Industry*, CAF, 2009.

[2] LOBOVIKOV, M., BALL, L., GUARDIA, M., RUSSO, L., `World Bamboo Resources: A Thematic Study Prepared in the Framework of the Global Forest Resources Assessment`, *Non-Wood Forest Products*, Issue 18. 73p., 2007.

[3] ZAIDAH, Z., `Case study as a research method`, *Jurnal Kemanusiaan*, bil.9, 2007.

Non-Conventional Materials and Technologies – NOCMAT for XXI Century   Materials Research Forum LLC
Materials Research Proceedings 7 (2018) 148-166   doi: http://dx.doi.org/10.21741/9781945291838-15

[4] SEIXAS, M., RIPPER J.L., GHAVAMI, K., ´Prefabricated Bamboo Structure and Textile Canvas Pavilions`, *Journal of the International Association for Shell and Spatial Structures*, 2016.

[5] MARY, W., KENMOCHI, C., COMETTI, N., LEAL, P., ´Avaliação de estrutura de bambu como elemento construtivo para casa de vegetação`, *Engenharia Agricola*, 2007. https://doi.org/10.1590/S0100-69162007000100003

[6] SHARMA, B., GHAVAMI, K., HARRIES, K., ´Performance Based Design of Bamboo Structures – Mungpoo Frame Test`, *International Wood Products Journal*, Vol 2, Issue 1, 2011.

[7] PRADO, G., ESTEVES, R., RAMIRES, M., BEGOSSI, A. ´The Caiçaras and the Atlantic Forest coast: Insights on their resilience`, *BioScience*, p.189 - 196, Vol. 4, N° 3, 2015.

[8] Wikipedia: Atlantic Rainforest, Southeast Brazil; https://en.wikipedia.org/wiki/tropical_rainforest_climate.

[9] LAUBER, W., ´Tropical architecture. Sustainable and humane building in Africa, Latin America and South-East Asia`. *Prestel*, Munich, 199p., p.85 'Fundamentals of climatically appropriate building and relevant design principals', 2005.

[10] Climate Data; https://en.climate-data.org/location/34847.

[11] KRAUTHEIM, M., PASEL, R., PFEIFFER, S., SCHULTZ-GRANBERG, J. ´City and wind. Climate as an architectural instrument`. *Dom Publishers*, Berlin, 201p., p 120 'designing a comfortable breeze', 2014.

[12] HIDALGO, O., ´The Gift of the Gods`. Bogotá, 2003, 553p.

[13] HEINSDORFF, M., ´The bamboo architecture. Design with nature`. *Hirmer Verlag*, Munich, 2011, 208p.

[14] Demets PhD 2006: Doctoraatsverdediging 29/06 *TU/e*: 'Naar een nieuwe houtskeletbouwwijze'

[15] Climate Data; https://en.climate-data.org/location/34846.

[16] Wikipedia: Oceanic climate; https://en.wikipedia.org/wiki/oceanic_climate.

[17] SNIJDERS C.J., GOUT, M., ´De gulden snede`, *Synthese*, 87p., 2007.

[18] SASSU, M., ANDREINI, M., DE FALCO, A., GIRESINI, L., ´Bamboo Trusses with Low Cost and High Ductility Joints`, *Open Journal of Civil Engineering*, Vol. 2, p. 229-234, 2012. https://doi.org/10.4236/ojce.2012.24030

[19] MINKE, G., ´Building with bamboo. Design and technology of a Sustainable architecture`. *Birkhauser,* Basel, 158p., 2012.

[20] MENTENS, J., RAES, D., HERMY, M., ´Green roofs as a tool for solving the rainwater runoff problem in the urbanized 21st century? `, *Landscape and Urban Planning,* Vol. 77, Issue 3, p. 217-226, 2006. https://doi.org/10.1016/j.landurbplan.2005.02.010

[21] NSR-10 Colombian Building Code. ´Estructuras de madera y estructuras de Guadua – Capitulo G.1 Requisitos Generales`. 156p., 2009.

[22] MOUTON, S., ALLACKER, K., GHAVAMI, K., VERSCHURE, H., ´Durability analysis of a bamboo community center, Brazil`, 2017.

Non-Conventional Materials and Technologies – NOCMAT for XXI Century   Materials Research Forum LLC
Materials Research Proceedings 7 (2018) 167-179   doi: http://dx.doi.org/10.21741/9781945291838-16

Structural behavior of Two-Floor Housing Made with Frames Braced with Prefabricated Bamboo Guadua Panels

C. Takeuchi[a*], E. Ayala[b], C. Castillo[c]

[a*] Universidad Nacional de Colombia; cptakeuchi@unal.edu.co

[b] Universidad Nacional de Colombia; eaayalat@unal.edu.co

[c] Universidad Nacional de Colombia; camacastillocar@unal.edu.co

Abstract. In the present research, the structural behaviour of two-store frames stiffened with **the structural system of Columns and Prefabricated Panels Frames (PCPP)** was evaluated. **The system is composed of columns in bamboo** *Guadua angustifolia* **Kunth anchored and assembled in-situ with prefabricated bamboo guadua panels, including beams.** Their performance under horizontal and vertical loads was studied. Then, with this structural system, two houses were designed: one for social interest housing and the other one for upper-middle-income housing. Three frames with a different geometric configuration were tested: two braced frames and one non-braced acting as a control frame. The aim was to find efficient systems in terms of displacements. All the frames were loaded until a maximum deck displacement of 0.57 m. Two frames were tested under monotonic horizontal load and the third one was tested with loading and unloading cycles. Additionally, mechanical characterization tests of the bamboo were done for the frames simulation in the structural analysis software SAP2000. Subsequently, the simulation was implemented in the two types of housing mentioned above. Based on the tests results, the frames with braced panels accomplish their structural function. This type of frames stiffens and controls the displacements of the structure with horizontal loads in an efficient way, when the structural system is still in an elastic range. However, due to the premature damage of the brace connections, the frames lose stiffness. At the end of each test, the frames presented remnant strain but always kept their structural integrity without collapse. According to numerical calibration, it is concluded that this system can be implemented in different architectural configuration housing, employing different types of panels. Some of these panels are required to resist vertical loads and the others, with diagonal elements, provide the required stiffness against horizontal load. The compliance of the design allowable stresses and the drift in elastic range according to the normative were verified.

Keywords: Bamboo Guadua Frames, Bamboo Prefabricated Panels, Drift, Bamboo Housing

Structural system of columns and prefabricated panels frames (PCPP)
The structural system studied in this research consists in a palafitos system of pre-assembled panels and columns in bamboo *Guadua angustifolia* Kunth. The system is composed of panels and continuous columns anchored in-situ from the foundation to the roof, called henceforth as *"Joining columns"*. The panels have beams on their upper part formed by two chords separated by vertical posts. The connections used between the elements of **the structural system of Columns and Prefabricated Panels Frames** PCPP had fish mouth cuts.

Most of the structure of the housing project is made or pre-assembled in the workshop and moved to construction for lifting and assembly. The type of support used for this housing consists in a reinforced concrete pedestal, which has a metal tube of 2.0" diameter and anchor hooks in bars of 3/8" diameter. The *Joining column* and foundation beams of the structure are anchored in this

Non-Conventional Materials and Technologies – NOCMAT for XXI Century   Materials Research Forum LLC
Materials Research Proceedings 7 (2018) 167-179   doi: http://dx.doi.org/10.21741/9781945291838-16

pedestal. The details of the reinforced concrete pedestal used in the foundation are presented in PHOTOGRAPH 1 (a).

The structural system is palafitos where only the pedestals make contact with the land, bringing constructive advantages for its easy execution, low cost and time saving. It does not need large volumes of excavation or filling and it is a friendly system with the environment, affecting the ground as little as possible. FIGURE 1 (b) presents the palafitos system.

(a)   (b)

FIGURE 1 - A) FOUNDATION PEDESTAL. B) PALAFITOS SYSTEM SOURCE: ARME IDEAS EN GUADUA.

Each column of the structure is composed of a Joining column plus the vertical elements of each panel as can be appreciated in FIGURE 2. Once all panels of a floor have been located in position, a 3/8 "diameter bar is inserted inside the top and bottom chord of the beams through all the perimeter of the house. This bar is threaded in the ends and tightened by nuts to compress the bamboo structure and form compression rings on each floor. This work is called "*Beam sewing* ".

FIGURE 2 - CORNER COLUMN IN HOUSING WITH PCPP STRUCTURAL SYSTEM. SOURCE: ARME IDEAS EN GUADUA.

Geometric configurations of prefabricated panels

The panels studied in this research cover the complete span of the frame on each floor. The dimensions of the panels depended on the physical space available in the laboratory. Three geometries of panels were selected. FIGURE 3 presents the geometric configurations and the dimensions of the panels considered in this research for analysis and loading tests.

Panels B and C correspond to braced panels; the panel C presents eccentric bracing, offering the possibility of locating some architectural elements like doors or windows, inside it. The panel B is concentrically braced and, according to preliminary simulations, presented the greatest stiffness

Non-Conventional Materials and Technologies – NOCMAT for XXI Century   Materials Research Forum LLC
Materials Research Proceedings 7 (2018) 167-179   doi: http://dx.doi.org/10.21741/9781945291838-16

under horizontal loads, employing an efficient quantity of material for its manufacture. Panel A or non-braced panel was considered for comparative purposes to observe the stiffness degree provided by braces in these types of prefabricated panels.

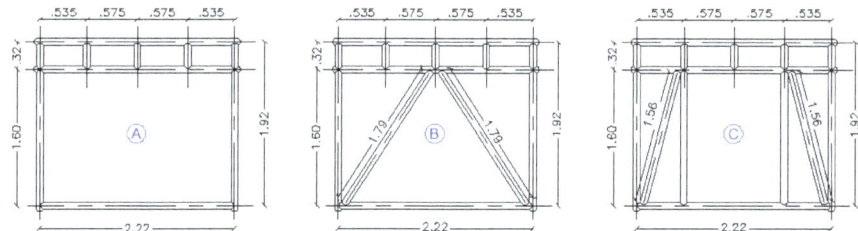

FIGURE 3 - *GEOMETRIC CONFIGURATIONS OF PREFABRICATED PANELS TO TEST.*
Source: Authors

Manufacture of panels
The connection of the panel culms was made by canes and bolts using threaded bars of 3/8" diameter. A detail of this connection is presented in FIGURE 4, where the bolt location must be always behind the stem diaphragm. Additionally, all the fish mouths of the panels were confined using strengthening hoop. The guaduas were filled with mortar in the connections zones to prevent crushing.

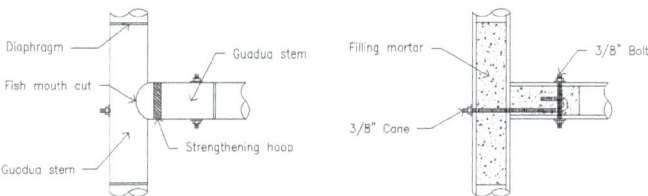

FIGURE 4 – *CANE-BOLT CONNECTION. Source: Authors*

Panel-Frame Assembly
Two ways to connect the panels with each other and also with the *Joining columns* were performed: one with the corner fish mouths of each panel and another with the sewed beam. From the fitting of the fish mouths, the panels are connected to each other vertically and horizontally to the *Joining columns*. On the other hand, it is possible to adjust the panels and columns adequately with the sewed beam.

Monotonic loading pcpp tests
Two tests, corresponding to the frames with panels A and C, were performed under monotonic load. For these tests, a metallic anchorage platform was built in the laboratory to guarantee the correct embedment of the frames and control the lifting and sliding during the tests. For the PCPP tests, readings of applied load and displacements were taken only during the loading process. Displacements were recorded in 8 different points which are indicated in FIGURE 3 (a): points 1 to 4

correspond to horizontal displacements in roof and deck, while points 5 to 8 correspond to displacements in the bases.

Test 1- PCPP with non-braced panels

The first PCPP was the frame without braces, using only panels A. For this test, a maximum horizontal roof load of 1222 kg was obtained. From that point it was not possible to apply greater load because the actuator piston reached its maximum displacement. For the maximum load applied, an average roof displacement of 0.57 m was achieved.

Types of failures presented

In general, the failures presented in the frame were minor. The main ones were rips in some stems and in certain fish mouths of the panels. The maximum roof displacement achieved was mainly possible due to the flexibility of the material. At the end of the test, the permanent displacement of the frame was recorded, with a mean value of 0.26 m on roof. In FIGURE 5 (c) the achieved displacement is illustrated.

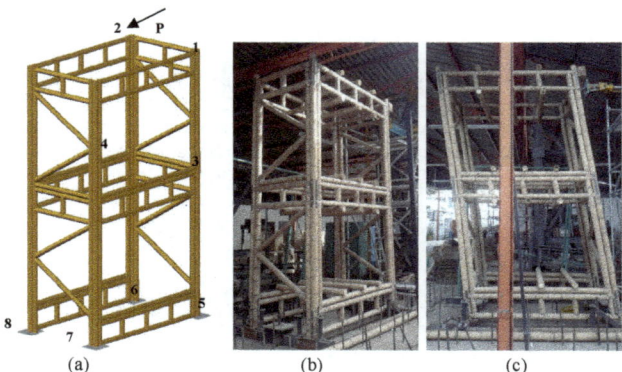

(a)   (b)   (c)

FIGURE 5 - *A) DISPLACEMENT MEASUREMENT POINTS AT THE PCPP. B) PCPP WITHOUT BRACES. C) MAXIMUM DISPLACEMENT AT THE END OF THE TEST. Source: Authors*

Test 2- PCPP with eccentric braced panels

The second PCPP tested was the frame with eccentric diagonals, using panels C. A maximum roof load of 978 kg was achieved reaching a maximum roof displacement of 0.57 m that corresponds to the maximum displacement of the actuator piston.

Types of failures presented

It was evinced that mortar filling of the stems were not carried out properly and failed by crushing in the connection zones of the compression diagonals the stems. The diagonals subjected to traction forces presented typical failures. The most frequent one was the disconnection of the cane from the respective bolt (see FIGURE 4). FIGURE 6 (b) presents this failure mode. There was a different type of failure where the cane of the traction diagonal kept its location but due to the lack of filling of the stem where the brace is connected, punching was generated by the nut and finally caused a crushing in the stem as can be appreciated in FIGURE 6 (c).

Non-Conventional Materials and Technologies – NOCMAT for XXI Century   Materials Research Forum LLC
Materials Research Proceedings **7** (2018) 167-179   doi: http://dx.doi.org/10.21741/9781945291838-16

(a)   (b)   (c)

FIGURE 6 *- A) PCPP WITH ECCENTRIC BRACES. B) DISCONNECTION OF CANE IN DIAGONAL TO TRACTION. C) LOWER VIEW OF THE CRUSHING IN STEM BY PUNCHING OF THE NUT. Source: Authors*

Loading and unloading pcpp test

The last PCPP tested was the frame with panels B. This methodology of load was aimed to evaluate the loss of stiffness that undergoes the frame when is subjected to loading and unloading cycles.

PCPP test with concentric braced panels

For this test, the frame was loaded and unloaded three times (3 cycles). Readings were recorded only during the loading process. Due to the failure presented in the diagonal connections of the eccentric frame, two actions were implemented. First the filling activity was improved in the near areas of the diagonal connections. Second, certain tensile diagonals were reinforced by the addition of bolts to avoid its disconnection. The maximum load achieved in all the cycles was 1372 kg with a roof displacement of 0.47 m. The maximum roof displacement reached at the end of the test was 0.57 m.

Types of failures presented

The failures presented appeared in the last load cycles where the horizontal force applied on the frame was greater than 1000 kg. One of them was the shear failure in the connection bolt of a tensile diagonal. This failure was due to the force applied by the associated cane, as shown in FIGURE 7 (b). Another failure mode was the opening of the cane hook, causing the diagonal disconnection. This event was presented in two of the four tensile braces and is illustrated in FIGURE 7 (c). It is important to mention that the two tensile diagonals that were reinforced with additional bolts did not show any type of damage.

(a)   (b)   (c)

FIGURE 7 *- A) PCPP WITH CONCENTRIC BRACES. B) SHEAR FAILURE OF THE BOLT PLACED AT THE TOP END OF THE DIAGONAL. C) DEFORMED AND OPENING OF THE DIAGONAL CANE HOOK. Source: Authors.*

Non-Conventional Materials and Technologies – NOCMAT for XXI Century   Materials Research Forum LLC
Materials Research Proceedings **7** (2018) 167-179   doi: http://dx.doi.org/10.21741/9781945291838-16

Experimental graphics

Graphic 1 presents the comparison of the curves obtained from the three tests in the roof displacements. The curves correspond to the envelope of the cycles. It is evident the difference in stiffness of the braced frames curves with respect to the non-braced frame (brown curve) in an initial zone of the tests where all the frames were still in their elastic range. This verifies that the panel braces accomplish its function of stiffening the PCPP. The PCPP curve with concentric braces presents a longer stiffness than the eccentric PCPP (blue curve). This happens because the filling of the stems was better executed during assembly and the short length of the brace connection canes prevents their disconnection. At the end of the tests, the three frames were loaded until a maximum roof displacement of 0.57 m. The maximum loads achieved in the tests were very similar to each other. Values were 11.9 kN (1221 kg), 9.6 kN (977 kg) and 13.5 kN (1372 kg) for the non-braced, the eccentric and the concentric frame, respectively, with remnant roof displacements of 0.26 m, 0.25 m and 0.39 m.

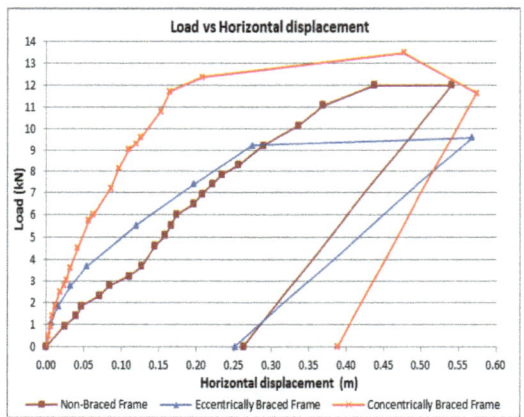

FIGURE 8 – COMPARATIVE CURVE: LOAD VS HORIZONTAL ROOF DISPLACEMENT.

Source: Authors

Numerical simulation

The calibration of each PCPP was performed in the elastic range and developed in the software SAP2000 V 14.0.0. Then, all parameters were implemented in housing simulation and their behaviour was verified.

Materials

It was considered that the bamboo guadua behaves as an isotropic material, thus a single value for the elastic modulus was used. This value was determined in the mechanical characterization tests performed to a set of 30 samples were an average of **14.183 MPa** was obtained. The Poisson´s ratio used was found by Luna, Lozano, & Takeuchi [1], who established an average value of **0.35**. The

Non-Conventional Materials and Technologies – NOCMAT for XXI Century   Materials Research Forum LLC
Materials Research Proceedings 7 (2018) 167-179   doi: http://dx.doi.org/10.21741/9781945291838-16

chosen specific weight was **8 KN/m³** as recommended in NSR-10 [2]. The mechanical properties of mortar and steel were taken from the literature and manufacture catalogues

Frame elements
Frame elements were used to simulate columns, beams, joists, rib lath (the vein of the steel mesh), bolts and canes. It is important to highlight that continuity in elements was respected without including joints in every intersection. Two kinds of elements were used for area elements. Shell Membrane element was used to simulate the floor plate because it does not contribute to resist bending moments, allowing loads to be transmitted completely to the structural system. On the other hand, Shell-Thin element was employed for mortar walls because it omits shear strain.

Simulation calibration
The calibration was made according to the load-displacement curves obtained from each test, using the average roof displacement as homogenous parameter. The displacement precision used for the simulations was of millimeter. At each simulation, the horizontal force was applied by means of point loads to the roof nodes.

PCPP with non-braced panels
The calibration of this test aimed to establish the value of the internal spring *(Partial Fixity)* of the fish mouth connection for the ends of all the elements that compose the PCPP. For this, a stiffness value was assigned yo the spring constant with the purpose of decreasing the rotation restriction around the two axes (axis 2 and 3) of frame elements. The value found for this spring was **6.9 kN-m/rad**. With these values an average roof and deck displacement of 0.2083 m and 0.1108 m, respectively, was achieved.

PCPP with eccentric braced panels
Using the spring value found for the PCPP without braces, the behaviour of the eccentric frame was simulated. The eccentric diagonals were simulated with *Partial Fixity* at their ends using the same spring value found previously. Additionally it was necessary to simulate the canes of these as *Frame* elements pinned at their ends to achieve the test displacement. The calibration parameter was the value of the cane rotation spring whose value was set at **0.017 kN-m/rad** for both ends. With these values an average roof and deck displacement of 0.0065 m and 0.0035 m, respectively, was achieved.

PCPP with concentric braced panels
The concentric frame was simulated in the same way as the eccentric. The experimental displacement, was achieved with a spring value for the canes of **0.013 kN-m/rad** obtaining an average roof and deck displacement of 0.0079 m and 0.0043 m, respectively.

Comparison of simulations
In order to determine the contribution of the braces in the panels and in the structural system, the same magnitude of horizontal force was applied to the three models. For this case, the load magnitude used was 100 kg. TABLE 1 shows the average roof displacements recorded in the simulations of each PCPP.

TABLE 1 – COMPARATIVE ROOF DISPLACEMENTS FOR PCPP CALIBRATED MODELS.

PCPP	$\Delta_{avg.}$ roof (m)
Without braces	0.0296
With eccentric braces	0.0058
With concentric braces	0.0056

Source: Authors

The similarity of displacements of the eccentric frame and the concentric one is due to the fact that the points used for each calibration are both on the same slope as shown in the initial section of Figure 8. Comparing the displacement of the braced frames with respect to the non-braced, a decrease of displacements is observed. This indicates that the braces in the system stiffen the structure. The displacement of the braced frames was reduced in 80% with respect to the PCPP without braces.

Loads

The numerical simulation only contemplates dead, live and seismic loads. Keeping in mind that this is a general case, these loads were considered as the most relevant and pertinent for this kind of buildings. In TABLE 2 vertical loads are resumed.

TABLE 2 – VERTICAL LOADS SUMMARY (KN/M^2).

DEAD LOAD	
Facades and partitions (Residential)	
Facades and partitions	2,0
Floor covered	1,4
LIVE LOAD	
(Residential)	
Rooms and halls	1,8
Roof	0,5

Source: NSR-10 Title B, Tables B.3.4.3-1, B.4.2.1-1 and B.4.2.1-2.

In this case, a structural system evaluation is needed. Therefore, the least favourable conditions must be employed to demonstrate that the system would be appropriate for every condition or location. In order to guarantee the foregoing, a high seismicity territory (Region 9) and a soil type E were chosen. According to the normative NSR-10, the guadua elements must be checked under admissible stresses method and the energy dissipation coefficient of guadua structures is $R_o= 2.0$.

Housing simulation

To evaluate the suitability of PCPP structural system, this system was simulated in two different kind of housing: a social interest housing (SIH) and an upper-middle-income housing (UMIH). The implementation in both cases will study its behaviour in different scenarios. To determine if the housing was adequate, 4 points were evaluated: allowable stresses in guadua elements, maximum drift structure, yield stresses in steel elements and maximum compression stresses in mortar.

Social interest housing (SIH)

"Given the magnitude of the housing shortage, we won't solve this problem unless we add people's own resources and building capacity to that of governments and market. That is why we thought of implementing an **OPEN SYSTEM** able to channel all the available forces at play. In that way people will be part of the solution and not part of the problem. On the other hand, it is a fact that available resources are not enough. In order to face scarcity, we propose a principle of **INCREMENTALITY**." [3]. Thereupon, the most important SIH alternatives are presented and 4 items, mentioned before, are checked.

174

Non-Conventional Materials and Technologies – NOCMAT for XXI Century   Materials Research Forum LLC
Materials Research Proceedings **7** (2018) 167-179   doi: http://dx.doi.org/10.21741/9781945291838-16

Alternative 1: Non-braced panel frames

TABLE 3 – *SIH ALTERNATIVE 1 – SUMMARY TABLE.*

SUMMARY	
ALLOWABLE GUADUA STRESSES	SI
Maximum tension (Mpa)	7.64
Maximum compression (Mpa)	8.05
DRIFT	NO
Maximum drift	18.61%
STEEL ELEMENTS STRESSES	SI
Maximum tension (Mpa)	57.35
MORTAR STRESSES	N.A

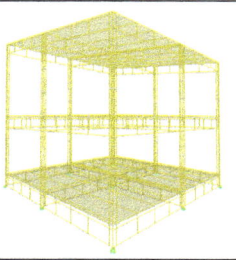

Source: Authors

This alternative does not consider mortar stresses and steel elements are only conformed by the bolts, because it is only be constituted by non-braced panels. Although the system fulfils two of three evaluation criteria, it presents serious inconvenient with drift control, reaching a maximum drift higher than 18% (TABLE 3).

Alternative 2: Eccentric braced panel frames in the perimeter
As can be appreciated in TABLE 4 the diminution in drift is significant. However, besides the maximum displacement is 7 times less than the first alternative, it is not enough, considering that maximum allowable drift is 1.5%. On the other hand, it is evinced an increment of guadua elements stresses, but without exceeding allowable stresses. Additionally, it was registered a maximum steel element stress higher than 400MPa. Those stresses were presented in cane elements, showing equivalent results than the tests, where some diagonal connections reach yields stresses inducing their permanent deformation and subsequently disconnection.

TABLE 4 – SIH ALTERNATIVE 2 – SUMMARY TABLE.

SUMMARY	
ALLOWABLE GUADUA STRESSES	SI
Maximum tension (Mpa)	11.52
Maximum compression (Mpa)	11.85
DRIFT	NO
Maximum drift	2.62%
STEEL ELEMENTS STRESSES	NO
Maximum tension (Mpa)	414.58
MORTAR STRESSES	N.A

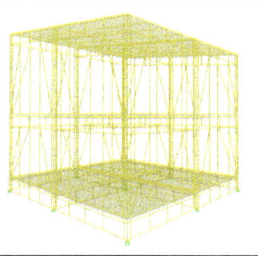

Source: Authors

Alternative 3: Eccentric braced panel frames and Bahareque Encementado (wall with bamboo guadua frames plastered with mortar) in the perimeter

As it is shown, the system is mixed. It is defined by frames with diagonals and *Bahareque Encementado* walls. The bracing is just in the perimeter and the housing is completely free inside, allowing a high architectural flexibility. Below, in TABLE 5, is presented a summary for each evaluation criteria.

TABLE 5 – SIH ALTERNATIVE 3 – SUMMARY TABLE.

SUMMARY		
ALLOWABLE GUADUA STRESSES	SI	
Maximum tension (Mpa)	7.34	
Maximum compression (Mpa)	7.59	
DRIFT	SI	
Maximum drift	0.28%	
STEEL ELEMENTS STRESSES	SI	
Maximum tension (Mpa)	83.60	
MORTAR STRESSES	SI	

Source: Authors

It is evinced that this alternative accomplishes all evaluation criteria; reaching maximum guadua stresses lower than 8 MPa, a maximum drift much smaller than allowed, a maximum steel element stress lower than 206 MPa (yield stress in SAE 306), and a maximum mortar stress lower than 10.5 MPa, corresponding to the maximum mortar compression stresses multiplied by a reduction factor. As it was shown in the tests, the canes are the steel elements that are suppressed under higher stresses. For this reason, it was necessary to use 1/2" rods and not the conventional of 3/8" diameter.

Upper-middle-income housing (UMIH)
To demonstrate the applicability of the PCPP system under every architectural configuration and upper-middle-income housing was selected. Due to having irregularities in top and lateral view and following the recommendations presented in the "*Manual de construcción de vivienda sismo resistente de viviendas en Bahareque Encementado*" [4], it was decided to divide the structure in regular zones and subsequently, in the construction, employs joints to assembly all the housing. To present the results, it was assigned a nomenclature for each zone. It can be appreciated in the Figure 9.

Alternative 1: Eccentric braced panel frames and Bahareque Encementado
As it was used in SIH, in the principal module and in 3.1m tower, it was employed a mixed structural system, which is composed by guadua frames with diagonals and *Bahareque encementado* walls. However, in the principal module, it was necessary to implement *Bahareque encementado* walls inside the housing, but allowing all the access (doors and windows), which are described in the architectonical blueprints. On the other hand, the first floor module and 4m tower do not need *Bahareque encementado* walls and the structural system only use the PCPP system.

Non-Conventional Materials and Technologies – NOCMAT for XXI Century   Materials Research Forum LLC
Materials Research Proceedings 7 (2018) 167-179   doi: http://dx.doi.org/10.21741/9781945291838-16

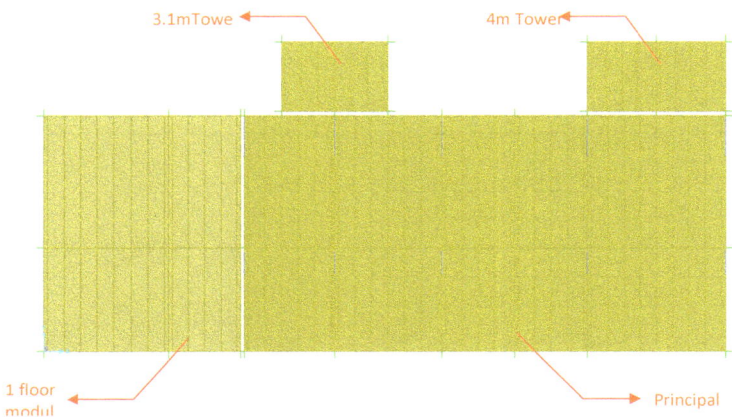

FIGURE 9 - *REGULAR ZONES NOMENCLATURE. SOURCE: AUTHORS*

Table 6 shows that four evaluation criteria are accomplish, demonstrating that PCPP system can be implemented in architectural configurations more complex than SIH housing and used for different purposes.

TABLE 6 – *UMIH ALTERNATIVE 1 – SUMMARY TABLE.*

SUMMARY	
ALLOWABLE GUADUA STRESSES	SI
Maximum tension (Mpa)	16.47
Maximum compression (Mpa)	10.29
DRIFT	SI
1 FLOOR MODULE	
Maximum drift	1.37%
3.1m TOWER	
Maximum drift	0.00%
4.1m TOWER	
Maximum drift	1.26%
PRINCIPAL	
Maximum drift	0.11%
STEEL ELEMENT'S STRESSES	SI
Maximum tension (Mpa)	170.45
MORTAR STRESSES	SI

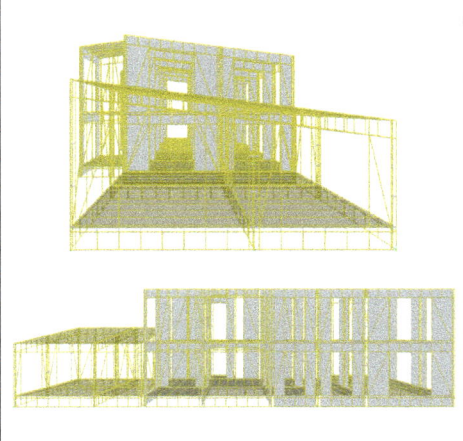

Source: Authors

Non-Conventional Materials and Technologies – NOCMAT for XXI Century   Materials Research Forum LLC
Materials Research Proceedings 7 (2018) 167-179   doi: http://dx.doi.org/10.21741/9781945291838-16

Alternative 2: Eccentric braced panel frames and Bahareque Encementado
Because alternative 1 accomplishes loosely all requirements, it was decided to check a second alternative with less bracing (TABLE 7), especially in the principal module which is the more complex.

***TABLE** 7 – UMIH ALTERNATIVE 2 – SUMMARY TABLE.*

SUMMARY	
ALLOWABLE GUADUA STRESSES	SI
Maximum tension (Mpa)	16.62
Maximum compression (Mpa)	10.82
DRIFT	SI
1 FLOOR MODULE	
Maximum drift	1.37%
3.1m TOWER	
Maximum drift	0.00%
4.1m TOWER	
Maximum drift	1.26%
PRINCIPAL	
Maximum drift	0.24%
STEEL ELEMENT'S STRESSES	NO
Maximum tension (Mpa)	391.67
MORTAR STRESSES	SI

Source: Authors

Although the mortar does not present a high stress increment, it is evinced that the reduction in braced panels produces an increment in steel element stresses, especially in those which conform the rib lath. Those stresses are much larger than the allowable, generating creep in the rib lath and tears in the mortar subsequently. For this reason, the first alternative was chosen.

Conclusions
The PCPP structural system presents ease and fast constructive process, being a system totally prefabricated in the workshop and assembled in the field, fact that was verified with the assembly of the frames tested.

The PCPP system kept its structural integrity during all tests, presenting remnant strain but not collapse.

According to the numerical simulation of PCPP with non-braced panels, the connection between bamboo guadua elements using *"fish mouth"* cuts has an approximate stiffness of 6.9 kN-m/rad.

In an elastic range, the numerical simulations carried out on braced frames reduce the displacement under horizontal loads by 80%.

The PCPP system has inconvenient facts about the drift control, but it is appropriate in allowable stresses control for guadua elements.

The currently PCPP system cannot be implemented without the use of mortar in some panels.

The PCPP system can work alone and control the drift, provided that diagonals have an inclination near to 45°.

Housing systems, which uses uniquely non-braced panels are unsuited because they have problems in both, drift control and allowable stresses.

Non-Conventional Materials and Technologies – NOCMAT for XXI Century   Materials Research Forum LLC
Materials Research Proceedings 7 (2018) 167-179   doi: http://dx.doi.org/10.21741/9781945291838-16

The use of mortar is very useful in drift control.

The use of rib lath is necessary to resist the tension stresses presented in the walls and avoid the appearance of tears in the mortar.

The orientation and location of diagonal´s connections represent important effects in the displacements of the system and drift control.

The actual connection design generates important stresses in the canes, inducing the premature failure of the connection and the system.

The displacements under vertical loads (dead and live load) are small and with normal conditions of load do not represent inconvenient for the system.

Recommendations

For numerical simulation, it was used other elements (rib lath and mortar) which were not part of experimental stage. It is recommended to make some test including all the materials and identifying how they work ensemble.

References

[1] Luna, P., Lozano, J., & Takeuchi, C. (2014). Experimental Determination of Characteristics values for Guadua Angustifolia. *Maderas. Ciencia y Tecnología, 16* (1), 77–92.

[2] Reglamento Colombiano de Construcción Sismo Resistente - NSR-10, Title G-Structures of Wood and Structures of Guadua,

[3] Aravena, A., & Iacobelli, A. (2016). ELEMENTAL: *Manual de vivienda incremental y diseño participativo* (Hatje Cant). New York.

[4] Asociacion Colombiana de Ingenieria Sismica (AIS). (2014). Manual de construcción sismo resistente de viviendas en bahareque encementado.

[5] Rivera, J. F. (2008). Comportamiento estructural de pórticos en guadua arriostrados mediante diagonales en guadua. Universidad Nacional de Colombia.

Non-Conventional Materials and Technologies – NOCMAT for XXI Century   Materials Research Forum LLC
Materials Research Proceedings 7 (2018) 180-189   doi: http://dx.doi.org/10.21741/9781945291838-17

Bamboo Reinforced Concrete Beams for Precast Slab

Thomas Lage Gonçalves[a*], Luís Eustáquio Moreira[a]

[a] Departamento de Engenharia de Estruturas, EEUFMG, Brasil

tomaslagon@hotmail.com; luis@dees.ufmg.br*

Abstract. The precast slabs are one of the most commonly used elements in Brazilian construction industry. They are made out of pre-fabricated beams which is composed of a steel truss fixed in a rectangular base of concrete, and this way commercialized. These easily produced beams are equally spaced and arranged to make the slabs. Afterwards, the gaps between then will receive fill blocks, which may be made out of Styrofoam, or ceramic. These blocks are permanent forms. Over this system a capping steel mesh is placed and then a concrete layer of 3 to 5 centimeters thick is set. This way the steel apex of the steel truss is merged into the concrete mass leading to the precast slab, a monolithic system easy to manufacture, low weight and high safety level, a plate ribbed by the beams. This paper presents the mechanical tests and the construction techniques that allowed replacing the steel bars by bamboo. The three points bending flexural test were performed in 2 prototypes beams, one of them made of steel and the other made of bamboo. Also characterization tests were run: - concrete compression and bamboo tension. The results were very effective, leading to the possibility of replacement of the steel by bamboo. The replacement of steel bars by bamboo bars on concrete precast slabs means using 2,9 kgf of bamboo instead of 4 kgf of steel in 1 square meter of slab. In 100 m², 400 kgf of steel could be saved.

Keywords: Bamboo, Reinforced Concrete, Slab, Bending Mechanical Tests, Sustainability

Introduction

The population growth rate increases in an exponential progression, just like the consumption of raw materials. The main consume is for housing. Therefore the environment impacts rise, mostly due to pressure for iron ores. The use of renewable resources like bamboo is one measure that helps saving natural resources and contributes for the sustainable development. Bamboo's plantations with structural features in degraded regions or poor quality soil may even improve soil characteristics; generate material for many facilities like furniture, housing, glues-laminate sheets and paper. It also release great amount of oxygen and fix carbonic gas while building the polymers of the bamboo. The polymers gather together and constitute the cellulose and lignin of the bamboo.

The great capacity of fixing carbonic gas into its cells makes the bamboo an important agent against the greenhouse effect. It mitigates the growth of CO_2 concentration, generating carbon credits while producing the gas of life, the oxygen in O_2 form. The use of bamboo in a global level also contributes for the sustainable development in order to rise the lifetime of the iron ores reserves. Urgent policies concerning the sustainable development should be adopted. The consumerism also grows in addition to the population's increase, since it is adopted by many nations as an incentive policy to industrial development. The economic growth makes no sense when it comes to Earth's finite reserves. The consumerism clearly points to the other way round the sustainability. Brazil is a land with very favorable conditions to the cultivation of bamboos, if planted the correct species for our conditions, such as *Guadua, Dendrocalamus giganteus, Bambusa tuldoides* and *Phyllostachys aurea* - and subtropical *–Phyllostachys pubescens* and also *Phyllostachys aurea*.

The precast beans reinforced with bamboo are the theme of this present investigation. The easy manufacturing process allied to the safety and satisfactory mechanical performance makes it an

Non-Conventional Materials and Technologies – NOCMAT for XXI Century   Materials Research Forum LLC
Materials Research Proceedings **7** (2018) 180-189   doi: http://dx.doi.org/10.21741/9781945291838-17

interesting product. As the ribs of these slabs are the preformed joints, the proper functioning of the beams relies on the adequate manufacture of these joints. The main intention of this paper is to reinforce the use of bamboo as a construction material. Bydeveloping a bamboo truss prototype we could put it under performance tests and confirm its total viability to replace steel truss. Through this paper it's faced all the constructions stages in details as well as the material characterization tests.

Materials and methods
The concrete recipe used in the manufacture beans process is the same used in a conventional bean industry in the city of Sete Lagoas, Minas Gerais State, Brazil. The research covers the following script:

- Concrete characterization tests;
- Bamboo *Phyllostachys pubescens* characterization tests;
- Bamboo slats and prototype's manufacture;
- Prototype's test;
- Outcome analysis;

Concrete Characterization
The concrete recipe dosed in volume was 1,5:2:1 (stone 0:sand:cement). Due to the small size of the grains, the dimensions of the specimens follow the standard for mortar tests ABNT (NBR 7215) [1].Compression tests were run with 6 cylindrical proof-bodies measuring 50 mm in diameter, 150 mm in length, Figure 1. A clip gauge measures the strain. We got an average modulus of elasticity of $7,9 \pm 0,4$ GPa and average tensile strength of $12,6 \pm 1,03$ MPa, with a maximum value of 14,76 MPa.

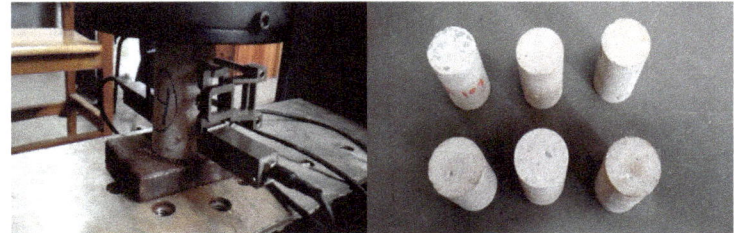

FIGURE 1 - CONCRETE COMPRESSION TESTS.

As a result there are the curves stress versus strain in Figure 2. The curves represent the three cycles of increase/decrease pressure and the final loading until the failure. The concrete presents a linear behavior until the tensile of 6*MPa*.

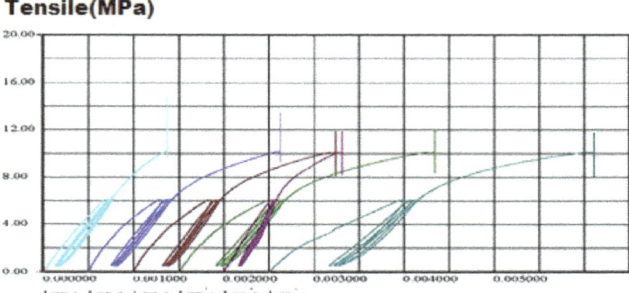

FIGURE 2 - *CURVES TENSILE X DEFORMATION OF THE CONCRETE'S COMPRESSION*

Bamboo characterization

The ironmongery was developed with bamboo of the species *Phyllostachys pubescens*, popularly known as "bamboo mossô". This type of bamboo is the most used in China and is planted commercially in a small scale in the states of Rio de Janeiro and São Paulo. The rectilinear form, high biomass production, average diameter of 10 cm and high resistance are the great attractions of this species. The 6 specimens were prepared for tensile stress according to ISO / DIS 22156 [2], Figure 3.The lot had a 8.8% moisture content, and the specimens had an average modulus of elasticity of 15,9± 1,27 *GPa* and average tensile strength of 166 ± 24,9 *MPa* , Figure 4. This bamboo specie is a material with high mechanical strength and flexibility, which makes it to be known as "the green steel".

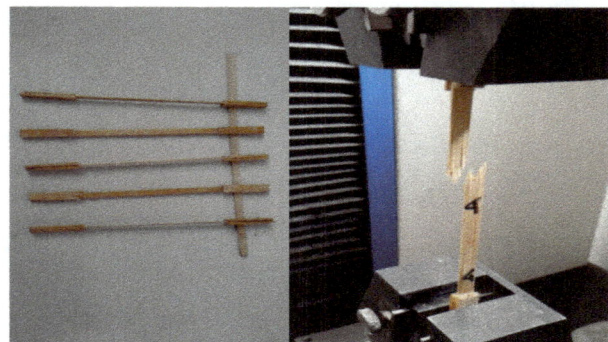

FIGURE 3 - *BAMBOO TENSION TEST*

Non-Conventional Materials and Technologies – NOCMAT for XXI Century   Materials Research Forum LLC
Materials Research Proceedings **7** (2018) 180-189   doi: http://dx.doi.org/10.21741/9781945291838-17

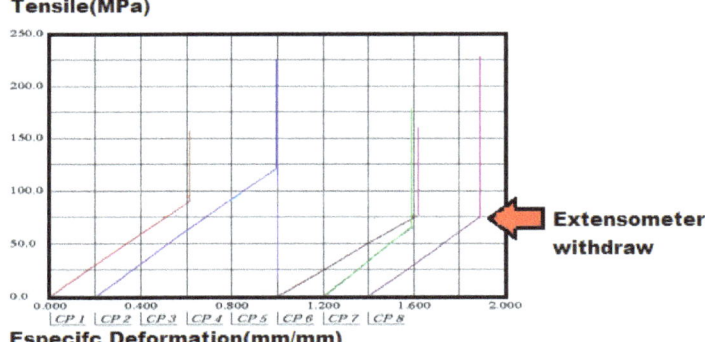

FIGURE 4 - *STRESS – STRAIN FOR BAMBOO IN TENSION*

Bamboo truss and design process

Differently from wood the bamboo can be used as strips, with small transversal sections. This is because of the perfect alignment of the bamboo fibers, Figure 5. Because of this, bamboo has been object from researches using it as concrete reinforcement, G h a v a m i [3]. The extraction of bamboo's thin cables it is a developed process over China. They have simple radial scissors machines where the bamboo tube is pushed in. On the present work the strips were withdraw with a machete.

FIGURE 5 - *HANDMADE WITHDRAW OF THE BAMBOO STRIPS AND SURFACE REGULARIZATION*

The result is a perfect straight strip with 6 meter long, Figure 6.

FIGURE 6 - *LINEAR BAMBOO STRIPS.*

The great innovation of this investigation is beyond the purpose of fabricate precast slabs. The technique of developing the ironmongery in truss by three independent plans was fundamental for the accomplishment of the task, Figure 7. The independent plans were made with the strips correctly dimensioned and connected with 2 *mm* wire and then connected in the same way. By a handmade procedure (it could be an industrial procedure as well), we got strips with transversal section in average of 10*mm*× 4*mm*.

FIGURE 7 - *A) NAKED BAMBOO TRUSS - 3 INDEPENDENT PLANS   B) BAMBOO TRUSS CORRUGATED BY SAND.*

In Figure 7 there are the steps of manufacturing the bamboo space truss used as ironmongery. The three plans connected were polished with a steel brusher number 36 and gather together with wire. In order to reach its corrugation, the lattice was painted with *Sikadur 32 gel* and then sprayed with sand. These steps work waterproofing the bamboo against the concrete curing water and also increasing significantly the adherence between the reinforcement and the concrete, Navarro[4]; In Figure 8a), there is the prototype of precast beam. In figure 8b), there are the forms from the process of concreting the bamboo prototype and the conventional control group.

FIGURE 8 - *A) TRUSS MADE WITH BAMBOO STRIPS; B) CONCRETE FORMWORK*

The number of bamboo bars was based on areas' equivalence, adjusting the difference from the elasticity modules by area, since there is $\frac{E_s}{E_b} \cong 12{,}6$. Ghavami [3] recommends that the bamboo reinforcement section should be from 3% of concrete section. In fact, the ratio in area equal to 12,6 was not assumed.

The three point bending flexural test
Two trusses were developed with the same transversal section and the same concrete recipe. The first one was reinforced with steel as it is commonly done. The second was reinforced with corrugated bamboo truss. Both were tested under the three points bending test, with 1.2 *m* of span as indicated on Figure 9.

FIGURE 9 - *BENDING OF THE BAMBOO BEAM - A) CRUSHING OF THE CONCRETE*

The test was run with a universal machine EMIC-INSTRON DL 30 ton. The instrumentation for the test was a loading cell and a displacement transducer. On Figure 10 there are Load × Vertical Displacement for the steel reinforced bean. The longitudinal bars of the lattice system consists in two diameters of 4,2 *mm* in tension on the lower part as well as the diagonal bars and one 6 *mm* longitudinal bars in the top of the truss.

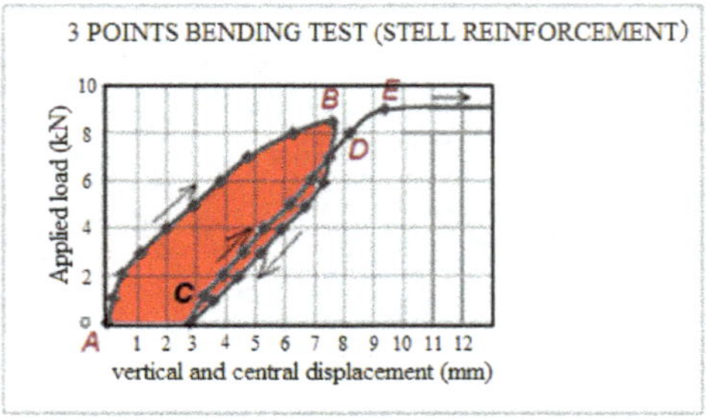

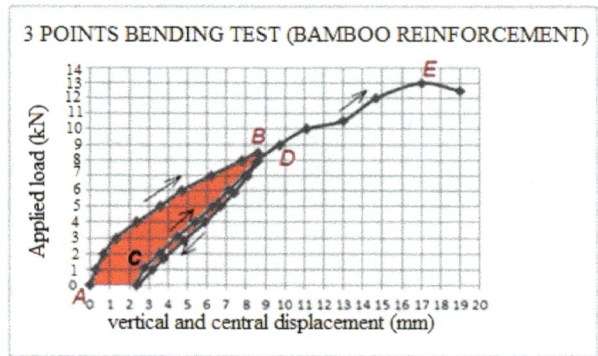

FIGURE10 - *LOAD × DISPLACEMENT δ FOR THE STEEL BEAM.*

In Figure 11 there is the curve Load versus Vertical displacement for the bamboo specimen beam. It achieves 12,5 *kN* before breaking for crush the concrete on the top area, as shown in Figure 9a). After the beam was unload until 8,5kN- the ultimate load for the steel bean - and then carried again. From the load of 9 *kN*, point D, the concrete started to get a plastic aspect. The longitudinal reinforcement is made by inferior bars with (10×4) *mm²* and 2 superior bars with the same size, Figure 12.

FIGURE11 - *LOAD × DISPLACEMENT δ FOR THE BAMBOO BEAM.*

Results analysis
From the straight-line segments $\overline{CD}$ of Figures 10 and 11, the bending stiffness *EI* is obtained to the beams, given by the plot of Eq. (1) of the maximum deflection in 3 points flexural test.
Despising its own weight, there is the Eq.(1):

$$P = \frac{48EI\delta}{l^3} \tag{1}$$

Non-Conventional Materials and Technologies – NOCMAT for XXI Century   Materials Research Forum LLC
Materials Research Proceedings 7 (2018) 180-189   doi: http://dx.doi.org/10.21741/9781945291838-17

Assuming tanθ as the slope of the axis extract from the plot of the Eq. (1), the flexural stiffness for the beans is given by

$$EI = \frac{l^3 tan\theta}{48} \tag{2}$$

Since the tested beans are 1200mm long, it was obtained respectively tanθ = 13,7 for the steel bean and tanθ= 11,5 for the bamboo one. Therefore for the steel bean it was gotten E_aI = 4,93 $\times 10^8 kNmm^2$ and E_bI = 4,14 $\times$ $10^8 kNmm^2$ for the bamboo one. The steel bean came out to be 19% more rigid under bending than the bamboo one. Increasing in 19% the area of the bamboo reinforcement under tension could compensate this, which is a small value and there wouldn't bring any construction problems of excessive vicinity of the reinforcement bars. The steel area under tension in the reference bean is equivalent of 2 longitudinal bars of 4,2 mm each, which results in 0,28 cm^2. Since the reasons of the modules are 12,6, the positive reinforcement of bamboo should be from 3,5 cm^2. Actually it was placed 4 bars of 0,4cm^2 each, resulting in 1,6 cm^2 or 46 % of the reinforcement required for the same final stiffness. The difference of only 19% in the bending stiffness EI can be linked to the non-homogeneity of the bamboo batch from where the sticks were taken; or the fact of the transversal section of concrete in the bean reinforced with bamboo was slightly superior than the concrete area in the steel specimen, as can be seen on Figures 12 e 13.

It is possible to observe that in the steel bean, the steel reaches the yield limit before the critical crushing of the concrete, for the load of 8,5 kN, while the limit of the bamboo joist was the crushing of the concrete under compression. Despite the low modulus of elasticity, the strength of the bamboo is very high. Thus, the area of positive bamboo reinforcement, from 1.6 cm2 would withstand a mean tension load equal to 16,6 × 1.6 = 26,56 kN while positive steel reinforcements would withstand 60 × 0,277 = 16, 3 kN for C60 rebar as used.

Resistant diagrams can be proposed within linear limits of concrete behavior, since stirrups do not confine the compressed portion, which is not justified considering the parable rectangle diagram. On the other hand, the resistant diagram in the limit load becomes hard to obtain, since from the compressive stress of 6 MPa, Figure 2, the concrete already presents an accentuated nonlinear behavior, reaching average rupture stress of 12 , 6 MPa.

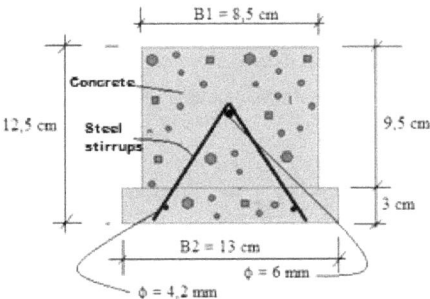

FIGURE 12 - STEEL BEAM DIMENSIONS.

Non-Conventional Materials and Technologies – NOCMAT for XXI Century   Materials Research Forum LLC
Materials Research Proceedings 7 (2018) 180-189   doi: http://dx.doi.org/10.21741/9781945291838-17

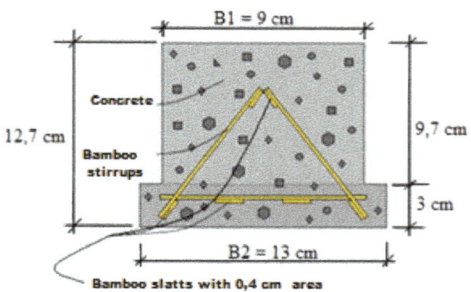

FIGURE 13 - BAMBOO BEAM DIMENSIONS.

Up to the 8 kN load the steel and bamboo beams have very similar behavior and present maximum displacements of 6,3 mm for the bamboo beam and 5,3 mm for the steel one, a deflection 19% larger due to the smaller bending stiffness of the bamboo beam..

The maximum strain in the lower bamboo reinforcement can be obtained by dividing the tension of 166 MPa by the average value of the modulus of elasticity obtained from the tension tests, $\varepsilon_b = \frac{166}{15600} = 1\%$. This means that bamboo is working in an elastic zone, because the rupture strain can reach up to 2 % in tension - brittle rupture.

Considering as safety factor 2 over the maximum load on the beam and considering that the load limit is 9 kN, therefore the safety load of 4,5kN is obtained. Considering a 3 m long bean and comparing the bending moment of the 4,5kN load in the prototype, with the moment of a load distributed in 3 m we have

$$\frac{q \times 3^2}{8} = \frac{4,5 \times 1,2}{4} \therefore q = 1,2 \frac{kN}{m}$$

Assuming this bean is being used as a dormitory floor in which the ABNT (NBR 6120) [5] standard is applied, there is a recommended accidental load of qk = $1,5 \frac{kN}{m^2}$. Assuming the beans are placed 0,4 m from each other, there is a distributed load of $0,6\frac{kN}{m}$. Adding it with the bean's own weight ($0,29 \frac{kN}{m}$), there is $(0,29+0,6)\times1,4 = 1,13 < 1,2 \rightarrow$ ok.

From the success of this experiment it's possible to imply the bamboo precast slab's success. By changing steel to bamboo there is a global reduction of weight, since 4 *kgf* of steel are substituted by 2,9 *kgf* of bamboo. This means that in a 100 m^2 slab, 400 *kgf* of steel would be saved and substituted by 290 *kgf* of bamboo.

Conclusion

The bamboo reinforced concrete beams was possible due to the innovation process of the bamboo ironmongery through independent plans, an idea which can be applied for any concrete beams and columns with rectangular section armed with bamboos. The conventional beam was 19% more rigid than the bamboo one. Increasing 19% of the positive area of the bamboo reinforcement could compensate this.

Non-Conventional Materials and Technologies – NOCMAT for XXI Century   Materials Research Forum LLC
Materials Research Proceedings 7 (2018) 180-189   doi: http://dx.doi.org/10.21741/9781945291838-17

One of the tension test specimens, Figure 4, had yield strength of 225 MPa. Only very well-prepared specimens are able to fail by tension. These specimens must be no more than 2 mm thick. The bamboo does not have a yield level in tension. It has a brittle fracture. The coating of the bamboo reinforcement with gel paint and $2mm$ diameter sand sprayed over is fundamental to maintain the dimensional stability of the bar and to increase the adhesion with the concrete, according to Navarro [4]. Its adhesion to concrete is 4 times higher than steel's reinforcements, if the bars are benefited this way. The epoxy 32 gel resin may be replaced by another resin with lower price, if it still holds such a good mechanical and physical performance. This is fundamental to reduce the final cost of the bamboo truss. Bamboo reinforcement can be manufactured with simple tools like knives and saws and can also be obtained by machines.

References

[1] ABN (NBR 7215). Cimento Portland– Determinação da Resistência à compressão, 1996.

[2] ABNT(ISO/DIS 22156) - Projeto de Estruturas de Bambu – Procedimento, 2016.

[3] Ghavami, K. Bamboo as reinforcement in structural concrete elements. Cement& Concrete Composites 27, 637-649. www.sciencedirect.com, 2005. https://doi.org/10.1016/j.cemconcomp.2004.06.002

[4] Navarro, E.H.A. Bambu – Material "High Tech" como reforço em concreto; PhD thesis; Pontifícia Universidade Católica do Rio de Janeiro, RJ, 2011.

[5] ABNT (NBR 6120). Cargas para o Cálculo de Estruturas de Edificações,1980.

Construction of a Sustainable Multi-Level Bamboo Structure

Chaaruchandra Korde[a*], Puspanjali Das[b], Roger West[a], Sudhakar Puttagunta[c]

[a] Trinity Collage Dublin, Ireland; kordec@tcd.ie*, rwest@tcd.ie

[b] Green Bam InfraTech Pvt. Ltd., India; puspanjali.iitd@gmail.com

[c] Haritha Ecological Institute, India; sudhakar.puttagunta@gmail.com

Abstract. This paper describes an innovative construction technology used in the construction of a multi-level bamboo structure built in New Delhi, India. It will demonstrate the capability of building three storey structures with bamboo without using nut and bolt components. A total of 1400 bamboo poles are used in the course of construction of this structure. The structure has four levels comprising over 70m², including a third floor as a terrace. The structure is built on a bamboo beam and column network which is integrated with non-load bearing bamboo wall panels. All the upper floors and walls are built using bamboo and then finished with cement mortar internally for walls and a concrete screed for flooring. It is then finished externally with a weather-proof coating and internally with paints. This paper describes the complete construction process undertaken during the building of the structure. The fact that these bamboo components sequestrate an estimated 12 tons of CO_2 further demonstrates the necessity for systematically researching bamboo as one of nature's structural gifts to utilize it for all possible rural, urban and agricultural infrastructures.

Keywords: Bamboo, Multi-Level Structure, Sustainability, Carbon Sequestration

Introduction

Bamboo is a natural grass, known to the people of India since antiquity, where India contains almost 45% of the world's bamboo forests, with one-third of its growing stock in the North-Eastern States [1]. The annual trade in bamboo products is estimated at US $ 15 billion which brings out the necessity for a special focus on research into developing applications. These products don't include the bamboo being used for construction due to an absence of adequate standards.

These are the times when India is anticipating major construction projects while being strongly committing to the global community in reducing its carbon footprint, thereby addressing one of the challenges of climate change. India's resource utilisation, such as for sand, stone, brick, cement and steel, is going to create extra stress on the environment and its ecology in fulfilling the natural aspirations of its people. Also, it has been strongly emphasised that the construction industry has a pincer role in terms of its contribution to global warming as it seems to contribute 32 % of global final energy consumption [2]. It is thus the right time to explore new materials as options for creating sustainable infrastructural solutions. Bamboo is one such material having a promising potential for achieving green and highly sustainable infrastructural development.

There is a keen awareness of increased utilisation of bamboo for construction at the present time across the world. A significant number of applications of bamboo in construction is available in which bamboo is specified for construction of concrete composite beams [3], arches [4], columns, slabs, trusses, frames, wall panels, double layer grid systems, domes, masts, bridges, pavilions, etc. A study of traditional houses in the Assam and Mizoram State reveals that houses with bamboo can last as long as 40 years [5], which is also the case in other parts of North Eastern India. Furthermore, it is documented that houses built with bamboo have not reported any significant damage during earthquakes [6–8], something to which certain parts of India are prone. This resilience of bamboo to dynamic excitation brings out the hidden and scientifically under-explored potential of this naturally growing, indigenous and affordable material.

Non-Conventional Materials and Technologies – NOCMAT for XXI Century   Materials Research Forum LLC
Materials Research Proceedings 7 (2018) 190-198   doi: http://dx.doi.org/10.21741/9781945291838-18

Scope of paper

The scope of this paper is to describe the stage-wise construction process during the building of a multi-level bamboo structure, namely a guardhouse at the entrance to a substantial property in New Delhi. The researchers have selected this particular aspect of the project from a comprehensive study which documented the complete architectural and structural design, construction and maintenance of this multi-level bamboo structure. The prime objective here is towards documenting the construction process involved at various stages and thus demonstrating the potential for multi-level bamboo structures.

Construction process

This section covers the complete stage-wise construction process of building the structure. It has various stages starting from sourcing and treating the bamboo, the foundation works, the fabrication of beams, columns, wall panels and floors, and the integration of the ground, first mezzanine floors and roof into an integral composite structure. The construction of the roof was also a challenging activity because its geometry comprised a hyperbolic paraboloid of dimension 11 m x 8 m, prefabricated on the ground and placed on the roof. The structure was finished with the necessary finishing materials and handed over to the client, a government agency. The details of the complete process of construction are described hereunder.

Sourcing and treatment of bamboo

The bamboo used for construction of the structure is sourced from Madhya Pradesh, India. There are three species of bamboo used, two of which are *Dendrocalamus strictus* and *Bambusa bambos* and the third one is not known. The first two are treated under a pressurized treatment plant, as shown in Figure 1(a) and the third one is treated on site as shown in Figure 1(b). The on-site treatment facility is reviewed by the officials for confirmation of compliance with the specified treatment process as seen from the figure.

(a)   (b)

FIGURE 1 – PRESERTATIVE TREATMENT OF BAMBOO (a) PRESSURE TREATMENT (B) ON SITE TREATMENT

Foundation of structure

The terrain on which the structure is built is partially rocky with a bearing capacity of around 200 kN/ m^2. Figures 2 (a and b) show the original porta-cabin structure for housing four guards. The site which is trapezoidal with length 8m and either end widths of 1.5m and 3.5 m. The excavation of the site for the foundations resulted in some boulders which had to be chipped out. As seen from figure 3a, the pre-fabricated bamboo beam-column frame is installed in place and cast in situ with a concrete foundation. The subsequent frames are lined up and integrated with wall panels as shown in figure 3(b). The bamboo columns are embedded into the concrete foundation to create adequate

Non-Conventional Materials and Technologies – NOCMAT for XXI Century   Materials Research Forum LLC
Materials Research Proceedings 7 (2018) 190-198   doi: http://dx.doi.org/10.21741/9781945291838-18

anchorage with the foundation. After casting the footings for the columns, the concrete plinth beams are poured, connecting each column in a 2-D grid in orthogonal directions. This will ensure proper integration of the structure at the plinth level. After curing of the footings and plinth beams, the subsequent internal pockets are filled with compacted building waste and the concrete floor is poured and finished. It is ensured that the finished floor level of the structure remains well above the ground level so as to protect the columns from splashing during heavy rainfall and to prevent any accumulation of water in the nearby area.

Skeleton, floors and walls

The column – beam frames erected with the footings are integrated with wall panels. The complete structure thus gets properly integrated with the column, beam and wall panel system using bamboo pins and an interlocking mechanism. It is ensured that no gaps are left in between the walls and columns. The beams are all checked to be at the same levels so as to receive a flat horizontal bamboo floor which are also integrated with the beams using bamboo pins. In order to erect the first floor, whole bamboo culms are laid along the length of the structure in a first layer and then subsequently they are laid in the orthogonal direction. These are projected on each side of the structure to develop cantilevering weather sheds. This two layer system creates a rigid diaphragm for the complete framed structure thus integrating it at first floor level (Figure 3(b)).

(a)

(b)

FIGURE 2 - *(a) ORIGINAL PORTACABIN BARRACKS FOR FOUR GUARDS (b) CLEARED SITE VIEW*

(a)

(b)

FIGURE 3 - *(a) BEAM COLUMN FRAME (b) GROUND FLOOR CONSTRUCTION*

After the first floor is constructed in place, the columns are extended to receive the second floor columns and beams (Figure 4(a)). A similar process as the one for the ground floor is followed for the first floor and mezzanine floor column-wall panel integration as shown in Figure 4(b). Here

Non-Conventional Materials and Technologies – NOCMAT for XXI Century   Materials Research Forum LLC
Materials Research Proceedings **7** (2018) 190-198   doi: http://dx.doi.org/10.21741/9781945291838-18

again two layers of bamboo in orthogonal planes are used to create a rigid floor diaphragm for the mezzanine floor. The top level of the structure is then integrated with two layers of bamboo runners which run around the perimeter of the structure. This is done to tie the top level of the columns at roof level with rigid members to ensure that it behaves in an integrated manner. Another purpose of this kind of integration is to create a strong stiff frame to receive the bamboo-concrete roof. Thus the complete activity of integration of columns, beams, floors and wall panels reaching up to a height of 8 m at one end and 6.5 m at the other end (see Figure 4(b)) using bamboo is both innovative and challenging.

(a)   (b)

FIGURE 4 - _(a) FIRST FLOOR CONSTRUCTION   (b) MEZANNINE (SECOND) FLOOR_
construction

Erection of roof

Another of the most challenging aspects of this construction project is the fabrication and erection of 11m long and 8m maximum width bamboo hyperbolic paraboloid roof using 200 bamboo poles. Since it would be very difficult to build such an intricately shaped roof in situ above the structure (which is 8 to 6m tall), it was decided to build it on the ground. The roof upon fabrication on the ground is shown in Figure 5(a). This roof was fabricated using single bamboo poles in a two layer grid system with each layer running almost orthogonal to each other as the geometry dictates. The next challenge was to lift the constructed roof from ground level and place it on the roof level. This particularly becomes challenging due to its span and the flexibility of the bamboo. Also, since one end is narrow and the other is broad it is particularly difficult to lift the same uniformly without inducing additional tortional stresses. The approximate centre of gravity points were derived and two cranes used during the erection process (Figure 5(b)). These two cranes are placed at opposite ends, with the first one lifting the narrow end and moving in reverse mode, while the other lifts the broad end and moves in a forward mode. The track to be traversed is a C-shaped path from the site of fabrication to the site of erection, which was covered with trees. However, the complete job was executed by very skilled crane operators and finally the roof was placed on the structure (Figures 5(c) and (d)). Upon placing, it is then covered with half split bamboo which are spread across the roof with its outer surface facing inwards giving a nice smooth surface at the bottom of the roof and the inner part exposed to the sky as shown in Figures 6(a) and (b).

Architectural concepts and actual structure

The architectural concept, as shown in Figure 7(a), was developed at an early stage, keeping in mind the technology and methodology which was going to be used for the erection of the structure. The structure upon completion (Figure 7(b)) resembles very closely the one planned at the initial stage. The principal deviation in the final built structure compared to the one conceptualised is a reduction in the extent of the roof overhang. This was necessary because upon design of the structure for wind forces, it was realized that such long cantilevers at the roof level will affect the stability of structure under strong winds.   The vicinity, comprising open parkland with low density plantations, particularly invites strong winds resulting in significant lateral and uplift forces on the structure. Upon completion of the bamboo work, the structure is finished with a concrete screed on the intermediate floor, mezzanine floor and on the roof. The walls are plastered using cement mortar internally, later painted, to give a smooth finish. The final architectural view and different views are shown in Figure 8 (a) and (b). The structure is then painted with a weather proof coating externally as shown in Figure 9 (b).

(a)   (b)

(c)   (d)

FIGURE 5 - *(A) 11M X 8 M BAMBOO HYPER PARABOLA   (B) TWO CRANES ROOF LIFTING (C) PLACING OF ROOF (D) LONG DISTANCE VIEW*

Non-Conventional Materials and Technologies – NOCMAT for XXI Century   Materials Research Forum LLC
Materials Research Proceedings **7** (2018) 190-198   doi: http://dx.doi.org/10.21741/9781945291838-18

(a)   (b)

***FIGURE   6** - (a) BAMBOO FOR RECEIVING CONCRETE(b) COMPLETION OF BAMBOO ROOF WORK*

Some environmental consideratons

The primary material used in building the structure, as can be seen from the figures, is bamboo. One hectare of bamboo plantation sequesters up to 62 tons of CO_2/year, whereas one hectare of young normal wooded forest sequesters 15 tons of CO_2/year [9]. The average bamboo culm production rate in well-managed plantations is of the order of 10 tons per hectare per year, so by selecting bamboo as the primary material for construction, a considerable amount of sequestering of CO_2 can be achieved. In the present building, which uses nearly 1400 poles of bamboo, about 12 tons of bamboo is used which sequestrates more than 12 tons of CO_2 in total for the bamboo's complete growth. Thus, the uses of such a building will result in the creation of a long term carbon sinks for the life cycle of the structure, which is at least 20 years. Further, the additional impact of soil conservation, increased ground recharge due to greater water percolation under bamboo (owing to higher leaf litter), a continuous green foot print (unlike trees when cut down) and the impact on flora and fauna are certainly positive environmental factors in the choice of bamboo.

(a)   (b)

***FIGURE 7 -** (a) ARCHITECTURAL CONCEPTS (b) FINAL CONSTRUCTION ON SITE – ENTRANCE VIEW*

(a)   (b)

FIGURE 8 - (a) FINAL CONSTRUCTION ON SITE – FRONT VIEW (b) CLOSE VIEW

Socio-economic impact
With land becoming an expensive resource and with a significant increase in the world's population, both housing and employment generation are a few of the key socio-economic challenges. The increase in land use is one of the main factors in increase in global warming and the challenge to generate employment at a local level is increasingly intractable. In this scenario, the present innovative technique of building bamboo-based multi-level houses will surely have a far reaching impact. Without affecting the soil, unlike brick mining, the bamboo plantations will rather enrich the land with its leaf litter and create widespread employment opportunities right from nursery management, plantation management to harvesting, processing at local levels to use on site. Also, a meagre amount of steel and cement mortar (which can also be replaced by mud, cow dung and husk plaster) will result in mostly local consumption of goods and create new employment opportunities. Further, it will enable the villages to become suppliers of finished products, thus enhancing the village economy and truly achieving a de-centralized production system.

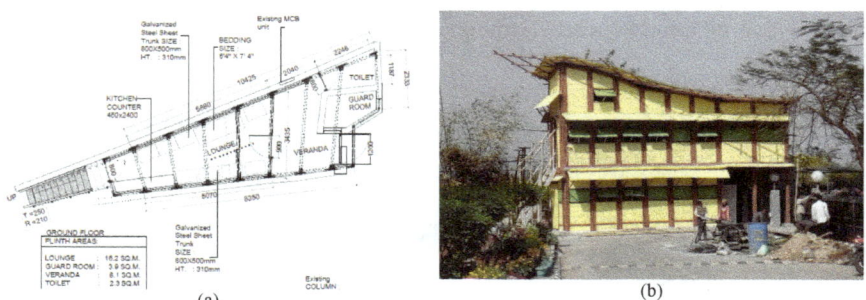

(a)   (b)

FIGURE 9 - FINAL AS-BUILT MULTI-LEVEL BAMBOO STRUCTURE (a) GROUND FLOOR PLAN (b) ACTUAL STRUCTURE

Non-Conventional Materials and Technologies – NOCMAT for XXI Century   Materials Research Forum LLC
Materials Research Proceedings 7 (2018) 190-198   doi: http://dx.doi.org/10.21741/9781945291838-18

Conclusions

The design and construction of this bamboo structure was indeed very challenging, particularly for two reasons i) this is the first of its kind being built with bamboo and ii) the complete bamboo superstructure from ground to the roof level hardly uses a single nut or bolt or, for that matter, not even any steel rods or wires in its construction. The steel reinforcement is used only for column foundations and plinth beams whereas nuts and bolts are used only at discrete locations for engaging the roof. This is significantly innovative in terms of its green index and the amount of carbon sequestration being undertaken. Further, with this type of construction, it strongly hints to a possibility of building such structures both as carbon sinks as well as utilising sustainable materials in construction. The researchers as engineers had to use their ingenuity and experience of working with bamboo to realize every component of the structure and its joinery. Thus this paper clearly demonstrates the potential of using bamboo as a green alternative for construction of an attractive strong durable building (see Figure 9b). This establishes the potential for the use of bamboo structurally in various sectors of infrastructure such as cost effective housing, ancillary infrastructures such as car parking shelters, bus stands, stand-alone latrines, shops, halls, etc in urban areas, rural housing and agricultural infrastructure such as cattle shelters, greenhouses, vegetable stores, solar dryers, etc. This also emphasises the benefit of more focussed research on bamboo based on practical engineering principles.

Acknowledgments

The authors acknowledge the funding support from Madhya Pradesh State Bamboo Mission (MPSBM) and Madhyanchal Bhawan, New Delhi for allowing the construction on their premises. Also, the architectural design and support of Ms. Revathi Kamath of Kamath Design Studio is acknowledged.

References

[1] Pandey B, Tripathi YC and Hazarika P, A hand book of propagation, cultivation and management of bamboo, Van Vigyan Kendra, Rain Forest Research Institute (ICFRE), Jorhat, Assam, 2008.

[2] Lucon O, Urge-Vorsatz D, Zain Ahmed A, Akbari H, Bertoldi P, Cabeza, LF, NEyre N, Gadgil A, L. Harvey DD, Jiang Y, Liphoto E, Mirasgedis S,Murakami S, Parikh J, Pyke C and M. V. Vilarino MV. Buildings. In: Climate Change 2014: Mitigation of Climate Change. Contribution of Working Group III to the Fifth Assessment Report of the Intergovernmental Panel on Climate Change [Edenhofer, Pichs-Madruga OR, Sokona Y, Farahani E, Kadner S, Seyboth K, Adler A, Baum I, Brunner Eickemeier SP, Kriemann Savolainen BJ, Schlomer von Stechow SC, Zwickel T and Minx JC (eds.)]. Cambridge University Press, Cambridge, United Kingdom and New York, USA, 2014.

[3] Ghavami K., Bamboo as reinforcement in structural concrete elements. Journal of Cement & Concrete Composites. 2005; 27: 637-649. https://doi.org/10.1016/j.cemconcomp.2004.06.002

[4] Korde C, West R, Gupta A and Puttagunta S. Laterally restrained bamboo concrete composite arch under uniformly distributed loading, Journal of Structural Engineering. 2014; 141, 3: B4014005.

[5] Das P, Korde C, Sudhakar P and Satya S. Traditional Bamboo Houses of North-Eastern Region: A Field Study of Assam and Mizoram. 2012; Key Engineering Materials: 517: 197-202. https://doi.org/10.4028/www.scientific.net/KEM.517.197

Non-Conventional Materials and Technologies – NOCMAT for XXI Century   Materials Research Forum LLC
Materials Research Proceedings 7 (2018) 190-198   doi: http://dx.doi.org/10.21741/9781945291838-18

[6] Jain SK, Murty CVR, Chandak N, Seeber L, and Jain NK. The September 29, 1993, M6.4 Killari, Maharashtra Earthquake in Central India. 1994: EERI Special Earthquake Report, EERI Newsletter, 28(1).

[7] Kaushik H, Dasgupta K, Sahoo D and Kharel G. Reconnaissance report Sikkim earthquake 2006, NICEE, IIT Kanpur.

[8] Murty CVR, Rai D, Jain SK, Kaushik H, Mondal G and Dash S. Performance of Structures in the Andaman and Nicobar Islands (India) during the December 2004 Great Sumatra Earthquake and Indian Ocean Tsunami. 2006: Earthquake Spectra; 22(S3): 21–54.

[9] Janssen J. Designing and Building with Bamboo. 2000, INBAR, Technical Report No. 20; 130-133.

Non-Conventional Materials and Technologies – NOCMAT for XXI Century   Materials Research Forum LLC
Materials Research Proceedings 7 (2018) 199-213   doi: http://dx.doi.org/10.21741/9781945291838-19

Evaluation Tools for R&D Projects Sponsored by the Brazilian National Network for Bamboo's Research and Development – Redebambu/Br

Ohayon Pierre [a*], Ghavami Khosrow [b]

[a*] Departamento de Contabilidade / Faculdade de Administração e Ciências Contábeis / Universidade Federal do Rio de Janeiro - FACC/UFRJ, Brazil; pohayon@facc.ufrj.br

[b] Departamento de Engenharia Civil / Pontifícia Universidade Católica do Rio de Janeiro (CIV/PUC-Rio), Brazil; ghavami@puc-rio.br

Abstract. Since the 70ties local energy saving materials, cement composites reinforced with vegetable fibers, bamboo as well as renovated ancient technologies started to be investigated by scientists and researchers in order to substitute industrialized materials which are highly polluting and high energy demanding in their production. Although proved technically and scientifically that the newly developed *non-conventional materials and technologies* (NOCMAT) were superior to the conventional industrialized materials they have not been used in large-scale projects. There is an intense on-going search in Brazil for non-polluting materials which consume little energy in their production and/or utilization. Even with the accumulation of technical data concerning the developed materials and structural elements obtained from the research programs, they are not systematically used in large scale in civil construction. Therefore, a systematic and methodological evaluation framework is needed. In this paper a short description of the materials and structural elements using bamboo are given. Then the evaluation tools for the successful implementation of the results in large NOCMAT projects is discussed considering those Research and Development (R&D) projects sponsored by Redebambu/BR – the recent network created in Brazil, which applies the national policy to encourage the bamboo´s handling and sustainable planting. Five important outcomes reflected into the NOCMAT R&D projects such as Efficiency, Effectiveness, Impact, Relevance and Sustainability and its pertinent indicators are presented. Additionally, four relevant dimensions, specifically, (1) Political, Strategic and Normative; (2) Organizational; (3) Allocation and Management of Resources; and (4) Technical, Scientific and Economic evaluation dimensions are considered and discussed.

Keywords: Evaluation Tools, NOCMAT R&D Projects, Redebambu/BR

Introduction

Since the 70ties local energy saving materials such as rice husk ash, soil-vegetable fibers, cement composites reinforced with vegetable fibers, bamboo as well as renovated ancient technologies started to be investigated by scientists and researchers in order to substitute industrialized materials which are highly polluting and high energy demanding in their production. Although proved technically and scientifically that the newly developed *non-conventional materials and technologies* (NOCMAT) were superior to the conventional industrialized materials they have not been used in large-scale projects. The NOCMAT findings could contribute immensely to the pursuit of sustainable development, which is a major global issue especially in developing countries including Brazil. These countries are in urgent need for adequate housing for an ever-increasing population. The present energy crisis in the world due to the extreme low level of water available over the last decades besides the indiscriminate industrial growth based on the program from industrialized nations' interferences has caused increasing concern about managing the energy resources still available besides the environmental degradation. There is an intense on-going search for non-

Non-Conventional Materials and Technologies – NOCMAT for XXI Century   Materials Research Forum LLC
Materials Research Proceedings 7 (2018) 199-213   doi: http://dx.doi.org/10.21741/9781945291838-19

polluting materials, which consume little energy in their production and/or utilization. Attention of researchers has turned to materials such as vegetable fibers including bamboo, soil, mining, industrial and agricultural wastes for engineering applications. New cements using all types of wastes are being developed and used for the production of composites reinforced with vegetable fibers around the world in a global effort to find a substitute for health hazardous asbestos cement which is prohibited by law in industrialized countries and used in most of developing countries with low cost.

To overcome the serious construction problem in Brazil and other developing countries in the world, several successful research programs have been carried out since 1979 at PUC-Rio. The results are being propagated through the ABMTENC (Brazilian Association of the Science of Non-Conventional Materials and Technologies) and implemented in other universities using indigenously available local materials such as bamboo, vegetables fibers, soil, quick lime, and cement mortars in the production of new structural elements such as bamboo space structures, corrugated sheets made of cement mortar composites reinforced with bamboo pulp, sisal and coconut fibers, soil-fibers composite for load bearing walls and concrete elements reinforced with bamboo [1]. Although with the accumulation of technical data concerning the developed materials and structural elements obtained from the research programs, they are not systematically used in large scale in civil construction. To reduce this barrier, a systematic and methodological selection and evaluation framework including a pertinent set of indicators is needed. In this paper a short description of the materials and structural elements using bamboo and vegetable fibers are given. Then the methodology for the selection and evaluation framework for the successful implementation of the results in large NOCMAT projects is discussed within the Redebambu/BR – the very recent network created in Brazil, which applies the national policy to encourage the bamboo´s handling and sustainable planting.

Bamboo and composites reinforced with vegetable fibers

In civil engineering the development and application of materials, with low cost and reduced energy consumption has turned into an actual basic requirement. The industrialized materials, so called conventional, mobilize vast financial resources, consume an enormous amount of energy and require centralized processing. In consequence of this, among other effects, activities are suppressed in rural areas or even in small towns, and non-renewable materials are wasted and causing permanent pollution. In this sense, it becomes obvious that ecological materials satisfy some fundamental requirements such as minimization of energy consumption, conservation of non-renewable natural resources, reduction of pollution and maintaining a healthy environment [2, 3].

Bamboo presents a tremendous economic potential, as it reaches its full growth in only a few months and its maximum mechanical resistance in a few years, besides the fact that it occurs in abundance in tropical and sub-tropical regions of the globe. The energy necessary to produce $1m^3$ per unit of stress projected in practice for materials commonly used in civil construction has been compared with that of bamboo. It was found that for steel it is necessary to spend 50 times more energy than for bamboo. The use of bamboo is attractive as a substitute for steel, especially when considering the relation between tensile strength and specific weight of bamboo, which is six times greater than that for steel. Although used intensively in South America by the natives for centuries, European colonizers did not know the potentials of bamboo. Systematic studies have been carried out on bamboo for more than two decades in order to develop methodologies for its application in space structures and as reinforcement in concrete considering their safety and durability. Bamboo to be used in a large scale as an economically viable material in engineering and with the possibility of its industrialization requires, at R&D of NOCMAT projects level, a systematic and logical framework for planning and evaluation in view of a sustainable technological development.

200

Non-Conventional Materials and Technologies – NOCMAT for XXI Century   Materials Research Forum LLC
Materials Research Proceedings 7 (2018) 199-213   doi: http://dx.doi.org/10.21741/9781945291838-19

Evaluation functions

Science and technology have contributed largely in the last three decades to the economic development without considering adequately different social classes. The intensive R&D activities in the rapidly growing areas of ST&I (Science, Technology and Innovation) such as new high resistance cements, steel, petrochemical derived materials, among others have not given the opportunity to less developed nations to cut the vicious circle which maintained them technologically dependent on industrialized countries. The *Science, Technology & Innovation Green Book* of the Brazilian Ministry of S&T, presented for discussion on July 2001, brings new challenges for the next ten years with its priorities notably related to low cost energy materials and technologies which are ecologically acceptable. It indicates that one of the main "bottle-necks" in terms of information, which restricts seriously the proper ST&I planning and decision making process, is the production of pertinent indicators [4, 5].

In order to overcome these difficulties, six new interrelated key functions of technological resources management, characterizing the *what, why, when, where, how* and *who* for strategic and operational applications, should be considered [6]. They have not been systematically regarded for the assessment of innovative projects related to the use of locally available materials in abundance and appropriate technologies, in developing countries [7, 8, 9]. These strategic functions are: *to carry out an inventory* of technological resources (available technologies, expertise and skills); *to evaluate* technological resources, their strengths and weaknesses, and their economic potential; *to optimize* (make the best use of technological resources); *to enrich* technological resources through investigation, acquisition, alliance, research, development, improvement, innovation, renewal and replacement, as well as to further develop human technological expertise and skills by recruiting, training and team building; *to watch* developments in the scientific, technological and competitive environment employing an appropriate *technological vigilance and intelligence system*; *to protect* technological resources by safeguarding intellectual property, and by preserving human expertise and skills.

The accelerated rhythm in which the results of the research on NOCMAT are being introduced into a society, principally used to conventional materials and technologies imported from industrialized countries and not sufficiently prepared to receive them, create new economic, financial, administrative, organizational and human resources problems. Specifically NOCMAT projects which benefited from an unconditional enthusiasm by researchers are seen by the community as suspicious not because of their "few" results but of their "any" results obtained. To show the reliability and durability of the newly developed materials and technologies, in addition to the results obtained in the laboratories, large scale constructions should be built and permanently monitored, requiring higher and continued investments from sponsoring agencies and private organizations. Therefore, interest to establish rational framework integrating scientific institutions and sponsoring agencies for research programs, which are directed to social, economic and technological advancement is increasing [10, 11, 12, 13, 14, 15, 16].

NOCMAT R&D Projects Evaluation Capacity

Project is understood as a set of actions, performed in a coordinated way by a temporary organization, in which necessary inputs are allocated for, in a given period, achieving a specific goal [17]. There are over the last decades numerous conceptualizations for research project and technology development - R&D [18, 19, 20].

Several relevant aspects are considered in the management of R&D projects, namely: (i) project team; (ii) project life cycle; (iii) organizational climate project and environmental conditions.

Evaluation in most developing countries is becoming an important tool for the management of technology, as a necessary link between R&D and society needs. A key feature of a successful S&T organization is the ability to learn from past experience and react to market or client responses.

Selection and Evaluation capacities can play an important role in influencing policy analysis and formulation; improving resource allocation and budgetary process; improving investment programs and projects, examining fundamental missions. However, in these countries the adequate use of feedback in formulating projects, programs and policies and allocating resources is only incipient. Sensitivity to public criticism and the fear of political fallout from selection and evaluation findings are inhibiting factors. Many social appropriate technologies still lack the essential requirements of effective selection and evaluation. The quality of information and access to it is often insufficient, mechanisms for feedback into the decision making process are weak and a culture of accountability by using pertinent indicators is not firmly applied.

The barriers in the selection and evaluation of NOCMAT are mainly high cost of their procedures and lack of interest and commitment to the selection and evaluation functions at the political level; feedback mechanisms for applying selection and evaluation findings; more attention given to preparing and appraising programs and projects than to evaluating their performance on completion; involvement of institutional and national staff in selecting and evaluating externally financed programs and projects; attention to the quality of information; objectivity and independence in conducting selection and evaluation; access to the research result on low-cost energy materials and technologies; trained staff [21]. In addition, most experts receive their education in industrialized countries and are not necessarily aware of local conditions and local solutions for a sustainable program. These experts could even damage or hinder the development of the project. In state and federal sponsoring agencies, selection and evaluation does not have high priority for major reasons such as: little effective methods for selection and evaluation; lack of incentive for future productivity and limited freedom of action.

NOCMAT R&D Projects Framework Matrix

The R&D of NOCMAT project framework is being recommended by IAEA - International Atomic Energy Agency based on models provided by World Bank and UNIDO - United Nations Industrial Development Organization among other technical assistance agencies [22]. It is a tool which helps to think through all aspects of a project and analyzes its "logic" in order to ensure a good quality proposal which: (i) responds to a real need; (ii) produces significant economic or social impact; and (iii) demonstrates high potential for sustainability through strong commitment of the government and social groups concerned. The project framework matrix presented in Table 1 points out the most important aspects of the project. The first column constitutes the *project design* and work plan; the second column *project's performance indicators* outlines the objectively verifiable performance indicators used to judge "success"; the third column indicates the *means of verification* of the indicators and the fourth column notes the *main assumptions* that may influence the course of the project [23].

The *project's design* contains the five specific elements (input; activities; project outputs; specific objective; development/overall objective) linked to each other by logical cause-effect relationships. The *overall objective* is the highest level result to which the project should contribute directly or indirectly. It explains why it would be necessary to carry out the project. The *specific project objective* explains what is expected to be achieved through the use of project outputs, and for whom (end user or target group), if successfully completed in a given time.

Project outputs are the immediate results that can be guaranteed as a consequence of project activities carried out during implementation. *Project activities* are the actions taken or planned to transform the inputs into outputs: how the project intends to produce the outputs. *Project inputs* are the products, services, or resources (financial, human, materials etc.) needed (necessary and sufficient) to carry out activities of the project.

The *project's performance indicators* are the signals that allow the measurement of achievement of the main design elements and must provide quantifiable and verifiable evidence of the progress made towards the objective. Performance indicators describe and specify what is expected to be obtained through the use of the outputs by the direct recipients, in terms of: quantity (how much, how many); quality (how good or how well); target group (for whom); time and location (by when and where).

TABLE 1 - PROJECT FRAMEWORK MATRIX FOR NOCMAT

	Project Design Elements	Verifiable Indicators	Means of Verification	Main Assumptions
To contribute	Development/overall objective: National capacities are established for low cost housing.	Number of new local housing with cost-effectiveness.	National and international documents.	Environmental and safety regulations are assumed
To Achieve	Specific objective: Capabilities to enable NOCMAT participation in housing programs.	% of effective cost reduction in implemented housing using NOCMAT.	Surveys related to the implementation of housing and low cost energy programs using NOCMAT.	Social/economic adherence to the housing program implementation is assured.
To Produce	Project outputs: Servicing facilities and training for local use of appropriate materials and technologies.	New techniques adopted for local uses, time required for program implementation, % of NOCMAT used.	Follow-up of the reports, comparing different programs results and laboratory results.	Technical patterns and costs are maintained; contract negotiations are successful.
	Activities: Construction of laboratories, training events, technical tests and evaluations; expertise on NOCMAT.	Education of human resources specializing in appropriate materials and technologies.	Project schedule and costs accomplishment.	Human/financial resources are available, effective training are provided.
Provided	Inputs: Experts' advice, laboratories, equipments and space.	Time, quantity, quality of specific resources.	Progress reports and national/international comparisons for NOCMAT projects.	Inputs are provided just in scheduled chronogram.

Source (adapted): [23, pp. 6].

Means of verification indicate the sources of information necessary to verify the accomplishment of the indicators and should include the information which is to be made available, in what form, by whom and when. Baseline data, implementation records and progress reports are necessary to monitor and evaluate the achievement of the project's objective.

Main assumptions should point out conditions that ought to exist for the project to succeed but which are outside the direct control of the project management. They are positive conditions that are logically necessary, for example, for the activities to lead to the outputs. The likelihood/probabilities of these assumptions should be analyzed at the formulation stage and monitored throughout implementation, as it is a decisive factor for taking corrective actions or modifying the work plan.

NOCMAT R&D Projects Outcomes Evaluation

For the elaboration of a NOCMAT project different proposal should be examined. Evaluation should take into account five important outcomes, reflected into the NOCMAT project framework matrix: efficiency, effectiveness, impact, relevance and sustainability [24, 25].

Efficiency considers how well inputs are converted into outputs for example: the percentage of non-conventional materials included in products/objects and economic efficiency measured by energy and costs reductions, safety and life-cycle improvements.

Effectiveness defines the extent to which the innovative project is likely to achieve its main objectives. It should reveal the effective changes observed by government action and relative advantages of the innovation to the users, improvements in warranty, reduction in complaints, new norms created or adopted for constructions, reduction of construction deficit.

Impact is related to the longer-term effects on the problem situation or need which relates to the Development/Overall Objective. It should present the economic and social advantages pointing out changes in local culture, increased applications for constructions, pollution reduction, waste reduction, number of inhabitants benefited by the project implementation, increased Human Development Index by NOCMAT utilization, improved dissemination of technical information.

Relevant projects are those, which are necessary to solve the problems/needs, and their outputs are necessary to achieve the objective through direct and verifiable cause-effect relationships. *Relevance* should indicate: the degree of urgency and rate of growth of private and public funds allocated for housing programs using non-conventional materials and technologies; improved state of art in selected scientific disciplines and contributions to solve technical problems; professional learning; adaptability of NOCMAT application for construction in specific environmental and socioeconomic conditions.

Sustainability refers to the extent to which the improved situation (as resulted from the achieved objective) can be maintained by users in their own way. It is linked to the local availability of funds for continued operation, maintenance of equipment and re-training of staff, and to the institutional and managerial capabilities. It should point out the scientific and technological prestige using appropriate indicators, local/regional and international cooperation, financial saving and new contracts with state sponsoring agencies.

Main dimensions and indicators for nocmat r&d projects sponsored by redebambu/br

The Brazilian Redebambu/BR Network and its R&D Sponsored Projects

The Ministry of Science, Technology and Innovation (MCTI) of Brazil and its National Council for Scientific and Technological Development (CNPq) launched the call for R&D projects N° 66/2013 to select proposals for financial support able to contribute significantly for the scientific and technological development and innovation in the country, specifically for structuring the National Network for Research and Development of Bamboo namely called Redebambu/BR.

The Redebambu/BR network supports the implementation of PNMCB - National Policy to Encourage Handling, Sustainable Planting and Preservation of Bamboo which has been created by the Law 12.484/2011. The Redebambu/BR network has representation in five Brazilian regions through the Regional Centers. These are "diffusion technology centers for strengthening the culture of bamboo in the country, offering technology solutions to all levels of the bamboo´s production chain, from the family farmer to the processing industry" [26].

The activities of Redebambu/BR network are managed by a Steering Committee composed by the heads managers of the six Regional Centers, a representative of the MCTI and a representative of the industrial sector.

The sponsored projects cover one or more research areas within the main Thematic Area entitled "Chain development bottlenecks in production of bamboo" which applies to large knowledge areas such as Engineering, Biology, Agriculture or subareas such as Management of Research and Technology, among others. The motivation to develop a methodological framework, for scientific and technological activities involving this main Thematic Area covering 17 sub-themes[1] comes essentially from the new priorities of the brazilian Government in meeting the technological demands concerning the implementation of PNMCB.

It is therefore suggested a model for evaluating the activities of scientific and technological development within the Redebambu/BR network. The evaluation system being proposed should be able to still achieve the following specific objectives:

(i) Measurement of the degree of implementation of the objectives and strategic goals for Redebambu/BR network under the national policy PNMCB and its projects adherence to international trends in scientific and technological research which were submitted and supported by the CNPq/MCTI sponsor agency;

(ii) Sorting and analysis of relevant results from the point of view of the scientific and technological development policy, based on the goals of PNMCB, S&T sponsor agencies and Redebambu/BR network which constitutes an "operator" of the supported projects;

(iii) Registration of all possible divergences between expected results and actual results, analyzing its causes and implications;

(iv) Balance of projects performance, those supported under the main Thematic Area and its 17 sub-themes;

(v) Evaluation of the direct impacts of research activities within the Redebambu/BR network (those relative to their more immediate goals) and indirect impacts of supported projects regarding policies and involved institutions;

(vi) Production of a summary of statistical data covering all aspects of the supported projects and their implementation;

(vii) Construction of a system of indicators to analyze and monitor the Redebambu/BR´s projects.

Six research projects were selected by the CNPq/MCTI from the following Brazilian research centers/Universities:

(1) Faculty of Animal Science and Food Engineering/University of São Paulo (State of São Paulo) and Department of Civil Engineering/Pontifical Catholic University of Rio de Janeiro (State of Rio de Janeiro);

(2) Department of Plant Science-Agricultural Sciences Center/Federal University of Santa Catarina (State of Santa Catarina);

---

[1] (1) Identification of native and exotic species; (2) propagation of bamboo´s species; (3) bamboo´s propagation and cultivation for temperate climate; (4) Production of reinforced concrete, activated coal and glued laminated bamboo ("plyboo"); (5) Planting for large-scale biomass and cellulose; (6) Production and manufacturing of bamboo buds; (7) Cultivation of tropical bamboo and use in landscaping; (8) Structural application of bamboo; (9) Sustainable handling of *Guadua sp* forest; (10) Identification of species with greater economic potential; (11) Technologies Demonstration for energetic use of bamboo; (12) Application of bamboo for restoration of degraded lands; (13) Strategy for introducing the culture of bamboo in family farming; (14) Technology transfer for production and planting of seedlings; (15) Introduction of new species of bamboo in Brazil; (16) Use of bamboo for high performance composite materials; (17) Production of raw materials for cosmetics and food industry.

(3) Department of Organic and Inorganic Chemistry/Federal University of Ceará (State of Ceará);
(4) Thematic Area of Biotechnology-EMBRAPA Genetic Resources/EMBRAPA - Brazilian Agricultural Research Enterprise (Brasília/ Federal District);
(5) Department of Forestry Engineering-Center for Agricultural Sciences/Federal University of Viçosa (State of Minas Gerais);
(6) Department of Rural Engineering-School of Agronomy/Federal University of Goiás (State of Goiás).

Each project should be developed originally in 24 months beginning in January 2014. The amount of financial resources approved by the CNPq/MCTI varied between 800,000.00 and 1,300,00.00 Reais *per* project (ie between 320 thousand and 520 thousand US Dollars). The total amount of estimated resources in the Call for R&D sponsored Projects has been 6 million Reais (ie about 2,4 million US Dollars). The actual economic crisis in Brazil impacted on grants and the project's completion are delayed.

Main Dimensions of Evaluation Framework
The functional dimensions suggested facilitate the construction of an evaluation tool which considers the most relevant aspects of the project its social impacts. Considering the NOCMAT R&D project as an open system, its dynamics is determined by an internal process of learning and the influences of the external environment to which it is concerned. Each one receives specific inputs generating results that directly affect each respective domain. They interact at all levels of the project activity and influence the behavior of all actors involved such as executing institutions and project teams. These main Dimensions are [27, 28, 29, 30, 31, 32, 33, 34, 35, 36, 37]:

Political, Strategic and Normative Dimension, including the processes which generate and manage the policies, strategies, norms and guidelines and involves the design and operationalization of decision making at the project level.

Organizational Dimension, comprising the organizational processes which result in operational project monitoring, within the formal rules highlighted by the political, strategic and normative Dimension; among the structures considered include all instances of the projects' institution and official agencies (Treasury, R&D sponsors, etc.).

Scientific-Technical and Economic Dimension, comprising the technical, scientific and economic results obtained from the institutional project executor (universities, research institutes, etc.).

Allocation and Resource Management Dimension, including the processes responsible from its origin in the sponsor institution until its final destination, its executing institutions and their projects, through all intermediate organizational levels. This Dimension also comprises the processes responsible for the management and use of project's management information system, and interacts closely with all other Dimensions: for example, with the Technical-Scientific one, for the project's monitoring; with the Organizational one for aspects related to project's design; and with the Political-Strategic-Normative Dimension for project's impacts and effectiveness.

These four Dimensions (Figure 1) interact at all levels of its programmatic action (annual and multi-annual operational plans) and influence the behavior of all final actors involved (executor institutions ie, institutional operators of projects executed by different teams engaged) and which produce impacts on the external environment (notably the productive and educational sectors) within their respective domains. In its turn, the external environment (economic, political, social) influences the project's development.

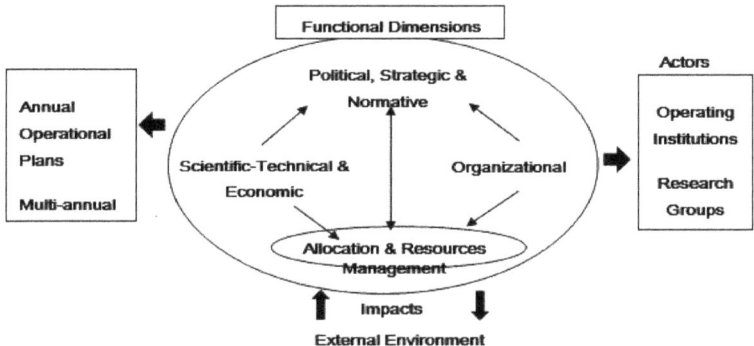

FIGURE 1 – *DIMENSIONS FOR NOCMAT R&D PROJECTS EVALUATION SPONSORED BY REDEBAMBU/BR*

TABLE 2 – *NOCMAT R&D PROJECT EVALUATION FOR SCIENTIFIC-TECHNICAL AND ECONOMIC DIMENSION WITHIN THE REDEBAMBU/BR NETWORK*

Performance Indicators	Project Data		Context Data	
	Planned	**Realized**	**Planned**	**Realized**
Inputs **Indicators:** • Research Groups • Involved Institutions • Applied Resources				
Outputs **Indicators:** • Human Resources Formation • Products and Processes Developed • Patents required • Publications	**Data Sources:**			
Impacts **Indicators:** • Licensed Patents • University-Industry Collaborations	**Institutional Project Operators** **Projects Sponsored** **Official Organisms Involved** **Sponsor Agencies**			
Economic **Indicators:** • New employments generated • Equipment's Modernization • New markets Penetration				

Finally, for the assessment of each Dimension, it is required to relativize, whenever possible, the performance indicators of the project not only with what has been *planned*, but also with those of its larger *context*.

Thus, for example, within the *Scientific-Technical and Economic* Dimension (see Table 2), it is important to know and compare the results reached for the formation of human resources in projects with the results achieved by other similar initiatives, in this same research area (for example, those initiatives coming from self-effort of the executor institution using its own resources or with support from funding agencies). This would permit to know how important is the project in thematic areas where it operates in relation to other government (and possibly private) actions in these same areas, taking into account both the results achieved by the research projects submitted/sponsored.

Selected Indicators for NOCMAT R&D Projects

The model adopted consists of five main indicators groups, namely: (i) resources ("*inputs*"); (ii) dynamics of R&D activities ("*Process*"); (iii) direct results ("*Direct outputs*"); results effectively used, and impacts [38]. Each group consists of a set of specific indicators, reliable in its measurement and relevant for decision making.

Table 3 presents examples of indicators derived from the literature [39, 40, 41] and few propositions of this study that could be added and therefore better assess the NOCMAT R&D Projects.

Indicators should be applied consistently throughout the evaluation process, such as: proposal application and selection stage; ongoing evaluation and monitoring stage; final evaluation stage. The last stage is probably the most important element of the overall continuous evaluation and monitoring process in terms of assessing the achievement of objectives of NOCMAT projects and thus encompasses aspects of networking, dissemination and outputs. The final set of measurement indicators should include: (i) *additionality* (improvement of R&D to main objective); (ii) outcomes and achievements; (iii) quality; (iv) benefits; (v) continuity and; (vi) gaps (areas of research competence not covered or not represented by representatives [42, 43].

Strategic objectives which express the specific ones with NOCMAT refer to: reduction of pollution shown by the increased use of NOCMAT; biodiversity protection through the number of families adopting NOCMAT; improvement of the quality of life of the population given by the number of dwellings using NOCMAT and the increased *per capita* of local revenue; organization/community and institutional participation reflected by the number of Non-Governmental Organizations, NGO, and public institutions and by the number of NOCMAT courses promoted [44].

Sustainable Development is aimed at meeting the material needs of the present generations without compromising the ability of the future generations to meet their needs [45]. At the upstream stage of the sustainable technological development, indicators of *prevention* are pointed out such as: less energy inputs for raw materials measured by: (i) reduce use of natural resources not renewable; (ii) reduce ecological disturbances; (iii) reduce noise and vibration and emissions. At the process stage, *containment* is considered, still pollution is being generated and has to be controlled. At the downstream stage, *utilization* of NOCMAT is focused, assuring less wastes generation, energy savings and financial resources savings for needy populations [46].

TABLE 3 - SELECTED INDICATORS FOR NOCMAT R&D PROJECTS WITHIN THE REDEBAMBU/BR NETWORK

Indicators Group	Sub-group	Selected Indicators
Resources ("*inputs*")	- Human Resources	Rating of researchers qualification by academic title
		Rating of researchers among overall project members
	- Financial Resources	Total financial amount for R&D Project
		Total amount allocated per researcher
	- Infrastructure Resources and Physical Space	Number of scientific equipment by project
		Laboratory availability
	- Information Resources	Information Center availability
		Digital database availability
	- Organizational Resources	Number of developed thematic researches
		Number of laboratories involved in the project
Dynamics of R&D Activities ("*Process*")	- Management	Time of experience in research
		Number of projects finalized and aborted
Direct Results ("*Direct Outputs*")	- Education	Number of master and PhD degrees accomplished by students involved in the project
	- Diffusion	Number of new students engaged in the Project with scientific vocation
		Number of new projects with NOCMAT popularization
Results effectively used	- Physical	Number of Projects implemented / commercialized/ patentes deposited
	- Bibliometric	Citation index of project researchers
	- Outcomes	Average financial benefits generated by projects
		Average financial benefits generated by courses, seminars and other events originated by project knowledge
Impacts	- For Science	Number of new papers and different scientific and technical publications on NOCMAT
	- For Political Relevance	Amount of national budget from the Ministry of ST&I for NOCMAT activities
	- For Commercialization	Number of new contracts established with industries
		Average contracts duration
	- For Education and Training	Number of new undergraduate and graduate courses oriented to NOCMAT
	- For Environment	Waste reduction by using NOCMAT

Source: Prepared by the authors.

Concluding remarks

The understanding of NOCMAT and its sustainability in constructions has undergone changes over the years. First attention was given to the issue of non-renewable resources and how to reduce their impact on the environment. Now emphasis is placed on more technical issues such as bamboo and composites reinforced with vegetable fibers for the construction components and technologies with energy related design concepts. In this paper, it is shown that the implementation of NOCMAT could well succeed by applying the selection and evaluation framework for R&D projects, which should be

Non-Conventional Materials and Technologies – NOCMAT for XXI Century   Materials Research Forum LLC
Materials Research Proceedings 7 (2018) 199-213   doi: http://dx.doi.org/10.21741/9781945291838-19

seen by all concerned parties as a way to learn and improve their integrated action in a systematic form. This is possible when the efforts are made to build social, political, economic, environmental, cultural and technical indicators for innovative projects, which are designed to serve the real needs of decision makers at local, state and federal level.

Acknowledgements

The authors would like to thank ABMTENC Association which made available all the NOCMAT Conferences data and the 17th International Conference on Non-Conventional Materials and Technologies – Construction Materials & Technologies for Sustainability (17th NOCMAT 2017), November 26-29, 2017, Mérida, México organizers for their valuable effort and proceedings for the discussions, which form the basis of the meeting. Also, financial support given by FAPERJ and FACC/UFRJ are appreciated.

References

[1] Ghavami, K. Bamboo as reinforcement in structural concrete elements. *Cement and Concrete Composites*, Elsevier, v. 27, n. 6, p. 637-649, July 2005.
https://doi.org/10.1016/j.cemconcomp.2004.06.002

[2] United Nations. The Habitat Agenda: Chapter IV: C. Sustainable human settlements development in an urbanizing world, 14 June 1996. Available at: <http://www.un-documents.net/ha-4c.htm>. Access in: January 16, 2017.

[3] Ghavami, K. Application of bamboo as a Low-Cost Energy material in Civil Engineering. In: Third CIB/RILEM Symposium - Materials for Low Income Housing, Mexico City. *Proceedings*. 1989, p. 526-536.

[4] Brasil. Ministério da Ciência e Tecnologia *Livro Verde* – O debate necessário: Ciência, Tecnologia, Inovação – Desafios para a Sociedade Brasileira. Brasília: MCT/Academia Brasileira de Ciências, July 2001.

[5] Albuquerque, M. E. E.; Bonacelli, M. B. M.; Weigel, P. A questão ambiental e a contribuição dos institutos de pesquisa à geração de tecnologias ambientalmente sustentáveis. In: *Parcerias Estratégicas*, Brasília/DF, v. 15, n. 30, p. 9-24, January 2010.

[6] Kuhlman, S. *Evaluation as a source of 'strategic intelligence*. In: Shapira, P.; Kuhlman, S. Learning from Science and Technology Policy Evaluation – Experiences from the United States and Europe. Northampton/MA: Edward Elgar, 2003. (Chapter 18, p. 352-379).

[7] Wickremasinghe, S. I.; Gupta, V. K. *Science & Technology Policy and Indicators for Development* – Perspectives from Developing Countries. Dehli: Daya Publishing House, 2008.

[8] Bellen, H. M. Van. *Indicadores de Sustentabilidade* – uma análise comparativa. 2ª Ed., São Paulo: FGV Editora, 2006.

[9] Morin, J.; Rafferty, P. J. *The six key functions of technological resources management.* Miami/Florida-USA: Institute of Industrial Engineering. Proc. of the Second International Conference on Management of Technology, Feb. 28–March 2, 1990, p. 621-627.

[10] Soltman, C.; Stucki, T.; Woerter, M. *The performance effect of environmental innovations.* KOF Working Papers, n. 330. Zurich: Swiss Federal Institute of Technology, February 2013.

[11]Edler, J.; Berger, M.; Dinges, M.; Gök, A. The practice of evaluation in innovation policy in Europe. *Research Evaluation*, Oxford, v. 21, n. 3, p. 167-182, July 2012.
https://doi.org/10.1093/reseval/rvs014

[12]Edler, J.; Georghiou, K.; Blind, E.; Uyarra, E. Evaluating the demand side: New challenges for evaluation. *Research Evaluation*, Oxford, v. 21, n. 1, p. 33-47, March 2012.
https://doi.org/10.1093/reseval/rvr002

[13] Clark, W.; Crutzen, P.; Schellnhuber, H. Science for Global Sustainability: Toward a New Paradigm. *CID Working Paper* n. 120. Science, Environment and Development Group, Center for International Development Cambridge: Harvard University, 2005.

[14] Sistema Integrado de Información sobre Investigación Científica, Desarrollo Tecnológico e Innovación. Available at: <**http://www.siicyt.gob.mx**>. Access in: January 16, 2017.

[15] Dale, R. Evaluating Development Programmes and projects. 2nd Ed., London: Sage Publications, 2004.

[16] Hong, H. D.; Boden, M. R&D programme Evaluation: theory and practice – a Comparative Analysis of Large Scale R&D Programme Evaluation. Surrey/UK: Ashgate Publishing, 2003.

[17] Ellis, l. Introduction to Evaluating R&D Process Management. In: *Evaluation of R&D Processes*: Effectiveness through Measurements. Norwood/MA: Artech House, 1997.

[18] Geisler, E. *The Metrics of Science and Technology*. Westport/CT: Quorum Books, 2000.

[19]Cleland, D. I.; Ireland, L. R. O Gerenciamento de Projetos. In: 2. ed. *Gerenciamento de Projetos*. Rio de Janeiro: LTC, 2007. (Chapter 1).

[20] Clifford, F. G.; Larson, E. W. Gerenciamento de projetos moderno. In: *Gerenciamento de projetos – o processo gerencial*. 4a. ed. São Paulo: McGraw-Hill, 2009. (Chapter 1).

[21] The World Bank Building Evaluation Capacity. Washington: The World Bank/Operations Evaluation Department. *Lessons & Practices*, n. 4, p. 1-11, 1994.

[22] International Atomic Energy Agency *Planning and Designing IAEA Technical Co-Operation Projects: Guidelines*. Vienna/Austria: IAEA, Department of Technical Co-Operation. June 1997.

[23] Knowlton, L. W.; Phillips, C. C. *The Logic Model Guidebook* – Better Strategies for Great Results. Thousand Oaks/Ca: Sage, 2009.

[24] Chen, H-T. *Evaluation Outcomes*. In: Practical Program Evaluation. London: Sage Publications, 2005. (Chapter 9, p. 195-229). https://doi.org/10.4135/9781412985444.n10

[25] Holvoet, N.; Renard, R. *Desk Screening of Development Projects: Is It Effective?* In: Stern, E. Evaluation Research Methods. London: Sage Publications, 2005. (Vol. 4, Chapter 60, p. 87-107).

[26] Brasil. Conselho Nacional de Desenvolvimento Científico e Tecnológico. *Chamada MCTI/Ação Transversal/CNPq N° 66/2013*. Available at:
<file:///C:/Users/Pierre/Downloads/Chamada+66-2013.pdf>. Access on: Oct. 15, 2014, p. 15.

[27] Champagne, F.; Hartz, Z.; Brouselle, A.; Contandriopoulos, A.-P. *A Apreciação Normativa*. In: Avaliação – conceitos e métodos. Rio de Janeiro: Fiocruz, 2011. (Chapter 4, p. 77-94).

[28] Champagne, F.; Brouselle, A.; Contandriopoulos, A.-P.; Hartz, Z. *A Análise Estratégica*. In: Avaliação – conceitos e métodos. Rio de Janeiro: Fiocruz, 2011. (Chapter 5, p. 95-104).

[29] Farand, L. *A Análise da Produção*. In: Avaliação – conceitos e métodos. Rio de Janeiro: Fiocruz, 2011. (Chapter 7, p. 115-158).

[30] Champagne, F.; Brouselle, A.; Contandriopoulos, A.-P.; Hartz, Z. *A Análise dos Efeitos*. In: Avaliação – conceitos e métodos. Rio de Janeiro: Fiocruz, 2011. (Chapter 8, p. 159-182).

[31] Brouselle, A.; Lachaine, J.; Contandriopoulos, A.-P. *A Avaliação Econômica*. In: Avaliação – conceitos e métodos. Rio de Janeiro: Fiocruz, 2011. (Chapter 9, p. 183-216).

[32] Brouselle, A.; Lachaine, J.; Contandriopoulos, A.-P. *A Análise da Implantação*. In: Avaliação – conceitos e métodos. Rio de Janeiro: Fiocruz, 2011. (Chapter 10, p. 217-238).

[33] Geisler, E. *Science and Technology, The Economy, and Society*. In: Creating Value with Science and Technology. Westport: Quorum Books, 2001. (Part IV, p. 167-315).

[34] Moraes, L. A. F. de; Ohayon, P.; Ghavami, K. Application of Non-Conventional Materials: Evaluation Criteria for Environmental Conservation in Brazil. *Key Engineering Materials*, Trans Tech Publications, Switzerland, v. 517, p. 20-26, 2012.

[35] Ohayon, P.; Rosenberg, G. Análise dos *indicadores de ciência, tecnologia e inovação no âmbito da Fundação Oswaldo Cruz - Fiocruz*. Artigo publicado na RSP – Revista do Serviço Público/ENAP, Brasília/DF, v. 65, n. 2, p. 297-319, July/September 2014.

[36] Ohayon, P.; Barreiros, D. S.; Ghavami, K. Science and Technology Observatory for "NOCMAT" in Brazil: Role and Proposed Framework. *Key Engineering Materials*, Zurich/Switzerland, Trans Tech Publications, v. 600, p. 399-412, 2014.

[37] Ohayon, P.; Ghavami, K.; Jesuz, K. *Specifications of Building Environmental Evaluation Methods within Redebambu Network in Brazil*. In: Proceedings of the NOCMAT 2015 - 16th Nonconventional Building Materials and Technologies International Conference 2015 – Construction for Sustainability - Green Materials & Technologies. August 10-13, 2015, Winnipeg/Canadá.

[38] Ohayon, P. *Modelo Integrado de Indicadores de Ciência, Tecnologia e Inovação no Estado do Rio de Janeiro*. Rio de Janeiro: UFRJ, 2007. (Research Project sponsored by CNPq/MCTI-Brazil, Edital Universal 2004, Vol. 2, p. 2-24).

[39] Kusek, J. Z.; Rist, R. *Step 3: Selecting Key Performance Indicators to Monitor Outcomes*. In: Ten Steps to a Results-Based Monitoring and Evaluation System. Washington, D.C.: The World Bank, 2004. (Chapter 3, p. 65-79). https://doi.org/10.1596/0-8213-5823-5

[40] Franceschini, F.; Galetto, M.; Maisano, D. *Management by Measurement* – Designing Key Indicators and Performance Measurement Systems. Torino: Springer, 2010.

[41] Parmenter, D. *Key Performance Indicators* – Developing, Implementing, and Using Winning KPIs. 2nd Ed., Hoboken: John Wiley & Sons, 2010.

Non-Conventional Materials and Technologies – NOCMAT for XXI Century   Materials Research Forum LLC
Materials Research Proceedings 7 (2018) 199-213   doi: http://dx.doi.org/10.21741/9781945291838-19

[42] Ministério da Ciência, Tecnologia e Inovação - MCTI. *Plano Anual de Monitoramento e Avaliação*. Comissão Permanente de Monitoramento e Avaliação. Brasília/DF: ASCAV - Assessoria de Acompanhamento e Avaliação das Atividades Finalísticas/Secretaria Executiva, December 2015.

[43] TECHNOPOLIS GROUP & MIOIR. *Evaluation of Innovation Activities*: Guidance on Methods and Practices. Study funded by the European Commission. Brussels: Directorate for Regional Policy/European Union, 2012.

[44] Brasil. Ministério do Meio Ambiente. *Relatório sobre a Aplicação Preliminar dos Indicadores do Projeto AMA para Monitoramento do PPG7*. Brasília: MMA/ Secretaria de Coordenação da Amazônia. Projeto Piloto para Proteção das Florestas Tropicais do Brasil – PPG7. Projeto: Apoio ao Monitoramento e Análise – AMA. March 2001.

[45] UNEP Bergen Ministerial Declaration of Sustainable Development in the ECE Region. Bergen/Norway, 14-15 May, 1990, *Industry and environment*, v. 13, n. 2, p. 54-56, April/June 1990.

[46] Mullick, A. K. *Role of Cement and Concrete in Sustainable Societal Development*. 1st International Conference on Concrete & Development, Tehran/Iran, April 30–May 2, 2001, p. 573-582.

Bibliometric Analysis of Scientific and Technical Papers within NOCMAT 1984-2015 International Conferences

Ohayon Pierre[a]*, Sharafi Rad Ali[b], Ghavami Khosrow[c], Siqueira Cristiana Pinheiro Machado de[d]

[a]* Departamento de Contabilidade/ Faculdade de Administração e Ciências Contábeis/Universidade Federal do Rio de Janeiro - FACC/UFRJ, Brazil; pohayon@facc.ufrj.br

[b] Departamento de Engenharia Industrial/ Pontifícia Universidade Católica do Rio de Janeiro (DEI/PUC-Rio), Brazil; a.sharafii@gmail.com

[c] Departamento de Engenharia Civil/ Pontifícia Universidade Católica do Rio de Janeiro (CIV/PUC-Rio), Brazil; ghavami@puc-rio.br

[d] Departamento de Biblioteconomia e Gestão de Unidades de Informação / Faculdade de Administração e Ciências Contábeis/Universidade Federal do Rio de Janeiro - FACC/UFRJ, Brazil; cristiana.siq@gmail.com

Abstract. The last International Conference on Non-Conventional Materials and Technologies held in Canada in 2015 presented more than 130 new articles among different NOCMAT themes, showing that there is a great number of scientific and technical studies being developed. 1,399 papers have been presented along the 15 International Conferences, since the first one held in Brazil, in 1984. The series of NOCMAT have proven to be a leading forum where scholars, governmental and non-governmental agencies, practitioners exchange innovations of low energy cement technologies, new ecological materials and systems such as bamboo and natural fibers. The accelerated rhythm in which the results of the research on NOCMAT are applied in practice is not at a desired speed. This is principally due to the conventional materials and technologies imported from industrialized countries which are dominating the economy, financial, administrative, and human resources in developing countries. NOCMAT projects which benefited from an unconditional enthusiasm by researchers are seen by the community as suspicious not because of their "few" results but of their "any" results obtained. To show the reliability and durability of the newly developed materials and technologies, in addition to the results obtained in the laboratories, large scale constructions should be built and permanently monitored, requiring higher and continued investments from sponsoring agencies and private organizations. The general objective of this article is to analyze the betweenness of researches presented within all the thematic areas of the NOCMAT Conferences. It presents a bibliometric analysis of the network created, mapping the actors (nodes) and its relations (edges) by participating authors, involved Institutions; participating Countries and Year, and NOCMAT Themes, along the last 31 years.

Keywords: NOCMAT Themes, International Conferences, Papers, Bibliometric Analysis

Introduction – NOCMAT research context
The understanding of sustainability in building construction has undergone changes over the years. First attention of specialists was directed towards the topic of limited resources, especially energy, and its impact on the natural environment. Now emphasis is placed on technical issues such as materials, building components, construction technologies and energy-related design concepts as well as on non-technical issues such as economic and social sustainability.

Non-Conventional Materials and Technologies – NOCMAT for XXI Century   Materials Research Forum LLC
Materials Research Proceedings 7 (2018) 214-233   doi: http://dx.doi.org/10.21741/9781945291838-20

The pursuit of sustainable development as defined in the Gro Harlem Brundtland Report [1] as "development that meets the needs of the present without compromising the ability of future generations to meet their own needs" has become a major issue when trying to meet the challenge of providing proper housing for an increasing world population. Recently, Pierre Papon emphasized the necessity to take time because progress is not as fast as we think and also developments are becoming heavier [2].

To increase understanding of sustainable materials, also known as Nonconventional Materials and Technologies (NOCMAT) using organic materials, which are used either alone or as reinforcement in different types of matrices such as soil, cement and polymers, many research programs have been carried out all over the globe. In recent years, there has been an increase in research into the use of natural and non-conventional construction materials, which have suitable physical and mechanical properties for structural applications, and are desirable as ecological alternatives to more commonly used industrialized materials.

The study of non-conventional materials and technologies (NOCMAT) for construction began in the 1970's. This field of research has continued to grow in the decades since. The increased use of locally available natural and waste materials in construction can promote environmental sustainability and aid in the eradication of extreme poverty.

Many of the world's most poverty-stricken people live in remote locations, typically rich in natural resources. In many of these areas, the use of steel and reinforced concrete in construction has become a symbol of economic status. These materials are imported for use when local, natural materials can also be used to create structures that can successfully meet the intended need at a lower cost and environmental impact (considering precautions are taken to prevent deforestation or depletion of other resources).

Organizations like ABMTENC - Brazilian Association of Non-Conventional Materials and Technologies [3] at the Pontifical Catholic University of Rio de Janeiro and the INBAR - International Network for Bamboo and Rattan strive to achieve sustainable development and poverty alleviation by promoting the use of these two natural resources Bamboo and Rattan [4].

Since the development of industrialized materials, including steel and concrete, natural materials have been abandoned for their structural use. While the mechanical properties of natural materials are less predictable than those of the industrial materials, the natural materials are readily available in many rural locations and can be used with significantly less processing. The availability of steel and cement, on the other hand, depend on the presence of factories and an abundant feedstock supply to create these materials, as well as adequate infrastructure to permit transportation to the locations for their use.

Now, with energy consumption and sustainability becoming increasingly important issues in the construction industry, industrialized materials are not always optimal when there are many natural materials that have excellent structural properties, are renewable, and require minimal energy input to be construction-ready. Therefore, as the progress of the civil engineering industry relies on the continued development of different materials in construction, non-conventional materials (NOCMAT) should be considered in the search for the low-energy structural materials of the future.

ABMTENC´S contributions to NOCMAT international conferences

The Brazilian Association of Non-Conventional Materials and Technologies – ABMTENC was founded on 29th of March 1996, at the Pontifical Catholic University of Rio de Janeiro – PUC-Rio, by professor Ghavami and the rector of the University, the Vice Rector Jesus Hortal Sanches S.J., the previous Rector - Father Laércio Moura S.J., and the deans and scientists, engineers, architects and many other professionals, who are interested to the study and development of the science of non-conventional materials and technologies.

Once professionals, organizations and institutions in Brazil have been brought together to share their common goals of assisting science and developing non-conventional materials and technologies, ABMTENC intends to develop research and the expand the knowledge of this field, as well as encouraging communication between scientist and universities, institutions and non-governmental organizations throughout the inside of country and abroad, also. ABMTENC intent to share knowledge of these sciences through the publication of books and magazines, as well as promoting meetings, congresses, conferences, courses and techno-scientific reunions.

ABMTENC's success depends fundamentally on the interest and participation of its members, individuals and enterprises that want to share advances in research into non-conventional technologies and materials in many different fields of application such as: *Bamboo, environment, sanitation, bau-biologie, composites materials with vegetal fibers, earth architecture* [3].

NOCMAT international conferences
The former International Conference on Non-Conventional Materials and Technologies (IC-NOCMAT) has been realized in Winnipeg, Canada in August 2015, coinciding with the 31th anniversary of NOCMAT.

This IC-NOCMAT follows previous successful events in Pirassununga, State of São Paulo/Brazil (2014), João Pessoa/Paraíba-Brazil (2013), Changsha, Hunan/China (2011), Cairo/ Egypt (2010), Bath/England (2009), Cali/Colombia (2008), Maceió/Alagoas-Brazil (2007), Salvador/Bahia-Brazil (2006), Rio de Janeiro/RJ-Brazil (2005), Pirassununga/ São Paulo-Brazil (2004), João Pessoa/Paraíba-Brazil (2003), Hanoi/Vietnam (2002), Bhubanewar/India (1997) after the first NOCMAT 1984 held in Rio de Janeiro/Brazil.

Study Objetive and Methodology
The aim of this study is to map and analyze thematic trends of scientific and technical production within the NOCMAT conferences systematizing 1399 papers published in them. Since its first edition in 1984 to his last in 2015 are highlighted: (1) the production of published articles and authorship; (2) the chronology of the thematic areas; (3) the geographical distribution by country; (4) the institutional origin; (5) axes and thematic trends. The technical themes of knowledge application considering the thematic areas identified in each NOCMAT conference were pointed out for trend analysis.

It is included in the field of studies on *State of the Art*[1] [5] that aims to understand the evolution and trends of scientific and technological knowledge produced on certain topics or areas of expertise. Articles published in conferences and magazines coming from dissertations, doctoral theses or even activities of research and development (R&D) are one of the main forms of scientific communication for knowledge in construction. This field permits to examine national policies, programs and strategies that aim to foster notably open access (OA) and discuss how OA policies are monitored and enforced [6, 7]. This knowledge capital oriented to practical applications has been studied by many intellectuals in order to assess high quality processes concerning research and technological development such as accountability, productivity and measurement, absorptive capacity, intellectual property, among many others [8, 9, 10, 11, 12, 13].

Bibliometric Tools for Scientific and Technical Paper's Analysis
Bibliometrics is a branch of science that quantitatively studies the processes of scientific communication, *"[...] with the purpose of explaining the dynamics of science and technology from indicators of outputs and inputs."* [14]. Initially known as statistical bibliography, it was popularized

---

[1]"The term '*state of the art*' refers to the highest level of general development, as of a device, technique, or scientific field achieved at a particular time. It also refers to such a level of development reached at any particular time as a result of the common methodologies employed at the time" [5].

in 1969 from an article written by Pritchard [15], although originally created by Paul Otlet in 1934 [16].

Since its origins, it has had two major concerns: scientific production and search for immediate practical benefits for libraries, initially developed for measuring the number of editions and copies of books, the number of words contained in them, the space occupied on bookshelves and usage statistics. Over time, it has been oriented for the study of other formats of bibliographic production (periodical papers and other types of documents) pointing out authors' productivity, scientific research directions and networks, social collaborations, and citations analysis.

The analysis of scientific networks brings about important indicators such as network density (calculated by the number of relations existing by the number of possible relations in a network), degree of centrality (indicates the number of connections an actor has), index of centralization when an actor occupies a central position in the network and the others have to pass through it), degree of intermediation (possibility that an actor has to facilitate the communication between pairs of actors) and degree of proximity (ability of an actor to connect to all the other actors in the network) [17].

Different computer programs were developed for constructing and viewing bibliometric maps. More specifically, for the authors [18]:

"VOSviewer pays special attention to the graphical representation of bibliometric maps. The functionality of VOSviewer is especially useful for displaying large bibliometric maps in an easy-to-interpret way. VOSviewer's ability to handle large maps is demonstrated by using the program to construct and display a co-citation map of viewing bibliometric 5,000 major scientific journals".

Bibliometric data generate indicators that allow a global view of research development, providing relevant information in different scales, from the broadest (country) to the most specific (institution). In this way, it is possible to estimate a tendency of growth or decrease in many areas of research, providing a better decision making. The collectivization and internationalization of research is a growing trend verified through bibliometric mappings.

Data and analysis
In this topic are presented the level of production covering all the 14 NOCMAT conferences realized in different countries since 1984 to 2015, this one being its last edition. All the data were extracted from CD-ROMs and Books of abstracts/full papers and made available by ABMTENC Association.

Mapping papers by the participating authors
NOCMAT 1984 - the first of the series held in Rio de Janeiro/Brazil, began with a single overall theme entitled "Low-Cost and Energy Saving construction Materials". Since the second NOCMAT held in Bhubaneswar/India in 1997 until now, emerged the term *"Non-Conventional"* in the Conference's title.

Table 1, Figures 1 and 2 present the distribution of all the papers by year (NOCMAT Conference serie), country of the authors, and number of authors per paper (1 to more than 6).

The data collected totalize 1399 papers for all the NOCMAT Conference series. Figure 1 shows that 52% of the papers (721 papers) are published by 2 or 3 authors and, 89% of the papers (1251 papers) are published by more than one author. Figure 2 indicates a significant growth of the number of papers presented from 1984 to 2015.

Mapping papers by the involved centers of NOCMAT knowledge (Institutions)
Tables 2 and Figure 3 present the distribution of all the papers by involved Institution in all NOCMAT Conference series from 1984 to 2015.

Non-Conventional Materials and Technologies – NOCMAT for XXI Century   Materials Research Forum LLC
Materials Research Proceedings 7 (2018) 214-233   doi: http://dx.doi.org/10.21741/9781945291838-20

TABLE 1 - NUMBER OF PAPERS PER PARTICIPATING AUTHORS

Year	City	Country	Participating Authors							Total Papers
			1	2	3	4	5	6	>6	
1984	Rio de Janeiro	Brazil	31	16	11	4	1	-	1	64
1997	Bhubaneswar	India	8	5	7	3	-	1	0	24
2002	Hanoi	Vietnam	8	22	16	10	4	2	1	63
2003	João Pessoa	Brazil	3	13	24	11	6	1	1	59
2004	Pirassununga	Brazil	3	13	18	13	3	5	2	57
2005	Rio de Janeiro	Brazil	6	21	17	18	6	4	1	73
2006	Salvador	Brazil	5	23	34	7	12	3	2	86
2007	Maceió	Brazil	4	23	18	16	12	4	1	78
2008	Cali	Colombia	9	35	24	16	19	9	1	113
2009	Bath	UK	19	30	23	27	13	6	0	118
2010	Cairo	Egypt	16	26	34	23	14	5	1	119
2011	Hunan	China	6	28	44	39	13	6	3	139
2013	João Pessoa	Brazil	18	27	43	20	10	13	7	138
2014	Pirassununga	Brazil	2	30	33	26	18	18	9	136
2015	Winnipeg	Canada	10	28	35	21	24	8	6	132
			148	340	381	254	155	85	36	1399
			11%	24%	27%	18%	11%	6%	3%	100%

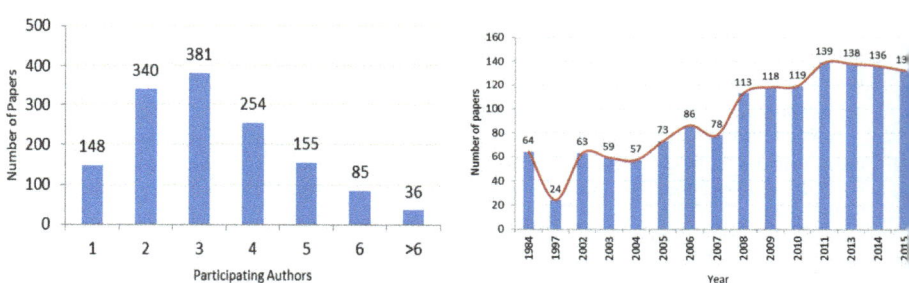

FIGURE 1: *NUMBER OF PAPERS PER PARTICIPATING AUTHORS* **FIGURE 2:** *NUMBER OF PAPERS PER YEAR*

The data from Table 2 and Figure 3 show that about 36.4% of the papers (509 papers among a total of 1399) are published by 2 and 3 institutions and about 38.5% of the papers (538 papers) are published by more than one institution in all NOCMAT Conference series. The level of multiple authorship increased indicating a greater integration of human resources and institutions.

Non-Conventional Materials and Technologies – NOCMAT for XXI Century   Materials Research Forum LLC
Materials Research Proceedings 7 (2018) 214-233   doi: http://dx.doi.org/10.21741/9781945291838-20

TABLE 2 - *NUMBER OF PAPERS PER INSTITUTIONS*

Year	City	Location	Participating Institutions						Total Papers
			1	2	3	4	5	6	
1984	Rio de Janeiro	Brazil	49	11	3	1	-	-	64
1997	Bhubaneswar	India	17	7	-	-	-	-	24
2002	Hanoi	Vietnam	41	17	5	-	-	-	63
2003	João Pessoa/ Paraíba	Brazil	36	18	2	2	1	-	59
2004	Pirassununga/ Brasil	Brazil	36	20	1	-	-	-	57
2005	Rio de Janeiro/ Brasil	Brazil	46	21	4	1	1	-	73
2006	Salvador/ Brasil	Brazil	62	21	2	-	1	-	86
2007	Maceió	Brazil	46	26	5	-	1	-	78
2008	Cali	Colombia	61	43	8	1	-	-	113
2009	Bath	UK	64	38	14	2	-	-	118
2010	Cairo	Egypt	70	28	20	1	-	-	119
2011	Hunan Changsha	China	88	40	10	1	-	-	139
2013	João Pessoa	Brazil	84	33	16	5	-	-	138
2014	Pirassununga	Brazil	78	39	12	6	-	1	136
2015	Winnipeg	Canada	83	37	8	4	-	-	132
Total papers			861	399	110	24	4	1	**1399**
Percentage (%)			61.5	28.5	7.9	1.7	0.3	0.1	100.0

FIGURE 3 - *NUMBER OF PAPERS PER PARTICIPATING INSTITUTIONS*

Table 3 and Figure 4 show the number of Institutions which participated in all NOCMAT Conference series.

Non-Conventional Materials and Technologies – NOCMAT for XXI Century   Materials Research Forum LLC
Materials Research Proceedings 7 (2018) 214-233   doi: http://dx.doi.org/10.21741/9781945291838-20

TABLE 3 - *NUMBER OF INSTITUTIONS PER YEAR AND HIGHER NUMBER OF PAPERS PRESENTED BY AN INSTITUTION*

Year	Institutions	Institution with highest n° of papers		n° Papers	Country
1984	67	LEHIGH	Lehigh Univ	4	USA
1997	21	PUC-RIO	Pontifícia Universidade Católica do Rio de Janeiro	5	Brazil
2002	53	HANYANG	Hanyang University	6	Korea
2003	48	PUC-RIO	Pontifícia Universidade Católica do Rio de Janeiro	9	Brazil
2004	38	USP	Universidade de Sao Paulo	8	Brazil
2005	58	CEFET-PR	Centro Federal de Educação Tecnológica do Paraná	8	Brazil
2006	53	UFPB	Universidade Federal da Paraíba	14	Brazil
2007	55	UFAL	UFAUniv Federal de Alagoas	8	Brazil
2008	82	UNIVALLE	Universidad del Valle	16	Colombia
2009	122	BATH	University of Bath	19	UK
2010	68	HBRC	Housing and Building National Research Center	16	Egypt
2011	88	HNU	Hunan University	17	China
2013	108	UFPB	Universidade Federal da Paraíba	14	Brazil
2014	81	USP	Universidade de Sao Paulo	22	Brazil
2015	81	PUC-RIO	Pontifícia Universidade Católica do Rio de Janeiro	22	Brazil

Table 3 shows that the number of institutions participating in NOCMAT Conference series increased, in a very significant growth, from 21, its lowest level (in 1997/India) to 122, its highest level (in 2009/UK).

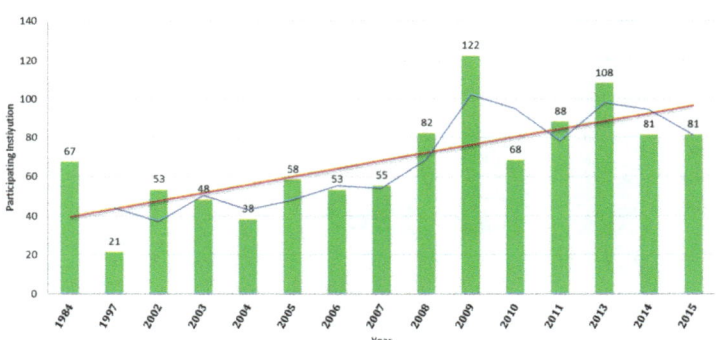

FIGURE 4 - *NUMBER OF INSTITUTIONS PER YEAR*

Figure 4 shows an increasing number of institutions participating to the NOCMAT Conferences from 1984 to 2015. Table 3 shows that the highest number of papers by institutions is increasing from 1984 (4 papers from LEHIGH) to 2015 (22 papers from PUC-RIO).

The data collected reveals that more than 600 Institutions have participated in all NOCMAT Conference series. In average, 69 Institutions have participated in all of them.

TABLE 4 - NUMBER OF INSTITUTIONS PARTICIPATING IN ALL NOCMAT CONFERENCE SERIES BY EACH COUNTRY

Item	Countries	N^o of Countries	N^o of Institutions	%
1	Brazil	1	142	23.4%
2	UK	1	53	8.7%
3	China	1	46	7.6%
4	USA	1	43	7.1%
5	Colombia	1	34	5.6%
6	India	1	26	4.3%
7	Mexico	1	19	3.1%
8	Argentina	1	18	3.0%
9	Egypt	1	16	2.6%
10	Canada	1	15	2.5%
11	France	1	15	2.5%
12	Japan	1	15	2.5%
13	Italy, Spain, Algeria	3	35	5.8%
14	Australia, Iran, Germany, Portugal	4	34	5.6%
15	Korea, Romania, Vietnam, Cuba, New Zealand, Nigeria, Peru	7	33	5.4%
16	Poland, South Africa, Thailand, Ghana, Libya, Malaysia, Venezuela, Belgium, Chile, Ecuador, Greece, Nepal	12	35	5.8%
17	Pakistan, Sweden, The Netherlands, Turkey, Bolivia, Botswana, Cypre, Denmark, Ethiopia, Guatemala, Jordan, Kuwait, Mozambique, Paraguay, Philippines, Qatar, Russia, Senegal, Singapore, Switzerland, Taiwan, Tanzania, UM	22	27	4.5%
Total		60	606	100%

Table 4 shows that among all institutions that participated in all the NOCMAT Conference series, Brazil is in first position with 142 participating institutions. One of its important impacts is the increasing number of call for R&D projects sponsored by the Brazilian Ministry of Science, Technology and Innovation [19].

Mapping papers within NOCMAT Conferences by participating Countries and Year

Table 5, Table 6, Figures 5 and 6 present the distribution of all the papers by participating country and year of NOCMAT Conference series.

The data from Table 5 show that 145 papers are published by authors from 2 countries and 166 papers among 1399 are published by more than one country in all NOCMAT Conference series.

60^2 countries have participated in all NOCMAT Conference series.

Figure 5 shows significant growth of the number of papers engaging authors of two countries or more. For NOCMAT Conferences held in Brazil and India, respectively in 1984 and 1997, just 3 papers were presented by authors from 2 countries. In 2015, the number increased to 15.

Researchers of different countries are being involved in common projects and engaging resources from multiple sponsor agencies, increasing engagements at the international level.

Table 6 (which indicates more specifically the location of each NOCMAT Conference) and Figure 6 show that NOCMAT Conference 2009 held in United Kingdom (University of Bath) integrated 30 countries, the highest registered in all NOCMAT Conference series. In average, 18 countries have participated in all NOCMAT Conference series.

TABLE 5 - NUMBER OF PAPERS PER PARTICIPATING COUNTRIES

Year	City	Location	Participating Countries/ Number of Papers				Total papers	%
			1	2	3	2 or 3		
1984	Rio de Janeiro	Brazil	61	3	-	3	64	4.6%
1997	Bhubaneswar	India	21	3	-	3	24	1.7%
2002	Hanoi	Vietnam	54	7	2	9	63	4.5%
2003	João Pessoa	Brazil	52	6	1	7	59	4.2%
2004	Pirassununga	Brazil	52	5	-	5	57	4.1%
2005	Rio de Janeiro	Brazil	66	6	1	7	73	5.2%
2006	Salvador	Brazil	81	5	-	5	86	6.1%
2007	Maceió	Brazil	64	12	2	14	78	5.6%
2008	Cali	Colombia	96	12	5	17	113	8.1%
2009	Bath	UK	101	15	2	17	118	8.4%
2010	Cairo	Egipt	104	14	1	15	119	8.5%
2011	Hunan Changsha	China	126	12	1	13	139	9.9%
2013	João Pessoa	Brazil	123	14	1	15	138	9.9%
2014	Pirassununga	Brazil	118	16	2	18	136	9.7%
2015	Winnipeg	Canada	114	15	3	18	132	9.4%
Total papers			1233	145	21	166	1399	100.0%

---

[2] (1) Algeria, (2) Argentina, (3) Australia, (4) Belgium, (5) Bolivia, (6) Botswana, (7) Brazil, (8) Canada, (9) Chile, (10) China, (11) Colombia, (12) Cuba, (13) Cypress, (14) Denmark, (15) Ecuador, (16) Egypt, (17) Ethiopia, (18) France, (19) Germany, (20) Ghana, (21) Greece, (22) Guatemala, (23) India, (24) Iran, (25) Italy, (26) Japan, (27) Jordan, (28) Korea, (29) Kuwait, (30) Libya, (31) Malaysia, (32) Mexico, (33) Mozambique, (34) Nepal, (35) New Zealand, (36) Nigeria, (37) Pakistan, (38) Paraguay, (39) Peru, (40) Philippines, (41) Poland, (42) Portugal, (43) Qatar, (44) Romania, (45) Russia, (46) Senegal, (47) Singapore, (48) South Africa, (49) Spain, (50) Sweden, (51) Switzerland, (52) Taiwan, (53) Tanzania, (54) Thailand, (55) The Netherlands, (56) Turkey, (57) UK, (58) USA, (59) Venezuela, (60) Vietnam.

Non-Conventional Materials and Technologies – NOCMAT for XXI Century　　Materials Research Forum LLC
Materials Research Proceedings 7 (2018) 214-233　　　　doi: http://dx.doi.org/10.21741/9781945291838-20

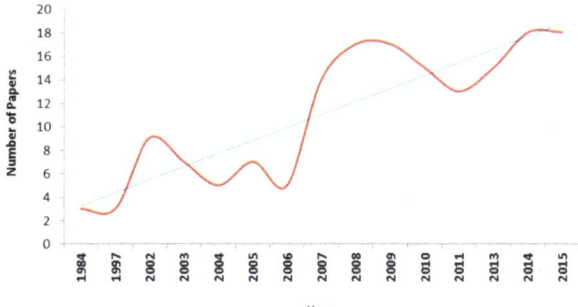

FIGURE 5 - *NUMBER OF PAPERS FOR TWO PARTICIPATING COUNTRIES PER YEAR*

TABLE 6 - *NUMBER OF COUNTRIES PER YEAR*

Year	City	Location	Number of Countries
1984	Rio de Janeiro	Brazil	17
1997	Bhubaneswar	India	9
2002	Hanoi	Vietnam	21
2003	João Pessoa	Brazil	17
2004	Pirassununga	Brazil	11
2005	Rio de Janeiro	Brazil	17
2006	Salvador	Brazil	12
2007	Maceió	Brazil	16
2008	Cali	Colombia	17
2009	Bath	UK	30
2010	Cairo	Egypt	15
2011	Hunan Changsha	China	16
2013	João Pessoa	Brazil	27
2014	Pirassununga	Brazil	19
2015	Winnipeg	Canada	21

Non-Conventional Materials and Technologies – NOCMAT for XXI Century   Materials Research Forum LLC
Materials Research Proceedings 7 (2018) 214-233   doi: http://dx.doi.org/10.21741/9781945291838-20

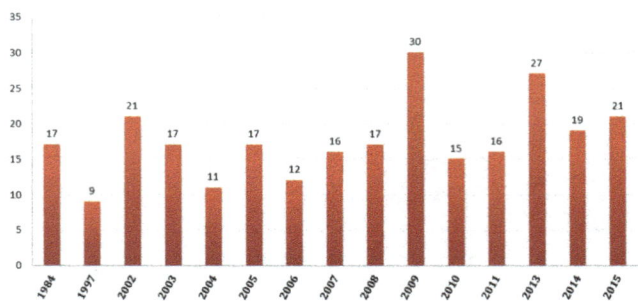

FIGURE 6 - NUMBER OF COUNTRIES PER YEAR

Mapping papers within NOCMAT Conferences by Themes

Tables 7, 8, 9 and 10, Figures 7, 8 and 9 present the distribution of all the papers included in different Themes.

Table 7 shows the number of papers presented by each NOCMAT Conference and by different themes. It can be observed in this Table that NOCMAT Conferences of 1997, 2003, 2004, 2005 and 2006 were not included. The reason is that the proceedings of these one didn´t associated papers to themes.

Table 8 shows the number of Themes increased from 1 (in 1984/Rio de Janeiro-Brazil) to 14 (in 2014/Cairo-Egypt) and average is 7 Themes for all NOCMAT Conference series. It is observed that the numbers of knowledge areas are increasing and being included in new Themes along the NOCMATs.

Regarding the number of papers presented in each Theme, Tables 7 and 9 show those being most focused by the authors in each NOCMATs. For instance "Inorganic Matrix Composites" with 44 papers is the most focused Theme in NOCMAT Conference 2008/Cali-Colombia.

TABLE 7 - THEMES PER NOCMAT CONFERENCES SERIE

Themes List	N° of Papers
1984	64
- Low Cost and Energy Saving Construction Materials	64
1997	24
- Not reviewed	24
2002	63
- Bamboo, Soil and Waste Materials	30
- Concrete and Composite Materials	33
2003	59
- Not reviewed	59
2004	57
- Not reviewed	57

2005	73
- Not reviewed	73
2006	86
- Not reviewed	86
2007	78
- Bamboo as construction material	10
- Composites materials	21
- Durability aspects of non-conventional materials	3
- Environmentally friendly, energy efficient construction	3
- Non-conventional materials and technologies management	2
- Other topics of interest	3
- Technology applied for low costing houses	7
- Vegetable fibers and soil construction	7
- Waste materials in building construction	22
2008	113
- Bamboo, Guadua and Others	14
- Inorganic Matrix Composites	44
- Organic Matrix Composites	15
- Residues and Subproducts Valorization	8
- Sustainable Technologies	16
- Techniques and Characterization	16
2009	118
- Bamboo	8
- Case studies	18
- Cement, lime and concrete materials	20
- Earth building	21
- Innovative masonry	11
- Natural fibers and fiber composites	11
- Recycled materials	6
- Straw bale	7
- Timber	16
2010	119
- Advances in research methodologies and materials testing	15
- Applications in improving building safety through materials and constructional development	7
- Construction materials and technologies to deliver affordable housing	13
- Construction materials and technologies to reduce climate change	2

Non-Conventional Materials and Technologies – NOCMAT for XXI Century   Materials Research Forum LLC
Materials Research Proceedings 7 (2018) 214-233   doi: http://dx.doi.org/10.21741/9781945291838-20

- Durability and performance of construction materials	12
- Environmentally friendly, energy efficient construction	16
- Infrastructures systems and materials	6
- Innovations in development of low carbon materials and technologies	8
- Life cycle assessment of materials	3
- Low cost housing concepts, prototypes and applications	5
- Miscellaneous	1
- Recycling of industrial, agricultural and urban waste stream materials	21
- Seismic engineering resistance, flood mitigation and disaster prevention	5
- Standards and guidelines	5
2011	139
- Bamboo materials	17
- Bamboo structures	19
- Case study of non-conventional buildings	7
- Cements and concretes	16
- Low carbon, energy saving design and practice	39
- Natural fibers and materials	8
- Non conventional mortars, renders and earth building	4
- Recycled materials	21
- Timber and masonry	8
2013	138
- Advances in research methodologies and material testing	8
- Bamboo and wood	22
- Cement, mortar and concrete	31
- Construction materials and technologies to deliver affordable housing and low energy	15
- Earthen Materials and Constructions	19
- Life-cycle assessment of materials and durability	3
- Natural fibers and materials	20
- Recycling of industrial, agricultural, and urban waste stream materials	13
- Standards, guidelines and policy issues	7

2014	136
- Advances in research methodologies and material testing	15
- Applications in improving welfare and comfort	8
- Construction materials and technologies for sustainability, energy, efficiency, affordable housing	16
- Durability, life cycles and performance of building materials	9
- Earth architecture and construction	11
- Natural fibers and plant-based materials (bamboo, straw, plant fiber)	22
- Non conventional and innovating research and technology	16
- Non-conventional structures	6
- Recycling of industrial materials, agricultural and urban waste	33
2015	132
- Academic reviews, Regulations & Feasibility Studies	6
- Bamboo	27
- Concrete / Earth	1
- Concrete / Earth	6
- Concrete / Mortar	26
- Concrete Aggregates	9
- Earth / Soil	14
- Fiber characterization, Panelized Straw Bale construction	7
- Material properties and behaviours	14
- Microscopic binder studies + Mycelium insulation	14
- Roofs, Rainwater, Cooling	8
Total number of papers	**1399**

Non-Conventional Materials and Technologies – NOCMAT for XXI Century   Materials Research Forum LLC
Materials Research Proceedings **7** (2018) 214-233   doi: http://dx.doi.org/10.21741/9781945291838-20

TABLE 8 - NUMBER OF THEMES PER YEAR

Year	1984	2002	2007	2008	2009	2010	2011	2013	2014	2015
N° of themes	1	2	9	6	9	14	9	9	9	11

TABLE 9 - THEMES WITH MOST FOCUSED IN EACH NOCMAT CONFERENCE

Year	Theme
1984	*Low Cost and Energy Saving Construction Materials*
2002	*Concrete and Composite Materials*
2007	*Waste materials in building construction*
2008	*Inorganic Matrix Composites*
2009	*Earth building*
2010	*Recycling of industrial, agricultural and urban waste stream materials*
2011	*Low carbon, energy saving design and practice*
2013	*Cement, mortar and concrete*
2014	*Recycling of industrial materials, agricultural and urban waste*
2015	*Bamboo*

TABLE 10 - NUMBER OF PAPERS PER MAIN AND SUB-GROUPS OF THEMES

Main and Sub-groups	N° of Papers
Bamboo and Natural fibers	**196**
Bamboo	127
Natural fibers	62
Vegetable fibers	7
Cement, Composite and Concrete Materials	**201**
Cement, Composite and Concrete Materials	201
Construction Material	**347**
Building	30
Climate	2
Earth	55
Housing	162
Masonry	15
Straw bale	9
Technologies	24
Timber	19
Durability	31
Energy	**22**
Energy	22
General	**57**
Case Study	21
Miscellaneous	6
Residues	9
Seismic	5
Technologies	16

Material and Testing	87
Infrastructures	8
Life cycle	6
Low carbon material	11
Testing	62
n-conventional materials	60
Building	9
Case Study	7
Housing	1
Technology	20
Structures	13
Durability	6
Management	4
Recycling	109
Recycling	109
Safety	7
Safety	7
Standards	14
Standards	14
Others	299
Others	299
Total number of papers	**1399**

Table 10 and Figures 7, 8 and 9 were created by the authors of this paper. Data collected were categorized in Main and Sub-groups of Themes.

Table 10 and Figure 7 show a strong presence of papers related to the Sub-themes *"Construction Materials"* with 347 papers, *"Cement, Composite and Concrete Materials"* with 201 papers, and *"Bamboo and Natural fibers"* with 196 papers among a total of 1100 papers.

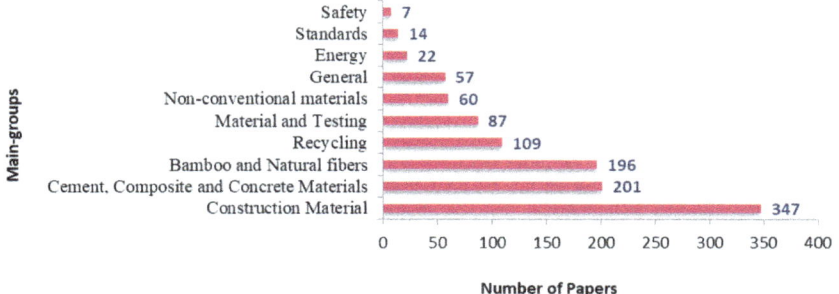

FIGURE 7 - *NUMBER OF PAPERS PER MAIN GROUP OF THEMES*

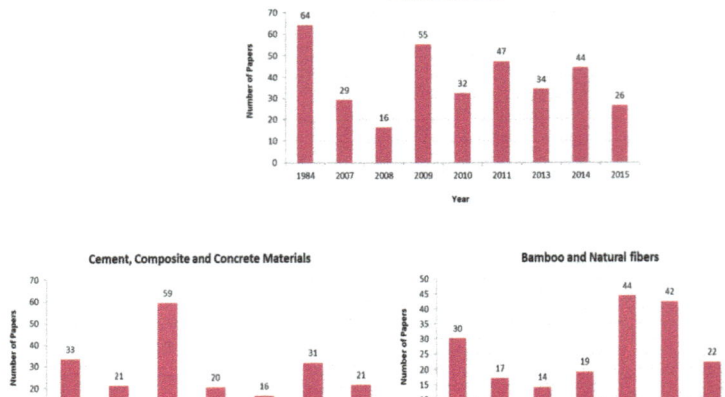

FIGURE 8 - *NUMBER OF PAPERS FOR 3 MAIN GROUPS PER NOCMAT CONFERENCE SERIES*

Figure 8 presents the concentration of papers in the 3 main groups of Themes by NOCMAT Conference series. For *"Construction Material"* Theme, the highest number of papers has been presented in 1984/Brazil (64 papers). For *"Cement, Composite and Concrete Materials"* Theme, the highest number of papers has been presented in 2008/Colombia (59 papers). For *"Bamboo and Natural Fibers"* Theme, the highest number of papers has been presented in 2011/China (44 papers).

Figure 9 presents all the papers in each Sub-groups of Themes. Among 27 sub-groups of Themes, the five more expressive cover 662 papers (ie, 60% of a total papers). Sub-group *"Climate"* presents one of the lowest numbers of papers.

Final remarks

This study mapped 1399 papers along 31 years and 15 NOCMAT Conferences showing regularity with increasing production in a wide range of NOCMAT sub-groups of Themes. This varied interest is of great importance, because since the 1st NOCMAT Conference in 1984, they become a core subject in scientific and technological policies aiming at a more environmentally friendly and sustainable construction, and also allowing a new "NOCMAT culture" within an increasing number of countries and public and private institutions notably of education and research. Universities are creating multidisciplinary courses integrating Engineering with Management, Economy, Biology, Agriculture Sciences among others traditional knowledge areas. R&D projects integrate more and more researchers specialized in these areas.

These transformations sponsored especially by research funding agencies with increasing grants and scholarships for NOCMAT Themes in civil engineering, led to a greater appreciation of studies and research results presented in different editions of these NOCMAT Conferences. Thus, guidelines

Non-Conventional Materials and Technologies – NOCMAT for XXI Century   Materials Research Forum LLC
Materials Research Proceedings 7 (2018) 214-233   doi: http://dx.doi.org/10.21741/9781945291838-20

for studies that integrate different areas of knowledge, such as nanotechnology for civil engineering, with significant implications for academic backgrounds to be created, gain increased importance.

In turn, thematic application areas identified in this study, negligible as compared to others, permit to identify thematic gaps that require more attention from now on, notably a greater and deep sense of accountability.

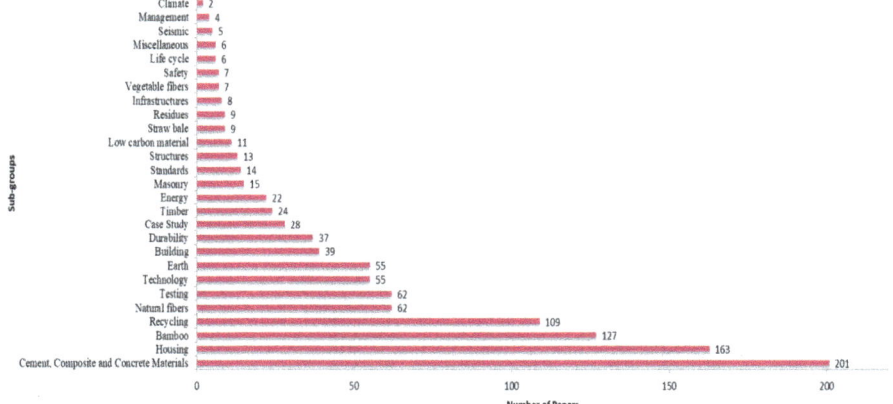

FIGURE 9 - *NUMBER OF PAPERS PER SUB-GROUPS OF THEMES*

For the future NOCMAT Conferences some improvements are suggested. Among existing data structure, it is recommended a database and integrated information system mainly on authors, academic level and area, countries and institutions, and themes. The database related to each submitted paper should consider one or more different defined themes. It also should be considered a defined application area regarding the papers content by the Organizer Committee for each submitted paper into the database.

Suggestions for future studies in the field of studies on *State of Art* should map and analyze authors' network regarding the institutions and countries, and also, themes centrality and closeness. This mapping should be continued to recognize the new themes and areas trends, their increasing or stagnation.

A Science and Technology Observatory for "NOCMAT" not yet existing should be created. This one should have the function of "observing", "tracking", "anticipating" and "monitoring" the development of knowledge areas with a "keen" eye on critical issues, notably those presented in NOCMAT Conferences [20]. As a central issue for a transition to create the NOCMAT Observatory, institutions concerned with these materials and technologies should be initially involved. Their role and success for the creation of a pertinent NOCMAT Observatory depends fundamentally on the interest and participation of its members, individuals and enterprises, and more specifically, in an initial step, the IC-NOCMAT (International Committee). This paper shows that 606 institutions engaged, 60 countries involved in NOCMAT Conference series along 31 years of experience justify such an integrative initiative.

Finally, for this engagement new fundamental values for NOCMAT Conferences serie should be considered, those presented by ABMTENC: *"The rapidly changing world is facing many threats and serious challenges. The glory of modern industrialization has been in doubt with ample facts of environmental deterioration problems. Yet, Development and survival are still the central issues facing a great portion of mankind. The world needs new role models of development and industrialization that are characterized by environmentally friendliness and sustainability. Construction industry is in the forefront facing today's challenges. Providing durable, efficient and effective infrastructure system, affordable housing, clean water, energy, communication and sanitation is fundamental to economic prosperity, social justice, political stability and quality of Life"*.

Acknowledgements
The authors would like to thank ABMTENC Association which made available all the NOCMAT Conferences data and the 17th International Conference on Non-Conventional Materials and Technologies – Construction Materials & Technologies for Sustainability (17th NOCMAT 2017), November 26-29, 2017, Mérida, México organizers for their valuable effort and proceedings for the discussions, which form the basis of the meeting. Also, financial support given by FAPERJ and FACC/UFRJ are appreciated.

References

[1] UNITED NATIONS. *Our Common Future*. Report of the World Commission on Environment and Development. Oslo, March 20, 1987. Available at: <http://www.un-documents.net/our-common-future.pdf>. Access in: April 10, 2017.

[2] PAPON, P. *FUTUR*: les avancées technologiques 2025, 2050, 2100. Paris: Le Monde, 2013. (Special Edition, February-April, 2013).

[3] ASSOCIATION OF NON-CONVENTIONAL MATERIALS AND TECHNOLOGIES. Available at: <http://www.abmtenc.civ.puc-rio.br/17nocmat>. Access in: April 10, 2017.

[4] INTERNATIONAL NETWORK FOR BAMBOO AND RATTAN. Available at: <http://www.inbar.int/#1>. Access in: April 10, 2017.

[5] WIKIPEDIA.ORG State of the art. Available at:<https://en.wikipedia.org/wiki/State_of_the_art>. Access in: May 15, 2017.

[6] ARCHAMBAULT, E.; CARUSO, J.; NICOL, A (2014) *State-of-art analysis of OA strategies to peer-review publications*. Science Metrix, October 22, 2014. (Document: RTD-B6-PP-2011-2: Study to develop a set of indicators to measure open access Deliverable D.2.1, 2014 Update, Version 5b. Available at: <http://science-metrix.com/sites/default/files/science-metrix/publications/d_2.1_sm_ec_dg-rtd_oa_policies_in_the_era_update_v05p_0.pdf>. Access in: May 15, 2017.

[7] Côté, G.; Roberge, G.; Archambault, E. Bibliometrics and Patent Indicators for the Science and Engineering Indicators 2016 — Comparison of 2016 Bibliometric Indicators to 2014 Indicators. Science Metrix, January 14, 2016. Available at: <http://science-metrix.com/sites/default/files/science-metrix/publications/science-metrix_comparison_of_2016_bibliometric_indicators_to_2014_indicators.pdf>. Access in: May 15, 2017.

[8] Weber, K. M.; Hemmelskamp, J. *Towards Environmental Innovation Systems*. Heidelberg: Springer, 2005. https://doi.org/10.1007/b138889

[9] Kibert, Ch. J. *Sustainable Construction* – Green Building Design and Delivery. 4th Ed. Hoboken: Wiley, 2016.

[10] Moraes, L. A. F. de; Ohayon, P.; Ghavami, K. Application of Non-Conventional Materials: Evaluation Criteria for Environmental Conservation in Brazil. *Key Engineering Materials*, Trans Tech Publications, Switzerland, v. 517, p. 20-26, 2012.

[11] Harries, K. A.; Sharma, B. *Nonconventional and Vernacular Construction Materials* – Caracterisation, Properties and Applications. Cambridge: Woodhead Publishing/Elsevier, 2016.

[12] Huang, A. L.; Chapman, R. E.; Butry, D. *Metrics and Tools for Measuring Construction Productivity*: Technical and Emprirical Considerations. Gaithersburg/ Maryland: NIST – National Institute for Standards and Technology, September 2009.

[13] *A Policy Study on the Sustainable Use of Construction Materials*. OECD: Paris, September 11, 2015. (Working Party on Resource Productivity and Waste, ENV/EPOC/WPRPW(2014)4/FINAL).

[14] Canchumani, R. M. L. *Scientific Domains in UFRJ*: mapping of fields of knowledge. Thesis (PhD in Information Science) – Federal University of Rio de Janeiro, School of Communication, Brazilian Institute for Information in Science and technology. Graduate Program in Information Science, Rio de Janeiro, 2015, 185p.

[15] Pritchard, A. A Statistical Bibliography or Bibliometrics? *Journal of Documentation*, Bingley/UK, v. 25, n. 4, p. 348-349, January 1969.

[16] Otlet, P. M. G. *Traité de Documentation* – le livre sur le livre - théorie et pratique. Bruxelles: Editiones Mundoneum – Palais Mondial, 1934.

[17] Vasudevan, R. K.; Ziatdinov, M.; Chen, C.; Kalinin, S. V. Analysis citation networks as a new tool for scientific research. *Materials Research Society*, Warrendale/PA, v. 41, p. 1009-1015, December 2016.

[18] Eck, N. J. van; Waltman, L. Software survey: VOSviewer, a computer program for bibliometric mapping. *Scientometrics*, Budapest, v. 84, n. 2, p. 523-538, August 2010. https://doi.org/10.1007/s11192-009-0146-3

[19] Ohayon, P.; Ghavami, K.; Jesuz, K. *Specifications of Building Environmental Evaluation Methods for within Redebambu Network in Brazil*. In: NOCMAT 2015 - 16th Nonconventional Building Materials and Technologies International Conference 2015 – Construction for Sustainability - Green Materials & Technologies. August 10-13, 2015, Winniped/Canada.

[20] Ohayon, Pierre; Barreiros, D. S.; Ghavami, Khosrow. *Science and Technology Observatory for "NOCMAT" in Brazil: Role and Proposed Framework*. Key Engineering Materials, v. 600, p. 399-412, 2014. Trans Tech Publications, Switzerland, doi: 10.4028/www.scientific.net/KEM.600.399. https://doi.org/10.4028/www.scientific.net/KEM.600.399

Non-Conventional Materials and Technologies – NOCMAT for XXI Century   Materials Research Forum LLC
Materials Research Proceedings 7 (2018) 234-246   doi: http://dx.doi.org/10.21741/9781945291838-21

Rural School in Santiago Del Estero- Argentina, Bioclimatic Conditioning and Evaluation of Thermal, Energy and Economic Efficiency

Gabriela Giuliano [a], Beatriz Garzón [b*]

[a] Architect and Professor in Secondary Education Professional Technical Modality, Doctoral Student in Doctorate in Sciences in the Area of Renewable Energies, Doctoral Fellowship of CONICET- MinCyT, ITA, CESPER, FCEyT- UNSE, Member of IAP Projects, SCAIT, UNT;. Address: Belgrano Avenue (south) N° 1912, Santiago del Estero, Argentina, PC: 4200, PN: +54-0385-4509560, gm.giuliano@gmail.com

[b*] Architect and Doctor in Sciences in the Area of Renewable Energies; Specialist in Management and Technological Linkage; Independent Researcher of CONICET; Director of of IAP Projects, SCAIT, UNT; Professor - Researcher Level II of MinCyT-MEN; Adjunct Professor of the Chair of Environmental Conditioning II, IAA, FAU, UNT, Member ABMTENC. Address: Néstor Kirchener Avenue N° 1900, San Miguel de Tucumán, Argentina, PC: 4000, PN: +54 0381-4364093, bgarzon@gmail.com

Abstract. The present work has as object of study a rural state school located in the town of Vilmer in Santiago del Estero, province in the northwest of Argentina, with the purpose of contributing to improve the user`s comfort conditions of this educational habitat. The article is part of a technical advisory agreement, with the relevant state institution General Direc torate of Architecture (GDA), responsible for the production of school buildings in the province. The results obtained were: 1) Definition of geographical and climatic conditions, 2) Strategies determination and bioenvironmental patterns selection, 3) Architectonic-technological analysis of the prototype for the bioclimatic adjustment, 4) Energetic evaluation with labeling and its comparison with regulated values and proposals for improvement, 5) Evaluation of construction costs of the both prototypes. In conclusion, it can be seen that in energy efficiency labeling is possible to scale from the lower category -less efficient- "H" -in the case of the conventional state prototype, to an optimal level "C" in the improved school prototype. This level was possible, through the consideration of the following aspects: determining the best orientations for the sunshine, incorporating protections to the carpentries, modifying the components that define the envelope and reinforcing the thermal insulation. The proposals for the improvement of the original prototype are economically viable: their execution involves only a 6% of over cost, respect to the total construction price of a conventional school prototype.

Keywords: Rural School, Bioclimatic Adequacy, Thermal- Energetic Efficiency

Introduction

Several International research groups (Almeida et. al. [1]) have evaluated environmental quality: thermal, acoustic, and indoor air in school buildings; where they found that cognitive performance of students increases in conditions of comfort. In Argentina, we examined power consumption in school institutions in different geographical-climatic areas and the impact of technological improvements (San Juan et. al. [2]); design studies of sun protection of windows were made at schools in the province of Mendoza (Pattini, et. al.) [3]. In La Pampa, schools with environmentally conscious design (DAC) were projected and their energy performance was estimated by computer programs (Marcilese, et. al. [4]), bioclimatic schools were also implemented, with control of thermal and energetic behavior by monitoring in different seasons of the year (Filippín, et. al. [5]). In Santiago del

Non-Conventional Materials and Technologies – NOCMAT for XXI Century   Materials Research Forum LLC
Materials Research Proceedings 7 (2018) 234-246   doi: http://dx.doi.org/10.21741/9781945291838-21

Estero, so far there was no history of school buildings with bioclimatic and thermal-energetic efficiency. Therefore, this work aims to make contributions to this line, proposing improvements in conditions of thermal comfort and energy saving in schools in Santiago; the same is part of an agreement of mutual collaboration of work with the pertaining state institution "Direction General de Arquitectura" (DGA), responsible for the production of school buildings in the province.

Methodology

Geographical and climatic conditions of the area were considered, general strategies were determined and specific guidelines were selected, for the bioenvironmental area where the rural school under study is located, of IRAM standard N ° 11.603 [6] "Bioenvironmental classification of the Argentine Republic". The architectural-technological analysis of the state prototype was carried out and its setting were raised, through the application of bioclimatic adequacy guidelines. Calculations of "K" values of traditional and proposed closures were carried out, comparing them with the admissible from the IRAM standard N ° 11.605 [6] of "Maximum values of thermal transmittance". Values of thermal loads of cooling "QR"; and cooling volumetric coefficients "Gr" of the improved state prototypes, corroborating with values of admissible of IRAM "GR" N ° 11.659 [6] "saving of energy in refrigeration" standard. Values of loss of heat "G" with admissible for both prototypes were checked and thermal loads of annual heating "Q" was calculated, according to IRAM standard N ° 11.604 [6] of "energy saving in heating". The thermal and energetic evaluation was completed, labelling in both situations, according with the IRAM standard N ° 11.900 [6] for "Label energy efficiency of heating for buildings". Finally, the cost of m2 of construction was estimated, of both school buildings.

Objetive

Through the analysis of the architectural and technological characteristics of a prototype of rural school of state production in Santiago del Estero, it is considered to formulate proposals for improvement by generating a school prototype with bioclimatic recommendations and thermal - energetic adjusting in relation to standardised values and to determine the difference of construction costs for both prototypes.

Results

The project of the secondary school in Vilmer, located in the town that bears the same name, in the Department Robles, located towards the centre of the province, next to the capital of Santiago del Estero, distanced only by 12 km (Figure 1). The climate is characterized as "semi-arid, dry and steppe" (Köppen, [7]), with a very warm summer, high temperatures and rainfall between the months of October to March; in contrast, the winter season is dry, and low temperatures are recorded (see table 1 and 2). The prevailing winds are "North-East" orientated from October to February and with South direction, from June to September.

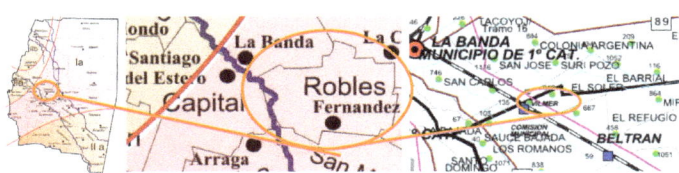

FIGURE 1– SANTIAGO DEL ESTERO, DEPARTAMENT OF ROBLES, LOCALITY OF VILMER

Non-Conventional Materials and Technologies – NOCMAT for XXI Century   Materials Research Forum LLC
Materials Research Proceedings 7 (2018) 234-246   doi: http://dx.doi.org/10.21741/9781945291838-21

TABLE 1 – *CLIMATIC DATA (NATIONAL METEOROLOGICAL SERVICE, (8).*

SUMMER DATA (DECEMBER)			WINTER DATA (JULY)		
(T. Mín. Med; HR Máx.)	(T Med; HR Med)	(T. Max. Med; HR. Mín.)	(T. Mín. Med, HR Máx.)	(T Med; HR Med)	(T. Max. Med.; HR. Mín.)
20.1 °C ; 74%	26.7 °C ; 64%	33.8 °C; 51%	5.6 °C ; 76%	12.4 °C ; 68%	20.5 °C ; 55%

TABLE 2 – *GEOGRAPHICAL DATA OF THE AREA (IRAM N° 11.603 STANDARD, [6])*

Bioenvironmental Area	Geographical Area	Latitude	Longitude	Altitude
Ia zone: very warm Thermal extent up to 14°C	Locality: Vilmer Departamento: Robles	27°47'15"S	64°09'11"O	182 m.s.n.m

This village belongs to the bioenvironmental zone "I": very warm and subarea "a": temperature range higher than 14 ° C, depending on the bioenvironmental classification of the Argentine Republic (Standard IRAM N ° 11.603, [6]).

Determination of bioenvironmental strategies and guidelines

Bioclimatic strategies were determined through "Olgyay and psychrometric" diagrams (Figures 2 and 3). From monthly climatic data of the locality under analysis: values of temperature and relative humidity, maximum, medium and minimal average values; segments on both charts have been traced. Of the methodologies used there were obtained and considered those with higher percentages in relation to their departure from the comfort zone (Garzón, [9]).

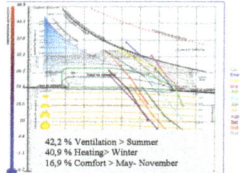

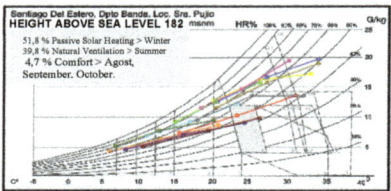

FIGURE 2 – *OLGYAY DIAGRAMAM*   **FIGURE 3** – *PSICROMETRIC DIAGRAMAM*

In the diagram of Olgyay (Figure 2) there appear the heating and ventilation strategies for the summer and winter periods, they have upcoming percentages: 42.2% and 40.9% respectively and the comfort zone is only 16.9%. This shows that most part of the year, it is outside the comfort zone, being necessary to adopt general strategies and specific design guidelines that allow to provide maximum internal comfort with minimum cost of conventional energy. After reading the psychometric chart (Figure 3) the obtained percentages of passive solar heating are 51,8%, natural ventilation 39,8%, comfort with 4,7% and mechanic ventilation of 3,6%, in which the most significant to be considered were the passive solar heating and natural cooling, for the summer and winter periods respectively. Hereunder the selected bioenvironmental patterns are described, according to each priority strategy:

-Passive solar heating strategies: it is applied in the winter period; in this way the design of the building should encourage the uptake, accumulation and distribution of heat. In this case, it is chosen as appropriate guidelines of "Direct gain through windows" and the "Protection of openings" to control heat loss.

-Natural cooling strategy: used in summer period and increases the feeling of comfort to influence people; likewise, enables structural cooling. The guideline selected in this case is "Cross-ventilation".

Architectural-technologic analysis and improvement of the prototype by bioclimatic setting
The school building project was introduced to the front of the field (Figure 4 and 5). A "linear" scheme with growth projection can be seen in "L". Its main axis corresponds to Northwest-Southeast (NO - SE) orientations and the secondary axis with the direction to the northeast - southwest (NE - SO).

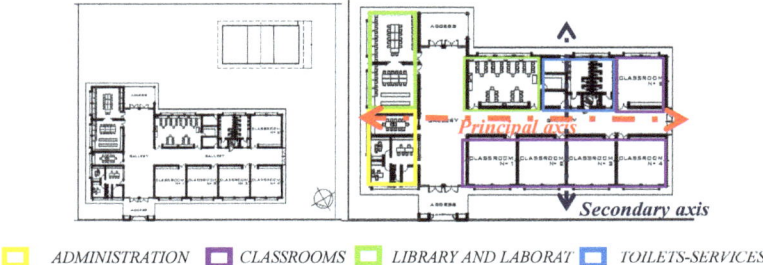

ADMINISTRATION   CLASSROOMS   LIBRARY AND LABORAT   TOILETS-SERVICES

FIGURE 4 – LOCATION IN THE FIELD PICTURE 5 – PLANT WITH FUNCTIONAL ZONATION

Recommendations according to the standard IRAM N ° 11.603
It indicates that for a very warm bioclimatic area named "Ia", the beneficial sunlight orientations are "NW - N - NE and SW -S - SE". Being the most disadvantaged of the year the summer season, where minimum exposure is suggested. They thermal orientations are shown below in Figure 6: optimum, regular and unfavorable orientations.

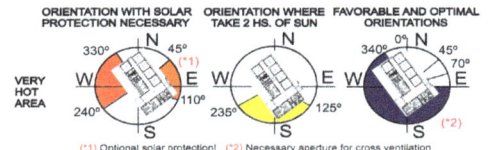

FIGURE 6 – RECOMMENDED ORIENTATIONS AREA I –VERY WARM- SUBAREA A (IRAM 11.603, [6])

The state prototype shows that there are more surface exposed to the NO-N-NE and SO-S-SE orientations which are considered "beneficial", and less surface exposed to the E -O orientations, that are considered "unfavourable". At the plant, the North is oriented at 45 °, thus 4 of 5 classrooms are oriented towards NE-N - NO, and only 1 of them is oriented toward the NO - SO.

Proposal for the protection from the sunstroke with vegetation
In the project of the improved prototype, it was raised to place trees not only on the sidewalk and the perimeter field by its esthetic and visual value and to provide privacy but above all for its bioclimatic value that already cast shadow on the area (Figures 7 and 8).

Non-Conventional Materials and Technologies – NOCMAT for XXI Century   Materials Research Forum LLC
Materials Research Proceedings **7** (2018) 234-246   doi: http://dx.doi.org/10.21741/9781945291838-21

FIGURE 7 – PLANT WITH VEGETATION

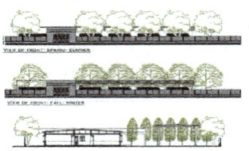

FIGURE 8 – VIEW AND CUT OF THE IMPROVED PROTOTYPE

The selection of the species is important, taking into account the form and characteristics of the tree. For this case, two species were suggested, advisable for the climate of the province, both of deciduous and rapid growth leaves: 1) Alamo "Palo jabón – no plateado": has a cup with shape of column that is suitable to safeguard the S-SE-SO orientations. 2) Fresno "No Americano": its cup has rounded form that is useful to preserve the NO-N-NE-E-O orientations (Figure 9). Lapacho, Paraíso, Tipa and Mora are other recommended tree species, but they should be placed away from any construction because its roots overgrow invasively.

FIGURE 9 – CUT A-A- VEGETAL PROTECTION FROM THE INCOMING SOLAR RADIATION

Verification of the fixed protection of the state prototype and the redesign of the improved prototype. The openings of the state prototype are aluminum with anti-vandalism mesh outside and sun shield fixed coffered ceiling H °A °, identical in all their facades. It does not have protections of lattices or roller shades (Figure 10).

Three orientations for the analysis of the spans were defined:
- ☐(SO)-laboratory of science and technology, health care, classroom 5.
- ☐(NE) classrooms 1, 2, 3 and 4.
- ☐(SE) management - staff room, storage, library, and computer lab.

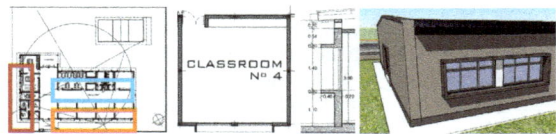

FIGURE 10 – PLANT OF OPENINGS AND CLASSROOM PLANT, CUT TYPE AND SUN SHADE

The woodwork protection of the state prototype is a coffered ceiling, consisting of fixed screens: two vertical and two horizontal 0.40 m wide and 0.10 m thick, which frame two windows per

Non-Conventional Materials and Technologies – NOCMAT for XXI Century   Materials Research Forum LLC
Materials Research Proceedings 7 (2018) 234-246   doi: http://dx.doi.org/10.21741/9781945291838-21

classroom and is applied in the same way in all fronts to protect different orientations. This was collated by means of the "diagram of Vision of sky vault"; when superimposing it on the "Solar path diagram" (matching with each considered orientation axis) periods of solar obstruction were generated, both in months and hours. According to the analysis carried out through the methodology described above and the applied study of sunlight through the "Sketch Up" program, that allows geolocation of the 3D model, and orientation of the school size correctly according to solar north; it was concluded that the sun shield did not verify for the orientations "NE-SE- SO" because the obstruction period required for the unfavorable months of spring and summer school hours (8.00 am to 18.00 pm) was minimal. Whereupon it was proposed the redesign the sun shade adapting it to its profile to the different needs of protection of the different orientations. In Figure 10, you can see the original sun shade in 3D identical for each facade and in figures 11, 12 and 13, the proposals of the different sun shades for its adaptation to the needs of each orientation.

The sun shades were redesigned to improve the protection of the sunstroke of the prototype, incorporating variants of vertical and horizontal intermediate screens by slightly modifying the original provision coinciding with the divisions of the carpentry, intending to optimize the result of protection. For Northeast orientation (NE) is suggested to place the sun shades of 0,60 m overhang at the top with two intermediate horizontal slats in a way to protect of the incident solar radiation more efficiently (Figure 11).

FIGURE 11 – FACADE NORTHEAST OF CLASSROOMS 1 TO 4 AND SUN SHADE CUT

In South-East orientation (SE) it is recommended to use the sun shade of the type formwork panel without intermediate lamas since the upper horizontal blade of 0,60 m cantilever is sufficient protection for the period of unfavorable sunstroke (Figure 12).

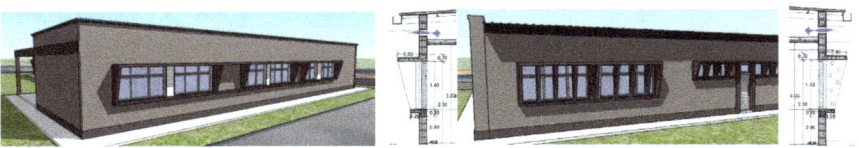

FIGURE 12 – 13 - FACADE SOUTHEAST AND SUN SHADE CUT- FACADE SOUTHWEST AND SUN SHADE CUT

In Southwest orientation (SO) it was proposed to place the sun shades with intermediate vertical slats within the concrete caisson, and to facilitate the ventilation of the hot air in summer was raised circular holes (Figure 13).

Propositions to improve ventilation

Non-Conventional Materials and Technologies – NOCMAT for XXI Century   Materials Research Forum LLC
Materials Research Proceedings 7 (2018) 234-246   doi: http://dx.doi.org/10.21741/9781945291838-21

The arrangement of the openings in classrooms of the state prototype coincides with the direction of the prevailing winds since four classrooms have openings that point between NE-N - NO, and in the fifth classroom windows are facing the S-SW. However the ventilation is not crusade (Figures 14 and 15) because this type of prototype does not allow bilateral ventilation. Therefore, in an improved prototype it is suggested to place openings in top of the cabinets and above the door of the classroom (which in general remains closed by acoustic issues and does not provide ventilation); both swing type which can be permanently open, and thus generate bilateral ventilation with permanent air exchange and efficient due to a "fireplace" effect (Figures 16 and 17).

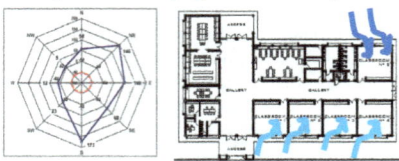

FIGURE 14 – 15 - DIRECTION WINDS (KM/H)– STATE PROTOTYPE: UNILATERAL VENTILATION OF CLASSROOMS

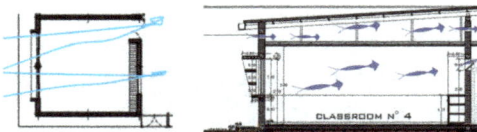

FIGURE 16 – 17 - IMPROVED PROTOTYPE: BILATERAL OR CROSSED VENTILATION – CHIMNEY EFFECT

Technological characteristics of the envelope
The state prototype was designed to function in traditional system of external vertical walls of solid ceramic thick of 0.18 m thick, load-bearing and non-bearing, with mixed structure: B) in the area of classrooms, health care, kitchen and laboratory of science and technology, the structure is load-bearing and chained seismic-resistant wall, whit a lightweight and metal cover roof structure, with suspended poly vinyl chloride (PVC) fire retardant ceiling. A) in the area of secretariat, administration and computer lab, raised in a timely manner, with columns, beams and slab of reinforced concrete with what is expected if it were necessary to enlarge, the upper floor (Figures 18 and 19).

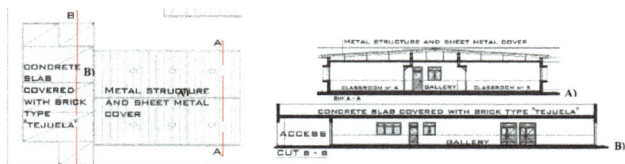

FIGURE 18 – PLANT OF THE ROOF OF THE SCHOOL   PICTURE 19 – CUTS A-A AND B-B OF THE AREAS

Energetic evaluation of the state prototype and the improved prototype labeled in

Non-Conventional Materials and Technologies – NOCMAT for XXI Century   Materials Research Forum LLC
Materials Research Proceedings **7** (2018) 234-246   doi: http://dx.doi.org/10.21741/9781945291838-21

comparison to standard values

Calculation of "K" thermal transmittance (W/m2. K)

They were calculated values of thermal transmittance of the constructive enclosures of walls and ceilings of the state prototype and its respective proposals for improvement (charts 3 and 4), for the bioenvironmental area Ia - "very warm" condition of summer (for being the worst), through the use of the "TRANS-Q-E" program (Garzón et al., [10]). The resulting values were compared, in relation to the maximum values of thermal transmittance of the walls and ceilings permissible for this season, setting three different levels of thermal comfort conditions recommended: A: Recommended, B: Medium and C: Minimum (Standard IRAM N° 11.605, [6]).

TABLE 3 – VALUES FROM THE WALLS IN THE SUMMER

Exterior enclosure	K_{cal}	$K_{máx.\ Adm}$	Conclusion
Exterior wall of hollow ceramic brick (LCH) thickness=0,18 m + indoor and outdoor plaster thickness=0, 02 m each, total thickness of 0,22m.	1,53	C:1.80; B: 1.10	It validates levels C and B
Exterior improved wall (LCH 0.12 m+ insulation 0,05 m + LCH 0,18 m).	0,44	C:1,80; B: 1,10; a:0,45	It validates levels C, B and A

TABLE 4 – VALUES OF DE K_{CAL} Y $K_{MÁX.\ ADM}$. $(W/M^2.K)$ FROM THE ROOF IN THE SUMMER (IRAM N ° 11.605, [6])

Exterior enclosure	K_{cal}	$K_{max\ adm}$	Conclusion
Roof 1: Reinforced concrete slab with subfloor of pearlite of expanded polystyrene (E.P.) thickness: 0,10 m.	0,86	C: 0,72	It does not validate the level C
Improved Roof 1: lightened slabs with subfloor and insulation of expanded polystyrene (E.P.) thickness: 0,10 m.	0,38	C: 0,72; B:0,45	It validates levels C and B
Roof 2: Metal rough ceiling and insulation of glass wool of 0,038 m of thickness with unvented attic.	0,68	C: 0,72	It validates level C
Improved roof 2: metal + aluminum film with rough ceiling and insulation of glass wool of 0,075 m of thickness with vented attic.	0,44	C: 0,72; B: 0,45	It validates levels C and B

In charts 3 and 4, types of walls and ceilings are explained, which are usually used in traditional construction and in the state production of schools. You can see that the calculated thermal transmittance values compared to the regulated, are not verified or only verified in level C and in the case of the tile ceiling, the minimum comfort moisture for the most unfavorable situation of summer it is not checked, in the climatic zone under study. Therefore, alternatives for optimization of the enclosures were proposed, where "K" values have been improved, by verifying in level B and even A, in the case of the enhanced alternative of the wall. It is suggested to use for the case in analysis of the secondary school of Vilmer, these last options where the thermal transmittance coefficients are lower, using double wall for the classrooms area, and in area of ceiling tile it is suggested especially to strengthen the thermal insulation with expanded polystyrene; finally, where the structure and cover are metal, it is convenient to place double insulation: in structure a film of aluminum, a suspended ceiling (forged), a good thickness of thermal insulation, and for air chamber (attic), providing grids for the renewal of the air since this allows to decrease the transmission of heat to the environment (Figure 20).

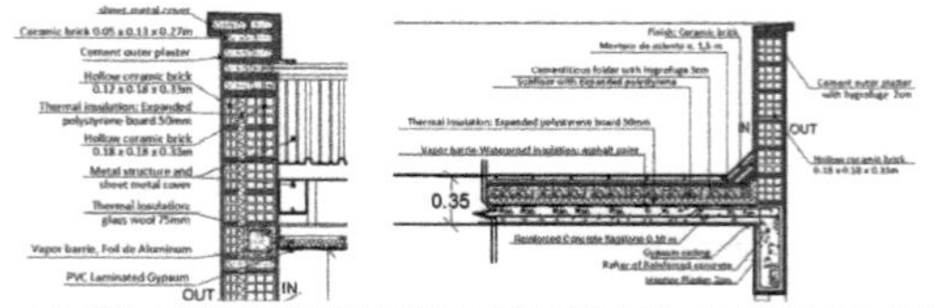

FIGURE 20 – WALL AND ROOF DETAILS OF THE IMPROVED PROTOTYPE

Calculation of volumetric coefficient of cooling GR and thermal load of refrigeration QR
Through the program "TRANS-Q-E" (Garzon et. al., [10]) the volumetric coefficient of cooling (GR) and the heat load of refrigeration (QR) were calculated by analyzing the two prototypes again in comparison. Where the improved prototype presented a slight improvement in the reduction of thermal load (W); however, in both cases it was evident the need for active systems of mechanical ventilation (table 5); by not complying with the value of GR admissible stipulated by the standard (IRAM N ° 11.659, [6]). This shows the summer climate harshness and the verification of the need for implementation of the strategy determined by psychometric chart: Mechanical ventilation 3,6% (Figure 4). To reduce the thermal load of cooling (QR), it was proposed: 1) to improve the level of thermal insulation of the envelope, in order to reduce profits by conduction, 2) to adjust sun protection to reduce profits by sunlight and 3) to control lighting system to reduce profits by sensible heat from the inside.

TABLE 5 – VOLUMETRIC COEFICIENT OF COOLING OF GR CALCULATION AND ADMISIBLE GRadm

State Prototype		Improved Prototype	
Cooling thermal load (W)	14.173,79	Cooling thermal load (W)	8585,22
Cooled volmun of the classroom (m^3)	155,52	Cooled volmun of the classroom (m^3)	155,52
Calculation coefficient $G_{R\ (W/m^3)}$	90,95	Calculation coefficient $G_{R\ (W/m^3)}$	55,20
Admissible coefficient $G_{R\ adm\ (W/m^3)}$	41,00	Admissible coefficient $G_{R\ adm\ (W/m^3)}$	41,00
$G_R > G_{Radm}$	Does not validate	$G_R > G_{Radm}$	Does not validate

Calculation of volumetric coefficient of heat loss "Gcal"
The standard (IRAM N° 11.604, [6]) of thermal insulation of buildings, allows to evaluate them in order to promote energy saving, depending on the volumetric coefficient of loss of heat (Gcal), which takes into account the losses through opaque enclosures, contact with the ground and local air renewals. It can be seen in the summary in chart 6, that the values obtained from the calculation of the state prototype do not verify, while the improved prototype got optimize coefficient considered in relation to the permissible value for the standard of reference, due to the adjustment proposed in the increase of the thermal resistance of the enclosure components: walls, ceilings and windows.

TABLE 6 – COMPARATIVE VALUES OF GCAL Y GADM IN BOTH PROTOTYPES

STATE PROTOTYPE		IMPROVED PROTOTYPE	
G= volumetric coefficient bof heat loss	W/ m³.k	G= volumetric coefficient bof heat loss	W/ m³.k
G $_{cal=}$	2,50	G $_{cal=}$	1,66
G $_{adm}$ (IRAM 11.604)=	2,09	G $_{adm}$ (IRAM 11.604)=	2,09
G $_{cal}$ > **G** $_{adm}$	It does not validate	**G** $_{cal}$ < **G** $_{adm}$	It validates

Labelling of energetic efficiency

By using the calculation of energy labelling program "EtiquEArq" (Garzón et al, [11]), the level of energy efficiency of the envelopes of the state prototype and the improved prototype was determined, capable of being heated according to the thermal transmittance of the envelope, in relation to parameters that are specified in standard (IRAM N ° 11.900, [6]). The same sets as design interior temperature of 20° C; and specifies eight kinds of energy efficiency, according to the variation of average temperature (ζm), between the inside surface of the envelope and the design interior temperature in Celsius degrees. The state prototype was classified in class "H", the lower efficiency; while the improved prototype with its efficiency could adjust until the third ladder of class "C" of the green group, the highest or optimum efficiency (Figures 21 and 22).

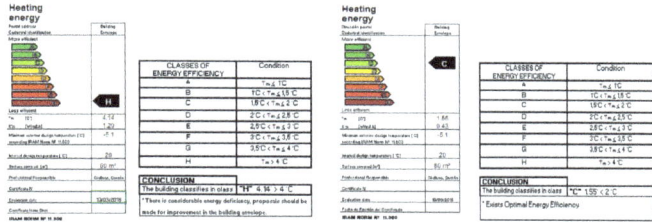

FIGURE 21 – STATE PROTOTYPE LABEL   FIGURE 22 – IMPROVED PROTOTYPE LABEL

Assessment of construction costs of both prototypes

A comparative table of construction costs ($/ m2) of the state prototype and the improved state prototype was made, corresponding to the month of March of the year 2016. The prices of budget template was ceded by the secretary of studies and projects of the general direction of architecture of Santiago del Estero, in the framework of the Convention of collaborative work DGA-CONICET. In chart 7, it can be observed from a total of 72 items, the analysis of 7 differential items with respect to the state prototype, corresponding to the improved prototype: 1) sun shades and H ° A° screens, 2) double wall in the classroom area, 3) and 4) ceiling slab and covered with insulation supplement 5) and 6) roof and ceiling insulation surcharge 7) carpentry to improve ventilation cross with chimney effect in classrooms. It was calculated and collated the cost of each item in relation to the percentage of divergence that represents the final cost of the amount of each work: the state prototype and the improved prototype; finding that it differs in the sum of $442,775.16; and that it represents only the 6% increase over the total of 100%. Whereupon, the construction cost difference is not significant and it will be amortized over the useful life of the building through energy savings and the gaining of benefits from improvements in the conditions of their future users hygrothermal comfort

TABLE 7 – VALUES OF CONSTRUCTION COSTS FOR THE PROJECTS OF THE STATE PROTOTYPE AND THE

N°	PROTOT YPE	DIFFERENTIATED OF THE STATE PROTOTYPE AND THE IMPROVED PROTOTYPE	QUANTITY (UNIT) M^2/M^3	PRICE ($ AR)	TOTAL PRICE ($ AR)	COST DIFFER ENCE (%)	(%) OF THE ÍTEM OF 100%
1	State	Without reinforced concrete sun shades and screens	0,00 m²	0,00	0,00		
1	Improved	With reinforced concrete (R° C°) sun shades and screens	30,62 m²	8035,04	246.057,03	100	3,14
2	State	Muro de ladrillo cerámico hueco de 0,18, sector Aulas	62,71 m²	469,70	29.454,89		0,40
2	Improved	Double wall (e=0,18 m y 0,12 m) + insulation of E.P. 0,05 m	62,71 m²	833,45	52.265,65	77	0,67
3	State	Slab roof of reinforced concrete (H° A°)	79,72 m³	8035,04	640.537,32		8,65
4	Improved	Covered with ceramic tiles and charcoal (C.)	400,27m²	302,54	121.097,69		1,64
3	State	Slab roof of reinforced concrete (R° C°)	79,72 m³	8035,04	640.537,32		8,17
4	Improved	Cover with insulation of EPS, subfloor de EPS, tile C.	400,27m²	510,63	204.389,87	69	2,61
5	State	Metal roof, insulation with polyethylene foam y aluminium film of 10 mm	612 m²	665,01	406.896,12		5,50
6	State	Rough ceiling of flame retardant PVC	528,55m²	397,77	210.240,54		2,68
5	Improved	Metal roof, insulation with polyethylene foam y aluminium film of 10 mm	612 m²	665,01	406.896,12	0	5,50
6	Improved	Rough ceiling of flame retardant PVC with glass wool 75mm.	528,55m²	478,40	252.857,36	20,27	3,22
7	State	Without woodwork with crossed ventilation- chinmey	0,00 m²	0,00	0,00		0
7	Improved	Without woodwork with crossed ventilation-chinmey	12,66 m²	3791,34	47.998,36	100	0,61
					7.401.513,34		100
		TOTAL WORK OF STATE PROTOTYPE					
					7.844.288,50	**5,98**	100
		TOTAL WORK IMPROVED PROTOTYPE					

Conclusion

Within the framework of the agreement of cooperative work DGA - CONICET, it was held an advisory report considering the described technical conditions of the project of state school prototype designed by the direction of architecture of Santiago del Estero; in that report, it became noted that it was evidenced the need to propose ways of new solutions that allow to optimize the thermal comfort for future users of the school establishment. When defining the bioclimatic strategies considered for the bioenvironmental zone "Ia - very warm", it was possible to select among the design guidelines that were considered as recommended for a production of the school habitat with bioclimatic adaptation to the context, that respond better to the thermal-energetic, ventilation, technological needs, among others. It was proposed to keep the architectural-morphological project, redesigning only in its enclosure the technological provisions, in way to allow the rational use of conventional energy. Differentiated protections were designed to each orientation of the openings and it was recommended to shade with vegetation of deciduous leaves carefully selected by their characteristics. At its technological resolution level, the ceiling tile from the state school building did not verify the minimum level "C" of the calculation of heat transmission "K" (W/m2. K), while the structure roof and metal cover and walls checked only at the minimum level for the standard IRAM 11.605-1996. With the setting in the choice and disposition of the constructive elements in the envelope of the improved prototype, reduction was achieved in the value of "K" calculation; that allowed to verify the "B - average" level for both ceilings and "A- recommended" for the suggested double wall; as indicated by the standard IRAM 11.605 reference values. With regard to the labelling of energy efficiency, in the case of the improved prototype the third level "C - optimal" was reached, optimizing five positions from the lower hierarchy "H-deficient", measured in the state school prototype; According to the values set by standard IRAM 11.900. The new enclosures that were

Non-Conventional Materials and Technologies – NOCMAT for XXI Century   Materials Research Forum LLC
Materials Research Proceedings 7 (2018) 234-246   doi: http://dx.doi.org/10.21741/9781945291838-21

studied up to level of construction details, were formed in order to reduce the losses and gains of heat in the different seasons of the year. All differential items for the proposals made in the improved prototype, with respect to the original budget of the project of the state prototype were analyzed in comparison and the divergent costs were calculated by incorporating it and/or modifying them. Resulting from the analysis, a negligible extra cost in the execution of the work of the improved prototype, which it will be fully amortized in energy savings in heating and the reduction of the use of active systems for cooling. This work allowed to devise a set of guidelines, which may be applied in the design of new school buildings as the analyzed case; or to delve into improvements for future reforms of school buildings in general in the province of Santiago del Estero. In this way, the contribution helps the purpose of optimizing the conditions of comfort in the livability and therefore the quality of life in school communities.

Recognition

We would like to thank to the General Direction of Architecture of Santiago del Estero, for allowing the agreement of collaborative work between DGA and CONICET; and also the members of the area of the Secretariat of studies and projects, to provide the mapping and the budget of the executive project of the secondary school of Vilmer, for the study and development of this work. This work, allowed to contribute to adjust the original state prototype so that it includes items of improvements in project, specification and budget for tender of the work and its future construction.

References

[1] Almeida R.; Freitas V.; Delgado J. School Buldings Rehabilitations. Indoor Environmental Quality and Enclosure Optimization. Springer, ISBN 978-3-319-15359-9. p 3, 2015.

[2] San Juan G.; Hoses S.; Gonzalez D.; Piñeyro J. Evaluación Energética e Incidencia de Mejoras Tecnológicas en Tipologías Escolares Bonaerenses. AVERMA. Volumen 4. ISSN 0329-5184, 2000.

[3] Pattini A.; Villalba A.: Córica L.; Ferrón L.; del Rosso R. (2009). Elementos de control de luz solar directa en fachadas vidriadas de edificios no residenciales de ciudad oasis. Rediseño para aulas. Avances en Energías Renovables y Medio Ambiente ISSN 0329-5184, Vol. 13, 2009.

[4] Marsilesi M., D.; Crowther, J.; Análisis simplificado de la eficiencia energética de una escuela rural en La pampa- Argentina- a lo largo de su ciclo de vida. Revista Hábitat Sustentable; Volumen 3, No. 1, pp. 3-14, 2013.

[5] Filippín C.; Bescochea A.; Gorozurreta, J. Comportamiento Higrotérmico y Energético de la Escuela Bioclimática de Catriló en la Provincia de La Pampa. AVERMA. Volumen 5. ISSN 0329-5184, 2001.

[6] Normas del Instituto Argentino de Normalización y Certificación: Norma IRAM N° 11603, 1996. Acondicionamiento térmico de edificios. Clasificación bioambiental de la República Argentina. Norma IRAM N°11601: Aislamiento térmico de edificios, métodos de cálculo, 2002. Norma IRAM N° 11605: Acondicionamiento térmico de edificios, condiciones de habitabilidad en edificios, 1996. Norma IRAM N° 11604: Verificación de sus condiciones higrotérmicas. Ahorro de energía en calefacción, 2001. Norma IRAM N° 11659: Ahorro de energía en refrigeración, 2007. Norma IRAM N° 11900: Etiqueta de eficiencia energética de calefacción para edificios, 2010.

Non-Conventional Materials and Technologies – NOCMAT for XXI Century   Materials Research Forum LLC
Materials Research Proceedings 7 (2018) 234-246   doi: http://dx.doi.org/10.21741/9781945291838-21

[7] Kôppen Das geographische system der klimate. Berlin Verlang von Grâvuder Borntraeger, seite C13, 1936.

[8] Servicio Meteorológico Nacional. Estadísticas Climatológicas Período 2005- 2015. Buenos Aires, Argentina, 2015.

[9] Garzón, B. Determinación de Estrategias Bioclimáticas para Localidades Rurales de Tucumán, Argentina. FAU-SeCyT, UNT – CONICET. 2006.

[10] Garzón, B.; Mendonca C. TRANS-Q-E. Programa calculador de transmitancia, cargas térmicas de calefacción y refrigeración y consumos energéticos (según Normas IRAM N° 11.601, 11.605, 11.604 y 11.659). 2012.

[11] Garzón, B.; Giuliano G. EtiquEArq. Programa de Cálculo del Etiquetado de Edificios. Determinación del Nivel de Eficiencia Energética de Calefacción (según Norma IRAM N° 11.900). 2015.

Non-Conventional Materials and Technologies – NOCMAT for XXI Century   Materials Research Forum LLC
Materials Research Proceedings 7 (2018) 247-256   doi: http://dx.doi.org/10.21741/9781945291838-22

Traditional and Alternative Masonry for the Acoustic Improvement of Architectural Spaces

Paterlini Leonardo[a], Garzón Beatriz[b]

[a] Architect, Doctoral Student in "Doctorate in Architecture" at the "FAPyD de la UNR", Doctoral Scolarship CONICET- MINCyT, FAU-SCAIT, member of IAP Projects. Address: Av. Kirchner 1900, San Miguel de Tucumán, Argentina +54-381-4364093. paterlinileonardo@gmail.com

[b] Architect and Doctor in Sciences in the Area of Renewable Energies. Specialist in Management and Technological Linkage. Independent Researcher of CONICET. Member of FAU-SCAIT, UNT project. Director of FAU-SCAIT, UNT projects. Professor-Investigator Level II of MINCyT-MEN. Adjunct Professor of the Chair of Environmental Conditioning II, IAA, FAU, UNT. ABMTENC Member. Kirchner Avenue No 1900, San Miguel de Tucumán, Argentina, PC 4000, PN +54 0381-4364093. bgarzon@gmail.com

Abstract. This paper aims to qualify and quantify the properties, in general, and acoustic isolation, in particular, of new construction systems. These systems use mix of traditional and alternative masonry materials in Tucumán, Argentina. Its goals are: a) to investigate, identify and catalog alternative masonry; b) to check their uses in mix construction systems with traditional masonry available in the local context, c) to evaluate these systems according to: its construction process, its energy cost, its economical price its physical, technological and sound insulation properties; d) to compare both of them. A "methodological strategy of "Participatory Action Research" was carried out to reach the objectives. The results achieved were: a) a good deal of alternative masonry were found, many of them are used as it raw material, urban and industry wastes; b) the identified masonry can be used in new constructive mix systems and they can achieve a good acoustic isolation; c) these new construction systems are able to use by the population. The conclusions obtained were: a) it is necessary to generate population and industry awareness; b) proposal masonry and systems need to get use in the construction market. c) In addition, of the acoustical benefits for users, these systems contribute to achieve sustainable and healthy habitat.

Keywords: Masonry, Acoustic Isolation, Evaluation

Introduction

Currently, there are many investigation which are developing different kind materials of waste (organic, solid urban, industry, etc.). They reuse and build different kind masonry, with different characteristics, many of this are being study still.

Numerous types of masonry, which today have become "traditional" type of construction methods, are sold in the metropolitan area of San Miguel de Tucumán.

We have to say, studies on acoustics isolation many times are insufficient and it's the spirit of this work to study, compare and to develop construction improved systems with this new kind masonry made of waste.

Methodology

The next job is part of the qualitative methodology as regards procedures enabling a building of knowledge that takes place on the basis of concepts. The concepts are that allow the reduction of complexity and it is through the establishment of relationships between these concepts are generated by the internal coherence of the scientific product.

Research is applied and has experimental involves observation, manipulation and registration of the variables (dependent, independent, involved, etc.) which affect an object of study. It's a collection of research designs used manipulation and controlled tests to understand the causal processes. In general, one or more variables are manipulated to determine their effect on a dependent variable. It is registered in participatory action research, as the user arises as the subject of action and not study, understanding that he can achieve the transformation of their habitat.

Objectives

The general objective of this research is to evaluate the acoustic behavior of traditional and alternative masonry and to develop an improved construction system with better acoustic isolation from these.

From this main objective we get individual ones. a) To identify alternative masonry developed from reuse or recycling wastes made in the north of Argentina. b) To compare with the traditional masonry made in the same area. c) To evaluate acoustic isolation of each and compare them. d) To develop improved system using these materials and to evaluate their acoustic isolation and compare them.

Evaluation of traditional and alternative masonry
Current state of situation

Currently, there are many investigations in the north Argentina. These investigations are developing different kind objects from raw materials. Some, are building masonries and try to get them into the commercial market. This "new alternative" materials have some improved properties than traditional, for example, they reduce ambient contaminants, they use residual material from industries, and some cases, their price is no further than traditional.

Tucumán city, is in the north Argentina and its main industry is from the sugar bowl. Constructive tradition is based material like the common ceramic brick, hollow ceramic brick, concrete block and "adobe" brick (which is a clay brick naturally air dry). Most of the buildings are made of these material and they use an earthquake resistant structure mixture with masonries.

We have to say that in Tucumán, the tradition is above technology and its innovations. This causes difficulty to promote new masonries and its improved properties.

This paper aims to highlight the sound isolation capabilities of these new alternative masonry, compare them with traditional and to study some now systems configuration to get an improved technology with the best sound isolation. Also, to show their low environmental impact, low use of resources and the possibility of getting into the commercial market.

Traditional masonry

TABLE 1 – TRADITIONAL MASONRY USED IN TUCUMÁN

		MATERIAL	MEDIDAS			DENSIDAD	PESO UNIT.	MASA	APLICACIÓN
			e (m)	h (m)	l (m)	kg/m3	kg	kg/m2	
CONVENCIONALES		LADRILLO MACIZO COMÚN	0.13	0.05	0.27	1600-1800	3.15	208	
		LADRILLO HUECO	0.12	0.18	0.33	1300-1500	10.69	156	
		LADRILLO DE ADOBE	0.15	0.10	0.30	1 6 0 0	7.2	240	
		BLOQUE HORMIGÓN	0.19	0.19	0.39	1 7 6 6	24.86	335.54	

Alternative masonry

TABLE 2 – ALTERNATIVE MASONRY IN THE AREA

		MATERIAL	MEDIDAS			DENSIDAD	PESO UNIT.	MASA	APLICACIÓN
			e (m)	h (m)	l (m)	kg/m3	kg	kg/m2	
NO CONVENCIONALES		LADRILLO DE PUZOLANA VEGETAL COMPRIMIDA (V C P)	0.13	0.07	0.27	1 8 0 0	4.42	234	
		LADRILLO DE PET	0.125	0.055	0.26	1 1 5 0	2.05	143.75	
		BLOQUE DE PET	0.205	0.20	0.40	1 1 5 0	18.86	235.75	
		HORMIGÓN CELULAR CURADO EN AUTOCLAVE	0.15	0.25	0.50	680	12.75	102	
		LADRILLO DE SUELOCEMENTO MACIZO	0.15	0.05	0.27	1 8 0 0	3.64	270	

Acoustic reduction comparison between traditional and alternative masonry using their commercial thickness.

From the calculation method of the "index of reducing acoustic r", where multiply the specific gravity on the surface of the element in study, in this case a Grouting theoretical of 1 meter by one meter of every masonry in study for comparison. The following table is made taking into account the

Non-Conventional Materials and Technologies – NOCMAT for XXI Century   Materials Research Forum LLC
Materials Research Proceedings **7** (2018) 247-256   doi: http://dx.doi.org/10.21741/9781945291838-22

different geometric properties of every masonry, which, depending on its production process and thus its size yields the following result.

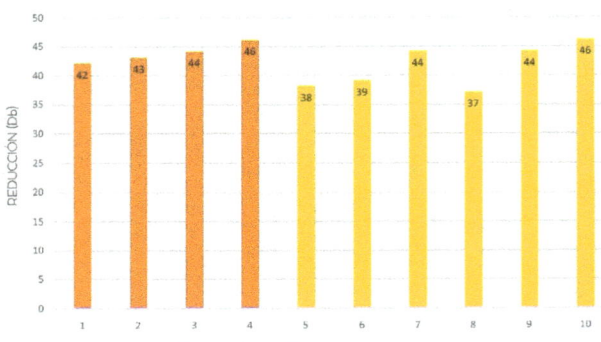

FIGURE 1 – ACOUSTICAL COMPARISON. TRADITIONAL AND ALTERNATIVE MASONRY

References: 1: Common ceramic brick masonry Pe=1600 kg/m3; **2:** Hollow ceramic brick masonry Pe= 1300kg/m3; **3:** "Adobe" Brick Masonry Pe= 1600kg/m3; **4:** Concrete block masonry Pe=1766kg/m3; **5:** Vegetal compress puzolana brick Pe= 1050 kg/m3; **6:** "PET" brick masonry Pe= 1150kg/m3; **7:** "PET" block masonry Pe= 1150 kg/m3; **8:** Autoclaved cellular concrete block Pe= 680 kg/m3; **9:** Solid cement floor brick masonry Pe= 1800kg/m3; **10:** TAPIAL: Pe= 2000 kg/m3.

Energetic const on the manufacturating process

It is important to know at the time of the choice of a system of masonry, and probably the most important to the user, its price. It should be mentioned that the masonry made with alternative masonry are not significantly more expensive than traditional ones, but the acoustic benefits of insulation, as well as also the environmental are noteworthy.

TABLE 3 – ENERGETIC COST IN MANUFACTURATION PROCESS

	MATERIAL	PRICE PER UNIT (S/UNIDAD)	PRICE PER SQUARE METER+LABOR ($/M2)		MATERIAL	PRICE PER UNIT (S/UNIDAD)	PRICE PER SQUARE METER+LABOR ($/M2)
CONVENTIONAL	COMMON CERAMIC BRICK	$2 c/u	$ 200 x m2	UNCONVENTIONAL	CELLULAR CONCRETE	$40 c/u	$430 x m2
	HOLLOW CERAMIC BRICK	$8 c/u	$240 x m2		PET BRICK	$3 c/u	$250 x m2
	ADOBE BRICK	$ 1,20 c/u	$140 x m2		VEGETAL COMPRESS PUZOLANA BRICK	$1,50 c/u	$150 x m2
	CONCRETE BRICK	$12 c/u	$320 x m2		SOLID CEMENT FLOOR BIRCK	$3,50 c/u	$275 x m2
					TAPIAL	-	$230 x m2
					PET BLOCK	$7 c/u	$220 x m2

Acoustic isolation evaluation of proposal systems
Cellular concrete with plaster in both faces

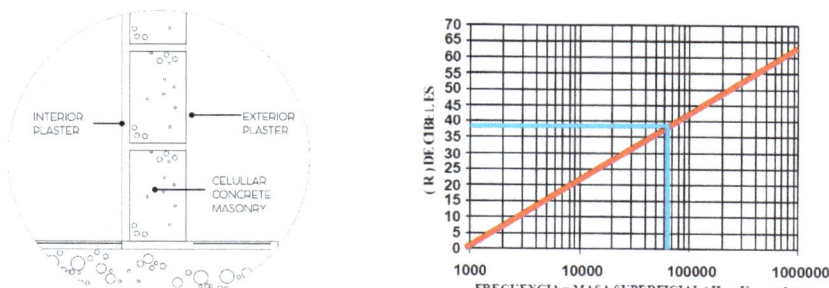

FIGURE 2 – *CELLULAR CONCRETE WITH PLASTER MASONRY - DETAIL ACOUSTIC ISOLATION*

This constructive system was made of celluar concrete with plaster in both faces o the masonry. The isolation reduction you can get with this masonry technology is 38 dB from exterior to interior. This kind of masonry is not structural independent, so it´s needed to be built with a earthquake resistant squeleton.

Puzolana brick with interior plaster plus air chamber and PET brick masonry

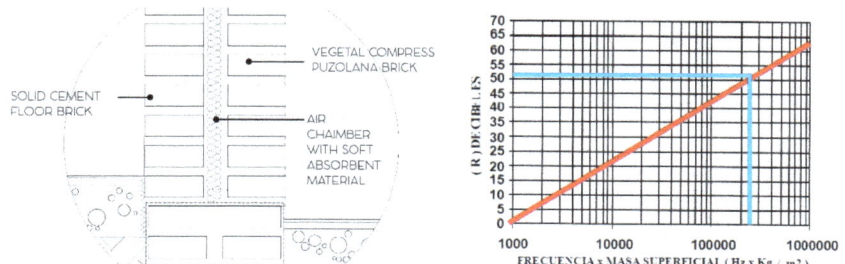

FIGURE 3 – *PUZOLANA BRICK WITH PLASTER AND AIR CHAMBER - DETAIL ACOUSTIC ISOLATION*

This masonry was made at the exterior with vegetal compress puzolana bricks and interior with a solid cement soil brick. In between there´s an air chamber with soft absorbent material. The isolation reduction you can get with this technology is above 50 dB from exterior to interior. This kind of masonry is not structural independent, so it´s needed to be biult with a earthquake resistant squeleton.

PET brick with air chamber and vegetal compress puzolana brick with interior plaster.

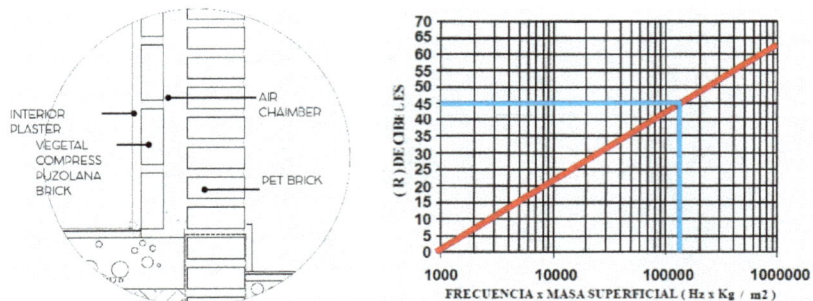

FIGURE 4 – PET AND VEGETAL COMPRES PUZOLANA BRICK MASONRY - DETAIL ACOUSTIC ISOLATION

This masonry was made of PET brick at the exterior and interior of vegetal compressed puzolana bricks. In between there is an air chamber to improve acoustic and thermic isolation. One can get an isolation of 45 dB from side to side. This kind of masonry is not structural independent, so it is needed to be built with a earthquake resistant squeleton in some areas.

PET brick within compact soil

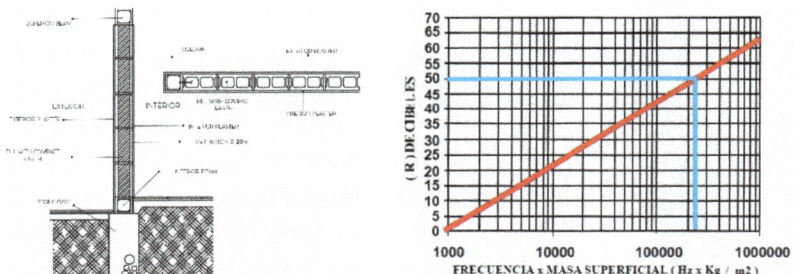

FIGURE 5 – PET BRICK WITHIN COMPACT SOIL - DETAIL ACOUSTIC ISOLATION

This constructive system was made with PET bricks (0.20*0.20*0.40 meters), with cement and sand adhesive (0.02 meters). Also has been fill in with compact soil in its interior. It has plaster in both faces. It´s thin constructive system but with a high acoustical reduction performance. It has the benefit that you can use the soil from basement digging to fill in the masonry.

Non-Conventional Materials and Technologies – NOCMAT for XXI Century   Materials Research Forum LLC
Materials Research Proceedings **7** (2018) 247-256   doi: http://dx.doi.org/10.21741/9781945291838-22

Double ashes brick masonry within compact soil

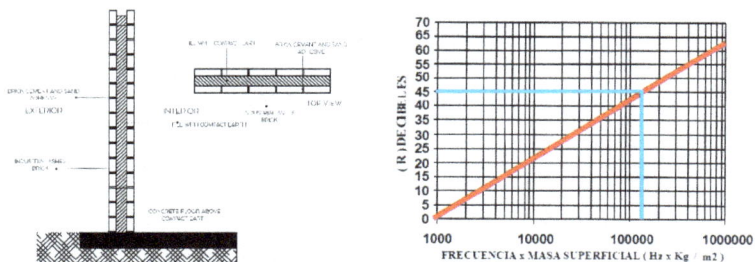

FIGURE 6 – *DOUBLE ASHES BRICK WITHIN COMPACT SOIL MASONRY - DETAIL ACOUSTIC ISOLATION*

Constructive system made by to sides of ashes bricks whitin compact soil. The ashes brick were 0.07 meters thick each and the compact soil was 0.10 m. The idea of this masonry is to get an high acoustical reduction with a low thickness. The reduction you get from it is 45 db.

Soil Cement with a cork chamber and interior PET brick

Constructive system made with 2 kind masonries. One is from compact soil and cement bricks (0.30 m). The other was made with PET brick at the interior (0.05 m thickness). In between there is a cork lamella (0.02 m). It had plaster only in its interior. The reduction you can get is over 50 dB.

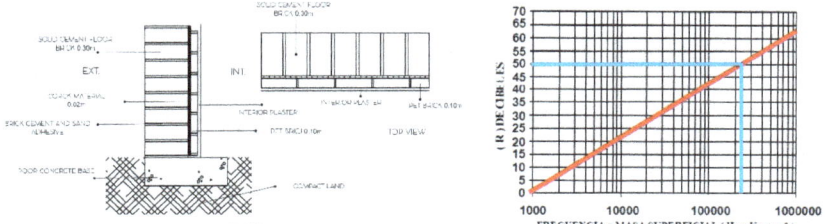

FIGURE 7 – *SOIL CEMENT WITH A CORK CHAMBER AND PET BRICK MASONRY- DETAIL ACOUSTIC ISOLATION*

ICF brick plus PET brick

It is a constructive system made with an ICF (0.12 m) and PET bricks masonry (0.05 m). It had plaster both sided. This is a low thickness system but it had a good acoustical reduction. You can get an acoustic isolation near 50 dB.

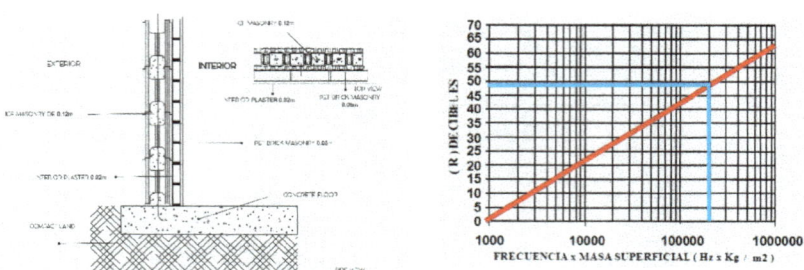

FIGURE 8 – ICF AND PET BRICK MASONRY - DETAIL ACOUSTIC ISOLATION

Cellular concrete and ashes brick masonry

This system was made by cellular concrete masonry (0.17 m) with exterior plaster and ashes brick masonry to the interior (0.07 m) without plaster. This system had a high thickness compared to its low acoustical reduction. This is because all its components have a low specific weight. Anyway, you can get almost 50 dB reduction with it.

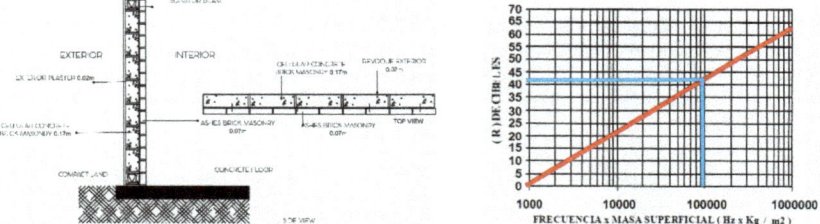

FIGURE 9 – CELLULAR CONCRETE AND ASHES BRICK MASONRY - DETAIL ACOUSTIC ISOLATION

Conclusions

After the investigation we achieve the following conclusions. All traditional masonry in general have a good behavior as insulation acoustic if being used correctly. The difference it´s in production cost. This should be noted before choosing any kind of masonry technology.

The alternative masonry that stand out for their behavior as acoustic insulation are "PET" brick masonry; solid cement soil brick masonry. Firstly this is because their own weight in direct relation to their geometric characteristics (directly related to the isolation ability of materials).

Vegetable compress puzolana brick materials; "PET" brick masonry; "PET" block masonry have a good response as acoustic insulation, but they are thin, so their geometric characteristics decrease its acoustic response. You could evaluate to enlarge the masonry to achieve a competitive acoustic insulation.

All the alternative masonry have a low energy cost, except Cellular concrete; while the traditional ones have a high energy cost except Adobe and cinder block.

Price-wise, the alternative are presented as proposals, generally of lower cost than the traditional ones.

Non-Conventional Materials and Technologies – NOCMAT for XXI Century   Materials Research Forum LLC
Materials Research Proceedings 7 (2018) 247-256   doi: http://dx.doi.org/10.21741/9781945291838-22

In short, we are in presence of masonry that:

- use alternative raw materials and improve habitat conditions,
- have an appropriate acoustic behavior,
- use low resources for their production,
- have an appropriate price, population affordable.

However, this new technologies are not market popular, maybe because there is and important pressure for the existent companies not to let this new technologies start competing. Therefore, here we have an important challenge to face in the future.

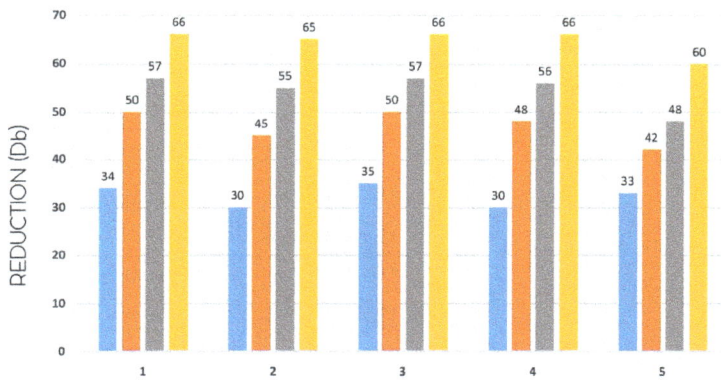

FIGURE 10 – ACOUSTIC ISOLATION REDUCTION COMPARISON BETWEEN CONSTRUCTIVE SYSTEMS

References: 1- PET block within compact soil. 2- Double ashes masonry filled with compact soil. 3: Soil-Cement masonry plus PET brick masonry within cork chamber. 4- ICF brick and PET masonry. 5- Cellular concrete and ashes brick masonry.

References

[1] Mariane Krause. "La investigación cualitativa. Un campo de posibilidades y desafíos". Revista Temas de Educación N°7. pp. 19-39. 1995.

[2] Rafael Francisco Mellace; Carlos Eduardo Alderete; Stella Maris Latina Lucìa Elizabeth Arias; Mirta Eufemia Sosa; Irene Cecilia Ferreyra. "Contrucción del centro regional de investigaciones de arquitectura de tierra cruda (Criatic)- FAU UNT. Criatic.

[3] Leonardo Paterlini, Beatriz Garzón. "Evaluation and comparison of acoustic insulation of traditional and alternative masonry in Tucumán, Argentina. 22nd International Congress on Aoustis. ICA 2016. Buenos Aires. Argentina.

[4] Baschuk, B; Dimarco, S.: "Manual de acústica para Arquitectos". Buenos Aires: Espacio Editora.

[5] Burk, W. "Manual de medidas de acústica para el control de ruidos". Barcelona: Blume.

[6] Vechiatti, N.; Iasi, F.; Muzzio, D. "Medición y estimación del aislamiento acústico en edificios". VI Congreso Iberoamericano de Acústica FIA 2008, Buenos Aires, 2008.

[7] Miguel Iglesias Ortiz-Quintana. Composites con Reciclados para la mejora del aislamiento acústico. Universidad Politécnica de Valencia. Escuela Politécnica Superior de Gandía. Máster en Ingeniería Acústica. Tesis de Master. España, 2010.

[8] Eva E. Porta. Nuevas Pantallas Acústicas a partir de Materiales Reciclados. Escuela Politécnica Superior de Gandía. Máster en Ingeniería Acústica. Tesis de Master. España, 2010.

[9] Rosana Gaggino. Ladrillos y placas prefabricadas con plásticos reciclados aptos para la autoconstrucción. Revista INVI, agosto, año/vol. 23, número 063. Santiago, Chile. 2008.

Non-Conventional Materials and Technologies – NOCMAT for XXI Century   Materials Research Forum LLC
Materials Research Proceedings 7 (2018) 257-264   doi: http://dx.doi.org/10.21741/9781945291838-23

Use of Alkaline Activated Cements from Residues for Soil Stabilization

Cosa J. [a], Soriano L. [a], Borrachero M.V. [a], Payá J. [a], Monzó J*. [a]

[a] Instituto de Ciencia y Tecnología del Hormigón, Universitat Politècnica de València (ICITECH-UPV); jcosa@hotmail.es, lousomar@upvnet.upv.es, vborrachero@cst.upv.es, jjpaya@cst.upv.es, jmmonzo@cst.upv.es*

Abstract. In recent decades, Portland cement (OPC) production has grown significantly as a result of economic and population growth. However, the cement industry is classified as a highly polluting sector and of great environmental impact. The production of one tonne of OPC requires the exploitation of a high volume of raw materials (mainly limestone and clay) and the emission of one tonne of CO_2 and other polluting gases (NOx and SOx). The OPC is the most used binder in soil stabilization, one of the alternatives with less environmental impact, is the use of so-called alkaline activated cements (AAC) and / or geopolymers. In the research work, is used a fluid catalytic cracking catalyst (FCC) residue as a precursor and a mixture of rice straw ash (RSA) and sodium hydroxide as activator. Reducing the economic and environmental cost, making it viable in developing countries. To obtain the RSA, a burner has been designed and built in which the rice straw is transformed into RSA. During the burning process, the different burning zones and their temperatures were studied. The objective of these measurements is to obtain the optimum quality RSA to synthesis of the alkaline activator in the geopolymerization reaction. The AAC was used for soil stabilization, the compressive strengths were obtained for ages between 7 and 90 days. Soils were stabilized with OPC and AAC, and the results were compared, being higher the results with OPC. However, the compressive strengths obtained with the AAC were sufficient for the stabilization of the soils.

Keywords: Sustainable Construction Materials, Waste Reuse, Alkali Activated Cement, Soil Stabilization

Introduction

Is called stabilized soil to the homogeneous mixture of a soil with a binder and eventually water, which a suitably compacted to acquires sufficient strength and stiffness to support the load for which it has been designed. The commonly used binders are OPC and lime. There are several applications in civil engineering and also in architecture, such as road stabilization or building blocks for housing.

There are numerous studies about the use of cement or lime as soil stabilizers from the last century to the present [1-3], therefore, it is a field well studied and characterized. But in the last decades, researchers are focusing their efforts on using alternative building materials to Portland cement (OPC), to improve above all the environmental impact that involves the OPC manufacture.

Within this type of alternative materials, we can find so-called alkaline activated cements or geopolymers. These materials are based on the reaction of a silicoaluminous nature source with a solution of high alkalinity (activator). These materials have very good mechanical performance and in many fields can be an ideal substitute for OPC [4-6]. There are not many studies that use geopolymers as soil stabilizers, however, the studies that are emerging show very good results comparable to those obtained by OPC [7-9].In the field of soil stabilizers, the use of geopolymers may has an economic disadvantage, since some times its use it requires an additional cost because the activator, this can make the final mixture more expensive. This problem can be solved by using as much waste as possible in the preparation of the geopolymer making it economically viable and environmentally friendly.

There are several research groups worldwide reusing residues to try to obtain a sodium silicate alternative to commercial [10-12]. For this, residues rich in silicon oxide are sought that are capable of reacting with the sodium hydroxide, forming sodium silicate. Among these studied residues is the rice husk ash (RHA), a residue that has obtained very positive results in studies with soils stabilized with geopolymers [13].

The present article wants to explore the possibility of using RSA in this type of mixtures based on the results of the RHA. The straw will be burned in a burner designed and built at the Universitat Politècnica de València (UPV), in order to obtain ashes of high reactivity.

Experimental
Materials
As is mentioned above, it is possible to stabilize soils with cement or geopolymer, using different materials that will be described briefly below. A OPC was used for cement stabilization, its oxides composition is shown in Table 1. In the geopolymer-stabilized soils the fluid catalytic cracking catalyst (FCC) residue is used as precursor material, its composition is shown in Table 1. To activate said material there are two preparation routes. The first route is the use of an activator solution formed by NaOH / Na_2SiO_3. The second one is the use of the RSA as a source of silica, which is reacted with NaOH for the formation of the sodium silicate.

The composition of the RSA is also shown in Table 1. It should be noted that the percentage of silica in the RSA exceeds 50%, a positive data for its use as a source of silica. This percentage is lower than the RHA ($\cong$90%) which is the material previously studied in this type of mixtures, so its effectiveness is expected to be lower.

The FCC was supplied by the company OMYA Clariana and the RSA is obtained in the burner designed and located in the Universitat Politècnica de València (UPV). The FCC is factory milled having an average diameter of 18 microns while the RSA needs to be milled before use. For this, a Roller 1 jars mill is used where an average diameter of 40 microns is obtained. Subsequently, a ball mill model Nannetti Speedy1 is used where its average diameter is reduced to 13 microns.

TABLE 1 – COMPOSITION IN PERCENTAGE OF OXIDES OF THE MATERIALS USED MEASURED BY X-RAY FLUORESCENCE * LOSS ON IGNITION

	SiO_2	Al_2O_3	Fe_2O_3	CaO	MgO	SO_3	K_2O	Na_2O	P_2O_5	TiO_2	Cl^-	Others	LOI*
OPC	20,80	4,60	4,80	65,60	1,20	1,70	1,00	0,07	-	-	-	-	2,02
FCC	47,76	49,26	0,60	0,11	0,17	0,02	0,02	0,31	0,01	1,22	-	-	0,51
RSA	52,44	0,47	0,17	8,01	2,71	2,26	12,05	0,89	2,58	-	3,52	0,29	14,60

Methodology
An important part of investigation development about the use of RSA has been the design of the burner for RSA obtaining. The burner construction has been carried out in the facilities of the UPV, with an approximate capacity of 3m³ of biomass. The burner consists of concrete blocks forming a ring, with holes opening between blocks to supply oxygen to the combustion, a metal sheet cover composed of two parts is arranged in order to facilitate the fumes, preventing in turn, the output of particles from the combustion. At one meter in height, two metal plates are placed, one to house the biomass and the other as a door, both perforated to allow the entrance of oxygen. Figure 1 shows the burner.

Non-Conventional Materials and Technologies – NOCMAT for XXI Century   Materials Research Forum LLC
Materials Research Proceedings 7 (2018) 257-264   doi: http://dx.doi.org/10.21741/9781945291838-23

 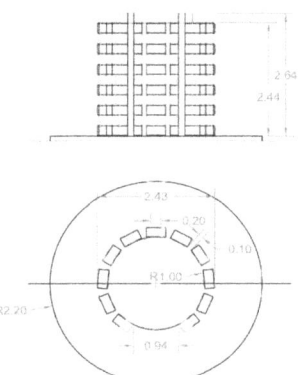

FIGURE 1 - BIOMASS COMBUSTION ENCLOSURE.

The temperature reached and the combustion time depend on the calorific value determined by the biomass, the amount of oxygen that exists between the biomass fibers, and the oxygen supplied by the different burner openings.

The RSA is a material obtained by own means, and it has been characterized by the techniques of X-ray diffraction, thermogravimetric analysis and scanning electron microscopy of field emission. Also, OPC, FCC and RSA have been characterized by X-ray fluorescence to determine their chemical composition.

The X-ray fluorescence kit is a Philips Magix Pro model, equipped with rhodium tube and beryllium window.The equipment used for the X-ray diffraction analysis is a Brucker AXS D8 Advance, the diffractogram being recorded for the interval 2Θ between 5° and 70°, with a pitch angle of 0.02 and an accumulation time of two seconds.

For the thermogravimetric analysis, was used a Mettler Toledo TGA 850 module, the sample was heated from 35 to 1000 ° C using an alumina crucible and heating at a rate of 20 ° C / min in an air atmosphere.

The equipment used for field emission scanning electron microscopy was a ULTRA 55-ZEISS equipment, and the samples were carbon coated for analysis.

As for the soil used, a sample of 25 kg has been taken and its homogenization is carried out by means of the quartet technique. After, we selecting the fraction that passes through the opening sieve 4mm and proceed to its drying at 60°C. A thermogravimetry test was carried out on the soil used from 35 to 1000 ° C to determine the nature of the soil. Figure 2 shows the TG and DTG soil curves, observing a double peak in the DTG curve starting at 700°C, indicating the presence of dolomite in the chosen soil.

Non-Conventional Materials and Technologies – NOCMAT for XXI Century   Materials Research Forum LLC
Materials Research Proceedings 7 (2018) 257-264   doi: http://dx.doi.org/10.21741/9781945291838-23

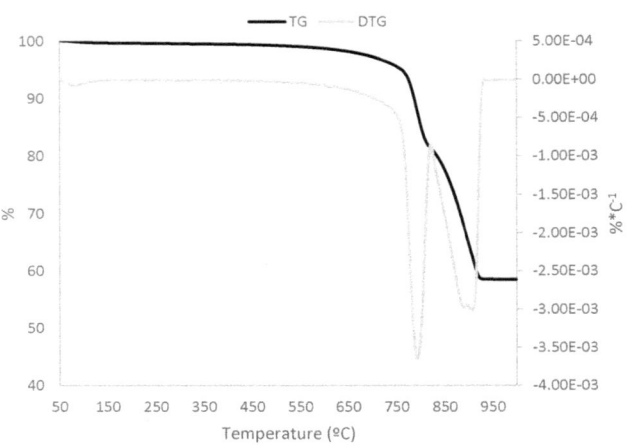

FIGURE 2 - TG AND DTG CURVE (SOIL)

From the optimum dry density obtained by mini harvard, soil kneading was done with different types of stabilizers, to make a comparison of their resistance to compression at ages between 7 and 90 days.

First, soil-water is compacted to know its compressive strength by its self, without other stabilizers, secondly, soil with OPC is stabilized as a pattern, because is the most used stabilizer. Subsequently, the program was marked for stabilization with alkaline activation cements, taking as precursor the FCC and the two types of activator previously indicated. The NaOH/ Na_2SiO_3 solution is composed of 60.8% Na_2SiO_3, 26% H_2O and 13.22% NaOH. While the RSA activator solution contains 56.80% H_2O, 17.04% NaOH and 26.16% RSA, to achieve the dissolution of the silica present in the RSA all the materials are mixed in a thermostated vessel and held for 24 hours.

Both in soil mixtures containing OPC and in the geopolymer-containing samples, these materials are used in 10% by weight with respect to the weight of soil, this percentage is within the usual percentages for soil stabilization (7-12%). The samples are kneaded together with the soil for one and a half minutes. After kneading, the entire kneading is pocketed to prevent the mixture from drying out, the mixture is divided into 24 plastic bags that are used to fill the mold designed to obtain cubic specimens of 40x40x40mm.

In the mold designed for this purpose, a dynamic compaction is applied, lowering a mass of 1.5 kg to 20 cm in height and performing 19 movements per layer (3 layers per cube). The specimens obtained are cured at an average temperature of 19 ° C. and a relative humidity of 65%.

To obtain the compressive strength of the specimens an Instron Model 270 press is used.

Results and discussion
Rice straw ash (RSA) characterization
The analysis of the characterization of the RSA is started by studying the obtained results by X-ray diffraction. As can be seen in Figure 3, the combustion of the RSA has produced an ash with a high amorphous character, such as observed with the deviation of the baseline between 15-35°. As crystalline phases the presence of Silvina (KCl), Calcite ($CaCO_3$), Arcanite (K_2SO_4) and Quartz

(SiO_2) are observed. The presence of potassium compounds accounts for about 12% of K_2O obtained by X-ray fluorescence (Table 1). The presence of calcite may be due to soil contamination at the time of collection.

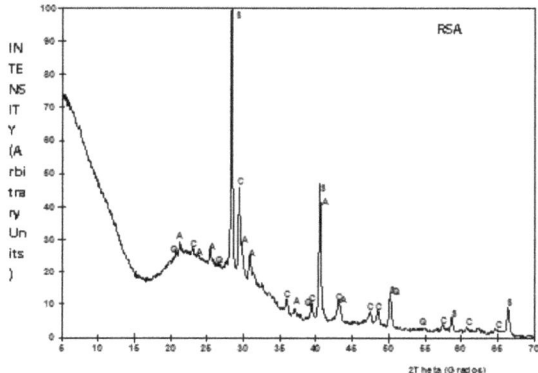

FIGURE 3 - *DIFFRACTOGRAM CORRESPONDING TO THE RSA. S: SYLVITE (KCL), C:*
CALCITE (CACO3), A: ARCANITE (K2SO4), Q: QUARTZ (SIO2)

Figure 4 shows the TG and DTG curves obtained when heating the straw ash to 1000°C, as can be seen in the TG curve there is a continuous loss of mass throughout the interval, obtaining a percentage of mass loss around of 9%. The DTG curve shows us two major mass losses, one in the range of 450°C to 550°C due to incomplete straw burning, and another loss from the 850°C that may be due to the carbonates present in the sample probably due to soil contamination.

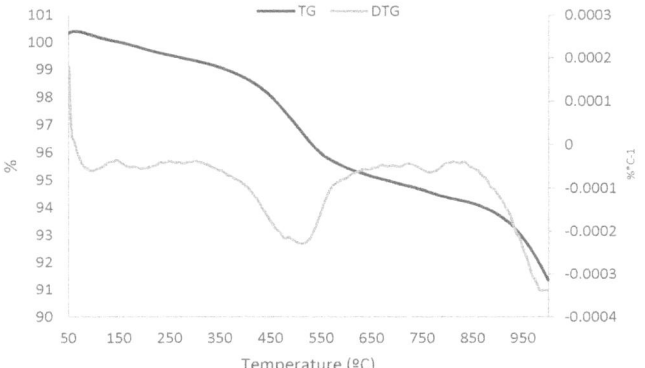

FIGURE 4 - *TG AND DTG CURVE (RSA)*

Finally, on micrographs obtained by scanning electron microscopy of field emission, are shown a sample without milling to observe more clearly the structure of the straw after its combustion. In the micrographs the presence of phytoliths is observed, formations rich in silicon oxide that have conserved its structure in spite of the combustion of the same one.

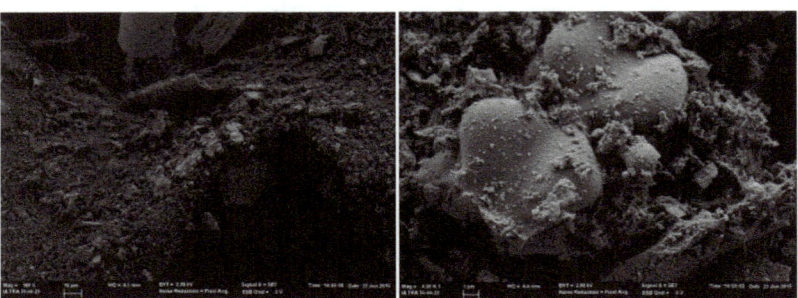

FIGURE 5 - UNMILLED RSA MICROGRAPHS

Compressive strengths (stabilized soils)
The compressive strengths study was carried out to four curing ages to observe the evolution of the different soil stabilization systems. Results are plotted in figure 6.

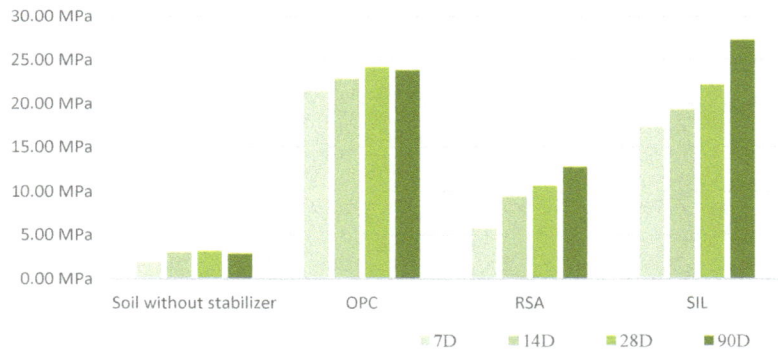

FIGURE 6 - COMPRESSIVE STRENGTHS IN MPA FOR SOILS STABILIZED AT 7, 14, 28 AND 90 DAYS OF CURING

There are several behaviours in the different evaluated samples that distinguish between them. It can be observed that: In the case of soil without cementitious stabilizer, we obtain results of compressive strength in the range of approximately between 2 - 3MPa.

By adding 10% of OPC, we obtain a maximum of compressive strength of 24 MPa, much higher than the soil without stabilizer.

Non-Conventional Materials and Technologies – NOCMAT for XXI Century   Materials Research Forum LLC
Materials Research Proceedings 7 (2018) 257-264   doi: http://dx.doi.org/10.21741/9781945291838-23

In the case of stabilizing with a 10% FCC geopolymer activated with a NaOH/ Na_2SiO_3 mixture (SIL), a maximum compressive strength of 27.2MPa is obtained; these samples show a greater evolution of resistances with time from 7 to 90 days.

In the case of FCC geopolymer stabilizing, with RSA activator a compressive strength of 12.7MPa at 90 days of curing time is obtained. It is lower than obtained by the cement samples and the NaOH/ Na_2SiO_3 mixture geopolymer, but much higher than the required for these kind of samples.

Conclusions

The present study has demonstrated the feasibility of using soil stabilization systems alternative to OPC using wastes. The use of geopolymers for soil stabilization using as a precursor the catalytic cracking catalyst residue gives lower results than the cement samples, but higher than not stabilized soils samples. The use of straw ash as a source of silica in the preparation of the activator reflect lower compressive strength results less than the use of the commercial silicate but is a very positive result, considering that the percentage of silicon oxide thereof is 52.4%. However, the compressive strengths achieved are sufficient for their application in stabilizing soils for their use in road pavements.

Acknowledgements

To the Ministry of Economy and Competitiveness of the Government of Spain, for the support granted to the APLIGEO project (BIA2015 70107-R), which includes this work research. And to PAVASAL for its collaboration and supply of soil. To the Institute of Science and Technology of Concrete, of the Universitat Politècnica de Valencia for ceding its facilities for the development of the experimental part.

References

[1] Khadka B., Shakya M. Comparative compressive strength of stabilized an un-stabilized rammed earth. Materials and Structures, vol 9, n° 49, pp 3945-3955, 2016.

[2] Alrubaye A.J., Hassan M., Fattah M-Y. Stabilization of soft kaolin clay with silica fume and lime. International Journal of Geotechnical Engineering, vol 11, n°1, pp 90-96, 2017.

[3] Rios S., Viana da Fonseca A., Baudet B. Effect of the porosity/cement ratio on the compression of cemented soil. Journal of Geotechnical and Geoenvironmental Engineering, vol 138, n°11, pp 1422-1426, 2012.

[4] Marinkovic S., Dragas J., Ignjatovic J., Tosic N. Environmental assessment of green concretes for structured use. Journal of Cleaner Production, vol 154, pp 633-649, 2017.

[5] Tashima M.M., Akasaki J.L., Castaldelli V.N., Soriano L., Monzó J., Borrachero M. V, Payá J. New geopolymeric binder base on fluid catalytic cracking catalyst residue (FCC). Materials Letters, vol 80, pp 50-52, 2012.

[6] Huiskes DMA., Keulen A, Yu QL., Brouwers HJH. Design and performance evaluation of ultra-lightweight geopolymer concrete. Materials and Design, vol 89, pp 516-526, 2016.

[7] Zhang M., Guo H., El-Korchi T., Zhang G. Tao M. Experimental feasibility study of geopolymer as the next-generation soil stabilizer. Construction and Building Materials, vol 47, pp 1468-1478, 2013.

[8] Rios S., Cristelo N., Viana da Fonseca A., Ferreira C. Stiffness behaviour of soil stabilized with alkali-activated fly ash from small to large strains. International Journal of Geomechanics, vol 17, no° 3, pp 1-12, 2017.

[9] Zhang M., Zhao M.X., Zhang G.P., Nowak P., Coen A., Tao M.J. Calcium-free geopolymer as a stabilizer for sulfate-rich soils. Applied Clay Science, vol 108, pp 199-207, 2015.

[10] Bouzón N., Payá J., Borrachero M.V., Soriano L., Tashima M.M., Monzó J. Refluxed rice husk ash/NaOH suspension for preparing alkali activated binders. Materials Letters, vol 115, pp 72–74, 2014.

[11] Mejía J.M., Mejía de Gutiérrez R., Montes C. Rice husk ash and spent diatomaceous earth as a source of silica to fabricate a geopolymeric binary binder. Journal of Cleaner Production, vol 118, pp 133-139, 2016.

[12] Puertas F., Torres-Carrasco M. Use of glass waste as an activator in the preparation of alkali-activated slag. Mechanical strength and paste characterization. Cement and Concrete Research, vol 57, pp 95-104, 2014.

[13] Alamán M. Estudio para la estabilización de bloques de tierra mediante la utilización de geopolímeros a partir de residuos. Aplicación para viviendas de bajo coste en Barranquilla (Colombia). Proyecto Final de Carrera. Universitat Politècnica de València, 2014.

Non-Conventional Materials and Technologies – NOCMAT for XXI Century   Materials Research Forum LLC
Materials Research Proceedings 7 (2018) 265-274   doi: http://dx.doi.org/10.21741/9781945291838-24

Construction with Earth and Sustainability: Analysis in the Light of Public Housing Policies

Matías Ortega [a], Beatriz Garzón [b*]

[a] FAU-UNT, SCAIT, CONICET; mateduortega@hotmail.com

[b*] FAU-UNT, SCAIT, CONICET; bgarzon@gmail.com

Abstract. Social housing public policies are the main tools that Latin American States have to reduce high housing deficit. This problem requires to be considered along with global warming, since thinking in the construction industry as energy-efficient it is imperative in the contemporary world. Earth construction is among the most interesting alternatives from this angle. This paper examines the experiences of public policies that allow the construction of homes made with earth in the province of Tucumán, in north-west of Argentina, with the purpose of approaching this reality and to examine the relation between these state programs of design and production of social habitat with the non-conventional technologies that incorporate the ground as an input for the construction. At the same time, this work aim to know the potentials for the development of these policies on a larger scale. The methodology used is research with participatory action (IAP), and the correlational study and of case is presented. The developed activities were analysed taking into consideration the history and documentation of the institutions involved, observation and analysis of sites, community and institutional approach, survey of housing and their contexts, data registration and its systematization to achieve results and conclusions. The results suggest that there are existing potentials for the development of public policies that implement the earthen construction on a larger scale, since the institutional, technological and socio-cultural identity precedents show it.

Keywords: Public Policies, Social Habitat, Earthen Construction

Introduction

We are witnessing today, and for several years, a process of very profound changes. A process of change of paradigms that challenge us to rethink the world that surrounds us. The crisis of civilization that we live in, according to several authors, is caused by the nature of our development, globally. With the crisis of the petroleum in the 70's, it has been gaining awareness of the value of minimizing energy consumption, since the resources that produce it are finite. For Garzón [1], "it is necessary to begin to change the focus and deal with our problems from a different perspective, a bioenvironmental perspective of the world and reality". Thus the energy-saving variable has been gaining a central role in discussions on contemporary development.

During the last decade, in Argentina this topic began to be institutionally discussed. In this way it is necessary to highlight the national programme of rational and efficient energy (PRONUREE), passed by decree 140/2007. This declares that the rational and efficient use of the energy is topic of interest and a national priority and establishing energy efficiency as a permanent activity of medium-to-long term. It is also defined as an indispensable component of energy policy and the preservation of the environment.

For these reasons, it is imperative to carry out actions that allow to optimize energy consumption, a scenario of limited resources and where environmental concern is growing. According to Chevez et al. [2], energy saving and the efficient use of energy are fundamental pillars to deal with these problems seriously and there are many states that have adopted it as a policy, including the country but with emerging results.

Non-Conventional Materials and Technologies – NOCMAT for XXI Century   Materials Research Forum LLC
Materials Research Proceedings 7 (2018) 265-274   doi: http://dx.doi.org/10.21741/9781945291838-24

Energy efficiency in social housing

In this framework, the construction industry is strongly crossed, since it generates one of the greater impacts in our territories due to the very high levels of energy consumed in the production, use and destruction of buildings. So, it is necessary to advance in the awareness raising of the use of building alternative technologies and to achieve energetic efficiency in buildings, particularly in housing. The management of sustainable technology and the use of passive thermal-energetic conditioning systems are subjected to numerous studies and design and construction practices. As Bracco affirms et al. [3], his survey, registration, analysis and application are making their mark in a set of "good practices" that should expand their space of recognition and dissemination in the environment of research and the professional practice of architecture, and thus to lay groundwork for a constant improvement.

It is interesting to note the GEF 4861 program, "Assessment of energy efficiency and renewable energy in the design, construction and operation of public housing", which also serves as illustrative inter-institutional coordination. The objective of the project is to establish technological and regulatory guidelines for the construction of social housing with energetic efficiency and renewable energy across the country, aimed at reducing demand for energy consumers and reducing emissions Greenhouse Gases (GHG). Measurement, registration, transfer and processing of data arising from the monitoring will allow to calculate different indicators associated with the result of the incorporation of energetic efficiency (EE) and renewable energy (ER), applicable in the social housing measures. Simultaneously with the attainment of this objective, the conditions of hygrothermal comfort in which occupants are in the interior of their homes will be evaluated, since consumption in heating and cooling are linked with the above-mentioned conditions.

It is possible to observe limits between the promotion of sustainable development and the strategies to achieve it, when public policies are implemented to reduce the housing deficit, that is to say, in the production of social housing. In the construction industry, there is an incipient level of actions on the issue. It is obvious the need for the state to promote a sustainable construction industry through the dynamics of the federal institutions of promotion of the construction of social housing.

Considerations on sustainability and energetic efficiency (EE), in earthen construction.

Earth construction and sustainability are closely linked. This is due to the low environmental impact that this technology has with respect to others conventionally used and dominant in the market.

As Bestraten et al. [4] affirms, the analysis of the ground of the place where a work will be implemented may suggest a technique to be used, and even improved its composition and granulometry; Thus the impact of waste and the cost of transport in materials is reduced, with a focus on the total amount of the work.

Working with techniques that incorporate the ground is more energetically efficient since it consumes less energy in the production of the constituent elements. According to Minke [5], what is needed to produce a material or a constructive element is considered as the primary energy content (CPE), and the ground along with other natural products has a lower CPE. Garzón [6], following MacKillop, affirms that the energy required to manufacture adobe is 13 kcal/unit, while a compressed earth block (CEB), stabilized with 10 percent of cement requires 94 kcal/unit; while the brick will need 379 kcal/unit and a concrete block will need 3830 kcal/unit. In the case of the compressed earth blocks (CEB), it also takes a very low percentage of water.

Energetic efficiency is also observed in the use and maintenance of the buildings due to the low conductivity of the earth; for example, the conductivity of the adobe is 37.5% of the baked brick. According to Garzón [6], the coefficient of thermal conductivity of adobe is 0.25 w/m ° C, while the brick of 0.85 w/m °c and the concrete is 1.50 w/m ° C.

The masonry on ground make those buildings that best incorporate them in terms of the resulting hygrothermal comfort since they retard the transfer of heat and regulate the percentage of humidity in

Non-Conventional Materials and Technologies – NOCMAT for XXI Century   Materials Research Forum LLC
Materials Research Proceedings 7 (2018) 265-274   doi: http://dx.doi.org/10.21741/9781945291838-24

the indoor environment, collaborating with the general welfare of people. Generated indoor conditions reduce or even avoid the use of air conditioning systems, which saves energy and money.

Another variable considered with respect to the sustainability of the materials or building elements is determined by the amount of their emissions of Greenhouse Gases (GHG), e.g. carbon dioxide (CO2), which are released in their production.

TABLE 1 – EMISSION OF C02 OF THE DIFFERENT MATERIALS ACCORDING TO BESTRATEN

Material	Density	Emissions per kg	Emissions por m3
Rammed earth (not stabilized)	2.200 kg/m^3	0,004 kg CO_2/kg	9,7 CO_2/m^3
Adobe	1.200 kg/m^3	0,06 kg CO_2/kg	74 CO_2/m^3
Mass concrete in site	2.360 kg/m^3	0,14 kg CO_2/kg	320 CO_2/m^3
Prefabricated concrete, 2 % of steal	2.500 kg/m^3	0,18 kg CO_2/kg	455 CO_2/m^3
Solid brick wall	1.600 kg/m^3	0,19 kg CO_2/kg	301 CO_2/m^3
Hollow brick wall	670 kg/m^3	0,14 kg CO_2/kg	95 CO_2/m^3

Earth construction in housing policies.

The rediscovery of the ground material applied to construction poses a profound change in the classic western paradigm of development, as it categorizes a constructive material of traditional techno-science, almost forgotten by the technological innovations of the 19th and 20th centuries. Since the ' 60s, the value of the land for construction has been recovered, since that was a context in which criticism towards modernity was generalized and vernacular values and local identities were rescued.

If we understand that the homes are the primary cells of the urban fabric, buildings on land would constitute human settlements very different from those who currently promote housing policies in general.

Human settlements contribute to improve social, economic and physical development at the same time constituting the indicator which measures the quality of such development. Therefore, they play a central role in economic and social policy and the management of the interaction between the built environment and the natural environment. In the durable valorisation of those settlements, two currents of thought converge relating to the management of human activity. One is focused on the development goals and the other intends to achieve these goals without compromising vital systems of the planet or endanger the interests of future generations. Only sustainable development is possible thanks to a rational and sensible management of all aspects of human settlements. Compared to others, the earth is a material that allows to build low-cost houses at the time which possesses excellent qualities. Moreover, for Carazas [7], the land allows to achieve an architecture that responds to the current needs.

It is interesting here to approach the local experiences of public policies that allow or promote earth construction for the implementation of social interest housing.

Objectives

To present the reality of earthen construction from public housing policies in the province of Tucumán, Argentina. To examine the relations between these public investments programmes to generate housing and technological production of social habitat that has to the ground as one of the used building materials, and meet the existing potential for development of housing in a larger-scale land policies, promoting sustainable design and production.

Non-Conventional Materials and Technologies – NOCMAT for XXI Century   Materials Research Forum LLC
Materials Research Proceedings 7 (2018) 265-274   doi: http://dx.doi.org/10.21741/9781945291838-24

Methodology
The methodology used is the participatory action research (IAP), since, according to Garzón [6], it seeks to know and influence the housing policy based on participation and reflection joint with beneficiary families and institutions involved, in order to improve them. The purpose of this method is to know for joint action and the transformation of reality.

At the same time, it is a descriptive, heuristic and inductive case study. Characteristic of this method is the in-depth study of cases, understanding it according to what Barrio del Castillo et al. [8] affirms, as "a system limited by the limits required by the object of study, but framed in the global context where it occurs". In this way, is intended to be a tool to analyse a set of certain houses, which will be used to gain a deep understanding of a system defined by specific social policies and their natural and cultural environments.

It is also, correlational, since it consists in the search for some kind of relationship between two or more variables and to what extent the variation of one of the variables affects the other. Thus, it seeks to identify the relations between the technologies used and the modes of production of the habitat. The variables considered are defined along two axes: housing policies and forms of production of habitat and employed construction techniques and supplementary facilities.

The activities were related to the analysis of precedents and the documentation provided by the institutions involved to the observation and analysis of sites, community and institutional approach through interviews, survey of housing and their contexts, the registry of data through sheets, chart, planimetries, photographs, etc. and its systematization to achieve results and conclusions.

Definition of the study area
Earth construction is present in the Argentinian architectural tradition, as it is in the continent and a great part of the world. According to Tomasi [9], "the data of the national population census of 2001 show us the vitality of these building techniques, with 2.54 percent of homes in the country with adobe walls (246.959 units)", and argues that the provinces of the northwest of Argentina have a percentage above the average. Salta has 10.45 percent, Jujuy the 14.11% and Catamarca the 17.47%. There are constructive logics that incorporate earth and which are present in the cultural identities of several localities in the region.

For Garzón [6], in rural areas of Tucumán, these percentages can amount to 87% in Colalao del Valle, if we consider the presence of an enclosure of adobe; on the plain, in the town of Balderrama, the presence of adobe is reduced to 18%, but the technique of cane and mud reaches a 72%.

For this study, it is important to know how to influence public policy in the generation of domestic architecture on land in the province of Tucumán. In relation to this, we identified dwellings distributed throughout the provincial territory.

These homes are located in areas different bioenvironments, as characterized by the standard IRAM 11.603: areas IIa, IIb, IIIa.

Zone	Departments
IIa	Chicligasta, Graneros, Juan Bautista Alberdi, La Cocha, Monteros y Rio Chico
IIb	Burruyacú, Capital, Cruz Alta, Famaillá, Leales, Tafi Viejo, Trancas, Lules y Yerba Buena
IIIa	Tafi del Valle

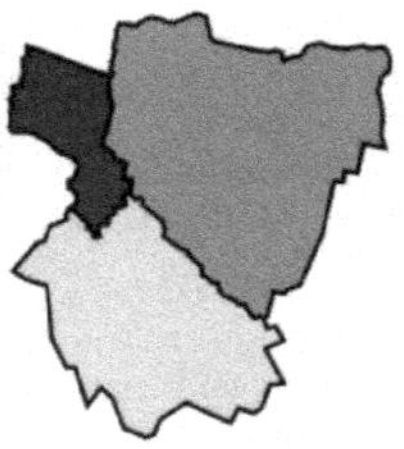

FIGURE 1 – BIOENVIRONMENTAL AREAS IN THE PROVINCE OF TUCUMÁN

Non-Conventional Materials and Technologies – NOCMAT for XXI Century   Materials Research Forum LLC
Materials Research Proceedings 7 (2018) 265-274   doi: http://dx.doi.org/10.21741/9781945291838-24

Results

The variables identified between the observed houses, those that were produced through public housing policies throughout the provincial territory, allow to identify potential on conditions of social organization, both productive and administrative and of management, such as traditional and emerging ideas about modes of living and producing such habitat and the technological means to do so.

Involved public policies

It is understood that housing policies are those social policies that build human habitats as a strategy of inclusion and social cohesion. In this sense, for Barreto et al [10], "the housing as a policy problem lies mainly in the difficulty that have millions of people to have access to decent housing; in other words, in social conditions that impede such access. Fundamentally it is, then, a problem of accessibility and, therefore, the housing issue is part of the social and housing policy is a social policy. (...) Access to the various services and housing submarkets is closely linked to the belonging of the families at different levels of the social structure, either through its hierarchy of income as well as by cultural factors."

In the province of Tucumán, dwellings were distinguished according to the various implemented programmes. Five specific channels were identified, from which it was possible to build with earth.

ProCreAr

It is the bicentennial of Argentine credit program, and it became "an inclusive and transparent vehicle for access to housing for large sectors of the population throughout the country. Also, it encouraged employment in construction and related industries." According to Tomasi [9], it is known to have been constituted as the program model of federal character for the access to mortgage credit of the Argentine middle class, and is characterized by low interest rates, being very attractive for a large percentage of employees who wanted to access to homeownership.

In the province of Tucumán, this program was observed in IIb and IIIa, areas in the northern half of the province and, even though its construction demands might slightly complicate design decisions, it has been an opportunity to build with innovative techniques. Homes built using the cimbra technique or lightened earth formwork were made with funding from this program, with a work tracking methodology that allowed to meet the construction needs.

FIGURE 2 – *HOUSE BUILT WITH THE CIMBRA TECHNIQUE IN TAFÍ DEL VALLE*

ProMeVi

Among the federal social housing programs implemented in the province and that they are relevant to this study is Programa Federal Mejor Vivir (Federal Program Better Living), housing improvement program, known as PROMEVI. According to their official website [11], "the program is designed for completion, expansion / renovation of housing of all family group that need their current home to be

Non-Conventional Materials and Technologies – NOCMAT for XXI Century   Materials Research Forum LLC
Materials Research Proceedings 7 (2018) 265-274   doi: http://dx.doi.org/10.21741/9781945291838-24

completed and/or improved, having started the construction of their only house through their own effort, and that do not have access to conventional forms of credit."

Prototypes homologated by the Provincial Institute of Housing and Urban Development of Tucumán (IPVyDU), have been built in the bioenvironmental zone IIIa, corresponding to the Department of Tafi del Valle, in the Valley of Tafi and in the Calchaquí valleys. It is of great value the coordination with socio-economic organizations such as cooperatives of construction, with much presence in the forms that the industry is taking in the province of Tucumán. The cooperatives Los Zazos, Casas Viejas, Los Molles, Ampimpa, Muso Cascanakuy, and Sinchy Huasi, have been the protagonists of the construction of these homes.

At the same time, in the locality of Colalao del Valle, the co-author has participated in the coordination of an experience of intersectional and inter-institutional articulation where conducted participatory design and the co-direction of the execution by self-management assisted homeownership in adobe and covered with mud, through PROMEVI and CONICET; according to Garzón [12], this allowed to have a greater surface in the prototype, with a bioclimatic architectural disposal and use of solar energy systems, trombe wall, systems for heating of water, etc.

FIGURE 3 – HOUSE BULT IN COLALAO DEL VALLE

ProMHIB

It is the federal program of improving housing and basic infrastructure and depends on the direction of programs for emergency and the direction of improvement programs and basic infrastructure. Officially [13], "it seeks to promote the development and the improvement of the conditions of the habitat, the house and social infrastructure of households with unsatisfied basic needs NBI, and vulnerable groups in situations of emergency, risk or marginality, located in small towns, spots, rural areas or Aboriginal communities. "The program funds, on a non-refundable basis, the acquisition of materials for the construction, improvement or completion of houses, and the construction of community equipment, for example: multi-purpose rooms, first aid rooms, nurseries, among other community infrastructure."

Its objective is to improve the conditions of the habitat of vulnerable population groups or in a situation of emergency, risk or marginality, facilitating access to basic housing, the completion of those that are recoverable, to basic community infrastructure. It also seeks to strengthen the capacities of subsistence and self-management of the vulnerable groups in the coverage of unmet basic needs (NBI), developing and strengthening the social, productive, technological and labor organization of villagers and intermediate beneficiary associations.

Like the previous one, observed only in zone IIIa.

Non-Conventional Materials and Technologies – NOCMAT for XXI Century   Materials Research Forum LLC
Materials Research Proceedings **7** (2018) 265-274   doi: http://dx.doi.org/10.21741/9781945291838-24

ProMat
It is Provision of Materials, a sub programme of the provincial capital Pedro Fernando Riera program. It is characterized by being destined for the most vulnerable, those sectors that do not have access to credits. Methodologically, they differ from the previous two due to those are articulated with local cooperatives, while these work through self-construction and mutual help. From the magazine of the National Housing Council or CONAVI [14], this program allows the use of characteristic materials from different areas where it is deployed, since it seeks to build decent housing at the lowest possible cost.

Although it was applied throughout the territory of the province, the cases that incorporate adobe are located mainly in zone IIIa, corresponding to the intermontainous valleys and product of the strong determinant of constructive local identity; while cases located in areas IIa and IIb were identified also the use of compressed earth blocks.

Rural Housing and Indigenous Peoples
They are methodologically similar to the previous one, also working through self-construction. In this case the funds are national. Like the ProMat, it seeks to contain the population more sensitive to economic problems, improving their living conditions, trying to avoid their migration to urban centres.

Building systems used
There were technologies that respond to local building traditions and innovations.

Adobe
This technique is traditional in vernacular architecture of the Argentine Northwest, and in the province of Tucumán are distributed throughout the province, mainly emphasizing its use in the valleys, while, according to Garzón [6], on the plain is also observed the use of ground-concrete blocks and lattices of cane and mud. This technique is historically rooted in the region, it was able to also adapt to new practices in construction and industrial materials.

It is usually manufactured with measures of 10 cm x 20 cm x 30 cm, since it is frequent to see it used not as a supporting structure, but as the closing in specific structures of beams and reinforced concrete columns, being the section of these columns of 20 cm x 20 cm.

Compressed Earth Blocks (CEB)
These masonry were used in zones IIa and IIb, in developments created to relocate families located on the edge of provincial routes. They were carried out according to the methodology of self-construction and mutual help, using machines Cinva-Ram that belong to the Provincial Institute of Housing and Urban Development (IPVyDU), at a variable rate of between 200 and 300 blocks per day.

Cimbra
For Gatti and Mirkin [15], is a technique that "seeks practicality, economy and comfort". Presented at the 8th Ibero-American seminar of architecture and construction (SIACOT), and 2nd Argentine seminar on the subject, held in June 2009 in Tucuman, this technique has been developed and installed in the western valleys, although we also observe it in places like Yerba buena or El Siambón, in the area of IIb.

Product of an innovation which sought a constructive system allowing to develop projects of self-construction assisted with not skilled labor, using simple technologies which are incorporated to popular culture.

Non-Conventional Materials and Technologies – NOCMAT for XXI Century   Materials Research Forum LLC
Materials Research Proceedings 7 (2018) 265-274   doi: http://dx.doi.org/10.21741/9781945291838-24

As it can be seen in figure 4, once the seismic-resistant structure is defined with reticulated columns and beams, window panels are installed, through high resistance plastic strip subject to the structural elements and which define a mesh of 30 cm x 30 cm. This functions as an armor of distribution, absorbs the bending stresses by earthquake and contains "the mixture of soil cement plastic at the time of filling".

***FIGURE 4** – HOUSE UNDER CONSTRUCTION (Credits to: www.arquimaster.com.ar [16])*

Lightened earth formwork
This technique was observed in one case, located in the town of Las Talitas, area IIb. It consists of adding to land a porous additive (straw), to lighten the mixture and increase its thermal insulation. It was used as the closing between wooden frames, as shown in figure 5. According to Minke [5], "the advantage of mud soaked with straw can be the low cost of materials and the possibility of using it without the need for tools and specialized machinery. That is why it is especially appropriate for self-construction".

FIGURE 5 – HOUSE IN LAS TALITAS

Equipment and services
Depending on the program, and hence the final recipients, there are differences with respect to the used facilities. In general, they all have determined the services of potable water, electricity and sewage. There is a case where a collection system of rainwater and the variable discharge sewage resolved with infiltration bed for the irrigation of the land where is implanted, minimizing the risk of contamination of phreatic joins.

Non-Conventional Materials and Technologies – NOCMAT for XXI Century   Materials Research Forum LLC
Materials Research Proceedings 7 (2018) 265-274   doi: http://dx.doi.org/10.21741/9781945291838-24

Conclusions

The great potential that this province has for the development of an architecture in Earth according to the demands of the society can be inferred from the observed. It is imperative to move forward in setting policy frameworks that allow the free deployment of these building techniques.

There are many cooperatives of construction which, in articulation with the housing institutions in the province, as well as infrastructure and urban development institutions manage to install local construction practices, historically practiced, although the passage of time and the development of industrial materials are conditioning them. This vernacular architecture is the result of an intimate relationship between man and his environment, and reflects their ways of living, in a direct way, delivering items that indicate the emergence of an alternative architecture that is strongly installed in urban centres.

So it is very interesting the articulation between these local ways of living and the State programs. This is a result on the one hand the economic organization forms of social sectors, such as cooperatives, which condition the housing policies and adapt them to its traditions and identity construction. On the other hand, the vocation of some of these programs surveyed to work with methodologies that incorporate self-construction and mutual help, shows the potential to be replicated on a larger scale.

Technical-constructive and theoretical innovations were observed at the same time, trying to bring the benefits of industry and prefabricated materials with earth construction, as it is the cimbra technique, strongly installed in some places, while there is the presence of the permaculture stream that is leaving footprint also in the imaginary and constructive practices.

Housing and programs that allowed its execution are thus valuable precedent to motivate further impetus from the use of ground as building material in the province of Tucumán.

Acknowledgements

The authors thank the families that generously allowed them to enter their homes. The architects García Villar, Gatti and Guzman for opening the doors of their houses and illustrate the techniques used and their experiences with the Pro.Cre.Ar. We would like to thank the architect Carrizo for providing information about her experience in the management of the programmes Pro.Mat and Rural housing, and indigenous peoples. To those responsible and technical staff of the Program Federal Integration Partner Community of the Provincial Institute of housing and urban development of Tucumán (IPVyDU), and of the office of auditing and Management Control of the province of Tucumán by providing the required documentation.

References

[1] Garzón, B. (2007). Patrimonio doméstico rural, adecuación ambiental y tecnologías tradicionales: el caso de Tucumán, Argentina. In: Construcción con tierra 3. Buenos Aires: CIHE - SI - FADU - UBA y IAA - SI - FADU - UBA. p. 49-50.

[2] Chevez, P., Martini, I., Discoli, C. (2013). Construcción de escenarios urbano-energéticos a partir de medidas de eficiencia en el sector residencial. ASADES.

[3] Bracco, M., Angiolini, S., Jerez, L., Pacharoni,A., Sánchez, G., Tambussi, R., Avalos, P., Gatani, M. (2010). Verificación de pautas de diseño sustentable en una vivienda serrana en Córdoba. ASADES. http://www.cricyt.edu.ar/asades/modulos/averma/trabajos/2010/2010-t005-a005.pdf

[4] Bestraten, S., Hormías, E. y Altemir, A. (2010). Construcción con tierra en el siglo XXI. In: Informes de la Construcción. Barcelona, p. 18.

[5] Minke, G. (2013). Manual de construcción con tierra. Bariloche: BRC Ediciones.

[6] Garzón, B. (2005). Variables bioclimáticas y uso de la energía en viviendas espontáneas y oficiales de interés social: análisis y propuestas. Disertación en Los edificios bioclimáticos en los países de Iberoamérica. Red iberoamericana para el uso de energías renovables y diseño bioclimático en viviendas y edificios de interés social. San Martín de los Andes, Neuquén, Argentina, p. 117

[7] Carazas Aedo, W. (2001). Vivienda urbana popular de adobe en el Cusco, PERÚ. Asentamientos humanos y medio sociocultural, Organización de las Naciones Unidas para la Educación, la Ciencia y la Cultura, UNESCO.

[8] Barrio del Castillo, Gonzalez Gimenez, Padín Moreno, Peral Sánchez, Sánchez Mohedano y Tarín López (2017). El estudio de casos. Disponible en www.uam.es

[9] Tomasi, J. Pro.Cre.Ar y el adobe (2014). In: Página 12. Disponible en < www.pagina12.com.ar >

[10] Barreto, M. A. y Lentini, M. compiladores (2015). Hacia una política integral del hábitat. Buenos Aires: Café de las Ciudades.

[11] Promevi (2017). Programa Federal de Mejoramiento de Viviendas. Disponible en <http://www.vivienda.gob.ar/mejorvivir/descripcion.html>

[12] Garzón, B. (2015). Vivienda rural sustentable en Tucumán, Argentina. VIII Congreso de Vivienda Rural. Santiago del Estero, Argentina.

[13] Promhib (2017). Programa Federal de Mejoramiento Habitacional e Infraestructura Básica. Disponible en <http://www.vivienda.gob.ar/promhib/descripcion.html>

[14] Consejo Nacional de la Vivienda (CONAVI, 2004). Tucumán. Obras en la provincia.

[15] Gatti, B.; Mirkin, G. (2010). Sistema abierto para autoconstrucción. In: Construcción con tierra 4. Buenos Aires, Argentina: CIHE y IAA (SI-FADU-UBA).

[16] Disponible en: <http://www.arquimaster.com.ar/notas/metodo_cimbra.htm>

Non-Conventional Materials and Technologies – NOCMAT for XXI Century   Materials Research Forum LLC
Materials Research Proceedings 7 (2018) 275-284   doi: http://dx.doi.org/10.21741/9781945291838-25

Cumulative Impact Energy Absorption of Sandwich Panels with Foam Cores and Flax FRP Facings

Dillon Betts [a], Pedram Sadeghian [a, *], Amir Fam [b]

[a] Dalhousie University, Canada; dillonbetts@dal.ca, pedram.sadeghian@dal.ca*

[b] Queen's University, Canada; amir.fam@queensu.ca

Abstract. In this paper, the cumulative energy absorption of sandwich panels constructed of foam cores and flax fiber-reinforced bio-based polymer (FFRP) faces is studied. Nine sandwich panels were constructed and tested under varying amounts of impact energy using a drop weight test. The parameters of the study were facing thickness and foam core density. Three different facing thicknesses (one, two, and three layers of a bidirectional flax fabric with a nominal unit weight of 400 kg/m^2) and three different core densities (32 kg/m^3, 64 kg/m^3, and 96 kg/m^3) were used. An accelerometer was placed on the bottom face of each specimen and on the drop weight. The data from the accelerometers is used to determine each specimen's fundamental frequency after each impact to determine the amount of damage in the specimen. To calculate curvature, strain gauges were placed at the center of the top and bottom face at mid-span. Two string-type displacement gauges were placed at mid-span below the specimen, to accurately measure the displacement during the impact event. The data sampling rate was 25 kHz. The information obtained is used to determine the ability of these panels to absorb the energy from multiple impact events. It also helps to establish a fundamental understanding of the behavior of the panels under impact for future research. To date all specimens have been fabricated and one of the specimens has been tested. The results are presented in this paper and more data will be available for presentation during the conference.

Keywords: Flax, Fiber, Sandwich Panel, Impact, Energy

Introduction
Where lightweight and insulation efficiencies are required, sandwich panels are a popular choice of building material. Sandwich panels are typically constructed of two structural faces separated by a lightweight core with high insulative properties. The separation of the faces provides a large moment of inertia to resist bending [1]. A common choice for facing materials are fiber-reinforced polymers (FRPs) because of their high specific strength. Because failure of sandwich panels is often initiated in the weaker core material [2,3], there is an option to use lower strength, yet environmentally friendly facing materials, such as FRPs made with renewable constituents, such as flax fibers and bio-based resins.

Sandwich panels constructed with synthetic faces and foam cores have been studied under impact loading [4–6]. Panels constructed with FFRP faces have been studied in flexural applications [2,3,7,8] and axial loading applications [9]. However, there is a gap in the available research which is the behavior of sandwich panels with FFRP faces under impact loading. In this study, the cumulative energy absorption of sandwich panels with FFRP faces and foam cores is examined. The research in this paper will serve as a starting point to further investigations on this topic.

Experimental program

Test Matrix
Nine sandwich beam specimens were fabricated for testing under a drop weight impact with increasing energy until failure. The parameters of the tests were facing thickness (1, 2 and 3 layers of

bidirectional flax fabric) and core density (32, 64 and 96 kg/m^3). The test matrix is shown in Table 1. In this paper, the behavior of specimen 2FL-P400 is examined.

TABLE 1 – TEST MATRIX

No.	Specimen I.D.	Number of Layers	Core Density (kg/m^3)
1	1FL-P200	1	32
2	2FL-P200	2	32
3	3FL-P200	3	32
4	1FL-P400	1	64
5	2FL-P400	2	64
6	3FL-P400	3	64
7	1FL-P600	1	96
8	2FL-P600	2	96
9	3FL-P600	3	96

Note: FL = Flax Layers, PYYY = Core Type (P200 = 32kg/m^3, P400 = 64 kg/m^3, P600 = 96 kg/m^3)

***FIGURE 1** – SPECIMEN FABRICATION: (A) PLAIN FOAM SECTION (B) APPLICATION OF FLAX LAYER (C) APPLICATION OF BIO-BASED EPOXY (D) APPLICATION OF PARCHMENT PAPER (E) CURING (F) REMOVAL OF PARCHMENT PAPER (G) FINISHED SAMPLES*

Non-Conventional Materials and Technologies – NOCMAT for XXI Century   Materials Research Forum LLC
Materials Research Proceedings **7** (2018) 275-284   doi: http://dx.doi.org/10.21741/9781945291838-25

Materials and Fabrication

The specimens used in this study were fabricated as a part of a larger study including specimens used in a study on the static flexural behavior of the panels [8]. The facings were made of a bidirectional flax fabric with a nominal areal weight of 400 g/m^2 (manufacturer: Composites Evolution, Chesterfield, UK) and a bio-based epoxy resin (manufacturer: Entropy Resins, Hayward, CA, USA).

The cores used were closed cell polyisocyanurate foams (manufacturer: Elliott Company, Indianapolis, IN, USA). Figure 1 shows the typical procedure for manufacturing each sandwich panel. The panels were made in 610 mm (weft fabric direction) by 1220 mm (warp fabric direction) sections and were cut down to the specimen size of 150 mm by 1200 mm after curing. A more detailed description of the materials and fabrication is available in the former study by Betts et al. [8], where similar specimens were tested under monotonic loading.

Test Setup and Instrumentation

Each specimen was tested using a drop weight with a mass of 10.413 kg at increasing heights until failure. The test set-up is shown in Figure 2. The span length of each test specimen was 1117 mm. Strain gauges were installed at mid-width of the specimen, at mid-span, on both the top and bottom faces. Two string-type displacement sensors were located beneath the specimen at mid-span, 20 mm in, from each edge of the specimen. The average result from the two displacement sensors was taken as the total displacement of the specimen throughout the impact event. An accelerometer was placed on the bottom of each specimen at mid-span and an additional accelerometer was placed on the drop weight.

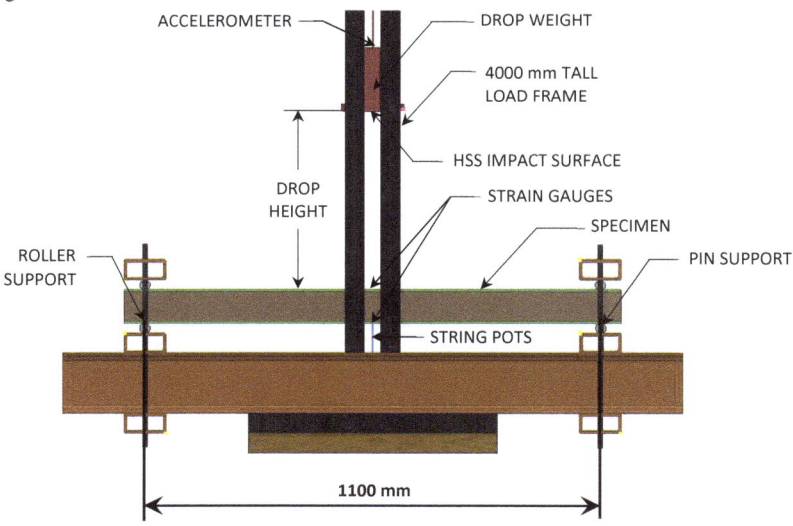

FIGURE 2 – TEST SET-UP

Results and discussions

The specimen 2FL-P400 was subjected to impacts from heights of 300 mm, 600 mm, 900 mm, 1200 mm and 1500 mm by a drop weight with a total mass of 10.413 kg. Figure 3 shows the impact event captured by a high-speed camera in three stages: before impact, maximum deflection and after impact.

Non-Conventional Materials and Technologies – NOCMAT for XXI Century   Materials Research Forum LLC
Materials Research Proceedings **7** (2018) 275-284   doi: http://dx.doi.org/10.21741/9781945291838-25

FIGURE 3 – IMPACT EVENT FROM A DROP HEIGHT OF 1200 MM (A) BEFORE IMPACT (B) MAXIMUM DEFLECTION   (C) AFTER IMPACT

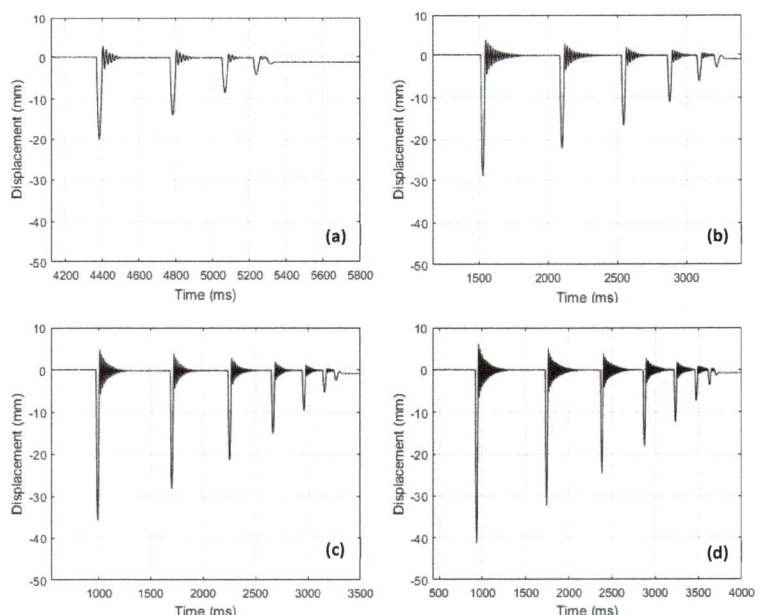

FIGURE 4 – DEFLECTION VS. TIME (A) 300 MM DROP (B) 600 MM DROP (C) 900 MM DROP (D) 1200 MM DROP

Non-Conventional Materials and Technologies – NOCMAT for XXI Century   Materials Research Forum LLC
Materials Research Proceedings 7 (2018) 275-284   doi: http://dx.doi.org/10.21741/9781945291838-25

Figure 4 shows the measured deflection after a moving average filter was applied to the data. The moving average filter was used to eliminate signal noise. After completing the moving average, the filtered data was plotted alongside the raw data to verify that the average did not affect the actual deflection data. The filtered deflection data was used to determine the specimen stiffness, damped period and damping coefficient.

The damped period of the specimen was determined by finding the average time elapsed between the crests and troughs of the deflection curve while the specimen was experiencing free vibration (i.e. while the weight and specimen were not touching). The damping ratio, ξ, was also calculated using the deflection data assuming that it was less than 10%, using Eqn. 1.

$$\xi = \frac{1}{2\pi p} \ln\left(\frac{u_n}{u_{n+p}}\right) \tag{1}$$

Where p is the number of full periods between the peak deflections, u_n and u_{n+p}, as shown in Figure 5.

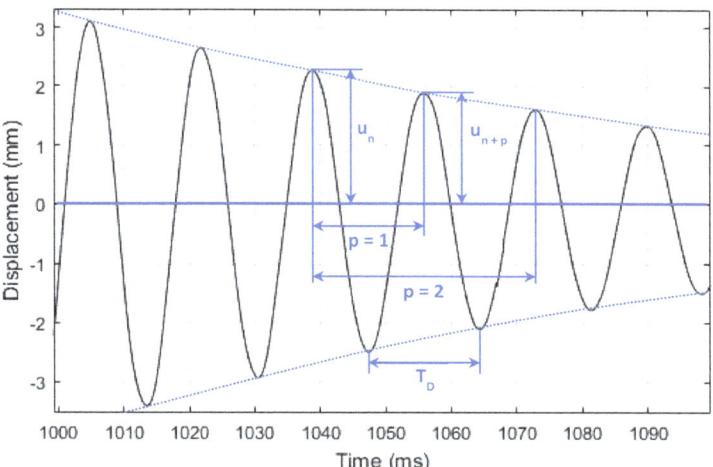

FIGURE 5 – *DAMPING OF SANDWICH PANEL UNDER FREE VIBRATION*

The natural angular frequency, ω_n, could then be calculated by Eqn. 2 assuming the damped and undamped periods are the same (i.e. $T_n = T_D$), which is valid if $\xi < 10\%$.

$$\omega_n = \frac{2\pi}{T_n} \tag{2}$$

The specimen stiffness, K, could then be calculated using Eqn. 3 as follows:

$$K = \frac{\omega^2 mL}{2} \tag{3}$$

Non-Conventional Materials and Technologies – NOCMAT for XXI Century   Materials Research Forum LLC
Materials Research Proceedings 7 (2018) 275-284   doi: http://dx.doi.org/10.21741/9781945291838-25

Where m = specimen mass / length and L = span length. Each peak deflection after the first is a rebound impact. The figure shows that after the drop from 300 mm, the drop weight rebounded three times. After the 600 mm drop initial impact, the drop weight rebounded four times. When dropped from the heights of 900 mm and 1200 mm, the drop weight rebounded five times.

Figure 6 shows the strain-time plot for each impact drop height other than the 1500 mm. By comparing the maximum strains from each drop height and comparing to the load-strain diagrams from the static tests [8] of the previous study, we can determine the equivalent static load to cause the strains in this specimen. Looking at the maximum strain in the tensile (bottom) face from the plots in Figure 6 and comparing to the previous study, the 300 mm drop is approximately equivalent to a static load of 2 kN, a 600 mm drop is equivalent to 2.5 kN, a 900 mm drop is equivalent to 3 kN, and a 1200 mm drop is equivalent to 3.2 kN. This same procedure can be completed by comparing the maximum deflections from Figure 5 to obtain similar results. The maximum facing strain in the previous study [8] was approximately 0.009 mm/mm, which is similar to the maximum strain of 0.0091 mm/mm in the current study, as shown in Figure 6d and Table 2. Therefore, future tests can be tailored to reach failure energy by predicting approximate specimen behavior based on the static test data. However, it should be noted that due to the different strain rates in static and impact tests, the material properties can vary. In future tests, the effect of strain rate on the material properties will be determined.

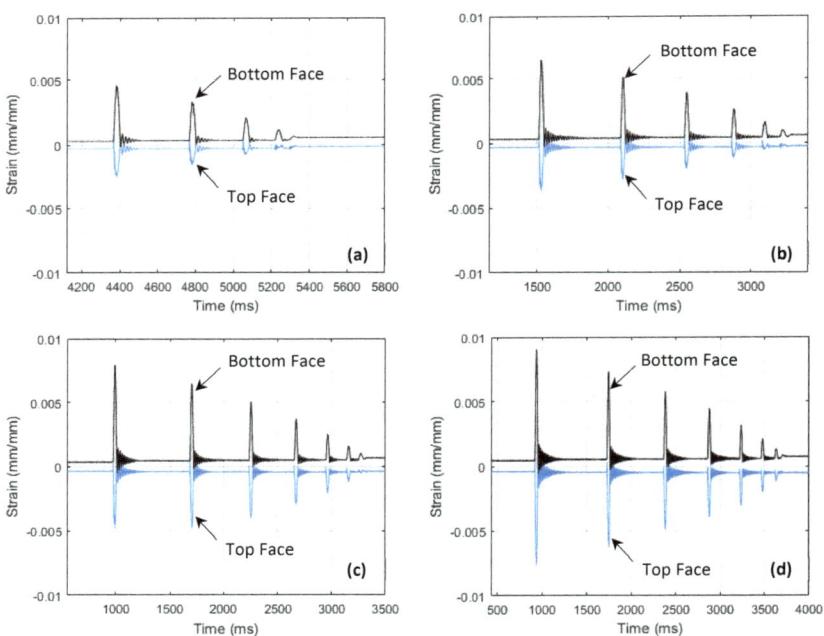

***FIGURE 6** – STRAIN VS. TIME (A) 300 MM DROP (B) 600 MM DROP (C) 900 MM DROP (D) 1200 MM DROP*

Non-Conventional Materials and Technologies – NOCMAT for XXI Century   Materials Research Forum LLC
Materials Research Proceedings 7 (2018) 275-284   doi: http://dx.doi.org/10.21741/9781945291838-25

Table 2 shows the test results of the specimen at each drop height. At a drop height of 1200 mm, the energy is 122.58 N-m. Upon examining the load-deflection curve of the static test performed by Betts et al. [8], the energy absorption of the static test specimen was found to be approximately 75 N-m . This shows that the energy absorption of the impact test was higher than that of the static test. The specimen experienced ultimate failure at an impact energy of 153 N-m (drop height of 1500 mm). This indicates that the failure energy is between 123 N-m and 153 N-m. The failure is shown in Figure 7. The specimen failed simultaneously in tensile rupture and shear failure. The shear-type failure symmetrically about the drop weight impact as shown in Figure 7a.

FIGURE 7 – SPECIMEN FAILURE AT DROP HEIGHT OF 1500 MM (A) DURING IMPACT (B) AFTER IMPACT (C) DETAIL (D) DETAIL

Figure 8a shows the effect of impact energy on the calculated stiffness of the specimen. It shows that the stiffness is relatively constant until the 1500 mm drop which caused failure. This is indicative that there was no damage in the specimen before the final loading. This agrees with the test observations as no visible damaged was seen during the testing before failure. If damage were present in the specimen after an impact event, it would be expected that the stiffness of the member would decrease. Figure 8b shows the damping ratio as calculated from the free vibration of the first

Non-Conventional Materials and Technologies – NOCMAT for XXI Century   Materials Research Forum LLC
Materials Research Proceedings 7 (2018) 275-284   doi: http://dx.doi.org/10.21741/9781945291838-25

impact from each energy level. It shows that after the first hit there is a reduction in the damping ratio of approximately 50%, after which the damping ratio stays within 3 to 4 %. As the damping ratio is always under 10%, it verifies the earlier assumption that the natural period can be approximated as the damped period, T_D.

As expected, Figure 8c, shows that as the impact energy increases, the maximum deflection also increases. Figure 8d similarly shows the increase in the strain on the top and bottom faces. Figure 8d also shows that the bottom face strain is approximately 0.002 mm/mm larger than the top face strain. However, as shown by Betts et al. (2017), the neutral axis of the specimens is located close to the midplane [8] and the strains should be opposite but approximately equal in magnitude. It is hypothesized that this difference is caused by the HSS impact surface shown in Figure 2. As there is a strain gauge in the center of the top face, a 25 mm diameter hole was placed in the middle of the HSS impact surface, to avoid damage to the strain gauge. This could have potentially caused local tension in this area of the face, thereby reducing the strain gauge reading of the bending strain in the face.

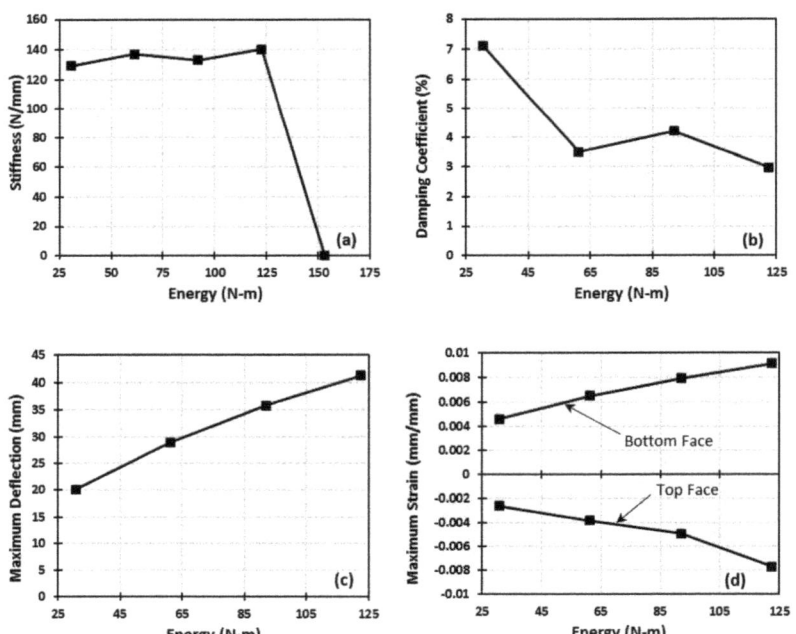

FIGURE 8 – EFFECT OF INPUT ENERGY ON SPECIMEN (A) STIFFNESS (B) DAMPING COEFFICIENT (C) MAXIMUM DISPLACEMENT (D) MAXIMUM FACING STRAIN

Conclusions

In this study, a sandwich specimen constructed of two-ply FFRP facings and a 64 kg/m^3 foam core was tested under varying impact energies. The specimen was not visibly damaged after impact events of up to 122 N-m, but it failed at an impact energy of 153 N-m. As expected, the specimen deflection and facing strains increased with impact energy. However, the stiffness was relatively constant until failure. Based on the results of the tests in this study, the following conclusions were made:

- The maximum cumulative energy absorption on of specimen 2FL-P400 is between 123 and 155 joules and failed simultaneously by tensile face rupture and shear.
- Both maximum deflection and facing strain increased with impact energy;
- There was no obvious damage present before failure;
- And the sandwich panels can absorb more energy from an impact event than they can under quasi-static loading.

These tests are a part of an ongoing research and helped to provide insight into the general behavior of these structures as well as to provide an understanding of future testing requirements, such as:

- the use of smaller energy increments to capture the damage initiation energy and actual failure energy;
- testing of all specimen types to determine effect of the facing thickness and core density on energy absorption and failure mode
- development of a numerical model to predict impact behavior of sandwich specimens

Acknowledgement

The authors would like to thank Brandon Fillmore, Jesse Keane, Brian Kennedy for their assistance in the lab. The authors would also like to acknowledge and thank Bioindustrial Innovation Canada (BIC), Queen's University, and Dalhousie University for their in kind and financial support.

References

[1] Allen HG. Analysis and Structural Design of Sandwich Panels. Oxford, UK: Pergamon Press; 1969.

[2] Sadeghian P, Hristozov D, Wroblewski L. Experimental and analytical behavior of sandwich composite beams: Comparison of natural and synthetic materials. J Sandw Struct Mater 2016:1–21.

[3] Fam A, Sharaf T, Sadeghian P. Fiber element model of sandwich panels with soft cores and composite skins in bending considering large shear deformations and localized skin wrinkling. J Eng Mech 2016;142:1–14. https://doi.org/10.1061/(ASCE)EM.1943-7889.0001062

[4] Atas C, Potoglu U. The Effect of Face-Sheet Thickness on Low-Velocity Impact Response of Sandwich Composites with Foam Cores. J Sandw Struct Mater 2016;18:215–28. https://doi.org/10.1177/1099636215613775

[5] Schubel PM, Luo J-J, Daniel IM. Low velocity impact behavior of composite sandwich panels. Compos Part A Appl Sci Manuf 2005;36:1389–96. https://doi.org/10.1016/j.compositesa.2004.11.014

Non-Conventional Materials and Technologies – NOCMAT for XXI Century   Materials Research Forum LLC
Materials Research Proceedings 7 (2018) 275-284   doi: http://dx.doi.org/10.21741/9781945291838-25

[6] Torre L, Kenny JM. Impact testing and simulation of composite sandwich structures for civil transportation. Compos Struct 2000;50:257–67. https://doi.org/10.1016/S0263-8223(00)00101-X

[7] Mak K, Fam A. Bio Resins and Bio Fibers for FRP Applications in Structural Engineering Applications. 7th Intnernational Conf. Adv. Compos. Mater. Bridg. Struct., Vancouver, BC, Canada: 2016, p. 1–6.

[8] Betts D, Sadeghian P, Fam A. Structural Behaviour of Sandwich Panels Constructed of Foam Cores and Flax FRP Facings. CSCE Annu. Conf. Vancouver, BC, Cananda, 2017, p. 1–10.

[9] Codyre L, Mak K, Fam A. Flexural and axial behaviour of sandwich panels with forced polymer skins and various foam core densities 2016.

Non-Conventional Materials and Technologies – NOCMAT for XXI Century   Materials Research Forum LLC
Materials Research Proceedings 7 (2018) 285-294   doi: http://dx.doi.org/10.21741/9781945291838-26

Green Sandwich Composites Fabricated from Flax FRP Facings and Corrugated Cardboard Cores

Aidan McCracken [a], Pedram Sadeghian [a, *]

[a] Dalhousie University, Canada; Aidan.McCracken@dal.ca, Pedram.Sadeghian@dal.ca*

Abstract. Composite sandwich beams and panels made of fiber-reinforced polymer (FRP) and lightweight, low-density core materials have been shown to be effective in reducing weight and increasing strength and stiffness in a variety of structural applications. The FRP skins resist the tensile and compressive stresses as a result of flexure, similar to the action of the flanges on an I-Beam, while the core resists shear stresses, provides insulation and increases the distance between skins resulting in a higher moment of inertia. In this study, sandwich panels made of green materials are studied. Namely, flax fibers and partial bio-based epoxy were used for the FRP skin and three flute varieties of corrugated cardboard with bulk densities of 127, 138 and 170 kg/m^3 were used for the core. A total of 30 small-scale sandwich beam specimens were manufactured across six unique beam varieties with dimensions of 50 mm in width, 25 mm in depth, and 200 and 350 mm in length (150 mm and 300 mm spans) to be tested under four-point bending up to failure. This is an ongoing research and so far 6 of the sandwich beams have been tested and the results are presented in this paper. The load-deflection behavior, load-strain behavior and moment-curvature behavior as well as the strength and stiffness of the sandwich beam specimens were analyzed. Overall, the flax FRP and cardboard sandwiches displayed promising structural behavior and may be considered as a viable, green option for the fabrication of sandwich composite panels. More results will be presented during the conference.

Keywords: Green, Flax Fiber Composite, Bio-Based Polymer, Cardboard, Sandwich

Introduction

The abundance of structural sandwich panels is growing as civil engineers look to improve the structural efficiency of building materials. Comprised of two high-strength facesheets to resist tensile and compressive stresses of bending as well as a low-density core, sandwich panels are often favoured due to their light weight and high moment of inertia [1]. In addition to separating the two facesheets, the core provides strength to resist transverse and longitudinal shear stresses and may also provide greatly improved thermal insulation [2]. In order to be more environmentally-concious, building materials must be reevaluated to determine how they can become more sustainable and have a smaller environmental impact during production. This will limit waste and pollution in the process of constructing and mnaintaining buildings and infrastructure.

Although synthetic fiber-reinforced polymer (FRP) composites , such as glass FRP or carbon FRP, are often used for the facesheets of sandwich beams, the concept of using natural fibers, such as flax or hemp, has also been explored [3][4]. Although the natural fibers have a lower strength than their synthetic counterparts, it has been showed that this may be acceptable since the core strength is what often governs the failure of the beams [5][6]. Additionally, natural fibers have many economic and environmental advantages compared to synthetic fibers [7]. Thus, flax FRP facesheets represent a viable structural option for sandwich beams and are a more environmentally-friendly choice than synthentically produced fibers.

Many different core materials have been explored for use in composite sandwich beams and panels. Core materials that are commonly studied include low-density foam and plastic or metal honeycombs [8][9]. In order to present a more sustainable option, this study will use corrugated cardboard as the core material. According to the Paper and Paperboard Packaging Environmental Council (PPEC), approximately 85% of corrugated cardboard in Canada is recycled and new cardboard is produced with nearly 100% recycled materials [10]. Along with being 100% biodegradable, corrugated cardboard is a very sutainable as it can be repuposed and produces very little waste. Although studies have been conducted on bio-based sandwich composites [11], recycled corrugated cardboard has not been explicitly studied in the context of a sandwich beam with natural fibres and natural epoxy.

In this paper, flax FRP facesheets is combined with corrugated cardboard cores to manufacture sandwich beams. In addition to these materials, the beams were cured using a non-toxic and organic epoxy. As a result, the sandwich beams produced were constructed using entirely green materials. The aim of the study is to analyze and evaluate the structural performance of corrugated cardboard and flax FRP composite sandwich beams. Although flax has previously studied for use in sandwich beams, the combination of flax FRP with cardboard has yet to be analyzed. This combination of materials represents a structural panel that has a minimal impact on the environment as corrugated cardboard is readily available and composed almost entirely of recycled material.

Experimental program

Test Matrix
In total, 30 flax FRP and corrugated cardboard sandwich beams were fabricated to be tested in four-point bending. All specimens were constructed using one layer of flax FRP skin on either side and a corrugated cardboard core with a thickness of approximately 25 mm. The variables being tested were span length as well as the flute of the corrugated cardboard. Two span lengths, 150 mm and 300 mm, as well as three cardboard flutes, B, C and BC, were tested. More information concerning the flutes can be found in the following section, *Material Properties*. A complete summary of this study's test matrix is shown in Table 1. Note that five identical specimens were manufactured and tested per case. All specimens are identified with a specimen ID which follows the format X-SY where X identifies the cardboard flute, S stands for span and Y iodentifies the specimens test span in mm. For example, the specimen ID B-S150 designates a flax FRP and caerdboard sandwich beam constructed using B flute cardboard with a test span length of 150 mm.

TABLE 1 - TEST MATRIX

Case #	Specimen ID	Flute	Span (mm)
1	B-S150	B	150
2	B-S300	B	300
3	C-S150	C	150
4	C-S300	C	300
5	BC-S150	BC	150
6	BC-S300	BC	300

Material Properties
As previously mentioned, three unique flutes were used in the fabrication of the sandwich beams: B, C and BC. Cardboard flutes are standard in international packing and are identified with a single capital letter. Each flute has a different nominal thickness and density. Table 2 compares the approximate measured dimensions of each flute in this study. The density measurements were taken after the flute layers had been combined into a core for the specimens. Thus, this density reflects the actual density of the core, including the small amount of adhesive used to combine the layers of cardboard.

Non-Conventional Materials and Technologies – NOCMAT for XXI Century   Materials Research Forum LLC
Materials Research Proceedings 7 (2018) 285-294   doi: http://dx.doi.org/10.21741/9781945291838-26

TABLE 2 - FLUTE COMPARISON

Flute	Thickness (mm)	Flutes per Meter	Density (kg/m^3)
B	2.8	160	170
C	4.0	120	127
BC	6.6	Mix	138

Figure 1 shows a visual comparison between the flutes with both a photo of the flutes as a part of a core as well as a 2D side-view schematic.

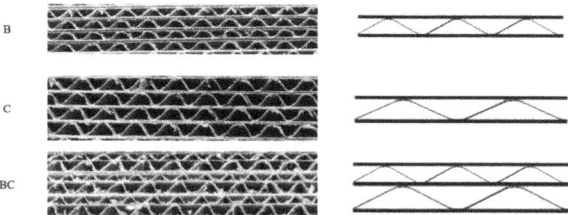

FIGURE 1 - VISUAL FLUTE COMPARISON

For the flax FRP skins, a unidirectional flax fabric with a reported aerial weight of 275 g/m^2 (gsm) was used. In terms of epoxy, Super Sap ONE was used, which is a bio-based epoxy with a reported tensile strength, modulus and elongation of approximately 53.23 MPa, 2.65 GPa and 6 %, respectively. Betts et al. [12] condcuted a study on the tensile properties of flax FRP composites manufactured using the same unidirectional flax fabric and three different epoxies. For the flax FRP samples tested with the bio-based Super Sap ONE epoxy, the average tensile strength and inital modulus were reported to be 198.0 ± 9.3 MPa and 17.09 ± 0.63 GPa, respectively. A secondary modulus was reported as 11.93 ± 0.39 GPa as it was found flax FRPs display an approximately bi-linear mechanical behavior.

Specimen Fabrication

The first step in the fabrication of the sandwich beams was to construct the cardboard cores. To do this, strips of cardboard (manufacturer: Maritime Paper, Dartmouth, NS, Canada) approximastely 25 mm in with were cut from larger panels using a straight edge and a sharp blade. The two span lengths being tested were 150 and 300 mm, thus strips were cut to lengths of 200 and 350 mm to provide an overhang of approximately 25 mm on each end of the specimen. To bond the strips together, a small amount of Tri-Tex Tribond P-1031 adhesive was used. This adhesive was provided by Maritime Paper and is the same used in the manufacturing of corrugated cardboard.. The number of strips in the core varied per flute as all cores were manufactured to have an approximate width of 50 mm. Figure 2 shows the fabrication process of the cardboard cores.

Non-Conventional Materials and Technologies – NOCMAT for XXI Century   Materials Research Forum LLC
Materials Research Proceedings 7 (2018) 285-294   doi: http://dx.doi.org/10.21741/9781945291838-26

FIGURE 2 - *CARDBOARD CORE FABRICATION: (A) CUTTING; (B) APPLYING AHDESIVE; (C) COMBINING INTO ONE CORE; AND (D) COMPLETED CORES FOR 150 MM SPAN.*

FIGURE 3 - *FABRICATION: (A) APPLYING EPOXY; (B) SATURATING FLAX FABRIC; (C) PLACING CARDBOARD CORES ON SATURATED FABRIC; (D) FIRST SIDE COMPLETE; AND (E) BOTH SIDES COMPLETE.*

Once the cardboard cores were completed, the flax FRP skins were applied using the standard wet lay-up method. Sheets of flax fabrix aqpproximatelty 300 mm in width and either 200 or 350 mm in length were pre-cut before the mixing of the epoxy. A sheet of parchment paper was put on the bottom surface and a layer of epoxy was applied. Next, a sheet of flax fabric was applied to the epoxy, then the top side of the fabric was saturated with another layer of epoxy. Each of the five cores per case was placed on the saturated sheet of flax. A piece of particle board was placed on top of the cores while the bottom layer of flax FRP cured. Once the first side of had cured, this process was repearted for applying the flax FRP skin to the other side of the cores. This method allowed for

Non-Conventional Materials and Technologies – NOCMAT for XXI Century   Materials Research Forum LLC
Materials Research Proceedings **7** (2018) 285-294   doi: http://dx.doi.org/10.21741/9781945291838-26

the curing FRP to always be below the cardboard core to help ensure that unwanted resin did not seep down into the cardboard. Figure 3 shows the application process of the second side of flax FRP.

Applying larger sheets of flax fabric allowed for a quicker fabrication process. Once both sides had fully cured, a bandsaw was used to cut the beams to their approximate width of 50 mm and a rotary sander was used to smooth the edges of the flax composite and ensure it was in line with the sides of the core. A completed sandwich beam is shown in Figure 4.

FIGURE 4 – COMPLETED BEAMS.

Test Setup

All specimens were tested under four-point bending with a loading span proportional to the supporting spans of 150 and 300 mm. As per ASTM D7249 [13] and D7250 [14], the loading span (L) was to be equal to (2/11) of the supporting span (S). A schematic of the four-point bending set up is shown in Figure 5 where S is the supporting span, L is the loading span and P is the applied load.

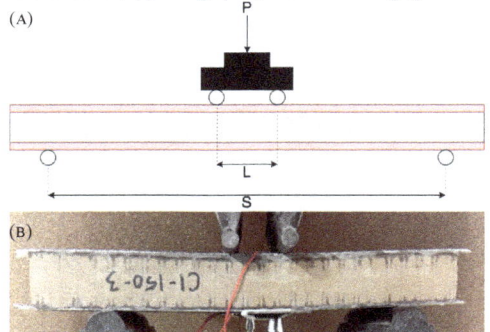

FIGURE 5 – (A) FOUR-POINT BENDING SCHEMATIC; AND (B) SPECIMEN READY FOR TESTING.

Non-Conventional Materials and Technologies – NOCMAT for XXI Century   Materials Research Forum LLC
Materials Research Proceedings 7 (2018) 285-294   doi: http://dx.doi.org/10.21741/9781945291838-26

In terms of instrumentation, a strain gauge was applied on either side of the sandwich beam, centered in the longitudinal direction to measure he tensile and compressive strains. Additionally, two linear potentiometers were setup in the middle of the beam's span to measure an average mid-span deflection. These values, along with the applied load, measured every 0.1 seconds, were collected for data processing. All tests were completed using a 100 kN universal testing machine and were displacement controlled using a fixed rate of 2 mm/min. Figure 6 shows photos of the sandwich beams before and after testing.

Results and discussion
A summary of the test results as well as the modes of failure for the C-S150 and C-S300 specimens is shown below in Table 3.

TABLE 3 - TEST RESULTS

ID	Peak Load (N)		Initial Stiffness (N/mm)		Deflection at peak (mm)		Peak moment (N-m)		Curvature at peak (1/km)		Failure Mode
	AVG	SD	AVG	SD	AVG	SD	AVG	SD	AVG	SD	
C-S150	3985	297	2209	144	3.37	0.25	122.3	9.1	953	188	Vertical crushing
C-S300	1715	201	387	27	9.73	1.05	105.2	12.4	1168	208	Longitudinal crushing

NOTE: AVG = AVERAGE; SD = STANDARD DEVIATION

Failure Modes
As expected, the failure of the core was the initial source of failure in both the 150 and 300 mm span sandwich beams. Due to their higher stiffness, the 150 mm span specimens did not flex very much, only deflecting and average of 3.37 mm at peak load. All three tested 150 mm specimens failed by vertical crushing of the core due to transverse shear stresses. This was followed by indentation of the top layer of flax. However, the 300 mm span samples reached a significantly lower peak load and failed by longitudinal crushing of the core due to bending and longitudinal shear stresses. Once the corrugated cardboard had begun crushing longitudinally, this created a noticeable increase in compressional strain on the top of the beam, which caused the flax FRP to rupture after the peak load. This result is somewhat expected, as corrugated cardboard is designed to have strength in the vertical direction to resist crushing. Images of these two failure modes can be seen in Figure 7.

FIGURE 7 – *FAILURE COMPARISON: (A) VERTICAL CRUSHING; (B) LONGITUDINAL CRUSHING; AND (C) DETAIL OF COMPRESSIONAL RUPTURE.*

Non-Conventional Materials and Technologies – NOCMAT for XXI Century   Materials Research Forum LLC
Materials Research Proceedings 7 (2018) 285-294   doi: http://dx.doi.org/10.21741/9781945291838-26

Load-Strain Behavior

As previously mentioned, the 300 mm span samples experienced much larger compressive strain compared to tensile strain. This was caused by the longitudinal crushing of the cardboard near the top facesheet of the sandwich. The 150 mm span samples experienced comparable tensole and compressive strains until the vertical crushing of the cardboard core. Graphs comparing the load-strain behavior of the C-S150 and C-S300 specimens is shown in Figure 8.

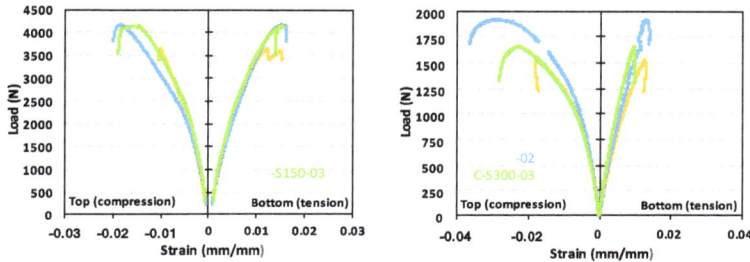

FIGURE 8 – *LOAD-STRAIN GRAPHS: (A) C-S150; AND (B) C-S300.*

Moment-Curvature Behavior

As expected, the moment curvature behavior was similar between the two spans that were compared. Graphs comparing the moment-curvature behavior of the C-S150 and C-S300 specimens is shown in Figure 9. Initial flexural stiffness and the shear rigidity of the core will be discussed further in a following section.

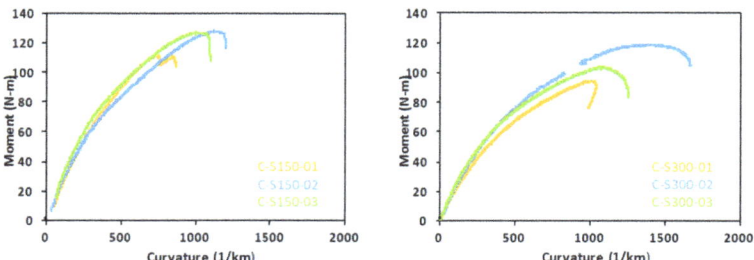

FIGURE 9 – *MOMENT-CURVATURE GRAPHS: (A) C-S150; AND (B) C-S300.*

Load-Deflection Behavior

A Graphs comparing the load-deflection behavior of the C-S150 and C-S300 specimens is shown in Figure 10. Typical failure for the 150 mm specimens was transverse shear failure of the core. Considering the 300 mm specimens, typical failure was longitudinal crushing of the core under compressive normal stress which was followed by crushing of the top facesheet.

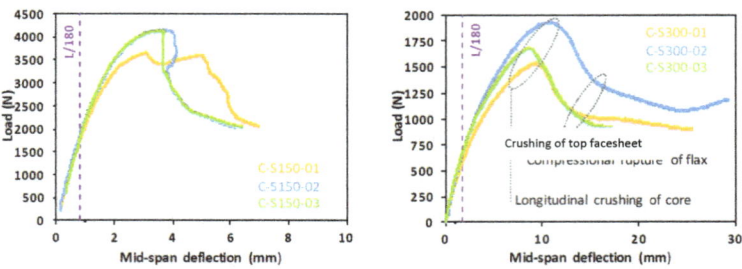

FIGURE 10 – *LOAD-DEFLECTION GRAPHS: (A) C-S150; AND (B) C-S300.*

Flexural Stiffness

Flexural stiffness (D) of the sandwich beams was calculated based on moment-curvature behaviour. Additionally, by comparing the initial stiffness of two span lengths, flexural stiffness (D) and transverse shear rigidity (U) can be calculated by the equation below where K is initial stiffness in N/mm, S is the span length in mm and L is the loading span in mm [6] [14] as follows:

$$K_i \frac{(2S_i^3 - 3SL_i^2 + L_i^3)}{96D} + K_i \frac{(S_i - L_i)}{4U} = 1 \tag{1}$$

where $i = 1$ denotes the parameters to the short-span specimens, and $i = 2$ to the long-span specimens. The first term in the equation is related to flexural deformation and the second term to shear deformation. Combining the equations for each span length and simplifying gives:

$$D = \frac{\alpha_2 - \frac{\alpha_1 \delta_2}{\delta_1}}{96(\frac{1}{K_2} - \frac{\delta_2}{\delta_1 K_1})} \tag{2}$$

$$U = \frac{\delta_2 - \frac{\alpha_2 \delta_1}{\alpha_1}}{4(\frac{1}{K_2} - \frac{\alpha_2}{\alpha_1 K_1})} \tag{3}$$

where

$$\alpha_i = 2S_i^3 - 3S_i L_i^2 + L_i^3 \tag{4}$$

$$\delta_i = S_i - L_i \tag{5}$$

Table 4 shows the calculated values for D and U as well as an experimental value of D based on moment-curvature behavior. As the tests move forward this table will be complete. The results will be also compared with similar sandwich beams with alternative synthetic materials.

TABLE 4 - FLEXURAL PROPERTIES

Flute	D (N-m^2) [Calculated]	D (N-m^2) [Average based on curvature]	U (kN)
C	239.19	225.13	178.0

Conclusion

In this study, flax FRP facesheets and three different flutes of corrugated cardboard cores, namely B, C and BC, were used to manufacture composite sandwich beams with two different span lengths of 150 and 300 mm. This is an ongoing research and currently only the sandwich beams containing the C-flute core have been tested under four-point bending. Compared to the 300 mm span samples which failed at an average load of 1715 N, the 150 mm span samples failed at an average load of 3985 N. Corrugated cardboard displayed impressive strength against transverse shear, however it was not as strong under compressive normal stress in the longitudinal direction. Once the remainder of specimens are tested, a more comprehensive understanding of how corrugated cardboard performs as a core will be developed. Although more research must be conducted, the all-natural flax FRP and corrugated cardboard sandwich beams displayed encouraging structural behaviour and may prove to be a sustainable and structurally efficient building material.

Acknowledgement

The authors of this paper would like to acknowledge the efforts of Dillon Betts, MASc Student at Dalhousie University, who aided in the manufacturing and provided guidance throughout the project as well as Dalhousie's technicians, Jesse Keane and Brian Kennedy, who helped immensly with the setup and testing process. In addition, the authors acknowledge the National Science and Engineering Research Council of Canada (NSERC) for the Undergraduate Student Research Award (USRA) for the first author, Dalhousie University for supplemental funding through the Innovation-Themed Undergraduate Research Funding, and Maritime Paper (Dartmouth, NS, Canada) for providing the corrugated cardboard and adhesive.

References

[1] Reis EM, Rizkalla SH. Material characteristics of 3-D FRP sandwich panels. Construction and Building Materials. 2008 Jun 30;22(6):1009-18. https://doi.org/10.1016/j.conbuildmat.2007.03.023

[2] Allen HG. Analysis and design of structural sandwich panels: the commonwealth and international library: structures and solid body mechanics division. Elsevier; 2013 Oct 22.

[3] Manalo A, Aravinthan T, Fam A, Benmokrane B. State-of-the-art review on FRP sandwich systems for lightweight civil infrastructure. Journal of Composites for Construction. 2016 Jul 5;21(1):04016068. https://doi.org/10.1061/(ASCE)CC.1943-5614.0000729

[4] Mallaiah S, Sharma KV, Krishna M. Development and comparative studies of bio-based and synthetic fiber based sandwich structures. Int J Soft Compos Eng. 2012;2:332-5.

[5] Betts D, Sadeghian P, Fam A. Structural Behaviour of Sandwich Panels Constructed of Foam Cores and Flax FRP Facings. Canadian Society for Civil Engineering (CSCE) Annual Conference, May 31 – Jun. 3, 2017, Vancouver, BC, Canada.

Non-Conventional Materials and Technologies – NOCMAT for XXI Century   Materials Research Forum LLC
Materials Research Proceedings 7 (2018) 285-294   doi: http://dx.doi.org/10.21741/9781945291838-26

[6] Sadeghian P, Hristozov D, Wroblewski L. Experimental and analytical behavior of sandwich composite beams: Comparison of natural and synthetic materials. Journal of Sandwich Structures & Materials. 2016 May 31:1099636216649891.

[7] Wambua P, Ivens J, Verpoest I. Natural fibres: can they replace glass in fibre reinforced plastics?. composites science and technology. 2003 Jul 31;63(9):1259-64.

[8] Gupta N, Woldesenbet E. Characterization of flexural properties of syntactic foam core sandwich composites and effect of density variation. Journal of composite materials. 2005 Dec;39(24):2197-212. https://doi.org/10.1177/0021998305052037

[9] Wadley HN, Fleck NA, Evans AG. Fabrication and structural performance of periodic cellular metal sandwich structures. Composites Science and Technology. 2003 Dec 31;63(16):2331-43. https://doi.org/10.1016/S0266-3538(03)00266-5

[10] Paper and Paperboard Packaging Environmental Council. Factsheet 03-2017. http://www.ppec-paper.com/pdfFiles/factsheets/2017/Packaging/FS03-2017.pdf

[11] Dweib MA, Hu B, O'donnell A, Shenton HW, Wool RP. All natural composite sandwich beams for structural applications. Composite structures. 2004 Feb 29;63(2):147-57. https://doi.org/10.1016/S0263-8223(03)00143-0

[12] Betts D, Sadeghian P, Fam A. Tensile Properties of Flax FRP Composites. In 6th Asia-Pacific Conference on FRP in Structures (APFIS 2017) 2017 Jul 19. 6th Asia-Pacific Conference on FRP in Structures, Singapore.

[13] ASTM D7249. Standard test method for facing properties of sandwich constructions by long beam flexure. West Conshohocken, PA, USA: ASTM International, 2012.

[14] ASTM D7250. Standard practice for determining sandwich beam flexural and shear stiffness. West Conshohocken, PA, USA: ASTM International, 2012.

Non-Conventional Materials and Technologies – NOCMAT for XXI Century   Materials Research Forum LLC
Materials Research Proceedings 7 (2018) 295-312   doi: http://dx.doi.org/10.21741/9781945291838-27

Physical-Mechanical Behavior of Metakaolin Based Geopolymer Systems Reinforced with Stainless Steel Fibers

Liuski Roger Caballero[a,*], Maria D. M. Paiva[a], Eduardo de M. R. Fairbairn[a], Romildo D. T. Filho[a]

[a] Federal University of Rio de Janeiro - UFRJ, Brazil; liuskiroger@gmail.com, ninanatal@gmail.com, eduardo@coc.ufrj.br, toledo@coc.ufrj.br

Abstract. Geopolymers are a promising alternative to ordinary Portland cement (OPC) binders in the manufacture of concrete, as their synthesis generates much less greenhouse gas emissions. Geopolymer binders are subject to drying shrinkage, which can be controlled by decreasing the water to cement ratio and adding inert materials to produce mortars. These systems are also brittle in nature and, therefore, it is important to investigate the fiber reinforcement efficiency under tensile loading. This study is based on typical metakaolin-based geopolymer pastes and mortars, activated with sodium silicate and sodium hydroxide. The aim of the study is to evaluate two metakaolin batches with different contents of crystalline phases, the addition of a fixed natural sand content and different contents of stainless steel fibers and their effect on the physical-mechanical performance of the composites. The geopolymer networks were confirmed by XRD and FTIR characterizations. Microstructure results showed that the geopolymer composites formed a dense and uniform base matrix, with no microvoids and very little Na_2CO_3 efflorescence. Metakaolin-based geopolymer pastes and mortars presented typical mechanical properties, with uniaxial compressive strengths ranging from 33 to 50 MPa and flexural tensile strength ranging from 3.4 to 10.0 MPa. Finally, the geopolymer binders exhibit similar physical-mechanical performances, despite the crystalline content of the metakaolin. Pure and reinforced geopolymer systems display overall superior mechanical performance, when compared to OPC matrices, especially regarding the maximum strain at failure, suggesting that these systems are suitable to replace OPC systems in structural applications.

Keywords: Geopolymer, Geopolymer Mortar, Fiber-Reinforced Geopolymer Composite, Metakaolin, Mechanical Properties

Introduction

The search for alternative materials to conventional concrete has grown remarkably nowadays, and is still an important challenge faced by the concrete industry. The initial industry motivation was to reduce the consumption of Portland cement, focusing on reducing the environmental impact due to CO2 emissions. Currently, the goal is also to seek solutions that enable the development of concretes, mortars and pastes with green footprint and also higher mechanical and durability performances. Geopolymer binders appear as a new class of sustainable materials that fit this gap [1-8].

Geopolymer, geopolymer binder or inorganic polymers are in general defined as a solid and stable material, identified by Si-O-Al sialate bonds that are formed by the alkali activation of an aluminum-silicate precursor source [1-8]. This precursor can be a calcium silicate, as the ones used in the alkali-activation of Portland cement clinkers, or an aluminum-silicate-rich precursor, such as a metallurgical slag, a natural pozzolan, a fly ash or a bottom ash [2]. The geopolymer defining characteristic is that the binding phase comprises an alkali aluminosilicate gel, with aluminum and silicon linked in a three-dimensional tetrahedral gel framework that is relatively resistant to water dissolution [3, 4, 8].

Geopolymers are a promising alternative to ordinary Portland cement (OPC) binders in the manufacture of concrete, as their synthesis generate a much smaller amount of greenhouse gas emissions, when compared to the OPC binders. Due to that, the replacement of OPC was the primary technological application of geopolymers. Furthermore, since they can be manufactured from waste materials and their engineering properties are, in some cases, superior to those obtained with the Portland cement, their use enables different technology applications, depending on the raw materials and the activators adopted in their synthesis [5]. This technology is receiving much attention worldwide, but so far very little application [6].

Metakaolin is one of the most commonly used aluminosilicates sources to produce geopolymers, and it is obtained from the kaolin calcination at a controlled temperature range between 600 and 900°C, where the adopted calcination temperature and time ranges depend on the clay structure [7]. This is one of the most used precursor materials for obtaining geopolymers [1, 8-15].

Geopolymers are subject to drying shrinkage, a phenomenon that occurs to various materials that transition from liquid to solid states. Due to volume reduction, drying cracks are generally observed and attributed to evaporation of unbound water from the binder. This mechanism can be accelerated by curing temperature, heating exposition or when the binder is not cured on a humid and sealed environment. For OPC and geopolymer systems, the drying shrinkage is governed by the volume fraction of solids in the paste, and can be controlled by decreasing the water to cement ratio and by adding inert materials, such as sand, to produce mortars. Some authors have outlined the sand content and aluminosilicate sources in which acceptable drying shrinkage and maximum mechanical performance are achieved [15, 16, 17].

Differently from OPC, geopolymeric binders produce water from its reaction, therefore their thermal stability is affected. Typically, there is around 15% of unbound water loss, followed by a small shrinkage, in an endothermic reaction at 100-200°C and there is further irreversible shrinkage around 800°C, when the metakaolin-based geopolymer network experiment a phase change, becoming dimensionally stable. [10-11, 13-14].

Geopolymers are brittle, in nature, as are OPC systems. Therefore, many authors have devoted time to study the reinforcement efficiency of various micro-fibers on those systems under tensile loading. The addition of steel and stainless fibers in the millimeter scale is a well-known topic for OPC system, but very few studies have evaluated this fiber size reinforcement effect in geopolymer systems [17-18]. It was noticed that steel fibers of 22 mm length have strong bonding with the geopolymer paste, since it tends to behave as a hydrophilic material, which results in significant improvement in energy absorption and flexural strength of the correspondent composite [19].

The present study is based on typical metakaolin-based geopolymer pastes and mortars, activated with a sodium silicate and a fixed 10M sodium hydroxide solution. The aim of the study is to evaluate two metakaolin batches (MK1 and MK2), with different content of crystalline phases, the addition of a fixed 40% (v%/v%) natural sand content and the reinforcement compatibility of different contents (1.2 to 2.4 (v%/v%) of stainless steel fibers in the physical-mechanical performance of these geopolymer compounds.

Materials and methods

The raw materials chemical composition, source and function are summarized on Table 1. Specific gravity, specific surface area and particle size characterization results are summarized on Table 2.

Raw material powders for characterization were obtained by milling the samples until obtaining a particle size passing a 106µm sieve and, when needed, previously grounding them in an agate mortar.

Chemical compositions were determined by X-Ray Fluorescence (XRF) semi quantitative analysis using a XRF-Shimadzu model EDX-720, with a 3kW tube and Rh target and samples were analyzed under vacuum.

Specific gravities were determined in a Gas Helium Pycnometer Micromeritics AccuPyc 1340, with the samples previously dried on an oven at 40°C for 2 hours, and calculating an average of five readings.

Specific surface areas were determined using the BET Micromeritics ASAP 2020 Plus Analyzer, calculating an average of three readings. The samples were heated from 0 to 150°C at 10°C/min for 60 minutes to remove sample humidity and, afterwards, a heat treatment from 150 to 300°C at 5°C/min for 800 minutes was conducted.

Particle size distributions were obtained using the Malvern MasterSizer 2000 laser granulometer, using deionized water as a dispersion agent.

The sodium silicate chemical composition and specific gravity of $1.58g/cm^3$ were provided by Diatom. Sodium hydroxide pellets (NaOH P.A.-97%) were supplied by VETEC.

Natural Sand was sieved to 0-2mm prior to use. Its specific gravity is $2.43g/cm^3$, determined using the ABNT NBR NM 52:2003 standard [20].

TABLE 1 – RAW MATERIALS CHEMICAL COMPOSITION AND SUMMARY INFORMATION.

Oxide (%)	MK1	MK2	Na_2SiO_3	Sand
Al_2O_3	48.74	41.69	-	11.74
SiO_2	45.01	51.85	33	80.71
Fe_2O_3	2.74	1.91	-	0.71
TiO_2	1.52	1.38	-	0.42
SO_3	0.99	1.09	-	1.76
K_2O	0.63	1.89	-	3.739
CaO	-	-	-	0.815
Na_2O	-	-	15	-
ZrO_2	0.016	0.039	-	0.051
BaO	0.27	-	-	-
Cr_2O_3	0.02	0.01	-	-
Other	0.064	0.141	-	0.025
H_2O	-	-	52	-
Description	Metacaulim HP Ultra	Metacaulim HP Ultra	R2252	Natural Sand
Source	Metacaulim do Brasil	Metacaulim do Brasil	Diatom	UFRJ
Function	Al and Si source	Al and Si source	Sodium silicate solution	Inert filler

Corrugated stainless steel fibers (Figure 1) were characterized by [21] and provided by Di Martino Metallurgical Industries Ltda. They have a 0.51mm diameter, 25mm length, $7.85g/cm^3$ specific gravity and were used as micro-reinforcement for the geopolymeric mortars.

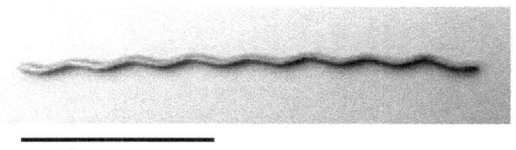

10mm

FIGURE 1 – STAINLESS STEEL FIBER USED IN THIS STUDY.

Non-Conventional Materials and Technologies – NOCMAT for XXI Century   Materials Research Forum LLC
Materials Research Proceedings 7 (2018) 295-312   doi: http://dx.doi.org/10.21741/9781945291838-27

The mineralogical analysis of the raw materials and the geopolymers admixtures was performed by powder X-ray diffraction (XRD), using a Bruker-AXS D4 diffractometer with Co Kα radiation. The samples were step-scanned at 0.02° 2θ and integrated at the step time of 1.83 s/step.

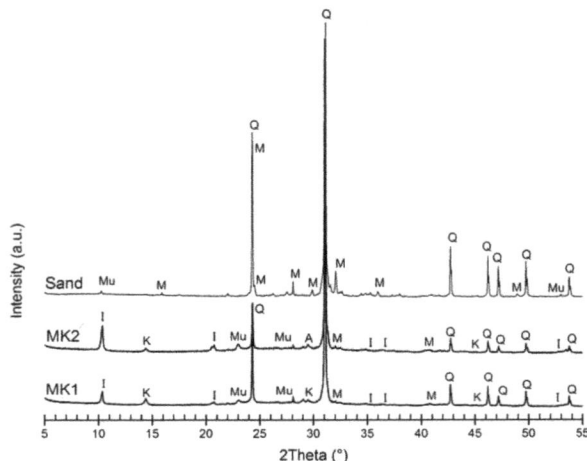

FIGURE 2 – *X-RAY DIFFRACTOGRAM OF RAW MATERIALS. K: KAOLINITE*
(Al$_2$Si$_2$O$_5$(OH)); M: MICROCLINE INTERMEDIATE 1 (KAlSi$_3$O$_8$); MU: MUSCOVITE
(KAl$_2$(AlSi$_3$O$_{10}$)(F,OH)$_2$; A:ANATASE (TiO$_2$); I: ILLITE
((K,H$_3$O)(Al,Mg,Fe)$_2$(Si,Al)$_4$O$_{10}$[(OH)$_2$,(H$_2$O)]); Q: QUARTZ (SiO$_2$).

As shown in Figure 2, quartz (SiO2) is the major crystalline phase contaminant present in both MK1 and MK2. Other minor crystalline contaminants are kaolinite, microcline intermediate 1, muscovite, anatase and illite. MK1 and MK2 amorphous contents, presented in Table 2, were determined using Rietveld quantitative amorphous content analysis (RQACA) [12]. It was found that MK1 has greater amorphous content than MK2. The differences of amorphicity could be attributed to variations in the kaolin source composition and calcination parameters adopted to convert kaolin sources to MK1 and MK2. The natural sand is composed mainly by quartz, with microcline and muscovite as minor crystalline constituents.

TABLE 2 – *RAW MATERIALS PHYSICAL PROPERTIES*

Material	Specific gravity (g/cm^3)	Surface area (m^2/g)	Amorphous content (%)	d$_{10}$ (μm)	d$_{50}$ (μm)	d$_{90}$ (μm)
MK1	2.65	14,50	75	3.368	24.935	61.242
MK2	2.81	14,04	47	3.061	17.729	53.013

Non-Conventional Materials and Technologies – NOCMAT for XXI Century   Materials Research Forum LLC
Materials Research Proceedings 7 (2018) 295-312   doi: http://dx.doi.org/10.21741/9781945291838-27

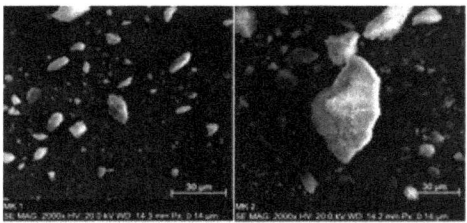

FIGURE 3 – SEM MICROGRAPHS OF (LEFT) MK1 AND (RIGHT) MK2.

Scanning electron micrograph (SEM) images of raw materials and fractured surfaces of pastes and mortars were obtained using a FEI Quanta 400 microscope, to elucidate the microstructural characteristics of the metakaolin and corresponding geopolymer compounds obtained. Samples were coated with 20 nm of gold. MK1 and MK2 SEM images indicates mainly irregularly shaped particles and some agglomerations, with particle dimensions in the microscale range, as shown in Figure 3.

Fourier transform infrared (FTIR) spectra of MK1, MK2, geopolymer pastes and mortars were collected in a VARIAN 3100 spectrometer Model-Excalibur in absorbance mode. Sample pellets were prepared using the normal KBr procedure, dried in an oven at 60°C overnight, and pressed before spectra were taken. The analyses were performed in the spectral range of 4000 to 400 cm-1, with a resolution of 4 cm-1 and the number of scans was 60 [11, 14].

Experimental

Geopolymer Systems Synthesis and Mixing

Two pure geopolymer matrices (G1 and G2) were, respectively, formulated from MK1 and MK2 and considering the molar ratios summarized in Table 3 chosen from previous studies [8, 10-14].

TABLE 3 – MOLAR RATIOS ADOPTED FOR DESIGNING THE GEOPOLYMER MATRICES.

Mixture	SiO_2/Al_2O_3 (%wt)	SiO_2/Na_2O (%wt)	H_2O/Na_2O (%wt)	liquid/solid (wt)
G1	2.45	4.21	16.50	1.25
G2	3.0	3.68	16.41	1.14

The synthesis of geopolymers was carried out by first dissolving sodium hydroxide pellets in deionized water, obtaining a solution with concentration of 10M. Sodium silicate solution was then added to the previous solution and the final activation solution was allowed to cool down to room temperature (~22°C).

The geopolymers binders were mixed according to the API RP 10B-2-2013 [22] and ABNT NBR 9831-2006 [23] standards, in a waring blender 24CB10C. Minor adjustments in mixing procedure was conducted to keep the same mixing energy required by [22, 23], in order to prepare 750mL paste volume. The activation solution was poured into the mixer vessel, the metakaolin was added gradually within 19 seconds at 4000rpm. The speed was increased to 12000rpm and kept for 44s to obtain a homogeneous paste and totalizing the mixing time as 63 seconds.

Mortars AG1 and AG2 were respectively produced from the above-mentioned pure matrices, with a fixed sand content of 40% (v%/v%) and reinforced with 1.2, 1.8 and 2.4% (v%/v%) of stainless steel fibers. Both sand and fiber contents were calculated based on the total volume of the mixtures.

Non-Conventional Materials and Technologies – NOCMAT for XXI Century   Materials Research Forum LLC
Materials Research Proceedings 7 (2018) 295-312   doi: http://dx.doi.org/10.21741/9781945291838-27

The geopolymer mortars were mixed according to the ASTM C305-14 [24] standard in a planetary mixer Soiltest CT-345 with minor adjustments, as follows. The solid materials were weighed and homogenized in a plastic bag. Then, they were placed and homogenized in the mixer at about 60-80rpm speed for 1 minute. After the first minute, the activator solution was slowly added. Upon completion of 2.5 minutes, the speed was changed to the 300rpm mode completing the total mixing time of 4 minutes.

To produce the reinforced mortars, the fibers were added to fresh AG1 and AG2 after completing the mixing time, as described above. The fibers were carefully added during 1 minute at the about 60-80rpm speed of mixer, and blended for a further 2 minutes at 60-80 speed, completing the total time of mixing of 7 minutes. The reinforced mortars were named, according to the content of fibers, as AG1-F-1.2%, AG1-F-1.8%, AG1-F-2.4% and analogously for AG2 matrices. The pure geopolymers G1 and G2, as well as their respective geopolymer mortars and reinforced mortars, may also be referred in the text as Geopolymer system 1 and 2, respectively.

The setting times at room temperature of geopolymer binders were obtained using ABNT NBR 16607-2017 standard [25].

Geopolymer Systems Molding and Curing

Cylindrical specimens of fresh pastes, mortars and reinforced mortars, with 50 mm (diameter) x 100 mm (height) were cast in two layers, vibrating each layer for 60 seconds to remove entrained air. Afterwards, the samples cured in water bath at $22\pm2°C$, for 7 days, prior to testing under uniaxial compressive loads.

For the four points bending tests, the same methodology of casting described above was followed, using open prismatic specimens 40x40x160 mm, which were cured in a humid chamber at 100% relative humidity for the same period. For reinforced matrices, the vibration time was decreased to 15 seconds per layer to prevent fiber segregation.

Geopolymer Samples Testing

The unconfined compressive strengths of geopolymers and mortar samples were determined following ABNT NBR 5739-2007 [26] using a 200kN Wykeham Farrance press. The loading velocity was 0.1 mm min-1. The lateral strain was calculated from stroke measurements during the tests with two LVDTs. The stiffness (Young's modulus, E) was calculated from the stress-strain curves, using the secant modulus from two points in the linear elastic section, the first corresponding to 40% of the failure stress, and the second corresponding to a strain of 50μS (Equation 3-1). Flexural strength of composites and geopolymer binders were conducted using a four-point-bending fixture according to ABNT NBR 12142-2010 [27], with load rate of 0.3 mm min-1. The specimen's deflection was measured during the tests with one LVDT. Each reported value corresponds to the average of 3 measurements, for both compressive and flexural tests.

$$E = \frac{\sigma_{c2} - \sigma_{c1}}{\varepsilon_{c2} - \varepsilon_{c1}} \qquad (1)$$

After testing, some fragments of the tested specimens were separated for microstructural analysis by SEM in fracture surfaces, as well as chemical and mineralogical characterization by XRD and FTIR. The fragments were milled until obtaining a particle size passing 106μm. Immediately, the geopolymerization reaction was stopped using a solution of alcohol/acetone (1:1%v/%v), as recommended by different investigators [28, 29].

Non-Conventional Materials and Technologies – NOCMAT for XXI Century   Materials Research Forum LLC
Materials Research Proceedings **7** (2018) 295-312   doi: http://dx.doi.org/10.21741/9781945291838-27

Results and discussions

Mechanical Properties

Stress-strain curves for Systems 1 and 2 are displayed in Figure 4. All curves exhibit a well-defined elastic regime and reinforced mortars also display post-peak yielding.

Compressive strength at failure (f_c), elastic moduli (E) and maximum lateral strain (ε_m) of the systems are summarized in Table 4 and Table 5. For each system, compressive strengths are higher for the geopolymer pastes and similar for the mortars and reinforced mortars. Geopolymer G1 shows higher compressive strength and failure strain compared to G2, but the mortars AG1 and AG2 and their reinforced derivatives have similar strengths.

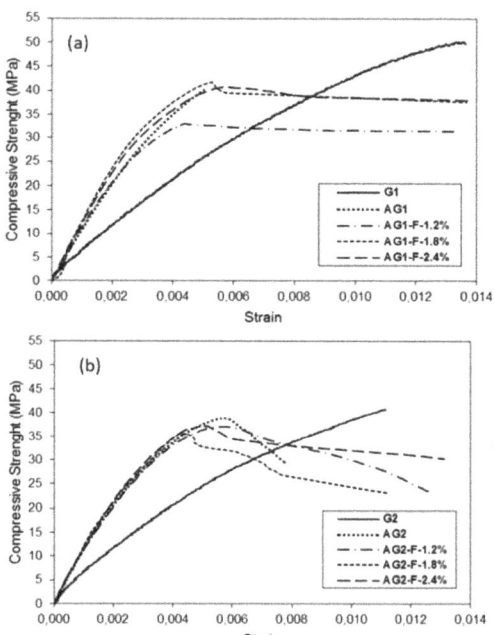

FIGURE 4 – STRESS-STRAIN CURVES FOR THE GEOPOLYMER BINDERS (A) SYSTEM 1 E (B) SYSTEM 2.

TABLE 4 – COMPRESSION MECHANICAL PARAMETERS FOR THE GEOPOLYMER BINDERS.

Mixture	f_c (MPa)	ε_m	E (GPa)
G1	50.22±0.99	0.013±0.001	6.32±0.02
G2	38.86±1.47	0.010±0.001	7.22±0.12

As shown in Table 4, systems based on geopolymer G2 are stiffer than their correspondents based on G1. Mortars are significantly stiffer than their base geopolymers, due to the high sand content and its stiffness, but the difference between AG1 and AG2 is much smaller. Addition of fibers causes a small increase in stiffness.

TABLE 5 – *COMPRESSION MECHANICAL PARAMETERS FOR THE GEOPOLYMER MORTARS.*

Mixture	f_c (MPa)	ε_m	E (GPa)
AG1	39.04±0.30	0.005±0.0002	10.59±0.02
AG1-F-1.2%	34.17±1.60	0.004±0.0005	10.29±0.13
AG1-F-1.8%	41.45±0.40	0.005±0.0002	12.56±0.57
AG1-F-2.4%	39.55±2.40	0.006±0.0003	12.03±0.18
AG2	39.49±0.68	0.005±0.0010	10.83±0.23
AG2-F-1.2%	36.44±0.70	0.005±0.0010	13.30±0.69
AG2-F-1.8%	33.29±2.93	0.004±0.0007	12.14±0.53
AG2-F-2.4%	36.11±2.49	0.004±0.0009	12.37±0.96

Flexural strength results of pure geopolymers are summarized in Table 6 and shown in Figure 5. It was found that the addition of sand increases the flexural strength of the geopolymer matrices from 3.5 MPa to 4.0–4.8 MPa. Fiber reinforcement increases the strength with increasing volume fraction, up to 10 MPa for a content of 2.4%. Similar to the compressive strength results, in general geopolymer systems based on G1 have higher flexural strength than the corresponding formulations based on G2. Mortars summary flexural mechanical behavior and parameters are presented, in Figures 6-8, and Table 7.

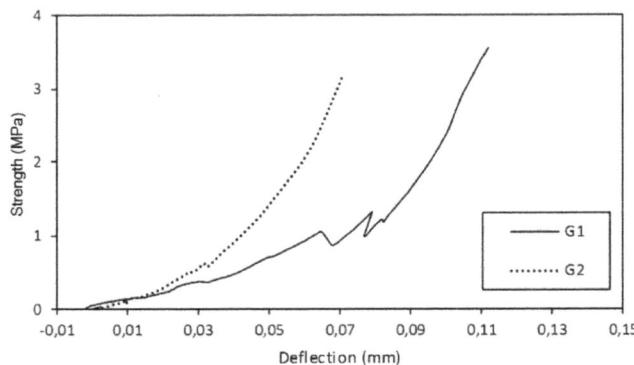

FIGURE 5 – *STRESS-STRAIN CURVES FOR THE GEOPOLYMER BINDERS.*

Non-Conventional Materials and Technologies – NOCMAT for XXI Century   Materials Research Forum LLC
Materials Research Proceedings 7 (2018) 295-312   doi: http://dx.doi.org/10.21741/9781945291838-27

TABLE 6 – FLEXURAL MECHANICAL PARAMETERS FOR THE GEOPOLYMER BINDERS.

Mixture	σ_f (MPa)	δ_m (mm)
G1	3.47±0.65	0.11±0.02
G2	3.50±1.00	0.06±0.01

TABLE 7 – FLEXURAL MECHANICAL PARAMETERS FOR THE GEOPOLYMER MORTARS.

Mixture	σ_f (MPa)	δ_m (mm)
AG1	4.75±0.43	0.004±0.001
AG1-F-1.2%	6.39±0.70	0.16±0.01
AG1-F-1.8%	9.27±0.30	0.44±0.04
AG1-F-2.4%	9.99±0.69	0.16±0.02
AG2	4.02±0.30	0.08±0.01
AG2-F-1.2%	7.57±1.15	0.37±0.09
AG2-F-1.8%	8.54±0.10	0.60±0.09
AG2-F-2.4%	9.87±0.42	0.32±0.10

Previous studies [10, 11, 14, 30, 31] report a strong correlation between the Si/Al molar ratio and the compressive strength, with higher silica contents leading to higher strengths. Based on this, it would be expected that G2 (Si/Al = 3) should have higher strength than G1 (Si/Al = 2.45). However, G1 was is based on MK1, which has a larger amorphous content compared to MK2. The amount of alkali soluble material is a critical quantity for geopolymer formation [32], even though not all amorphous material is accessible by the alkali solution [33]. In this specific case, the reactivity of the precursor had a greater effect than the molar ratio.

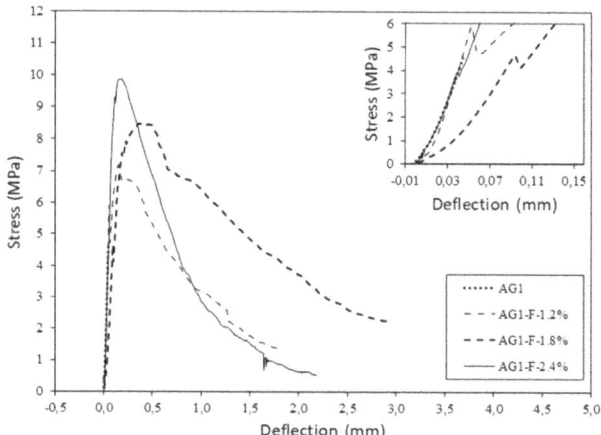

FIGURE 6– STRESS-STRAIN CURVES FOR THE AG₁ GEOPOLYMER MORTAR.

Non-Conventional Materials and Technologies – NOCMAT for XXI Century   Materials Research Forum LLC
Materials Research Proceedings **7** (2018) 295-312   doi: http://dx.doi.org/10.21741/9781945291838-27

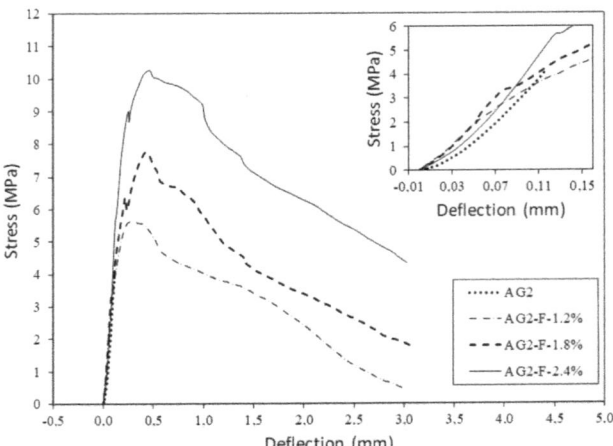

FIGURE 7 – STRESS-STRAIN CURVES FOR THE AG₂ GEOPOLYMER MORTAR.

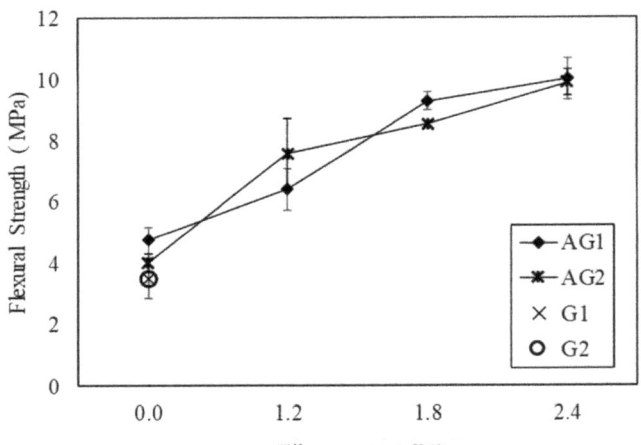

Fiber content (%)

*FIGURE 8 – FLEXURAL MECHANICAL PERFORMANCE OF THE GEOPOLYMER
SYSTEMS WITH AND WITHOUT FIBER REINFORCEMENT.*

Geopolymer binders and mortars characterization

XRD diffractograms (Figures 9 and 10) show a typical broad amorphous hump around 25-40° 2θ, less pronounced in G2 and AG2, due to its higher crystalline content, compared to G1 and AG1. This suggests a higher degree of geopolymerization and that a purer geopolymer binder was obtained for G1, with 90% of amorphous content found by RQACA. This may explain the higher compressive strength of G1, compared to G2. This typical broad hump is mentioned by [8, 10-13, 34-36], as the characteristic signature of amorphous geopolymers. Quartz (SiO_2) is the major crystalline phase contaminant present in geopolymer binders and mortars. Other minor crystalline contaminants are kaolinite, microcline intermediate 1, muscovite and anatase, which originate from the metakaolin precursors. The remaining crystalline phases present in geopolymer binders and mortars are not involved in the geopolymerization reaction, but are present like inactive fillers [37, 38].

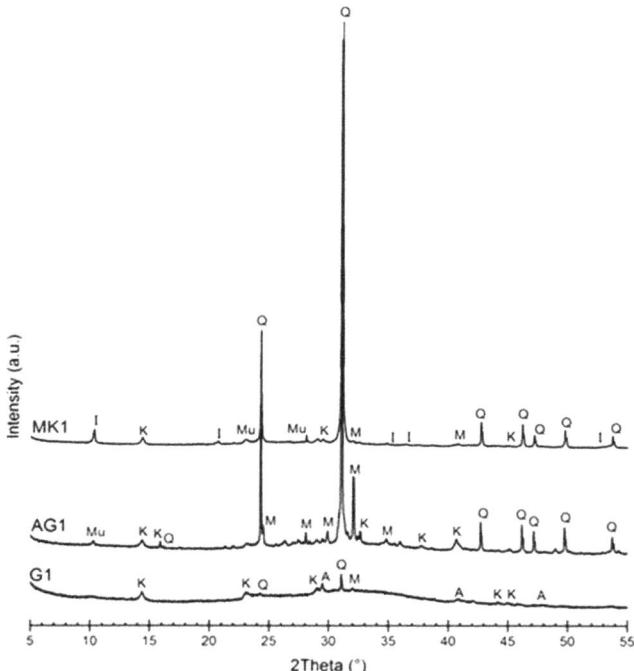

FIGURE 9 *– X-RAY DIFFRACTOGRAM OF* system 1*. K: KAOLINITE ($Al_2Si_2O_5(OH)$); M:*
MICROCLINE INTERMEDIATE 1 ($KAlSi_3O_8$); MU: MUSCOVITE ($KAl_2(AlSi_3O_{10})(F,OH)_2$;
A:ANATASE (TiO_2); Q: QUARTZ (SiO_2); I: ILLITE
$((K,H_3O)(Al,Mg,Fe)_2(Si,Al)_4O_{10}[(OH)_2,(H_2O)])$.

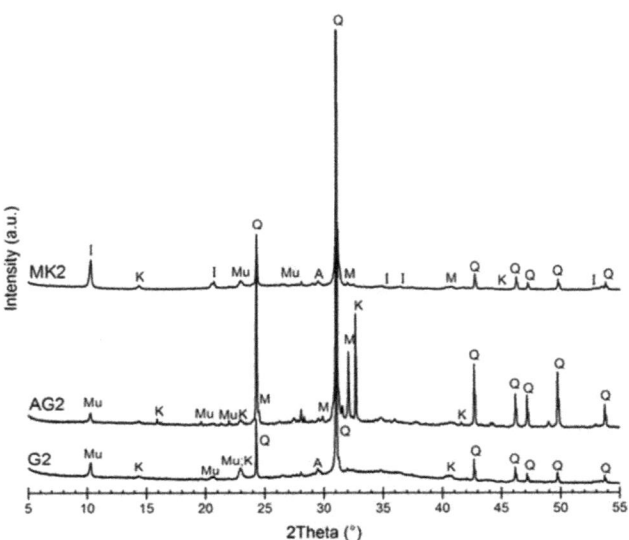

FIGURE 10 – *X-RAY DIFFRACTOGRAM OF* SYSTEM 2. *K: KAOLINITE (Al₂Si₂O₅(OH)); M: MICROCLINE INTERMEDIATE 1 (KAlSi₃O₈); MU: MUSCOVITE (KAl₂(AlSi₃O₁₀)(F,OH)₂; A:ANATASE (TiO₂); Q: QUARTZ (SiO₂); I: ILLITE ((K,H₃O)(Al,Mg,Fe)₂(Si,Al)₄O₁₀[(OH)₂,(H₂O)]).*

FTIR spectra (Figures 11 and 12) present the main usual bands, described as follows: below 700cm-1, related to the asymmetric vibration of the Si-O-(Si, Al) bonds, around 1000cm-1, related to Si-O-Si bonds and, around 1600cm-1 and 3400cm-1, attributed respectively to stretching vibrations (H-O) and deformations (H-O-H) of absorbed or structural water molecules present in the geopolymer porosity [11, 14, 36, 37].

While in the unreacted MK1 and MK2 the Si-O-Si band appears at 1037 cm-1 and 1076 cm-1, in the activated samples it shifts to lower wave numbers in both systems studied (between 1016 and 1010 cm-1). This shift occurs due to the formation of aluminosilicate gel, indicating the condensation of Si-O and SiO4 tetrahedra and AlO4 tetrahedra in the geopolymer network [36, 38, 39]. The characteristic metakaolin Si-O-Al band at 801 cm−1, after the geopolymerization, is replaced by several weaker bands in the range from 600 to 800 cm−1 [39]. The band at about 460 to 600 cm-1 is due to Si-O bending vibration [39]. The smaller peak at 2926 cm-1 is related to C-O bonds. The lower peak intensities found for system 2 are attributed to the higher MK2 crystalline content compared to system 1, which agrees with the XRD analysis.

Non-Conventional Materials and Technologies – NOCMAT for XXI Century   Materials Research Forum LLC
Materials Research Proceedings 7 (2018) 295-312   doi: http://dx.doi.org/10.21741/9781945291838-27

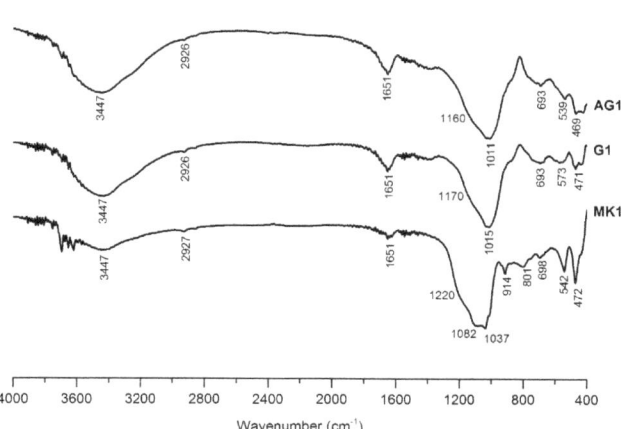

FIGURE 11 – FTIR SPECTRA OF SYSTEM 1.

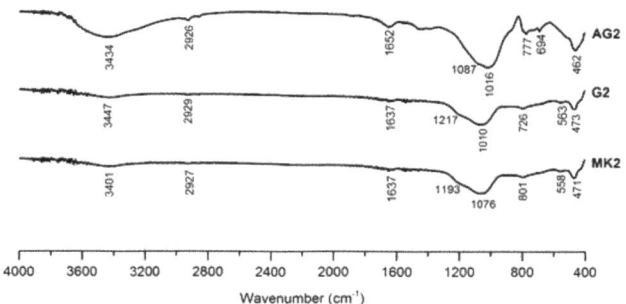

FIGURE 12 – FTIR SPECTRA OF SYSTEM 2.

Microstructure images from fracture surfaces (Figures 13 and 14) showed that geopolymer systems obtained from MK1 and MK2 formed a dense and uniform base matrix, with no microvoids and very little efflorescence in the form of NaOH and Na_2CO_3 micro-crystals. This was confirmed by EDX analysis.

Figure 13a shows a dense gel and uniform base matrix with no microvoids, with some small crystals deposited on the surface, suggesting that G1 is mainly composed of pure geopolymer binder. Needle crystal structures, also found in other studies [12, 34], were identified as NaOH or Na_2CO_3 microcrystals. G1 has a high H_2O/Na_2O ratio, which increases the tendency of alkaline leaching out phenomena on the surface, due to the weakly bound Na+ ions in the nanostructure of geopolymer gel [40], which can be leached almost completely, with no significant compromise of the binder compressive strength [41]. Na+ ions may also diffuse to the surface, reacting with atmospheric CO_2

and forming visible efflorescence, mainly composed of $Na_2CO_3 \cdot nH_2O$ and $NaHCO_3$, due to alkali carbonation [34, 42].

SEM images of G2 (Figures 13c and 13d) exhibit a dense matrix, similar to that found in G1, in which some unreacted particles and were identified. Some laminar structures are deposited on the surface of the sample and trapped in the base matrix. EDX analysis of these laminar structures confirmed the presence of the crystalline contaminant phases detected by XRD in MK2 and consequently in G2. Figure 14 shows SEM images of geopolymer mortars. It can be clearly appreciated that both matrices, like the pure geopolymers, are mainly formed by a dense and uniform matrix. As verified for pure geopolymers, AG2 shows higher amounts of contamination than AG1, and the amount of laminar structures found in AG2 is also higher when compared to G2. These overall observations regarding microstructure are in accordance with well-formed geopolymers described by [42].

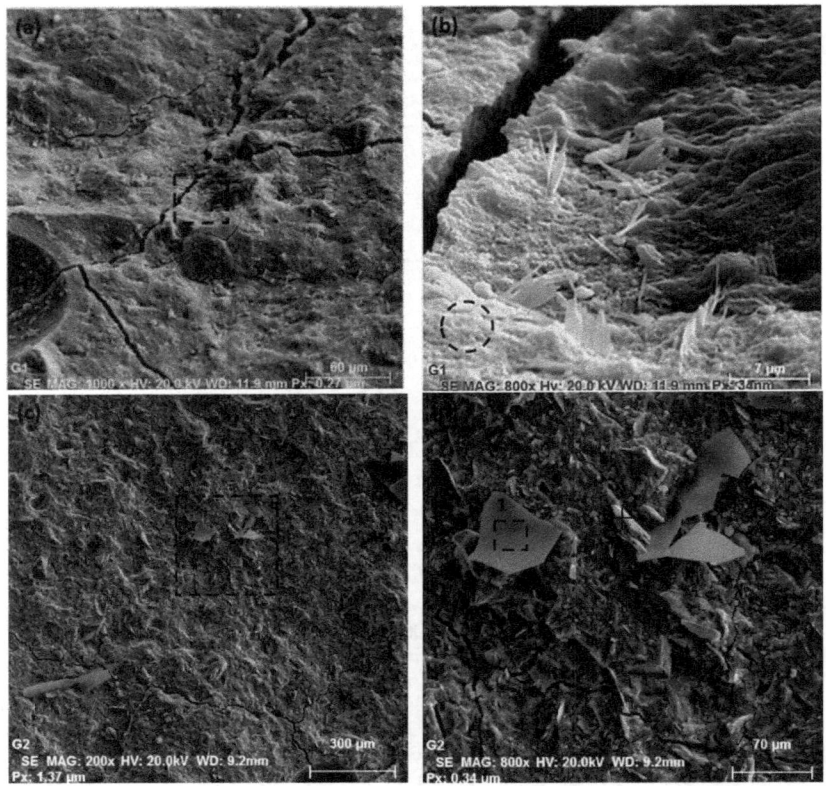

***FIGURE 13** – SEM MICROGRAPHS: (A) OVERVIEW OF G1 MATRIX; (B) ZOOMED VIEW OF G1; (C) OVERVIEW OF G2 MATRIX; (D) ZOOMED VIEW OF G2.*

Non-Conventional Materials and Technologies – NOCMAT for XXI Century   Materials Research Forum LLC
Materials Research Proceedings 7 (2018) 295-312   doi: http://dx.doi.org/10.21741/9781945291838-27

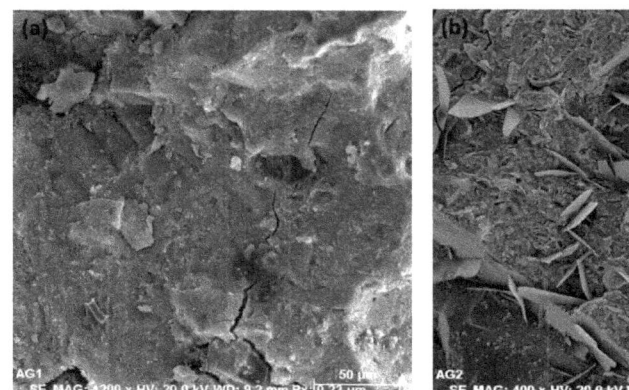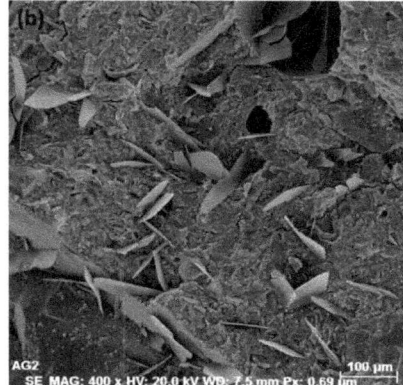

FIGURE 14 – SEM MICROGRAPHS: OVERVIEW OF AG1 (A) AND AG2 (B) MORTARS.

The geopolymer networks were confirmed by XRD and FTIR characterizations. Microstructure results showed that the geopolymer composites formed a dense and uniform base matrix, with no microvoids and very little Na_2CO_3 efflorescence. Metakaolin-based geopolymer pastes and mortars presented typical mechanical properties, with uniaxial compressive strengths ranging from 33 to 50 MPa and flexural tensile strength ranging from 3.4 to 10.0 MPa. Finally, the geopolymer binders exhibit similar physical-mechanical performances, despite the crystalline content of the metakaolin. Pure and reinforced geopolymer systems display overall superior mechanical performance, when compared to OPC matrices, especially regarding the maximum strain at failure, suggesting that these systems are suitable to replace OPC systems in structural applications.

Conclusion
Composite geopolymer systems were designed with the addition of sand and stainless-steel fibers to geopolymer matrices obtained by alkaline activation of two different metakaolin batches with different crystalline contents.

The metakaolin-based geopolymer pastes G1 and G2 presented a mechanical performance comparable to systems found in the literature, with uniaxial compressive strengths ranging from 38-50 MPa and stiffness around 7 GPa. The paste G1 produced with MK1, which has a greater amorphous content and higher reactivity, was significantly stronger that G2, which is based on the more crystalline precursor MK2.

Mortars AG1 and AG2, containing a fixed amount of sand, displayed lower strength and significantly higher stiffness than the base geopolymer matrices. The resulting properties were less sensitive to the quality of the precursor.

Increasing amounts of fiber increased significantly the flexural strength, from 3.4 MPa to 10 MPa, demonstrating the reinforcement efficiency of this type of fiber and its compatibility with the matrices. In uniaxial compression tests, the addition of fibers also increased significantly the maximum strain, overcoming the brittleness of the base mortars.

The formation of amorphous geopolymer networks was confirmed by XRD and FTIR characterizations. Diffraction results demonstrate also that crystalline contaminants in the precursor are found mostly unreacted in the final product. These contaminants can be clearly seen in SEM

Non-Conventional Materials and Technologies – NOCMAT for XXI Century   Materials Research Forum LLC
Materials Research Proceedings 7 (2018) 295-312   doi: http://dx.doi.org/10.21741/9781945291838-27

images of G2 and AG2 and confirmed by EDX analysis, while they are mostly absent in images of G1 and AG1. Overall, SEM images showed that the geopolymer composites formed a dense and uniform base matrix, with no microvoids and very little Na_2CO_3 efflorescence.

References

[1] J. L. Provis and J. S. J. Van Deventer, Geopolymers: structures, processing, properties and industrial applications. Elsevier, 2009. https://doi.org/10.1533/9781845696382

[2] J. L. Provis and J. S. J. Van Deventer, Alkali avtivated materials. 2014.

[3] K. J. D. MacKenzie, "What are these things called geopolymers? A physicochemical perspective," Ceram. Trans., 2003.

[4] C. a Rees, J. L. Provis, G. C. Lukey, and J. S. J. van Deventer, "Attenuated total reflectance fourier transform infrared analysis of fly ash geopolymer gel aging.," Langmuir, vol. 23, no. 15, pp. 8170–8179, 2007. https://doi.org/10.1021/la700713g

[5] P. Duxson, A. Fernández-Jiménez, J. L. Provis, G. C. Lukey, A. Palomo, and J. S. J. Van Deventer, "Geopolymer technology: The current state of the art," J. Mater. Sci., vol. 42, no. 9, pp. 2917– 2933, 2007. https://doi.org/10.1007/s10853-006-0637-z

[6] L. Struble and J. K. Hicks, Geopolymer Binder Systems, vol. 1566 STP. 2014.

[7] Mota, A.M.P. et al. "Análise das Propriedades do estado endurecido de concretos empregando metacaulim e aditivo superplastificante.," in 48° Congresso Brasileiro do ConcretoIBRACON, 2006., 2006.

[8] Davidovits, J., "Geopolymers : Inorganic Polymeric New Materials," J. Therm. Anal., vol. 37, pp. 1633–1656, 1991. https://doi.org/10.1007/BF01912193

[9] J. L. Susan A. Bernal, Erich D. Rodríguez, Ruby Mejía de Gutiérrez, Marisol Gordillo, Provis, "Mechanical and thermal characterization of geopolymers based on silicate-activated metakaolin/slag blends," J. Mater. Sci., vol. 46, no. 16, pp. 5477– 5486, 2011. https://doi.org/10.1007/s10853-011-5490-z

[10] Barbosa, V.F.F et al. "Thermal behavior of inorganic polymers and composites derived from sodium polysialate". Materials Research Bulletin vol. 38, pp.319-331, 2003. https://doi.org/10.1016/S0025-5408(02)01022-X

[11] Paiva, M. D. M. "Otimização e análise mecânica de pastas geopoliméricas para uso em poços sujeitos à injeção cíclica de vapor," D.Sc. Dissertation. Natal (RN, Brasil): UFRN, 2008.

[12] L. Reig, M. M. Tashima, M. V. Borrachero, J. Monzó, C. R. Cheeseman, and J. Payá, "Properties and microstructure of alkali activated red clay brick waste," Constr. Build. Mater., vol. 43, pp. 98–106, 2013. https://doi.org/10.1016/j.conbuildmat.2013.01.031

[13] Davidovits, J. Geopolymers based on natural and synthetic metakaolin - A critical review. Materials Today 2016. Available https://www.materialstoday.com/polymers-soft-materials/features/geopolymers-natural-and-synthetic-metakaolin/

[14] M. Meftah, W. Oueslati, N. Chorfi, and A. Ben Haj Amara, "Intrinsic parameters involved in the synthesis of metakaolin based geopolymer: Microstructure analysis," J. Alloys Compd., vol. 688, pp. 946–956, 2016. https://doi.org/10.1016/j.jallcom.2016.07.297

Non-Conventional Materials and Technologies – NOCMAT for XXI Century   Materials Research Forum LLC
Materials Research Proceedings 7 (2018) 295-312   doi: http://dx.doi.org/10.21741/9781945291838-27

[15] Tahri, W. "Shrinkage and mechanical performance of geopolymeric mortars based on calcined Tunisian clay". Journal of Chemistry and Materials Research, vol. 4, pp. 6-11, 2013

[16] T. T. Zhang et al., "Control of Drying Shrinkage of Magnesium Silicate Hydrate Gel Cements", Key Engineering Materials, Vol. 709, pp. 109-113, 2016. https://doi.org/10.4028/www.scientific.net/KEM.709.109

[17] Davidovits, J. Reinforced Geopolymer Composites: A critical review. Materials Today 2016. Available from: https://www.materialstoday.com/polymers-soft-materials/features/reinforced-geopolymer-composites-a-critical-review/

[18] Pelisser, F. e tal. "Micromechanical characterization of metakaolin-based geopolymers" Construction and Building Materials, vol. 49, pp.547-553, 2013. https://doi.org/10.1016/j.conbuildmat.2013.08.081

[19] Ranjbar, N. et al "Mechanisms of interfacial bond in steel and polypropylene fiber reinforced geopolymer composites". Composites Science and Technology, vol. 122, pp.73-81, 2016. https://doi.org/10.1016/j.compscitech.2015.11.009

[20] NBR NM 52, "Agregado miúdo - Determinação de massa específica e massa específica aparente," 2009.

[21] M. Jorivaldo, "Refratários de Elevada Tenacidade para Uso em Aplicações Críticas na Indústria do Refino de Petróleo", D.Sc. Dissertation. Rio de Janeiro (Brazil): UFRJ, 2012.

[22] API RP 10B-2-2013: Recommended practice for testing well cements, American Petroleum Institute, 2nd ed. 2013.

[23] ABNT, "ABNT NBR 9831-2006, Cimento Portland destinado à cimentação de poços petrolíferos - Requisitos e métodos de ensaio.

[24] ASTM C305-14, Standard Practice for Mechanical Mixing of Hydraulic Cement Pastes and Mortars of Plastic Consistency, ASTM International, West Conshohocken, PA, 2014

[25] ABNT NBR 16607-2017, Cimento Portland – Determinação dos Tempos de Pega. ABNT.

[26] ABNT, "ABNT NBR5739- 2007, Concreto – Ensaios de compressão e corpos-de-prova cilíndricos," p. 5739, 2007.

[27] R. Curti and A. Vaquero, "Concreto – Determinação da resistência à tração na flexão de corpos de prova prismáticos," 2010.

[28] N. SAIKIA, A. USAMI, S. KATO, and T. KOJIMA, "Hydration Behaviour of Ecocement in Presence of Metakaolin," Resour. Process., vol. 51, no. 1, pp. 35–41, 2004. https://doi.org/10.4144/rpsj.51.35

[29] A. S. Taha, H. Eldidamony, S. A. Aboelenein, and H. A. Amer, "PHYSICOCHEMICAL PROPERTIES OF SUPERSULFATED CEMENT PASTES," Zement-Kalk-Gips, vol. 34, no. 6, pp. 315– 317, 1981.

[30] P. Duxson, "Geopolymer precursor design," Geopolymers Struct. Process. Prop. Ind. Appl., vol. 37, 2009.

Non-Conventional Materials and Technologies – NOCMAT for XXI Century　　Materials Research Forum LLC
Materials Research Proceedings 7 (2018) 295-312　　　　doi: http://dx.doi.org/10.21741/9781945291838-27

[31] P. Duxson, S. W. Mallicoat, G. C. Lukey, W. M. Kriven, and J. S. J. Van Deventer, "The effect of alkali and Si/Al ratio on the development of mechanical properties of metakaolin-based geopolymers," Colloids Surfaces A Physicochem. Eng. Asp., vol. 292, no. 1, pp. 8–20, 2007. https://doi.org/10.1016/j.colsurfa.2006.05.044

[32] A. Fernández-Jiménez, J. Y. Pastor, A. Martín, and A. Palomo, "High-Temperature Resistance in Alkali-Activated Cement," J. Am. Ceram. Soc., vol. 93, no. 10, pp. 3411–3417, 2010. https://doi.org/10.1111/j.1551-2916.2010.03887.x

[33] L. Vickers, A. van Riessen, and W. D. A. Rickard, "Precursors and Additives for Geopolymer Synthesis," in Fire-Resistant Geopolymers, Springer, 2015, pp. 17–37.

[34] J. He, J. Zhang, Y. Yu, and G. Zhang, "The strength and microstructure of two geopolymers derived from metakaolin and red mud-fly ash admixture: a comparative study," Constr. Build. Mater., vol. 30, pp. 80–91, 2012. https://doi.org/10.1016/j.conbuildmat.2011.12.011

[35] Z. Li and S. Liu, "Influence of slag as additive on compressive strength of fly ash-based geopolymer," J. Mater. Civ. Eng., vol. 19, no. 6, pp. 470–474, 2007. https://doi.org/10.1061/(ASCE)0899-1561(2007)19:6(470)

[36] X. Guo, H. Shi, and W. A. Dick, "Compressive strength and microstructural characteristics of class C fly ash geopolymer," Cem. Concr. Compos., vol. 32, no. 2, pp. 142–147, 2010. https://doi.org/10.1016/j.cemconcomp.2009.11.003

[37] G. Zhang, J. He, and R. Gambrell, "Synthesis, Characterization, and Mechanical Properties of Red Mud-Based Geopolymers," Transp. Res. Rec. J. Transp. Res. Board, vol. 2167, pp. 1–9, Dec. 2010. https://doi.org/10.3141/2167-01

[38] S. Zhang, K. Gong, and J. Lu, "Novel modification method for inorganic geopolymer by using water soluble organic polymers," Mater. Lett., vol. 58, no. 7, pp. 1292–1296, 2004. https://doi.org/10.1016/j.matlet.2003.07.051

[39] C. Ferone, G. Roviello, F. Colangelo, R. Cioffi, and O. Tarallo, "Novel hybrid organic-geopolymer materials," Appl. Clay Sci., vol. 73, pp. 42–50, 2013. https://doi.org/10.1016/j.clay.2012.11.001

[40] M. Amer, A. Abdullah, A. Ali, and N. Farzadnia, "Characterization of mechanical and microstructural properties of palm oil fuel ash geopolymer cement paste," Constr. Build. Mater., vol. 65, pp. 592–603, 2014. https://doi.org/10.1016/j.conbuildmat.2014.05.031

[41] F. Škvara, V. Šmilauer, P. Hlaváček, L. Kopecký, and Z. Cilova, "A weak alkali bond in (N, K)–A–S–H gels: evidence from leaching and modeling," Ceramics-Silikaty, vol. 56, no. 4, p. 9, 2012.

[42] P. Duxson, J. L. Provis, G. C. Lukey, S. W. Mallicoat, W. M. Kriven, and J. S. J. Van Deventer, "Understanding the relationship between geopolymer composition, microstructure and mechanical properties," Colloids Surfaces A Physicochem. Eng. Asp., vol. 269, no. 1–3, pp. 47–58, 2005. https://doi.org/10.1016/j.colsurfa.2005.06.060

Non-Conventional Materials and Technologies – NOCMAT for XXI Century   Materials Research Forum LLC
Materials Research Proceedings 7 (2018) 313-324   doi: http://dx.doi.org/10.21741/9781945291838-28

Confined Bamboo Guadua Laminate – CBGL

Alfonso Cruz [a], Caori Takeuchi [b,]

[a] Universidad Nacional de Colombia, Colombia; alfonsocrg@gmail.com

[b] Universidad Nacional de Colombia, Colombia: cptakeuchit@unal.edu.co

Abstract. The constructive technique called Confined Bamboo Guadua Laminate (CBGL) is an improvement of the mechanical properties of wood with low specifications, by means of the reinforcement of laminates of Bamboo *Guadua angustifolia* Kunth, in a composite section. The CBGL is a composite section of pieces of softwood that confine laminates of guadua. Thus, low mechanical specification wood can be used for structural constructions with the CBGL technique. Compression parallel fiber tests were carried out for wood, bamboo, *Guadua angustifolia* Kunth laminates and for the composite section of wood and bamboo guadua laminates, to demonstrate the advantages of the CBGL construction technique. The density was determined for each material. The elasticity modulus (E), the maximum stress ($\sigma_{máx}$) and the proportionality limit stress (σ_{LP}) were obtained with the results of stress–strain graph tests. The experimental data of the composite section were compared with theoretical and numerical data analysis, demonstrating the efficiency of the CBGL constructive technique. Wood from reforestation and fast-growing processes, with the CBGL construction technique, can accomplish structural requirements and become an alternative to hardwood with high mechanical resistance, but longer growing time. In this way, not only the selection processes of structural elements for constructions in wood are favored, but also, the environment.

Keywords: Bamboo, Wood, Laminated, Elastic Modulus, Ecoconstruction

Introduction

The cost of reforested wood is relatively low, because it is cut at an early age. However, generally it has low density, therefore, its resistance and stiffness are low compared to other wood. Regarding to laminated bamboo, different studies have found that its compressive strength is high, with values between 57.9 MPa and 62.7 MPa for bamboo *Mosso* laminated [1], 54.2MPa and 69.6 MPa for bamboo *Phyllostachys* laminated, 42.5MPa and 58.4 MPa for bamboo *Dendrocalamus* laminated [2] and 54.8 MPa and 71.2 Mpa for laminates of bamboo *Guadua angustifolia* [3]. However, its cost is quite high due to its manufacturing process that involves bamboo cutting, striping, drying, laminating (whit slats dimensions limited by the thickness of the wall and the diameter of the culm), glue applying and pressing. A viable alternative from the point of view of resistance and cost is the use of composite sections of reforested wood and bamboo laminates. Reyes and Rayo [4] have tested beams of a large rectangular composite section of pine in the interior and boards of guadua laminates located in the upper and lower side. They found that those beams behaved better than normal pine beams. Cruz [5] also studied the behavior of composite sections of pine in the interior and boards of bamboo guadua in the exterior under compression. However, the external boards of guadua tended to open, because they were elaborated with sheets of bamboo that only have longitudinal fibers.

The guadua laminate can be confined by softwood to prevent premature failures. In this way, the laminate could reach its entire compression capacity and, also, the resistance can be increased under confined conditions. Then, this document contains the results of compression parallel fiber tests of Confined Bamboo Guadua Laminate (CBGL).

Non-Conventional Materials and Technologies – NOCMAT for XXI Century   Materials Research Forum LLC
Materials Research Proceedings 7 (2018) 313-324   doi: http://dx.doi.org/10.21741/9781945291838-28

Methodology

This section explains the methodology of laboratory tests and numerical analysis of pine samples, bamboo guadua laminates and CBGL under compression parallel fiber.

Materials

Bamboo guadua

Bamboo guaduas used in this research were extracted from a natural bamboo guadua forest in the department of Cundinamarca (Colombia). The bamboo culms were cut to a length of about 240 cm and cured on site. Then, the culm was cut in strips of about 0.022 x 0.005 x 1.20m approximately and finally it was dried in a greenhouse for a period of 15 days. Afterwards, the strips were brushed on both sides by means of a mechanical woodworking brush. Figure 1(a) shows the approximate dimensions of the strips. The boards were made of 0.11 x 0.50 x 1.20m and randomly joined into four (4) groups as described in Table 1 and Figure 1(b). The guadua boards were used to produce the laminated guadua and the composite section test pieces.

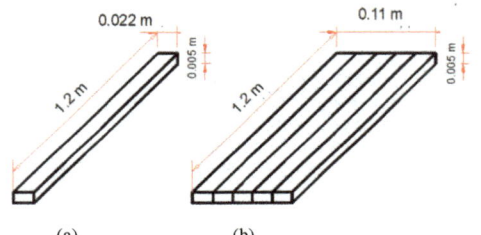

(a)   (b)

FIGURE 1 - *APPROXIMATE DIMENSIONS OF GUADUA LAMINATES. (A) STRIPS. (B) BOARD.*

Source: Authors

TABLE 1 - *AMOUNTS OF BASE MATERIAL GUADUA ANGUSTIFOLIA KUNTH*

	Group 1	Group 2	Group 3	Group 4
Strips: 0.022x0.005x1.20m.	30	30	40	40
Boards: 0.11x0.005x1.20m.	6	6	8	8
Laminated guadua specimens	1A, 1B, 1C	2A, 2B, 2C	3A, 3B, 3C	4A, 4B, 4C

Source: Authors

Radiata pine.

One ledge of commercial radiata pine 3.2 x 14 x 490 cm was used.

Laboratory assemble and testing

Pressing process

There were two stages of pressing: first, bamboo laminated boards and second, blocks of guadua laminates with sheets oriented in one direction, or blocks of composite sections with pine and bamboo laminates. The applied pressure was between 0.6MPa and 0.7MPa in normal temperature conditions for a minimum time of 12 hours.

Non-Conventional Materials and Technologies – NOCMAT for XXI Century   Materials Research Forum LLC
Materials Research Proceedings 7 (2018) 313-324   doi: http://dx.doi.org/10.21741/9781945291838-28

Glue
Melamine formaldehyde urea (MUF 1242 resin and catalyst) was used.

Number of test specimens
The number of test specimens was:

- Fourteen (14) pine specimens enumerated consecutively from one (1) to fourteen (14). The average dimensions of the test pieces were 39.17 x 38.76 x 82.34 mm.
- Twelve (12) specimens of bamboo guadua laminates listed as indicated in **Table 2.1.** The average dimensions of the specimens were 43.92 x 41.91 x 81.27 mm.
- Fourteen (14) specimens enumerated consecutively from one (1) to fourteen (14). The specimens had average dimensions of 54.87 x 36.35 x 91.08 mm and were formed by two pieces of pine of average dimensions 36.35 x 22.08 x 91.08 mm in the exterior and one piece of bamboo laminate of average dimensions 36.35 x 10.71 x 91.08 mm.

Strain gages were used in the specimens identified in Table 2 and the position is showed in Figure 2. All the test pieces were measured using a calibrator of 0.01mm precision and weighed with a digital balance of 0.01 gr precision. The application speed of compression load was 0.01 mm/s.

TABLE 2 - INSTRUMENTED SPECIMENS

Bamboo guadua Specimens	Radiata pine Specimens	Composite section Specimens
1A, 1B, 2A, 3A, 4A	2, 4, 10, 12	1, 2, 3, 4, 5, 6, 7, 8, 9, 10.

Source: Authors

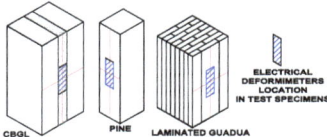

FIGURE 2 - APPROXIMATE DIMENSIONS OF GUADUA LAMINATES. (A) STRIPS. (B) BOARD. Source: Authors

Results
The modulus of elasticity of each instrumented specimen was calculated with the slope of linear regression of the stress–strain graph between 25% and 60% of the maximum load.

Bamboo guadua laminates
The test results of the bamboo guadua laminates, hereinafter GPLG, are shown in Figure 3, Figure 4 and Table 3, corresponding to Graph σ_c vs Δ/L, graph σ_c vs ε and the summary table of results, respectively. In these figures, σ_c is the compression stress parallel fiber (MPa), Δ is the displacement between the universal machine plates (m), L is the length of the specimen (m), ε is the strain (mm / mm), E_g is the modulus of elasticity of guadua (MPa) and γ_g is the guadua density (g/cm^3).

Non-Conventional Materials and Technologies – NOCMAT for XXI Century   Materials Research Forum LLC
Materials Research Proceedings 7 (2018) 313-324   doi: http://dx.doi.org/10.21741/9781945291838-28

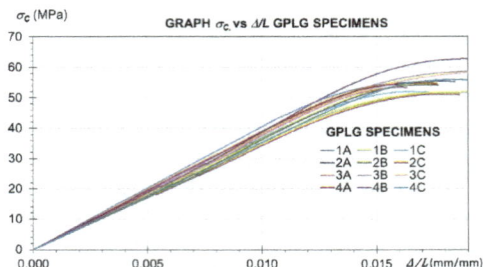

FIGURE 3 - *GRAPH σ_C VS Δ/L GUADUA SPECIMENS (GPLG). Source: Authors*

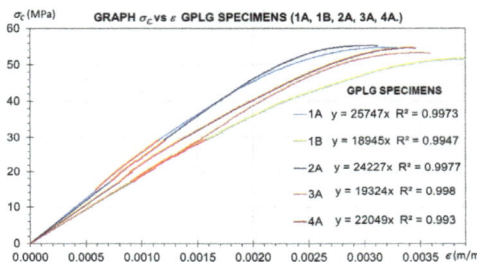

FIGURE 4 - *GRAPH σ_C VS ε OF THE GUADUA INSTRUMENTED SPECIMENS (GPLG) (1A, 1B, 2A, 3A, 4A.) Source: Authors*

TABLE 3 - *RESULTS OF $\sigma_{máx}$, σ_{LP}, MODULUS OF ELASTICITY (**E**), COEFFICIENTS OF LINEAR REGRESSIONS AND DENSITY IN SPECIMENS OF GPLG*

GPLG Specimens	$\sigma_{máx.}$ MPa	σ_{LP} MPa	Modulus of elasticity		Y_g g/cm³
			E_g (MPa)	Correlation coefficient	
1A	54.80	32.88	25747	0.9973	0.70
1B	51.63	30.98	18945	0.9947	0.69
1C	51.96	31.17			0.67
2A	55.33	33.20	24227	0.9977	0.70
2B	54.17	32.50			0.69
2C	51.08	30.65			0.70
3A	53.30	31.98	19324	0.9980	0.70
3B	58.47	35.08			0.72
3C	58.26	34.96			0.73
4A	54.70	32.82	22049	0.9930	0.75
4B	62.70	37.62			0.75
4C	55.91	33.55			0.71
Average	55.19	33.12as	22058	0.99614	0.71
s	3.19	1.91	2664		0.02
c.v.	0.058	0.058	0.121		0.035

Non-Conventional Materials and Technologies – NOCMAT for XXI Century   Materials Research Forum LLC
Materials Research Proceedings **7** (2018) 313-324   doi: http://dx.doi.org/10.21741/9781945291838-28

Five (5) specimens showed a detachment of the external fibers during the test (1A, 1B, 1C, 3C and 4B). In other cases, crushing occurs at one end of the specimen, as presented in six (6) test specimens: 2B, 2C, 3B, 4A, 4C and 3A. Finally, a case of diagonal crack failure was obtained in the middle section of the specimen 2A (Figure 5).

| 1A | 4C | 2A |

FIGURE 5 - *EXAMPLES OF FAILURES IN GUADUA SPECIMENS (GPLG). Source: Authors*

The detachment of external fibers in guadua laminates is a characteristic situation that has been corroborated by other authors in their respective tests.

The modulus of elasticity and the maximum compression stress parallel fiber of guadua laminates obtained in this research are in the ranges of the values reported by other authors, as shown in Table 4.

TABLE 4 - *COMPARISON OF MODULUS OF ELASTICITY AND MAXIMUM COMPRESSION STRESS OF GPLG WITH OF BAMBOO LAMINATES OBTANINED BY OTHER AUTHORS*

AUTHOR	$\sigma_{máx}$ MPa	E_g (MPa)	Y_g (g/cm³)
Takeuchi: [3]	65.70	30044	0.80
López and Correal: [6]	48.00	19137	0.72
Li and other: [1]	60.90	9361	*
Correal and other: [7]	62.00	32271	0.72
Verma and other: [8]	31.90	11700	*
Barreto:[9]	47.07	*	*
Author: [5]	55.19	22058	0.71

* Value not evaluated

Source: Authors

Radiata Pine
The test results of the pine specimens are shown in Figure 6, Figure 7 and Table 5, corresponding to Graph σ_c vs Δ/L, graph σ_c vs ε and the summary table of results, respectively. In these graphs, E_p is the modulus of elasticity of radiata pine (MPa) and Y_p is the pine density (g/cm³).

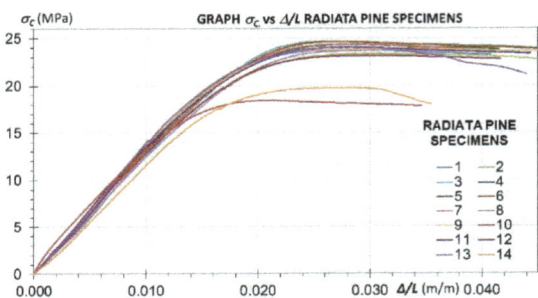

FIGURE 6 - *GRAPH σ_c VS ΔL RADIATA PINE SPECIMENS. Source: Authors*

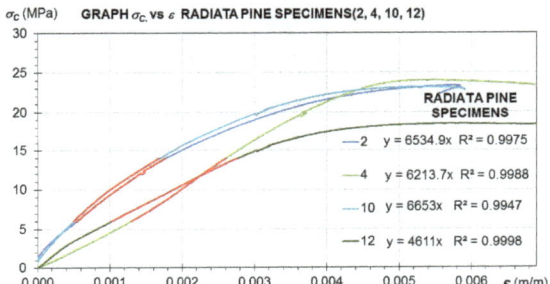

FIGURE 7 - *GRAPH σ_c VS ε RADIATA PINE INSTRUMENTED SPECIMENS (2, 4, 10, 12)*

Source: Author.

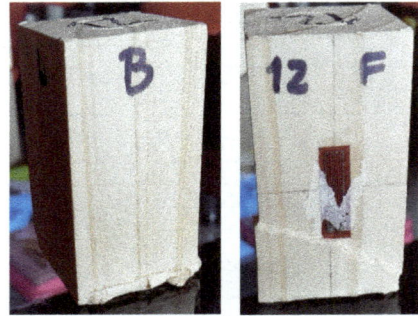

FIGURE 8 - *FAILURES AT THE EXTREMES AND DIAGONAL FISSURES IN RADIATA PINE. Source: Author.*

Non-Conventional Materials and Technologies – NOCMAT for XXI Century   Materials Research Forum LLC
Materials Research Proceedings 7 (2018) 313-324   doi: http://dx.doi.org/10.21741/9781945291838-28

TABLE 5 - *RESULTS OF $\sigma_{máx}$, σ_{LP}, MODULUS OF ELASTICITY (E), LINEAR REGRESSION COEFFICIENTS AND DENSITY IN RADIATA PINE SPECIMENS*

Radiata Pine Specimens	$\sigma_{máx.}$ MPa	σ_{LP} MPa	Modulus of elasticity		γ_p g/cm^3
			E_p (MPa)	Correlation coefficient	
1	23.81	14.28			0.40
2	24.64	14.79	6535	0.9975	0.39
3	23.34	14.01			0.39
4	24.01	14.41	6214	0.9988	0.39
5	24.34	14.61			0.40
6	24.63	14.78			0.40
7	24.37	14.62			0.39
8	23.73	14.24			0.41
9	23.93	14.36			0.40
10	23.18	13.91	6653	0.9947	0.39
11	24.05	14.43			0.39
12	18.41	11.05	4611	0.9998	0.37
13	24.04	14.43			0.40
14	19.74	11.85			0.48
Average	23.30	13.98	6003	0.9977	0.40
s	1.79	1.07	820		0.02
c.v.	0.077	0.077	0.137		0.059

Source: Author.

The pine specimens presented homogeneity in the results. The test piece No.12 presented less resistance than other specimens, however, it was not ruled out, since all the elements came from the same shelf and the simple and composite section were made with them.

The radiata pine specimens presented failures at both ends and diagonal fissures as seen in Figure 8.

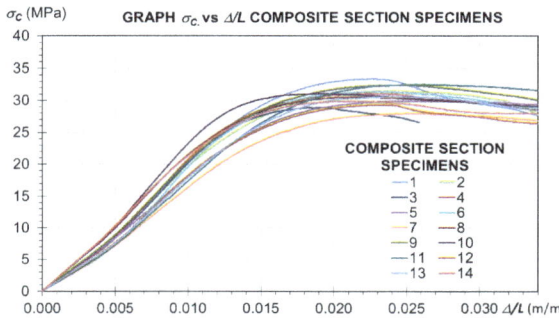

FIGURE 9 - *GRAPH σ_c. VS Δ/L COMPOSITE SECTION SPECIMENS. Source: Author.*

CBGL composite section.
The test results of the composite section are shown in Figure 9, Figure 10 and Table 6, corresponding to Graph σ_c. vs Δ/L, Graph σ_c. vs ε and the summary table of results, respectively.

Non-Conventional Materials and Technologies – NOCMAT for XXI Century   Materials Research Forum LLC
Materials Research Proceedings 7 (2018) 313-324   doi: http://dx.doi.org/10.21741/9781945291838-28

In these figures, E_{SC} is the modulus of elasticity of the composite section (MPa) and y_{SC} is the density composite section (g/cm^3).

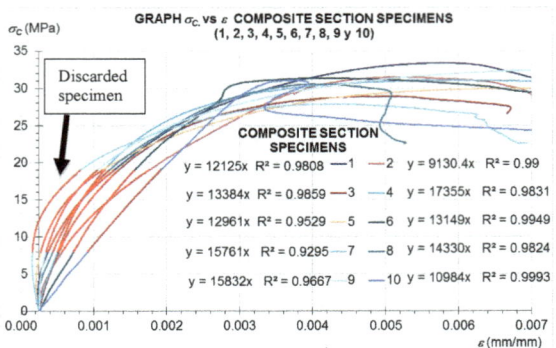

FIGURE 10 - *GRAPH σ_c VS ε COMPOSITE SECTION INSTRUMENTED SPECIMENS (1, 2, 3, 4, 5, 6, 7, 8, 9 Y 10)*

Source: Author.

TABLE 6 - *RESISTANCE, MODULUS OF ELASTICITY IN SPECIMENS OF COMPOSITE SECTION*

Composite Section Specimens	$\sigma_{máx}$ MPa	σ_{LP} MPa	Modulus of elasticity		y_{SC} g/cm^3
			E_{SC} (MPa)	Correlation coefficient	
1	33.34	20.00	12125	0.9808	0.50
2	31.45	18.87	9130	0.9900	0.48
3	28.88	17.33	13384	0.9859	0.47
4	31.04	18.63	17355	0.9831	0.46
5	29.90	17.94	12961	0.9529	0.48
6	31.31	18.78	13149	0.9949	0.49
7*	27.86*	16.72*	15761*	0.9295*	0.46*
8	30.53	18.32	14330	0.9824	0.46
9	32.41	19.44	15832	0.9667	0.50
10	31.05	18.63	10984	0.9993	0.46
11	32.48	19.49			0.49
12	29.26	17.56			0.45
13	30.55	18.33			0.47
14	29.83	17.90			0.47
Average	30.92	18.55	13250	0.98	0.47
s	1.25	0.75	2313	0.01	0.02
c.v.	0.041	0.041	0.14	0.014	0.034

* Data discarted for atypical behavior in **Figure 11**.

Source: Author

In the composite section there are not delamination or notorious failures of one component over the others, since the elements of this composite section worked as a unit. (See Figure 11).

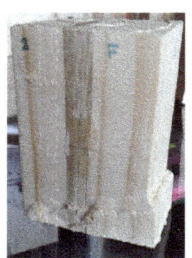

FIGURE 11 - TYPICAL FAULT IN COMPOSITE SECTION (SPECIMEN 2). Source: Authors

As shown in Table 7, the dispersion of the results obtained separately for guadua and pine is greater than in the composite section. The laminate elements decrease the dispersion of the results in the compressive stress parallel fiber test. However, the dispersion of the modulus of elasticity was greater in the composite section. (See Table 8)

TABLE 7 - MAXIMUM COMPRESSIVE STRESS PARALLEL FIBER ($\sigma_{máx}$) IN ALL ELEMENTS

		min (MPa)	máx (MPa)	s (MPa)	Average (MPa)	CV %
$\sigma_{máx}$	GPLG	51.08	62.70	3.19	55.19	5.78
	PINE	18.41	24.64	1.79	23.30	7.68
	COMPOSITE SECTION	28.88	33.34	1.25	30.92	4.04

Source: Authors

TABLE 8 - MODULUS OF ELASTICITY (E) COMPRESSION PARALLEL FIBER IN ALL THE ELEMENTS

	min (MPa)	máx (MPa)	s (MPa)	Average (MPa)	CV %
E_g	18945	25747	2664	22058	12.08
E_p	4611	6653	820	6003	13.65
E_{SC}	9130	17355	2313	13250	17.45

Source: Authors

Theoretical calculation of the modulus of elasticity of the composite section.
The theoretical module of elasticity of the composite section is determined with the percentage of the its materials: the laminated bamboo guadua and the pine. In Table 9, the percentages of each material are shown.

TABLE 9 - PERCENTAGES OF MATERIALS IN COMPOSITE SECTION

	PINE	GPLG
A (mm²)	1604.78	388.12
%	80.52%	19.48%

Source: Authors

Non-Conventional Materials and Technologies – NOCMAT for XXI Century   Materials Research Forum LLC
Materials Research Proceedings 7 (2018) 313-324   doi: http://dx.doi.org/10.21741/9781945291838-28

The experimental elasticity modulus under compression parallel fiber, obtained in the composite sections, presented intermediate values in their compounds, as shown in Table 10.

TABLE 10 - *EXPERIMENTAL AVERAGE OF MODULES OF COMPRESSION PARALLEL FIBER ELASTICITY.*

E_g (MPa)	E_p (MPa)	E_{SC} (MPa)
22058	6003	13250

Source: Authors

The theoretical value depends on the percentages of pine and guadua in the composite section. Using Equation (1), E_{SC} is equal to 9131 MPa.

$$E_{sc} = 80.52\%E_p + 19.48\%E_g \qquad \text{Equation (1).}$$

The experimental result of the modulus of elasticity of the composite material is higher (13250 MPa) than the theoretical expected (9131 MPa). Consequently, the stiffness has been increased, not only in the wood , but also in the laminated guadua, because it was confined. The composite section with pine (80%) and guadua (20%) in the interior showed a good behavior. The modules of experimental elasticity were higher than the theoretical ones, approaching a combination of 55% of pine and 45% of guadua.

The maximum theoretical resistance can be obtained multiplying the maximum resistance of each material by its percentage of area in the composite section:

$$\sigma_{máx} \text{ (theoretical)} = 19.48\%*55.19 + 80.52\%*23.30 = \textbf{29.51 MPa}$$

This result is positive, because the experimental result wasslightly higher (30.71MPa). The two materials have big mechanical differences. However, they behave well from the proportionality limit up to the maximum of their resistances. This demonstrates that the composite sectional was improved in 4.1%, compared to the theoretical results.

These results are summarized in Figure 12. The graph shows the σ_{LP} and the $\sigma_{máx}$ of each material: laminated bamboo guadua, pine, theoretical and experimental composite section, with their corresponding deformations. The deformation corresponding to the σ_{LP} of the experimental composite section draws a vertical that intersects the curve of the pine and the laminated guadua. The distance of the line, between the point of the laminated guadua and the intersection with the pine, constitutes the difference σ_g-σ_p. With this dimension and the percentages of each material in the composite section, we obtained the point and the slope of the theoretical composite section.

Non-Conventional Materials and Technologies – NOCMAT for XXI Century   Materials Research Forum LLC
Materials Research Proceedings 7 (2018) 313-324   doi: http://dx.doi.org/10.21741/9781945291838-28

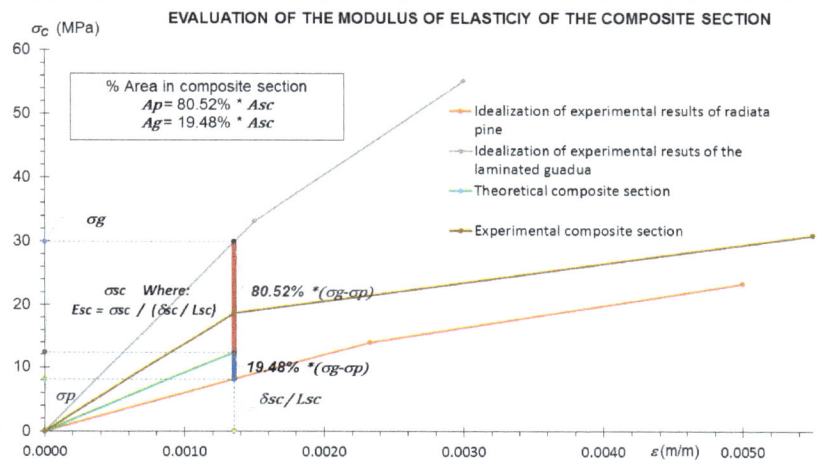

FIGURE 12 - *SUMMARY OF RESULTS. Source: Author.*

ε = Unitary deformation (mm / mm).
σ_c = Compression stress parallel fiber (MPa).
σ_{sc} = Compression stress parallel fiber in the composite section (MPa).
σ_p = Compression stress parallel radiata pine fiber in the composite section (MPa).
σ_g = Compression stress parallel guadua fiber in the composite section (MPa).
δ_p = Pine deformation by parallel compression stress fiber (mm).
δ_g = Guadua deformation by parallel compression stress fiber (mm).
δ_{sc} = Composite section deformation of parallel compression stress fiber (mm).
A = Cross-section area of the specimens (mm²).
A_p = Cross-section area of pine in the composite section (mm²).
A_g = Cross-sectional area of the guadua in the composite section (mm²).
A_{sc} = Cross-sectional area of composite section specimens (mm²).
E_{sc} = Modulus of elasticity of the composite section (MPa).
L_{sc} = Length of the composite section (m).

Conclusions

The compression strength of the composite section with 20% of laminated bamboo guadua and 80% of pine is in average 30.88 MPa , which contrasts the 23.3 MPa of the pine.

The modulus of elasticity of the composite section with 20% of laminated bamboo guadua and 80% of pine is in average 13250 MPa, which contrasts the 6003 MPa of the pine.

The tendency of transverse failure presented by the guadua laminates under longitudinal compressive stresses is controlled with the confinement given by the softwood. Additionally, when analyzing graphs σ_c vs ε, it is observed that lower deformation in the guadua is required to reach the elastic limit, than in wood with low mechanical specifications. This means that, in a composite section, the guadua cross first the elastic limit than the wood. Then the guadua keeps resisting the compression stresses in a plastic range, because the wood needs more deformation to reach the elastic limit. Finally, when the guadua is confined, it transmits triaxial compressive stresses that

Non-Conventional Materials and Technologies – NOCMAT for XXI Century   Materials Research Forum LLC
Materials Research Proceedings 7 (2018) 313-324   doi: http://dx.doi.org/10.21741/9781945291838-28

improve the mechanical properties of the composite section. Thus, the CBGL is a structural alternative to implement softwoods with low densities ($0.40g/cm^3$) compared to hardwood with densities closer to one.

References

[1] H. T. Li, Q. S. Zhang, D. S. Huang, and A. J. Deeks, "Compressive performance of laminated bamboo.," *Compos. Part B Eng.*, vol. 54, no. 1, pp. 319–328, 2013. https://doi.org/10.1016/j.compositesb.2013.05.035

[2] M. C. Yeh and Y. L. Lin, "Finger joint performance of structural laminated bamboo member," *J. Wood Sci.*, vol. 58, no. 2, pp. 120–127, 2012. https://doi.org/10.1007/s10086-011-1233-7

[3] C. P. Takeuchi, "Caracterización mecánica del bambú guadua Laminado para uso estructural," Universidad Nacional de Colombia., 2014.

[4] E. M. Reyes Ramirez and G. Rayo Morales, "Comportamiento a Flexión de Vigas de Madera de Gran Longuitud de Sección Compuesta Reforzadas con Láminas Pegadas Prensadas de Guadua 'Angustifolia Kunth'.," Universidad Militar Nueva Granada, 2014.

[5] A. Cruz Guzmán, "Módulo de elasticidad a compresión de un material compuesto de madera en su núcleo y guadua laminada pegada en el exterior," 2017.

[6] L. F. López and J. F. Correal, "Exploratory study of the glued laminated bamboo guadua angustifolia as a structural material," *Estud. Explor. los laminados bambú guadua angustifolia como Mater. estructural*, vol. 11, no. 3, pp. 171–182, 2009.

[7] J. F. Correal, J. S. Echeverry, F. Ramírez, and L. E. Yamín, "Experimental evaluation of physical and mechanical properties of Glued Laminated Guadua angustifolia Kunth," *Constr. Build. Mater.*, vol. 73, pp. 105–112, 2014. https://doi.org/10.1016/j.conbuildmat.2014.09.056

[8] C. S. Verma, N. K. Sharma, V. M. Chariar, S. Maheshwari, and M. K. Hada, "Comparative study of mechanical properties of bamboo laminae and their laminates with woods and wood based composites," *Compos. Part B Eng.*, vol. 60, pp. 523–530, 2014. https://doi.org/10.1016/j.compositesb.2013.12.061

[9] W. M. Barreto Castillo, "Evaluación de guadua laminada pegada aplicada a propuesta de reticulado plano," 2003.

Non-Conventional Materials and Technologies – NOCMAT for XXI Century   Materials Research Forum LLC
Materials Research Proceedings 7 (2018) 325-331   doi: http://dx.doi.org/10.21741/9781945291838-29

A Short View on the Novel uses of Geopolymer Metakaolin based to Control Air and Water Pollution

A. Manzano-Ramírez[1a*], J. Ramón-Gasca[2], H. R. Guzman-Carrillo[1b], J. L. Reyez-Araiza[3] and Mauricio Mondragon-Figueroa[1c]

[1a*] CINVESTAV-Querétaro, Qro. México, amanzano@cinvesta.mx

[1b] CINVESTAV-Querétaro, Qro. México, hr_guzmanc@hotmail.com

[1c] CINVESTAV-Querétaro, Qro. México, mmondragon@cinstav.mx

[2] Universidad de Guanajuato, Campus Celaya-Salvatierra, Celaya, Guanajuato, México, ragatsi99@yahoo.com

[3] DIPFI, Facultad de Ingeniería, Universidad Autónoma de Querétaro, C. U. Cerro de las Campanas, Centro, Querétaro, Qro. C.P. 76010, México. reyesaraiza@yahoo.com.mx

Abstract. Results obtained on the funtionalizacion of inorganic polymers metakaolin based are presented. Geopolymers are not only cementitious material instead it will be shown how semiconductors, at nanometric scale, can be added as admixture compounds or incorporated in the aluminosilicates structures by ion exchange. In addition, the efficiency on adsorption and photodegradation of cationic organic compounds in aqueous suspention is described based on the kinetic evaluation by first and second order reaction.

Keywords: Geopolymer, Semiconductors, Photodegradation, Kinetic

Introduction

Geopolymer is a broadly termed for "inorganic polymer" (1) .In the synthesis of geopolymer, the chemical reaction may consist of the following steps: (i) dissolution of Si and Al atoms from the source material through theaction of hydroxide ions, (ii) transportation, (iii) orientationor , (iv) condensation of precursor ions into monomers.(2) setting or polycondensation/polymerisation ofmonomers into polymeric structures (3, 4).

Geopolymer can be used as a binder instead of portland cement paste, to produce concrete. Concretes based on fly ash geopolymers have been synthesized and characterized [5, 6–8]. In addition, geopolimers have gained, in the past fifteen years, a major interest so that some of their applications are in cements and high tech materials [9] fire protection [10], immobilizations of waste and toxic materials [11] and radioactive waste encapsulation [12].

On the other hand, heterogeneous photocatalysis has been growing rapidly in the past three decades, this uses the light and a semiconductor material so that light-induced redox reactivity which may be initiated by the incident light.

Hence, in the present work are presented the results obtainend when an "inorganic polymer" metakaolin based , semiconductors at nanometric scale, are added as admixture compounds or incorporated in the aluminosilicates structures by ion exchange . In addition, the kinetic evaluation on adsorption and photodegradation of cationic organic compounds in aqueous suspention is described based on first and second order reaction.

Materials and methods

Commercial metakaolin was Metamax from BASF Corporation with the following chemical composition obtained from X-ray fluorescence 51.55% SiO2, 44.78% Al2O3,0.48% Fe2O3, and others. Sodium hydroxide and sodium silicate were purchased from SIDESA-Corporation México.

Non-Conventional Materials and Technologies – NOCMAT for XXI Century   Materials Research Forum LLC
Materials Research Proceedings 7 (2018) 325-331   doi: http://dx.doi.org/10.21741/9781945291838-29

Whereas kaolin from Tisayuca , Hidalgo, México was used. The chemical composition (weight percentage) of kaolin was determined by X-ray fluorescence, i.e: 73.19% SiO_2, 24.80% Al_2O_3, 1.26% SO_2, and others.

TiO_2 as a Anatase with an average size of 300 nm was obtained from Xinhai whereas the molar concentration of 1.0 of zinc acetate dehydrate was used to obtain nanospheres of ZnO with an average diameter of 38 nm.

Geopolymer preparation

The temperature at which complete dehydroxylation of the kaolin and formation of metakaolinite was determined by DTA analysis to take place at 540°C with a heating time of 120 min . Hence the alcinations of kaolin was carried out at 560°C for two hours. The conversion of the kaolinite to metakaolinite was confirmed by XRD , thereafter geopolymer samples were prepared by mechanically mixing stoichiometric amounts of metakaolin, sodium hydroxide, distilled water and sodium silicate (Na_2O/SiO_2 wt. ratio: 0.31) as well as TiO_2 particles with an average size of 300 nm, at two different weight percent concentrations i.e. 20 and 50 and cured at 90°C. The mechanical mixing was followed by 15 min of vibration to obtaine an homogenous slurry which was poured into cylindrical acrylic mould. The mixture constituents were formulated to follow the molar oxide ratios: SiO_2/Al_2O_3 = 3.3, Na_2O/SiO_2 = 0.25, Na_2O/Al_2O_3 = 0.4883 and H_2O/Na_2O = 13.73.

The slurry was dried for 2 hrs at 40°C and cured in a laboratory oven at 90°C for 24h. After cooling, geopolymers were cut with a diamond disc to produce samples of 1.0371± 0.059 g weight. When commercial metakaolin was used, this was mixed directly with the same molar ratios indicated above, dried and cured at the same time and temperature.

In the case of metakaolin from BASF Corporation, this was mixed directly with the same molar ratios indicated above, dried and cured at the same time and temperature.

Each set of samples was named as it is shown in Table 1.

TABLE 1 - WEIGHT PER CENT OF TIO₂ PARTICLES IN EACH SET OF SAMPLES

Sample TiO₂ particles wt%	Geo_cal	Geo_cMK	Geo 50-50TiO₂	Geo 80-20TiO₂
	-----	------	50	20

To determine the adsorption equilibrium time, experiment was conducted, firstly using 0.01 g of nano-particles of either ZnO or TiO_2 and secondly circular geopolymer sample (12.5 mm diameter 2.6 mm thickness) were placed within the MB solution. The degradation in dark (adsorption) was performed during 105 minutes

Photocatalytic degradation of MB

UV spectrometric techniques were used to study the photocatalytic degradation of MB. The UV-Vis calibration curve for aqueous MB was obtained using three different concentration of methylene blue , i.e 4.5×10^{-6} , 8.5×10^{-6} , 1.0×10^{-5} mg/L MB at the natural pH of the solution. The experiments were conducted under an UVA lamp (340 nm) , 14 w.

From a stock solution 8.36 ml were drawn to prepare 100 ml of a solution of methylene blue at a concentration of 8.5×10^{-6} M which then was poured into a 200 ml glass beaker (pyrex) with 0.01 g. of solid particles with nano-scale dimension of either ZnO or TiO_2 , circular geopolymer sample (12.5 mm diameter 2.6 mm thickness) thereafter the glass beaker was placed into the UV lamp system with a magnetic stirrer, Figure 1 . The distance between the liquid surface and the lamp was 17 cm.

Non-Conventional Materials and Technologies – NOCMAT for XXI Century   Materials Research Forum LLC
Materials Research Proceedings **7** (2018) 325-331   doi: http://dx.doi.org/10.21741/9781945291838-29

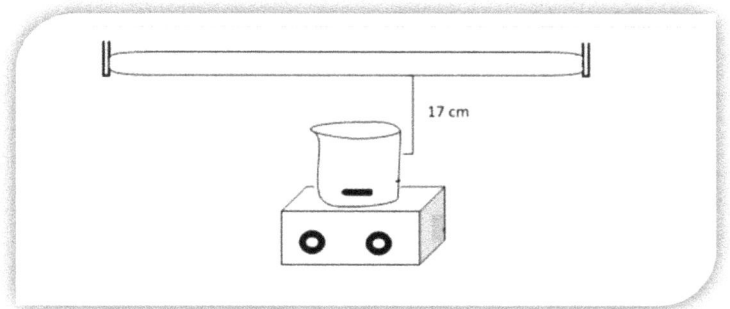

*FIGURE 1 - EXPERIMENTAL SET UP FOR THE ADSORPTION AND
PHOTODEGRADATION OF MB.*

Results and discussion

In a heterogeneous catalysis, the adsorption is a main factor that has been considered before any chemical reaction arises. Hence in the present work the adsorption equilibrium time was shown to be at 30 mins, Table 2. This effect was clearly observed for the ZnO particles, effect ascribed to the size of the nano-spheres ZnO particles, 38 nm, while in the case of TiO$_2$ an average variation within +/- 0.03 was observed.

TABLE 2 - ADSORPTION EQUILIBRIUM TIME

Time (min)	Absorbance stability	
	ZnO	TiO2
0	0.5799 *	0.5799 *
15	0.4690	0.4733
30	**0.4808**	**0.4021**
45	0.4733	0.4997
60	0.4710	0.4632
75	0.4737	0.5048
90	0.4907	0.4917
105	0.4768	0.4514

Figure 2 (a) shows the UV-Vis absorption spectrum obtained for MB solutions prepared at concentrations of 1.0 x 10^{-5} , 8.5 x 10^{-6} and 4.5 x 10^{-6} mg/L the calibration . In addition the absorbance at 664nm showed a linear dependence with MB concentration.

Non-Conventional Materials and Technologies – NOCMAT for XXI Century   Materials Research Forum LLC
Materials Research Proceedings 7 (2018) 325-331   doi: http://dx.doi.org/10.21741/9781945291838-29

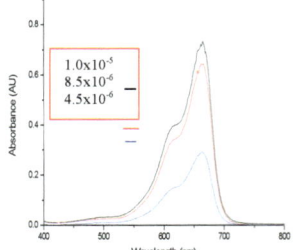

CONCENTRATION mg/L	ABSORBANCE
1.0×10^{-5}	0.73350
8.5×10^{-6}	0.64351
4.5×10^{-6}	0.29002

FIGURE 2 - (A) UV-VIS ABSORPTION SPECTRUM FOR MB SOLUTIONS PREPARE AT CONCENTRATIONS OF 1.0 X 10^{-5} , 8.5 X 10^{-6} AND 4.5 X 10^{-6} mg/L , 1(B) LINEAR DEPENDENCE WITH MB CONCENTRATION

The spectra obtained shows a strong absorbance at 664 nm and a shoulder at 664 nm and a shoulder at about 610 nm indicating that the MB solution contain free monomers (MB^+) and free dimmers $(MB^+)_2$, respectively [14]. In all the cases, after 5 mins of exposure, the MB solutions present very broad absorbance bands.

On the other hand, the degradation efficiency of MB was determined using the fallowing equation:

$$\eta = \left[\frac{A_0 - A_t}{A_0}\right] x \ 100 \tag{1}$$

where η is the removal effiency , A_0 is the initial absorbency at time 0 , A_t is the absorbency at time t after the UVA lamp was swiched on Table 3.

TABLE 3 - REMOVAL EFFICIENCY OF MB AFTER IRRADIATION

Illumination Time (min)	Degradation efficiency η (%)	
	ZnO	TiO$_2$
0	0.0	0.0
30	28.6	19.8
45	32.6	52.2
60	40.1	64.5
75	46.9	67.2
90	53.4	71.8
105	63.1	76.9

Non-Conventional Materials and Technologies – NOCMAT for XXI Century   Materials Research Forum LLC
Materials Research Proceedings 7 (2018) 325-331   doi: http://dx.doi.org/10.21741/9781945291838-29

On this, it may be observed how the TiO_2 particles show the greatest removal efficiency, this may implies that the degradation of MB strongly depend on the adsorption capacity of the photocatalyst since the photocatalitic redox reactions occur on the surface of the former.

Table 4 shows the removal rates in dark of both semiconductors while Table 5 shows the removal rates under UVA radiation:

TABLE 4 - *COMPARISON OF THE REMOVAL RATES OF ZNO AND TIO₂ IN DARK*

Conditions	First order kinetics parameter	Second order kinetics parameters		
	R^2	k_2 (g/(mg min))	q_e (mg/g)	R^2
ZnO without light	0.148	4.403	0.483	0.999
TiO_2 without light	0.001	7.156	0.472	0.988

TABLE 5 - *COMPARISON OF THE REMOVAL RATES OF ZNO AND TIO₂ UNDER UVA*

Conditions	First order kinetics parameter	Second order kinetics parameters		
	R^2	k_2 (g/(mg min))	q_e (mg/g)	R^2
ZnO with light	0.468	0.294	---	0.955
TiO_2 with light	0.366	0.362	----	0.931

Based on these results the degradation kinetics of MB may be considered to correspond to the second order rate equation. Hence it may be suggested that the reaction depends on the concentrations of the demethylated intermediates formed (azure B and A) from MB degradation under UV-A irradiation [15].

Figure 3 shows the degradation of MB for geopolymer samples in aqueous solution at different conditions, without and with illumination:

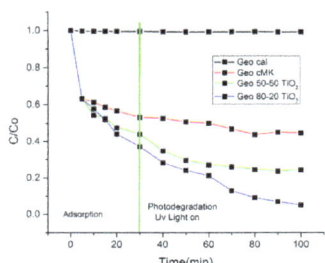

FIGURE 3 - *DEGRADATION OF MB FOR GEOPOLYMER SAMPLES AT DIFFERENT CONDITIONS*

On this it may observed how the highest effect is shown by the geopolymer with 20 % of TiO_2, effect that may be ascribed to a better distribution of the TiO_2 particles in the geopolymer matrix , Figure 4 , compared to the clusters observed in the geopolymer-50%TiO_2 Figure 5, clusters that may avoid an effective photo-activation of the semiconductor particles.

Non-Conventional Materials and Technologies – NOCMAT for XXI Century   Materials Research Forum LLC
Materials Research Proceedings 7 (2018) 325-331   doi: http://dx.doi.org/10.21741/9781945291838-29

FIGURE 4 - *TiO$_2$ PARTICLES WELL DISTRUSTED IN THE GEOPOLYMER-20%TiO$_2$*

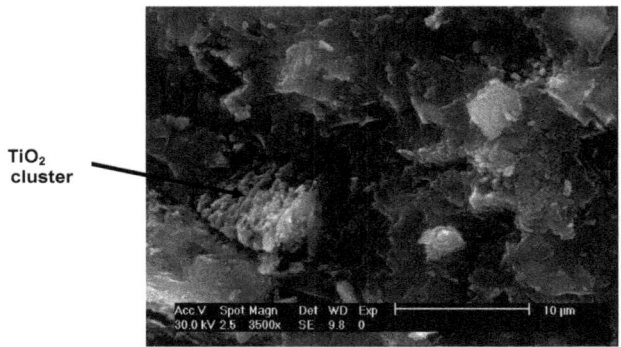

FIGURE 5 - *TiO$_2$ CLUSTERS OBSERVED IN THE GEOPOLYMER-50%TiO$_2$*

Conclusions

Results have shown how the geopolymer with 20 and 50% TiO$_2$ is the must be effective for the degradation of MB in solution exposed to UV$_A$ light.In addition the removal efficiency of MB on the surface of the geopolymer with the semiconductor particles agrees with the second order reaction kinetics effect that may suggested that the oxidation of MB on the surface along the reduction of O$_2$ by the iluminated semiconductor suggests the reaction depends on the concentrations of hydroxyl and hydroperoxy radicals as well as on the demethylated intermediates formed (azure B and A) .

Acknowledgments

M.Mondragon-Figueroa thanks to CONACYT for scholarship as well as LITRA_CINVESTAV for providing the infrastructure.

Non-Conventional Materials and Technologies – NOCMAT for XXI Century   Materials Research Forum LLC
Materials Research Proceedings 7 (2018) 325-331   doi: http://dx.doi.org/10.21741/9781945291838-29

References

[1] Barbosa, V.F.F. and MacKenzie, K.J.D., Thermal Behavior or Inorganic Geopolymers and Composites Derived from Sodium Polysialate, *Materials Research Bulletin*, 2003, vol. 38, no. 2, pp. 319–331. https://doi.org/10.1016/S0025-5408(02)01022-X

[2] Powers, T.C., Structure and Pysical Properties of Hardened Portland Cement Paste, *J. Am. Ceram. Soc.*, 1958, vol. 61, pp. 1–5. https://doi.org/10.1111/j.1151-2916.1958.tb13494.x

[3] Davidovits, J., Chemistry of Geopolymeric Systems Terminology, In: Proceedings of geopolymer, international conference, France, 1999.

[4] Xu, H. and Deventer, J. S. J. V., The Geopolymerisationof Alumino-Silicate Minerals, *International Journal ofMineral Processing*, 2000, vol. 59, no. 3, pp. 247–266.

[5] Hardjito, D., Wallah, S.E., Sumajoaw, D.M.J., and Rangan, B.V., Brief Review of Development of Geopolymer Concrete, George Hoof Symposium, American Concrete Institute, Las Vegas, USA, 2004.

[6] Hardjito, D. and Rangan, B.V., Development and Properties of Low-Calcium Fly Ash-Based Geopolymer Concrete, Research Report GC 1, Fac. of Engineering, Curtin University of Technology, Perth, Australia, 2005

[7] Chindaprasirt, P., Chareerat, T., and Sirivivatnanon, V.,Workability and Strength of Coarse High Calcium Fly Ash Geopolymer, *Cem. Concr. Compos.*, 2007, vol. 29, pp. 224–229. https://doi.org/10.1016/j.cemconcomp.2006.11.002

[8] Palomo, A, Fernández-Jimenez, A, Lopez-Hombrados,C., and Lleyda, J.L., Railway Sleepers Made of Alcali Activated Fly Ash Concrete Revista, *Ingenieria de Constructión*, 2007, vol. 22, pp. 75–80.

[9] R.Cioffi, L.Maffucci and L.Santoro, Optimization of geopolymer synthesis by calcinations and polycondensation of a kaolinitic residue., Resources Conservation & Recycling. 40 (2003), 27-38. https://doi.org/10.1016/S0921-3449(03)00023-5

[10] Giancaspro, J., Balaguru, P., Lyon, R., Fire protection of flammable materials utilizing geopolymers. SAMPE J, v. 40, Issue 5, p.42-49, 2004.

[11] Zhang, J.G., Provis, J.L., Feng, D.W., Van Deventer, J.S.J. Geopolymers for immobilization of Cr6+, Cd2+, and Pb2+. Journal of Hazardous Materials, v. 157, Issue 2-3, p. 587-598, 2008. https://doi.org/10.1016/j.jhazmat.2008.01.053

[12] Pereira, C.F., Luna Y., Querol X., Antenucci D., Vale J., Waste stabilization/solidification of an electric arc furnace dust using fly ash-based geopolymers. Fuel Journal, v. 88, Issue 7, p. 1185-1193, 2009. https://doi.org/10.1016/j.fuel.2008.01.021

[13] James J. Beaudoin et.al. (2011) Cement & Concrete Composites 33, pag. 246-250.

[14] Colling G Joseph, Yun Hin Taufiq-Yap, Gianluca Li Puma, Kogularama Sanmugam and Hye Shane Quek, Photocatalytic degradation of cationic dye simulated wastewater using four radation sources , UVA, UVB, UVC and solar lamp of identical power output, Desalination and water treatment, 57 (2016) 7976-7967. https://doi.org/10.1080/19443994.2015.1063463

Non-Conventional Materials and Technologies – NOCMAT for XXI Century   Materials Research Forum LLC
Materials Research Proceedings 7 (2018) 332-340   doi: http://dx.doi.org/10.21741/9781945291838-30

Dry Etching Plasma Applied to *Guadua Angustifolia* Bamboo Fibers: Influence on their Mechanical Properties and Surface Appearance

P. Luna [a,*], J. Lizarazo-Marriaga [a], A. Mariño [b]

[a] Department of Civil and Agricultural Engineering, Universidad Nacional de Colombia. Bogotá – Colombia; plunat@unal.edu.co*, jmlizarazom@unal.edu.co

[b] Department of Physics, Universidad Nacional de Colombia. Bogotá – Colombia; amarinoca@unal.edu.co

Abstract. Natural fibers are becoming a valuable resource for composite industry. The main disadvantage of using them as reinforcement of polymeric matrices is their low compatibility with common matrices, resulting in a poor mechanical behavior. Plasma treatments applied to fibers could be used in order to increase the bonding between composite phases. This paper shows the results of a research aimed to study the effect of the dry etching plasma on natural fibers; for this, *Guadua angustifolia* bamboo fibers were used. The influence of etching time on the fiber's tensile strength and the surface appearance was evaluated. Results showed that guadua fibers could be treated using dry etching plasma without a strength decreasing but with a significant modification on their surface appearance.

Keywords: Guadua Angustifolia Bamboo, Natural Fibers, Mechanical Properties, Dry Etching Plasma

Introduction

Composite materials are formed by two phases, the reinforcement material (discontinue phase) and the matrix (continuous phase). Their mechanical behavior depends on the mechanical properties of its constituents, distribution and reinforcement length, among other characteristics. However, the interaction between constituents plays an essential role on composite performance; the physicochemical interactions between composite components are the responsible of stress transfer from the matrix to the reinforcement [1–4].

The use of natural fibers as reinforcement of polymeric matrices has multiple advantages as high mechanical properties per unit of weight and the cost of manufacturing per unit of volume is fairly low, among others [5–7]. Indeed, according to different authors [8–12], the use of natural fibers in composite industry has increase in last two decades. Nevertheless, the potential of natural fibers in composite industry has not been completely profited, due to the low chemical compatibility between components. This behavior is consequence of the hydrophilic nature of natural fibers and hydrophobic nature of commonly used polymeric matrices, leading in an inadequate mechanical performance of composite material due to the low adherence between composite phases [13–15] [16].

According to literature, there are two main procedures in order to overcome the limitations of using natural fibers as reinforcement of polymeric matrices. Both procedures are focused on the modification of physico-chemical properties of composite components; the first one is aimed on the matrices, and the second one on the fibers. The second procedure is commonly used for industrial applications [2]. The modification of physico-chemical properties of fibers can be achieve by using coupling agents [17–19], making a graft polymerization of monomers compatible with the polymer matrix [2,20] or using plasma treatments [21–23].

Non-Conventional Materials and Technologies – NOCMAT for XXI Century   Materials Research Forum LLC
Materials Research Proceedings 7 (2018) 332-340   doi: http://dx.doi.org/10.21741/9781945291838-30

In general terms, plasma consists in the ionization of a gas or gasses; this ionization is reached by applying an electric field to the gas [24–27]. Plasmas are composed by ions and electrons, and according to its internal temperature can be classified as hot (thermal equilibrium) and cold plasmas (non-thermal equilibrium) [28–31]. Due to most materials present changes on their microstructural composition and degradation at high temperatures, in material science are commonly used cold plasmas. Cold plasmas can be generated using low (10-4 to 10-2 kPa) or atmospheric pressure; in both cases, the ionization is started and maintained by using direct current (DC), radio frequency (RF) or microwave (MW) power, with or without an additional electric (bias) or magnetic field [31]. The main advantage of using cold plasma treatments is that the alterations do not affect the bulk properties of the material [22,32–34].

From a general point of view, cold-low pressure plasmas fulfill three main purposes: functionalization of surfaces, deposition of thin films, and etching [20,24,30]. Etching processes are focused on removing material from the surface, which can be attained using a purely chemical process (wet etching), or using a physical approach (dry etching). Dry etching can be achieved by physical sputtering, chemical reaction or ion-assisted mechanism. By using a physical sputtering procedure, ions in the plasma transfer significant amounts of energy and momentum to the substrate, causing the atoms removal [35].

On this research was studied the effects of using dry etching plasma for treating Guadua angustifolia (guadua) bamboo fibers. The influence of the treatment on fibers' tensile strength and surface appearance were studied. All treatments were made using the same energy and treatment time was variable. This research explores this plasma technique as improver of interfacial properties of polymeric composite materials.

Materials and experimental procedure

Fiber Extraction

Guadua fibers were obtained using a chemical-mechanical process, which is shows schematically on Figure 1. Culms between 3 and 6 years old were used for obtaining fibers, and they were only extracted from the bottom part of the plant. Initially, the culms were divided in longitudinal strips, which were polished to remove the protuberances at nodes and the external layer. The objective of the chemical process is to soften the material, which was achieved by the immersion of polished strips during 3 hours in a sodium hydroxide solution, with a concentration of 2.5% previously heated at 80°C; the temperature was maintained during the immersion time. After immersion time, the strips were washed deeply using tap water. To obtaining the fibers, was used a mechanical process using the machine described in [36]. All fibers were carefully washed using tap water, and dried at laboratory temperature. The average length and cross-sectional area of obtained fibers were 10cm and 0.034mm^2, respectively.

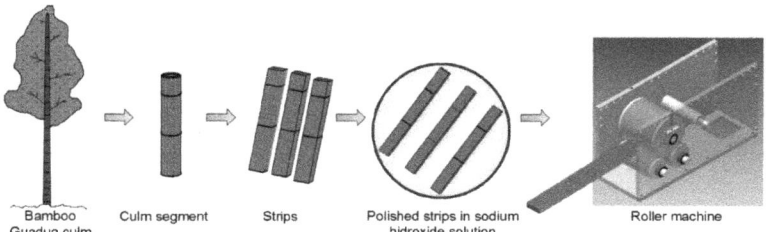

Bamboo   Culm segment   Strips   Polished strips in sodium   Roller machine
Guadua culm   hidroxide solution

FIGURE 1 - GUADUA FIBERS EXTRACTION PROCESS

Non-Conventional Materials and Technologies – NOCMAT for XXI Century   Materials Research Forum LLC
Materials Research Proceedings 7 (2018) 332-340   doi: http://dx.doi.org/10.21741/9781945291838-30

Dry Etching Plasma Treatments

Dry etching plasma treatment was carried out using a DC sputtering (etching) system. The gas used for plasma generation was Argon (Ar). Guadua fibers were exposed to ion bombardment for different times; the exposure times were 400, 1000, 1500 and 2000s. All treatments were carried out using an average current of 30 ± 3 mA and a working pressure of 10^{-2} kPa.

Tensile Tests

Through tensile tests was studied the effect of treating guadua fibers using dry etching plasma technique. All tests were performed following the guidelines of ASTM 1557-14, and using a load rate of 1.5mm/min. Before mechanical tests, all fibers were placed on paper frames, as shows Figure 2, working with a gage length of 40mm, to avoid deviation of axial load due to incorrect positioning of the fiber on testing grips. After setting the testing grips, the paper frame was carefully cut.

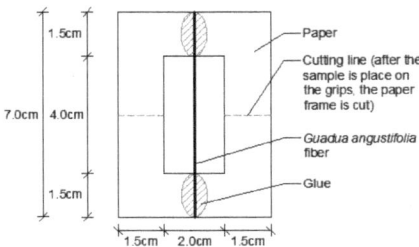

FIGURE 2 - PAPER FRAME USED FOR FIBERS' MECHANICAL TESTS

In order to obtained results that represent the mechanical behavior of guadua fibers, for each treatment time was tested 16 replicates. This sample size was calculated in function of an expected error of 20MPa on the determination of fiber strength, with a confidence level of 90%.

The tensile strength of each fiber, σ_t, was calculated using Eqn. 1, where P_{max} indicates the failure load of each fiber and A gives the cross-sectional area. After tests, the cross-sectional area of each fiber was measured at the fracture plane with high accuracy taking micrographs through a metallographic microscope Leco500 and a 100X magnification. Prior the micrographs were taken, the fibers were put in a transparent resin and afterward, the software ImageJ used to calculate the actual measurement. This procedure ensured that the perpendicular axis of the image cross-section matched the observer's eyes axis, which avoided optical distortions.

$$\sigma_t = \frac{P_{max}}{A} \tag{1}$$

In a previous research [37], was established the equipment's system compliance (C_s) according to the ASTM1557-14. Thereby, the fiber elongation, Δl, can be determined with Eqn. 2, where ΔL is cross-head displacement during the fiber test and F the applied force.

$$\Delta l = \Delta L - C_s F \tag{2}$$

Non-Conventional Materials and Technologies – NOCMAT for XXI Century   Materials Research Forum LLC
Materials Research Proceedings **7** (2018) 332-340   doi: http://dx.doi.org/10.21741/9781945291838-30

The modulus of elasticity (E) of each fiber was calculated as the slope, between 30% and 60% of maximum σ_t, of stress-strain curves as shown in Figure 3. The fiber's strain, ε , was calculated using equation (3), where 40 corresponds to the gage length used on the paper frame (Figure 2).

$$\varepsilon = \frac{\Delta l}{40} \tag{3}$$

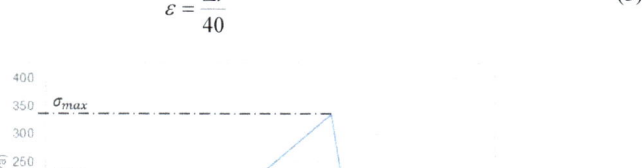

FIGURE 3 - *STRESS-STRAIN CURVE OBTAINED FROM EXPERIMENTAL MEASUREMENTS*

Morphological Analysis

In order to assess the influence of etching treatment on surface appearance of fibers, treated and nontreated samples were observed using a scanning electron microscope Tescan Vega 3 SB, working in secondary electron mode.

Results and discussion

Tensile Tests

Figure 4 and Figure 5 show the results in terms of tensile strength and modulus of elasticity, respectively. The values for 0s of treatment time, correspond to the results for untreated fibers. A statistical analysis was performed to identify significant differences among experimental results for different treatment times. Initially, using a Shapiro-Wilk test, it was verified whether each data set fulfilled a normal distribution. The homoscedasticity among data sets was verified using a Levene test. Due to each data set satisfy the criteria of normality and homo-elasticity, an ANOVA analysis was applied, using a significance level of 0.05.

Non-Conventional Materials and Technologies – NOCMAT for XXI Century   Materials Research Forum LLC
Materials Research Proceedings 7 (2018) 332-340   doi: http://dx.doi.org/10.21741/9781945291838-30

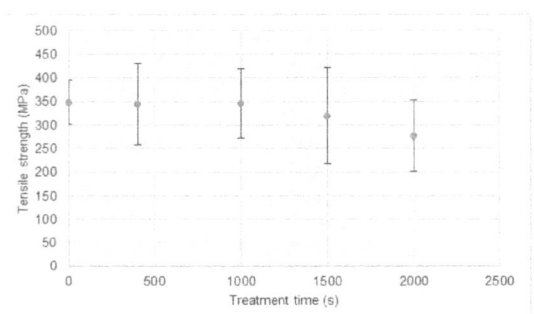

FIGURE 4 - *TENSILE STRENGTH FOR EACH TREATMENT TIME*

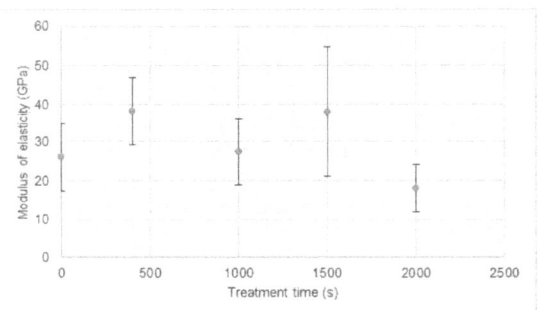

FIGURE 5 - *MODULUS OF ELASTICITY FOR EACH TREATMENT TIME*

The ANOVA analysis made for the tensile strength indicates that there are no significant differences among treatment times applied. Researchers who have evaluated the influence of dry etching plasma on tensile strength of different fibers have also concluded that there are no alterations on the mechanical property [22,23]. On the other hand, modulus of elasticity has changes with treatment applied; however, the obtained results do not permit to establish a trend for this behavior. There is no substantial differences among modulus of elasticity calculated for untreated fibers and treated during 1000 and 2000s; moreover, there are no significant differences between E values obtained for fibers treated during 500 and 1500s.

SEM images
Figure 6 shows the SEM micrographs for guadua fibers. Treated fibers exhibit a rough and coarser surface, even with the formation of some cracks. This change in the surface is attributed to etching effect due to ion bombardment. Similar results were found in other research works [38–42]. Untreated fiber (Figure 6a) shows a mild surface with some continuous, slight ridges and irregularities generated by the presence of non-cellulosic materials (hemicellulose and lignin constituents). Some little-isolated microcracks begin to be seen on the surface of fibers treated during 400s (Figure 6b), which increase for higher treatment times (Figure 6c and Figure 6d). For fibers

treated during 2000s, those microcracks coalesce together and the surface tends to become continuous (Figure 6e).

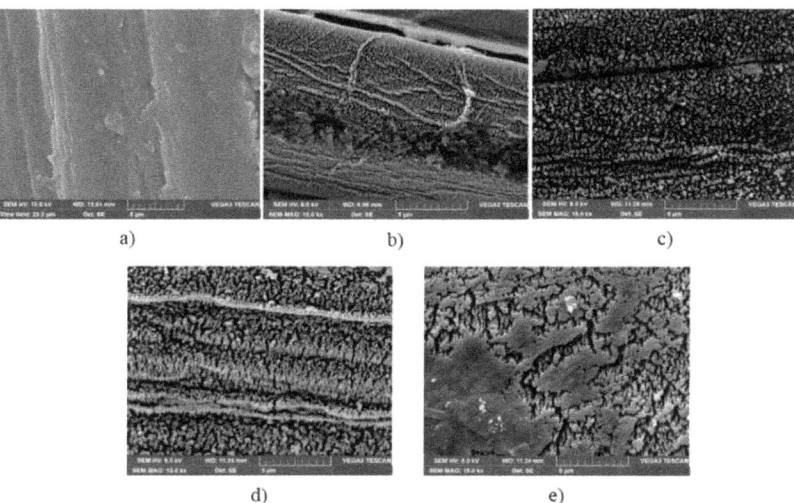

a)   b)   c)

d)   e)

FIGURE 6 - *SEM MICROGRAPHS FOR GUADUA FIBERS: A) WITHOUT TREATMENT, AND TREATED DURING B) 400S, C) 1000S, D) 1500S AND E) 2000S*

Conclusions

From the statistical analysis made for experimental results, it can be concluded that treating guadua fibers with dry etching plasma does not change the fibers' tensile strength. On the other hand, the treatment alters the fibers' modulus of elasticity; however, obtained results do not permit to establish a trend for this behavior.

Results obtained through SEM images suggest that treating guadua fibers with dry etching plasma could increase the bonding between composite phases, because the treatment causes rough and coarser fiber surfaces. This could lead in an increment of the mechanical grip between the reinforcement and the polymer matrix.

From mechanical results and SEM images it can be concluded that the coalesence effect does not affect the tensile resistance of treated fibers.

The results of this research will be examined with depth detail during the second stage of the research, which aims the development of a composite material with a polyester matrix reinforced by guadua fibers.

Acknowledgments

The authors acknowledge the support provided by Colciencias (funding 6172), and Universidad Nacional de Colombia (funding DIB 34835).

References

[1] Daniel I, Ishai O. Engineering mechanics of composite materials. Primera Ed. Oxford University Press; 1994.

[2] Valadez A. Efecto del tratamiento superficial de fibras de Henequén sobre la resistencia interfacial fibra-matriz y en las propiedades efectivas de materiales compuestos termoplásticos. Universidad Autónoma Metropolitana, México D.F; 1999.

[3] Hodgkinson JM. Mechanical testing of advanced fibre composites. Primera ed. TJ International; 2000.

[4] Agarwal B, Broutman L, Chandrashekhara K. Analysis and performance of fiber composites. Tercera Ed. John Wiley & Sons, Inc; 2006.

[5] Tapia C, Paredes C, Simbaña A, Bermúdez J. Aplicación de las Fibras Naturales en el Desarrollo de Materiales Compuestos y como Biomasa. Rev Tecnológica ESPOL. 2006;19:113–20.

[6] Taj S, Munawar M, Khan S. Natural fiber-reinforced polymer composites. Proc Pakistan Acad Sci. 2007;44(2):129–44.

[7] Spear M. Natural Composites in Construction. Welsh Composites Center; 2009. p. 1–9.

[8] Bledzki A, Reihmane S, Gassan J. Properties and modification methods for vegetable fibers for natural fiber composites. J Appl Polym Sci. 1996 Feb 22;59(8):1329–36. https://doi.org/10.1002/(SICI)1097-4628(19960222)59:8%3C1329::AID-APP17%3E3.0.CO;2-0

[9] Bledzki A, Gassan J. Composites reinforced with cellulose based fibres. Prog Polym Sci. 1999;24:221–74. https://doi.org/10.1016/S0079-6700(98)00018-5

[10]   Rijswijk K, Brouwer WD, Beukers A. Application of Natural Fibre Composites in the Development of Rural Societies. 2001.

[11]   Mohanty AK, Misra M, Drzal LT. Sustainable Bio-Composites from Renewable Resources : Opportunities and Challenges in the Green Materials World. J Polym Environ. 2002;10:19–26. https://doi.org/10.1023/A:1021013921916

[12]   Faruk O, Bledzki AK, Fink H-P, Sain M. Progress Report on Natural Fiber Reinforced Composites. Macromol Mater Eng. 2013 Jun 19;n/a-n/a.

[13]   Barkoula NM, Alcock B, Cabrera NO, Peijs T. Fatigue properties of highly oriented polypropylene tapes and all-polypropylene composites. Polym Polym Compos. 2008;16(2):101–13. https://doi.org/10.1177/096739110801600203

[14]   Kushwaha PK, Kumar R. Studies on performance of acrylonitrile-pretreated bamboo-reinforced thermosetting resin composites. J Reinf Plast Compos [Internet]. 2010 May 27 [cited 2013 Jul 17];29(9):1347–52. Available from: http://jrp.sagepub.com/cgi/doi/10.1177/0731684409103701

[15]   Song W, Zhao F, Yu X, Wang C, Wei W, Zhang S. Interfacial characterization and optimal preparation of novel bamboo plastic composite engineering materials. BioResources. 2015;10:5049–70. https://doi.org/10.15376/biores.10.3.5049-5070

[16]   Khan Z, Yousif BF, Islam MM. Fracture behaviour of bamboo fiber reinforced epoxy composites. Compos Part B Eng [Internet]. 2017;116:186–99. Available from: https://doi.org/10.1016/j.compositesb.2017.02.015

[17]   Kalia S, Kaith B, Kaur I. Pretreatments of Natural Fibers and their Application as Reinforcing Material in Polymer Composites — A Review. Polym Eng Sci. 2009; https://doi.org/10.1002/pen.21328

[18]   Araújo JR, Waldman WR, De Paoli MA. Thermal properties of high density polyethylene composites with natural fibres: Coupling agent effect. Polym Degrad Stab. 2008 Oct;93(10):1770–5. https://doi.org/10.1016/j.polymdegradstab.2008.07.021

[19]   Kwon H-J, Sunthornvarabhas J, Park J-W, Lee J-H, Kim H-J, Piyachomkwan K, et al. Tensile properties of Kenaf Fiber and Corn Husk FlourReinforced Poly(lactic acid) Hybrid Bio-Composites: Role of Aspect Ratio of Natural Fibers. Compos Part B Eng [Internet]. 2013 Aug [cited 2013 Aug 27]; Available from: http://linkinghub.elsevier.com/retrieve/pii/S1359836813004113

[20]   Li R, Ye L, Mai Y. Application of plasma technologies in fibre-reinforced polymer composites: a review of recent developments. Compos Part A Appl Sci Manuf [Internet]. 1997;28(1997):73–86. Available from: https://doi.org/10.1016/S1359-835X(96)00097-8

[21]   Rodríguez L, Fangueiro R, Orrego C. Efecto de tratamientos químicos y de plasma DBD en las propiedades de fibras del seudotallo del plátano. Rev Latinoam Metal y Mater. 2015;35(2):1–10.

[22]   Barra BN, Santos SF, Bergo PVA, Alves C, Ghavami K, Savastano H. Residual sisal fibers treated by methane cold plasma discharge for potential application in cement based material. Ind Crops Prod. 2015;77:691–702. https://doi.org/10.1016/j.indcrop.2015.07.052

[23]   Luna P, Lizarazo-Marriaga J, Mariño A. Guadua angustifolia bamboo fibers as reinforcement of polymeric matrices: An exploratory study. Constr Build Mater [Internet]. 2016;116:93–7. Available from: http://dx.doi.org/10.1016/j.conbuildmat.2016.04.139

[24]   Albella JM, editor. Láminas delgadas y recubrimientos: preparación, propiedades y aplicaciones. Madrid, España: Consejo Superior de Investigaciones Científicas; 2003.

[25]   Costa THC, Feitor MC, Alves CJ, Freire PB, de Bezzera CM. Effects of gas composition during plasma modification of polyester fabrics. J Mater Process Technol. 2006;173:40–3.

[26]   Sparavigna A. Plasma treatment advantages for textiles. Cornell Univ Libr. 2008;

[27]   Morshed M, Alam M, Daniels S. Plasma Treatment of Natural Jute Fibre by RIE 80 plus Plasma Tool. Plasma Sci Technol. 2010;325. https://doi.org/10.1088/1009-0630/12/3/16

[28]   Bogaerts A, Neyts E, Gijbels R, Van der Mullen J. Gas discharge plasmas and their applications. Spectrochim Acta - Part B At Spectrosc. 2002;57(4):609–58. https://doi.org/10.1016/S0584-8547(01)00406-2

[29]   Morent R, De Geyter N, Desmet T, Dubruel P, Leys C. Plasma surface modification of biodegradable polymers: A review. Plasma Process Polym. 2011;8(3):171–90. https://doi.org/10.1002/ppap.201000153

Non-Conventional Materials and Technologies – NOCMAT for XXI Century   Materials Research Forum LLC
Materials Research Proceedings 7 (2018) 332-340   doi: http://dx.doi.org/10.21741/9781945291838-30

[30]   Denes FS, Manolache S. Macromolecular plasma-chemistry: An emerging field of polymer science. Prog Polym Sci. 2004;29(8):815–85. https://doi.org/10.1016/j.progpolymsci.2004.05.001

[31]   Tendero C, Tixier C, Tristant P, Desmaison J, Leprince P. Atmospheric pressure plasmas: A review. Spectrochim Acta - Part B At Spectrosc. 2006;61(1):2–30. https://doi.org/10.1016/j.sab.2005.10.003

[32]   Carlsson CMG, Stroem G. Reduction and Oxidation of Cellulose Surfaces by Means of Cold-Plasma. Langmuir [Internet]. 1991;7(11):2492–7. Available from: http://pubs.acs.org/doi/abs/10.1021/la00059a016

[33]   Costa THC, Feitor MC, Alves Junior C, Bezerra CM. Caracterização de filmes de poliéster modificados por plasma de O2 a baixa pressão. Rev Matéria. 2008;13(1):65–76.

[34]   de Oliveira M, Reis H, Pereira JC, de Brito CL, Rodrigues M, Alves C. O Uso Do Plasma De Nitrogênio Para Modificação Superficial Em Membranas De Quitosana. Rev Bras Inovação Tecnológica em Saúde. 2010;1–15.

[35]   Flamm DL, Herb GK. Chapter 1: Plasma Etching Technology—An Overview. In: Plasma Etching [Internet]. Academic Press, Inc.; 1989. p. 1–89. Available from: https://doi.org/10.1016/B978-0-08-092446-5.50006-8

[36]   Luna P, Lizarazo-Marriaga J. An extraction methodology of Guadua angustifolia bamboo fibers. In: 6th Amazon & Pacific Green Materials Congress and Sustainable Construction Materials LAT-RILEM Conference. 2016.

[37]   Luna P, Lizarazo-Marriaga J, Mariño A. Preliminary study on the compatibilization techniques of natural fibers as reinforcement of polymeric matrices. In: Sustainable Construction Materials and Technologies (SCMT4). 2016.

[38]   Kan CW, Chan K, Yuen CWM, Miao MH. Surface properties of low-temperature plasma treated wool fabrics. J Mater Process Technol [Internet]. 1998;83(1–3):180–4. Available from: https://doi.org/10.1016/S0924-0136(98)00060-0

[39]   Yip J, Chan K, Sin KM, Lau KS. Low temperature plasma-treated nylon fabrics. J Mater Process Technol. 2002;123(1):5–12. https://doi.org/10.1016/S0924-0136(02)00024-9

[40]   Sun D, Stylios GK. Fabric surface properties affected by low temperature plasma treatment. J Mater Process Technol. 2006;173(2):172–7. https://doi.org/10.1016/j.jmatprotec.2005.11.022

[41]   Cheng SY, Yuen CWM, Kan CW, Cheuk KKL, Daoud WA, Lam PL, et al. Influence of atmospheric pressure plasma treatment on various fibrous materials: Performance properties and surface adhesion analysis. Vacuum [Internet]. 2010;84(12):1466–70. Available from: http://dx.doi.org/10.1016/j.vacuum.2010.01.012

[42]   Amirou S, Zerizer A, Haddadou I, Merlin A. Effects of corona discharge treatment on the mechanical properties of biocomposites from polylactic acid and Algerian date palm fibres. Sci Res Essays. 2013;8(21):946–52.

Non-Conventional Materials and Technologies – NOCMAT for XXI Century   Materials Research Forum LLC
Materials Research Proceedings 7 (2018) 341-348   doi: http://dx.doi.org/10.21741/9781945291838-31

Effect of Production Variables on the Properties of Cement Bonded Flake Board from *Polyalthia longifolia* (Sonn.)Thw. Wood

Owoyemi J. M. [a], and Oyeleye I.O [a]

[a,] Department of Forestry and Technology, Federal University of Technology, Akure. P.M.B. 704, Akure, Ondo State, Nigeria

jacobmayowa@yahoo.com, onigbinde2012@gmail.com

Abstract. The need to source for other woody materials for particle board like spent ornamental is on the increase due to reduction in the supply of timber from natural forest. Therefore, this study was aim at evaluating the effect of flake size and mixing ratio on the physical and mechanical properties of cement bonded particleboards produced from *Polyalthia longifolia* wood residues, a readily available and underutilized tree species. The production variables flake sizes of 19.85 mm, 19 mm and 6.35 mm and wood-cement ratios of 1:1, 2:1, and 3:1. The boards produced were subjected to physical properties tests such as water absorption, thickness swelling and linear expansion at 24, 48 and 72 hours. Water absorption, thickness swelling and linear expansion increased with decrease in flake size and mixing ratio. For the mechanical properties, the flake size of 19 mm had the highest strength as the mixing ratio increases while FS of 6.35 mm had the lowest strength. Flake size and mixing ratio had significant effects on the properties of the boards produced ($P>0.5$). Hence the study affirmed the suitability of spent *Polyalthia longifolia* flakes for Cement Bonded Board which can be used for both interior and exterior applications in building.

Introduction

Research interests in wood cement bonded boards are on the increase in developing countries. The interest in wood cement bonded boards can be attributed to the compatibility of wood to cement as well as the simplicity and availability of technology for board production [1]. With advances in technology and increase in the global population, the demand for wood in the forest product industry has grown over the years. In addition, the application of wood in new areas had also caused a significant pressure on the current standing forest resources. This has generated the necessity for people in the forest industry and scientists studying in this field to find alternative biomasses or raw materials.

Alternative fibers such as agro fibers will play an important role in the wood fiber supply and demand map of the future [2]. However there are little or no information on the use of residue from ornamental trees for other purposes especially particle board production. Particleboard is mainly use in the construction industry, bracing walls and flooring and other structural applications [3]. Wood cement board is a versatile material suitable for interior and exterior use for core and low-cost housing construction. It can be moulded into any form and shape to meet specific end use and has resistant to fire, water, rot, termites, insects and fungi attack. The production of flake board from *Polyalthia longifolia* using cement as binder may provide an alternative to the use of natural source from the natural forest and other agricultural wastes materials. *P. longifolia* planted as ornamental trees around houses have grown to large size with the roots causing structural problems on buildings and are now cut down. The proceeds are processed into chips and flakes to produce value added material. One way to explore the economic benefit of wood residues is to incorporate the wood residues into cement bonded particle board production as this will reduce the environmental hazard emanating from wood waste burning but will also contribute to economic growth of the country and reduce pressure on trees from forest [4].

Non-Conventional Materials and Technologies – NOCMAT for XXI Century   Materials Research Forum LLC
Materials Research Proceedings 7 (2018) 341-348   doi: http://dx.doi.org/10.21741/9781945291838-31

Materials and methods
Material Procurement and Preparation
Polyalthia longifolia wood flakes were obtained from felled ornamental trees at School of Agriculture and Agricultural Technology, Federal University of Technology, Akure. The inorganic binder used (Portland cement) was bought from fresh consignment at a cement shop and also the additive (calcium chloride) was added. The flakes were treated with hot water at temperature of 80 °C and were air dried in controlled ambient temperature for two weeks to attain moisture content of 12% approximately prior to use. The flakes were separated into three different flake sizes of (FS1) 19.85 mm, (FS2) 19 mm and (FS3) 6.35 mm.

Mat Formation
The dimension of cement bonded flake boards (CBFB) produced were 8 x 350 x 350 mm. Portland cement and sawdust were weighed based on the mixing ratio of about 70:30% ratio which was then compounded. The amount of water used was calculated using the formula by Simatupang (1979), which is expressed as:

$$Wt = W (0.30\text{-}MC) + 0.60C.$$

Where: Wt= Weight of water (g), W= Wood dry weight (g),
MC= Moisture Content (%) and C= Cement weight (g).

The stock was hand formed in a wooden mould of 350 x 350 mm and placed on a metal caul plate covered with polythene sheet; plywood plate was used to pre-press the formed mat and covered with polythene sheet before the top metal caul plate was placed on it. Three replicates were prepared. Thereafter, the boards were trimmed, cut into specimen sizes and subjected to tests in accordance to modified ASTDM 1037 [5] standard. Water absorption (WA) and linear expansion at 24, 48 and 72 hours, Thickness swelling (TS) were determined manually while 195 x 50 mm specimen size was used for modulus of rupture (MOR) and modulus of elasticity (MOE) in accordance to ASTM [6] Standard Method for particle board test.

Statistical analysis
A 3 by 3 factorial in Completely Randomized Design was used. The main factors considered were; flake size (FS1, FS2 and FS3), and mixing ratio (1:1, 2:1, 3:1). Duncan Multiple Range Test was used at 95% probability level to test the significance of treatment means.

Results and discussion
Effect of Production Variables on Water Absorption
Figure 1 revealed that water absorption after 24hr soaking which ranged from 34.87± 1.33% to 19.33±1.36% for FS1 at MR 1:1 having the highest water absorption value and FS3 at MR 3:1 having the least water absorption value. For FS1, the value ranged between 34.87± 1.33% to 27.28±0.58%, while for FS2, the value ranged between 34.56±5.04%, to 25.18± 4.51%. FS3, the value was between 29.77.5±0.81 to 19.33±1.36% for 24 hr. Water absorption after 48 hr ranged from 40.34 ±4.01% to 29.32±2.32% for FS1, 36.49±1.32% to 28.23± 2.00 % of FS2 and 34.04± 2.12% to 22.33±1.59% of FS3 while water absorption after 72 hr ranged from 42.76±1.55% to 32.77±6.01% of FS1, 38.01±0.46% to 29.45±7.78% of FS2 and 37.04 ± 4.04% to 25.59±5.13% of FS3. The board water absorption properties decreased with increase in cement binder mixing ratio and decreased with flake size at 24, 48 and 72 hr.

Non-Conventional Materials and Technologies – NOCMAT for XXI Century   Materials Research Forum LLC
Materials Research Proceedings 7 (2018) 341-348   doi: http://dx.doi.org/10.21741/9781945291838-31

The result of analysis of variance (ANOVA) in Table 2 showed that Water Absorption is significantly affected by the flake size, mixing ratio, and their interaction. Tables 1 showed that there was significant difference on board production between FS1 mm to FS3, except for the soaking at 72 hr which did not have a significant difference (P<0.05) between FS2 and FS3. The result also showed that mixing ratio has significant effect (P>5%) on board and it ranked the board produced at MR3:1 as the best and MR1:1 as the weakest. The trend of water absorption (WA) in response to water intake at the different flake sizes to mixing ratio is shown in Figure 1.

The water absorption observed in this study showed higher values with increase in flake size and decrease in cement content in term of mixing ratio. These observations agree with the findings of [4, 7] that increase in cement content caused improvement in the physical properties of the board. This result showed a downward trend as the cement/wood ratio increases. The lowest water absorption and thickness swelling was achieved at the highest mixing ratio. The board with the highest water absorption was obtained from the board produced from the highest flake size and this is due to the ability of the higher flakes size to absorb water. The observed higher percentage of water absorption could be attributed to the applied force and presence of void spaces in the boards which allowed water intake as reported by Ajayi et al., [8].

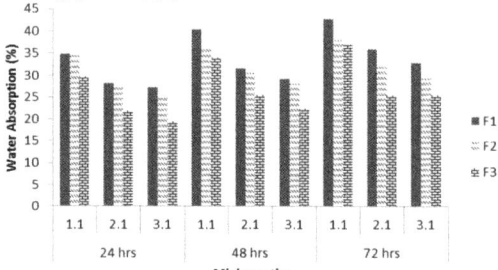

FIGURE 1- EFFECT OF FLAKE SIZE, DURATION AND MIXING RATIO ON WATER ABSORPTION

TABLE 1- DUNCAN MULTIPLE RANGE TEST OF FLAKE SIZE AN MIXING RATIO FOR PHYSICAL PROPERTIES

Flake Size	WA 24	WA 48	WA72	T.S 24	T.S 48	T.S 72	L.E24	L.E 48	L.E72
1	34.56[a]	36.16[a]	37.844[a]	6.20[a]	8.36[a]	10.19[b]	0.02[a]	0.04[a]	0.04[a]
2	27.39[b]	29.17[b]	29.71[b]	7.84[a]	10.61[a]	13.45[a]	0.44[a]	0.05[a]	0.05[a]
3	25.18[c]	26.71[c]	29.57[b]	5.77[a]	8.38[a]	11.45[ab]	0.03[a]	0.03[a]	0.03[a]
Mixin g Ratio									
1:1	30.35[b]	31.88[b]	32.42[b]	6.40[ab]	9.28[ab]	11.52[b]	0.31[a]	0..31[a]	0.31[a]
2:1	23.64[c]	25.36[c]	28.00[c]	8.11[a]	10.73[a]	13.93[a]	0.17[a]	0.02[a]	0.02[a]
3:1	33.46[a]	35.05[a]	37.01[a]	7.20[b]	7.20[b]	9.42[b]	0.44[a]	0.67[a]	0.07[a]

Alphabets with the same letter along the column shows that there is no significant difference

Non-Conventional Materials and Technologies – NOCMAT for XXI Century   Materials Research Forum LLC
Materials Research Proceedings 7 (2018) 341-348   doi: http://dx.doi.org/10.21741/9781945291838-31

TABLE 2 - ANALYSIS OF VARIANCE FOR THE FLAKE SIZE AND MIXING RATIO ON PHYSICAL PROPERTIES OF THE BOARDS

Source	Df	W.A 24 (%)	W.A 48 (%)	W.A 72 (%)	T.S 24 (%)	T.S 48 (%)	T.S72 (%)	L.E24 (%)	L.E48 (%)	L.E72 (%)
FS	2	71.07*	92.10*	27.87*	3.89*	5.26*	8.12*	1.00*	2.66*	2.66*
MR	2	74.39*	94.66*	31.04*	2.10*	2.93*	4.55*	0.47^{ns}	0.22^{ns}	0.22^{ns}
MR*FS	4	0.78*	1.45*	1.68*	1.02*	1.72*	2.02*	2.19*	1.51*	1.51*
Error	43									
Total	51									

*Significant (p 0.05)   ns – not significant

Effect of Production Variables on the Thickness Swelling on Properties of the Cement Bonded Flake Boards

The values (Figure 2) for thickness swelling ranged from $10.09 \pm 3.58\%$ to $5.25 \pm 2.67\%$ for FS1, $8.77 \pm 2.32\%$ to $6.00 \pm 4.12\%$ for FS 2 and $8.19 \pm 2.00\%$ to $4.96 \pm 2.70\%$ for FS 3 for 24 hr soaking. $12.95 \pm 6.68\%$ to $7.54 \pm 1.23\%$ for FS 1, $12.58 \pm 1.91\%$ to $6.96 \pm 2.51\%$ for FS2 and $10.93 \pm 4.51\%$ to $6.62 \pm 2.08\%$ for FS3 at 48 hr. $16.34 \pm 5.61\%$ to $9.15 \pm 2.06\%$ for FS1, $15.58 \pm 1.45\%$ to $8.83 \pm 3.35\%$ for FS2 and $13.35 \pm 3.75\%$ to $8.38 \pm 1.43\%$ of FS3 for 72 hr. FS1 board had the highest thickness swelling. The thickness swelling of the board decreases with decrease in flake size and mixing ratio. The trend of thickness swelling response of board to water intake at the different flake sizes and mixing ratio is shown in Figure 2. The result of analysis of variance (ANOVA) in Table 2 shows that Thickness Swelling is significantly (P>0.05) affected by flake size, mixing ratio, and their interaction. DMRT (Table 1) revealed that there was no significant difference (P<0.05) on board production between FS1 to FS2, except for the soaking at 72 hr which showed a significant difference between FS2 and FS3 (P>0.05). The result also showed that mixing ratio has significant difference on board and it ranked the board produced at MR3:1 as the best and MR 1:1 as the weak. The result also showed that there is significant difference between mixing ratio MR 2:1 to mixing ratio MR 3:1. There was decrease in Thickness swelling as the mixing ratio increase and flake size decrease which could be due to less irregular void spaces in the board and the wood particles from lower wood-cement ratio are not encapsulated by cement which resulted in higher thickness swelling. This also conforms to the findings of [4]. While reduction in thickness swelling arises because of sufficient encapsulation of the wood particles at high cement-wood ratios and the minimal swelling of the small particles. [9-11].

Effect of Production Variables on the Linear Expansion of the Cement Bonded Flake Boards Produced

Figure 3 showed the values of the linear expansion of the boards ranged from 0.067 ± 0.00 to $0.00 \pm 0.00\%$ for FS1 after soaking, 0.067 ± 0.000 to $0.067 \pm 0.000\%$ for FS2 and 0.067 ± 0.000 to $0.000 \pm 0.000\%$ for FS3 at 24 hr. 0.067 ± 0.817 to $0.000 \pm 0.000\%$ for FS1, 0.067 ± 0.041 to $0.00 \pm 0.041\%$ for FS2 and 0.067 ± 0.052 to $0.00 \pm 0.098\%$ for FS 3 after 48 hr soaking while 72 hr soaking recorded 0.067 ± 0.082 to $0.000 \pm 0.000\%$ for FS 1, 0.067 ± 0.041 to $0.00 \pm 0.041\%$ for FS 2 and 0.067 ± 0.052 to $0.00 \pm 0.098\%$ for FS 3. Linear expansion was not affected by the flake size of the board. This trend was the same for linear expansion at 24, 48 and 72 hours. The result of analysis of variance (ANOVA) in Table 2 shows that Linear Expansion is not significantly affected by the

Non-Conventional Materials and Technologies – NOCMAT for XXI Century   Materials Research Forum LLC
Materials Research Proceedings 7 (2018) 341-348   doi: http://dx.doi.org/10.21741/9781945291838-31

flake sizes but significantly (P<0.05) affected by mixing ratio, and their interaction. DMRT (Table 1) showed that there was no significant difference on board between FS 1 to FS 3. The result also showed that mixing ratio had no significant difference (P<0.05) on the board (Table 1). Wood is an anisotropic material. It shrinks most in the tangential direction, about half as much across the rings (radially), and only slightly along the grain (longitudinally). Longitudinal shrinkage of wood (shrinkage parallel to the grain) is generally quite small. This study showed that the expansion was small and this could be due to the complete filling of the likely void spaces within the boards with cement and the flakes may be responsible for poor responses to water intake and board expansion in linear direction [12].

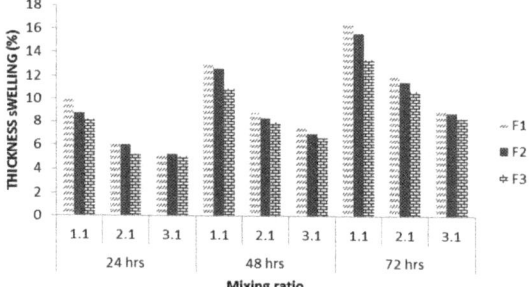

FIGURE 2 - *EFFECT OF FLAKE SIZE AND MIXING RATIO ON THICKNESS SWELLING*

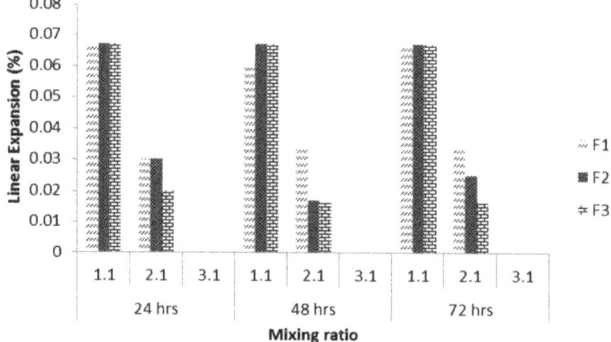

FIGURE 3 - *EFFECT OF FLAKE SIZE AND MIXING RATIO ON LINEAR EXPANSION*

Effect of Production Variables on Strength properties
The mean values of modulus of rupture of cement bonded board produced are presented in Figure 1. MOR ranged from 0.195 ± 0.380 to 1.662 ± 0.442 N/mm^2 from FS1 to FS3. Board produced from FS2 had highest strength property, while the lowest board in strength was produced at FS1 with MR

Non-Conventional Materials and Technologies – NOCMAT for XXI Century   Materials Research Forum LLC
Materials Research Proceedings 7 (2018) 341-348   doi: http://dx.doi.org/10.21741/9781945291838-31

of 1:1. The response of the boards to rupture showed that increase in the mixing ratio of cement to flakes showed increase in the modulus of rupture (Figure 4). FS2 had the highest modulus of rupture at the highest MR of 3:1 while modulus of elasticity (MOE) ranged from 900.84± 9.67 to 1007.63 ±23.22 for FS1. 1427.63 ±22.75 to 1950 ± 16.30 for FS2 and 1100.21±10.13 to 1395.16 ±16.10 for FS3 (Figure 5). From the data set it was observed that the highest strength property was produced at FS2. The results of Analysis of Variance presented showed that the strength properties (MOR and MOE) of the board is significantly affected by mixing ratio, flake size and the interaction between flake size and mixing ratio at 5% probability level. The DMRT analysis in Table 2 revealed significant different in boards between FS 1 and FS2, FS2 and FS3 but there was no significant difference between FS 1 and FS 3. This could implied that there is absence of void space in the board gives random distribution of the particles and this gave a well compact mat structure which enhanced the strength of the board [13],[4],[11],[14].

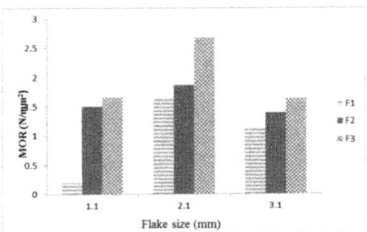

FIGURE 4- EFFECT OF FLAKE SIZE AND MIXING RATIO ON MODULUS OF RUPTURE

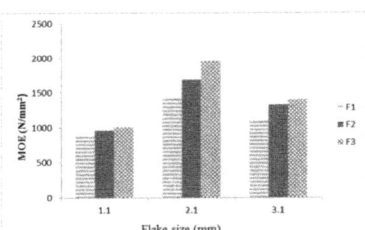

FIGURE 5- EFFECT OF FLAKE SIZE AND MIXING ON MODULUS OF ELASTICITY

TABLE 4 - DUNCAN MULTIPLE RANGE TEST OF FLAKE SIZE AND MIXING RATIO FOR MECHANICAL PROPERTIES

Flake Size	MOE	MOR	Mixing Ratio	MOE	MOR
FS1	955.97[b]	1.23[b]	1:1	1142.86[b]	1.18[b]
FS2	1684.67[a]	1.72[a]	2:1	1320.49[a]	1.35[b]
FS3	1273.77[ab]	1.61[ab]	3:1	1451.06[b]	2.03[a]

Alphabets with the same letter shows that there is no significant difference
Alphabets with different letter shows that there is significant difference

Non-Conventional Materials and Technologies – NOCMAT for XXI Century   Materials Research Forum LLC
Materials Research Proceedings 7 (2018) 341-348   doi: http://dx.doi.org/10.21741/9781945291838-31

TABLE 5 - *ANALYSIS OF VARIANCE FOR THE FLAKE SIZE AND MIXING RATIO ON MECHANICAL PROPERTIES OF THE BOARDS*

Source	Df	MOR	MOE
FS	2	29.62*	18.04*
MR	2	6.78*	7.69*
MR*FS	4	5.40*	8.47*
Error	43		
Total	51		

*Significant (p 0.05)   ns – not significant

Conclusion

This study affirmed that the use of ornamental plants like *P. longifolia* in the production of Cement Bonded Flake Board can serve as an alternative to forest trees as this could cause reduction in exploitation pressure on forest biodiversity. The board produced from *P. longifolia* had excellent physical and mechanical properties. There was a significant difference in the flake sizes for water absorption, thickness swelling except for linear expansion which showed no significant difference in the flake sizes and mixing ratios also showed a significant difference in the boards produced. The results of the study showed that increase in cement content lowers the physical properties of the boards. It was observed that the medium size flake exhibited the best effects in strength. From the quality point of view, FS2 ranked best followed by FS3 and FS1 respectively. These results indicated that increase in mixing ratio of cement to flakes resulted in the dimensional stability of the board.

Reference

[1] Owoyemi JM, Ogunrinde OS. Suitability of Newsprint and Kraft Papers as Materials for Cement Bonded Ceiling Board. Inter J. of Chem Mol, Nuclr, Mat and Met Eng. 2013; Vol: 7(9). PP 717-721

[2] Ibrahim B, Cengiz. G, and Hulya K. The manufacture of particle boards using sunflower stalks (*Helianthus annuusi*.) and poplar wood (*Populous alba* 1.) J. of composite mat. 2005; Vol.39 (5).PP 45-50.

[3] Erakhrumen AA, Areghan SE, Ogunleye MB, Larinde SL, Odeyale OO. Selected physico-mechanical properties of cement-bonded particleboard made from pine (Pinus caribaea M.) sawdust-coir (Cocos nucifera L.) mixture. Scientific Research and Essay. 2008 May 1; 3(5):197-203.

[4] Olufemi AS, Abiodun OD, Omajor O, and Paul FA. Evaluation of Cement –Bonded Particle Board produced from *Afzelia africana* wood residues. J. of Eng. Sci and Tech. 2012; Vol. 7(6).PP 732-743.

[5] ASTM D 1037-93 - Standard methods for evaluating properties of wood-based fiber and particle panel materials. American Society for Testing and Material, Philadelphia, PA, 1995.

[6] American Society for Testing and Materials. American Society for Testing and Materials. Annual book of ASTM standards. 100 Barr Harbor Dr., West Conshohocken, PA 19428, ASTM D570-98, reapproved in 2005;35-37.

[7] Moslemi AA, and Pfister, SC. The influence of cement/wood ratio and cement type on bending strength and dimensional stability of woodcement composite panels. Wood and Fiber Science, 1987; 19(2), 165-175.

[8] Ajayi, B. Olanike, O., Ayodele, E., and Ayorinde A. Assessment of Gmelina arborea sawdust-cement bonded rain water storage tank. Environmentalist. 2008, Vol 28. PP 123-127.

[9] Meneeis. C.H.S., Catro, V.G., and Souza, M.R. "Production and properties of a medium density wood-cement boards produced with Oriented Strands and Silica fume," Maderas: Ciencia technologia. 2001; Vol.9 (2), PP 105-116.

[10] Sadiku, N. Physico-mechanical characterization of wood plastic composite produced from some tropical wood species, Mtech project, department of forestry and wood technology, Akure. 2012.

[11] Karade, S.R. Irle, M. and Maher, K. Assessment of wood-cement compatibility: A new approach. Holzforschung (Int. J. of the Biol, Chem, Phys, and Tech. of Wood), 2003; Vol 57(6), PP 672-680.

[12] Ajayi. Reaction of cement-bonded composites from *Gmelina arborea* and *Leucaena leucocephala* to water treatment. Nig. J of Forestry, 2004,34(1-2), 125-131

[13] Frybort S, Raimund M, Alfred T, and Muller U. Cement bonded composites; A mechanical Review. BioRes. 2008, Vol 3(2): 602 - 626.

[14] Ma LF, Yamaguchi H, Pulido OR., Sasaki H, and Kawai S. Production and properties of orientated cement-bonded boards from sugi, in Wood-Cement Composites in the Asia-Pacific Region, ed P.D. Evans, ACIAR Proceedings No. 107. Australian Centre for International Agricultural Research, Canberra, 2000, PP 140-147.

Non-Conventional Materials and Technologies – NOCMAT for XXI Century   Materials Research Forum LLC
Materials Research Proceedings 7 (2018) 349-358   doi: http://dx.doi.org/10.21741/9781945291838-32

Flat Ring Flexure Test for Full-Culm Bamboo

Jelani Virgo[a], Richard Moran[b], Kent Harries[a,c,*], J.J. Garcia[b], Shawn Platt[a]

[a] Department of Civil and Environmental Engineering, University of Pittsburgh, USA

jjv32@pitt.edu, kharries@pitt.edu*, slp71@pitt.edu

[b] Escuela de Ingeniería Civil y Geomática, Universidad del Valle, Cali, Colombia

richard.moran@correounivalle.edu.co, josejgar@googlemail.com

[c] University of Bath, BRE Centre for Innovative Construction Materials, Bath, UK

Abstract. The development of a new simple test method suitable for assessing the tension strength perpendicular to the fibres (Mode I) of a bamboo culm – the flat ring flexure test – is presented. The proposed test places a short section of bamboo culm in through-cross section flexure, causing circumferential stresses at failure. The modulus of rupture at the failed section is a measure of the transverse tensile strength of the culm. The test is compression-based (indeed, in the field it can be run using free weights rather than a test machine) and uses a simple apparatus and specimen. The full culm specimen is symmetric and requires very little preparation. The study first investigated test parameters affecting results, thereby arriving at an appropriate and repeatable standard test method. The resulting test method is documented in a format consistent with ISO 22157 [1].

Keywords: Bamboo, Tension Perpendicular to Fibre, Test Method

Introduction

Standardisation of construction materials and practices serves both technical and social purposes. The objectives of a standard material test procedure, for instance, are to accurately determine a characteristic or design value of the material and to provide a common frame of reference for the user community. Data from such comparable tests can be compiled to obtain a more reliable understanding of a material's properties which can lead to the refinement of, and confidence in design values. This leads to broader acceptance of the material in the design community. Such acceptance, coupled with advocacy, can lead to broader social acceptance of previously marginalised vernacular construction methods.

Conventional construction materials such as steel and reinforced concrete were once nonconventional and unproven materials. Acceptance was achieved through decades of testing, analysis, and experience which evolved into standardised practices that continue to evolve. More recently, fibre reinforced polymer (FRP) composites, initially developed for aerospace applications, are being standardised for use in civil infrastructure and their use is burgeoning. Increasingly, focus is being placed on the standardisation of sustainable materials such as engineered natural fibre composites. Full culm bamboo, however, remains firmly in the vernacular.

Standard Test Methods for Bamboo

In 2004, the International Organisation for Standardisation (ISO) developed model standards for determining the mechanical properties of bamboo [2]. If the use of bamboo is limited to rural areas, ISO recognises established "experience from previous generations" as being an adequate basis for design [3]. However, if bamboo is to achieve its full potential as a sustainably obtained and utilised building material on an international scale, issues of the basis for design, prefabrication, industrialisation, finance and insurance of building projects, and export and import of materials all require some degree of standardization [4].

Non-Conventional Materials and Technologies – NOCMAT for XXI Century   Materials Research Forum LLC
Materials Research Proceedings 7 (2018) 349-358   doi: http://dx.doi.org/10.21741/9781945291838-32

The ISO test methods standard [2] includes tests for determining full-culm compressive strength, longitudinal tensile strength, longitudinal shear, and flexural capacity. These established tests provide a promising starting point for standardisation; however, they neglect important limit states such as longitudinal shear and connection-induced splitting. Splitting behaviour, in particular, has not been fully addressed and the need for additional work in this area was identified by Janssen [4, 5].

Assessing bamboo splitting behaviour

A dominant failure mode of bamboo is longitudinal splitting associated with bamboo carrying flexure, compression or tension loads; splitting is exacerbated by the use of simple bolted connection details common in some bamboo construction [6]. Janssen [5] describes the bending stresses in a culm as being characterised by the longitudinal compressive stress and transverse strain in the compression zone of the culm, with failure eventually occurring due to longitudinal splitting. This is ideally a Mode II[1] longitudinal shear failure; however, in the presence of perpendicular stresses, there is some Mode I component stress which significantly reduces the Mode II capacity. The longitudinal shear test [2], shown in Figure 1a, quantifies Mode II material behaviour, however, neglects the modest Mode I contribution which is believed to drive the splitting failure.

Richard et al. [7] demonstrate the effects of such mode mixity using longitudinal shear tests (pure Mode II), split pin tests (see below; pure Mode I) and culm bending tests of different spans resulting in different degrees of mode mixing. For two different species, a thin walled *P. edulis* and thick-walled *B. stenostachya*, the split pin tests resulted in Mode I capacities equal to only 18% of the Mode II capacity determined from the longitudinal shear tests. Beam tests having mixed mode behaviour exhibited shear capacities ranging from 40-70% of the Mode II capacity. A number of test methods have been developed to better understand bamboo splitting behaviour, as summarised in the following paragraphs.

The Split Pin Test [8], shown in Figure 1b, characterises the splitting capacity of bamboo using the Mode I critical stress intensity factor which provides a measure of the material's fracture toughness. A fracture mechanics approach was selected on the premise that this will normalise the quantification of material properties thereby reducing the scatter inherent in establishing mechanical properties of bamboo.

The Bolt Shear-out Test [6] is a variation of the split pin test developed to determine the behaviour of bolt-induced forces and assess their contribution to the eventual splitting behaviour of the bolted culm. Two distinct types of failures were documented depending on loading orientation highlighting the effects of mode mixity on bamboo splitting behaviour.

The Edge Bearing Test [9], shown in Figure 1c, was developed as a field-appropriate alternative to the more complex split-pin test. The failure mechanism of an edge bearing test specimen involves the formation of a pair of multi-pinned arches resulting from the hinges forming at the locations of maximum moment around the circumference of the culm section. From this behaviour, the culm wall bending properties may be determined [6, 10]. Specifically, the culm wall modulus of rupture is a measure of the transverse tension capacity of the culm wall and therefore the splitting behaviour.

The Concentric Annular Edge Bearing Test [11], shown in Figure 1d, was used to obtain greater resolution of through-culm-wall properties and to isolate the effect of the material property gradient. In this test, the culm was cut, using a water jet, into two or three concentric annular sections. Edge bearing test results for each "ring" provide an improved measure of through-thickness transverse properties than may be obtained from a single full-culm section although the test method was impractical for all but thick-walled species and cumbersome even then.

---

[1] Reference to Modes II and I are in relation to classical fracture mechanics in which Mode II refers to forces resulting in 'in-plane shear' and Mode I refers to perpendicular in-plane 'peeling' forces.

Non-Conventional Materials and Technologies – NOCMAT for XXI Century   Materials Research Forum LLC
Materials Research Proceedings 7 (2018) 349-358   doi: http://dx.doi.org/10.21741/9781945291838-32

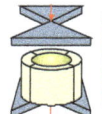    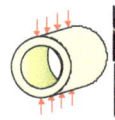 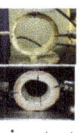 

a) longitudinal shear test [2]   b) split pin test [8]   c) edge bearing test [9]   d) concentric annular edge bearing test specimens [11]

FIGURE 1 - BAMBOO TEST METHODS TO ASSESS LONGITUDINAL BEHAVIOR

The split pin and edge bearing tests have been successfully adopted into the forthcoming revised ISO 22157 Standard [1]. The longitudinal shear test will also remain in this revised document. Nevertheless, these methods have their drawbacks. The longitudinal shear test captures only Mode II behaviour and therefore significantly over estimates splitting capacity. While the split pin test captures Mode I behaviour, it is a tension-based test requiring specialised apparatus and test specimen fabrication making it ill-suited as a field test. The edge bearing test, while simple to conduct, does not accurately reflect the nature of splitting failures. Additionally, test results are derived from simplified analysis and the test is not appropriate for thick-walled bamboo species.

Research significance
The present study reports the development of the flat ring flexure test, which is a new simple test method suitable for assessing Mode I behaviour of a bamboo culm. The test is compression-based (indeed, in the field it can be run using free weights rather than a test machine) and uses a simple apparatus and specimen. This developmental study investigates test parameters affecting results, thereby arriving at an appropriate and repeatable standard test method. The resulting test method is documented in a format consistent with ISO 22157 [1].

Proposed test method
The proposed flat ring flexure test places a short section of bamboo culm in through-cross section flexure. The modulus of rupture at the failed section is a measure of the transverse tensile strength of the culm. The test method utilises a full culm section requiring very little specimen preparation (cutting and sanding ends parallel) and results in a symmetric specimen in which only circumferential stresses are present.

A schematic view of the test arrangement is shown in Figure 2. In order to ensure alignment, a serrated-plate test arrangement with alignment pins was used as shown in Figure 3. The serrations provide accurate positions for the load and reaction rollers; span lengths and constant moment lengths may be selected in 10 mm increments (for the apparatus shown). This design ensures repeatability of tests and eliminates the need for fine measurements of each test set up.

The modulus of rupture, f_r, is calculated as:

$$f_r = 3Pa/(t_N + t_S)L^2 \qquad (1)$$

where P = total load applied to specimen

     a = shear span

     t_N and t_S = culm wall thicknesses at failure locations on either side of the culm

     L = length of culm section tested (flexural depth of specimen)

Non-Conventional Materials and Technologies – NOCMAT for XXI Century   Materials Research Forum LLC
Materials Research Proceedings **7** (2018) 349-358   doi: http://dx.doi.org/10.21741/9781945291838-32

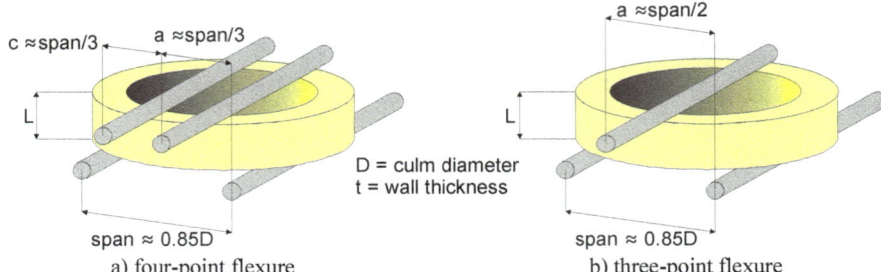

a) four-point flexure   b) three-point flexure

FIGURE 2 - FLAT RING FLEXURE TEST GEOMETRY

a) base plate with 100 mm diameter specimen

b) complete test frame with alignment pins in place

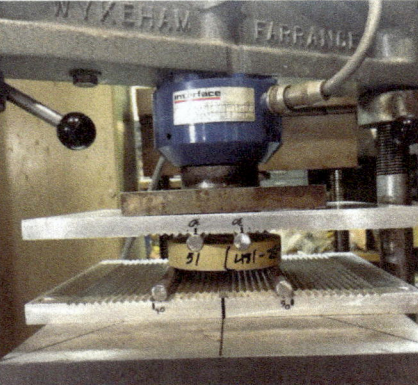

c) test being conducted in universal test machine
(shown 100 mm diameter span tested over 80 mm span with 25 mm shear span)

FIGURE 3 - FLAT RING FLEXURE TEST SET-UP

Pilot test programme

The objective of this pilot study was to establish standard test parameters for the proposed method. Specifically, appropriate test values for the specimen depth to diameter ratio, L/D and shear span to depth ratio, a/L. Other practical test parameters will also be discussed. The pilot programme used *P. edulis* culms having a diameter (D) ranging between approximately 50 and 115 mm. Specimen depth was selected to be nominally $L = 0.1D$, $0.2D$ or $0.3D$. The test spans were varied to some extent, targeting a span of approximately $0.85D$. Tests were conducted in four-point flexure with a constant moment region (c) and both shear spans (a) equal to approximately one third the span (Figure 2a). A limited number of smaller diameter culms were tested in three-point flexure (Figure 2b) for which the

Non-Conventional Materials and Technologies – NOCMAT for XXI Century   Materials Research Forum LLC
Materials Research Proceedings 7 (2018) 349-358   doi: http://dx.doi.org/10.21741/9781945291838-32

shear span (a) is one half the span. It is not practical to test smaller culm diameters using four-point flexure, especially considering the limitation of 10 mm increments for loading locations. The culm diameter D is taken as the diameter oriented parallel to the flexural test span. In some instances, the culms were slightly oval; these were tested such that D was the longer principal axis. Due to the symmetry of the test arrangement, this is not believed to effect results. All specimens were cut from Borax-treated culms having a moisture content of $10.5 \pm 1\%$. Table 1 summarises test arrangement and specimen geometries tested.

TABLE 1 - SUMMARY OF PILOT TEST PROGRAM: P. EDULIS SPECIMEN DIMENSIONS AND RESULTS

test	nominal L/D		D mm	D/t	L/D	a/L	failure locations		f_r (Eq. 1) MPa	
4 pt	0.1	avg.	117.2	9.2	0.100	2.66	n	19	avg.	12.30
		COV	0.034	0.046	0.074	0.105	CMR	15	COV	0.160
		Max	124.2	10.6	0.118	3.28	a	1	max	15.79
		Min	108.8	8.8	0.091	2.05	support	3	min	9.81
4 pt	0.2	avg.	99.5	10.5	0.204	1.28	n	21	avg.	14.01
		COV	0.070	0.127	0.056	0.088	CMR	15 + 2 outliers	COV	0.124
		max	114.7	12.4	0.222	1.45	a	2	max	16.86
		min	92.2	8.7	0.182	1.12	support	2	min	11.45
4 pt	0.3	avg.	102.6	10.8	0.305	1.04	n	20	avg.	15.63
		COV	0.065	0.142	0.028	0.093	CMR	8 + 1 outlier	COV	0.087
		max	114.6	12.4	0.315	1.21	a	0	max	18.25
		min	95.7	8.2	0.283	0.88	support	11	min	14.22
4 pt	0.2	avg.	49.9	7.45	0.215	0.933	n	12	avg.	18.24
		COV	0.015	0.086	0.021	0.027	CMR	12	COV	0.137
		max	51.2	8.11	0.222	0.977	a	0	max	21.77
		min	49.0	6.07	0.206	0.886	support	0	min	14.50
3 pt	0.2	avg.	49.1	7.52	0.220	1.86	n	9	avg.	18.87
		COV	0.036	0.087	0.046	0.022	midspan	9	COV	0.102
		max	50.5	8.11	0.241	1.90	a	0	max	22.30
		min	44.7	6.36	0.209	1.78	support	0	min	15.62

n = number of specimens
CMR = failing in constant moment region
a = failing in shear span
f_r is calculated based ONLY on specimens failing in CMR or at midspan; shaded failure modes are excluded
outliers noted are defined by ASTM E178 [12] (Grubbs method) at 99% significance level

Four-point flat ring flexure tests
72 four-point flexure specimens (Table 1, Figures 2a and 3) were tested in an initial evaluation of observed behaviour. Failures were classified as falling in i) the constant moment region (CMR), ii) the shear span (a), or iii) in the immediate vicinity of the support. A failure in the constant moment region is desired – additionally, this is the only failure mode for which the modulus for rupture may be calculated using Eq. 1 and without accounting for the in-plane curvature of the specimen. Distribution of failure types by specimen depth is shown in Table 1. Only specimens exhibiting a

failure in the constant moment region were considered in subsequent analysis. For those specimens failing in the CMR, outliers were determined using Grubb's Method at a 99% significance level [12]; these were also excluded from subsequent analyses.

It is seen in Table 1, that the deeper specimens (L/D = 0.3) exhibited fewer desirable failures in their constant moment regions (only 9 of 20 specimens). For specimens having a nominal depth to diameter ratio of L/D = 0.3 (actual values, $0.28 < L/D < 0.32$), the corresponding nominal shear span-to-depth ratio is a/L = 0.94 ($1.21 > a/L > 0.88$). It is hypothesised that such a short shear span led to arching action, rather than flexural behaviour and resulted in a disproportional number of failures in the support region (11 of 20 specimens) and an increase in observed capacity for those specimens that did fail as desired in the CMR. For this reason, it was concluded that these deeper specimens, being less reliable, should be excluded and that $L/D \leq 0.20$.

Modulus of rupture results are shown in Table 1 and plotted in Figure 4. Each grouping in Figure 4 can be shown (Table 2) to be statistically distinct from each other with confidence exceeding 97% (using double sided t-test). Thus, there is an increase in measured modulus of rupture with increasing L/D or decreasing a/L ratios. Once again, this is believed to be partially attributed to arching action in the deeper spans. Additionally, as is often observed, the smaller specimens (nominal diameter D = 50 mm) exhibited higher moduli of rupture than those having a larger diameter (nominal diameter D = 100 mm).

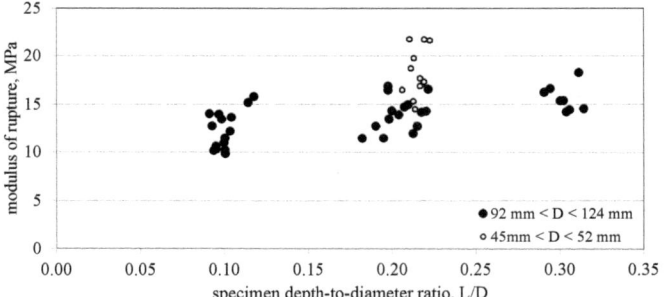

FIGURE 4 - *MODULUS OF RUPTURE VERSUS L/D FOR SPECIMENS TESTED IN FOUR-POINT FLEXURE*

TABLE 2 - *T TEST SIGNIFICANCE*

test			4 pt	4 pt	4 pt	4 pt
	D		100	100	100	50
		L/D	0.1	0.2	0.3	0.2
4 pt	100	0.2	0.018			
4 pt	100	0.3	0.000	0.025		
4 pt	50	0.2	0.000	0.000	0.008	
3 pt	50	0.2	na	na	Na	0.537

Non-Conventional Materials and Technologies – NOCMAT for XXI Century   Materials Research Forum LLC
Materials Research Proceedings 7 (2018) 349-358   doi: http://dx.doi.org/10.21741/9781945291838-32

Selection of L/D for testing

Based on the observations of the pilot programme, it is desirable to reduce L/D to the lowest value practical. Certainly, for L/D greater than 0.2, a disproportionate number of tests do not fail in the constant moment region. The proportion of unacceptable failures (i.e., "bad tests") for $L/D \leq 0.2$ is similar at about 20% of specimens.

In order to mitigate arching action, it is desirable to maximise a/L (thus minimising L/D), however this must be balanced with rational specimen dimensions. Particularly for smaller diameter culms, $L/D = 0.2$ is about as small a specimen as may be practically prepared. Thus the following recommendations are made: 1) $L/D \leq 0.20$; and 2) results from specimens having different L/D ratios should not be compared. As this test method is standardised, in order to mitigate the need for the second recommendation, the authors propose that $0.18 < L/D < 0.22$.

Three-point flat ring flexure results

Testing small diameter culms becomes impractical requiring four-point flexure tests and, even with $L/D = 0.20$ may result in relatively deep shear spans if the proposed test apparatus is used (requiring spans to multiples of 10 mm). Thus, a second series of specimens having a nominal culm diameter, $D = 50$ mm and $L/D = 0.2$ were conducted. Half of these tests were conducted in the four-point flexure condition as previously described (Figure 2a) and half in three-point flexure (Figure 2b). The shear span-to-depth ratio, a/L, can be made greater for the three-point test (almost doubled for the small culms tested here), thus the effects of arching action can be minimised. For a three point flexure test, a desirable failure falls within the distance L centred on the midspan load point. As is seen in Table 1, all smaller culms exhibited desirable failures and no outliers were identified. Furthermore, as indicated in Table 2, the modulus of rupture results are statistically indistinguishable from each other regardless of test set-up (t-test p = 0.54). Thus three-point flexure is equivalent to a four-point flexure tests provided (in both cases) the rupture occurs in the desired region: within the CMR for four-point flexure and within the distance L centered on the load point for three-point flexure. Results from these tests are shown in Figure 5. Nonetheless, the authors feel that four-point flexure, when it is practical, is preferred since the failure in the CMR will not be directly affected by the bearing of the load points. As this test method is standardised, the authors propose that four-point flexure be required for $D > 75$ mm but that three-point flexure be permitted for $D < 75$ mm.

Appendix A provides a proposed flat ring flexure test method written to be consistent with the current ISO test methods document (ISO 2017).

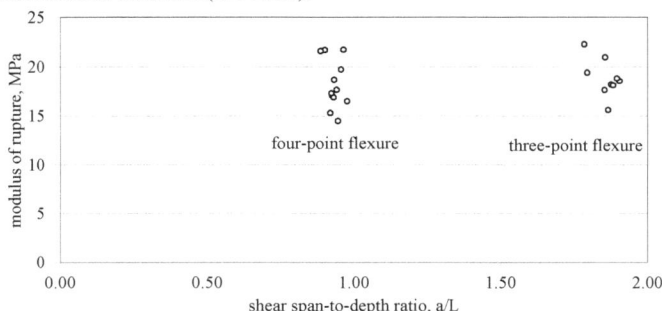

***FIGURE 5** - MODULUS OF RUPTURE OF NOMINAL D = 50 MM SPECIMENS HAVING L/D = 0.2 SUBJECT TO FOUR- AND THREE-POINT FLEXURE*

Non-Conventional Materials and Technologies – NOCMAT for XXI Century   Materials Research Forum LLC
Materials Research Proceedings 7 (2018) 349-358   doi: http://dx.doi.org/10.21741/9781945291838-32

Conclusions

The flat ring flexure test for assessing tension strength perpendicular to the fibres of full-culm bamboo specimens is proposed. Through a parametric study, and considering practical use of the test, specimen and loading parameters are determined for standardizing this test method. A proposed standard test – compatible with ISO 22157 [1] is presented in Appendix A.

The research team is presently establishing a database of flat ring flexure test results including data from at least four laboratories, covering (to date) ten species of bamboo. This data will be used to establish precision guidelines for the flat ring flexure test.

Appendix A

The following is a proposed test method written to be consistent with ISO 22157 [1]. This method is not presently included in, nor has it been considered for inclusion in ISO 22157 at the time of this writing.

A. Tension strength perpendicular to the fibres by flat ring flexure

This clause species a method for determining the tension strength perpendicular to the fibres on specimens form bamboo culms.

A.1 Apparatus

Test machine. Tests shall be carried out on a suitable testing machine capable of measuring compression load with a precision of at least 1%.

Flat ring flexure apparatus capable of applying four-point and/or third point bending loads to a flat ring bamboo specimen.

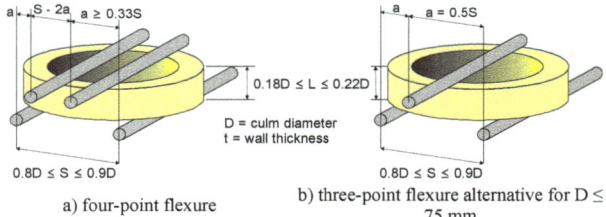

a) four-point flexure

b) three-point flexure alternative for D ≤ 75 mm

FIGURE A-1 – SCHEMATIC VIEW OF FLAT RING FLEXURE LOADING SETUP AND SPECIMEN.

FIGURE A-2 – VIEW OF EXAMPLE FLAT RING FLEXURE APPARATUS.

Rollers providing support and used to apply load shall have a minimum diameter of 10 mm and shall be continuously supported in the test machine. Rollers shall be aligned such that they are parallel and that their spacing remains constant during a test.

A.1.1 Test span, S

The flexural test span shall fall between 0.80 and 0.90 times the specimen diameter; that is, $0.8D \leq S \leq 0.9D$.

Non-Conventional Materials and Technologies – NOCMAT for XXI Century   Materials Research Forum LLC
Materials Research Proceedings 7 (2018) 349-358   doi: http://dx.doi.org/10.21741/9781945291838-32

A.1.2 Four point loading shear span, a, and constant moment regions

The loading rollers shall be arranged to provide a central 'constant moment region' and two equal shear spans. Each shear span shall be at least on third the flexural test span; that is, $a \geq 0.33S$.

A.1.3 Three point loading

For specimens having $D \leq 75$ mm, it is permissible to provide a single roller in the middle of the flexural test span, resulting in two equal shear spans of length $a = 0.5S$.

A.2 Preparation to test specimens

Test culms shall be selected according to [ISO 22157 Clause] **6**.

Flat ring flexure tests shall be made on specimens without a node. The length of the specimen shall fall between 0.18 and 0.22 times the diameter; that is, $0.18D \leq L \leq 0.22D$. The specimen ends shall be parallel.

A.3 Procedure

Measure the diameter of the specimen and select the axis for testing [typically this will be the larger of the two principle diameters of an oval specimen]. This axis will be assumed to intersect 0° and 180° around the culm circumference and be labelled N-S.

Measure the length, L, and wall thickness, t, at both the 90° (E) and 270° (W) locations.

The specimen shall be placed in the test apparatus so that its primary N-S axis is the flexural test span; that is, the N-S axis is perpendicular to the roller orientations.

A small load, not exceeding 1% of the expected failure load is initially applied to 'seat' the specimen in the apparatus.

Application of load shall comply with [ISO 22157 Clause] 5.2.

The following is ISO 22157 Clause 5.2; its inclusion here is in the interest of clarity.

5.2 Rate of load application

The rate of load application of the testing machine shall be selected such that failure is reached within (300 ± 120)s. Tests that fail in less than 30 seconds shall be removed from analysis. The load shall be applied continuously without interruption at the required rate throughout the test. For tests run in displacement control, the rate of traverse of the movable head of the testing machine shall be the free running or no-load speed of the head for mechanical drive type machines, and the loaded head speed for hydraulic or servo-hydraulic driven testing machines. The time to failure for each individual specimen shall be recorded in the test report.

The final reading of the maximum load, F_{ult}, at which the specimen fails shall be recorded. Following each test, obtain specimens for the determination of moisture content in accordance with [ISO 22157 Clause] 7.

A.4 Calculation and expression of results

The tension strength perpendicular to the fibre, f_r, shall be calculated from

$$f_r = 3F_{ult}a/(t_N + t_S)L^2 \qquad \text{Eq. 1}$$

where F_{ult} = total load applied to specimen
  a = shear span
  t_N and t_S = culm wall thicknesses at failure locations on either side of the culm
  L = length of culm section tested

A.5 Test report
The test report shall be in accordance with [ISO 22157 Clause] 5.4.

References

[1] International Organisation for Standardisation (ISO) (2017) CD22157, Bamboo structures – Determination of physical and mechanical properties of bamboo culms – Part 1: Test methods. Committee Document, Geneva.

[2] International Organisation for Standardisation (ISO) (2004a) ISO 22157-1:2004(E), Bamboo – Determination of Physical and Mechanical Properties – Part I: Requirements. Geneva.

[3] International Organisation for Standardisation (ISO) (2004b) *ISO 22156:2004(E), Bamboo – Structural Design.* Geneva.

[4] Janssen, J.A. (2005) International Standards for Bamboo as a Structural Material. *Structural Engineering International*, 15, 48-49. https://doi.org/10.2749/101686605777963288

[5] Janssen, J. (1981) Bamboo in Building Structures. *Doctoral Dissertation*, Eindhoven University of Technology, Netherlands.

[6] Sharma, B. (2010) Performance Based Design of Bamboo Structures. *Doctoral Dissertation*, University of Pittsburgh.

[7] Richard, M., Gottron, J., Harries, K.A. and Ghavami. K. (2017) Experimental Evaluation of Longitudinal Splitting of Bamboo Flexural Components, *ICE Structures and Buildings* Themed issue on bamboo in structures and buildings, 170 (4), 265-274.

[8] Mitch, D., Harries, K.A., and Sharma, B. (2010) Characterization of Splitting Behavior of Bamboo Culms. *ASCE Journal of Materials in Civil Engineering*, 22, 1195-1199. https://doi.org/10.1061/(ASCE)MT.1943-5533.0000120

[9] Sharma, B., Harries, K.A. and Ghavami, K. (2012) Methods of Determining Transverse Mechanical Properties of Full-Culm Bamboo, *Journal of Construction and Building Materials*, 38, 627-637. https://doi.org/10.1016/j.conbuildmat.2012.07.116

[10] Moran, R., Webb, K., Harries, K.A. and Garcia, J.J. (2017) Edge Bearing Tests to Characterize the Radial Gradation of Bamboo, *Journal of Construction and Building Materials*, 131, 574-584. https://doi.org/10.1016/j.conbuildmat.2016.11.106

[11] Sharma, B. and Harries, K.A. (2011) Effect of Fiber Gradation on the Edge Bearing Strength of Bamboo Culms. *Proceedings of the 13th International Conference on Non-conventional Materials and Technologies* (IC-NOCMAT 2011), Changsha, Hunan, China, September 2011.

[12] ASTM (2016) ASTM E178-16a Standard Practice for Dealing with Outlying Observations, ASTM International, West Conshohocken PA, USA

Non-Conventional Materials and Technologies – NOCMAT for XXI Century   Materials Research Forum LLC
Materials Research Proceedings 7 (2018) 359-365   doi: http://dx.doi.org/10.21741/9781945291838-33

A Bus Stop: Industrialized Construction in Glued Laminated Bamboo and Timber (Flex Material)

Marina de A. Patury, Jaime G. de Almeida, Vanda A. G. Zanoni*

ABSTRACT. In order to reduce impacts to the environment, the civil construction stands out as a strategic sector, as it enables the assimilation of cleaner technologies and more sustainable materials. Linked to the research project "Glued laminated bamboo and timber : development of a flexible material for the production of furniture", this research project sought to apply glued laminated bamboo and timber focusing on the industrialized production of a bus stop to be installed, experimentally, in Vila da Granja do Torto, Brasília - DF. This paper presents the process of producing a mockup to study its construction system in glued laminated bamboo and timber, aiming to contribute for its full-sized urban furniture. From the manufacturing and assembly of a mock up in the scale 1: 2.5, the objective was to understand the best ways and solutions for the industrialized production and in loco execution of the bus stop, including the improvement of the architectural design, its specifications and executive detailing. The evaluation of the assembly of the bus stop mockup provided needed information for the validation and feedback of the design process, as well as advances in the research for flex material applications in urban equipment of intensive use.

Key words: Bamboo, Lyptus, Glued Laminated, Flex Material, Design Process, Mockup

Introduction

The variety of applications of bamboo, in its natural or industrialized state, extends the viability for new uses and studies by making this material compatible with other building materials. The bamboos of the genus Dendrocalamus have thick stems and walls and also efficient mechanical properties. The Dendrocalamus asper presents an excellent growth rate (BERALDO; PEREIRA, 2007). Pereira (2001) states that the resistance to compression, traction and flexion in the asper's stem are greater than other species like Dendrocalamus, Guadua and Bambusa.

In addition, according to Américo (2009), bamboo can be easily processed, both manually and industrially. Timber, however, remains to these days as a product of great importance, due to the tradition of its uses, throughout history.

Lyptus® is a timber from renewable forests and is the result of the crossing of the species Eucalyptus grandis and Eucalyptus urophylla. Lyptus® is classified as noble timber and, because of its high density, is suitable for machining processes.

This work explores the potential of using a new material and is part of the project "lation of cleaner technologies and more sustainable materials. Linked to the research project "Glued laminated bamboo and timber: development of a flexible material for the production of furniture", coordinated by Professor Jaime Gonçalves de Almeida, the main objective of this research project was to develop a material composed of glued bamboo and timber (Glubam + Glulam), called flex material, in order to offer a differentiated and local industrial option to the furniture and urban production for the Federal District.

In this context, this article presents the development of digital models and mockups for studies of a constructive system with flex material, focusing on industrialized production, aiming to contribute to full-sized construction of a bus stop.

Non-Conventional Materials and Technologies – NOCMAT for XXI Century   Materials Research Forum LLC
Materials Research Proceedings 7 (2018) 359-365   doi: http://dx.doi.org/10.21741/9781945291838-33

A mock up is a scaled model of the object or its parts. According to Ramos (2005), the designs and models used throughout the design process are different types of recording of the design reasoning in progress. They also serve as a basis for rethinking or discussing other possibilities

Resulting from the stage in which the project is located. Ramos (2005) supports the idea that design and model have identical purposes. The characterization of both as a design process supports and promotes the creation in the initial stage. Moreover, by acting as consensual representations of the existing or desired reality, it allows the development of the established plans, which serve as verification of the proposals.

A mockup is a scaled model of the object or its parts. According to Ramos (2005), the designs and models used throughout the project process are different types of registry of the design reasoning in progress. They also serve as a basis for rethinking or discussing other possibilities resulting from the stage in which the project is located. Ramos (2005) supports the idea that design and model have identical purposes. The characterization of both as a design process supports and promotes the creation in the initial stage. Moreover, by acting as consensual representations of the existing or desired reality, it allows the development of the established plans, which serve as verification of the proposals.

From the manufacturing and assembly of a mock up in the scale 1: 2.5, this research aimed to understand the best ways and solutions for the industrialized production and *in loco* execution of a bus stop, including the improvement of the architectural design, its specification and executive detailing.

This work was based on the geometry generated by the architectonic party, designed by Professor Jaime Gonçalves de Almeida, through the designs of the executive project that allowed digital prototyping and modeling on a reduced scale. These stages propitiated the construction of the mockup that enabled the evaluation of the manufacturing processes of the pieces, dynamics of assembly of the construction system and adjustments in the technical drawings of the bus stop.

Materials and methods

The research project was carried out in four stages: 1) bibliographical review of the production processes of prefabricated building systems and, especially, those manufactured with timber and bamboo, thus supporting the planning of the manufacturing and assembly processes; 2) elaboration of the executive drawings and construction of virtual and physical mockups in scale 1:20; 3) production of components and prefabricated elements in glued laminated bamboo and timber for a 1: 2.5 scaled mockup; 4) evaluation of the production process of the mockup, during the manufacturing and assembly phases, identifying: difficulties encountered by the work team, team training dynamics; through the sizing of the work team and duration of tasks.

From the design drawings and basic architectural design, the executive project, specifications and constructive detailing were generated. They enabled the execution of the mockup in the 1: 2.5 scale. Figures 1, 2 and 3 present the first digital and physical models of the bus stop design.

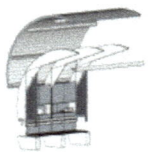  

FIGURE 1 -
DIGITAL MODEL OF
THE BUS STOP.

FIGURE 2 - DIGITAL MODEL
OF THE INSTALLED BUS
STOP IN VILA DA GRANJA
DO TORTO - BRASÍLIA - DF.

FIGURE 3 - PHYSICAL
MODEL OF THE BUS
STOP.

Subsequently, the assembly team was organized and trained; the preparation of the moulds was started, cutting and pressing the boards in Glubam and timber pieces. In this phase, the process of checking the bond strength of glued laminated beams and adapting the materials to the design geometry occurred. At the moment, it was tried to indicate divergences on the process of construction of the urban equipment in real scale and the 1:2.5 scaled mockup. The difficulties were analyzed and new solutions were developed to execute the model considering the obstacles that could be encountered when constructing the bus stop, in real scale, *in loco*. At the end of this step, the executing team prepared a record of the results of the developed experiments, identifying the strategies that worked and recommending those that proved more appropriate.

Results: the mockup production

The making of the mock up of the bus stop in the 1: 2,5 scale was done by the available materials for the production of the Glulam pieces. The Glulam boards (1.40 x 0.80 x 0.025 m) were glued with the PU glue (Figure 4), making it necessary to process them in the skid steer (Figure 5).

| *FIGURE 4 - GLULAM BOARD AFTER BONDING WITH PU GLUE.* | *FIGURE 5 - SKID STEER REMOVING PU GLUE LAYER.* | *FIGURE 6 - MOULDS FOR FLEX MATERIAL BEAMS PRODUCTION.* | *FIGURE 7 - CUTTING OF THE GLULAM BEAMS.* |

After passing through the air conditioning process, the timber samples (Figure 8) were cut on the 1: 2.5 scale according to the moulds of the pieces corresponding to the wooden beams of the bus stop design (Figure 9). After all pieces of the flex material (Glulam + Glubam) were made (Figure 9), the three beams in the bus stop were joined, glued and screwed to be compatible with one another by adjusting the posterior curvature on an uniform circumference by the sanding machine (Figures 10 and 11).

| *FIGURE 8 - FREIJÓ WOODEN BOARDS.* | *FIGURE 9 - FREIJÓ WOODEN PIECES FOR FLEX MATERIAL MAKING.* | *FIGURE 10 - GLULAM + GLUBAM BEAMS.* | *FIGURE 11 – COMPATIBILITY OF CIRCUMFERENCE OF FLEX MATERIAL BEAMS.* |

The niches of books in the bus stop (Figure 12) were made from 15 mm timber plywood, glued with PVA glue and fixed with the pneumatic pinwheel, and although they were made from the technical drawings on the mockup scale, it presented some differences in the compatibility with the already constructed beams. The adjustments have been made to the sanding machine. Subsequently, the beams of flex material received the cutouts for fitting the stringers (Figure 13 and 14).

  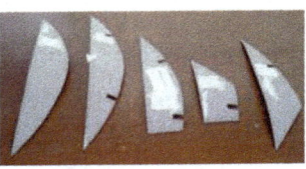

FIGURE 12 -	*FIGURE 13 - CUT*	*FIGURE 14 - PARANA PAPER*
ASSEMBLY OF	*OUT THE BEAMS*	*MOULDS FOR MARKING THE*
NICHES FOR	*TO FIT THE*	*BEAMS ON THE FLEX*
BOOKSHELVES.	*STRINGERS.*	*MATERIAL BEAMS.*

The assembly of the mock up consisted of the joining of flex beams, wooden stringers, book niches and flexible plywood cover. Firstly, the beams were connected parallel and aligned with the temporary fitting of the stringer (Figure 15). Next, the book niches were fixed to the frame with 40x30 mm bolts and the definitive groove of the stringers rectified the alignment (Figure 16).

FIGURE 15 - ALIGNMENT OF BEAMS AND	*FIGURE 16 - DEFINITIVE*
TEMPORARY FITTING OF STRINGERS.	*FITTING OF THE STRINGERS*
	AND VERIFICATION OF THE
	FIT OF THE BOOK NICHES.

The flexible 6mm plywood cover was installed with 40x30mm bolts and screws after stabilizing the main frame. Then, supported horizontally, the mock up was sanded and sealed with a mixture of equal proportions of thinner and marine varnish (Figure 17). The side panels of the bus stop were run in timber and cut into the industrial spit (Figure 18). The metallic frames that support the lateral structure of the bus stop were made in metallic profile T (Figure 19) and were installed with screws 40x30mm. In addition, the metal parts (Figure 20) designed to connect the metal frame to the flex beams were welded to the rear perimeter of the profile (Figure 21 and 22).

For the concreting of the support pillars that received the metallic frames (Figure 23), 12 mm plywood shapes were made. During the concreting process, so that the plumbing and alignment of the mock up billets were maintained, adjusted plywood moulds were used (Figure 24).

After 14 days, the pillars were unmounted and received gray acrylic painting (Figure 25). The metal parts were zarcon coated and painted with black matte spray paint (Figure 26). Then, the mockup base was made of plywood and received, on its underside, a metal bar, following the alignment of the pillars, to ensure the stability of the mockup base structure (Figure 27). Finally, after verifying the connections, alignments and plumbs, the mock up pillars were definitively fixed to the wooden base by screws (Figures 28, 29 and 30).

Non-Conventional Materials and Technologies – NOCMAT for XXI Century   Materials Research Forum LLC
Materials Research Proceedings **7** (2018) 359-365   doi: http://dx.doi.org/10.21741/9781945291838-33

FIGURE 17- DRYING OF THE ASSEMBLED AND SCREWED PARTS AFTER SEALING.

FIGURE 18 - FREIJÓ WOOD SIDE PANEL.

FIGURE 19- METALLIC FRAME IN T PROFILE AND SIDE PANEL TECHNICAL DRAWING.

FIGURE 20 - SIDE PANEL USED AS A WELDING ALIGNMENT BEACON FOR THE METAL FRAME.

FIGURE 21- METAL PART IN PROFILE COATED WITH ZARCON.

FIGURE 22- METAL PART IN WELDED T-PROFILE AND METAL PARTS CONNECTING THE BEAMS OF FLEX MATERIAL IN THE FRAME.

FIGURE 23 - ALIGNMENT OF METAL FRAMES FOR CONCRETING PROCESS.

FIGURE 24 - PROCESS OF CONCRETING THE METALLIC FRAMES IN THE FORMS OF MOCK UP SUPPORT PILLARS.

  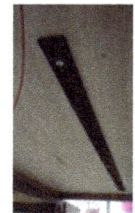

FIGURE 25 - UNMOULDED CONCRETE PILLAR COVERED IN MORTAR AND PAINTED IN GRAY.

FIGURE 26 - ALIGNMENT OF THE PILLARS FOR ANCHORING THEM ON THE BASE.

FIGURE 27 - METAL BAR FOR MOCKUP ANCHORAGE ON DEFINITIVE BASIS

Non-Conventional Materials and Technologies – NOCMAT for XXI Century   Materials Research Forum LLC
Materials Research Proceedings **7** (2018) 359-365   doi: http://dx.doi.org/10.21741/9781945291838-33

FIGURE 28 - *MOCKUP CONNECTIONS PLUG-INS VERIFICATION*

FIGURE 29 - *QUADRO METÁLICO ENCAIXADO NA VIGA DE MATERIAL FLEX.*

FIGURE 30 - *VERIFICATION OF THE ALIGNMENT OF THE SUPPORTING PILLARS AND METAL FRAMES.*

The last finishes were made with sanding machines. On the shelves of the niches were arranged some scenographic books (Figures 31 and 32).

FIGURE 31- FINISHED MOCKUP.

FIGURE 32 - SIDE VIEW OF THE FINISHED MOCKUP.

Discussion

The experimental application, in an innovative way, of the flex material in the execution of an experimental prototype for urban equipment allowed the verification of the technical variability of the constructive system and the production process, fabrication and assembly included, besides the technician, researchers, interns and involved professionals.

The results of the execution of the mock up generate modifications and enhancement in the architectonic project, especially in the system of fitting between parts, in the substitution of some materials, in the modifications of the panels more exposed to intense use and, also, in the sequence of mounting of the parts.

The panels of the bench seat and the side panels, initially projected in slats shaped like a *"muxarabi"*, were modified to be made, respectively, in wood plank and perforated metal frame, in order to guarantee the durability of the urban furniture subjected to intense use and climatic exposition conditions.

The installation of wood niches requires that these elements are independent of the structure and lateral panels, including to facilitate maintenance and possible substitution over it's lifespan.

The reduction on the peace sizes in the mock up confection on account of the 1:2,5 scale made precision difficult in some cuts and fittings, causing rework and necessity of addition of powder mixtures of bamboo and wood + PVA glue in the edges and corner of some pieces, what would be ineffective in a large scale fabrication and mounting *in lovo* of the real product. To the confection of

Non-Conventional Materials and Technologies – NOCMAT for XXI Century   Materials Research Forum LLC
Materials Research Proceedings 7 (2018) 359-365   doi: http://dx.doi.org/10.21741/9781945291838-33

the bus stop peaces, the bamboo utilized was Dendrocalamus asper. As for the wood, the Lyptus was replaced for the Freijó wood (*Cordia goeldiana*), for matters of research budget limits and material availability.

Conclusion

The experience showed that the execution of the mockup is a resource that effectively generates the project's improvement, verifies the efficacy of the constructive system and its production process, besides signalizing the aspects about the final product quality and its behavior on the intense everyday use.

The evaluation of the making of the mock up of the bus stop provided necessary information for the validation and feedback of the project process, besides advance in research for application of flex material in urban equipment.

So, the mounting of future prefabricated elements in real size and it's construction *in loco* can be realized in a more precise way, giving great importance to the insertion of prototyping in the process of designing.

From the didactic point of view, the bus stop is not just utilized as a design tool. By means of the confection of the bus stop it was possible to develop and experiment the constructive system, establish the relation between space and matter and still notice the visual, tactical and spatial aspects. That is, the process allowed the improvement and compatibilization of the project.

Acknowledgments

To Professor Jaime de Almeida for the general coordination of the Project of Scientific Initiation - PIBIC. To the Center for Research Applied to Bamboo (CPAB) for the availability of space and employees. To the Research Support Foundation (FAP) for funding the project. To the architect Atila Mendes for the monitoring and execution of the processes in the Workshop. To the Bamboo Workshop at Granja do Torto, to Jonatas Pereira and to the entire team for the manufacturing space and the mock up assembly. To the entire technical team and partners of the project "Glued laminated bamboo and timber: development of a flexible material for the production of furniture''.

References

[1] AMÉRICO, Leandro; Eco-Design e a utilização de materiais alternativos renováveis: o Bambu e sua inter-relação com o design. Anais do 2° Simpósio Brasileiro de Design Sustentável, São Paulo, 2009.

[2] PEREIRA, M. A. Bambu: espécies, características e aplicações. Bauru, São Paulo: Editora da UNESP, 2001.

[3] PEREIRA, Marco A. R.; BERALDO, Antonio L,. Bambu de corpo e alma. Canal, Bauru, 2007.

[4] RAMOS, F. G. V., Arquitetura: Os Planos De Propostas. Criação, Representação e Informação, II PROJETAR 2005 (FAU – Universidade Federal do Rio de Janeiro, Rio de Janeiro), 2005.

Non-Conventional Materials and Technologies – NOCMAT for XXI Century   Materials Research Forum LLC
Materials Research Proceedings 7 (2018) 366-372   doi: http://dx.doi.org/10.21741/9781945291838-34

Addition Effect of Polymer Residues Reinforced with Glass Fiber in Adobe as to Thermal Comfort

Rômulo Marçal Gandia[a] *, Andréa Aparecida Ribeiro Corrêa[b], Francisco Carlos Gomes[c], Yan Hideki Kawano[d], Sylvia Veiga Gruber Guffey[e] , Romeu Vidali Neto[f]

[a*] UFLA, Brasil; romagandia@gmail.com, [b] UFLA, Brasil; andrea.rcorrea@deg.ufla.br, [c] UFLA, Brasil; fcgomes@deg.ufla.br, [d] UFLA, Brasil; yanhideki@gmail.com, [e] UFLA, Brasil; sylviaveigagguffey@gmail.com, [f] UFLA, Brasil; romeubinneto@hotmail.com

Abstract. This works aimed to analyze and compare the thermal conductivity of traditional adobe and adobe using different fiberglass reinforced polymer (FRP) residue contents. A chamber made of MPD (Medium Density Particleboard) of sugarcane bagasse was used. The box is insulated to the outside and lined internally by insulation layers, one styrofoam, and a thermal blanket. A source of heat, incandescent lamp, in the lower part emits a constant heat that by measurements of thermocouples can be analyzed the incoming heat and that which passes through the sample. FRP is a widely used material which, in addition to generating a large volume of waste, also has a high degradation time. The FRP residue was pre-processed. Five treatments were analyzed: 0; 2.5; 5.0; 7.5 and 10.0% of FRP of dry soil mass. The equation K=P.E/ΔT was used to calculate the energy coefficient. The samples were submitted to tests for 3.33 hours each, corresponding to 1000 measurements, being measured at the frequency of 12 seconds. A solar radiation meter was used to measure the radiation of the lamp. Each treatment was submitted to 5 replicates, and the intermediate value was chosen. The thermal conductivities were: 0.86; 0.79; 0.74; 0.72 and 0.65W.m°C⁻¹ respectively for the treatments 0, 2.5, 5.0, 7.5 and 10.0%. It was concluded that the greater the addition of synthetic fiber residues in the adobe will decrease the thermal conductivity, thus improving its thermal comfort.

Keywords: Non-Conventional Materials, Sustainability, Residues, Thermal Conductivity

Introduction

The construction industry is a worldwide highlight in environmental pollution and energy expenditure. This occurs for example in the extraction of natural resources, production of steel and cement and ceramic burned materials. Faced with this finding regarding environmental and energy issues, an earth construction in Brazil is revitalized. The adobe, millennial masonry, is a typical example of the structures of production and execution of a global civil construction environment, aimed at efficient and sustainable engineering. Amongst the advantages, it is worth mentioning the non-incorporation of cement into production, a possibility of reuse and a possibility of the raw material being the soil of the local area. The energy consumption in the production of the adobe is much inferior when compared to the ceramic materials for the fence (SHUKLA) [15]. The consumption of water and up to 60 times in relation to ceramic materials (CORRÊA) [7].

The Glass Fiber Reinforced Polymer (FRP) has excellent mechanical properties and low density, which contributed to its high industrial growth. However, it generates a high volume of waste, 13 thousand tons per year (ORTH) [11]. The FPR residue has a very low degradability, and even when discarded in correct locations it takes about 300 years to decompose (KEMERICH) [9].

The use of natural or synthetic fibers in building composites improves their physical and mechanical properties. Corrêa et al. [8] Incorporated *Bambusa vulgaris vittata* particles with 6% in dry mass of the soil in the production of adobes and obtained an increase of up to 90% in the compressive strength.

Non-Conventional Materials and Technologies – NOCMAT for XXI Century   Materials Research Forum LLC
Materials Research Proceedings 7 (2018) 366-372   doi: http://dx.doi.org/10.21741/9781945291838-34

The thermal comfort, the adobe has favorable characteristics. Palme [12] comparing thermal efficiency in four construction models, adobe, clay, concrete blocks and wood concluded that earth buildings have a lower temperature variation, lower thermal conductivity and have a better thermal comfort when compared to the others models.

Research on building materials combined with sustainability has expanded. Materials produced from renewable sources or using agricultural or industrial waste and natural fibers are increasingly studied as to their physical and mechanical properties to be accepted by the standards

Ortiz [11] defines the term sustainability as increasing the quality of life, social, economic and environmental relations for present and future generations. Taking into consideration the construction sector and sustainability, the objective of this work is to analyze and compare the thermal behavior of the stabilized adobe with the FPR residue.

Materials And Methods

The soil sample was taken on the campus of the Federal University of Lavras - UFLA at a depth of 1.20 m, horizon B and free of organic matter. The soil was classified as Red Latosol according to the classification of Embrapa [16]. The result of the granulometric analysis was very clayey texture (600g kg -1). This was corrected with average sand at 600g kg -1 to stay in the ideal proportions for the adobe. In Table 1 the results of the physical tests of the soil in natura and corrected are presented.

TABLE 1 - RESULTS FOR THE PHYSICAL TESTS OF THE SOIL.

Tests		Natural soil	Modified soil	References
Granulometry (%)	Clay	66	41	NBR 7181/84[2]
	Silt	6	2	
	Sand	28	57	
Consistency limits (%)	LL	40,02	27,02	NBR 6459/84[4]
	LP	29,79	20,30	NBR 7180/84[1]
	LC	16,80	19,02	NBR 7183/82, adaptado[5]
Density (g cm⁻³)		1,98	2,37	NBR 9776/DNER-ME 093/94[3]

The FPR residue was processed, crushed and selected, by a crusher with a three-phase 7.5 CV motor using a 55 mm sieve. The particles of the generated residue were analyzed with Image J software according to Figure 1. Was selected 30 samples of random fibers having a mean size of 10.507 mm with a standard deviation of 0.439. Using the Chapman bottle method modified by NBR 9776/87 [3], it was possible to find the density of the FPR residue 1.25 g cm⁻³.

The composition of the adobe, produced with the soil corrected by sand, with PRFV residue was 0; 2.5; 5.0; 7.5 and 10.0% of dry soil mass and water with humidity control in the oven for 24 hours at 103 ± 2 ° C with three replicates compared to field tests. The amount of water to be added in the production was calculated between the liquidity limit and the plasticity limit of the soil according to the data of Table 1. In order to reach the ideal humidity, the mixtures were submitted to the "Fall of the Ball" tests (BARBOSA; GHAVAMI) [6] and "Vicat Test" (RUIZ; LUNA) [13]. Figure 2 shows the production steps and humidity tests.

The thermal conductivity test was made in a MDP (Medium Density Particleboard) chamber made of sugarcane bagasse coated with two layers: Styrofoam and thermal blanket for insulation from the external environment. The lower part consists of a heat source (incandescent lamp) connected to a thermostat that regulates the temperature by a thermocouple just above the lamp at 47.0 ° C. The system had 4 thermocouples: lamp temperature, ambient temperature, and temperature before and

Non-Conventional Materials and Technologies – NOCMAT for XXI Century   Materials Research Forum LLC
Materials Research Proceedings 7 (2018) 366-372   doi: http://dx.doi.org/10.21741/9781945291838-34

after traversing the sample. The entire system is connected to an open source microcontroller (Arduino) hardware development platform. To evaluate the possible dissipation of heat by other locations of the chambers, images were made with a Fluke infrared sensor, model TI55FT20 / 54 / 7.5, with an accuracy of ± 0,05°C (Figures 3 and 4).

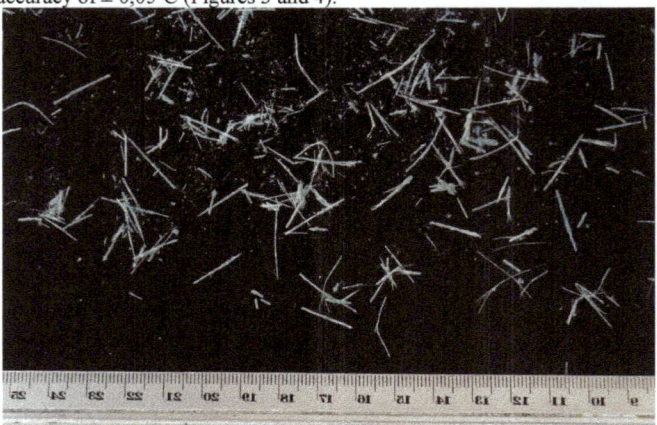

FIGURA 1 - PARTÍCULAS DO RESÍDUO DE PRFV PARA ANÁLISE DE DIMENSÃO MÉDIA.

FIGURE 2 – STEPS FOR ADOBE´S PRODUCTION.
Figure 2. Production stages and moisture tests: corrected soil + FPR residue (A), after addition of water and mixture with the hoe and trampling (B), homogenization in the "maromba" (C), confection and deformation of the adobe (D), adobe drying at 7 days of production (E), "Fall of the Ball" test (F) and Vicat Test (G).

The samples were taken from the retraction box and sawed in the dimensions of 7x7.5cm with a variable height between 2.5 and 3cm for each sample, according to each retraction. Each sample was placed in the thermal box for a predetermined time of 3.33 hours, value for 1000 readings and each reading of 12 seconds. Five replicates were made per treatment using intermediate sampling.

Non-Conventional Materials and Technologies – NOCMAT for XXI Century   Materials Research Forum LLC
Materials Research Proceedings **7** (2018) 366-372   doi: http://dx.doi.org/10.21741/9781945291838-34

For calculate the thermal conductivity, the following equation was used, proposed by Silva [14]:
$K = P.E / \Delta T$:

K – Thermal conductivity [W/m°C];
P – Radiation by area measurement [W/m²];
E – Thickness of the samples [m];
ΔT – Temperature variation [°C].

The radiation of the lamp was obtained by a solar radiation meter Intrutherm model MES-100. Five measurements were taken, averaging 207.34 W / m². The temperature variation (ΔT) was measured for 3.33 hours, just after the temperature stabilization curve.

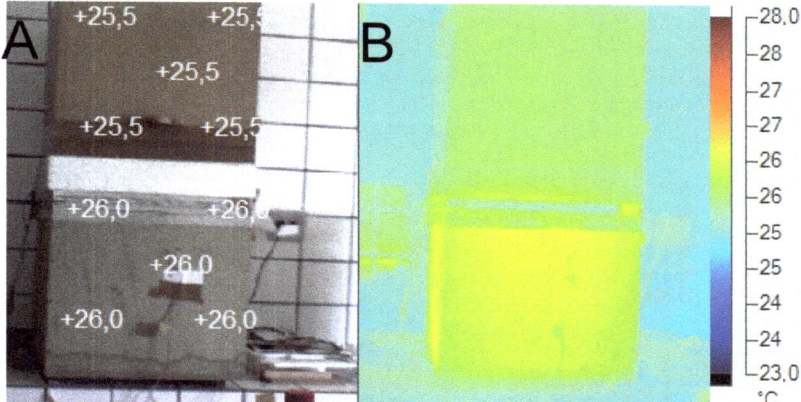

***FIGURE 3** - EXTERNAL VIEW OF THE THERMAL BOX AND ITS INFRARED IMAGE.*

Figure (A) shows the temperatures at the bottom, where the heat source and the top are allocated, where the sample is allocated. Figure (B) shows an infrared image of the thermal box and the temperature plot.

Results and discussion

Addition of the residue provided a temperature variation of 1.1 °C, Figure 5, of the control treatment for FV100. Figure 6 shows the thermal conductivity values of the adobes showing that the FV100 presents the lowest value when compared to the other treatments.

According to Mosquera [10] using two samples of adobes, one drought in an oven and another one with 1.67% humidity found values of 0.80 and 0.90 W / (m.K) respectively. Betting that the humidity in the adobe adversely influences thermal comfort and that the value found 0.86 W / (m ° C) is appropriate.

Non-Conventional Materials and Technologies – NOCMAT for XXI Century   Materials Research Forum LLC
Materials Research Proceedings 7 (2018) 366-372   doi: http://dx.doi.org/10.21741/9781945291838-34

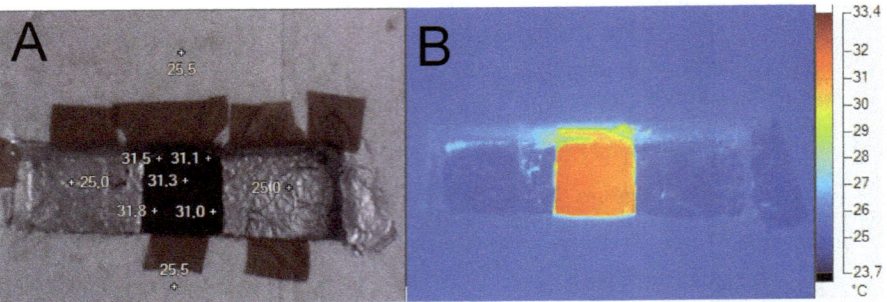

FIGURE 4 - *THERMAL BOX, TOP VIEW OF THE SAMPLE AND INFRARED IMAGE*

Figure (A) shows where the sample, the fence and its temperatures are inserted, showing that there is heat insulation. Figure (B) shows an infrared image of the thermal box and the temperature plot.

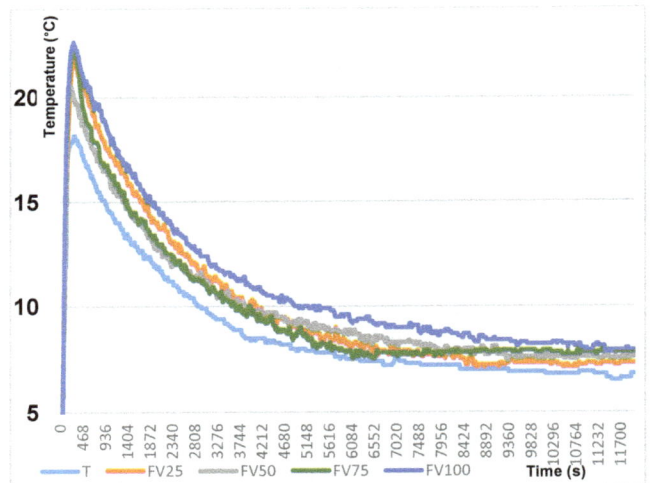

FIGURE 5 - *DIFFERENCES OF INPUT TEMPERATURES AND OUT OF TREATMENTS.*

Non-Conventional Materials and Technologies – NOCMAT for XXI Century   Materials Research Forum LLC
Materials Research Proceedings 7 (2018) 366-372   doi: http://dx.doi.org/10.21741/9781945291838-34

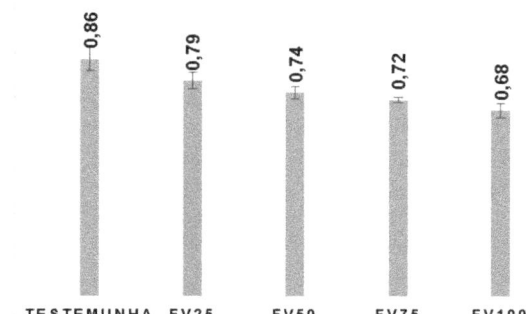

FIGURE 6 - *AVERAGE THERMAL CONDUCTIVITY OF TREATMENTS.*

Conclusion

The inclusion of the FRP residue in the adobe, besides providing a correct destination for this residue of high scale and low degradability, also provides an improvement in the thermal properties of the adobe.

According to the increase of the incorporation of FPR residue obtained a decrease in temperature and also decreases its thermal conductivity, thus improving thermal comfort.

References

[1] ABNT (1981) NBR-7180. Solo – Determinação do limite de plasticidade. Método de Ensaio. Associação Brasileira de Normas Técnicas, Rio de Janeiro, 03p.

[2] ABNT (1984) NBR-7181. Análise Granulométrica. Método de Ensaio. Associação Brasileira de Normas Técnicas, Rio de Janeiro, 13p.

[3] ABNT (1986) NBR-9776. Agregados - Determinação da massa específica de agregados miúdos por meio do frasco Chapman. Método de Ensaio. Associação Brasileira de Normas Técnicas, Rio de Janeiro, 03p.

[4] ABNT (2004) NBR-6459.Determinação do Limite de Liquidez. Método de Ensaio. Associação Brasileira de Normas Técnicas, Rio de Janeiro, 06p.

[5] ABNT (1982) NBR-7183. Determinação do Limite e Relação de Contração dos Solo. Método de Ensaio. Associação Brasileira de Normas Técnicas, Rio de Janeiro, 03p.4

[6] BARBOSA, NP, GHAVAMI, K. Terra Crua para Edificações. Materiais de Construção Civil e Princípios de Ciência e Engenharia de Materiais. São Paulo: Ibracon, 2007, 2.

[7] CORRÊA, AAR, et al. Avaliação das propriedades físicas e mecânicas do adobe (tijolo de terra crua). Diss. Universidade Federal de Lavras, 2006.

[8] CORRÊA, A. A. R., et al. Incorporation of bamboo particles and "synthetic termite saliva" in adobes. Construction and Building Materials, 2015, 98: 250-256. https://doi.org/10.1016/j.conbuildmat.2015.06.009

Non-Conventional Materials and Technologies – NOCMAT for XXI Century   Materials Research Forum LLC
Materials Research Proceedings 7 (2018) 366-372   doi: http://dx.doi.org/10.21741/9781945291838-34

[9] KEMERICH, PD da C, et al. Fibras de Vidro: Caracterização, disposição final e impactos ambientais gerados, Rev. Elet. *Gestão, Educação e Tecnologia Ambiental*, 2013, 10.10: 2112-2121.

[10] MOSQUERA, P, et al. Determination of the Thermal Conductivity in Adobe With Several Models. Journal of Heat Transfer, 2014, 136.3: 031303.

[11] ORTH, CM, BALDIN, N, ZANOTELLI, CT. Implicações do processo de fabricação do compósito plástico reforçado com fibra de vidro sobre o meio ambiente e a saúde do trabalhador: o caso da indústria automobilística. Revista Produção Online, Florianópolis, SC, v.12, n. 2, p. 537-556, 2012. https://doi.org/10.14488/1676-1901.v12i2.943

[12] PALME, M, GUERRA, J, ALFARO, S. Thermal performance of traditional and new concept houses in the ancient village of San Pedro de Atacama and surroundings. Sustainability, 2014, 6.6: 3321-3337. https://doi.org/10.3390/su6063321

[13] RUIZ HE, LUNA MA. Cartilla de pruebas de campo para selección de tierras en la fabricación de adobe. 1983.

[14] SILVA, A DA R. Estudo térmico e de materiais na construção de casas populares com blocos confeccionados a partir de um composto a base de cimento, gesso, eps e raspa de pneu., 2010. Universidade Federal do Rio Grande do Norte.

[15] SHUKLA, A, TIWARI, GN, SODHA, MS. Embodied energy analysis of adobe house. *Renewable Energy*, 2009, 34.3: 755-761. https://doi.org/10.1016/j.renene.2008.04.002

[16] SOLOS, Embrapa. Sistema brasileiro de classificação de solos. *Centro Nacional de Pesquisa de Solos: Rio de Janeiro*, 2013.

Non-Conventional Materials and Technologies – NOCMAT for XXI Century   Materials Research Forum LLC
Materials Research Proceedings 7 (2018) 373-382   doi: http://dx.doi.org/10.21741/9781945291838-35

The Influence of Hot Water Washing Cycles on Tensile Properties of Curauá Fiber

Bartosz Zukowski [a*], Flávio de Andrade Silva [b], Romildo Dias Toledo Filho [c]

[a, c] Department of Civil Engineering, COPPE, Universidade Federal do Rio de Janeiro, P.O. Box 68506 – 21941-972, Rio de Janeiro – RJ, Brazil. bzukowski87@gmail.com*, toledo@coc.ufrj.br

[b] Department of Civil Engineering, Pontifícia Universidade Católica do Rio de Janeiro (PUC-Rio), Rua Marques de São Vicente 225, 22453-900 - Rio de Janeiro - RJ, Brazil.

fsilva@puc-rio.br

Abstract. The work in hand presents the preliminary results of an experimental campaign on the influence of hot water treatment on the tensile strength of high performance natural curauá fibers and its dimensional changes. In this work we have studied the monotonic tensile behavior of natural and treated curauá fibers. Tensile tests were performed on a microforce testing system using four different gage lengths: 10, 20, 30 and 40mm. The cross-sectional area of the fiber was measured using scanning electron microscope (SEM) micrographs and image analysis. The measured Young's modulus was corrected for the machine compliance. Weibull statistics were used to quantify the degree of variability in fiber strength at the different gage lengths. Furthermore, the time length of the used hot water treatment was investigated. The fibers were divided into three groups: reference, 9 hours and 18 hours of treatment in hot water (80° C) with water being changed every 3 hours. This process was applied to remove the wax, impurities and plant residues from the surface of the fibers. The used treatment increased the tensile strength of the fibers and, after 18 hours, caused a notable cross-sectional area reduction.

Keywords: Curauá Fiber, Mechanical Properties, Fiber Treatment, SEM

Introduction

The curauá (*Ananas comosus* var. *erectofolius*) is a type of pineapple plant with a small fruit, less than 15 cm with low nutrition value, and the reason why it is mostly cultivated for the leaves. It has unarmed end straight erected leaves. The bromelain present in the leaves is used in medicine to heal wounds [1]. However, the most promising application is the fiber extracted from the leaves. The demand on curauá fiber has been growing since the fiber was used in automotive industry as an ecological alternative to glass fibers. Curauá is an easy cultivation plant, which needs hot and humid climate to grow, but is very susceptible to cold weather [2]. It has been seen that curauá can grow along the trees and can be used in reforestation areas [3]. In Brazil it is most cultivated in the North, the native area of curauá. With a low initial capital to start a plantation, curauá is considered as a possibility to improve the economic condition of agriculture states in Brazil [4].

There are various types of curauá, among them two types are most known: white and purple. The white has green-light leaves and the purple has reddish leaves [5]. The curauá fiber is classified as a high performance fiber among jute and sisal fibers. It has much better mechanical performance than piassava or coir fibers which are considered as low performance fibers and their average tensile strength is lower than 150 MPa [6]. High tensile strength and Young's modulus focus attention on curauá fiber as a possible reinforcement in cement composites [7].

This work presents the preliminary results on the study of curauá fiber tensile behavior under the influence of hot water treatment as a future reinforcement in cement composites with strain hardening behavior.

The Curauá fiber

The Curauá leaves are separated from the plant manually. Then the fibers are extracted by a mechanical process as illustrated in Figure 1. After extraction, the fibers are washed to remove residues and impurities. This residue is called mucilage (sap and fiber pieces) and is used as a nutrition component for animal or organic fertilizer. The washing process is important for fiber properties because the residues can cause fungus growth on fibers and reduce their resistance. The washing also unifies the color, making impossible to distinguish the difference between white and purple curauá. After washing the fibers are dried for 48 hours in natural conditions, and then tied into bundles ready to distribution for future use.

(a)   (b)   (c)

FIGURE 1 - CURAUÁ FIBER EXTRACTION, (A) MACHINE USED TO FIBER SEPARATION FROM LEAVES, (B) MUCILAGE (RESIDUE AFTER FIBER EXTRACTION, (C) CURAUÁ FIBER DRYING IN THE SUN.

The curauá leaf can contain up to 30 grams of fiber [8]. The quality of the fiber depends on many conditions: insolation, humidity, nutrients, and other plant species around. The leaf contains two types of fibers. The first (Figure 2), octagonal fiber, is located at the edges of the leaf and provides stiffness and strength. The second (Figure 2), arch or horseshoe fiber, also called the vascular fiber, is located in the middle of the leaf cross-section and is responsible for water and nutrients transportation [3]. The diameter of the curauá fiber vary, but most of the fibers range from 60 to 120μm.

Curauá leaf cross-section   500μm

FIGURE 2 - OCTAGONAL (AT THE EDGES) AND VASCULAR (IN THE MIDDLE) FIBER IN CURAUÁ LEAF CROSS-SECTION.

Non-Conventional Materials and Technologies – NOCMAT for XXI Century   Materials Research Forum LLC
Materials Research Proceedings 7 (2018) 373-382   doi: http://dx.doi.org/10.21741/9781945291838-35

The chemical composition of curauá fiber was verified on two sets of samples at the COPPE/UFRJ by calorimetric determination of hexuronic acids. The percentage values showed the amount of cellulose 58.8%, hemicellulose 23.8%, lignin 14.6%.

Tensile testing

The curauá fibers were investigated at different gage lengths (10, 20, 30 and 40 mm) after different time exposure (0, 9, 18 hours) into hot water (80° C). Before the test fibers were dried for 24 hours at 40° C. The preparation of the test was performed according to ASTM C1557 [9]. The tensile tests were performed in a microforce testing machine Tytron 250 at the strain rate of 0.00008 s^{-1}.

The compliance of the loading and gripping system was determined by obtaining the force versus displacement behavior of the fiber at various gage lengths (10, 20 and 30 mm) following the methodology used by Silva et al. [10] and Chawla et al. [11]. The total cross-head displacement during fiber testing δ_t can be expressed by:

$$\frac{\delta_t}{F} = \left[\frac{1}{EA}\right] l + c \qquad (1)$$

Where: c = the machine compliance, F = the applied force, E = the Young's modulus of the fiber and A = the cross-sectional area of the fiber.

Thus, a plot of δ_t/F versus gage length, l, will yield a straight line of slope $1/EA$ and intercept c, the compliance of the load train (Figure 3: Left).

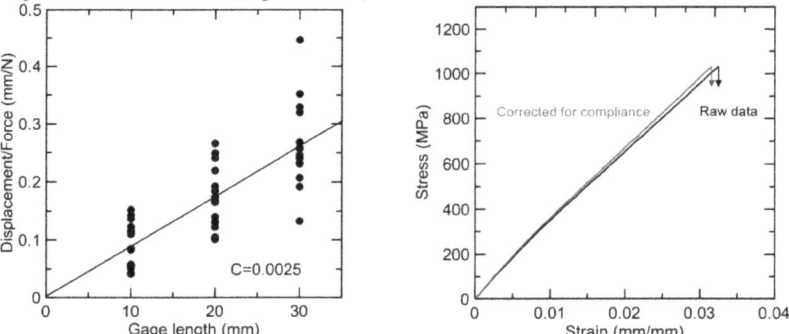

FIGURE 3 - DETERMINATION OF THE MACHINE COMPLIANCE (LEFT): DISPLACEMENT/FORCE VERSUS GAGE LENGTH. THE MACHINE COMPLIANCE, GIVEN BY THE INTERCEPT (C = 0.0025), WAS DETERMINED FROM THE PLOT. (RIGHT): STRESS-STRAIN BEHAVIOR OF A CURAUÁ FIBER TESTED AT 10 MM GAGE LENGTH, SHOWING THE RAW DATA COMPARED WITH THE DATA CORRECTED FOR COMPLIANCE.

Microstructural analysis

The fiber's microstructure was investigated using a SEM (TM 3000) at 25kV of accelerating voltage. The images obtained were processed using the software package ImageJ for measuring the cross-section of each fiber at the two external points of the sample. A contour line was interactively drawn to delineate the fiber cross-section and then the area was measured.

Non-Conventional Materials and Technologies – NOCMAT for XXI Century   Materials Research Forum LLC
Materials Research Proceedings **7** (2018) 373-382   doi: http://dx.doi.org/10.21741/9781945291838-35

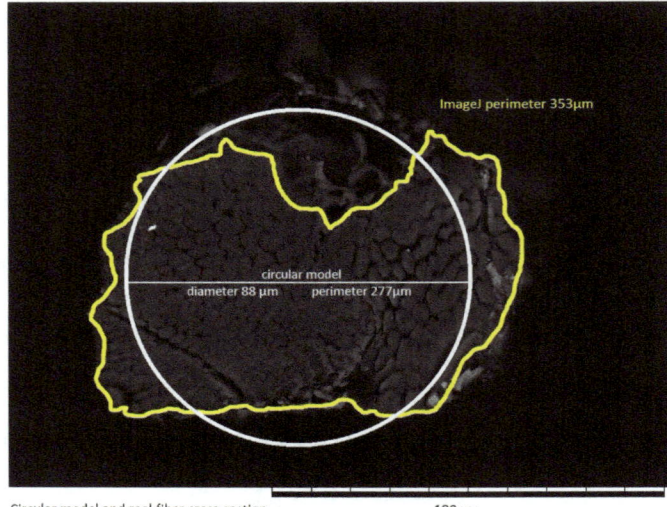

Circular model and real fiber cross-section   100 µm

FIGURE 4 - *CURAUÁ FIBER CROSS-SECTION AT SEM AND IMAGEJ ANALYSIS: YELLOW PRESENTS THE CONTOUR LINE, WHITE THE APPROXIMATION TO CIRCULAR MODEL DIAMETER.*

Results and discussion

Table 1, 2 and 3 present the tensile test results of the curauá fibers for the different gage lengths at different time exposure to hot water treatment called cycles. Each cycle contains 3 hours of washing followed by 24 hours of drying. The fibers were randomly chosen from a given batch and tested. Young's modulus was calculated in the elastic portion of the stress-strain curve and then corrected for compliance by measuring force versus displacement, at various gage lengths, using Eq (1)(Figure 4b). The gage length does not seem to influence Young's modulus of the fibers and the variability of modulus values are related to the microstructure of the curauá fibers and possible damages during the extraction process, as reported by Silva et al. [6]. However, the tensile strength seems to decrease with the increase in the gage length. For all groups of fibers tested the highest strength is for the shortest gage length. This can be caused due to the increased probability of fiber damage with the fiber length.

For untreated fibers the obtained average tensile strength values varied from 991 to 688 MPa with standard deviation oscillating around 150 MPa.

TABLE 1 - SUMMARY OF AVERAGE TENSILE TEST RESULTS AND STANDARD
DEVIATION FOR CURAUÁ FIBERS WITH NO EXPOSURE TO HOT WATER TREATMENT.

Fiber Length	Area	Diameter	Strain to failure	Force at rupture	Strength	Young's modulus	Weibull modulus
(mm)	(mm²)	(μm)	(%)	(N)	(MPa)	(GPa)	
10	0.00483 (0.00066)	78.21 (5.34)	3.54 (0.50)	4,78 (0.96)	991.18 (156.25)	27.33 (3.38)	4.61
20	0.00590 (0.00112)	85.92 (8.62)	2.57 (0.45)	4.44 (1.41)	735.19 (150.73)	28.71 (3.10)	5.69
30	0.00576 (0.00143)	84.55 (10.97)	2.15 (0.35)	3.93 (1.18)	688.83 (133.72)	32.55 (3.55)	3.55
40	0.00506 (0.00072)	79.87 (5.69)	2.22 (0.32)	4.54 (0.98)	928.51 (194.17)	41.67 4.73	2.74

Graphical representation of results and calculated average for curauá in nature (without hot water cycles) presents specimen results and calculated average (Figure 5).

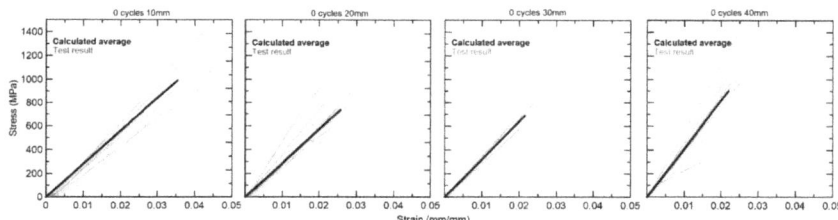

FIGURE 5 - GRAPHICAL REPRESENTATION OF TENSILE TEST RESULTS FOR CURAUÁ
FIBER WITHOUT TREATMENT.

The group of fibers after 9 hours of hot water treatment presented higher tensile strength ranging from 1328 to 895 MPa with standard deviation ranging from 117 to 300 MPa.

TABLE 2 - SUMMARY OF AVERAGE TENSILE TEST RESULTS AND STANDARD
DEVIATION FOR CURAUÁ FIBERS AFTER 9 HOURS OF EXPOSURE TO HOT WATER
TREATMENT.

Fiber length	Area	Diameter	Strain to failure	Force at rupture	Strength	Young's modulus	Weibull modulus
(mm)	(mm²)	(μm)	(%)	(N)	(MPa)	(GPa)	
10	0.00489 (0.00096)	78.33 (7.24)	4.44 (0.34)	6.12 (0.78)	1328.84 (300.54)	35.43 (5.81)	3.78
20	0.00461 (0.00113)	75.88 (8.88)	3.21 (0.38)	4.95 (1.05)	1096.72 (142.81)	35.92 (5.92)	6.33
30	0.00507 (0.00086)	79.99 (6.80)	2.80 (0.33)	5.10 (0.97)	1006.62 (117.73)	37.92 (5.43)	6.77
40	0.00572 (0.00098)	84.88 (7.15)	2.78 (0.21)	5.04 (0.87)	895.40 (138.85)	22.99 (3.25)	4.41

Graphical representation of results and calculated average for curauá after 3 cycles of hot water treatment are presented in Figure 6.

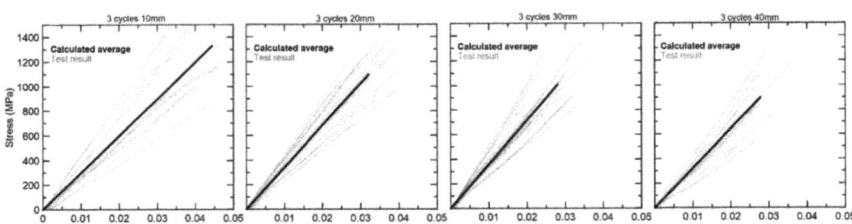

FIGURE 6 - GRAPHICAL REPRESENTATION OF TENSILE TEST RESULTS FOR CURAUÁ FIBER AFTER 9 HOURS OF HOT WATER TREATMENT.

The group of fibers after 18 hours of hot water treatment presented more uniform tensile strength between 1110 to 935 MPa with the smallest standard deviation values ranging from 64 to 108 MPa.

TABLE 3 - SUMMARY OF AVERAGE TENSILE TEST RESULTS AND STANDARD DEVIATION FOR CURAUÁ FIBERS AFTER 18 HOURS OF EXPOSURE TO HOT WATER TREATMENT.

Fiber length	Area	Diameter	Strain to failure	Force at rupture	Strength	Young's modulus	Weibull modulus
(mm)	(mm^2)	(µm)	(%)	(N)	(MPa)	(GPa)	
10	0.00402	70.18	3.31	4.44	1110.88	41.11	13.12
	(0.00123)	(10.53)	(0.36)	(1.30)	(64.10)	(4.98)	
20	0.00384	69.54	2.52	3.85	1003.72	44.26	8.41
	(0.00072)	(6.57)	(0.23)	(0.77)	(108.10)	(5.39)	
30	0.00326	63.80	2.40	2.95	920.65	39.45	7.56
	(0.00067)	(6.71)	(0.22)	(0.57)	(92.28)	(6.40)	
40	0.00397	70.43	2.18	3.65	935.24	44.13	7.50
	(0.00093)	(8.38)	(0.17)	(0.78)	(92.19)	(7.28)	

Graphical representation of results and calculated average for curauá fiber after 18 hours of exposure to hot water treatment (Figure 7).

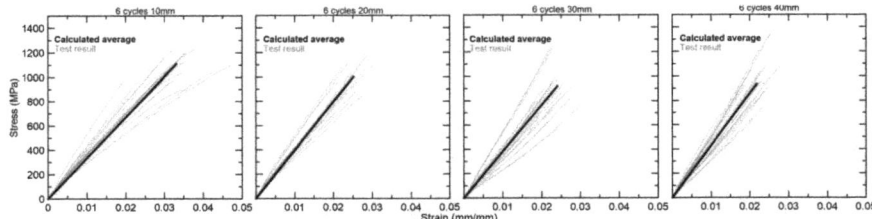

FIGURE 7 - GRAPHICAL REPRESENTATION OF TENSILE TEST RESULTS FOR CURAUÁ FIBER AFTER 18 HOURS OF HOT WATER TREATMENT.

The strain-to-failure seems to decrease when increasing the gage lengths with a significant difference between 10 and 20 mm gage lengths.

Non-Conventional Materials and Technologies – NOCMAT for XXI Century   Materials Research Forum LLC
Materials Research Proceedings 7 (2018) 373-382   doi: http://dx.doi.org/10.21741/9781945291838-35

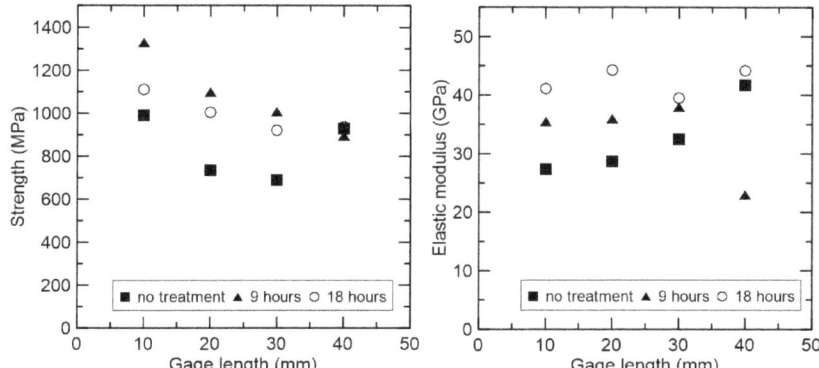

FIGURE 8 - *(LEFT) TENSILE BEHAVIOR OF THE CURAUÁ FIBER CORRELATED WITH GAGE LENGTH AND TIME OF HOT WATER TREATMENT. (RIGHT) YOUNG'S MODULUS OF THE CURAUÁ FIBER CORRELATED WITH THE GAGE LENGTH AND TIME OF HOT WATER TREATMENT.*

The tensile behavior of curauá fibers can be classified as a high performance fiber according to Silva [6]. The untreated curauá fibers presented strength higher than 600 MPa with the modulus greater than 27 GPa. The hot water treatment increased the value of the fiber tensile strength to more than 1000 MPa and Young's modulus to 35 GPa (and even to more than 40 for 18 hours of treatment) (Figure 8:Left). The increased strength seems to be correlated with the dimensional changes. The hot water treatment reduces the area of the fiber. This change can be shown as a diameter reduction of 6% for 9 hours and 15% after 18 hours in comparison to the untreated fiber (Figure 9). The dimensional change can be due to the wax and impurities removal and reduction of hemicellulose and lignin content (which will be investigated in the authors following work by describing the fiber chemical composition after the hot water treatment).

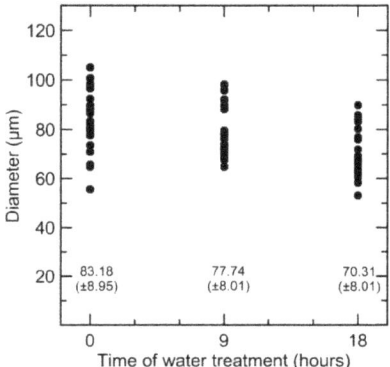

FIGURE 9 - *DIMENSIONAL CHANGES (DIAMETER) AS A FUNCTION OF THE HOT WATER TREATMENT TIME. MEAN VALUES WITH STANDARD DEVIATION (μM).*

Non-Conventional Materials and Technologies – NOCMAT for XXI Century   Materials Research Forum LLC
Materials Research Proceedings 7 (2018) 373-382   doi: http://dx.doi.org/10.21741/9781945291838-35

The application of Weibull distribution on the fibers tensile strength is described by several authors [10], [11]. In this study, the form presented by Silva et al. [10] was used. According to the Weibull analysis, the probability of survival of a fiber at a stress σ, is given by:

$$P(\sigma) = exp\left[-\left(\frac{\sigma}{\sigma_0}\right)^m\right] \tag{2}$$

Where:

σ = the fiber strength for a given probability of survival, m = the Weibull modulus, σ_0 = the characteristic strength, which corresponds to $P(\sigma) = 1/e = 0.37$.

The higher is value of m the lower is the variability in strength. Ranking of the fiber strengths is performed by using an estimator given by:

$$P(\sigma)_i = 1 - \frac{1}{N+1} \tag{3}$$

Where:

$P(\sigma)_i$ = the probability of survival corresponding to the i-th strength value;
N = the total number of fibers tested.

Substituting Eq. (3) into Eq. (2) yields:

$$\ln\ln\left[\frac{N+1}{N+1-i}\right] = m \ln\left(\frac{\sigma}{\sigma_0}\right) \tag{4}$$

Thus, a plot of $ln \, ln((N+1)/N+1-i))$ versus $ln(\sigma/\sigma_0)$ yields a straight line with slope of m. Figure 10.Left shows the variability of Weibull modulus among the group of fibers (after 9 hours of hot water treatment) for the different gage lengths. Figure 10.Right shows the variability of the Weibull modulus for three groups of curauá fibers which were verified. The Weibull modulus increased with the time exposure to hot water treatment showing the less variability of strength after treatment. According to Weibull modulus, 18 hours of hot water treatment reduced the variability of strength between the fibers.

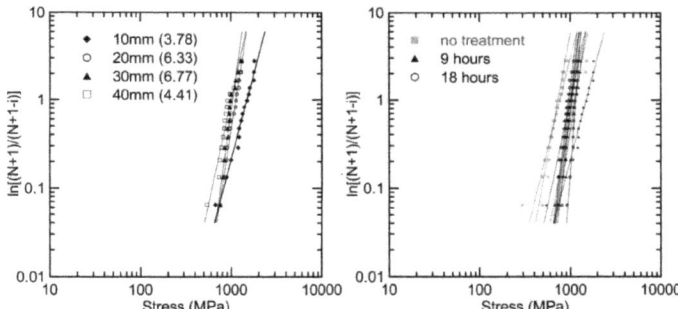

FIGURE 10 - *WEIBULL DISTRIBUTION OF THE (LEFT) CURAUÁ FIBER STRENGTH FOR DIFFERENT GAGE LENGTHS AFTER 9 HOURS OF HOT WATER TREATMENT. (RIGHT) CURAUÁ FIBER TENSILE STRENGTH FOR DIFFERENT GAGE LENGTHS AND DIFFERENT TIME OF HOT WATER TREATMENT.*

Conclusion

The curauá fiber is a high performance natural fiber. Its tensile strength presents higher values in comparison to other high performance natural fibers like Jute and Sisal. From the present work, the main findings are presented below:

Non-Conventional Materials and Technologies – NOCMAT for XXI Century   Materials Research Forum LLC
Materials Research Proceedings 7 (2018) 373-382   doi: http://dx.doi.org/10.21741/9781945291838-35

- The gage length has influenced the measured tensile strength of the fiber: the higher the gage length is the smaller the measured tensile strength.
- The hot water treatment reduced the area of the fibers.
- The hot water treatment decreased the data scatter, which is shown in a higher Weibull modulus representing the lower variability of measured strength.
- The hot water treatment improved the mechanical properties of the fibers encouraging their application as a reinforcement in cement based composites.

Acknowledgements

The authors gratefully acknowledge the Conselho Nacional de Desenvolvimento Cientifico e Tecnologico, CNPq, (Brazilian National Science Foundation), for financial support for this work.

References

[1] Silva T.A.L., Pereira Bresolin I.R.A., Campos-Takaki G.M., Aoyama H., Garrard I., Tambourgi E.B., (2014). In: *Extraction and preliminary characterization of bromelain from curaua (ananas erectifolius l.b.smith) purple and white*, Chemical Engineering Transactions, 37, 769-774 DOI: 10.3303/CET1437129.

[2] Reis, I.N.R. de S.; Lameira, O.A.; Cordeiro, I.M.C.C. (2004). In: *Efeito da adubação orgânica e de NPK no desenvolvimento do curauá (Ananas erectifolius)*. Congresso Brasileiro de Sistemas Agroflorestais, 5., 2004, Curitiba. Anais. Curitiba: Embrapa Florestas, 2004. p.332-334. (Embrapa Florestas. Documentos, 98).

[3] E.C.P. de Oliveira et al. (2008). In: *Leaf structure of curaua in different intensities of photosynthetically active radiation*. Pesquisa Agropecuária Brasileira, Brasília, DF, v.43, n.2, p. 163-169, fev. 2008.

[4] I.M. Castro Coimbra Cordeiro et al. (2009). In: Economical analysis of cultivation systems with Schizolobium parahyba var. amazonicum (Huber ex Ducke) Barneby (Parica) and Ananas comosus var. erectifolius (L. B. Smith) Coppus & Leal (Curaua) crop at Aurora do Pará, Brazil. Faculty of Agronomy Journal (LUZ). University of Zulia, Maracaibo, Venezuela, 2009, 26:p.243-265.

[5] Ledo, I. A. de M. (1967). In: *O cultivo do curauá no lago grande de Franca*. Belém: Banco da Amazönia S/A - BASA - 1967. 23 p.

[6] Silva F. de A., Toledo R.D., (2013). In: *The effect of fiber morphology on the tensile strength of natural fibers,* Journal of Materials Research and Technology, 2013;2(2):p 149-157. https://doi.org/10.1016/j.jmrt.2013.02.003

[7] d'Almeida, A. L. F. S., Melo Filho, J. A., & Toledo Filho, R. D. (2009). In: *Use of curauá fibers as reinforcement in cement composites. CHEMICAL ENGINEERING, 17*, 2009.

[8] Silva O. G. da, (2013). Produção e caracterização de compósitos à base de fibras de curauá, amido termoplástico e polietileno, utilizando-se a termografia, Belo Horizonte, 129p, Dissertação (Mestrado) - Escola de Design – UEMG, Universidade do Estado de Minas Gerais.

[9] ASTM. *ASTM C1557-03 Standard Test Method for Tensile Strength and Young's Modulus of Fibers* West Conshohocken, PA, USA: American Society for Testing and Materials; 2008.

Non-Conventional Materials and Technologies – NOCMAT for XXI Century   Materials Research Forum LLC
Materials Research Proceedings 7 (2018) 373-382   doi: http://dx.doi.org/10.21741/9781945291838-35

[10] Silva FA, Chawla N, Toledo Filho RD. (2008). In: *Tensile behavior of high performance (sisal) fibers*. Compos Sci Technol 2008;68:3438–43. https://doi.org/10.1016/j.compscitech.2008.10.001

[11] Chawla N, Kerr M, Chawla KK., (2005). In: *Monotonic and cyclic fatigue behavior of high-performance ceramic fibers*. J Am Ceram Soc 2005;88:101–8. https://doi.org/10.1111/j.1551-2916.2004.00007.x

Non-Conventional Materials and Technologies – NOCMAT for XXI Century   Materials Research Forum LLC
Materials Research Proceedings 7 (2018) 393-390   doi: http://dx.doi.org/10.21741/9781945291838-36

Influence of Water Hornification and Alkaline Treatment on the Stress-Strain Behaviour of Jute Fibers

Yasmim Gabriela dos Santos Mendonça [a*], Bartosz Zukowski [b], Romildo Dias Toledo Filho [c]

[a, b, c] Department of Civil Engineering, COPPE, Universidade Federal do Rio de Janeiro, P.O. Box 68506 – 21941-972, Rio de Janeiro – RJ, Brazil. yasmim.mend1@gmail.com*, bzukowski87@gmail.com, toledo@coc.ufrj.br

Abstract. The aim of this work is to present the influence of two treatments, namely water hornification, and alkaline treatment, on the stress-strain behaviour of jute fibers. A group of jute fibers was subjected to 5 cycles of wetting and drying. In a cycle, the fiber was immersed in water at 25 °C for 3 hours and dried for 16 hours at the temperature of 80°C. The other group was submitted to 1 cycle of alkaline treatment in a saturated solution of $Ca(OH)_2$ with water for 50 minutes and dried for 24 hours at the temperature of 40 °C. The tensile stress-strain behavior of the natural fibers was determined before and after each treatment. The effectiveness of the alkaline treatment was evaluated by tensile tests performed on a microforce testing system using three different gage lengths: 20, 30 and 40 mm. The cross-sectional area of the fiber was measured using scanning electron microscope (SEM) micrographs and image analysis. The Young´s modulus was measured for all the groups and gage lengths. Both applied treatments increased the tensile strength of the jute fibers in comparison with the untreated fibers.

Keywords: Jute, Water Hornification, Alkaline Treatment, Stress-Strain Behaviour, Scanning Electron Microscope

Introduction

Jute is a name of a fiber extracted from the plants in the genus *Corchorus*, from family *Tiliaceae*. About forty different species are known among *Corchorus* family, but just the *C.capsularis* (popularly known for "white jute") and *C.olitorius* are cultivated on a commercial scale [1].

Jute fibers are largely used in all the world mostly as packaging material, floor covering, insulation material, soil protection, handicrafts, among others [2]. Jute cultivation is predominantly annual. The cultivation process is laborious but requires relatively small amounts of fertilizers and pesticides. It can be carried out on small land. For these reasons, the production of jute is more concentrated in Bangladesh, India, Nepal, China, and Thailand than in other places. These countries correspond about 95% of the world production. India is the biggest producer and consumer of jute in the world. Bangladesh is the second biggest producer and the main exporter of fibers [3]. From the 1930's to the beginning of 1990 the production of fibers was dominant along the Amazon River, between Manaus and Santarem, in the region of lowlands. Japanese immigrants introduced the culture in the region to supply the demand for jute bags in southern Brazil, where the bags were used to the packaging of commodities, mainly the coffee. Changes in the packaging of goods, quality of the fiber, removal of tariffs on imported jute, contributed to jute market collapse in the 1990's. Despite this, the legacy of jute is still evident in the physical and social landscape of the region [4].

The mechanical properties of natural fibers present a lot of variation and depend on some factors: diameter of the fiber, structure, degree of polymerization, crystalline structure, origin (stem, leaf, fruit, and seed) and growth conditions [5]. The area of jute is smaller than the area of some other

Non-Conventional Materials and Technologies – NOCMAT for XXI Century   Materials Research Forum LLC
Materials Research Proceedings 7 (2018) 393-390   doi: http://dx.doi.org/10.21741/9781945291838-36

fibers as sisal and curauá. The tensile strength of the jute is less than curauá and sisal, but Young's modulus is not too distant from it (Table 1) [6].

TABLE 1 - PHYSICAL AND MECHANICAL CHARACTERISTICS OF UNTREATED NATURAL FIBERS (FERREIRA, 2016).

Fiber	Area	Force at rupture (N)	Tensile Strength	Strain to failure	Young's modulus
	(mm²)		(MPa)	%	(GPa)
Jute	0.004	1.05 (0.37)	249.23 (89.02)	0.60 (0.20)	43.90 (12.31)
Curauá	0.008	2.54 (0.46)	632.14 (138)	2.10 (1.00)	38.10 (18.01)
Sisal	0.023	10.28 (3.43)	447.20 (23.90)	3.00 (1.00)	19.28 (1.36)

The physical and mechanical properties of the fibers are influenced by their composition, mainly cellulose, hemicelluloses, and lignin. Higher tensile strengths and Young's modulus are obtained for fibers that contain more crystalline cellulose [7].

Fiber treatments can promote a better interaction between lignin and cellulose and hemicellulose, resulting in a stronger fiber, with a higher crystallinity and modulus. The natural fibers tensile strength increase with the application of hornification treatments, because of the change in the cellulose crystallinity and possible bonds created between different polymer chains in the microfibrils. The hornification treatment is presented as a simple solution to improve the mechanical behavior. It is possible that the cycles of wetting and drying can improve the mechanical properties of natural fibers, because of cross-section area reduction combined with higher dimension stability. Alkaline treatment promotes changing in cellulose chemical bonds, as well as the reduction of fiber-water absorption capacity. The jute is the fiber that presents a higher interaction with the calcium hydroxide. Jute contains cellulose of high crystallinity, which corroborates well with $Ca(OH)_2$. This chemical affinity is reported in the literature [6].

This work presents the preliminary results of the study of jute fibers tensile behavior under the influence of water hornification and alkaline treatment as a future reinforcement in cement composites with strain hardening behavior.

The jute fiber

The extraction occurs of the plant stem by a combination of the process based on the following steps: cutting, hardening, crushing, drying and classification [8].

The jute from Amazon region is combed and then cut in a guillotine.

The jute chemical composition was verified on two sets of samples at the COPPE/UFRJ by calorimetric determination of hexuronic acids. The percentage values showed the amount of cellulose 74.4%, hemicellulose 15.0%, lignin 8.4%.

The jute fiber treatments

The fibers were first separated into three groups, namely A, B and C. The group A refers to jute in nature (untreated). The group B refers to jute submitted to water hornification. The group C refers to jute submitted to alkaline treatment.

The fibers of the groups A, B and C were separated from the bundles and glued to the paper frames of gage lengths 20, 30 and 40 mm prior to tensile testing. Only the fibers of the groups B and C were subjected to a pretreatment. Pretreatment consists of washing the fibers in water at 80°C and remained submerged in it for 3 hours. Then the fibers were cooled at 40 °C for 5 hours.

Non-Conventional Materials and Technologies – NOCMAT for XXI Century   Materials Research Forum LLC
Materials Research Proceedings **7** (2018) 393-390   doi: http://dx.doi.org/10.21741/9781945291838-36

The purpose of the pretreatment is to wash the fibers to remove residues and impurities. The washing process is important for fiber properties and durability because the residues can cause fungus growth on fibers and reduce their resistance.

After the pretreatment, the fibers of the group B were subjected to 5 cycles of wetting and drying. In a cycle, the fibers were immersed in water at 25 °C for 3 hours and dried for 16 hours at the temperature of 80 °C.

The group C, after the pretreatment, was submitted to 1 cycle of alkaline treatment in a saturated solution of $Ca(OH)_2$ with water for 50 minutes, and then dried for 24 hours at the temperature of 40 °C.

Tensile testing
The jute fibers were investigated at different gage lengths (20, 30 and 40 mm). The groups A, B and C were tested. Before the test fibers were dried for 24 hours at 40 °C. The preparation of the test was performed according to ASTM C1557 [9]. The tensile tests were performed in a microforce testing machine Tytron 250 (Figure 1) at the strain rate of 1 mm/min.

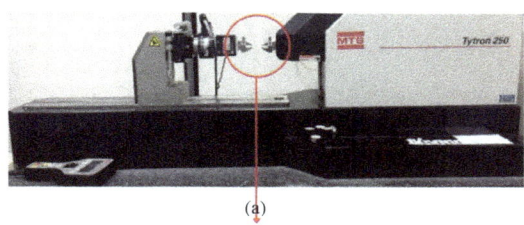

(a)

Specimen grips

(b)
FIGURE 1 - SETUP OF THE TENSILE ACCORDING TO ASTM C1557: (A) MICROFORCE TESTING MACHINE TYTRON 250, (B) DETAIL OF CLAMPS AND SPECIMEN.

The strength of each group of fibers was determinate by the Equation 1.

$$\sigma = \frac{F}{A} \hspace{5cm} (1)$$

The Young's modulus of each group of fibers was determinate by the Equation 2.

$$E = \frac{\sigma}{\varepsilon} \hspace{5cm} (2)$$

Where:

σ = strength;
F = the applied force;
A = the cross-sectional area of the fiber;
E = the Young's modulus of the fiber;
ε = strain in the elastic portion.

Microstructural analysis
The fiber's microstructure was investigated using a SEM (TM 3000) at 25kV of accelerating voltage. The images obtained were processed using the software package ImageJ for measuring the cross-section of each fiber. A contour line was interactively drawn to delineate the fiber cross-section and then the area was measured.

The visual analysis of fiber surface was carried at SEM. The fibers were covered with approximately 20nm layer of gold for before the visual investigation of treatments benefits under the microscope, as illustrate the Figure 2.

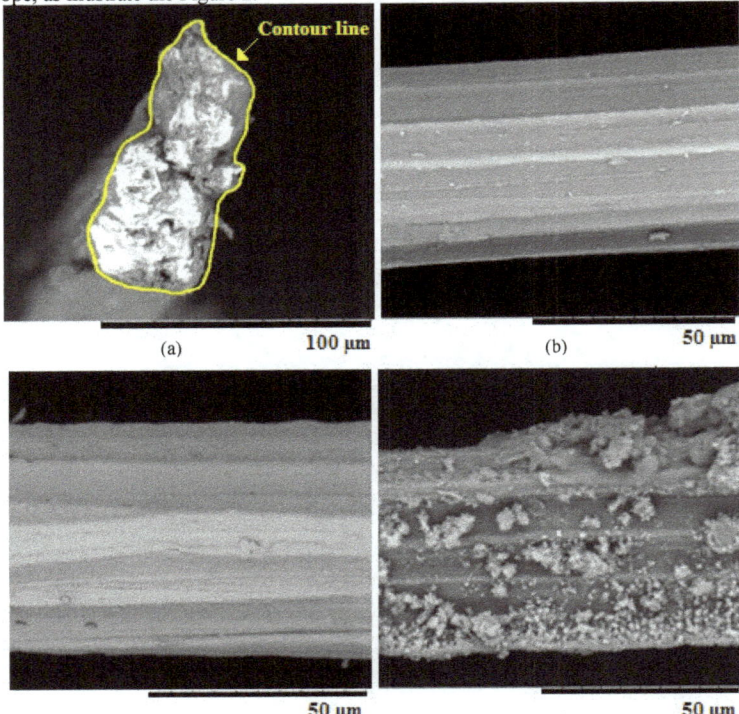

FIGURE 2 - MICROGRAPHS OF JUTE FIBERS AT SEM: (A) AREA CALCULATION USING IMAGEJ, (B) SURFACE OF UNTREATED FIBER, (C) SURFACE OF FIBER AFTER 5 CYCLES OF WETTING AND DRYING IN WATER, (D) SURFACE OF FIBER AFTER 1 CYCLE OF WETTING AND DRYING IN ALKALINE SOLUTION.

Non-Conventional Materials and Technologies – NOCMAT for XXI Century   Materials Research Forum LLC
Materials Research Proceedings 7 (2018) 393-390   doi: http://dx.doi.org/10.21741/9781945291838-36

Results and discussion

Table 2, 3 and 4 present the tensile test results of the jute fibers for the different gage lengths of three groups (A, B and C). The untreated fibers belong to the group A. The fibers submitted to 5 cycles of washing and drying belong to the group B. The fibers submitted to the 1 cycle of alkaline treatment belong to the group C.

The fibers were randomly chosen from a given batch and tested. Young's modulus was calculated in the elastic portion of the stress-strain curve.

The gage length does not seem to influence Young's modulus of the fibers and the variability of modulus values are related to the microstructure of the jute fibers and possible damages during the extraction process, as reported by Silva et al. [10]. However, the tensile strength seems to decrease with the increase in the gage length, as it happens in the groups B and C. For most groups of fibers tested the highest strength is for the shortest gage length. This can be attributed to the increased probability of fiber damage with the fiber length.

For untreated fibers (group A) the obtained average tensile strength values varied from 234 to 307 MPa with standard deviation oscillating around 111 to 73 MPa (Table 2).

TABLE 2 - SUMMARY OF AVERAGE TENSILE TEST RESULTS AND STANDARD DEVIATION FOR UNTREATED JUTE FIBERS (GROUP A).

Fiber length	N° of tested fibers	Area	Diameter		Strain to failure	Force at rupture	Strength	Young's modulus
			Major	Minor				
(mm)		(mm2)	(µm)		(%)	(N)	(MPa)	(GPa)
20	11	0.00366	95.32	50.01	1.06	0.89	244.34	22.17
		(0.00057)	(20.31)	(9.31)	(0.40)	(0.35)	(95.20)	(7.89)
30	8	0.00291	79.83	45.08	1.15	0.89	307.103	26.97
		(0.00171)	(27.43)	(27.43)	(0.25)	(0.21)	(72.67)	(5.93)
40	7	0.00423	101.91	51.67	0.99	0.99	234.31	22.10
		(0.00148)	(22.47)	(8.16)	(0.19)	(0.47)	(111.12)	(8.78)

Graphical representation of results and calculated average for untreated jute fibers presents specimen results and calculated average (Figure 3).

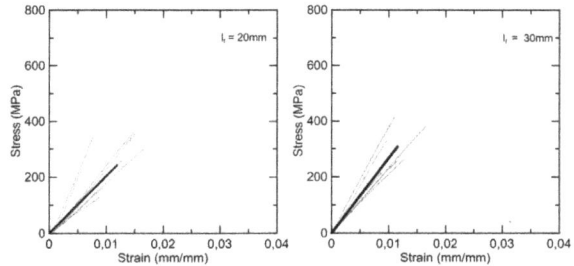

FIGURE 3 -GRAPHICAL REPRESENTATION OF TENSILE TEST RESULTS FOR UNTREATED JUTE FIBER (GROUP A).

Non-Conventional Materials and Technologies – NOCMAT for XXI Century   Materials Research Forum LLC
Materials Research Proceedings 7 (2018) 393-390   doi: http://dx.doi.org/10.21741/9781945291838-36

The fibers submitted to water hornification (group B) presented higher tensile strength than the group A varied from 276 to 535 MPa with a standard deviation ranging from 102 to 151 MPa (Table 3).

TABLE 3 - *SUMMARY OF AVERAGE TENSILE TEST RESULTS AND STANDARD DEVIATION FOR JUTE FIBERS SUBMITTED TO WATER HORNIFICATION (GROUP B).*

Fiber length	N° of tested fibers	Area	Diameter		Strain to failure	Force at rupture	Strength	Young's modulus
			Major	Minor				
(mm)		(mm²)	(μm)		(%)	(N)	(MPa)	(GPa)
20	8	0.00184	68.22	31.84	1.13	0.99	535.88	45.29
		(0.00130)	(25.60)	(9.32)	(0.35)	(0.28)	(151.01)	(7.71)
30	8	0.00299	104.33	37.56	0.79	0.83	276.41	35.16
		(0.00086)	(24.00)	(13.25)	(0.40)	(0.30)	(102.02)	(9.84)
40	6	0.00237	85.98	37.09	0.93	0.68	285.79	29.20
		(0.00056)	(35.53)	(8.97)	(0.22)	(0.18)	(74.74)	(5.57)

Graphical representation of results and calculated average for jute of the group B are presented in Figure 4.

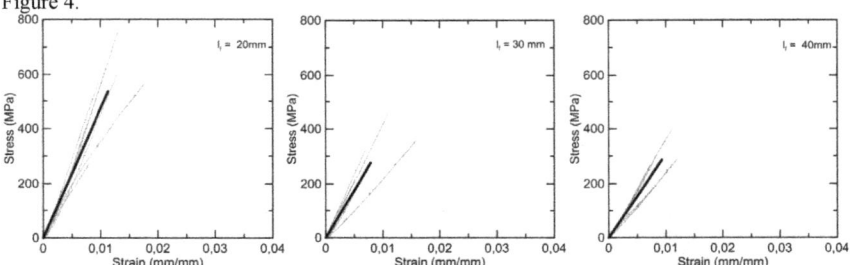

FIGURE 4 - *GRAPHICAL REPRESENTATION OF TENSILE TEST RESULTS FOR JUTE FIBER SUBMITTED TO WATER HORNIFICATION (GROUP B).*

The group of fibers after 1 cycle of alkaline treatment (group C) presented higher tensile strength than the groups A and B, between 475 to 690 MPa with the standard deviation values ranging from 174 to 74 MPa (Table 4).

TABLE 4 - *SUMMARY OF AVERAGE TENSILE TEST RESULTS AND STANDARD DEVIATION FOR JUTE FIBERS SUBMITTED TO ALKALINE TREATMENT (GROUP C).*

Fiber length	N° of tested fibers	Area	Diameter		Strain to failure	Force at rupture	Strength	Young's modulus
			Major	Minor				
(mm)		(mm²)	(μm)		(%)	(N)	(MPa)	(GPa)
20	7	0.00199	74.52	34.87	2.06	1.37	689.77	35.94
		(0.00050)	(14.72)	(10.97)	(0.77)	(0.15)	(74.80)	(10.83)
30	14	0.00242	77.12	37.31	1.21	1.16	480.80	38.27
		(0.00193)	(16.44)	(19.48)	(0.34)	(0.28)	(116.30)	(6.77)
40	6	0.00202	72.01	36.37	1.02	0.96	475.30	42.22
		(0.00069)	(11.32)	(12.00)	(0.30)	(0.35)	(174.37)	(9.27)

Graphical representation of results and calculated average for jute fiber after 1 cycle of alkaline treatment (Figure 5).

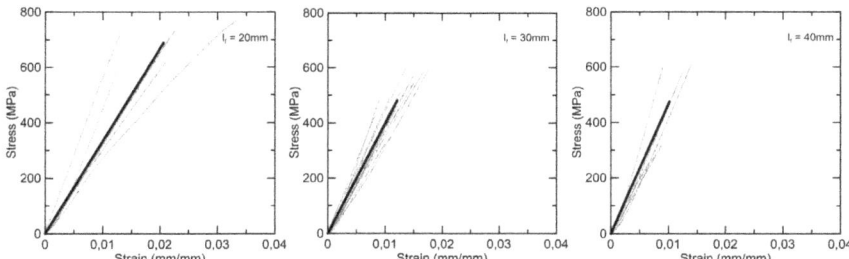

FIGURE 5 - *GRAPHICAL REPRESENTATION OF TENSILE TEST RESULTS FOR JUTE FIBER SUBMITTED ALKALINE TREATMENT (GROUP C).*

The tensile behavior of jute fibers can be classified as a high-performance fiber according to Silva [10]. The untreated jute fibers presented strength smaller than 307 MPa with the modulus greater than 22 GPa. However, the water hornification increased the value of the fiber tensile strength to more than 276 MPa and Young's modulus to more than 29 GPa. As well, the alkaline treatment increased the value of the fiber tensile strength to more than 475 MPa and Young's modulus to more than 35 GPa.

The increased strength seems to be correlated with the dimensional changes. The water hornification reduces the area of the fiber. The fiber group of alkaline treatment also reduced the area of the fiber. It can be because of the pretreatment with hot water made before the cycle of Ca $(OH)_2$. The dimensional change can be due to the wax and impurities removal and reduction of hemicellulose and lignin content.

The group C of fibers presented higher strain to failure and force at rupture than the groups A and B. The best results were for smaller lengths, 20 and 30 mm. While the fibers of 20 mm of the group C presented 2.06 % and 1.37 N of strain to failure and force at rupture respectively, the groups A and B presented 1.06 % and 0.89 N, and 1.13% and 0.99 N respectively, for the same length. The second higher force at rupture was presented by the fibers of 30mm of the group C, with 1.16N.

Acknowledgements
The authors gratefully acknowledge the Conselho Nacional de Desenvolvimento Cientifico e Tecnologico, CNPq, (Brazilian National Science Foundation), for financial support for this work.

Conclusion
The jute fiber is a high-performance natural fiber before submitted to a water hornification or alkaline treatment. From the present work the main findings are presented below:
- The gage length has influenced the measured tensile strength of the fiber: the higher the gage length is the smaller the measured tensile strength.
- The water hornification and the alkaline treatment reduced the area of the fibers.
- The water hornification and the alkaline treatment is presented as a simple solution to improve the mechanical behavior.
- The fibers submitted to alkaline treatment presented better results of tensile strength than the fibers submitted to water hornification, and both presented better results than fibers untreated (in nature).

Non-Conventional Materials and Technologies – NOCMAT for XXI Century   Materials Research Forum LLC
Materials Research Proceedings 7 (2018) 393-390   doi: http://dx.doi.org/10.21741/9781945291838-36

- The water hornification and the alkaline treatment improved the mechanical properties of the fibers encouraging their application as a reinforcement in cement-based composites.

References

[1] Singh, B., Gupta, M., Tarannum, H., Randhawa, A., 2011, "Natural Fiber- Based Composite Building Materials", in Cellulose Fibers: Bio and Nano Polymer Composites – Green Chemistry and Technology. Springer-Verlag. Edited by KALIA, S., KAITH, B. S., KAUR, I. https://doi.org/10.1007/978-3-642-17370-7_24

[2] Graupner, N., Müssig, J., 2010, "Technical Applications of Natural Fibres: An Overview", in Industrial Applications of Natural Fibres: Structure, Properties and Technical Applications. 1 ed., Wiley Series in Renewable Resources. Edited by JÖRG MÜSSIG.

[3] Rahman, S., 2010, "Jute - A Versatile Natural Fibre: Cultivation, Extraction and Processing", in Industrial Applications of Natural Fibres: Structure, Properties and Technical Applications. 1 ed., Wiley Series in Renewable Resources. Edited by JÖRG MÜSSIG.

[4] Winklerprins, A. M. G. A., 2006, "Jute Cultivation in the Lower Amazon, 1940– 1990: An Ethnographic Account from Santarém, Pará, Brazil", Journal of Historical Geography, v. 32, n. 4, pp. 818-838. https://doi.org/10.1016/j.jhg.2005.09.028

[5] Thomas, S., Paul, S. A., Pothan, L. A., Deepa, B., 2011, "Natural Fibres: Structure, Properties and Applications", in Cellulose Fibers: Bio and Nano Polymer Composites – Green Chemistry and Technology. Springer-Verlag. Edited by Kalia, S., Kaith, B. S., Kaur, I. https://doi.org/10.1007/978-3-642-17370-7_1

[6] Ferreira, S.R, Effect of surface treatments on the structure, mechanical, durability and bond behavior of vegetable fibers for cementitious composites. Tese de doutorado, Universidade Federal do Rio de Janeiro, 2016.

[7] Young, J., Mindness, S., Gray, R., & Bentur, A. (1998). The science and technology of civil engineering materials. Saddle River, NJ: Prentice Hall.

[8] Fidelis, M.E.A., Desenvolvimento e Caracterização Mecânica de Compósitos Cimentícios Têxteis Reforçados com Fibras de Juta. Tese de D.Sc., COPPE/UFRJ, Rio de Janeiro, RJ, Brasil, 2014.

[9] ASTM. ASTM C1557-03 Standard Test Method for Tensile Strength and Young's Modulus of Fibers West Conshohocken, PA, USA: American Society for Testing and Materials; 2008.

[10] Silva F. DE A., Toledo R.D., (2013). In: The effect of fiber morphology on the tensile strength of natural fibers, Journal of Materials Research and Technology, 2013;2(2): p 149-157. https://doi.org/10.1016/j.jmrt.2013.02.003

Non-Conventional Materials and Technologies – NOCMAT for XXI Century   Materials Research Forum LLC
Materials Research Proceedings 7 (2018) 391-402   doi: http://dx.doi.org/10.21741/9781945291838-37

Self-Supporting Bamboo Space Structure with Flexible Joints

Luís Eustáquio Moreira [a], Mario Seixas [b, c *], João Bina [b],
José Luiz Mendes Ripper [c]

[a] Universidade Federal de Minas Gerais, Brazil; luis@dees.ufmg.br

[b] Bambutec Design, Rio de Janeiro, Brazil; mario@bambutec.com.br

[c] Pontifícia Universidade Católica do Rio de Janeiro, Brazil; ripper@puc-rio.br

Abstract. This paper presents research results in the structural design and analysis of a self-supporting bamboo space structure. The developed structure presents a flexible connection system and a tensile structural behaviour. The modular frame of the architecture applied hinged lashed connections (HLC) in textile polyester ropes. The modular frame spans 15m width, 4m length and 7,5m high, using *Phyllostachys pubescens* bamboo culms. Nonlinear analysis of the structure under static loadings carried out using the Finite Element Method (FEM) through the SAP 2000 software. The analysis showed that loads induced by strong winds, overloads and self-weight are relatively low for the structural members and the developed connections. The results demonstrate that the self-supporting bamboo space structure meets the requirements of engineering design for safety. This analysis opens a series of another computational analysis calibrated with mechanical tests to determine natural frequencies and damping constant for the structure, demonstrating the potential to be used in earthquake regions.

Keywords: Bamboo, Self-Supporting, Space Structure, Flexible Joints, Engineering Analysis

Introduction

Building with tensile cables is an ancient tradition. Several vernacular designs and construction technologies developed worldwide employ cable structures and textile materials. Traditional dwellings in the desert of Gobi, such as the Mongolian Ger (Yurt), uses a lightweight mobile shelter built with timber, animal skin, ropes and remains updated in Asia until nowadays [1]. Indigenous from Africa and South America developed houses and useful objects, using straw and vegetal rods for structural applications. In Brazil, the *Asuriní* indigenous group from the *Tupi* ethnicity from Medium Xingu, keeps alive the culture of building with timber, straw and earth [2]. The major structure of the *Asuriní* indigenous village is the *Aketé*, measuring approximately 30m long, 12m wide and 7m high, which consists in the main ceremonial house of the village (Fig. 1). The *Aketé* structure is connected by lashed liana fibers and erected by bending timber rods over the structural timbering fixed on the ground, without the use of pins or nails. The *Aketé* roof employs *Piaçava* fiber layers (*Leopoldinia piassaba*) and consists in a bioclimatic roofing technique, protecting to the sun and rain and also allowing the passage of the air. In Brazil, several housing models developed by indigenous groups reflect their ability to build sustainable structures. In Maranhão state, vernacular shelters employing palm tree roofing apply the biodiversity forestry materials, as *Buriti* fibers (*Mauritia spp.*), with a similar joint technique based on lashed straw-based materials (Fig. 2). All over the Brazilian territory it can be observed housing techniques developed by hundreds of indigenous groups based on biomaterials weaving and lashing technologies. Several of them are still alive in Brazil in the present. It is time to recognize and understand the great contribution of the indigenous inhabitants of the South in subjective and objective fields of human being, their technologies and their full adaptability to the tropical climate.

Non-Conventional Materials and Technologies – NOCMAT for XXI Century   Materials Research Forum LLC
Materials Research Proceedings 7 (2018) 391-402   doi: http://dx.doi.org/10.21741/9781945291838-37

FIGURE 1 - INTERIOR VIEW OF THE MAJOR HOUSE OF THE ASURINÍ INDIGENOUS GROUP. PHOTO CREDITS: J. JANGOUX, 1979. FIGURE 2: VERNACULAR SHELTER USING TIMBER AND STRAW

Experimental bamboo structures

Research programs established the physical and mechanical properties of bamboo and its application in engineering. Researches were developed at PUC-Rio in the Civil and Environmental Engineering Department (since 1979), in the Arts & Design Department (since 1985) and at UFMG in the Structural Engineering Department (since 2005). Structural models inspired on the indigenous designs were studied. Several studies developed in the following themes: lightweight bamboo structures, bio-based materials, sustainable technologies and flexible connections [3]. Experimental programs investigated methods to joint bamboo bars combining materials such as wood, steel, cotton, resins and textile cables [4, 5, 6]. Investigations were realized about tensile, tensegrity, pantographic, polyhedral and geodesic structural systems using bamboo and natural fibers. Figure 3 presents a 2V bamboo geodesic structure inspired on the Buckminster Fuller geodesic dome geometry [7] employing nodal steel joints. The dome was developed and tested at Rio de Janeiro in the year of 1992 [8]. These nodal joints concentrated high load stresses in the structural members, favoring shear forces along the bamboo bars. The erection steps were the most critical processes of the dome assembly procedure, as the structural members needed to perform different geometrical positions during mounting steps and the nodal joints transmitted high torsion stresses along the structural members, favoring the generating of cracks in the bamboo bars. Assembly loadings are critical for bamboo structures according to the used connections.

(a)   (b)

(c)

FIGURE 3 - (A) BAMBOO STRUCTURE USING NODAL BOLTED JOINTS; (B) (C) DETAILS OF THE JOINTS DEVELOPED IN STEEL, AGAVE SISALANA ROPES AND BITUMEN

Non-Conventional Materials and Technologies – NOCMAT for XXI Century   Materials Research Forum LLC
Materials Research Proceedings 7 (2018) 391-402   doi: http://dx.doi.org/10.21741/9781945291838-37

In the year of 1999, a bamboo space structure developed using polypropylene textile cables with lashed timber tourniquets joints (Fig. 4). The space structure presented a simplified mounting procedure, avoiding torsion stresses on the bamboo bars. The structure built with flexible joints had an accessible mounting procedure, preserving the mechanical properties of the whole bamboo culms. This flexible joint technique allowed assembly and disassembly procedures with no damage occurring along the bamboo bars. On the other hand, it was observed displacements in the joints and also crushes at the end of the bars induced by diametrical compression bamboo bars stressed by the timber tourniquets, as bamboo bars were used in its cylindrical hollow raw state. Other prototypes were developed employing tourniquet lashed connections. These joints presented the potential to build with bamboo and textile cables with simplified techniques in areas of poor resources, generating non-pollutant and clean mounting processes, preserving the physical and mechanical properties of the bamboo bars.

(a)   (b)

***FIGURE 4** - (A) BAMBOO SPACE STRUCTURE CONNECTED WITH LASHED TEXTILE ROPES. (B) DETAIL OF THE JOINT DEVELOPED WITH TEXTILE ROPES AND TIMBER TOURNIQUETS*

The self-supporting bamboo space structure
The space structure was developed through form-finding physical scale models, to achieve the final architecture of the structure. Several scale models studied allowed design, detailing and planning the assembly procedure. Figure 5a presents a 1:25 scale model of the structure. Flexible materials such as bamboo, tensile membranes, textile ropes and steel cables were used in the development of the bamboo pavilion and its connection system, spanning an area of 15 x 23m and 7,50m height. A novel constructive system was developed combining a hinged bamboo structure and pantographic bamboo grids covered by PVC textile membranes, generating a tensile self-supporting space structure [9]. The structural members of the bamboo pavilion are portable and ultra-lightweight. The self-supporting bamboo structure presented deployable properties and can be assembled in 20 working days with the use of non-conventional materials and technologies. The first built bamboo pavilion erected in Rio de Janeiro in the year of 2013 presented a self-weight of 25 kN, i.e. 7.25 kgf/m^2 (Fig. 5b) [10].

Non-Conventional Materials and Technologies – NOCMAT for XXI Century   Materials Research Forum LLC
Materials Research Proceedings **7** (2018) 391-402   doi: http://dx.doi.org/10.21741/9781945291838-37

(a)   (b)

FIGURE 5 - *SELF-SUPPORTING BAMBOO STRUCTURE AND TEXTILE MEMBRANE PAVILION. (A) 1:25 SCALE MODEL. (B) 1:1 AS-BUILT MODEL*

The bamboo structure minimized the use of heavy foundations and 12 touch-down concrete anchors on the ground supported articulated bamboo bipods. The basic module of the developed space structure is a self-supporting bamboo frame connected with flexible joints using polyester ropes. The modular frame can be subdivided in 2 parts: the superior truss and the inferior bipods. The main structural members of the self-supporting frame are presented in figure 6.

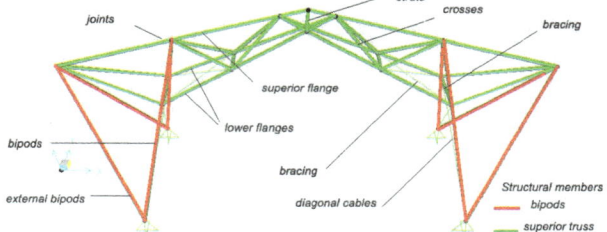

FIGURE 6 - *STRUCTURAL MEMBERS OF THE SELF-SUPPORTING BAMBOO FRAME WITH FLEXIBLE JOINTS*

Geometric optimization processes were developed to find a geometry with even less material and labor involved, generating a self-supporting bamboo frame model variant that was successfully employed during the assembly procedure (Fig. 7a). The optimized self-supporting frame presented in figure 7b continued to be a space truss, with small structural changes that does not change the main structural properties of the structure. The arrangement of the bars presented a slight variation of the first structural design, although the structural properties located near the ridge changed considerably, as the central flexible joint inserted therein disengages the bending moments of the two symmetrical parts of the frame (Fig. 7b). The structural design variation changed somewhat in the final aesthetics of the structure, however, we considered less important the imposition of an aesthetic, than the characteristics of a useful function. In order to understand what non-organizational variations on the structural design are, we propose that the original identity of the self-supporting bamboo frame is maintained and the geometric variation of the structural members of the frame do not change the identity of the structure. The structural members of the optimized geometry are presented in figure 7b.

Non-Conventional Materials and Technologies – NOCMAT for XXI Century   Materials Research Forum LLC
Materials Research Proceedings **7** (2018) 391-402   doi: http://dx.doi.org/10.21741/9781945291838-37

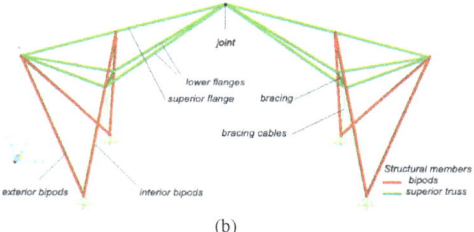

(a)   (b)

FIGURE 7 *- SELF-SUPPORTING BAMBOO SPACE STRUCTURE. (A) HINGED BAMBOO STRUCTURE DURING THE ASSEMBLY PROCEDURE. (B) STRUCTURAL MEMBERS OF THE OPTIMIZED SELF-SUPPORTING BAMBOO FRAME WITH FLEXIBLE JOINTS*

Textile-based flexible joints

Hinged lashed connections (HLC) in textile polyester ropes were developed as presented in figure 8a. Bamboo space structures designed with flexible joints presents simplicity of fabrication of the joints and allow smart assembly procedures, showing advantages in the mechanical operation of the structural members. Hinged lashed connections (HLC) are 4-degree of freedom joints, able to transmit assembly loadings to the structure without generating torsion stresses in the members, consisting in a great advantage in the assembly and operation of the bamboo space structures. HLC present an articulated property and a deployable mechanism for the design of adaptable structures, absorbing special assembly loadings. Instead, special assembly loadings can be critical for bamboo bolted structures.

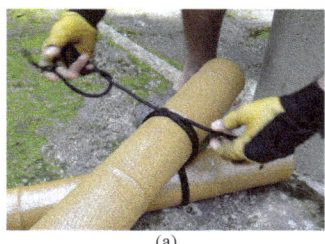

(a)   (b)

FIGURE 8 *- FLEXIBLE HINGED LASHED CONNECTION (HLC). (A) TEXTILE-BASED HLC IN POLYESTER ROPE. (B) HLC DURING THE ASSEMBLY PROCEDURE OF THE SPACE STRUCTURE*

The authors introduced the HLC for bamboo space structures based in the clove-hitch lashing [9]. Composite locking bandages were developed to avoid the sliding of the HLC as presented in figure 9a. The composite locking bandages consist in transverse reinforcements glued on the bamboo bars using cotton fabrics and castor-oil polymer bio-composites [11]. The HLC were braced using textile constraint ropes after the erection process, locking the rotation between bars and interrupting the degrees of freedom of the structural members. Figure 9b shows the bipod column with the higher loading for the structure submitted only to gravity actions.

Non-Conventional Materials and Technologies – NOCMAT for XXI Century   Materials Research Forum LLC
Materials Research Proceedings 7 (2018) 391-402   doi: http://dx.doi.org/10.21741/9781945291838-37

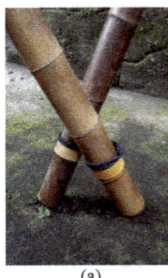 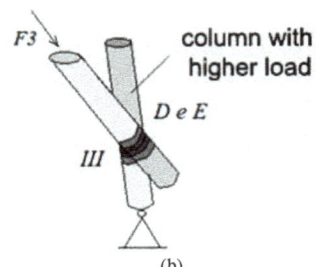

(a)   (b)

FIGURE 9 - ANALYSIS OF THE FLEXIBLE JOINTS. (A) HLC WITH COMPOSITE LOCKING BANDAGES GLUED ON THE BAMBOO BARS. (B) LOCAL ANALYSIS OF THE BRACED HLC APPLIED IN THE BIPOD JOINTS

Structural analysis

The previous experience of the authors in the design of bamboo lightweight structures is programmed for the object and its use in the environment. The present research outlined a working method based on experimental physical models and experimental computer models in interaction. In the present analysis, the structure was modeled as perfect trusses and eccentricities were not considered. The structural analysis used the Finite Element Method (FEM) SAP 2000 software [12]. The bamboo species *Phyllostachys pubescens* bars were sized with 90mm diameter and 9mm wall thickness. The frame dimensions are shown in Figure 10.

Wind loadings

Considering that the structure is built in the city of Belo Horizonte, Minas Gerais, Brazil and according to the isopleths mapped on the Brazilian map ABNT (NBR 6123), 1988 [13], the basic wind velocity v0=32m/s. Then, the characteristic speed of the wind is given by Eq.(1):

$$vk = v0 \times S_1 \times S_2 \times S_3 \quad (1)$$

The topographic factor S1=1 was used for a flat ground distant from hills and slopes. S2 is the roughness factor, function of the height "z" of the building, of the maximum dimensions of the frontal façades exposed to the wind load, divided into classes A, B and C; and also because of the obstacles that are located in the vicinity of the building, which would be the roughness proper, whose environments are divided into 5 categories, being category 1 for environment without obstacles and category 5 for environments with many high obstacles. For example, the maximum dimension of the plant or vertical building is 20m < 28m ≤ 50m ⇒ class B. The height of the building z = 6.0m > 5m. Category IV was adopted because of the average distance of 12m from the top of the obstacles.

Interpolating in ABNT Table 2 (NBR 6123), we obtain $S2 = 0,77$.

For buildings for commerce and industry with high factor of occupation has S3=1. Then, the characteristic wind speed will be

$$v_k = 32 \times 1 \times 0,77 \times 1 = 24,6 \; \frac{m}{s}$$

And the characteristic wind pressure in kN/m^2 will be

Non-Conventional Materials and Technologies – NOCMAT for XXI Century   Materials Research Forum LLC
Materials Research Proceedings **7** (2018) 391-402   doi: http://dx.doi.org/10.21741/9781945291838-37

$$q_k = 6{,}13 \times 10^{-4} v_k^2 = 6{,}13 \times 10^{-4} \times 24{,}6^2 = 0{,}37 \frac{kN}{m^2}$$

Aerodynamic coefficients for wind loading at 0° angle with the axis of the structure are presented in Figure 11a. The analysis considered by hypothesis, there is no risk of overturning during high wind loads and equal permeability is considered in all façades.

Then, the geometric dimensions of the structure are:
a = 28m (highest plant length)
b = 9,2m (lower plant length)
h = 4,21m (wall height)
z = 6,04m (maximum height of the building)

- $\frac{h}{b} = \frac{4{,}21}{9{,}20} = 0{,}46 < 0{,}5$

- $2 < \frac{a}{b} = \frac{28}{9{,}2} = 3{,}1 < 4\ m$

- $\frac{b}{3} = \frac{9{,}2}{3} = 3{,}07 ; \frac{a}{4} = \frac{28}{4} = 7{,}0\ m ; 2h = 8{,}42\ m$. Then, the depth of the higher wind suction load will be $p = 6.0$m

- Calculation of the angle $\theta = \tan^{-1}\left(\frac{6000-4211}{6647}\right) = 15{,}06°$

The depth **p** for wind suction at 90° is the smallest value between

$$2h = 2\times 4{,}21 = 8{,}42m \text{ and } \frac{b}{2} = \frac{9{,}2}{2} = 4{,}6\ m \therefore p = 4{,}6\ m$$

Thus, the aerodynamic coefficients for the wind loading at 90° angle with the axis of the structure are shown in Figure 11b.

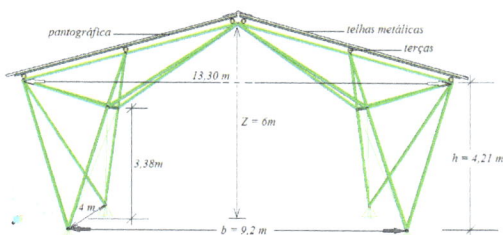

FIGURE 10 - *DIMENSIONS OF THE SELF-SUPPORTING BAMBOO FRAME*

Non-Conventional Materials and Technologies – NOCMAT for XXI Century   Materials Research Forum LLC
Materials Research Proceedings 7 (2018) 391-402   doi: http://dx.doi.org/10.21741/9781945291838-37

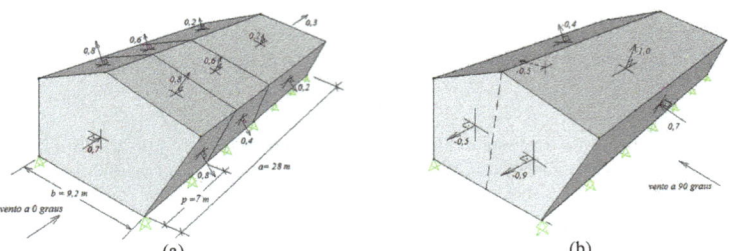

FIGURE 11 - *AERODYNAMIC COEFFICIENTS FOR WIND LOADING (A) WIND LOADING AT 0°. (B) WIND LOADING AT 90°*

The analysis considered the hypothesis of metal roof tiles weighting 0,05 kN/m^2 above the pantographic bamboo grid. The roof overload was 0,25 kN/m^2. The loads were combined according to the recommendations of the ABNT standard NBR8681 [14]. For normal loads, i.e. those that only have loads resulting from normal use (permanent loads, overloads and wind loads applied in the structure in the enduring situation of design and use) or the same as explain that the loading does not contain any special loading, or any loading corresponding to the assembly (these are considered special loads), nor any loading of exceptional nature or magnitude. The following combinations of loadings were considered for ultimate limit states, which by hypothesis would lead to the greatest stresses in the structure static loading. The analyses were realized applying 4 load combinations. The loading patterns were expressed in the equations as follows.

Loading hypotheses
Wind loads at 0° angle with the axis of the structure:
- *COMB1: $F_d = 1,4G + 1,4Q + 1,4Wc \times 0,6$* (live load as the main variable load)
- *COMB2: $F_d = 1,4G + 1,4Wc + 1,4 \times Q \times 0,7$* (wind pressure as the main variable load - not critical in relation to COMB1).
- *COMB3: $F_d = 1,4Wa - 0,9G$* (wind suction as the main variable load)

Wind loads at 90° angle with the axis of the structure:
- *COMB4: $F_d = 1,4Wa - 0,9G$* (wind suction as the main variable load)

As presented in the previous example, in these equations we have G for permanent loads, Q for live loads, Wc for wind pressure and Wa for wind suction. The unfavorable wind is considered as a resulting internal suction, an internal pressure coefficient $cpi = -0,3$; neglecting the coefficients of external suction. When we have equal permeability on all façades, the norm recommends $cpi = -0,3$, considering the overlap with the external effects given by the coefficients c_e. However, since the internal coefficients cpi can reach -0,9, depending on whether there are dominant openings due to accidents during high winds; it is recommended in practice to consider a $cpi = -0,3$ as the minimum result because of safety.

The probabilistic weighting coefficients of the actions are given by ABNT NBR8681, for normal, special and exceptional loads. The coefficient of combination of the secondary variable action is obtained from the same standard, $\psi 0 = 0,7$ for accidental loads in buildings with a high factor of occupation or presence of fixed equipment. Similarly, $\psi 0 = 0,6$ for dynamic wind pressure. Figure 12 presents the loads combined and the applied node loading of the combinations. The nodal stresses for each of the combinations are shown in Figure 12. COMB1 combination is critical to COMB2, reason why COMB2 was not used for the structural analysis.

Non-Conventional Materials and Technologies – NOCMAT for XXI Century   Materials Research Forum LLC
Materials Research Proceedings 7 (2018) 391-402   doi: http://dx.doi.org/10.21741/9781945291838-37

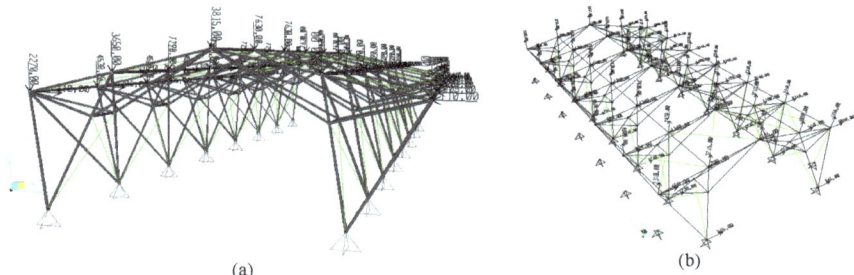

(a)

(b)

FIGURE 12 - (A) NODAL LOADS FOR COMB1 (N). (B) NODAL LOADS FOR COMB4 (N)

Wind loads at 0° for COMB1 and COMB3 applied in the structure are shown in figure 13.

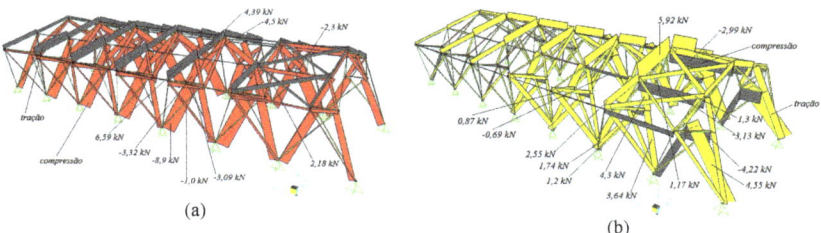

(a)

(b)

FIGURE 13 - (A) AXIAL LOADS FOR COMB1. (B) AXIAL LOADS FOR COMB3

Figure 14 presents the bending moments and shear forces in COMB4 for wind suction due only to the weight of the bars and not considered eccentricity in the bars, which means that they are totally negligible. Moment diagrams are not parabolic is a problem because the graphic output. The magnitudes of the analyzed applied loadings are consistent with the considered hypotheses.

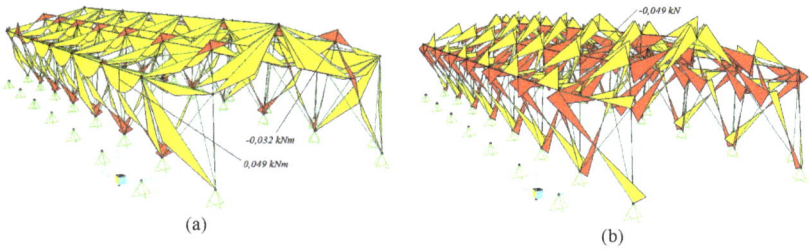

(a)

(b)

FIGURE 14 - (A) BENDING MOMENTS FOR COMB4. (B) SHEAR FORCES FOR COMB4

The axial forces for the combinations COMB1 and COMB3 are perfectly absorbable by the compressed members with the lengths of the designed bars. The higher compression forces of the upper flanges came from COMB4. In this case, the superior flanges and the bipods may require bracing for the reduction of buckling length, as showed during the analysis. Therefore, this

Non-Conventional Materials and Technologies – NOCMAT for XXI Century   Materials Research Forum LLC
Materials Research Proceedings **7** (2018) 391-402   doi: http://dx.doi.org/10.21741/9781945291838-37

consideration should adopt the structural design for the superior trusses and the bipods, as presented in figure 15.

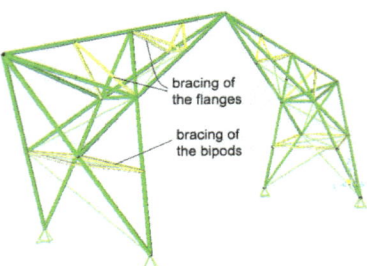

FIGURE 15 - OPTIMIZED SELF-SUPPORTING FRAME WITH BRACINGS IN THE SUPERIOR TRUSS AND IN THE BIPODS

The authors observe that the structure using this reduced number of continuous bars can only be safe using the developed flexible HLC in textile ropes or in bio-composite bandage joints [15]. Bolted connections would require interruption of bars and would need to use more bolts to absorb the stresses applied in the structural members. A single bolt used to the connection of the bars would not be sufficient for the transmission of the forces. Besides, the perforations that pass bolts are points of weakening and stress concentration in the bars, and certainly would increase the risk of collapse of the structure in earthquake regions.

Finally, for fixing the bipod column with the higher applied load for wind suction, figure 16 presents the map of the loads of the anchors reaction forces. In the supports, we will then have horizontal and vertical forces for wind suction. The higher loads for wind suction are presented in figure 16 and occurred for COMB4.

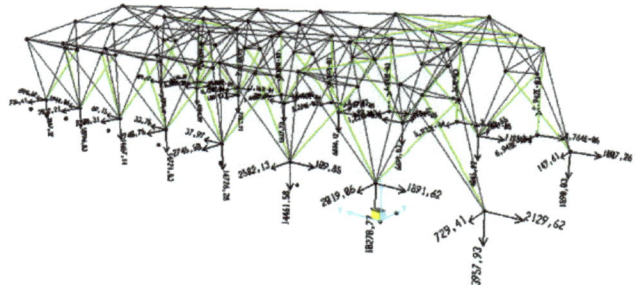

FIGURE 16 - ANCHORS REACTION FORCES (N) FOR COMB4.

Conclusions

The present paper introduced a self-supporting bamboo space structure for engineering applications. The observation and the monitoring of these novel structures in service will present a greater understanding and reliability of the designed ultra-lightweight structural system with flexible joints. The structural analysis shows that the forces are relatively low in the structure and do not represent any safety problem. The developed hinged lashed connection (HLC) have the advantage of not introduce torsion stresses in the bamboo bars. These joints also avoid the weakening of the bamboo bars with perforations, as in the case of bolted connections. Bolted joints in addition to local weakening, introduce moments of torsion in the bars through these perforated points of weakening, which is a structural disadvantage. The question of the self-supporting bamboo structures engineering design for safety is, therefore, not the difficulty or feasibility of the structural analysis, but a correct choice of the materials, the design and the detailing of the joints. A disadvantage would be the sliding of the connections along the bars, and the stability of the system due to the observed displacements of the joints. Mechanical tests of the connections and numerical modeling of the textile HLC were developed. The study considered the limits of resistance and local displacements of the joints, leaving for a further work the evaluation of the deformations and possible consequences for the global operation of the structure under dynamic analysis, since the flexible joints decrease the stiffness of the structure, thereby decreasing their natural frequencies of vibration.

Acknowledgments

The authors would like to thank the Fundação Carlos Chagas Filho de Amparo à Pesquisa do Estado do Rio de Janeiro FAPERJ for the support of this research project. Our special thanks are to the Emeritus Prof. Khosrow Ghavami from PUC-Rio for his advice in the present research.

References

[1] Khan, L.; Easton, B. Cobijo. Shelter Publications. Madrid: H. Blume Ediciones, 1979.

[2] Ribeiro, B. (org.); Ribeiro, D. (ed.) Suma etnológica brasileira. Updated edition of Handbook of South American Indians. Volume 2: Tecnologia indígena. Petrópolis: Ed. Vozes, 1987.

[3] Moreira, L.E.; Ripper, J.L.M. Jogo das Formas: Lógica do objeto natural. Rio de Janeiro: Nau Editora, 2014.

[4] Moreira, L.E.; Ripper, J.L.M. Métodos de ensino de design de produtos e sua aplicação às estruturas da engenharia civil. In: Congresso Brasileiro de Ensino de Engenharia COBENGE. Brasília, 2004.

[5] Ripper, J.L.M.; Moreira, L.E.; Silva, M.F. Desenvolvimento de estruturas auto-tensionadas de bambu no LOTDP. Agenda Pública: Drama Social. Faperj, Rio de Janeiro, RJ: 1999. P267-276.

[6] Seixas, M.A. Inserção social de arquiteturas temporárias de bambus e lonas têxteis utilizando tecnologias não-convencionais. Msc. Dissertation, Arts & Design Department. Rio de Janeiro: PUC-Rio, 2009.

[7] Fuller, R.B. Tensile-integrity structures US Patent 3063521 A, 1962.

[8] Ripper, J.L.M.; Moreira, L.E.; Ubésio, A. Cúpula Geodésica de Bambu. Proceedings of the V Encontro Brasileiro em Madeiras e em Estruturas de Madeira. Belo Horizonte, EEUFMG 1995. 12p.

[9] Seixas, M.A.; Ripper, J.L.M.; Ghavami, K. Prefabricated bamboo structure and textile canvas pavilions. Journal of the International Association for Shell and Spatial Structures, Vol. 57 n° 189, 2016, pp. 179-188. https://doi.org/10.20898/j.iass.2016.189.782

[10] Bambutec Design. www.bambutec.com.br (acessed on 22 September 2017).

Non-Conventional Materials and Technologies – NOCMAT for XXI Century   Materials Research Forum LLC

Materials Research Proceedings 7 (2018) 391-402   doi: http://dx.doi.org/10.21741/9781945291838-37

[11] Seixas, M.; Bina, J.; Stoffel, P.; Ripper, J.L.M.; Moreira, L.E.; Ghavami, K. Active bending and tensile pantographic bamboo hybrid amphitheater structure. J. IASS, Vol. 58 n° 193, 2017. https://doi.org/10.20898/j.iass.2017.193.872

[12] Computers and Structures Inc. CSI Analysis Reference Manual for SAP, 2015.

[13] ABNT. NBR 6123: Forças devidas ao vento em Edificações. 1988.

[14] ABNT. NBR 8681: Ações e Segurança nas Estruturas. 2003.

[15] Lanna, C.A.C.; Moreira, L.E. Treliças Planas de Bambu com Bioconexões Compósitas. In: II Congresso Luso Brasileiro de Materiais de Construção Sustentáveis, João Pessoa, PB, 2016.

Non-Conventional Materials and Technologies – NOCMAT for XXI Century   Materials Research Forum LLC
Materials Research Proceedings 7 (2018) 403-414   doi: http://dx.doi.org/10.21741/9781945291838-38

Living Form Finding of a Bamboo Modular Shelter Covered by Textile Composite

Sophie Madeleine Jung[a], Luís Eustáquio Moreira[b]*, Carlos Henrique Ferreira[b]

[a]Ecole Supérieure du Genie Urbain, EIVP, France; sophiejung12@gmail.com

[b]Escola de Engenharia da UFMG, Brazil; luis@dees.ufmg.br*, khenriquef3@gmail.com

Abstract. This paper presents the design of a modular bamboo icosahedron shelter covered by textile composites. The lightness of bamboo, the manufacturing techniques and the natural materials used, favor the finding of new constructive paradigms, as new connections and new coverings. The shelter is characterized by a spatial truss made of bamboo bars which corresponds to the icosahedron edges and a composite cotton shell as covering. Assembling connectors sustain the spatial geometry and the final connection wraps the first one through bandages of cotton fabric, hardened with polyurethane castor resin, named Cotton Composite Connection, or 3C connection, due to innovation. The truncated icosahedron consists of 15 equilateral triangular contours. The coverings consists of triangular cotton textile stretched over these bamboo triangles, and painted with castor polyurethane resin, named Cotton Composite Shell or 2CS shell, for also being an innovation. These considerations resulted in a resistant structure with minimum waste, in addition to low energy consumption and low environmental impact for manufacturing. The coupling of Form Finding and Living Design Methods, the Living Form Finding Method – L2F Method was the base of the development and it is explicit in the text. A structural analysis shows that forces and stresses are relatively low for the used materials conducting to a light, serviceability and safe shelter.

Keywords: Bamboo Modular Shelter, L2F Method, 3C Connections, 2CS Shell

Introduction
In these years we have been betting on a sustainable use of cultural products, including bamboo pipes in the Western culture of objects, we can distinguish 3 basic categories of bamboo structures:

> ➤ Bamboo Traditional Structures
> ➤ Bamboo Heavy Structures
> ➤ Bamboo Living Structures

The Bamboo Traditional Structures that we witness in the furniture or baskets and even in the work of building constructors are based on vernacular and primitive techniques. They are characterized by using natural fibers such as rattan or vines, for example, to tie the different segments of stems, or braids of slats and tapes, whose most beautiful buildings, which are true masterpieces, are found in Indonesia. They are also characterized by a careful elaboration of each constructive element and decorative forms that refer to another time or place. These solutions are incompatible with the timed industrial process of construction, and with the general workmanship already impoverished by the dismantling of constructive traditions and even by the elimination of apprentice workshops of the pre-industrial societies. In large Brazilian cities, handicraft has survived the margins of the industrial system, with their specific public spaces invaded by industrial products. Bamboo Heavy Structures are more recent stage in construction, initiated in mid-1970s of the twentieth century. They are characterized by bolted connections and ceramic tile roofing, highlighted by the significant architecture of the Colombian architect Vellez [1].

Non-Conventional Materials and Technologies – NOCMAT for XXI Century   Materials Research Forum LLC
Materials Research Proceedings 7 (2018) 403-414   doi: http://dx.doi.org/10.21741/9781945291838-38

Bamboo Living Structures are academic products of experimental studies of construction and Industrial Design of the Product. It differs from the others because more than the search for new constructive paradigms, they rationally manage the inputs of industrial societies. It is conditioned by the needs of damping the environmental impacts through the involvement of the engineers, architects, designers and builders with the local productive flow. The objective is to utilize all available resources in the generation of the utility objects.

Although they are handcrafted, Bamboo Living Structures differ from Bamboo Traditional Structures because they replace the props of these by principles of better physical and mechanical functioning of the structure in its interaction with the environment with a reliable technological development. New forms for bamboo structures are finding with a differentiated architecture. Bamboo as a tube with the favorable characteristics of being ultra-light, super resistant, in structures and in many objects substituting tubes of other materials, such as steel, aluminum and plastics. Bamboo is ultra-flexible, serviceable, renewable and workable by simple tools. Therefore, the taboo of a constructive purism is broken. Excessive details of traditional connections with fittings are greatly diminished even sometimes eliminated. Bamboos can be painted to be protected from UV rays and fungus, while the color stands out the unique and smart shape of the bamboos.

a) Tied spatial systems   b) Plane systems with 3C Connection
FIGURA 1 - FLEXIBLE BAMBOO CONNECTIONS

Bamboo Living Structures differ from Bamboo Heavy Structures by visible decrease in the number of bars and the type of coverage and closure. The first Bamboo Living Structures also appeared with bolted connections, but already with other concepts and reduction of fittings, Moreira and Ghavami [2]. The bolted connections, have been replaced by synthetic or cotton ropes, Seixas et al [3] as presented in Figure 1a and by Cotton Composite Connections – 3C connection – Lanna and Moreira [4] shown in Figure 1b. All these structures are characterized by the low weight and reduction of the excessive number of elements observed in the Heavy Structures in general, which makes them economical and easy to preserve.

The structure proposed in this research was conducted by the Living Design Method historically designed in LILD - Laboratory of Investigations in Living Design of PUC-Rio and in the LASE - Laboratory of Structural Systems of EEUFMG, which largely resembles the Form Finding Method of the Institute for Lightweight Structures from Stuttgart, Germany, to achieve optimum shape through the frequent participation of reduced physical models and prototypes, Otto [5].

Living Design and Form Finding Methods
The LIVING DESIGN and FORM FINDING methods are very similar in the procedure of creating the object from the direct dealing with materials in reduced physical models and in essence have a simple algorithm, similar to that proposed by the total quality programs, adapted here from Dynamics of the Knowledge, Silva e Ferreira [6]:

Non-Conventional Materials and Technologies – NOCMAT for XXI Century   Materials Research Forum LLC
Materials Research Proceedings **7** (2018) 403-414   doi: http://dx.doi.org/10.21741/9781945291838-38

The input in the algorithm of Figure 2, in the ideation rectangle, is done mainly with computational models, from the pilot scheme, (Figure 3). They are followed by physical models, which prevail over the computational models. So, the methods confer greater freedom of participation to the materials and accessories available in the market today. This is the goal: to be side by side with the materials, accessories, tools and machines throughout the development.

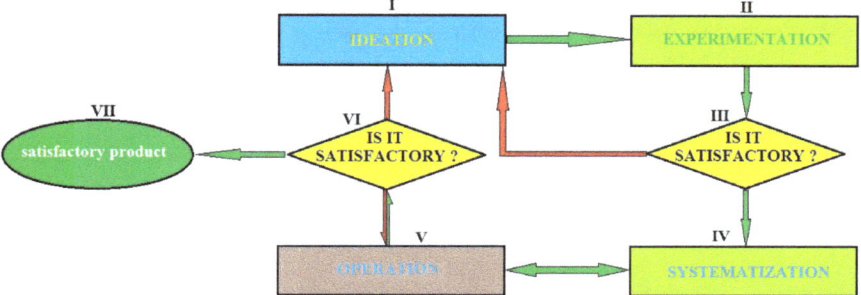

FIGURA 2 - *ELEMENTARY FLOW CHART OF THE FORM FINDING AND LIVING DESIGN METHODS*

The Figure 2 flowchart is complemented by flowchart of the ideation, Figure 3, for the solution of bamboo structures. The ideation product is the Pilot Scheme. These engineering assumptions and related analyzes are the Computational Experiments. Then the first entry in Figure 2, relative to the flow of both methods, is relative to Computational Experiments whose sufficient product will be a Structural System with dimensions and appropriate configuration. This step should not be extremely detailed, it is not an analysis of the object, because it has not yet been discovered in all its details, it is still in search of this object. Then, in possession of the Pilot Structural System, Figure 4, the Physical Experiments are initiated and so the flow chart in Figure 2 is repeated as often as necessary. The physical models are developing in stages - windows - and this segmentation of the productive process and its techniques extends the understanding of the object.

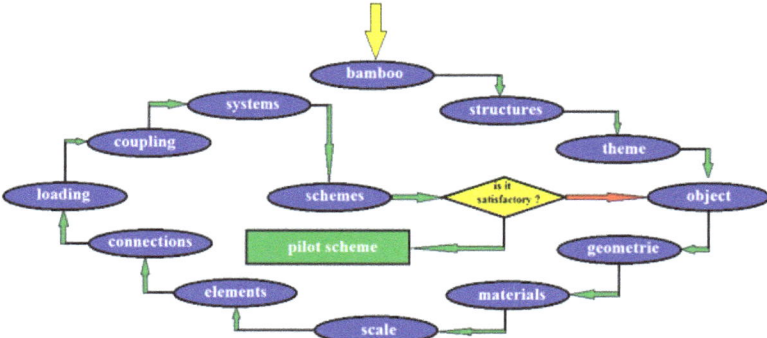

FIGURE 3 - *CIRCUIT OF IDEATION – THE GENERATION OF THE PILOT SCHEME*

Non-Conventional Materials and Technologies – NOCMAT for XXI Century   Materials Research Forum LLC
Materials Research Proceedings 7 (2018) 403-414   doi: http://dx.doi.org/10.21741/9781945291838-38

From the theme of developing a small shelter with bamboo bars, which could have urban applications such as magazine stands, bus stops, bungalows, among others, it was chosen the truncated icosahedrons structure by its favorable characteristics, Figure 4 :

- ✓ Engage a good internal volume
- ✓ Can be easily produced: 15 bars of equal length
- ✓ To allow the use of bamboos of small diameter, average diameters of 40 mm, abundant in the Brazilian rural areas.
- ✓ Admit different forms of covering (straw, membrane, shell, tiles)
- ✓ Because it is modular with faceted geometry of easy coupling to other modules.

Then, bars of 2,23 *m* in length, bamboo of the species *Bambusa tuldoides* were harvested already dried in unmanaged bamboo bush. Dry and non-attacked insect bars are rarely attacked later. Although the structural bamboos must be mature - more than 3 years in activity in bamboo bush -, there was no age control of the bamboos.

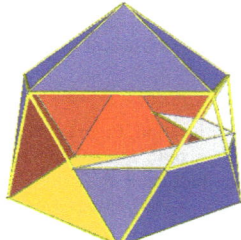

a)   Pilot scheme   b) Structural system - dimensions
FIGURE 4 - THEME – TRUNCATED ICOSAHEDRA SHELTER SEALED WITH TEXTILE COMPOSITE SHELL

First Computational Experiments
The length of the bars was set to give an entry height $Z = 1,9$ m. With the length of 2,23 m, the bamboos, which can be calculated from the diameter of the middle section and average wall thickness of the bar, ISO/DIS 22156 [7], have the following geometric properties:

- Cross-sectional area: $A \cong \pi \bar{D} t = \pi \times 4 \times 0,5 = 6,3 \ cm^2$
- Moment of inertia: $I \cong \pi \bar{R}^3 t = \pi \times 2^3 \times 0,5 = 12,57 \ cm^4$
- Slenderness of the bar: $\lambda = \dfrac{l_0}{\sqrt{\frac{I}{A}}} = \dfrac{223}{\sqrt{\frac{12,57}{6,3}}} = 158 < 250 \rightarrow ok!$

Since the mean modulus of elasticity of this species is 20 *GPa* under compression, each bar would have an Euler load given by

$$F_E = \frac{\pi^2 EI}{l_0^2} = \frac{\pi^2 \times 2000 \times 12,57}{223^2} = 4,99 \ kN$$

These bamboos can be selected with accidental imperfection of the axis equal to $e_a = \frac{l_0}{200} = \frac{223}{200} = 1,12 \ cm$. Let Eq. 1 give the total imperfection of the axis from the accidental imperfection.

$$\delta = \frac{e_a}{1 - \frac{P}{F_E}} \qquad (1)$$

From Eq. 1, for a maximum lateral displacement $\delta = 2e_a$, a force $P_d = \frac{F_E}{2} = 2,5 \ kN$ is obtained. The maximum compressive stresses in the middle of the bar would be equal to

$$\sigma_c = \frac{P_d}{\pi \overline{D} t} + \frac{P_d \delta_T D}{2I} = \frac{2500}{\pi \times 40 \times 5} + \frac{2500 \times 22,2 \times 50}{2 \times 12,6 \times 10^4} = 15,0 \ MPa < 22,4 \ MPa \ \rightarrow ok!$$

Evaluation of the pilot scheme (was it satisfactory?)-In each place that the shelter is placed there will be different wind loadings. Being ultra-light these structures, the own weight does not present problem. Similarly wind effects depend on where the shelter is placed. In this way one can continue the development and subsequently limit the conditions of use or even make some design changes for adequate absorption of the specific loads. So, the structure with the proposed dimensions is feasible.

Experiments on Reduced Physical Models
From the Pilot Scheme and computational pre-evaluation of the bars the Living Design and   Form Finding Methods turn to Physical Experiments in physical models, which become the organizing field where the whole game of discoveries of forms and techniques will take place, Moreira and Ripper [8] and who waves for the proper development of the product.

Reduced model assembler connector- Bamboos of 1 *cm* of average diameter of the species *Phyllostachys aurea* were selected for the reduced model. PET bottle segments were cut 6 *cm* long and 1,5 *cm* wide, after which they were fixed to each other by eyelets, Figure 5.

FIGURE 5 -ASSEMBLER CONNECTORS - PET TAPE CROSSPIECES WITH CENTRAL EYELETS

By folding the tapes into a tube and inserting them into the hollow of the bamboo ends, the architectural assembly of the icosahedrons is achieved, Figure 6. We can add that the length (6 *cm*) has been chosen to generate enough friction with bamboo wall, to avoid pull out effect especially during the montage of the structure.

Non-Conventional Materials and Technologies – NOCMAT for XXI Century   Materials Research Forum LLC
Materials Research Proceedings 7 (2018) 403-414   doi: http://dx.doi.org/10.21741/9781945291838-38

a) reduced model   b) bandage manufacture

FIGURE 6 - ASSEMBLER PET CONNECTORS SUPPORT THE 3D GEOMETRY

The idea is that the final connections are bandages of cotton fabrics saturated in polyurethane resin based on castor oil, the 3C connection. Then, 2 *cm* wide of fabric bands are immersed in resin for few minutes. The resin excess is removed and small rolls are made to facilitate the unwinding of the bandage around bamboos, Figure 6b). Some tension is given to the band as it is tightened around bamboos.

Evaluation of the model's mounting connector: (was it satisfactory?)-A low torsion resistance of the PVC strips was found, as expected, without however preventing the model from being mounted with relative ease and leading to the expected result. The prototype mounting connector may then follow the same crosshead principle but will have to be more resistant due to the larger weight and larger dimensions of the elements.

Reduced Model Closure

In this development, in continuation to the searches of Dias [9], the sealing with cotton composite shells was tried. A layer of square mesh cotton cloth was stretched over the model and then the fabric was painted with polyurethane resin both sides, avoiding excess and formation droplets and bubbles, Figure 7.

Evaluation of the shell (Was it satisfactory?) - The result was satisfactory. Because the bleached cotton fabric was well stretched over the structure the weight of the resin did not cause bulges. There is doubt about the prototype, if the weight of the resin will cause excessive bulging into the fabric, to the point of damaging the good functioning or object aesthetics.

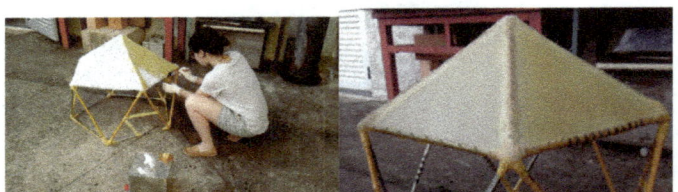

FIGURE 7 - FREE FORM FINDING OF THE 2CS SHELL

Computational Experiments for Prototype Assembly

The prototype mounting connectors are subject to the following forces due to own weight of the structure, Figure 8.

Non-Conventional Materials and Technologies – NOCMAT for XXI Century   Materials Research Forum LLC
Materials Research Proceedings 7 (2018) 403-414   doi: http://dx.doi.org/10.21741/9781945291838-38

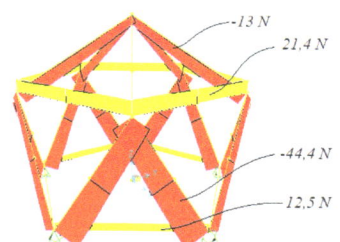

FIGURE 8 -*AXIAL MOUNTING FORCES*

A new connection of PET bottles was developed as a prototype assembly connector, after an experiment with green bamboo bars, which waved for further investigation in that direction, Figure 9a). However, as soon as the top apex of the prototype, Figure 9b) was started, with the thermal blower at a temperature sufficient to tighten the PET bottles, a problem occurred. Although for dried bamboos, the tension resistance perpendicular to the fibers was high and could reach 4 *MPa* on average, the temperature gradients and the geometrical bamboos irregularities gave rise to even greater stress gradients, cracking the bamboos.

a) Pilot experiment   b) Prototype experiment
FIGURE 9 - *TEST OF THE PROTOTYPE ASSEMBLAGE CONNECTOR*

With the elimination of the possibility of heating the PET without damage of dried bamboos, the idea arriving bamboo bars at the same point had to be aborted. That is, the Living Form Finding Method has led to the more economical solution like the one in which bamboos do not converge to the same point. This solution was developed in LILD and has been successfully applied in several types of innovative structures by BAMBUTEC - Rio, Figure 10. Following are important beacons placed at this moment of decision by the method:

- Bars reaching the same point make it very difficult to assemble and fabricate the bamboo space trusses, although they reduce the forces on the bars.
- Bars arriving at the same point necessarily lead to more expensive solutions because they require more sophisticated assembly connectors of the type developed here because they must absorb tension forces of assembly. Besides, these connectors will be lost elements within the composite connection.
- The holes that appear in the twirl at the meeting point of the bars, Figure 11, should allow the hand of the designer to pass through to make the composite connection, a requirement not observed in Figure 10, where the connections are only tied.

Non-Conventional Materials and Technologies – NOCMAT for XXI Century   Materials Research Forum LLC
Materials Research Proceedings 7 (2018) 403-414   doi: http://dx.doi.org/10.21741/9781945291838-38

FIGURE 10 - *BAMBUTEC BUNGALOWS WITH TIED CONNECTIONS AND WAXED COTTON MEMBRANE COVERS – CMC COVER*

The nodes in twirl, Figure 11, gives rise to bending moments and shear forces in the bars, however of negligible values. Figure 11 shows the new sequence of the Living Design: the mooring that in Figure 10 is the final solution becomes a connector of assembly with ropes of smaller diameter, just enough to arm the architecture, using a mooring known as Reliable Node, Moreira and Ripper [8]. This tie, besides giving a good grip to the bars, avoiding easy slip under their own weight, forms a kneecap between them, relieving the local mounting efforts. The tissue bands were cut 10 *cm* wide and 1 *m* long. Resin absorption of 450 g / m^2.

FIGURE 11 - *PROTOTYPE ASSEMBLING WITH LASHINGS*

Non-Conventional Materials and Technologies – NOCMAT for XXI Century   Materials Research Forum LLC
Materials Research Proceedings **7** (2018) 403-414   doi: http://dx.doi.org/10.21741/9781945291838-38

FIGURE 12 - COMPLETE PROTOTYPE WITH 3C CONNECTIONS AND 2CS SHELL AS COVERING

The Living Form Finding Method – L2F Method – led to a satisfactory prototype, as proposed in the beginning of the search. Then the engineering forecasts for the application context of the shelter.

Engineering forecasts
Two loading hypotheses were considered in order to estimate the applicant efforts in the in-service structure. An exceptional load corresponding to a load of P_k = 1 kN hanged on one of the vertices, Figure 13, and a normal loading with actions of its own weight, overload and winds with characteristic speed v_k = 18,5 m/s.

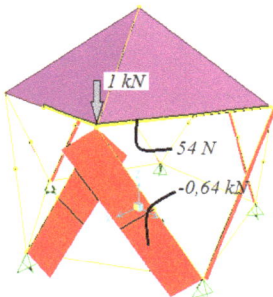

FIGURE 13 - AXIAL FORCES FOR A LOAD FK = *(1 KN + PERMANENT WEIGHT G_K)*

Figure 14 shows the resulting pressures estimated according to ABNT(NBR 6123) [10] considering the overlapping of the external aerodynamic coefficients with internal pressure coefficients of extreme values, *cpi* = +0.8 (internal pressure coefficient) and *cpi* = - 0.9 (internal suction coefficient). We have obtained the results for each case, in which in Figure 14b) a view was taken below the cover.

Non-Conventional Materials and Technologies – NOCMAT for XXI Century   Materials Research Forum LLC
Materials Research Proceedings **7** (2018) 403-414   doi: http://dx.doi.org/10.21741/9781945291838-38

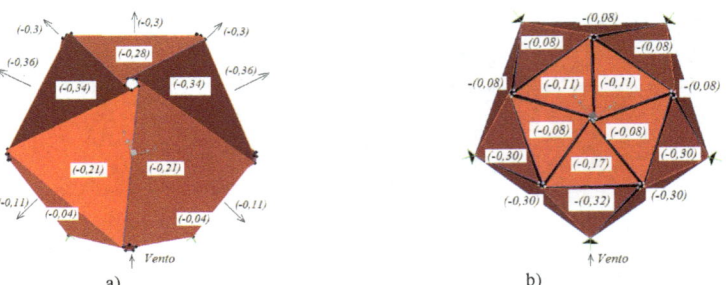

FIGURE 14 - *A) RESULTING PRESSURES* $\left(\frac{kN}{m^2}\right)$ *A)CE + CPI (+0,8) (EXTERNAL VIEW);*

B) CE + CPI = -0,9 (INTERNAL VIEW)

These loadings result in the axial forces of Figure 15a) and 15b) for each case as well as the resulting support reactions against wind that could drag and fly the shelter. The pulling forces are relatively low Fwk = - 0.74 kN ; and are easy to be counterbalance.

In all the analyzed loads, the maximum compression loads are still very low for the bamboo considered, as well as for the connections according to the results of Moreira and Lana [4], even if these values are multiplied by 1,4. The compressive forces remain below the maximum compression force F_d = 2,5 kN.

The structure should be properly anchored so that it is not drag out by the wind, but the structure do not suffer the slightest damage with the normal and exceptional loading considered.

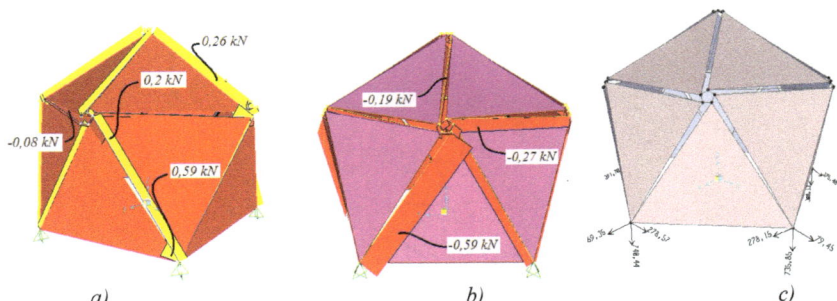

FIGURE 15 - *A) AND B) AXIAL FORCES F_{WK} (KN) CORRESPONDING TO WIND LOADING; AND C) SUPPORT REACTIONS TO DRAG WINDS*

Conclusion

With the Cotton Composite Connections - 3C - the small-diameter bamboo structures win the paradigm of pins and bolts, getting closer to the existing connections in nature where there are no pins. With the 3C connections, the Living Design Method of three-dimensional bamboo structures with small diameter is obtained without requiring precision of the bars lengths and jigs or sophisticated equipment and machines, making the manufacture of the most different cross-linked architectures, all of which fulfilling the requirements of resistance and usability of contemporary engineering. The 3C connections significantly decrease the relative displacements between the rods at the nodes, with respect to the rope lashings, precisely by the presence of the resin cementing the

Non-Conventional Materials and Technologies – NOCMAT for XXI Century   Materials Research Forum LLC
Materials Research Proceedings 7 (2018) 403-414   doi: http://dx.doi.org/10.21741/9781945291838-38

fabric and the mounting lashings. In the specific case of gyre connections, each bar is fixed at the end by two points, causing the boundary conditions of the bars to approach the crimping, with consequent increase of the compressive load limit. In this sense, all the gyre are performed in the same direction at each vertex of the structure, either clockwise or counterclockwise, it is the best solution, in the sense of contributing to generate a point of inflection in the compressed bar, with consequent increase of the resistance.

The upper apex hole allows the exit of hot air, which is accelerated in case of wind, since externally the wind causes suction. Rain inlet can be prevented by a hat made with the 2CS shell itself. Cotton Composite Shell or 2CS shell has been shown to be easy to use and is watertight, as investigated by Dias [9].

Additional mechanical experiments will contribute to improve the understanding of the global and local functioning of the structure and adjustment of the performed analyzes. However, no less important are the field experiments, for monitoring under adverse conditions of temperatures, winds and rains, where aging will be observed, possible proliferations of fungi and plagues; excessive exposure to UV rays, cracking of bamboos and tearing of the shell.

The *Form Finding* and *Living Design* Methods are similar in the search of the optimal solution through reduced physical models where the manufacturing and assembly techniques necessarily revert to the initial design orienting the optimum form within the physical conditions and available infrastructure in the production sites. However, the Living Design Method, because it opens up to structural elements with a natural, crude form, and does not require excessive precision, in coherence with the geometrical irregularities of the bamboo, takes a step beyond *Form Finding*, adding more freedom of play, opening a crack in the rigid logic of industrialized forms, allowing for paradigmatic changes like 3C connection and 2CS shell. The 3C connection explode the realization of complex reticulated architectures without the need for sophisticated equipment, crimping and machines. So, the *Living Form Finding Method*, or *L2F Method* is the more coherent name to express the research of the recent Bamboo Living Structures.

Acknowledgments

The authors are grateful to the Post Graduate Program of the Department of   Structures Engineering – POSSES of EEUFMG, by the supporting of part of the research. They also thank the company BAMBUTEC for photos of its original and awarded structures. The authors thank also to Khosrow Ghavami and Mario Augusto Seixas for theirs brilliant conceptual ideas.

References

[1] Velez, Simon: *Grow Your Own House – Simón Vélez and bamboo architecture*, Vitra Design Museum Publ., Weil and Rhein, Germany, 2000.

[2] Moreira, LE and Ghavami, K.: *Bamboo Space Structures*, Proceedings of the Fourth International Conference in Space Structures, Guildford, England, 1993.

[3] Seixas, M. A.; Ripper, J.L.M.; Ghavami, K.: *Prefabricated bamboo structure and textile canvas pavilions*. http://dx.doi.org/10.20898/j.iass.2016.189.782, v. 57, p. 179, 2016. https://doi.org/10.20898/j.iass.2016.189.782

[4] Lanna, C. A.C. e Moreira, L.E.: *Treliças Planas de Bambu com Bioconexões Compósitas*, Procedente do II Congresso Luso Brasileiro de Materiais de Construção Sustentáveis, João Pessoa, PB, Brasil, 2016.

[5] Otto, F.: *Lightweight Construction Natural Design* – Frei Otto Complete Works, Birkhäuser Basel-Boston-Berlin, 2010.

Non-Conventional Materials and Technologies – NOCMAT for XXI Century   Materials Research Forum LLC
Materials Research Proceedings 7 (2018) 403-414   doi: http://dx.doi.org/10.21741/9781945291838-38

[6] Silva, J.M and Ferreira, R.L. : *Compreendendo a Dinâmica do Conhecimento para Inovar em Gestão*, Proceedings of the XXIX Encontro Nacional de Engenharia de Produção, Salvador, Ba, 2009.

[7] ISO/DIS 22156: *Projeto de Estruturas de Bambu – procedimento*; Associação Brasileira de Normas Técnicas, Brasil, 2017.

[8] Moreira, L.E. and Ripper, J.L.M.: *Jogo das Formas – lógica do objeto natural*; Nau Editora, Rio de Janeiro, RJ, 2014.

[9] Dias, W. F.: Cascas compósitas de resina e tecido de algodão aplicadas ao desenvolvimento de sistemas construtivos, monografia de conclusão de curso, EEUFMG, Brasil, 2013.

[10] ABNT(NBR 6123): Associação Brasileira de Normas Técnicas (Forças devidas ao vento nas Edificações), Brasil, 1998.

Non-Conventional Materials and Technologies – NOCMAT for XXI Century   Materials Research Forum LLC
Materials Research Proceedings **7** (2018) 415-426   doi: http://dx.doi.org/10.21741/9781945291838-39

Bionanocomposite Bamboo: A Regioselective Impregnation with Silver Nanofillers for Antifungal Application

O. Pandoli[a*], Raquel S. Martins[a], Bernardo A. Barbosa[a], Karen L. G. De Toni[b], Sidnei Paciornik[c], Marcos H. P. Maurício[c], Renan M. C. Lima[c], Nikolas B. Padilha[c], Sonia Letichevsky[c], Roberto R. Avillez[c] and K. Ghavami[d]

[a]Chemistry Department, PUC-RIO, Rio de Janeiro, Brazil;

[b]Rio de Janeiro Botanical Garden, Rio de Janeiro, Brazil;

[c]Chemical and Materials Eng. Department, PUC-Rio, Rio de Janeiro, Brazil;

[d]Civil Eng. Department, PUC-Rio, Rio de Janeiro, Brazil;

* omarpandoli@puc-rio.br

ABSTRACT. Silver nanoparticles are promising nanofiller materials (Ag-NFs) for reinforcement and anti-bacterial effect in polymers and composites. In this investigation bamboo *Dendrocalamus giganteus Munro* specimens were impregnated using a colloidal solution of homemade Ag-NFs with the objective of filling up the vascular bundles vessels and improving its resistance to fungi attacks. The metal nanostructured materials with different charged organic ligands (trisodium citrate and chitosan) were used to fill up the internal microenvironment of bamboo. Chemical and physical characterization was made by DLS, UV-VIS, zeta-potential and STEM. The bamboo specimens (5x12x18mm) were submitted to 20 impregnation cycles through a vacuum system. A qualitative characterization of the metal aggregate depositions of impregnated bamboo specimens were analyzed with X-ray microtomography (μCT) revealing a gradient deposition of citrate-capped Ag-NFs into the parenchyma tissue with higher concentration at the outer part for bamboo, while the chitosan-capped Ag-NFs were deposited mainly in the vessel bundles. Both engineered bionanocomposite materials when exposed to air and humidity, after 14 months, the samples were free of fungal colonies, while colonization by the fungal hyphae were present in untreated bamboo specimens.

Keywords: Bamboo, Silver Nanofiller, Nanocomposite, Coating, Antifungical Activity

Introduction

A nanocomposite is a polymer with reinforced nanometric filler, such as, nanoparticles (NPs) or fibers. According to the guest nanofiller (NFs) material it is possible to enhance or add new properties to the host polymer such as: thermal stability, weather resistance, self-cleaning, fire-resistance, antibacterial and catalytic activities.(1)(2)   Bamboo is a natural nanostructured biopolymer made by polysaccharides, mainly, cellulose, lignin and hemicellulose. At the micro-level structure all the constituent polymers are organized into micro-fibrils that are self-organized in macro-fibrils around a straight microsized channels.(3) The microarray of lignin-cellulose parallel channels constitutes the internal vascular bundles system of bamboo with varied internal diameter from 50 to 200 μm, while the parenchymatic living cells are square and rectangular-shaped with variable volume of 1000-2000 μm^3. Amorphous and crystalline cellulose fibers mixed with lignin and hemicellulose, anatomically named sclerenchyma, are distributed anisotropically in a radial direction from the inner to outer section of the culm. Large presence of hydrogen-bonded and Van Der Waals Interactions between lignin molecules and cellulose microfibril, named lignin carbohydrate complex (LCC) (4) are responsible for the intrinsically mechanical strength and supports the vascular bundles system (pholem and metaxylema) to transport fluid and nutrients (figure 1).

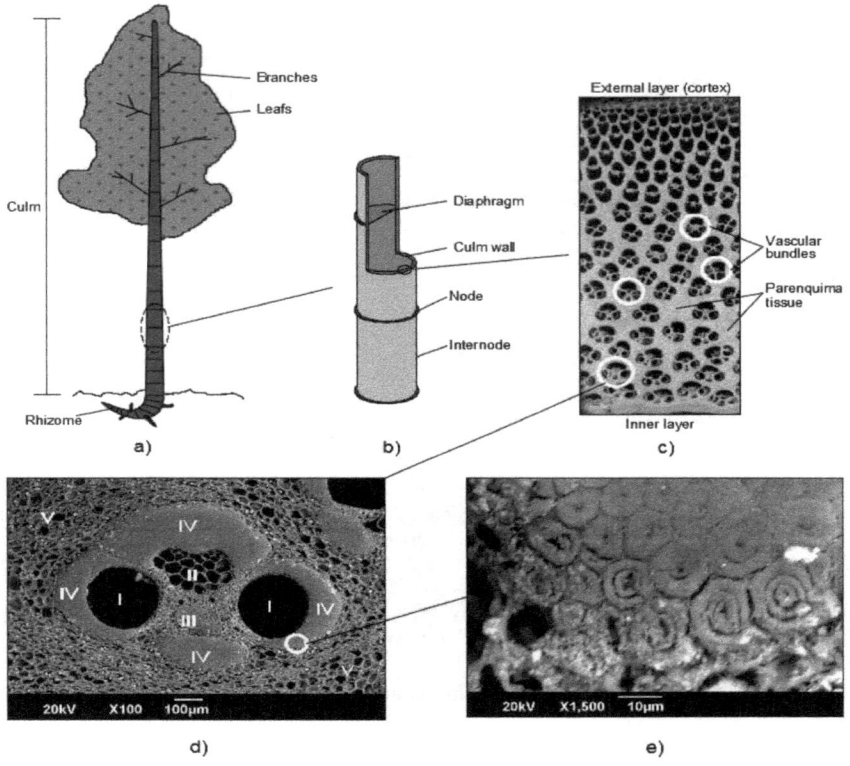

FIGURE 1 - BAMBOO STRUCTURE: A) PLANT PARTS; B) CULM PARTS; C) CULM WALL DISTRIBUTION; D) VASCULAR BUNDLE: (I) METAXYLEM VESSELS, (II) PROTOXYLEM, (III) PHLOEM, (IV) FIBER BUNDLES, (V) PARENCHYMA TISSUE, (E) AFM IMAGE OF ESCLERENCHYMA FIBER.

Bamboo is a giant grass plant which grows very fast, up to 30 cm/day, for this reason it is a cheap raw material used in engineering, furniture, textiles, paper, composite panels, decoration and as a building material.

The disadvantage of using bamboo is its low biostability and consequently its low durability over time. The high concentration of starch allows a natural biodegradation for microorganisms, fungus, molds and bacteria. Nanostructured materials have been already used for wood coating for different purposes and are already on the market: water proofing (clay, SiO_2, TiO_2); UV protection (SiO_2, TiO_2, ZnO, SiO_2, Fe_2O_3); biotic decay protection (Ag, Cu and ZnO); fire resistant (SiO_2, TiO_2 and

Non-Conventional Materials and Technologies – NOCMAT for XXI Century   Materials Research Forum LLC
Materials Research Proceedings **7** (2018) 415-426   doi: http://dx.doi.org/10.21741/9781945291838-39

clay); self-cleaning (TiO$_2$, ZnO); anti-scratching (Al$_2$O$_3$, SiO$_2$, TiO$_2$).(5–7) Jin et al. have used several nanoparticles for an external coated of bamboo timber to add new functional properties: superhydrophobicity and self-cleaning (TiO$_2$)(8); flame-retardant (TiO$_2$)(9); conductivity (Ag)(10) and mould-resistance (ZnO)(11).

Our recently paper, published in 2016 in *RCS Advances* was the first one which shows the possibility to create a new functional bionanocomposite based on bamboo specimen with internal impregnation with antimicrobial citrate-Ag-NFs.(12) Our proposal is to control the internal coating of natural microchannel array of bamboo by Ag-NFs with antifungal properties to finally increase the stability of engineered bio-composite. Herein, we show the possibility to deposit selectively different charged Ag-NFs into biological bamboo matrix for a biotic decay protection for almost 1and1/2 year against microorganisms.

Materials and Methods

Silver nitrate (AgNO$_3$, >99.9%pure), sodium borohydride (NaBH$_4$, >99%pure), chitosan and trisodiumcitrate (Na$_3$Citrate, >99.0%pure) were used as received from Sigma-Aldrich. All solutions were prepared using ultrapure water (resistivity of 18.2 MΩ cm^{-1}) obtained from a water purifier Milli-Q Gradient System A10, Millipore. Dendrocalamus giganteus bamboo culms, four years old, were obtained from FZEA-USP, Pirassununga-SP, Brazil. Specimens of different size were cut with an automatic precision cut-off machine, Miniton, from Struers.

Synthesis Procedures
Silver nanofiller colloidal solutions were synthesized with a glass chip microreactor of 100 μL internal volume and two syringe pumps. Aqueous stock solutions of AgNO$_3$ (10^{-2} mol L^{-1}) and Na$_3$Citrate (10^{-2} mol L^{-1}) were prepared dissolving the salts in purified water. A chitosan solution (10 g L^{-1}) in water and acetic acid (pH=5) was prepared dissolving slowly, step by step, small portions of polymer at 60˚C. NaBH$_4$ reducing agent solution, related with silver solution in 1.5:1 proportion, was used to promote the chemical reduction of silver ions to Ag-NFs. Ag-NF colloidal solutions were characterized by UV-VIS spectrophotometer (Perkin-Elmer, model Lambda 950), with a spectral range between 300 and 500 nm, Dynamic Light Scattering (DLS) and Zeta Potential (Horiba, Nano Particle Analyzer, SZ-100). For UV-vis, DLS and Zeta potential, the solution was diluted by 1:10 with pure water.

TSEM of Ag-NFs and SEM/EDS of Bamboo Specimens
A field emission scanning electron microscope (FEG-SEM) (JEOL, JSM-6701F) operating in the transmission mode (TSEM) at 30 kV with a work distance of 6.0 mm using the bright-field detector was used. Diameter distributions were inserted in the corresponding TSEM images with a mean value of 14.3 ± 3.6 nm for citrate-Ag-NFs and 4.0 ± 1.2 nm for chitosan-Ag-NFs. A scanning electron microscope (SEM) (JEOL, JSM-6510LV) and EDS system were also employed to analyze a longitudinal section of bamboo vascular bundle and Bamboo specimen impregnated with Chitosan-Ag-NFs.

Confocal Laser Scanning Microscopy
To obtain high-resolution optical images of the bamboo sections, previously described, a confocal laser scanning microscope (Leica TCS SPE) was used with laser channel at 488 and 405 nm to excite 0.01% auramine O with emission of 540–656 nm, and 0.1% calcofluor white with emission of 480–500 nm in order to detect lignin and cellulose, respectively.

Non-Conventional Materials and Technologies – NOCMAT for XXI Century   Materials Research Forum LLC
Materials Research Proceedings 7 (2018) 415-426   doi: http://dx.doi.org/10.21741/9781945291838-39

Bamboo Impregnation

18x12x5 mm specimens of *Dendrocalamus giganteus* Munro bamboo were placed into different vials with 4 mL of citrate-capped colloidal solution of silver nanoparticles (citrate-Ag-NFs) and chitosan-capped silver nanofiller (chitosan-Ag-NFs) for 20 impregnations cycles of 1 hour each under negative pressure through a vacuum system and then were left for 30 minutes drying in an oven at 60°C. For each impregnation cycle a fresh colloidal solution was used. Then the impregnated bamboo specimen was analyzed with X-ray microtomography.

X-ray microtomography (μCT)

A Zeiss Xradia 510 Versa microtomograph was employed to visualize and quantify the 3D structure of pure and impregnated bamboo. The typical microtomography operational parameters are compiled in Table 1. The μCT images were processed using FIJI/ImageJ free software enabling the identification of Ag-NF aggregates within the bamboo structure through a manual threshold and post-processing techniques. The image stack was then analyzed layer by layer to provide data from different axes: radial (from the external to the internal wall) and longitudinal (along the fiber direction). The 3D model was reconstructed using DragonFly software and analyzed. All measurements obtained were plotted using Origin software

.

TABLE 1. X-RAY MICROTOMOGRAPHY (μCT) OPERATIONAL PARAMETERS

	Bamboo specimens		
	With Citrate-Ag-NFs		With Chitosan-Ag-NFs
Lens	4.0x	4.0x	0.4x (macro)
Voxel (μm)	3.95	4.15	18.75
Voltage (kV)	80	50	80
Power (W)	7	4	7
Exposure time (s)	0.2	2	0.2
Number of projections	1601	1601	1601

Discussion

Confocal Laser Scanning Microscopy (CLSM) images in Figure 2 are shown with two selective fluorochromes: Calcofluor for cellulose (blu-cian) and Auromine O for lignin (green-yellow), respectively. Insight of Figure 2 (B and C), a transversal section of pure bamboo presents rich areas of lignin polymers into sclerenchyma tissues (fibers), while rich areas of cellulose are presented into the parachymatic living cells. The two polymers are spread out in both tissues. In Figure 3, longitudinal CLSM images of fibers (A) and living cells (B) show microsized and nanosized holes (500-600 nm), in fiber and parenchyma tissues, respectively. This information confirms the presence of a communication network between the principal holed vascular bundles and the parenchymatic living cells. The communication channels, from microsized vessels to nanosized holes present in living cell membranes, allow the nutrients to flow from the ground to the all matrix bamboo. This

Non-Conventional Materials and Technologies – NOCMAT for XXI Century   Materials Research Forum LLC
Materials Research Proceedings 7 (2018) 415-426   doi: http://dx.doi.org/10.21741/9781945291838-39

transportation channel system is our target to be coated with silver nanofiller and protect the bamboo matrix from the biotic decay.

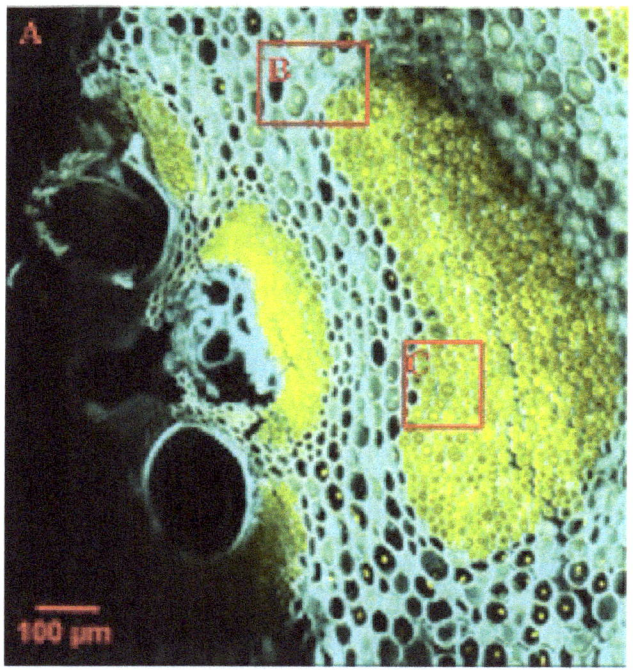

FIGURE 2. CONFOCAL LASER SCANNING MICROSCOPY (CLSM) IMAGES OF A PURE BAMBOO TRANSVERSAL SECTION WITH TWO FLUOROCHROMES: CALCOFLUOR FOR CELLULOSE (BLU-CIAN) AND AUROMINE O FOR LIGNIN (GREEN-YELLOW).

Silver Nanofiller was characterized by means of STEM microscopy and the corresponding diameter distributions are shown in Figure 4. Citrate-Ag-NFs and chitosan-Ag-NFs colloidal solutions present a mean diameter of 14.3 ± 3.6 nm and 4.0 ± 1.2 nm, respectively. Nevertheless, the chitosan-Ag-NFs have shown a smaller diameter distribution compared to citrate-Ag-NFs by means of STEM analysis, UV-VIS spectroscopy and DLS analysis give us information about the surface plasmon resonance (SPR) and hydrodynamic particle diameter, respectively, showing that positively charged chitosan polymer might stabilize clusters of Ag-NPs in the polymer matrix. This aggregation phenomena of Ag-NPs driven by chitosan polymer explains the red shift of the SPR band at 402 nm, compared to the 394 nm SPR band of citrate-capped Ag-NPs, as well as, the higher hydrodynamic diameter of 120 nm and higher zeta-potential value of +62.5 mV compared to citrate-Ag-NFs (Table 2).

Non-Conventional Materials and Technologies – NOCMAT for XXI Century   Materials Research Forum LLC
Materials Research Proceedings **7** (2018) 415-426   doi: http://dx.doi.org/10.21741/9781945291838-39

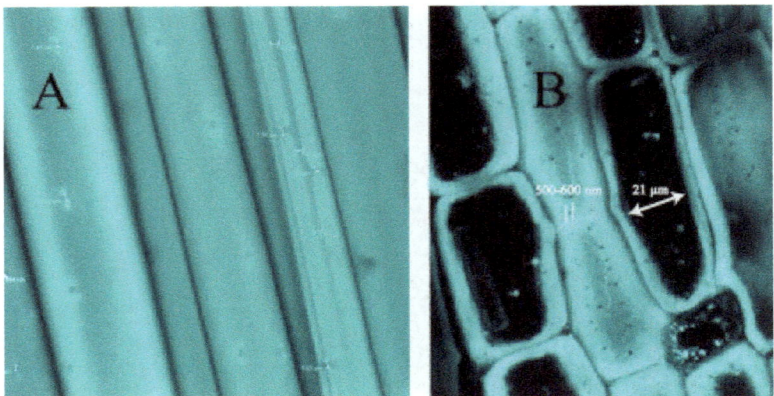

FIGURE 3. CONFOCAL LASER SCANNING MICROSCOPY IMAGES OF SCLERENCHYMA TISSUE (A) AND PARENCHYMA TISSUE (B) OF PURE BAMBOO, EVIDENCING THE LIGNIN THROUGH AURAMINE O.

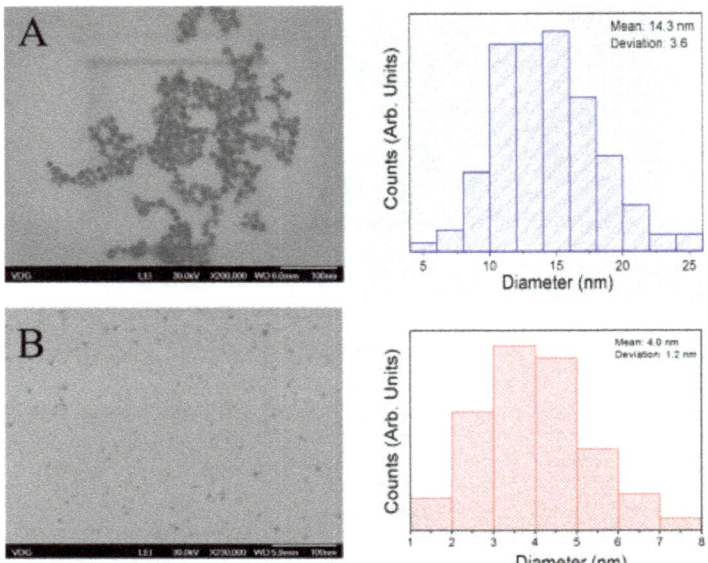

FIGURE 4. STEM IMAGES FOR (A) CITRATE-AG-NFS AND (B) CHITOSAN-AG-NFS AND CORRESPONDING DIAMETER DISTRIBUTION.

Non-Conventional Materials and Technologies – NOCMAT for XXI Century   Materials Research Forum LLC
Materials Research Proceedings **7** (2018) 415-426   doi: http://dx.doi.org/10.21741/9781945291838-39

TABLE 2. *DYNAMIC LIGHT SCATTERING (DLS) AND ZETA-POTENTIAL MEASUREMENTS FOR CITRATE-AG-NFS AND CHITOSAN-AG-NFs COLLOIDAL SOLUTION*

Ag-NFs	SPR Band	Hydrodynamic Particle diameter	Zeta Potential
Citrate-Ag-NFs	394 nm	17 nm	- 40.5 mV
Chitosan-Ag-NFs	402 nm	120 nm	+ 62.5 mV

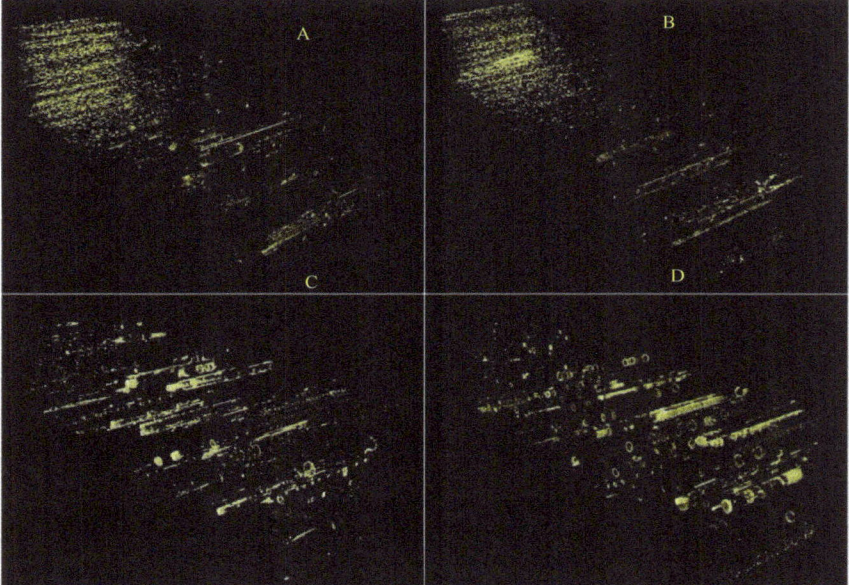

FIGURE 5. COMPARISON OF VOLUMETRIC DISTRIBUTION OF AG-NFS AGGREGATES: (A) EDGE CITRATE; (B) CENTRAL CITRATE; (C) EDGE CHITOSAN; (D) CENTRAL CHITOSAN.

The silver deposition in the bamboo matrix was analyzed by means of μCT and quantitative information of the silver clusters were extracted using digital processing of 3D images. A qualitative representation of all volumetric distributions of silver aggregates, shown in yellow color, is summarized in Figure 5. To simply the interpretation of the 3D μCT images, we take out the polymer bamboo matrix, keeping only the visualization of silver aggregate. Figures 5A-B visualize the edge and central bamboo specimens treated with citrate-Ag-NFs, with a higher concentration of metal deposition in the parenchymatic tissue and mainly in the external wall of bamboo.

Figures 5C-D visualize the edge and central bamboo specimens treated with chitosan-Ag-NFs, showing a high concentration of metal deposition in the microchannel array of bamboo.

Non-Conventional Materials and Technologies – NOCMAT for XXI Century   Materials Research Forum LLC
Materials Research Proceedings 7 (2018) 415-426   doi: http://dx.doi.org/10.21741/9781945291838-39

In table 3 a summary is given of the analytical results of the µCT 3D high resolution images of silver aggregates in bamboo specimens on different axis: radial (from the out to inner bamboo wall) and longitudinal (along the fiber and vascular vessels).

TABLE 3. RESUME OF 3D µCT HIGH RESOLUTION IMAGE ANALYSIS OF SILVER AGGREGATE ON BAMBOO SPECIMENS REGIONS ON DIFFERENT AXES: RADIAL AND LONGITUDINAL.

Nanofillers	Region	On Radial Axis		On Longitudinal Axis	
		n. aggregates /cm^2	Occupied Area (%)	n. aggregates /cm^2	Occupied Area (%)
Citrate-Ag-NFs	Edge	1492	0.15	1651	0.15
Citrate-Ag-NFs	Central	1657	0.17	1653	0.17
Chitosan-Ag-NFs	Edge	331	0.08	526	0.09
Chitosan-Ag-NFs	Central	809	0.23	984	0.25

The results, obtained on two differenta, are approximately the same, confirming an increase of reliability in the quantitative analyses. The difference in the number of aggregates can easily be explained by the difference in size of those aggregates. When we change the axis of analysis, a number of aggregates seemed to be independent showing themselves as a single unit. By comparison, the bamboo impregnated with chitosan-Ag-NFs, as seen on Figure 5C-D, is characterized by a metal aggregates deposition only in the internal wall of microchannel arrays. From the analysis of citrate-Ag-NFs metal aggregates through the tube's xylem tube's axis, along the overall bamboo sample (edge and central part), we found a percentage occupied area of around 0.16%. Note that there is a gap between 2500 µm and 4000 µm, which represents the gap between the edge and the central part of the sample analyzed. On the other hand, the chitosan-Ag-NFs metal aggregates in bamboo matrix present an increasing distribution from the edge to the central portion of the sample, from 0.08 to 0.25% respectively. A few presumptions can be taken to explain this behaviour. The negatively charged citrate-Ag-NFs colloidal solution pass through the vascular bundles, penetrate into the bamboo matrix until filling up some parenchymatic living cells, as shown in Figure 3. A 2D and 3D µCT image shows citrate-Ag-NFs aggregates with a random distribution mainly in the parenchyma tissue (Figure 6).

Non-Conventional Materials and Technologies – NOCMAT for XXI Century   Materials Research Forum LLC
Materials Research Proceedings 7 (2018) 415-426   doi: http://dx.doi.org/10.21741/9781945291838-39

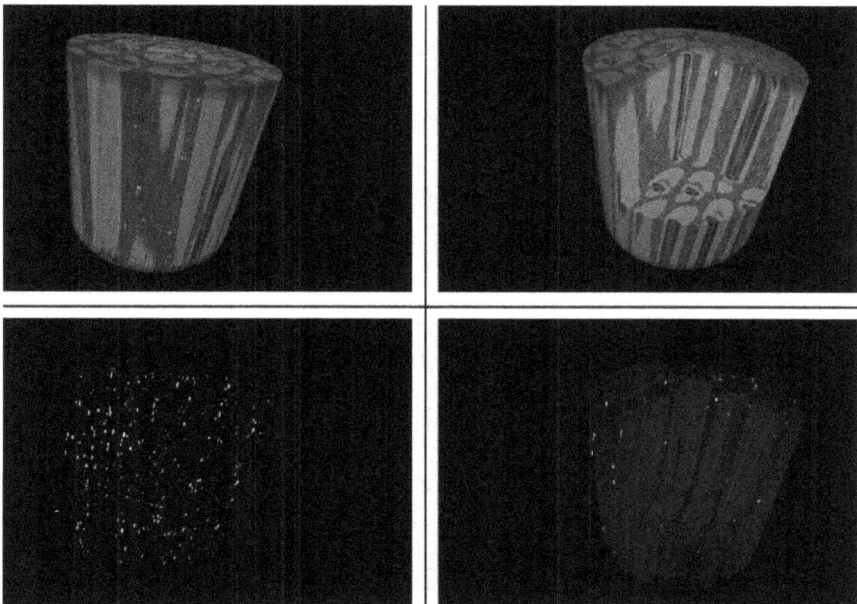

FIGURE 6. 2D and 3D µCT IMAGES WITH CITRATE-AG-NFS AGGREGATE DEPOSITED INTO THE PARANCHYMATIC LIVING CELLS.

On the other hand, the diffusion of chitosan-Ag-NFs colloidal solution in the vascular bundles is not uniform, observing a higher coating of internal channels' walls in the central portion of the bamboo specimen. On Figure 7, decreasing the resolution of µCT images with a macro objective lens of 0.4x, we can visualize the total bamboo size specimen (5x12x18 mm) with the formation of silver metal tube templated by the vascular bamboo vessels. The positively charged chitosan-Ag-NF might have a higher molecular affinity with the cellulose-lignin polymer constituent in the internal wall of vessels. After a primary deposition the following chitosan-Ag-NFs with a hydrodynamic diameter of 120 nm, which is higher when compared to the citrate-Ag-NFs, could block the holed internal wall of bamboo and become obstructed for the penetration to the parenchymatic tissue. A higher concentration was found in the central portion of the bamboo which might be due to the solvent evaporation process of the colloidal solution drag on the silver nanofiller from the external to the internal portion of the tubes. This process, repeated during the impregnation process (20 cycles), might have created this kind of cylindrical metal deposition inside the microchannel array. The selective distribution of chitosan-Ag-NFs has to be possible for the positively charged silver nanofillers, for the higher hydrodynamic and probably for the interfacial compatibility between the similar macromolecular structure of chitosan and cellulose.

Non-Conventional Materials and Technologies – NOCMAT for XXI Century   Materials Research Forum LLC
Materials Research Proceedings 7 (2018) 415-426   doi: http://dx.doi.org/10.21741/9781945291838-39

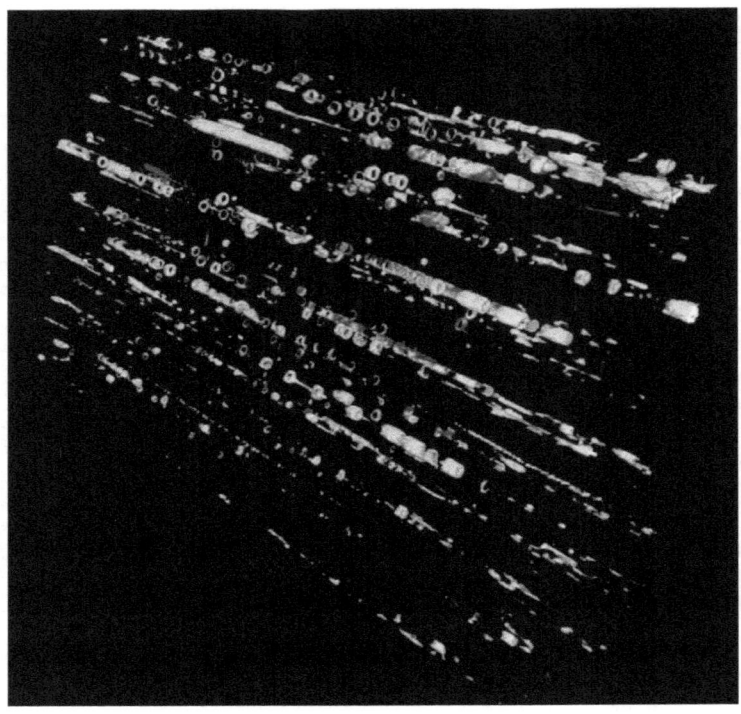

FIGURE 7. 3D µCT IMAGES OF CHITOSAN-AG-NFS INTO BAMBOO MATRIX, FOR EASY INTERPRETATION OF THE IMAGE THE BIOLOGICAL MATRIX OF BAMBOO WAS EXCLUDED DURING THE IMAGE PROCESSING;

Conclusions

Ag-NFs, capped with different anionic and cationic organic ligands, have been used to coat selectively the internal bamboo matrix. Ag-NFs stabilized with negative (sodium citrate) and positive (chitosan) charged organic ligand and hydrodynamic size diameter of 17 and 120 nm, respectively, were synthesized with a microfluidic system in a continuous flow. The self-sorting filling of hybrid metal biocomposite bamboo was driven for the electric charge, nanoparticles dimension of organic ligands-capped Ag-NFs and chemical affinity between organic capped agent and bamboo matrix formed mainly by cellulose-lignine polymer. Both Engineered bio-composite, obtained with different charged silver nanofiller, exposed to humidity and ambient conditions have not shown any biodegradation process for more than 1 year.

Non-Conventional Materials and Technologies – NOCMAT for XXI Century   Materials Research Forum LLC
Materials Research Proceedings 7 (2018) 415-426   doi: http://dx.doi.org/10.21741/9781945291838-39

References

1. Vaia RA, Wagner HD. Framework for nanocomposites. Mater Today [Internet]. Elsevier Ltd; 2004;7(11):32–7. Available from: http://dx.doi.org/10.1016/S1369-7021(04)00506-1

2. Šupová M, Martynková GS, Barabaszová K. Effect of Nanofillers Dispersion in Polymer Matrices: A Review. Sci Adv Mater [Internet]. 2011;3(1):1–25. Available from: http://openurl.ingenta.com/content/xref?genre=article&issn=1947-2935&volume=3&issue=1&spage=1

3. Siti S, Abdul HPS, Wan WO, Jawai M. Bamboo Based Biocomposites Material, Design and Applications. In: Materials Science - Advanced Topics [Internet]. InTech; 2013. Available from: http://www.intechopen.com/books/materials-science-advanced-topics/bamboo-based-biocomposites-material-design-and-applications

4. Youssefian S, Rahbar N. Molecular Origin of Strength and Stiffness in Bamboo Fibrils. Sci Rep [Internet]. Nature Publishing Group; 2015 Jun 8;5:11116. Available from: http://dx.doi.org/10.1038/srep11116

5. Nikolic M, Lawther JM, Sanadi AR. Use of nanofillers in wood coatings: a scientific review. J Coatings Technol Res [Internet]. Springer US; 2015 May 8;12(3):445–61. Available from: http://dx.doi.org/10.1007/s11998-015-9659-2

6. Marzi T. Nanostructured materials for protection and reinforcement of timber structures: A review and future challenges. Constr Build Mater [Internet]. Elsevier Ltd; 2015 Oct;97:119–30. Available from: http://dx.doi.org/10.1016/j.conbuildmat.2015.07.016

7. Fufa SM, Jelle BP, Hovde PJ, Rorvik PM. Coated wooden claddings and the influence of nanoparticles on the weathering performance. Prog Org Coatings. 2012;75(1–2):72–8.

8. Li J, Lu Y, Wu Z, Bao Y, Xiao R, Yu H, et al. Durable, self-cleaning and superhydrophobic bamboo timber surfaces based on TiO2 films combined with fluoroalkylsilane. Ceram Int [Internet]. 2016 Jun;42(8):9621–9. Available from: http://linkinghub.elsevier.com/retrieve/pii/S0272884216301791

9. Li J, Zheng H, Sun Q, Han S, Fan B, Yao Q, et al. Fabrication of superhydrophobic bamboo timber based on an anatase TiO 2 film for acid rain protection and flame retardancy. RSC Adv [Internet]. 2015;5(76):62265–72. Available from: http://xlink.rsc.org/?DOI=C5RA09643J

10. Jin C, Li J, Han S, Wang J, Yao Q, Sun Q. Silver mirror reaction as an approach to construct a durable, robust superhydrophobic surface of bamboo timber with high conductivity. J Alloys Compd [Internet]. Elsevier B.V.; 2015 Jun;635:300–6. Available from: http://dx.doi.org/10.1016/j.jallcom.2015.02.047

11. Yu Y, Jiang Z, Wang G, Tian G, Wang H, Song Y. Surface functionalization of bamboo with nanostructured ZnO. Wood Sci Technol [Internet]. 2012 Jul 18;46(4):781–90. Available from: http://link.springer.com/10.1007/s00226-011-0446-7

12. Pandoli O, Martins R dos S, Romani EC, Paciornik S, Maurício MHDP, Alves HDL, et al. Colloidal silver nanoparticles: an effective nano-filler material to prevent fungal proliferation in bamboo. RSC Adv [Internet]. 2016;6(100):98325–36. Available from: http://pubs.rsc.org/en/Content/ArticleLanding/2016/RA/C6RA12516F

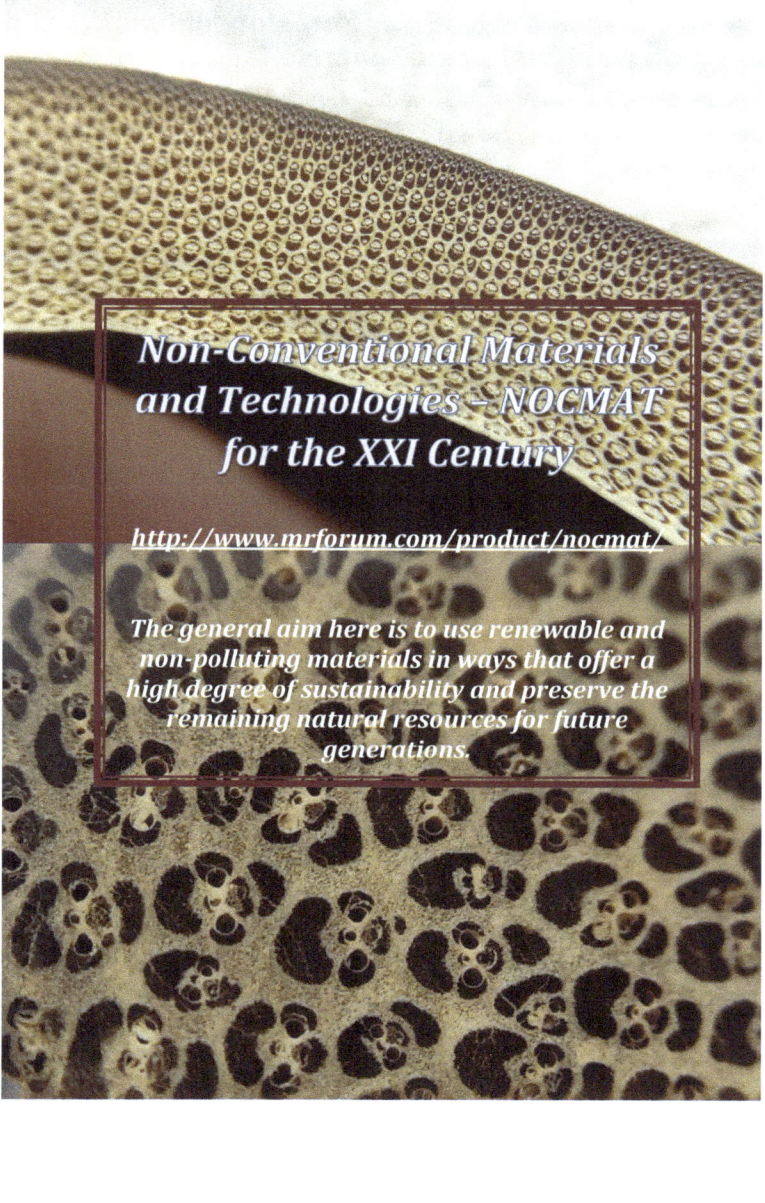

Non-Conventional Materials and Technologies – NOCMAT for the XXI Century

http://www.mrforum.com/product/nocmat/

The general aim here is to use renewable and non-polluting materials in ways that offer a high degree of sustainability and preserve the remaining natural resources for future generations.

Non-Conventional Materials and Technologies – NOCMAT for XXI Century   Materials Research Forum LLC
Materials Research Proceedings 7 (2018) 428-439   doi: http://dx.doi.org/10.21741/9781945291838-40

Performance Analysis of Corrosion Inhibitors in Pore Solution and Reinforced Concrete for Carbon Steel CA 50

Anilé Ossorio Domínguez [a], Marina Martins Mennucci [b], Denise Carpena Coitinho Dal Molin [c]

[a] Federal University of Rio de Janeiro, Department of Civil Engineering, Rio de Janeiro, RJ. e-mail: aniossoriobrasil@gmail.com

[b] University of São Paulo, Metallurgical and Materials Engineering Department, São Paulo, SP. e-mail: marinam@uol.com.br

[c] Federal University of Rio Grande do Sul, Department of Civil Engineering, Rio Grande do Sul, RS. e-mail: dmolin@ufrgs.br

Abstract. The aim of this work is to study the best performance in between sodium nitrite, sodium phosphate and ethanolamine as corrosion inhibitors of steel bars in the simulated concrete pore solution and then test it in the concrete. The effectiveness of these compounds as corrosion inhibitors was investigated by measuring the electrochemical impedance spectroscopy of steel bars immersed for 3 and 72 hours in solutions with and without chlorides. The results showed that higher performance was observed in the presence of the sodium nitrite. So, for the second stage of this work was chosen just sodium nitrite. Samples were taken of concrete reinforced with the addition of sodium nitrite and contaminated with chlorides. For this purpose, were selected two Brazilian types of cements CP-IV (pozzolan) and CP-V and three water/cement ratios (w/c 0.4; w/c 0.5; w/c 0.65). To simulate the marine environment aggression, chlorides accelerated tests were conducted. Even if one is liquid (pore solution) and another is solid (concrete), in both tests, the nitrite had increased efficiency with the exposure time.

Keywords: Corrosion Inhibitor, Corrosion, Chlorides, Pore Solution, Concrete

Introduction

The life cycle of reinforced concrete has a direct relationship with the corrosion of the reinforcing steel, once the corrosion begins, the kinetics of the process accelerates, being able to lead to economic losses in matenimentos or even to structural collapse in some cases.

Concrete suppliers have now reduced cement consumption, resulting in a large number of structures where the concrete used provides the specified strengths, but it does not provide the specifications of adequate durability. With consumption of up to 180 kg / m3, as it was used indiscriminately, it is difficult to achieve a service life of more than 30 years, resulting in concrete with a high porosity ([1], [2], apud [3]).

In this case, the use of additional protection methods that lead to an increase in the useful life cycle and to the reduction of economic losses caused by corrosion is initiated. Most standards and codes do not include any measurement other than coating, a / c ratio, ambient and Compressive Strength. There are some rules that superficially define alternative protection measures, but without placing the expected benefits and alternative conditions of application.

As the pathological manifestations related to corrosion increased. An interest in acquiring a greater understanding of corrosion and protection processes arises, with the presence of corrosion inhibiting additives in reinforced concrete. With the purpose of contributing to a reduction of the environmental impact, caused by expenses in maintenance and in some cases new works. [4]

In Brazil, after the 1990s, a gradual increase in the use of corrosion inhibiting additives began to increase the durability of reinforced concrete structures [5]

Non-Conventional Materials and Technologies – NOCMAT for XXI Century   Materials Research Forum LLC
Materials Research Proceedings 7 (2018) 428-439   doi: http://dx.doi.org/10.21741/9781945291838-40

In agreement by [6], the increase in the use of inhibitors in reinforced concrete is due to advantages such as a relatively low cost, easy application, besides being useful in prevention and repair when compared with other methods.

Looking for preventive measures of this pathological manifestation, the interest arises a deeper understanding of the processes of corrosion and protection in the presence of corrosion inhibiting additives in reinforced concrete.

The present work was divided in two phases, where from the studies in the first phase in the dissolution of the pore water is obtained information about the electrochemical behavior of the reinforcement in solutions that simulate the aqueous phase of the concrete. However, these tests do not contemplate the porous nature of the concrete which does the electrolyte in contact with the reinforcing steel not always continuous. [7]. In the present research two phases were carried out, in order to obtain a comparative analysis: in the synthetic solution that simulates pore water and in the reinforced concrete.

Experimental study

To perform the experimental program in the first phase, CA-50 S steel was used in the aqueous medium which simulates the pore water in the concrete, where the performance was evaluated in the presence of 3 inhibitors (sodium nitrite, phosphate sodium and ethanolamine). The second phase was performed in a real reinforced concrete, where only one inhibitor (sodium nitrite) was used and was added in the kneading step. The materials are characterized and the methods used in this research are described.

Materials and methods

The chemical composition of the carbon steel (CA-50S) used in this study is shown in Table 1 from the optical emission spectrometry[1].

TABLE 1 - CHEMICAL COMPOSITION OF CARBON STEEL (%)

Elenento	%
C	0,237
Si	0,195
Mn	0,773
P	0,0044
S	0,0474
Cr	0,0809
Mo	0,0083
Ni	0,0499
Al	0,0221
Co	<0,01
Cu	0,082
Nb	0,0055
Ti	0,032
V	<0,001
W	<0,01
Pb	<0,002
Sn	0,0027
Mg	<0,002
B	>0,132
Fe	<98,2

---

[1] The experiment was performed at the Laboratory of UFRGS (LAMEF)

Non-Conventional Materials and Technologies – NOCMAT for XXI Century   Materials Research Forum LLC
Materials Research Proceedings **7** (2018) 428-439   doi: http://dx.doi.org/10.21741/9781945291838-40

For the electrochemical impedance spectroscopy tests, in both working stages, the same reference electrode and electrode were used.

Reference electrode
For the study used the reference electrode of Ag / AgCl, the body of glass, simple junction, type of "banana", 4mm, with 1m cable brand Analyzer, mod. 3A11. This type of electrode has a large encapsulated Ag / AgCl crest reference element, which confers great stability over time.

Contra-electrode
A corrosion-resistant 304 series stainless steel plate, 4 mm thick, was used.

First phase
A CA-50S carbon steel (Table 1) rebar was sliced and use as the working electrode for the electrochemical measurements. In each slice was made the electric contact with a copper wire and that was placed in a mold and filled with epoxy resin. The exposed area of the working electrode was 0.312 cm². Samples were ground with silicon carbide paper up to No. 1000, rinsed with deionized water and degreased in acetone then dried.

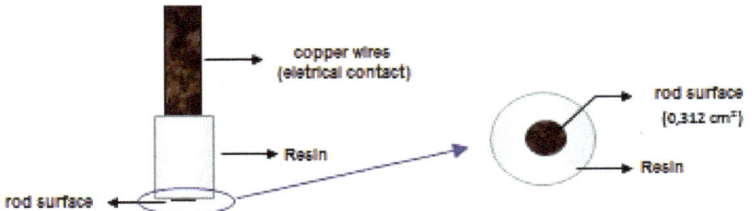

FIGURE 1 - *SCHEMATIC OF THE CA-50 S STEEL SAMPLES IN THE EPOXY RESIN FOR COLD CURE WITH THE ELECTRIC CONTACT.*

The reference solution was 2.8 wt% + NaOH 0.4 wt% + NaCl 3.5 wt%, to simulate the pore water of NaCl contaminated concrete (REF) (Mennucci et al., 2009). In this reference solution was added the corrosion inhibitors: in a solution I, sodium nitrite (2.0 wt%); In solution II, sodium phosphate (2.0 wt%) and in solution III, ethanolamine (2.0 wt%). Solutions I, II and III were used to comparing the inhibitory effect on corrosion resistance in the investigated carbon steel. All reagents used in the solutions are analytical grade. The exposure time was divided into two cycles 3 hours of exposure and then 72 hours of exposure.

Second phases
In the second phase, the mixtures of the different concretes were designed to reach three distinct levels of water-cement ratio (0.40, 0.50 and 0.65), representative of a rich, medium and poor screening, determined through the mix, for two types of cement (CP V ARI RS) and (CP-IV). For each a / c ratio, two mix were designed: one reference (without addition of inhibitor) and one with addition of inhibitor, substitution of 2% by mass of the cement in the design of the mixture by sodium nitrite powder.

Non-Conventional Materials and Technologies – NOCMAT for XXI Century   Materials Research Forum LLC
Materials Research Proceedings 7 (2018) 428-439   doi: http://dx.doi.org/10.21741/9781945291838-40

Cement
It was used the Portland cement of high initial resistance (CP-V ARI RS), (NBR 5733), compared to cement (CP-IV), which is a cement that contain a mass addition of pozzolan from 15% to 50%, according to NBR 5736. (ABCP, 2002).
Tables 2 and 3 show the chemical composition of each cement, according to the X-ray result performed[2]

***TABLE 2** - CHEMICAL COMPOSITION OF PORTLAND CEMENT CP-IV 32, OBTAINED BY X-RAY FLUORESCENCE (FRX)*

FRX	CaO	SiO_2	Al_2O_3	Fe_2O_3	MgO	SO_3	Na_2O	K_2O	CaO
%	37.39	36.06	9.86	5.58	1.39	2.27	0.098	1.78	37.39

***TABLE 1** - CHEMICAL COMPOSITION OF PORTLAND CEMENT CP-V ARI RS, OBTAINED BY X-RAY FLUORESCENCE (FRX)*

FRX	CaO	SiO_2	Al_2O_3	Fe_2O_3	MgO	SO_3	K_2O
%	62.46	17.96	3.09	4.54	1.26	4.74	1.19

Working electrode II
Samples of CA-50S steel 17 cm in length were used, which should be pickled with a dissolution of the acid base. The length will be divided into strips, leaving only 2 cm at the end of the steel bar to make electrical contact. All preparation of the working electrode was in accordance with UNE 83992-1: 2012

Aggressive angent
It was used de the 3% NaCl solution according to UNE 83992-1: 2012. This solution will be placed in the pool that will be attached to the concrete sample. The water used to make the solution is deionized water.

Corrosion inhibitor
At the end of phase I, the corrosion inhibitor to be used for the second phase was the sodium nitrite (N) concentration 2% as a function of the cement mass, it was placed at the time of the concrete kneading. As proposed by [8] that was the content that presented better compatibility between the cement properties and the use of the nitrite as corrosion inhibitor additive.

Mix desing
The design of the mixture was performed using the IPT / EPUSP method [9]. The optimal mortar content was obtained in 51% and the Slump set between 10-12 cm
From the mathematical equations of the dosage diagram and considering the great content of mortar they were determined traits unit mass. Table 5.6 shows each of the mixing designs used.
The concretes were shaped to reach three distinct water-cement ratio levels (0.40, 0.50 and 0.65), representative of a rich, medium and poor trace, determined through the dosage diagram. For each a / c ratio, two traces were molded: one reference (no addition of inhibitor) and another trace with inhibitor addition, with substitution rates of 2% by mass of the trace cement by sodium nitrite powder by the amount of Cement of the trace.

---

[2] The experiments were performed at the Laboratory of UFRGS (LACER)

TABLE 2 – MIX DESIGNS BASED ON UNIT WEIGHT

Cement	m	w/c	Mortar ratio %	Mix Unit	Slump cm	
					Without inhibitor	Wich inhibidor
CP-IV	3,5	0,40	51	1:1,32:2,23	8,5	9,5
	5,1	0,50	51	1:2,1:3,0	11	13,5
	7,1	0,65	51	1:3,12:3,95	12	15,5
CP-V ARI RS	3,6	0,40	51	1:1,32:2,24	9,5	15
	5,2	0,50	51	1:2,16:3,03	11	14
	7,64	0,65	51	1:3,40:4,24	9,0	10

To mold the samples we adapt our materials to the dimensions specified by the Spanish standard. The nominal dimensions were (14.5 x 7.5 x 7.5) cm, with a steel bar CA-50S embedded in the center of the sample, with a diameter of 6.3 mm, functioning as (working electrode), leaving only 2.5 cm of the same exposed to do The electrical contact. As shown in figure 2, it unifies the procedure

FIGURE 2 – PROCEDURE FOR MAKING SPECIMENS FOR EIE ASSAYS. (A) PREPARATION OF THE MOLD AND CONCRETING, (B) PREPARATION OF THE-POOLS FOR THE PLACEMENT OF THE AGGRESSIVE AGENT NACL, (C) FIXING THE POOL TO THE CONCRETE, (D) APPLYING THE INSULATION PAINT TO AVOID EVAPORATION WITH THE EXTERNAL ENVIRONMENT.

Fonte: OSSORIO,2014

Non-Conventional Materials and Technologies – NOCMAT for XXI Century   Materials Research Forum LLC
Materials Research Proceedings 7 (2018) 428-439   doi: http://dx.doi.org/10.21741/9781945291838-40

In the present work 12 mix designs were executed. For each ratio w / c, 9 samples with and without inhibitor, totaling 108 test specimens.

After the submerged cure up to the age of 28 days, the EIS test was carried out following the instructions of the Spanish standard (UNE 83992-1 EX) for the accelerated chloride test in the natural state. Some modifications of the specimens (dimensions, wetting cycles, and drying were reduced from 2 weeks to one), to adapt to our conditions.

Experimental results and discussions

Analysis results phase 1

Figure 3 shows the EIE diagrams in the pore water solution and in the chloride pore water solution (REF). It can be observed a reduction of 6% of the impedance in the electrolyte with chloride, comparing it with the electrolyte that simulates the pore water inside the concrete without contamination. This fact jeopardizes the rupture of the passive layer caused by the chloride ions at high concentrations, provoking the exposure of the metallic substrate. By showing that a well-crafted concrete does not allow the entry of aggressive agents and therefore, it will hardly result in corrosion.

Comparing the impedance at 10 mHz it is verified that the real component, attributed to the resistance of the system to the corrosive process, is much smaller for the system containing chloride (7k ohms) than the system with only pore water (107k ohms). In addition, the beginning of a second arc for the chloride system is observed, whereas only with pore water the first arc does not close, indicating that the kinetics in the case of pore water is smaller since it would require frequencies to close the first arc.

Normally the arc formed at higher frequencies is due to the process of charge transfer, the faster this process, the greater the frequency at which the resistance will appear.

The second arc is generally attributed to the process of species diffusion, which requires larger times to occur the process of transfer of charge. In general, the resistance associated with this process is called polarization resistance.

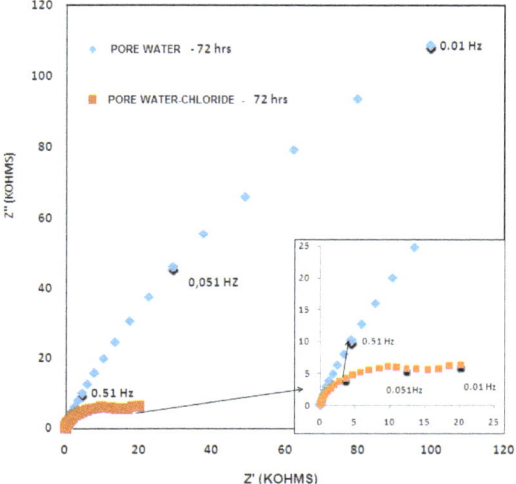

FIGURE 3 - *NYQUIST DIAGRAM FOR CARBON STEEL CA-50S IN THE PORE WATER SOLUTION AND THE PORE WATER SOLUTION WITH 3.5% (MASS) SODIUM CHLORIDE (REF) FOR 72 HOURS INMERSION.*

Non-Conventional Materials and Technologies – NOCMAT for XXI Century   Materials Research Forum LLC
Materials Research Proceedings 7 (2018) 428-439   doi: http://dx.doi.org/10.21741/9781945291838-40

These results are similar to those obtained by [3]; [7] also indicated the same decrease in impedance in a medium with the addition of chlorides in comparison to the medium that simulates pore water.

All solutions with corrosion inhibiting additives presented differences as a function of the exposure time.

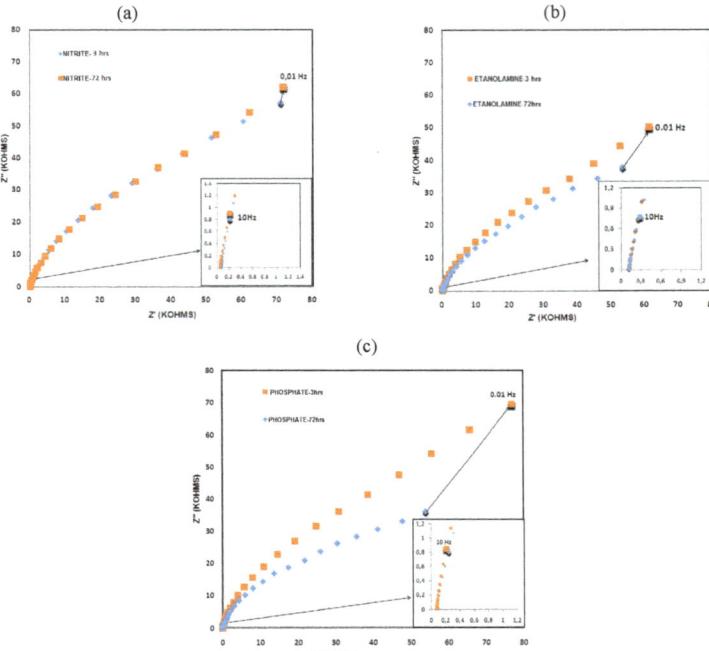

FIGURE 4 – *NYQUIST DIAGRAMS OF CA-50S SPECIMENS IMMERSED IN SOLUTION I (A), SOLUTION III (B) AND SOLUTION II (C). SHOWING THE BEST RESULTS WITH EXPOSURE TIME FOR NITRITE.*

Comparing the impedance value in 10 mHz for Nitrite (a), it is verified that improvement with the increase of the exposure time, reaching the highest values in the 72 hours. The real component, attributed to the system resistance to the corrosive process in the 72 hours, reached the best behavior arriving (65k ohms) when compared to 3 hours with (59k ohms).

In the case of Phosphate (b) comparing 10mHz, a decrease in impedance occurred with increasing exposure time, showing the best behavior at 3hrs with (73k ohms) when compared at 72hrs with (38k ohms). The actual component attributed to the system resistance to the corrosive process was shorter than 72 hrs. A similar behavior was observed with Ethanolamine, showing a higher impedance value at 3 hours of (53k ohms) than at 72 hrs with (38k ohms), reaching in the case of the 3hrs greater impedance changes at higher frequencies.

Non-Conventional Materials and Technologies – NOCMAT for XXI Century   Materials Research Forum LLC
Materials Research Proceedings **7** (2018) 428-439   doi: http://dx.doi.org/10.21741/9781945291838-40

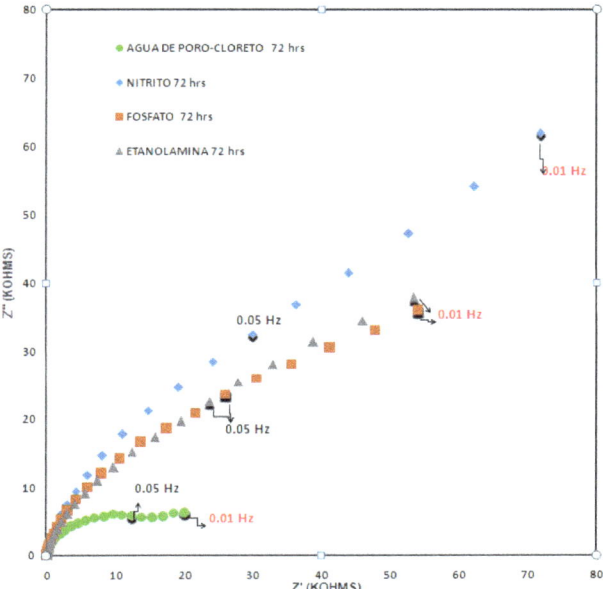

FIGURE 5 - *NYQUIST DIAGRAM FOR CA-50S CARBON STEEL IN SOLUTIONS WITH INHIBITORS: IN THE REFERENCE SOLUTION PORE WATER WITH CHLORIDE (PW-C/GREEN) NITRITE (N/BLUE), ETHANOLAMINE (EA/GREY), AND PHOSPHATE (F/ORANGE); ALL WITH 72 HOURS OF EXPOSURE.*

Comparing the impedance at 10 m Hz it is verified that the real and imaginary component, attributed to the system resistance to the corrosive process, is much smaller for the chloride-containing system (21k ohms) in relation to the system for each inhibitor studied. In the case of 10 m Hz analysis of the 3 inhibitors, the best behavior of the real and imaginary component attributed to the corrosion resistance of the system was nitrite (N) with (94k ohms), phosphate (F) with (65k ohms) Similar to the ethanolamine (EA) behavior with (64k ohms).

Analysis results phase II

In order to explain the performance of the inhibitors in the reinforced concrete in the presence of chlorides, EIA diagrams were followed in the solution for concretes with CP-IV cement and with CP-V with and without the presence of inhibitors after 3 exposure cycles. EIS was performed for all w / c ratios (0.4, 0.5, 0.65), with exposure time completing 70 days. The figure 6 shows the nyquits graph for concrete with CP-IV cement.

Non-Conventional Materials and Technologies – NOCMAT for XXI Century   Materials Research Forum LLC
Materials Research Proceedings **7** (2018) 428-439   doi: http://dx.doi.org/10.21741/9781945291838-40

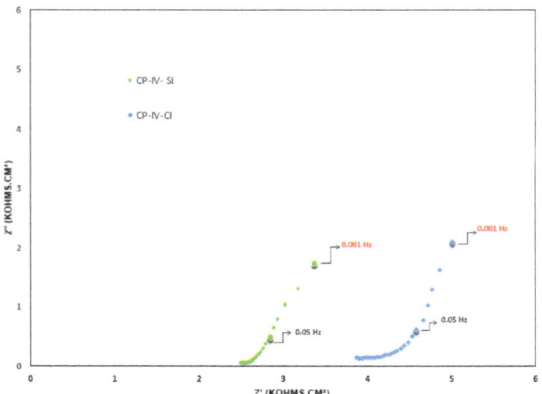

FIGURE 6 - NYQUIST PLOT FOR THE CONCRETE SYSTEM WITH CP-IV CEMENT IN THE W / C-0.65 RATIO IN THE THIRD CYCLE OF THE ACCELERATED IMMERSION AND DRYING PROCESS WITH THE PRESENCE OF CHLORIDE.

By comparing the impedance at 10 mHz associated with the species diffusion process to load transfer associated with system resistance, it needs readings at lower frequencies since the arc has not yet formed. It is verified that the real component, attributed to the system resistance to the corrosive process at the steel-corrosion product interface, is much smaller in the system without inhibitor (3.37 kΩ) than the system with inhibitor (5 kΩ), showing the Efficiency of the inhibitor [10].

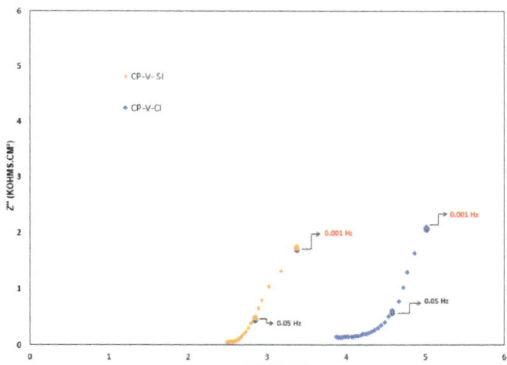

FIGURE 7 - NYQUIST DIAGRAMS FOR THE CONCRETE SYSTEM WITH CP-V ARI CEMENT IN THE W / C-0.65 RATIO WITH THE THIRD CYCLE OF THE ACCELERATED IMMERSION AND DRYING PROCESS WITH THE PRESENCE OF CHLORIDE.

It is also verified that the real component, attributed to the system resistance to the corrosive process at the interface between the steel and the corrosion product, is lower in the system without inhibitor (1.38 KΩ) than the system with inhibitor (1.47 kΩ).

Non-Conventional Materials and Technologies – NOCMAT for XXI Century   Materials Research Forum LLC
Materials Research Proceedings 7 (2018) 428-439   doi: http://dx.doi.org/10.21741/9781945291838-40

Next graphs are presented relating the efficiency of the inhibitor as a function of the cycles, directed to our main objective in the present research. The Nyquist graphs 8 and 9 for CP-IV and CP-V cements, with inhibitor for the first exposure cycle and the 3rd relative to w/ c 0.65.

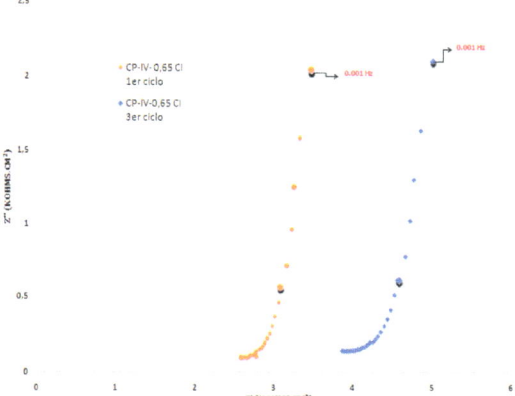

FIGURE 8 – *NYQUIST DIAGRAMS FOR THE CONCRETE SYSTEM WITH CP-IV ARI CEMENT IN THE RATIO W / C-0.65 FOR THE FIRST AND THIRD CYCLE OF THE IMMERSION AND ACCELERATED DRYING PROCESS WITH THE PRESENCE OF CHLORIDE.*

This is followed by showing the EIE graph for the CP-V ARI cement.

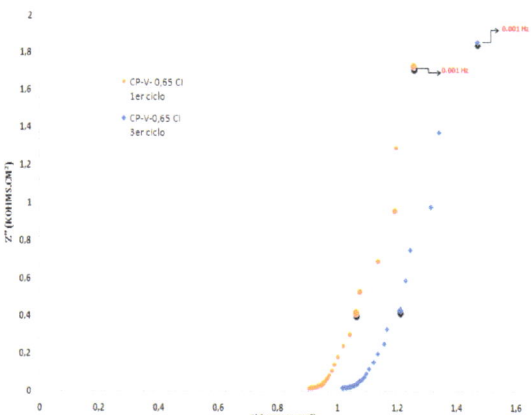

FIGURA 9 - *NYQUIST DIAGRAMS FOR THE CONCRETE SYSTEM WITH CP-V ARI CEMENT IN THE RATIO W / C-0.65 FOR THE FIRST AND THIRD CYCLE OF THE IMMERSION AND ACCELERATED DRYING PROCESS WITH THE PRESENCE OF CHLORIDE.*

Comparing the impedance at 10 mHz associated with the diffusion process of species to occur the transfer of load related to the resistance of the system, it is verified that the real component, attributed to the resistance of the system to the corrosive process at the interface steel-corrosion product, as it increases the exposure time from 1 cycle (24 days) to the 3rd cycle (70 days), increases the resistance of the system. For concrete with cement CP-IV 1st cycle was (3.46 KΩ) in relation to the system in the 3rd cycle of (5 kΩ), for the concrete with cement CP-V ARI 1st cycle was (1.25 KΩ) and in the 3rd cycle (1.47 KΩ). In this case, it did not present great differences with the passage of time which may be associated with the influence of the inhibitor according to the type of cement used in the concrete. Actually the system needs measurements at lower frequencies to determine the resistance of the transfer of loads. The following is a discussion of EIE results.

The nitrites react with the cement phases of the cement, so it is recommended to use them with low C_3A content. According to [11], it presents a thermogravimetric analysis, highlighting the three main bands of mass loss, especially in the first facsimile between 30°C and 370°C, referring to the decomposition of the C-S-H, gypsite and ethidite bonds [12]. By adding $NaNO_2$ (inhibitor) to concrete made with CP-V ARI cement which has high C_3A content, the Etringite formed consume the NO_2^- present in the system. Which will interfere with the efficiency of our inhibitor.

The author [7] carried out a transport analysis of the concentration of sodium nitrite in the concrete. It was classified as an anodic inhibitor, whose inhibitor group is the anion, where it was observed that, regardless of the initial concentration placed on the catholyte, from the 3rd day of the experiment it was possible to identify the presence of NO_3 (Nitrate) in the anolyte. Over time the amount of nitrate in the system is stabilized and the efficiency of the NO_2 (nitrite) is being slowed.

It was confirmed that Nitrite (NO_2) is the one that is transported to the anolith, undergoing an oxidation process whereby NO_3 is produced, causing a decrease of pH below 3 [7]. A further argument to explain the results obtained in our work for concrete with CP-V cement, where it should have shown a more resistive behavior, as in the case of a purer cement, should be more alkaline.

Based on the statistical results and the literature published by [13] and [3], the influence on the impedance results of the immersion cycles of the samples can be confirmed, confirming, in the case of nitrite, in The last measurement cycle in experimental phase I and phase II. Recalling that they are two totally different mechanisms since one is solid and the other liquid, in both experimental phases increased the resistance of the system in the measurement of the time of exposure, giving rise to a qualitative similarity of the behavior with the passage of time, Efficiency Of the (N).

Conclusions

From the impedance analysis, a better performance of the sodium nitrite inhibitor was obtained in relation to the other inhibitors analyzed (Phosphate and Ethanolamine)

It was observed that the concretes made with CP-V cement, where 2% concentrations of corrosion inhibitor additive were used in relation to the cement mass, the additive caused changes in the rheology, presenting fluidizing characteristics.

The Nitrite-based inhibitor additive had an increased efficiency in the last measurement cycle, both in experimental phase I and in phase II. Remembering that they are two totally different mechanisms since one is solid and the other liquid, in both experimental phases increased the resistance of the system in the measurement of the time of exposure. Resulting in a qualitative similarity of the behaviors.

A decrease in the impedance values occurred in the concrete where CP-V ARI cement was used when we related them to the concrete where CP-IV cement was used. This decrease in impedance may be related to the C_3A content of the cement, in addition to the need to perform impedance measurements at low frequencies and for longer periods

For the inhibitor Sodium Nitrite, both phases presented the same behavior, increasing their efficiency according to the cycles

Non-Conventional Materials and Technologies – NOCMAT for XXI Century   Materials Research Forum LLC
Materials Research Proceedings 7 (2018) 428-439   doi: http://dx.doi.org/10.21741/9781945291838-40

References

[1] Yiğiter, H., Yazıcı, H., & Aydın, S. (2007). Effects of cement type, water/cement ratio and cement content on sea water resistance of concrete. Building and environment, 42(4), 1770-1776. https://doi.org/10.1016/j.buildenv.2006.01.008

[2] A. C. Vasconcelos, " Inovações na engenharia do concreto, & quot; Revista Ibracon, 2002.

[3] M. M. Mennucci, " Avaliação da potencialidade de aditivos inibidores de corrosão do aço carbono CA-50 usado como armadura de estruturas de concreto, & quot; Dissertação de Mestrado, IPEN, São Paulo, 2006.

[4] Andrade, C; Alonso, C; Acha, M; Malric, B. Cement and Concrete Research 22, 869, 1992. https://doi.org/10.1016/0008-8846(92)90111-8

[5] BOLINA, C . et al. Inibidores de corrosão: Avaliação do desempenho frente à corrosão de armaduras induzida por cloretos e carbonatação em meio aquoso. ENGEVISTA, Brasil, V.15, p. 81-94, 2013.

[6] BOLINA, C. C. Inibidores de corrosão: Avaliação do desempenho frente à corrosão de armaduras induzida por carbonatação e cloretos. Tese de Doutorado. Dissertação (Mestrado em Engenharia Civil)–Universidade Federal de Goiás, Goiânia, 2008.

[7] MORENO, Mercedes Sánchez. Películas pasivas modificadas por el empleo de inhibidores de corrosión para la protección de armaduras: sistemas de prevención de la corrosión. Tese de Doutorado. Universidad Autónoma de Madrid, 2007.

[8] Lima, M. G.; Arvati Filho. A; HELENE, P. R.; Inibidores de corrosão: compatibilidade cimento-aditivo. EPUSP, 1996

[9] Helene, P. R., & Terzian, P. Manual de dosagem e controle do concreto. Pini, 1992.

[10] Andrade, C, Soler, L. and Nóvoa, X. R. Advances in electrochemical impedance measurement in reinforced concrete. MaterialsScienceForum, Trans. Tech. Publications, Switzerland, Vols. 192-194, pp 843-856. 1995.

[11] Hoppe Filho, J. Sistemas cimento, cinza volante e cal hidratada: mecanismo de hidratação, microestrutura e carbonatação de concreto (Doctoral dissertation, Universidade de São Paulo, 2008.

[12] Kosmatka S.H., Kerkhoff B. y Panarese W.C, Design and Control of Concrete Mixtures. 14th ed. Engineering Bulletin 001. Skokie, IL: Portland Cement Association, 358, 2002.

[13] Saura GÓMEZ, P. Inhibidores en el inicio y propagación del proceso de corrosión de las armaduras en el hormigón armado. Universidad de Alicante, 2011. ISBN 8469509608.

Non-Conventional Materials and Technologies – NOCMAT for XXI Century   Materials Research Forum LLC
Materials Research Proceedings 7 (2018) 440-448   doi: http://dx.doi.org/10.21741/9781945291838-41

Evaluation of the Mechanical Behavior of Composites Reinforced with Vegetal Fibers

Martha L. Sánchez [a,*], L.Y. Morales, J.D. Caicedo [a]

[a] Universidad Militar Nueva Granada, Colombia

martha.sanchez@unimilitar.edu.co*, luz.morales@unimilitar.edu.co,
asistente3.estructuras.sismica@unimilitar.edu.co

Abstract. At present, composites reinforced with vegetal fibers constitute one of the main areas of interest in the research and development of new materials. The chemical characteristics of most vegetable fibers and the fact of their being rich in hydroxyl groups determine their mechanical behavior and allow them to be used as reinforcement in polymeric composites, which use resins based on polyurethanes. This study analyzes the effect of fiber orientation and surface treatment on the mechanical performance of panels constructed using non-conventional materials made from fibers and resin of vegetable origin. To evaluate the effect of the orientation of the fibers on the performance of the composite, two fiber typologies were analyzed: long unidirectional fibers and randomly distributed short fibers. To improve the adhesion of the fibers to the matrix, a process of surface modification was carried out, using two types of treatment: chemical and plasma. As a matrix of the composite, the use of a material of vegetal origin is proposed: a polyurethane from *Higuerilla*. The fibers were obtained via the mechanical method. A pressing method was used for the construction of the panels. The physical characterization was based on the determination of the density, effective absorption, and dimensional stability. The mechanical characterization focused on the determination of the maximum tensile strength through compression and static bending tests. The results obtained demonstrate the viability of the use of vegetal fibers as reinforcement for natural composites.

Keywords: Vegetal Fibers, Mechanical Properties, Static Bending, Tensile, Compression

Introduction

The study of new materials denominated "bio-composites" has increased significantly in various branches of modern industry, mainly in civil engineering. Composites made from the combination of fibers and resin of vegetal origin exhibit good mechanical properties and low levels of toxicity, and they are biodegradable and can be recycled, which contributes to reducing the environmental impact generated by traditional materials [1-3].

Most vegetal fibers are less dense than the synthetic fibers traditionally used in manufactured composites [4]. This property, together with their mechanical characteristics, makes this reinforcement a viable option from the economic and technical point of view, mainly for those applications in which the use of light, resistant, and durable materials is required. Independent of the part of the plant (stem, leaves, seeds, fruits, etc.), the properties of the fibers may vary considerably [5]. It is for this reason that the selection of the fiber type suitable for a given application depends not only on its origin but also on parameters such as the diameter/length ratio, the degree of polymerization, and the crystalline structure [6].

There are several methods that allow the extraction of fibers from the stems of a plant. These methods can be classified into biological, mechanical, chemical, and physical methods [4]. The methods of mechanical extraction stand out for their speed, simplicity and economy. However, it is important to consider that depending on the origin of the fiber, surface degradation can occur, which may affect its performance as a reinforcement of the composite. On other hand, fibers obtained by mechanical means are generally thicker than fibers obtained with biological, physical, or chemical methods, which affects their diameter/length ratio [4].

Recent studies have shown that the surface texture of the fibers affects the mechanical performance of bio-composites. Through the application of techniques such as scanning electron microscopy (SEM) and atomic force microscopy (AFM), it has been possible to verify that the roughness of the fibers is a factor that favorably influences the adhesion between the fibers and the resin used as a matrix [7]. The presence of wax substances on the surface of the fibers prevents adequate impregnation of the resin, which leads to the development of a weak and poorly-bonded interface [8].

It has been found that with the application of physical or chemical modification methods on the surface of the fibers, it is possible to increase their roughness. Among the physical methods that can be used for the modification of the surface of vegetal fibers, treatment with plasma stands out. Recent studies have shown that regardless of the type of plasma used (argon, methane, etc.) and the duration and specific conditions of the treatment, it is possible to modify not only the roughness of the fibers but also their hydrophobicity [9–13].

One of the most frequently used chemical methods for fiber surface modification is treatment with sodium hydroxide solution. Recent results show that by subjecting a volume of fibers to an alkaline solution for a given period of time, it is possible to modify its crystallinity and partially remove the lignin and hemicellulose from the fibers. The application of this type of treatment causes modifications in the hydroxyl groups present in the amorphous regions of the material, improving its reaction with the binder. However, high solution concentrations or excessive treatment times can significantly deteriorate the fibers, resulting in loss of strength and stiffness [14–18].

In the present paper, the influence of the type of fibers, their orientation, and the effect of the surface treatment on the physical and mechanical properties of panels made with materials of vegetal origin is presented. In order to analyze the influence of the fiber type on the physical and mechanical properties of the bio-composite, two types of fibers were used: fibers of *Arundo Dónax* (AD) and fibers of bamboo of the species *Guadua Angustifolia Kunth* (GAK). To evaluate the effect of the length and orientation of the fibers on the performance of the material, two conditions were considered: randomly-distributed short bamboo fibers (FAD) and unidirectional long bamboo fibers (FLU). The determination of the effect of the surface treatment of the fibers was carried out on composites made with randomly-distributed bamboo fibers. Two treatments were analyzed: treatment with methane cold plasma and alkaline treatment. The determination of the physical and mechanical properties of the panels was carried out through laboratory tests of percentage of swelling, density, effective absorption, tensile, compression, and static bending.

Materials and methods

For the manufacturing process, fibers with an age between 1 and 2 years, extracted from the upper part of *Arundo Dónax* (AD) culms and fibers with an average age of three years extracted from the middle part of *Guadua Angustifolia Kunth* (GAK) bamboo culms were used. All the material used for obtaining fibers was immunized by immersion in borax and boric acid salts (according to the specifications of Colombian technical standard NTC 5301). A vegetable resin from *Higuerilla* oil was used as an adhesive. This resin consists of the combination of two components: a pre-polymer and a polyol. Its main physical properties are shown in Table 1.

TABLA 1- PHYSICAL PROPERTIES OF VEGETAL POLIURETHANE

Properties	Prepolymer	Polyol
Boilng point	190 °C	313 °C
Melting point	-14 °C	-10 °C
Density at room temperature	1.25 g/cm³	0.98 g/cm³

Non-Conventional Materials and Technologies – NOCMAT for XXI Century   Materials Research Forum LLC
Materials Research Proceedings 7 (2018) 440-448   doi: http://dx.doi.org/10.21741/9781945291838-41

For the extraction process, a mechanical crusher composed of a roller system coupled to a motor reducer was used (see Figure 1). This equipment allows the separation of long fiber bundles, which were cut according to the needs of their use. To evaluate the effect of the length and orientation of the fibers on the performance of the composite material, two test conditions were established: randomly distributed short fibers (FAD) and unidirectional long fibers (FLU). Lengths of 15 mm and 300 mm for the construction of panels using fibers extracted from GAK were selected. Due to the complexity of the process of extraction of AD fibers, only short fibers of 15 mm length were extracted.

FIGURE 1- *CRUSHING MACHINE*

For the surface modification of the fibers, two types of treatment were applied: physical and chemical. For the physical treatment, the fibers were subjected to the action of a methane cold plasma for 10 minutes. The working pressure was 27 Pa and the DC potential was -700 V, with a gas flow of 10 sccm and initial temperature of 18 °C and final temperature of 26 °C (see Figure 2). The chemical treatment consisted of the immersion of the material in a 10% sodium hydroxide solution for a period of 48 hours. This process was followed up with washing with distilled water and a drying process until reaching a constant mass value.

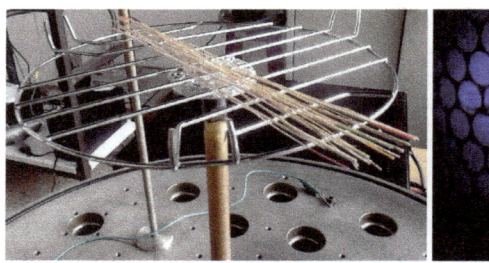 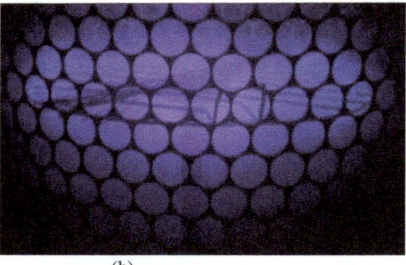

(a)   (b)

FIGURE 2 - *PLASMA TREATMENT: (A) FIBER INTO THE DEPOSITION CHAMBER, (B) FIBER UNDER METHANE PLASMA.*

Non-Conventional Materials and Technologies – NOCMAT for XXI Century   Materials Research Forum LLC
Materials Research Proceedings 7 (2018) 440-448   doi: http://dx.doi.org/10.21741/9781945291838-41

The dosing of the constituent materials was done with a steel mold of size 300 x 300 mm and average thickness 6 mm. In order to determine the environmental weight of the fibers, their density was determined according to the recommendations of Colombian technical standards. To establish the mixing proportions of the adhesive, at the suggestion of its supplier the use of a 1:1.5 ratio (pre-polymer-polyol) was selected. The use of these proportions allows a workability period of approximately 20 minutes before starting the drying process.

The manufacture of the panels was carried out using the method of manual molding and compaction by pressing at room temperature. This method consists of placing successive layers of fibers on a mold, doing the manual impregnation of the polyurethane, and compacting each layer in order to eliminate the air bubbles and at the same time contribute to the homogeneous distribution of the fibers and the resin in the mold. For the pressing of the panels, a hydraulic press was used, whose high rigidity frame is made up of four columns with a maximum load capacity of 100 tons, a digital indicator, a pressure transducer of 70 MPa, and a manual pump. The compression load was 18 tons, and the compaction time was 18 hours. Once removed from the mold, all panels were cured for 24 hours at room temperature. An image of the manufactured panels is shown in Figure 3.

The physical characterization of the composite was based on the determination of its density, effective absorption at 24 h, and percentage of swelling at 2 h and 24 h. The density was calculated according the specifications of ASTM D-2395 [19]. To estimate the percentage of effective absorption, the methodology described in ASTM D4442 [20] was employed. The percentage of swelling was calculated from the variation of thickness of square specimens of 50 mm in width after being submerged in water for periods of 2 h and 24 h.

(a)   (b)

FIGURA 3 - VEGETAL COMPOSITE PANELS: (A) RANDOMLY DISTRIBUTED SHORT FIBERS, (B) UNIDIRECTIONAL LONG FIBERS.

The mechanical evaluation was focused on the performance of tensile, compression, and static flexion tests. For the performance of the tests, the specifications of ASTM 1037 were followed [21].

Results and discussion
The physical properties of the untreated and chemically treated fibers are shown in Table 2.

TABLE 2 - PHYSICAL PROPERTIES OF VEGETAL FIBERS.

Properties	Untreated GAK	Treated GAK	Untreated AD	Treated AD
Moisture content (%)	8±0.9	9±4.0	8.4±2.4	8.8±2.1
Absorption (%)	63±1.3	45 ±2.4	71 ±6.5	58 ±7.1
Density (g/cm³)	0.8±0.04	0.7±0.2	0.62±0.01	0.59±0.01

From the results shown in Table 2, it is possible to note that with the application of an alkaline treatment, the lignin and other extractable substances that adhere to the surface of the fiber during the mechanical extraction are partially removed, which contributes to a reduction of approximately 30% of its absorption capacity and a reduction on the order of 10% of its density. Similar results were observed for the two types of fibers analyzed.

To evaluate the effect of the type of treatment on the physical properties of vegetal fibers, a treatment with methane cold plasma was carried out on GAK fibers of 15 mm in length. The results are shown in Table 3.

TABLE 3 - INFLUENCE OF TREATMENT ON THE PHYSICAL PROPERTIES OF FIBERS

Properties	Alkaline treatment	Methane cold plasma treatment
Moisture content (%)	9±4.0	9±1.40
Absorption (%)	45 ±2.4	28±6.05
Density (g/cm³)	0.7±0.2	0.6±0.03

Analyzing the results presented in Table 3, it can be verified that with the physical treatment the absorption capacity is reduced by 38% with respect to the value obtained for fibers treated with alkaline solution. At the same time, the reduction in density is approximately 15%. The results obtained affect not only the dosage of the fibers to be used in the construction of the panels, but also their weight and durability. The increased hydrophobicity of the fibers favors the development of a suitable fiber-matrix bond, contributes to the dimensional stability of the composite, and improves their performance, mainly in those applications in which they are exposed to wet conditions for long periods of time.

To evaluate the efficiency of the treatment with respect to the physical and mechanical properties of the bio-composites, panels with randomly-distributed bamboo short fibers were constructed. The surface of the fibers was treated following the two methodologies described above: alkaline treatment (AT) and treatment with methane cold plasma (MCPT). The results are presented in Table 4.

Non-Conventional Materials and Technologies – NOCMAT for XXI Century   Materials Research Forum LLC
Materials Research Proceedings 7 (2018) 440-448   doi: http://dx.doi.org/10.21741/9781945291838-41

TABLE 4 *- INFLUENCE OF THE TREATMENT ON THE PHYSICAL AND MECHANICAL*
PROPERTIES OF GAK PANELS

Properties	AT	MCPT
Density (g/cm^3)	1.08±0.04	0.89±0.02
Absorption (2 h) (%)	3.20±0.25	1.44±0.12
Absorption (24 h) (%)	23.11±2.45	6.04±0.11
Percentage of swelling (2 h) (%)	5.81±1.46	1.23±0.27
Percentage of swelling (24 h) (%)	11.85±3.34	4.05±0.67
Rupture Modulus (MPa)	27±0.70	35±0.50
Compression stress (MPa)	13±2.30	16±1.30

The results presented in Table 4 show that the value of the rupture modulus is modified with the treatment employed. The bending strength of panels made with fibers treated with methane cold plasma exhibited the best properties of flexural strength, 35 MPa, which is almost 20% greater than the stress of composites made with fibers treated with 10% of sodium hydroxide The increase in the values of the rupture modulus may be caused by bonding of the fiber with the vegetal resin, which improves the fiber-resin interaction. Similar results were obtained for the compression test.

In order to analyze the effect of fiber type on the physical and mechanical properties of panels made from materials of vegetal origin, panels using GAK fibers and AD fibers were constructed. The results are shown in Table 5. All fibers were treated with alkaline solution.

TABLE 5 *- INFLUENCE OF FIBER TYPE ON THE PROPERTIES OF COMPOSITES MADE*
FROM VEGETAL FIBERS

Properties	Type of fibers	
	GAK	AD
Density (g/cm^3)	1.02±0.03	1.03±0.03
Efective absorption (24 h) (%)	8.25±2.45	14±0.06
Percentage of swelling (2 h) (%)	6.60±1.46	8.99±1.7
Percentage of swelling (24 h) (%)	7.76±3.34	11.5±2.7
Tensile stress (MPa)	20.38±1.50	18.8±2.5
Modulus of elasticity (MPa)	1170±28	1890±45
Compression stress (MPa)	13±2.30	12±0.13

From the results presented in Table 5, it can be verified that panels made with GAK fibers are lighter and dimensionally stable. As can be seen, for panels made under the same conditions, there is a difference of about 30% in the result of the swelling percentage at two hours of immersion in water.

Non-Conventional Materials and Technologies – NOCMAT for XXI Century   Materials Research Forum LLC
Materials Research Proceedings 7 (2018) 440-448   doi: http://dx.doi.org/10.21741/9781945291838-41

This difference can be up to 40% at 24 hours of testing. Similar results were obtained when analyzing the effective absorption of both panels. Analyzing the mechanical properties, it can be seen that the tensile and compressive strength values of the two panels are similar (difference less than 10%); however, panels made with AD fibers exhibited a higher modulus of elasticity, showing a stiffness increase of approximately 38%.

In order to analyze the influence of fiber orientation on the physical and mechanical behavior of bio-composite panels, two types of reinforcement were considered: randomly distributed short fibers (FAD) and long unidirectional fibers (FLU). The results are shown in Table 6.

TABLE 6 - *INFLUENCE OF THE ORIENTATION OF THE FIBERS ON THE PROPERTIES OF COMPOSITES MADE WITH VEGETAL FIBERS*

Properties	Configuration of fibers	
	FAD	FLU
Density (g/cm^3)	1.02±0.03	1,11±0,04
Percentage of swelling (2 h) (%)	6.60±1.46	5.50±0.07
Percentage of swelling (24 h) (%)	7.76±3.34	6.70±0.04
Rupture modulus (MPa)	27±0.70	140.50 ±25
Modulus of flexural elasticity (MPa)	3700±26	14285 ±110
Compression stress (MPa)	13±2.30	13.90±1.50

The results shown in Table 6 allow verifying that on orienting the fibers in a single direction, there are no significant variations in the compressive strength of the composite. However, the results demonstrate that the use of unidirectional fibers not only improves the dimensional stability of the panel, but also considerably increases the static bending strength and the stiffness.

Conclusions

An analysis of the influence of the type and orientation of the fibers on the physical and mechanical properties of panels made with materials of vegetal origin was presented, evaluating the effect of the surface modification of the fibers on the performance of the bio-composites.

The results obtained indicate that independently of the type of treatment, it is possible to reduce the absorption capacity of the fibers, which contributes to improving the adhesion between the reinforcement and the matrix of the material and allows increasing the durability of the composite when it is exposed to humidity during prolonged periods of time.

From the present study, it can be concluded that factors such as the type of fiber, its interaction with the resin that acts as an adhesive, its orientation, and its size and surface condition are parameters that affect the mechanical behavior of the composite material, and for this reason they must be carefully analyzed during the conception, design, and manufacturing stages. In addition to the results presented in this paper, it is necessary to consider the effect of other variables that influence the performance of the bio-composites: the mixing percentages and the molding and compaction processes are variables which affect the properties and durability of these types of materials and require analysis in future studies.

Acknowlegment
This paper is a derivative product of the project (INV-ING-2392) financed by the Vice-rectory of Research of Universidad Militar Nueva Granada-validity (2017).

References

[1] Kumar N, Das D. Fibrous biocomposites from nettle (Girardinia diversifolia) and poly (lactic acid) fibers for automotive dashboard panel application. Composites Part B 2017:130:54-63. https://doi.org/10.1016/j.compositesb.2017.07.059

[2] Jacob M.J, Thomas S. Biofibers and biocomposites. Carbohydr Polym 2008:71:343-364. https://doi.org/10.1016/j.carbpol.2007.05.040

[3] Abel P, Lauter C, Gries T, Troester T. Textile composites in automotive industry. In: Carvalli V, Lomov SV, editors. Fatigues of textile composites. Amsterdam: Elsevier 2015: 383-401. https://doi.org/10.1016/B978-1-78242-281-5.00016-X

[4] Kalia S. et al., Cellulose Fibers: Bio- and Nano-Polymer Composites, Springer-Verlag Berlin Heidelberg 2011. https://doi.org/10.1007/978-3-642-17370-7

[5] Nevell T.P, Zeronian S.H.Cellulose chemistry and its applications. Wiley, New York, 1985.

[6] Toumis G.T. Structure, properties and utilization. Science and technology of wood.Van Nostrand Reinhold, New York, 1991, p 494.

[7] Pietak A, Korte S, Tan E, Downard A, Staiger M.P Atomic force microscopy characterization of the surface wettability of natural fibres. Appl Surf Sci 2007:253:3627–3635. https://doi.org/10.1016/j.apsusc.2006.07.082

[8] Tomczak F, Demetrio Sydenstricker T.H, Satyanarayana K.G et al. Studies on lignocellulosic fibres of Brazil Part II: morphology and properties of Brazilian coconut fibres. Compos Part A 2007:38:1710–1721. https://doi.org/10.1016/j.compositesa.2007.02.004

[9] Luna P., Lizarazo-Marriaga J., Mariño A. Guadua angustifolia bamboo fibers as reinforcement of polymeric matrices: An exploratory study. Construction and Building Materials 2016:116:93-97. https://doi.org/10.1016/j.conbuildmat.2016.04.139

[10] Shi T., Shao M., Zhang H., Yang Q., Shen X. Surface modification of porous poly(tetrafluoroethylene) film via cold plasma treatment. Applied Surface Science 2011:258(4):1474-1479. https://doi.org/10.1016/j.apsusc.2011.09.110

[11] Barra B.N., Santos S.F., Bergo P.V.A., Alves Jr. C., Ghavami K., Savastano Jr H.Residual sisal fibers treated by methane cold plasma discharge for potential application in cement based material. Industrial Crops and Products 77(2015):691-702. https://doi.org/10.1016/j.indcrop.2015.07.052

[12] Yuan X., Jayaraman K., Bhattacharyya D.Effects of plasma treatment in enhancing the performance of woodfibre-polypropylene composites, Composites Part A: Applied Science and Manufacturing 35(12) (2004):1363-1374. https://doi.org/10.1016/j.compositesa.2004.06.023

[13] Sever K., Erden S., Gülec H.A., Seki Y., Sarikanat M.Oxygen plasma treatments of jute fibers in improving the mechanical properties of jute/HDPE composites, Materials Chemistry and Physics 129(1)(2011):275-280. https://doi.org/10.1016/j.matchemphys.2011.04.001

Non-Conventional Materials and Technologies – NOCMAT for XXI Century   Materials Research Forum LLC
Materials Research Proceedings 7 (2018) 440-448   doi: http://dx.doi.org/10.21741/9781945291838-41

[14] Brígida A.I.S., Calado V.M.A., Gonçalves L.R.B., Coelho M.A.Z. Effect of chemical treatments on properties of green coconut fiber. Carbohydrate Polymers 79(2010):832–838. https://doi.org/10.1016/j.carbpol.2009.10.005

[15] Rokbi M., Osmania H., Imad A., Benseddiq N.Effect of Chemical treatment on Flexure Properties of Natural Fiber-reinforced Polyester Composite, Procedia Engineering 10(2011): 2092–2097. https://doi.org/10.1016/j.proeng.2011.04.346

[16] Akhtara M.N., Sulong A.B., Fadzly Radzi M.K., Ismail N.F., Raza M.R., Muhamad N., Khan M.A.Influence of alkaline treatment and fiber loading on the physical and mechanical properties of kenaf/polypropylene composites for variety of applications, Progress in Natural Science: Materials International 26(2016):657–664. https://doi.org/10.1016/j.pnsc.2016.12.004

[17] Orue A., Jauregi A., Unsuain U., Labidi J., Eceiza A., Arbelaiz A. The effect of alkaline and silane treatments on mechanical properties and breakage of sisal fibers and poly(lactic acid)/sisal fiber composites, Composites: Part A 84(2016):186–195. https://doi.org/10.1016/j.compositesa.2016.01.021

[18] Fiore V., Di Bella G., Valenza A. The effect of alkaline treatment on mechanical properties of kenaf fibers and their epoxy composites, Composites: Part B 68(2015):14–21. https://doi.org/10.1016/j.compositesb.2014.08.025

[19] ASTM D2395-14. Standard Test Methods for Density and Specific Gravity (Relative Density) of Wood and Wood-Based Materials. Annual Book of ASTM Standards, 2014.

[20] ASTM D4442. Standard test method for direct moisture content measurement of wood and wood-based materials. Annual Book of ASTM Standards, 2016.

[21] ASTM D1037. Standard Test Methods for Evaluating Properties of Wood-Base Fiber and Particle Panel Materials. Annual Book of ASTM Standards, 2012.

Non-Conventional Materials and Technologies – NOCMAT for XXI Century   Materials Research Forum LLC
Materials Research Proceedings 7 (2018) 449-456   doi: http://dx.doi.org/10.21741/9781945291838-42

Effect of Alkaline Hornification in Sisal Fibers on the Mechanical Behaviour

Santos, R. D. [a], Ferreira, S. R. [b], Santos, E. R. F. [a], Oliveira, G. E. [a], Silva, F. A. [c],
Souza Jr., F. G.[a], Toledo Filho, R. D. [a]

[a] Universidade Federal do Rio de Janeiro – UFRJ - COPPE, Brasil; renatadaniel@poli.ufrj.br, geizaesperandio@gmail.com, edinhorfs@gmail.com, toledo@coc.ufrj.br

[b] Universidade Federal de Lavras – UFLA, Brasil; ferreira.sr@hotmail.com

[c] Pontifícia Universidade Católica do Rio de Janeiro – PUC-Rio, Brasil; fsilva001@uol.com.br

Abstract. Aiming at a reduction in the water absorption capacity of lignocellulosic fibers, wetting and drying cycles are usually used in the paper and cellulose industry. This procedure stiffens the polymeric structure of the fiber-cells (process known as hornification) resulting in a higher dimensional stability. Several authors have proposed treatments in natural fibers, including hornification, that modifies the surface of the fibers and increase the mechanical behavior. The present study presents a comprehensive analysis of the influence of alkaline hornification with calcium hydroxide 0.7% (1 cycle) on the structure modification, mechanical response, durability performance and bond behavior of sisal fibers. The intrinsic changes on the fiber structure as well as their physical and chemical characteristics were evaluated through analytical techniques such as X-ray diffraction, Thermogravimetry, FTIR and Scanning Electronic Microscope, while their mechanical response was evaluated with direct tensile tests. The obtained results indicate that the hornification process changes the fiber properties, mainly morphological, physical and chemical properties, which improves their mechanical properties.

Keywords: Natural Fibers, Chemical Treatment, Alkaline Hornification

Introduction

Vegetables fibers are being used as reinforcement in cementitious composites not only for their reinforcement capability but also for other advantages, such as biodegradability, abundance, low cost, low health risk and the potential of economic development in the regions where they are cultivated. Nevertheless the use of vegetable fibers also presents some problems such as high water absorption and low durability in alkaline media and this can lead to fiber mineralization and low adhesion with cementitious matrices [1-3].

In order to overcome these problems some different strategies can be used, in an isolated or associated way, among them the most important are (i) the use of calcium hydroxide (CH) free matrices and (ii) the application of some chemical treatments such as acetylation, hornification, polymer impregnation, alkaline and thermal treatment [4-5].

The chemical treatments most widely used on vegetable fibers are acid hydrolysis and alkaline treatment. These treatments include the well-known delignification process. Vegetable fibers have also been subjected to modifications that accelerate their degradation, turning into monosaccharides. In the alkali treatments, the ester-linked molecules of the hemicellulose and other cell-wall components can be cleaved. This tends to increase the hydrophilicity and hence the solubility of the material. Alkali treatment of lignocellulosic substances disrupts the cell wall, dissolves hemicellulose and lignin by hydrolyzing acetic acid esters and by swelling cellulose, and decreases the crystallinity of cellulose. The biodegradability of the cell wall increases, due to the cleavage of the bonds between lignin and hemicellulose or between lignin and cellulose [6].

Non-Conventional Materials and Technologies – NOCMAT for XXI Century   Materials Research Forum LLC
Materials Research Proceedings 7 (2018) 449-456   doi: http://dx.doi.org/10.21741/9781945291838-42

The treatment of vegetable fibers with sodium hydroxide (NaOH) is widely used to modify the cellulosic molecular structure. It changes the orientation of highly packed crystalline cellulose order and forms an amorphous region. This provides more access for chemicals to penetrate. In the amorphous region, cellulose micro-molecules are separated, allowing increasing distances and the space for water molecules to infiltrate. Alkali sensitive hydroxyl (OH) groups, present among the molecules, are broken down, which then react with water molecules (HOH) and move out from the fiber structure. The remaining reactive molecules form fiber–cell–O–Na groups between the cellulose molecular chains. Due to this, hydrophilic hydroxyl groups are reduced and increase the fibers moisture resistance property. It also reduces a certain portion of hemicelluloses, lignin, pectin, wax and oil covering materials. As a result, the fiber surface becomes clean and more uniform due to the elimination of microvoids and thus the stress transfer capacity between the ultimate cells improves. In addition to this, it reduces fiber diameter and thereby increases the aspect ratio (length/diameter) of the fiber. If the alkali concentration and/or exposition are higher than the optimum, the excess delignification of the fiber can take place, which results in weakening or damaging the fibers. Treated fibers have lower lignin content, a partial reduction of wax and oil cover materials and distension of crystalline cellulose order [7].

In this study, sisal fibers were submitted to one cycle of soaking and drying in an alkaline hornification treatment with calcium hydroxide $Ca(OH)_2$ 0.7% w/v, to evaluate the impact on the fiber structure and the consequences in their stress-strain behavior. $Ca(OH)_2$ was selected because its presence in the matrix can be almost eliminated by the use of CH free matrices, but the long exposure, even in small concentration, causes severe damage to the vegetable fibers. Thus, the main objective of this study is to understand better the mechanism that $Ca(OH)_2$ modifies the vegetable fiber structure and how these changes influence the stress-strain behavior.

Materials and methods
Materials
Sisal fibers were obtained from the Associação de Desenvolvimento Sustentável e Solidário da Região Sisaleira (APAEB), town of Valente (BA), Brazil. The calcium hydroxide P.A. were supplied by Vetec. The used water was distillated in the own laboratory. All materials were used as received.

Vegetable Fibers Calcium hydroxide Hornification
Vegetable fibers were soaked in 0.7% w/v of $Ca(OH)_2$ solution under controlled temperature (21 ± 1°C) for 50 min. Literature [8] shows that soaking for periods from 30 to 60 min, in small alkali concentrations (0.5-1%) causes no degradation to the vegetable fibers. After that, the $Ca(OH)_2$ saturated fibers were dried in an air flow chamber at 40°C for 24h up to constant mass.

Vegetable Fibers Characterization
FTIR spectra were performed using a Perking Elmer spectrometer, model Frontier FT-IR/FIR, and ATR with a ZnSe crystal. The range measured was from 4000 to 600 cm^{-1}, with 4 cm^{-1} of resolution and 60 accumulated scans. XRD diffractograms were carried out using a Bruker, model D8 Focus X ray diffractometer with the FT (fixed time) method and CuKα radiation, with wavelength of 0.1542 nm. The used 2θ range was from 10 to 40° with angular steps equal to 0.05°/s and the tube voltage and current were equal to 30 kV and 35 mA, respectively. The crystalline degree was calculated by the Ruland method [9] [Eq. 1]:

$$Xc = \left(\frac{Ac}{Ac+Aa}\right)100 \qquad Eq(1)$$

Non-Conventional Materials and Technologies – NOCMAT for XXI Century   Materials Research Forum LLC
Materials Research Proceedings 7 (2018) 449-456   doi: http://dx.doi.org/10.21741/9781945291838-42

Where Xc is the crystalline degree, Ac is the crystalline area and Aa is the amorphous area. TGA analysis was performed using a TA instrument model SD 2960 with heating rate of 10°C/min under a nitrogen flow of 100 mL/min and temperature range from 35 to 800°C. The initial sample was around 10mg in Pt pan. The water absorption index (Iaw) was obtained from a methodology proposed by Toledo Filho [10], which is an adaptation from the one used by the paper industry [Eq. 2]:

$$I_{aw} = \frac{WF_{dri} - WF_{sat}}{WF_{dri}}$$

Eq(2)

Where WF_{dri} is the weight of dried fibers in an oven with about 40°C up to constant weight and WF_{sat} is the weight of the water saturated fibers, obtained by immersing the dried fiber in water for up to 36 hours at room temperature. The morphology of the fibers was determined using a scanning electronic microscope (SEM) from Hitachi model TM3000, under vacuum with a secondary electron detector from Everhart-Thornley – ETD and voltage of 15kV. The use of energy-dispersive X ray spectroscopy (EDX) is associated with SEM.

Vegetable Fiber Stress-Strain Behavior
The tensile strength tests of the fibers were carried out using an electromechanical Shimadzu AG-X100kN with a load cell of 1kN and a displacement rate of 0.1 mm/min. The fibers, with a length of 50 mm, were glued to a paper template, for a better alignment in the machine and a better gripping in the upper and lower jaws, in accordance with ASTM C1557 [11]. In order to calculate the tensile strength of the fibers, their diameters were measured by analyzing the images obtained from SEM.

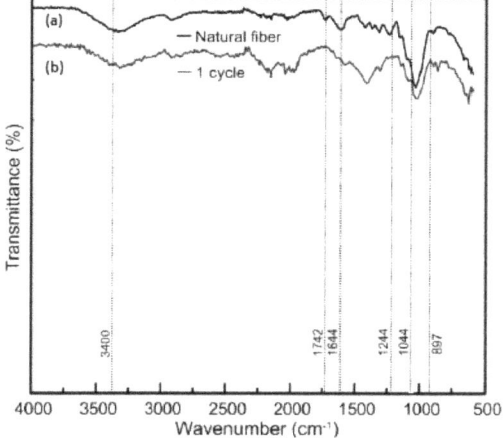

FIGURE 1 – FTIR SPECTRA OF SISAL FIBERS: (A) NATURAL; (B) 1 CYCLE CH.

Results and discussions
Figure 1 presents the FTIR spectra of sisal fibers before and after the alkaline hornification cycle. All spectra show same characteristic bands, typical for lignincellulosic materials. Stretching of –OH group can be seen around 3350 cm^{-1}, while the C–H stretching appears around 2920, 2850 and 1365 cm^{-1}. The characteristic band at 1742 cm^{-1} is related to coupled stretching of C=O and the C=C bonds. Bending mode of -OH group from water in the hemicelluloses is observed around 1644 cm^{-1},

Non-Conventional Materials and Technologies – NOCMAT for XXI Century   Materials Research Forum LLC
Materials Research Proceedings 7 (2018) 449-456   doi: http://dx.doi.org/10.21741/9781945291838-42

while the aromatic skeleton vibration of C-C bonds shows up at 1422 cm^{-1}. Scissoring deformation in the plane of the ring of O-H bond is at 1321 cm^{-1} and aromatic stretching of C=O is at 1244 cm^{-1}. The band at 1160 cm^{-1} is attributed to the C-O-C asymmetric stretching, and symmetric stretching of this bond appears at 1105 cm^{-1}. Stretching O-C-C of appears at 1165 cm^{-1} and 1036 cm^{-1}. The characteristic band at 897 is related to C-H scissoring deformation out-of-plane in the aromatic ring [12-14].

Hornification treatment caused a reducing in the intensity of all fibers spectra, mainly in the characteristic bands associated with the aromatic compounds, in this case, the lignin, which proves the considerable removal of this component from the fiber by the treatment.

Figure 2 shows the XRD of sisal fibers before and after the alkaline hornification cycle. It is possible to observe a typical diffraction pattern of ligniclellulosic materials, with 2 theta peaks at 16.6°, 22.5° and 34° attributed to (101), (002) and (040), respectively [15]. As expected the removal of lignin and hemicelluloses, which are amorphous, causes an increment in the crystalline, as noticed by the higher intensity in the peaks signal. The calculated crystalline degree is presented in the Table 1.

TABLE 1 – CALCULATED CRYSTALLINE DEGREE OF SISAL FIBER

Sisal Fiber Treatment	Crystalline Degree (%)
Natural	44.5 ± 2.2
1 cycle CH	46.7 ± 2.3

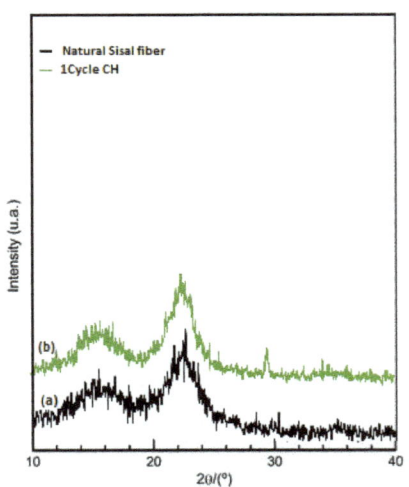

FIGURE 2 –XRD OF SISAL FIBERS: (A) NATURAL; (B) 1 CYCLE CH.

Figure 3 presents the results to TGA / DTG of the sisal fibers. The thermal degradation profile of this kind of fiber occurrs in three stages of weight loss: (i) evaporation of moisture, below 100°C, (ii) decomposition of hemicelluloses, from 250 to 350°C, and (iii) degradation of cellulose, from 325 to 400°C. The later one is major weight loss detected in this sample. Besides that, the lignin degradation

Non-Conventional Materials and Technologies – NOCMAT for XXI Century   Materials Research Forum LLC
Materials Research Proceedings **7** (2018) 449-456   doi: http://dx.doi.org/10.21741/9781945291838-42

occurs between 200°C and 600°C, with no evident degradation step, due to the way the lignin is spread in the cellulosic fiber structure [16]. It was noted that the hornification treatment reduces the amount of hemicelluloses present in the fiber.

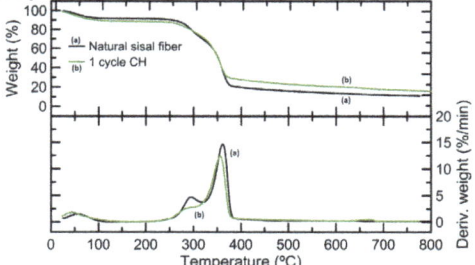

FIGURE 3 *–TGA / DTG OF SISAL FIBERS: (A) NATURAL; (B)1 CYCLE CH.*

Figure 4 presents the water absorption behavior of the tested vegetable fibers. Table 2 shows the value of water absorption. The sisal fibers achieved the maximum water absorption up to 5 minutes of soaking. In both conditions, before and after the treatment, the natural sisal fiber absorbed (193 ± 15) % of water while for 1 cycle CH sisal fiber the water absorption was (237 ± 25)%, respectively. Sisal fibers were able to absorb more water after 1 cycle the hornification. Therefore those fibers will take much more hornification cycles to achieve the dimensional stability.

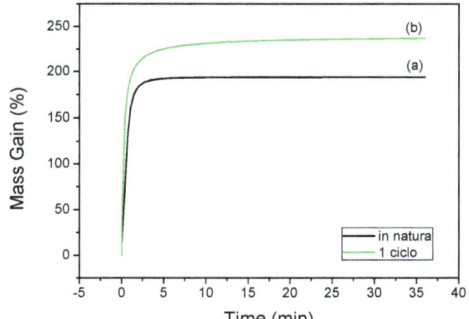

FIGURE 4 *–WATER ABSORPTION OF SISAL FIBERS: (A) NATURAL; (B) 1 CYCLE CH.*

TABLE 2 *– CALCULATED WATER ABSORPTION OF SISAL FIBER*

Sisal Fiber Treatment	Water Absorption (%)
Natural	193 ± 15
1 cycle CH	237 ± 25

Comparing the cross section of the fibers, before and after hornification treatment, shown in Figure 5, in this case an increase could be seen in the thickness of the secondary wall, like a swelling, causing a reduction of the fiber lumens. Figure 6 presents the side structure of sisal fibers, before and after the treatment. It is possible to see some $Ca(OH)_2$ deposition on the fiber surface. None significant change was observed in the fiber after only one cycle of alkaline hornification.

Non-Conventional Materials and Technologies – NOCMAT for XXI Century   Materials Research Forum LLC
Materials Research Proceedings 7 (2018) 449-456   doi: http://dx.doi.org/10.21741/9781945291838-42

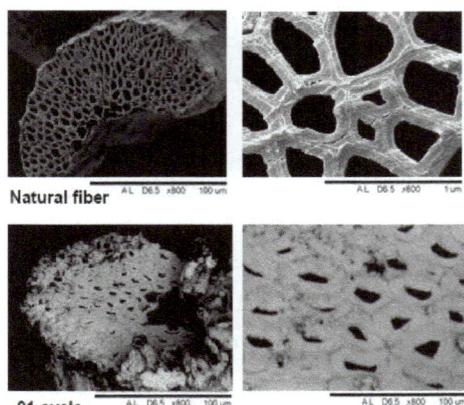

FIGURE 5 *–SEM IMAGES OF CROSS SECTION OF SISAL FIBERS: NATURAL AND 1 CYCLE CH.*

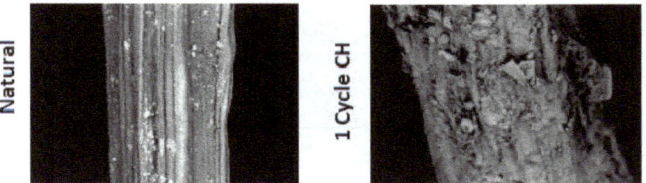

FIGURE 6 *–SEM IMAGES OF SIDE OF SISAL FIBERS: NATURAL AND 1 CYCLE CH.*

Stress-strain behavior curves of the sisal fibers are shown in Figure 7. Mechanical properties of these fibers, obtained from the curves presented in Figure 7, are presented in Table 3.

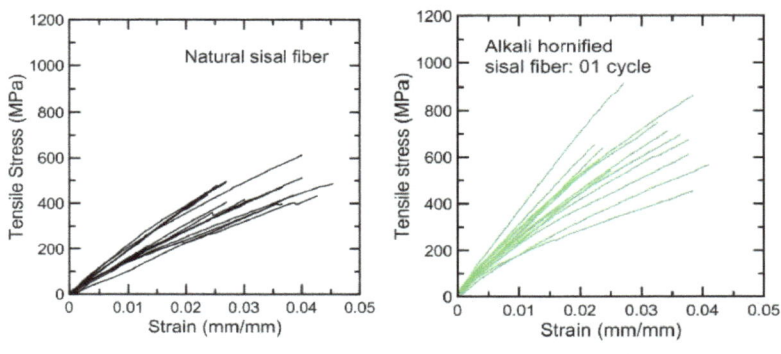

(a)   (b)

FIGURE 7 *– STRESS – STRAIN BEHAVIOR CURVES OF SISAL FIBERS: (A) NATURAL AND (B) 1 CYCLE CH.*

Non-Conventional Materials and Technologies – NOCMAT for XXI Century   Materials Research Forum LLC
Materials Research Proceedings 7 (2018) 449-456   doi: http://dx.doi.org/10.21741/9781945291838-42

TABLE 3 – *MECHANICAL PROPERTIES OF TESTED VEGETABLE FIBERS*

Treatment	Max Load (N)	Tensile Strength (MPa)	Tensile Strain (mm/mm)	Young's Modulus (GPa)
Natural	10.28 ± 3.43	453 ± 66	0.03 ± 0.01	20 ± 4
1 Cycle CH	13.36 ± 2.32	709 ± 103	0.03 ± 0.01	24 ± 2

It is possible to observe that the hornification treatment produced an increment in the mechanical resistance of fibers. This increment was 56% for sisal fiber in the tensile strength, and 20% for sisal fiber to the Young's modulus. As described in the literature [6-7] these kind of treatment removes partially the lignin and hemicelluloses from the fibers, leaving the cellulose, which is the crystalline phase and more resistant mechanically. Moreover, the lignin and hemicelluloses removal results in a reduction of fiber diameter, increasing their aspect ratio (length/diameter) and contributing to the improvement of the mechanical resistance.

Conclusion

Alkaline hornification treatment with calcium hydroxide, even with only one cycle of soaking and drying, proved to be very efficient to remove partially the lignin and hemicelluloses from the vegetable fibers. This kind of treatment is able to increase the crystalline degree of fibers, especially that of the jute fiber, which shows the highest crystalline degree increment. Besides that, alkaline hornification also promotes changes in the fiber morphology, with an increase in the thickness of the secondary fiber wall and reduction of the lumen. The water absorption by the vegetal fibers is also affected - after the hornification only curaua presented a decrease in water absorption. With regards to the mechanical resistance, the alkaline hornification treatment with only 1 cycle has improved significantly the mechanical resistance. The best result obtained was for the curaua fiber, which showed a significant increment in both tensile strength and Young's modulus.

Acknowledgments

The authors would like to thank to Conselho Nacional de Desenvolvimento Científico e Tecnológico (CNPq), Coordenação de Aperfeiçoamento de Pessoal de Nível Superior (CAPES), Financiadora de Estudos e Projetos (FINEP) and Fundação de Amparo à Pesquisa do Estado do Rio de Janeiro (FAPERJ) for financial support and scholarships.

References

[1] Onuaguluchi, O.; Banthia, N. Plant-based natural fibre reinforced cement composites: A review, Cem. Conc. Comp. 2016, 68, 96-108.

[2] Ferreira, S. R.; Lima, P. R. L.; Silva, F. A.; Toledo Filho, R. D. Effect of Sisal Fiber Hornification on the Fiber-Matrix Bonding Characteristics and Bending Behavior of Cement Based Composites, Key Eng. Mat. 2014, 600, 421-432. https://doi.org/10.4028/www.scientific.net/KEM.600.421

[3] Barra, B.; Paulo, B.; Alves Junior, C.; Savastano Junior, H.; Ghavami, K. Effects of Methane Cold Plasma in Sisal Fibers, Key Eng. Mat. 2012, 517, 458-468. https://doi.org/10.4028/www.scientific.net/KEM.517.458

[4] Beraldo, A. L.; Payá, J.; Monzó, J.M. Evaluation of Compatibility between Sugarcane Straw Particles and Portland Cement, *Key Eng. Mat. 2014*, 600, 250-255. https://doi.org/10.4028/www.scientific.net/KEM.600.250

[5] Hospodarova, V.; Singovszka, E.; Števulová, N. Characterization of Cellulosic Fibres Properties for their Using in Composites, Sol. Sta. Phen. 2016, 244, 146-152. https://doi.org/10.4028/www.scientific.net/SSP.244.146

[6] Arsène, M. A.; Bilba, K.; Savastano Jr, H.; Ghavami, K. Treatments of non-wood plant fibres used as reinforcement in composite materials, 2013. Mat. Res. 16(4): 903-923. https://doi.org/10.1590/S1516-14392013005000084

[7] Kabir, M. M.; Wang, H.; Lau, K. T.; Cardona, F. Chemical treatments on plant-based natural fibre reinforced polymer composites: An overview. Comp. Part B: Eng. 2012, 43 (7) 2883-2892. https://doi.org/10.1016/j.compositesb.2012.04.053

[8] Sasha P., Manna S, Chowdhury S. R., Sen R., Roy D., Adhikari B. Enhancement of tensile strength of lignocellulosic jute fibres by alkali-steam treatment. Bioresources Technology 2010;101:3182–3187. https://doi.org/10.1016/j.biortech.2009.12.010

[9] W. Ruland. X-ray Determination of Crystallinity and Diffuse Disorder Scattering, Acta Crystallogr. 1961, 14, 1180-1185. https://doi.org/10.1107/S0365110X61003429

[10] Toledo Filho, R. D.; England, G.L; Ghavami, K Comportamento em Compressão de Argamassas Reforçadas com Fibras Naturais - Parte A: Relação Tensão-Deformação Experimental e Processo de Fratura. *Rev. Bras. Eng. Agric. Amb. 1997* 01 (01), 79-88.

[11] ASTM C1557-14 Standard test method for tensile strength and Young's modulus of fibers ASTM International, West Conshohocken, PA 2014.

[12] Ahujaa, D.; Kaushika, A.; Chauhanb, G. S. Fractionation and physicochemical characterization of lignin from waste jute bags: Effect of process parameters on yield and thermal degradation. *Int. Jour. Biol. Macrom.* 2017, 139, 551-561. https://doi.org/10.1016/j.ijbiomac.2017.01.057

[13] Elias, E.; Costa, R.; Marques, F.; Oliveira, G.; Guo, Q.; Thomas, S.; Souza Jr, F. G. Oil-spill cleanup: The influence of acetylated curaua fibers on the oil-removal capability of magnetic composites. *Jour. Ap. Pol. Sci.* 2015, 132 (13), 41732.

[14] Silvertein, R. M.; Webster, F. X.; Indentificação Espectrometric de Compostos Orgânicos, 6ª ed., LTC, Rio de Janeiro 2000.

[15] Mazlita Y., H.V. Lee; S.B.A. Hamid Preparation of Cellulose Nanocrystals Bio-Polymer From AgroIndustrial Wastes: Separation and Characterization. *Polym. and Polym. Comp.* 2016, 24 (9) 719-728. https://doi.org/10.1177/096739111602400907

[16] Wei, J.; Meyer, C. Utilization of rice husk ash in green natural fiber-reinforced cement composites: Mitigating degradation of sisal fiber *Cem. Conc. Res.* 2016, 81, 84-111.

Non-Conventional Materials and Technologies – NOCMAT for XXI Century   Materials Research Forum LLC
Materials Research Proceedings 7 (2018) 457-461   doi: http://dx.doi.org/10.21741/9781945291838-43

Synthesis and Characterization of Nano Magnetite for Application in Cementitious Materials for Civil Construction

Leonardo Carreira Moren Scapim [a], Júnia Nunes de Paula [a, *], Luciana Patrícia Ferreira[a], Pedro Rodrigues de Almeida III[a], Sílvia Belloni Borges[a].

[a] Centro Federal de Educação Tecnológica de Minas Gerais – CEFET-MG, Brazil;
leo.moren17@gmail.com

junia@deii.cefetmg.br, lupiferreiracefetmg@gmail.com, pedroiii@yahoo.com.br,
silvia.bellonib@gmail.com

Abstract. The incorporation of nanoparticles in small amounts has been shown to have a significant influence on the properties of the materials. This can contribute to the creation of innovative and sustainable structures. The object of this research is the synthesis of Fe_3O_4 nano particles and the physical treatment of the chemical mixture for application in cementitious materials. The chemical synthesis was performed from the mixture of ferric chloride ($FeCl3$), potassium iodide (KI) and ammonia solution (25%), on temperature and controlled agitation. The addition of Nano Fe_3O_4 particles improves the cementitious materials microstructure, enhancing its durability and improving its mechanical properties.

Keywords: Nano Magnetite Synthesis, Cementitious Materials, Durability

Introduction

Nanotechnology deals with the production and application of physical, chemical and biological systems at scales ranging from few nanometers to submicron dimensions [1].

The use of nanoparticles in cement and concrete can lead to improvements in the nanostructure of building materials [2, 3].

Nanomaterials are known for showing physical and chemical properties that can lead to the development of more effective materials than the ones that are currently available [4].

These materials are considered unique because their size affects the behavior of the mixture in which they are added often providing significantly improved mechanical properties. In the nano scale, particles are able to form predictable grain-size materials of the same chemical composition. With the reduction in size, more atoms are located on the surface of a particle that, in addition benefits from the remarkable surface area of nano powders. This imparts a considerable change in surface energies and surface morphologies of the material. All these factors alter the basic properties and the chemical reactivity of nanomaterials [5].

Incorporation of nanoparticles in cement and cement-based concrete resulted in super mechanical properties, durable and desirable stress-strain behavior [6].

Magnetite is a mineral that in its natural formation comes from the reaction between ferrous oxide (FeO) and ferric oxide (Fe_2O_3), presenting an inverted spinel cubic structure, containing iron in the Fe^{+2} and Fe^{+3} oxidation states in the proportion 1 to 2, according to Figure 1 [7].

Synthesis, characterization and applications of iron oxide, nanocrystals have received tremendous attention in recent years due to their potentials for information storage device rotary shaft sealing, position sensing [8], as well as medical and pharmaceutical applications [9].

Fe_3O_4 nanoparticles in cementitious materials are able to act as nanofillers and recover the pore structure of the specimens by decreasing harmful pores to improve the water Fe3O4 nanoparticles in the cementitious materials as a foreign nucleation site could accelerate C-S-H gel formation as a result of increased crystalline $Ca(OH)_2$ amount especially as the early age of hydratation and hence increase the strengtht of the specimens permeability [10].

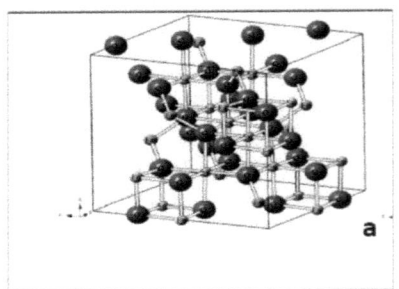

FIGURE 1- NANOMAGNETITE UNIT CELL.

The rate of the pozzolanic reaction is proportional to the amount of surface area available for reaction. This explain the enhanced mechanical performance of the cement mortar with nanoparticles. It was also found that nano magnetite exhibits a self-sensing of strain capability which can be used for structural health monitoring [11].

Several methods are known and recognized as eficient when synthesizing superparamagnetics nanoparticles. The co-precipitation synthesis method is one of the most used. The methodology used by [12] presented the possibility of the nanoparticles synthesis by means of a simple reaction and few reagents.

The objective of this research is to synthesize and characterize superparamagnetic nanoparticles for use in cementitious materials.

The importance of study is to enable, in the near future, the construction of structures in cementitious materials that self-monitor their deformation, what can increase their life cycle and consequently more sustainable.

Materials and methods
Materials
Table 1 presents the materials used in Fe_3O_4 nanoparticles co-precipitation synthesis.

TABLE 1 - MATERIALS USED IN CO-PRECIPITATION SYNTHESIS

Materials
$FeCl_3$ (Ferric Chloride)
KI (Potassium Iodide)
NH_4OH (Ammonia Hydroxide)25%
H_2O (Water)

Methods
The synthesis of the Fe_3O_4 nanoparticles by co-precipitation was performed by mixing the ferric chloride hexahydrate ($FeCl_3.6H_2O$) with the potassium iodide (KI) solution in a ratio of 3: 1 mol (Figure 2). At the beginning of the tests, the solution with the reagents becomes yellow due to the presence of Fe^{+2} and Fe^{+3} ions.

After mixing, the solution placed under constant rotation at room temperature to the equilibrium point, after the total precipitation of the iodine.

The solution filtered having the iodine retained in the filter (Figure 3).

Non-Conventional Materials and Technologies – NOCMAT for XXI Century   Materials Research Forum LLC
Materials Research Proceedings **7** (2018) 457-461   doi: http://dx.doi.org/10.21741/9781945291838-43

FIGURE 2 - *Solution A (FeCl3 + distilled water) and solution B (KI + distilled water)*

FIGURE 3 - *IODIDE RETAINED IN THE FILTER.*

In the resulting solution was added ammonium hydroxide (NH$_4$OH). The ammonia hydroxide added dropwise until complete precipitation of nano magnetite (Figure 4).

The precipitated magnetite filtered and dried at a temperature of 100 ° C in an oven for 16 hours.

After drying, the dry and agglomerated material produced is comminuted.

The present method was based on the methodology used by [12].

FIGURE 4 - *THE RESULTING SOLUTION*

For the chemical characterization of Fe$_3$O$_4$ nano particles an X-ray diffraction was performed.

The X-ray diffraction (XRD) is a technique used to acquire data on the crystalline structure of the material. To acquire the sample crystallization data, to identify and quantify the present phases, to identify the presence of amorphous material, to measure the size of the smallest crystal present in the structure and to define elements of the crystalline phase.

The microstructure of the material was analyzed by scanning electron microscopy (SEM).

459

Scanning Electron Microscopy (SEM) is a technique used for the microstructural visual analysis of materials, which has a nano metric resolution range.

To verify the paramagnetic behavior of the nanoparticle inside cement an amount of 0.1% was mixed the cement paste.

Results and discussions

Figure 5 shows the XRD of the synthesized nanoparticles and the standard XRD of the nano magnetite.

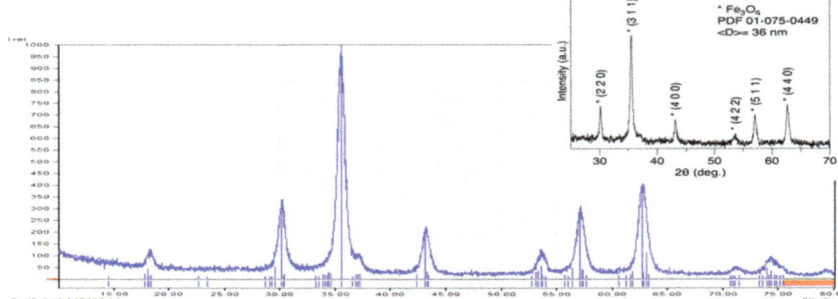

FIGURE 5 - *NANO PARTICLES SYNTHETIZED AND NANO MAGNETITE STANDARD XRX*

According to the presented result, the formation of a single phase is perceived. The X-ray diffraction peak characteristic of the interplanar distances of the Fe_3O_4 phase indexed according to the JCPDS-2003 crystal structure database.

The analysis of crystallite size performed by the Scherrer equation in relation to the main peak of the magnetite phase provided crystalline size of the order of 18.056nm.

Figure 6 shows the SEM of the nano magnetite synthesized by the co-precipitation method.

TM3000_8213   H   D6,3 x2,5k   30 um
CEFET-MG - DET

FIGURE 6 - *NANO MAGNETITE SEM IMAGES*

The SEM images presented grains with hexagons and octahedral forms, with heterogeneity of shape and size, characteristic of the formation of magnetite / maghemite phases, according to X-ray diffraction results. The particle size varied between micrometers and nanometers.

Non-Conventional Materials and Technologies – NOCMAT for XXI Century Materials Research Forum LLC
Materials Research Proceedings 7 (2018) 457-461 doi: http://dx.doi.org/10.21741/9781945291838-43

The cement paste molded with 0.1% nano magnetite addition presented para magnetic behavior when placed in the presence of magnetic field.

Conclusions

The synthesis by the co-precipitation process presented results compatible with the literature, presenting particles in micrometric and nano metric sizes with superparamagnetic properties.

The synthesized nanoparticles presented heterogeneity of shape and size.

The cement paste molded with 0.1% nano magnetite particles presented paramagnetic behavior when submitted of magnetic field. This suggests the possibility of producing cementitious materials with self-monitoring of deformations.

The results point to the need to improve the process allowing the production of superparamagnetic nanoparticles with greater homogeneity and low cost.

Acknowledgments

The authors thank FAPEMIG, CNPq and CEFET / MG.

References

[1] Bhushan B, editor. Handbook of nanotechnology, Springer; Berlim, 2004.

[2] El-Hosiny FI, El-Diasty F, El-Said HM, Ismail MIM. Hydration characteristics of admixed magnetite nanoparticles-cement pastes. Third International Conference on Nanotechnology in Construction (HBRC), Cairo, Egypt, 2011.

[3] Maile, Aiu. The chemistry and physics of nano-cement. Loyola

[4] Li, H., Yuan, J., Ou, J. Microstruture of cement mortar with nanoparticles, Compos: B Eng, 35(2), 185-9, 2004. https://doi.org/10.1016/S1359-8368(03)00052-0

[5] Cornelle Schwertmann. The iron oxides: Structures, Properties, Reactions and User. 1999

[6] Raj, K., Moskwistz, R. J. Magn. Mater, 85, 233, 1990. https://doi.org/10.1016/0304-8853(90)90058-X

[7] Xie, J., Chen, K., Lee, H.Y., Hsu, A. R., Peng, S., Chen, X., Sun, S. J. Am. Chem. Soc. 130, 7542, 2008. https://doi.org/10.1021/ja802003h

[8] Shouheng, S., Zeng, H. Size- Controlled Synthesis of Magnetite Nanoparticles. J. Am. Chem. Soc. 124, 8204-8205, 2002. https://doi.org/10.1021/ja026501x

[9] Kawashita, M., Kawamura, K. PMMA-based bone cements containing magnetite particles of hyperthermia of cancer. Acta Biomaterial. 6:3187-3192, 2010. https://doi.org/10.1016/j.actbio.2010.02.047

[10] Khoslaklagh, A., Nazari, A., Gholamreza, K. Effects of Fe_2O_3 Nanoparticles on Water Permeability and Stregth Assessmens of High Strength Self-Compacting Concrete. J. Mater. Sci, Technol. 28 (1), 73 – 82, 2012. https://doi.org/10.1016/S1005-0302(12)60026-7

[11] Li, X., Xiao, H., Ou, J. A study on the mechanical and pressure sensitive properties of cement mortar with nanophase materials. Cem Concr Res. 34(3): 435-8, 2004. https://doi.org/10.1016/j.cemconres.2003.08.025

[12] Mutasim, I. K. Co-precipitation in aqueous solution synthesis of magnetite nanoparticles using iron (III) salts as precursors. Arabian Journal of Chemistry, 279-284, 2015.

Non-Conventional Materials and Technologies – NOCMAT for XXI Century   Materials Research Forum LLC
Materials Research Proceedings 7 (2018) 462-469   doi: http://dx.doi.org/10.21741/9781945291838-44

Load Deflection Behaviour of Rammed Earth-Filled Polyvinyl Chloride Columns

Dick, K.J[a]* and Midence, L.[a]

[a]Biosystems Engineering, University of Manitoba, Winnipeg, Canada

Kristopher.Dick@umanitoba.ca*, lc_midence@hotmail.com

Abstract. This paper presents the results of a series of tests conducted on rammed-earth filled polyvinyl chloride (PVC) columns for use in structures in rural Honduras. While the use of PVC may not seem sustainable the project location in Honduras does not have an affordable source of timber for construction. Previous use of concrete-filled tubes led to the consideration to use rammed earth as a more sustainable, low cost option and provides a stay-in-place form. The tubing is available from production facilities in Central America.. 102 mm (4") PVC tubing with a length of 2440 mm was filled using 5 and 10% cement stabilized and non-stabilized rammed earth. A single piece of 10M rebar was positioned at the centre of each column. The specimens were then loaded in axial compression to evaluate the load-deflection behaviour. Test cylinders were prepared using the same mix designs and were tested to investigate the compressive strength. The rammed earth test cylinder specimens had a compressive strength ranging from 0.6 to 3.03 MPa. The PVC columns exhibited ultimate loads from 6.6kN to 28.1 kN. Using an out-of-plane deflection limit at mid-height of H/500 a recommended design value is presented. Based on the proposed end-use for these columns the capacity is more than adequate for the anticipated loads.

Keywords: Rammed Earth, Polyvinyl Chloride(PVC), Column Design, Compressive Strength, Filled Tubes

Introduction

Constructing the built environment is integral in the development of societies whether it is for housing, office buildings or other structures; however; with global climate change and a rise in material cost, many look towards alternatives to typical construction methods. Rammed earth is a construction practice that utilizes loose soil compacted inside formwork (Ciancio and Beckett, 2013) and has been implemented in many regions around the world including North Africa, Australia, North and South America, China and Europe (Maniatidis and Walker, 2003). Rammed earth construction promotes the use of locally available material, which reduces energy in transportation thus providing environmental and sustainable benefits (Ciancio and Beckett, 2013). However, this material has been primarily used for wall systems and is seldom incorporated into columns.

Research has been conducted on square and circular steel tubes filled with concrete providing a symbiotic relationship which increases the resistance in tension and compression of the system (Lehman and Roeder, 2012); however the steel is costly and erodes to weather conditions. The program Students for Sustainability (S4S) at the University of Manitoba, has implemented 4-inch polyvinyl chloride (PVC) columns and beams filled with concrete in their community projects within Honduras (Carter, 2013; Enns, 2015). The PVC tubing provides containment and a layer of protection for the concrete, it is readily available, easy to handle, and it is a cost-effective alternative to steel.

In an effort to make the columns implemented by S4S more sustainable, research and testing was conducted on column specimens that replaced the concrete in the previously tested design, with rammed earth. In order to obtain better results and provide alternatives to the columns, three sets of columns were designed with un-stabilized and stabilized (5% and 10% cement) rammed earth along with cylinder samples to portray the material behaviour independent from the PVC.

Materials and methods

For the rammed earth, soil with a particle size analysis of 45% sand, 35% silt and 20% clay characterized as loam soil textural class (Griffith, 2007) from Anola, Manitoba was used. The soil was initially sieved through a 12.7 mm screen to avoid any aggregate greater than 20 mm as is normally specified among practice (Maniatidis and Walker, 2003) and has been implemented in previous projects in Honduras (Enns, 2015; Carter, 2013).

A total of three test specimens were prepared for each of three mix designs that varied the cement stabilisation content for the cylinders and columns tested.

Mixing and Ramming Process

Mixing for the cylinders was done by hand rotating materials within a 5 gallon pale since a small amount of material was needed for each set of three. This also allowed getting a better sense for the mix consistency, and making sure the mix had obtained a water content considered optimum (Bui, Morel, Hans and Meunier, 2009). Table 1 shows the mix composition for the cylinders and Figure 1 shows the mixing and ramming process.

TABLE 1 - CYLINDER MIX COMPOSITION BY PROPORTIONS

Material	CEMENT STABILIZATION		
	0%	**5%**	**10%**
SOIL (%)	94.00	88.00	83.00
CEMENT (%)	-	5.00	10.00
WATER (%)	6.00	7.00	7.00

FIGURE 1 - CYLINDER MIX AND RAMMING PROCESS FROM LEFT TO RIGHT

The required consistency of the mix, for this purpose, was reached once the clay started to bind the materials in the mix, tested using a rolling by hand technique using palm and finger (Maniatidis and Walker, 2003). Upon acquiring the desired consistency of the mix, test cylinders were filled with two scoops of mix prior to ramming with a 2x2 piece of lumber, achieving around 63.50 mm (2.50in) per compacted layer.(Fig.1)

Mixing of the earth was done using a mortar mixer. The empty columns were set up in a vertical position and braced with a lumber framework attached to an existing steel framework at mid-height and at the top (Fig.2). The bottom plate was made of two plywood sheets with the top layer having 114.30 mm (4.50in.) holes to position the columns and the bottom layer had 14.28 mm (9/16in.) nuts in the center to hold the 10M ($^3/_8$in) rebar in place.

Non-Conventional Materials and Technologies – NOCMAT for XXI Century   Materials Research Forum LLC
Materials Research Proceedings 7 (2018) 462-469   doi: http://dx.doi.org/10.21741/9781945291838-44

TABLE 2 - COLUMN MIX COMPOSITION PROPORTIONS

Material	CEMENT STABILIZATION		
	0%	5%	10%
SOIL (%)	95.82	90.48	86.38
CEMENT (%)	-	4.37	8.55
WATER (%)	4.18	5.15	5.07

FIGURE 2 - (LEFT TO RIGHT) A) COLUMN SETUP; B) ABS FITTING USED FOR RAMMING; C)RAMMING OF 8FT COLUMNS; D) COLUMNS FINALIZED.

The ramming process was done by hand using an ABS pipefitting connected to a 1-1/2 diameter ABS tube (Fig. 2b). This simulates a method that is proposed for use in practice where no machinery or air compressors are available.

Curing Process
The specimen curing/drying process was done accordingly that typical of concrete curing since some of the test specimens were stabilized with cement; which specifies the material reaches 75% of its design strength in 7 days and 100% in 28 days.

The cylinders were weighed individually upon pouring and ramming to obtain their initial mass (m_i) and then stored on a shelf at room temperature for 28 days at which point they were again weighed to obtain their dry mass (m_f) The samples were initially left in the PVC mould for 3 days which was removed on the third day to allow more consistent curing. The moisture content was determined on dry basis. The average moisture content of the samples was average of 22% after 28 days.

The columns cured/dried for 28 days at room temperature of approximately 19°C; however these were contained within PVC through the entire curing/drying time thus limiting the curing process to any external ambient air through the column.

The average dry density obtained from the cylinder samples was 1804.63 kg/m³ while the columns had an average of 1967.78 kg/m³, both of which fall within the expected range of 1700-2200 kg/m³ (Maniatidis and Walker, 2003).

Testing
The axial compression tests of the cylinders were conducted using a universal test machine at the University of Manitoba (Fig 3) recording load (N) and displacement (mm). The load rate was set at a speed of 5mm/min. Two steel plates 152.4mm x 152.4mm (6in x 6in) in combination with a 6mm high density rubber matt between the plate and specimen were placed below and above each specimen to distribute the load as shown in Figure 3.

FIGURE 3 - UNIVERSAL TEST MACHINE, SAMPLE PLACEMENT AND TESTING

Upon multiple trials with stabilized and un-stabilized samples, the failure was consistent with the lower layer failing through the material crumbling apart with all three sets of cylinders. The top layers did not exhibit failure and remained solid even after testing.

The columns were tested using a load test frame at the Alternative Village at the University of Manitoba. Hinge connections at the two ends with 51 mm (2in) hitch balls held in position by a steel ring on the bottom and a bolted clip assembly were installed onto the frame to hold the column in position as seen in Figure 4. Two 19 mm ($^3/_4$in.) plywood guides were placed mid-height to restrain movement in one plane of lateral movement. Two linear potentiometers were installed onto the assembly to measure the axial displacement at the top of loading apparatus and lateral displacement mid-height.

FIGURE 4 - (LEFT TO RIGHT) TOP PLATE WITH BALL HINGE HELD BY A BOLTED CLIP; BOTTOM PLATE WITH BALL HINGE HELD IN PLACE BY STEEL RING; MID-SPAN PLYWOOD AND POTENTIOMETER FOR LATERAL DISPLACEMENT

FIGURE 5 - (LEFT TO RIGHT) A) COLUMN PLACEMENT INTO TEST FRAME; B) AXIAL LOADING OF COLUMN; C) MAXIMUM DISPLACEMENT OF COLUMN UPON FAILURE.

Non-Conventional Materials and Technologies – NOCMAT for XXI Century   Materials Research Forum LLC
Materials Research Proceedings **7** (2018) 462-469   doi: http://dx.doi.org/10.21741/9781945291838-44

Figure 5 contains images of a test specimen at various stages of loading. As load was increased the column gradually deflected laterally. For purposes of this test failure was defined when the maximum was reached and then dropped off substantially.

Results and Discussion

Cylinders

Cylinder dimensions were 101.6 mm (4-in) diameter and a 203.20 mm (8-in) height with an area of 0.01 m² and a volume of 1.65 x 10⁻³ m³. A total of 9 cylinders were tested. The maximum compressive strength in N/mm² (MPa) was determined for each of the specimens and is summarized in Table 3.

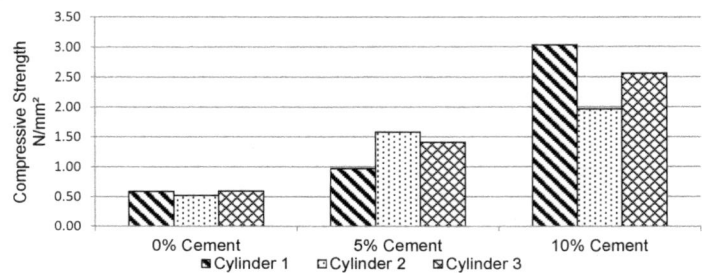

FIGURE 6 - COMPRESSIVE STRENGTH OF CYLINDER SAMPLES

TABLE 3 - MAXIMUM COMPRESSIVE STRENGTH FOR UN-STABILIZED AND
STABILIZED RAMMED EARTH CYLINDERS.

Stabilizer	Specimen	Maximum Displacement (mm)	Maximum Load (N)	Max. Compressive Strength N/mm² (MPa)	
				Individual	Average
0% Cement	Specimen E	10.16	4766.00	0.59	0.57
	Specimen F	7.48	4230.00	0.52	
	Specimen G	7.76	4837.00	0.60	
5% Cement	Specimen I	25.00	7869.00	0.97	1.32
	Specimen J	6.78	12810.00	1.58	
	Specimen K	10.61	11390.00	1.41	
10% Cement	Specimen L	13.11	24530.00	3.03	2.52
	Specimen M	11.16	15910.00	1.96	
	Specimen N	10.98	20710.00	2.56	

From data in Table 3 and Figure 6, the maximum compressive strength was recorded for the 10% cement stabilized samples with an average value of 2.52 N/mm² (326.25psi). This is approximately four times as much as the 0.57 N/mm² (82.65psi) by un-stabilized samples.

Columns

The column samples differed from the cylinders not only in length, but these were also contained within 4in PVC and had 10M ($^3/_8$in) re-bar full height at the centreline the column. Their dimensions were 101.6 mm diameter and a 2,438.40 mm height with an area of 0.01 m² and a volume of 0.02 m³. A total of 9 columns were tested similarly to the cylinders at a load rate of 0.25mm/min. Table 4 provides a summary of the maximum axial compressive load, mid-height out of plane deflection and a proposed design value.

TABLE 4 - *MAXIMUM COMPRESSIVE STRENGTH FOR UN-STABILIZED AND STABILIZED RAMMED EARTH REINFORCED 4IN. PVC COLUMNS*

Stabilizer	Specimen	Maximum Mid-Height Displacement (mm)	Maximum Load (N)	Coefficient of Variation	Load at Out of Plane Displacement of H/500 (N)	Proposed Design Load (N)
0% Cement	Specimen E	46.16	7,270.75	0.10	5,793.84	5,677.63
	Specimen F	57.03	8,351.11		7,751.07	
	Specimen G	58.83	6,616.84		3,487.98	
5% Cement	Specimen I	78.97	27,618.12	0.23	26,310.19	18,492.34
	Specimen J	77.72	19,163.74		16,227.91	
	Specimen K	59.05	16,214.44		12,938.93	
10% Cement	Specimen L	58.51	24,273.80	0.08	21,879.63	22,354.97
	Specimen M	73.44	28,056.55		24,474.28	
	Specimen N	76.17	23,612.42		20,711.00	

(1) **NOTE:** The proposed design load is based upon the average of the Load at H/500 for each stabilisation.

It is noted that *Specimen 1* in the 5% stabilized columns was the only specimen that reversed deflection when compared to the others Fig 7b. A proposed design value was based on a typical value for inter-storey drift of H/500 in building design used for wind and seismic loading. Thus the load corresponding to a deflection of H/500 was determined for all specimens. At a maximum load 26,579.29 N (6199.99) lbs the column reversed its deflection to the opposite direction. The value used for H/500 was that corresponding to a deflection of 4.88 mm after the maximum load was achieved.

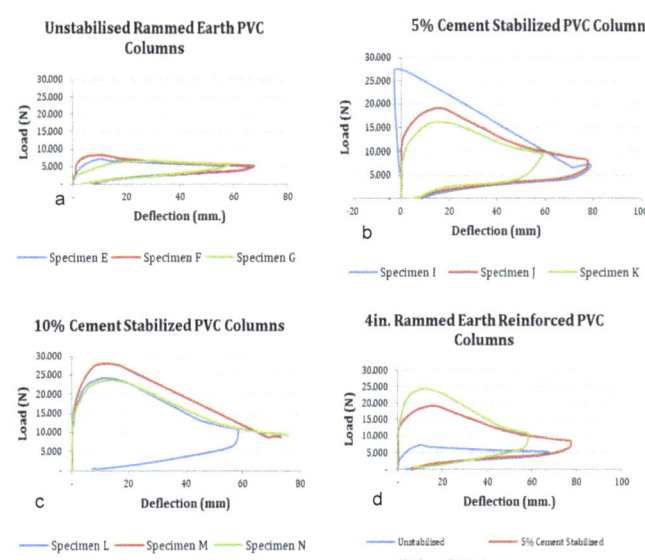

FIGURE 7 - *LOAD/DEFLECTION BEHAVIOUR OF 2440 MM (8FT) COLUMN SAMPLES*

Non-Conventional Materials and Technologies – NOCMAT for XXI Century   Materials Research Forum LLC
Materials Research Proceedings 7 (2018) 462-469   doi: http://dx.doi.org/10.21741/9781945291838-44

The proposed design out of plane deflection value obtained for these columns based on their height of 2.44m (96in) was 4.88mm (0.19in). From the data collected and the values calculated in Table 4, it is shown on average, by the time the column reached the design value deflection it had already achieved around 86% of its maximum compressive strength. This can also be seen in Figure 7, which shows the mid-height deflection of the columns versus applied load. As mentioned previously in the testing, the failure can be seen upon the sudden drop in loading. The 10% cement stabilized columns provided the highest compressive strength, with an average of 3.12 N/mm² (452.40psi), just over 3 times as much as the 0.91 N/mm² (131.95psi) provided by the un-stabilized rammed earth PVC columns.

While examining the load/deflection behaviour of the columns, it is interesting to notice the elasticity provided by the PVC containment of the rammed earth. On average, the columns recovered 88% based on deflection at maximum load. This could potentially present further study into the stress and strain relationship of the columns in order to obtain more accurate results on the elasticity of rammed earth reinforced PVC columns.

The behaviour of un-stabilized columns was different than the others. As load was applied lateral deflection was gradual. Stabilized, reinforced columns however, resisted the load until a sudden lateral movement occurred. It is postulated that this sudden drop in load corresponds to the point at which the bond strength between the earth-binder matrix and the rebar was exceeded.

Comments and Recommendations for Future Research

All specimens were built, cured and tested within a closed environment (The Alternative Village) at room temperature with no outdoor exposure. In practice, these columns would be built in place exposed to the sun, rain, weathering and freeze-thaw. These conditions may affect the column performance over the long term and should be evaluated either in-situ or in a test apparatus that can simulate these conditions.

The PVC used for this experiment had a 101.60mm (4in) inner diameter with a 6.10mm (0.24in) wall thickness. Depending upon the global location and manufacturer the thickness can vary. Testing with other PVC wall thicknesses is recommended to develop a relationship between design load and PVC dimensions.

For implementation purposes of the system in developing areas, the ramming process was done accordingly using as simple a mechanism possible, meaning no expensive tools or machinery. As the idea of the study was to provide an accessible yet effective construction practice, the ramming was done using a 31.75mm (1¼in) ABS pipe with a 76 mm (3") ABS reducing bushing attached to the bottom in order to maintain the rebar at the centre and also provide more consistent compaction all around. This provides a cost-effective and more accessible approach to ramming the earth within the columns.

As tests were being conducted on the reinforced stabilized columns, it was observed would exhibit a slight deflection until maximum load was reached upon which the column would jump to peak deflection while the un-stabilized columns slowly deflected. Theoretically, this behaviour is postulated be due to the bonding of the rammed earth and the 10M (³/₈in) rebar placed through the column. It is assumed that this behaviour is a function of the development length between the rebar and rammed earth.

Conclusions

Through research and testing conducted on stabilized and un-stabilized rammed earth reinforced polyvinyl (PVC) columns; the maximum compressive strength along with deflection and a calculated design value were determined. Cylinders were also tested to isolate the rammed earth and provide the characteristics and behaviour under compression loading. From the data collected for the columns, the 10% cement stabilized columns provided the highest resistance with an average maximum compressive strength of 3.12 MPa (452.40psi) and an average maximum deflection of 69.37mm (2.73in). An out-of-plane deflection limit of 4.88mm (0.19in) based on H/500 with a column height

Non-Conventional Materials and Technologies – NOCMAT for XXI Century   Materials Research Forum LLC
Materials Research Proceedings 7 (2018) 462-469   doi: http://dx.doi.org/10.21741/9781945291838-44

of 2.44m (96in) was used to determine a proposed design value. Based on this criterion a design load of 22.35kN (5018 lbs) for 10% stabilized reinforced columns is proposed. Similarly, for 5% cement stabilized and un-stabilized rammed earth columns a design load of 18.49kN (4157lbs) and 5.68kN (1,276 lbs) respectively is proposed. Comparing the average compressive strength between the cylinders and the columns, the increase of compressive strength provided by the enclosing of the PVC was, on average, 0.73MPa (105.85psi). Based on this test series the columns could be implemented into construction practice as long as it is designed accordingly to the environment and the design loads implicated in the location proposed.

Acknowledgements

The authors would like to thank The Alternative Village at the University of Manitoba that provided all of the funding, materials and equipment required for this study.

References

[1] Bui, Q., Morel, J. Hans, S. Meunier, N. 2009. Compression behaviour of non-industrial materials in civil engineering by three scale experiments: the case of rammed earth. Materials and Structures. 42: 1101-1116. https://doi.org/10.1617/s11527-008-9446-y

[2] Carter, J. 2013, Design and implementation of a sustainable housing system in Honduras. Unpublished M.Sc. thesis, Department of Biosystems Engineering, University of Manitoba, Winnipeg MB.

[3] Ciancio, D. and Beckett, C. 2013, Rammed earth: An overview of a sustainable construction material, University of Western Australia.

[4] Dawn E. Lehman and Charles W. Roeder, "Foundation Connections for Circular Concrete-Filled Tubes." Journal of Constructural Steel Reasearch 78, (2012): 212-225. https://doi.org/10.1016/j.jcsr.2012.07.001

[5] Enns, G. 2015. Design and Implementation of a Rammed Infill Adobe and Plastic Bottle Wall System in Honduras. Department of Biosystems Engineering, Unpublished M.Sc. thesis, University of Manitoba, Winnipeg MB.

[6] Griffiths, K. 2007. Physical Properties of an Anola Soil Based Cob Re-inforced with Chicken Feathers. Department of Biosystems Engineering, University of Manitoba, Winnipeg MB.

[7] Maniatidis, V. & Walker, P. 2003. A Review of Rammed Earth Construction.

[8] Morsier, Y. Our Experience with Rammed Earth: A Manual for Rammed Earth Building, Numbugga, New South Wales, Australia,

Non-Conventional Materials and Technologies – NOCMAT for XXI Century   Materials Research Forum LLC
Materials Research Proceedings 7 (2018) 470-481   doi: http://dx.doi.org/10.21741/9781945291838-45

Small Diameter Bamboo Plane Truss with 3C Connections

Cinthia Aparecida Carneiro Lana [a*], Luís Eustáquio Moreira [b]

[a]Uniaraxá -MG, Brasil; cinthialana@uniaraxa.edu.br

[b] Universidade Federal de Minas Gerais - UFMG, Brasil; luis@dees.efmg.br

Abstract. Small diameter bamboos have been little investigated in relation to the structural application, probably because they have relatively low moment of inertia and wall thickness compared to the *Dendrocalamus, Phyllostachys pubescens and Guadua* bamboos. However, such bamboo species exist in large numbers in Brazilian rural areas, both with pachymorphs rhizomes and with leptomorphs rhizomes. In this paper, bamboos of the *Bambusa tuldoides* specie, with an average diameter of 5 cm, harvested in the vicinity of the city of Santa Luzia - MG, were used for the development of plane trusses. By avoiding to connect the bars by bolts, a composite of cotton fabric with a single-component castor-based polyurethane resin was developed, named in this paper 3C connection, or Cotton Composite Connection. Two prototypes were tested, with 3 and 4 *m* of free span, respectively, along with the characterization of the materials used and the bond strength. The second prototype was able to withstand 114 times the own weight, with limit load equal to 16,9 *kN*. The shear stress at shear bond strength reached 5,3 *MPa* in tension tests and 4,86 *MPa* in compression tests. Bamboos present modulus of elasticity in compression and tension of 16,2 *GPa* and 20,7 *GPa* respectively. The tests showed that the prototypes have a predictable behavior, easily computationally simulated, with linear load versus displacement behavior for load levels close to the limit load. The 3C connection has a better performance than connection with bolts because do not splits bamboos under large displacements.

Keywords: Small Diameter Bamboos, Plane Truss, 3C Connection, Mechanical Tests

Introduction

The development of techniques for manufacturing and assembling objects of utility, such as bamboo structures, has been a priority of LILD - Laboratory of Free Design Research - in the Department of Arts and Design of PUC-Rio, more than 20 years ago. Like this laboratory, LASE - Laboratory of Structural Systems, of the Department of Structural Engineering of UFMG, work in the same line research about 8 years ago, where the structure and connections discussed in this paper were developed.

In these laboratories, different ideas are embodied in reduced physical models in an initial viability study. Immediately the models reveal the importance of the development of the technique in order to have a well solved solution in mechanical, spatial and constructive terms. Viewed as a system of objects, the different models can be arranged in time, to reveal an evolution of the technique [2]. It can be said that the technique walks with autonomy; a concrete solution inspires another one, and so on, discovering ways to solve the problems that arise. The 3C connection is inserted in a search for replacement of the bolts to connect some bamboo structures [3], not necessarily all. In this sense, the models developed in LILD, (Figure 1), can be placed as antecedents of this investigation. .

Non-Conventional Materials and Technologies – NOCMAT for XXI Century   Materials Research Forum LLC
Materials Research Proceedings **7** (2018) 470-481   doi: http://dx.doi.org/10.21741/9781945291838-45

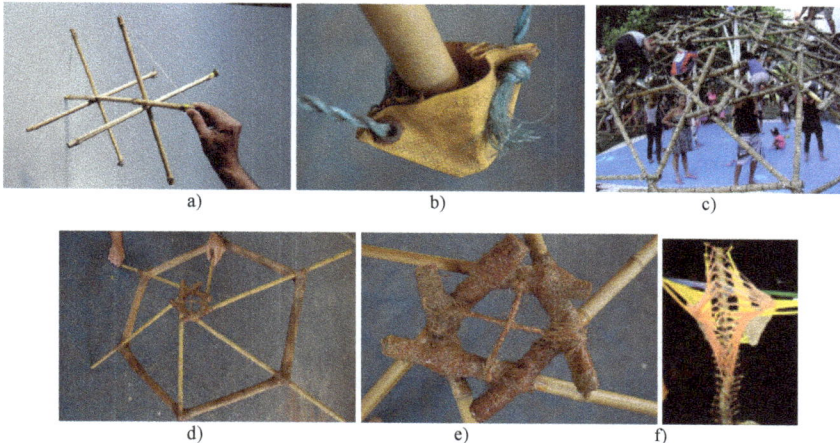

FIGURE 1 - *A) TENSEGRITY MODEL OF BARS AND CABLES; B) CAPUCHINS CONNECTION C) BAMBOO GEODESIC WITH FLEXIBLE TEXTILE JOINTS D, E) BAMBOO BARS TIED WITH J2C CONNECTION – JUTE COMPOSITE CONNECTION (PVA GLUE); F) TENSEGRITY MODEL OF THE VERTEBRAL COLUMN.*

Materials and methods

The 3C connection consists of the union of tubular bars of bamboo by composite rolls of cotton cloth that binds them internally and a bandage that connects them externally, (Figures 2 and 3). First, bands of cotton fabric are soaked in resin until they saturate. With the excess removed, for the connection of the elements of the upper and lower flanges, composite rolls are formed which are hoisted into the bars being connected by a small diameter rope (Figure 3).

FIGURE 2 - *INTERNAL CONNECTION*

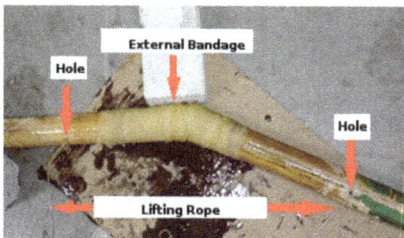

FIGURE 3 - *EXTERNAL CONNECTION*

Resistance of the bamboo-composite bonding

Shear strength of bonding under compression

The bonding between bamboo and the composite bandage was measured in compression and tension tests.

Shear strength of bonding under compression

The specimens were prepared according to Figure 4.

Non-Conventional Materials and Technologies – NOCMAT for XXI Century   Materials Research Forum LLC
Materials Research Proceedings 7 (2018) 470-481   doi: http://dx.doi.org/10.21741/9781945291838-45

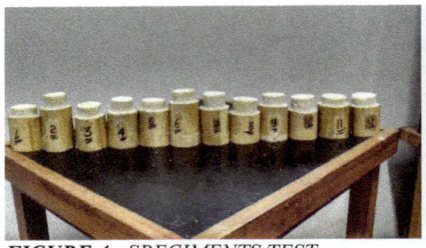

FIGURE 4 - *SPECIMENTS TEST*   **FIGURA 5** - *COMPRESSION SHEAR TEST*

The bonding technique consists of sanding the inner surface of the bamboo with thick steel sandpaper no. 35, causing circumferential and longitudinal roughness increasing the bonding surface. Good bonding technique also requires waiting 10 to 20 min for surface wetting and adhesive penetration before the introduction of the composite roll. Bonded surfaces should be allowed to stand until complete adhesive polymerization. Vibrations are not advisable. Three days after gluing, the specimens were tested (Figure 5). Six proof bodies were tested for each type of test. The results are shown in Table 1. The mean shear stress was 4,86 *MPa* with an mean bonding length of 25 *mm*.

TABLE 1 - *SHEAR RESISTANCE IN COMPRESSION*

Machine Numbering	Proof bodies	Maximum Force [N]	Rupture Stress [MPa]
CP1	1	14664	5,08
CP5	2	8270	2,79
CP7	3	9965	3,68
CP8	4	15396	5,52
CP9	5	15038	4,88
CP10	6	10390	4,70
CP11	7	15065	6,62
CP12	8	11718	5,05
CP13	9	13322	5,07
CP14	10	14715	5,13
CP15	11	13518	4,31
CP16	12	16845	5,53
			Mean ± Standard deviation 4,86±0,96

Tension bonding shear resistance
Test specimens were prepared according to Figure 6. Likewise, good bonding requires that the bamboos be previously painted and after a wetting time the composite is placed between the bamboo segments. The parts are clamped until the drying takes place. Care must be taken that the segments

Non-Conventional Materials and Technologies – NOCMAT for XXI Century   Materials Research Forum LLC
Materials Research Proceedings 7 (2018) 470-481   doi: http://dx.doi.org/10.21741/9781945291838-45

are in the same alignment so that no moment is introduced in the bonding zone. In Table 2 we have the results of the experiment whose average resistance was 5,35 *MPa*.

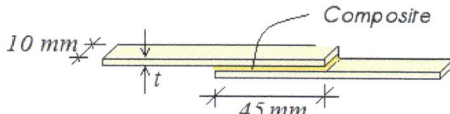

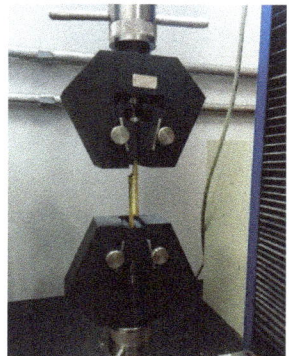

FIGURE 6 - *THEORETICAL DIMENSIONS – SHEAR PROOF BODY*   **FIGURE 7** - *TENSION SHEAR TEST*

TABLE 2 - *SHEAR RESISTANCE IN TENSION*

Proof bodies	Real bonding lenght [mm]	Real width [mm]	Rupture Stress [MPa]
1	46,5	11,6	4,81
2	45,8	16,3	5,60
3	43,7	14,0	5,87
4	44,2	14,3	5,14
5	45,5	11,3	4,26
6	43,9	10,9	6,44
Mean ± Standard Deviation			5,35 ± 0,78

Composite compression tests
Composite cylinders of rolled fabric (Figure 8) were tested according to Figure 9. PVC pipes with 45 mm internal diameter were used as cast.

FIGURE 8 - *COMPOSITE PROOF BODIES*

FIGURE 9 - *COMPOSITE COMPRESSION TEST*

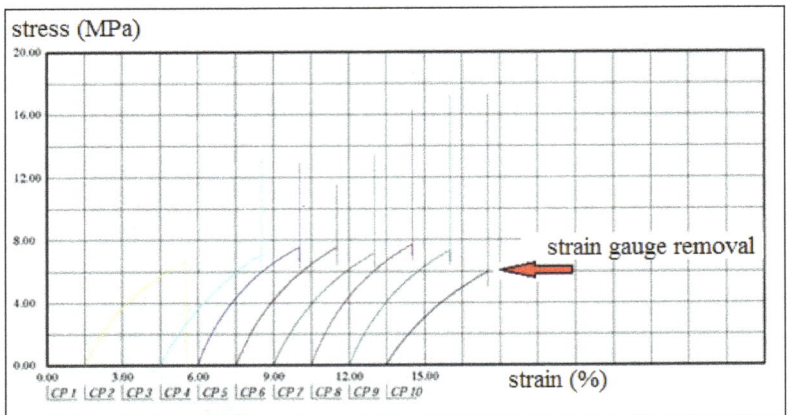

FIGURE 10 - *STRESS VERSUS STRAIN - COMPOSITE IN COMPRESSION*

The material presented a non-linear elastic behavior in compression. Results are in Table 3.

TABLE 3 - *COMPOSITE IN COMPRESSION – MECHANICAL PROPERTIES*

Machine Numbering	Speciments Test	Maximum Force [N]	Resistance Limit [MPa]	Deformation on strain gauge removal [%]	Modulus of Elasticity of the initial line [MPa]
CP1	CP1	18350,08	11,29	4,01	347,10
CP4	CP2	21881,98	13,23	4,02	324,15
CP5	CP3	20958,3	12,89	4,05	387,45
CP6	CP4	18937,88	11,65	4,03	348,60
CP7	CP5	22193,87	13,41	4,03	315,85
CP8	CP6	25972,53	16,33	4,03	361,00
CP9	CP7	27334,90	17,19	4,02	320,63
Mean Resistance± Standard Deviation 13,71 ±2,24				Mean Modulus± Standard Deviation 343,54 ± 25,62	
CP10	CP7	27540,25	17,32	4,03	217,09

Composite Tension Tests
Test specimens were prepared according to Figure 11, where the central part has 3 overlapping tissue layers, previously embedded in the resin until saturation, followed by excess removal. At the claw attachment site the specimens were increased from 2 layers of tissue to each side of the central tissues. The 3 cm wide specimens were cut with a ceramic saw. The rupture occurred in the central section in all specimens (Figure 12).

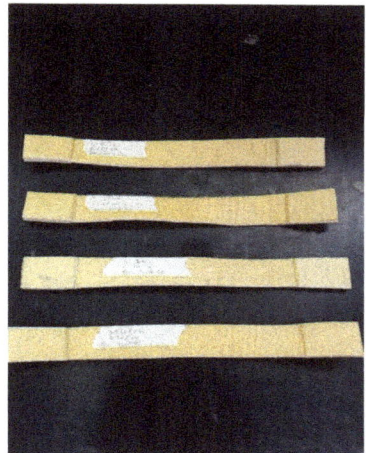

FIGURE 11 - PROOF BODIES FOR TENSION TESTS   *FIGURE 12 - TENSION TESTS*

Figure 13 shows the constitutive relationships of the composite in tensile strength. Again, a non-linear behavior, with a modulus of elasticity of the mean initial line of 841 *MPa* and a mean tensile stress of 12,9 *MPa* (Table 3), is observed.

TABLE 4 - COMPOSITE IN TENSION - MECHANICAL PROPERTIES

Machine Numbering	Proof bodies	Maximum Force [N]	Resistance Limit [MPa]	Strain (on strain gauge removal) [%]	Modulus of Elasticity of the initial line [MPa]
CP3	1	1140,63	12,6	3,5	1065,0
CP4	2	1145,77	15,1	6,64	946,0
CP5	3	1171,46	13,9	2,0	964,0
CP6	4	1108,09	13,9	1,6	712,0
CP7	5	1126,93	12,0	1,21	729,0
CP8	6	1116,65	12,4	1,61	760
CP9	7	940,25	10,1	1,61	713
Mean Resistance ±Standard Deviation: 12,9±1,62				Mean Modulus± Standard Deviation	841 ±146

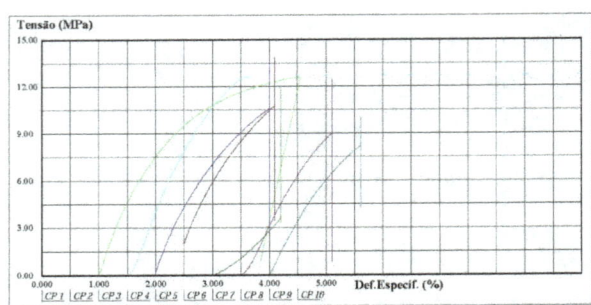

FIGURE 13 - STRESS VERSUS STRAIN - COMPOSITE IN TENSION

Prototypes Tests

Two prototypes were tested according to Figures 14 and 15.

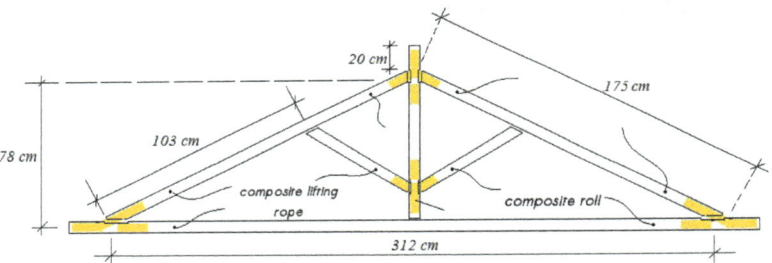

FIGURE 14 - PROTOTYPE 1- DESIGN AND DIMENSION.

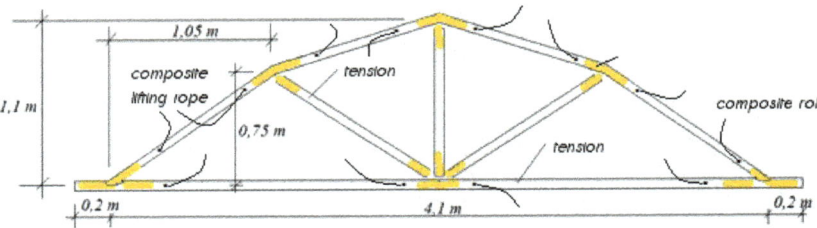

FIGURE 15 - PROTOTYPE 2- DESIGN AND DIMENSION.

Prototype 1 Mechanical Test

The loads were applied at the top node of the prototype, with manual control actuator, and load steps of 0,89 *kN*, recorded by a dynamometric ring. A displacement transducer -DT- measured the vertical displacement of the truss in the upper third of the central and vertical bar. The truss was bi-supported at the ends. The bracing of the upper and lower nodes was performed with polypropylene ropes to avoid lateral instability. (Figures 16 and 17).

Non-Conventional Materials and Technologies – NOCMAT for XXI Century   Materials Research Forum LLC
Materials Research Proceedings **7** (2018) 470-481   doi: http://dx.doi.org/10.21741/9781945291838-45

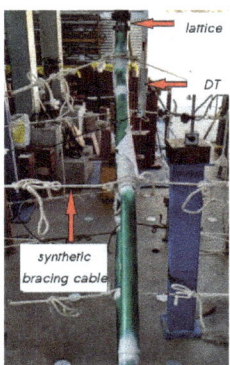

FIGURE16 - *PROTOTYPE 1 – FRONT VIEW*

FIGURE 17 - *PROTOTYPE 2:*
DEVICES

The loading and unloading versus vertical displacement cycles are shown in Figure 18. The failure occurred in the bottom flange for a maximum load of 11,2 kN (Figure 22).

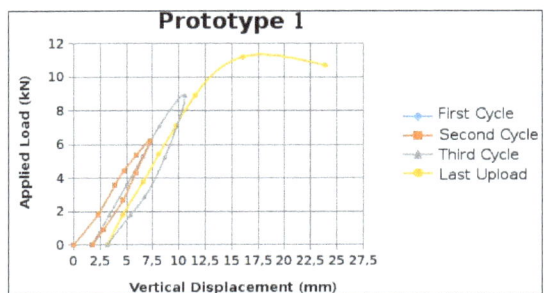

FIGURE 18 - *APPLIED LOAD VERSUS VERTICAL DISPLACEMENT – PROTOTYPE I*

Prototype II Mechanical Test
The bottom flange has an amendment near the center to verify the composite connection in tension. Similarly, the diagonals are now in tension due to the angles given to the upper flange. Vertical displacements were measured step by step in a centesimal clock (Figure 19). The bottom flange bracing is also fundamental for the stability of the structure, without which the central node would move perpendicular to the truss plane because the compression force.

Non-Conventional Materials and Technologies – NOCMAT for XXI Century   Materials Research Forum LLC
Materials Research Proceedings **7** (2018) 470-481   doi: http://dx.doi.org/10.21741/9781945291838-45

FIGURE 19 - POSITION OF THE COMPARATOR CLOCK   *FIGURE 20 - BRACINGS*

The loading and unloading versus vertical displacement cycles are shown in Figure 21. For the 16,91 *kN* load the upper flange buckled laterally despite the bracing. The symmetric nodes of the upper flange, lateral to the point of load application, moved horizontally in opposite directions, about 30 cm from the truss plane (Figure 23). Curiously, with the loading withdrawal, the truss returned to the plane configuration. Both bamboos and 3C connections do not suffer any visual damage after the large displacements caused by buckling.

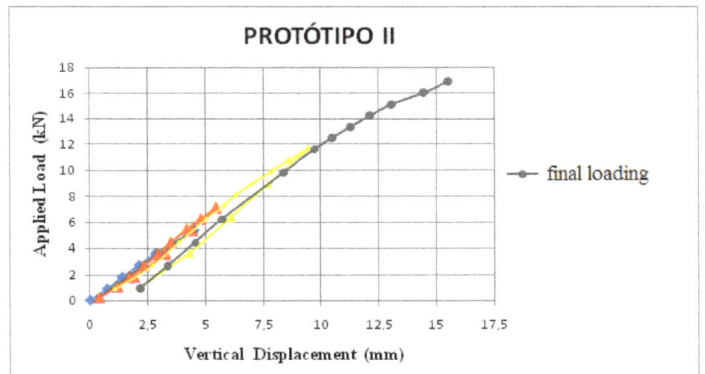

FIGURE 21 - APPLIED LOAD VERSUS VERTICAL DISPLACEMENT – PROTOTYPE 2

Figures 22 and 23 shows the limit conditions for each prototype. Prototype *I* was limited by failure connection and Prototype *II* by upper flange buckling.

Non-Conventional Materials and Technologies – NOCMAT for XXI Century   Materials Research Forum LLC
Materials Research Proceedings **7** (2018) 470-481   doi: http://dx.doi.org/10.21741/9781945291838-45

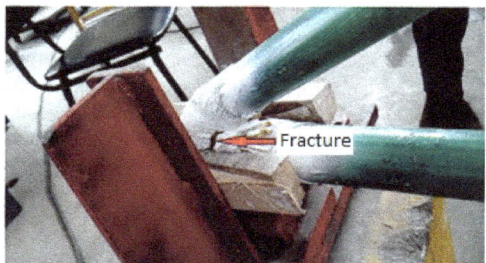 

***FIGURE 22** - PROTOTYPE 1 – CONNECTION FRACTURE*   ***FIGURE 23** - PROTOTYPE 2 -*
BUCKLING

Analysis of the results
Both prototypes showed a lineal relationship stress versus strain until high levels of loadings.

Prototype 1
For prototype *1* a load versus displacement lineal relationship was recorded up to the load of 8,9 *kN*. The lost of energy between cycles is the accommodation at nodes. The third cycle registered a negligible residual and vertical displacement equal to 3,2 *mm*. The nonlinearity presented for loads greater than 8,9 *kN* is associated with the second order effect of individual compressed bars. Bars are not perfectly straight and 3C connections have nonlinear behavior both in tension and compression. In Figure 24, we compare the final experimental curve with two different numerical simulations by SAP 2000 v14. The bamboos were simulated as straight pipe elements with diameter and wall thickness equal to the mean of the diameters and wall thicknesses of all bars, 5 *cm* and 6 *mm*, respectively; compression modulus of elasticity equal to 15,9 *GPa* and tension modulus of elasticity of 20,7 *GPa*, experimentally determined. In the first simulation the presence of the composites was not considered. In the second simulation, 8.6 mm of the end of each bar was considered for been a composite material. The composite modulus in compression, $Ecc = 344$ *MPa*, was adopted for this small segment (Table 2). The parallelism of the two straight lines shows that the simulation of the bamboo bars with composite terminals gives a precise result.

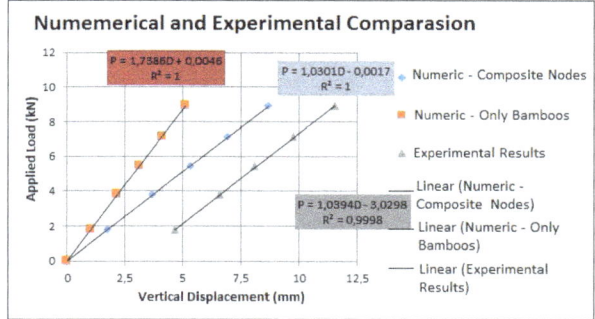

***FIGURE 24** - EXPERIMENTAL AND NUMERICAL ADJUSTMENTS – PROTOTYPE I*

Non-Conventional Materials and Technologies – NOCMAT for XXI Century   Materials Research Forum LLC
Materials Research Proceedings 7 (2018) 470-481   doi: http://dx.doi.org/10.21741/9781945291838-45

Prototype II

Different from Prototype *I*, the Prototype *II* has tensioned diagonal. As in Prototype *I*, the relationship between applied load versus vertical displacement was lineal up to the load of 13,35 *kN*. In other words, the second order effects arose for loads greater than 13,35 *kN*. A negligible residual and vertical displacement of 2 mm was also recorded in the third cycle discharge. The simulation by SAP 2000 v14 considered two cases, in the same way as in Prototype *I*. The geometric and mechanical properties described before were also adopted here. The ends of the diagonals under tension, a segment of 8,2 *mm* was considered with the modulus of elasticity of the composite in tension, or E_{tc} = 841 *MPa* (Table 3), and a perfect adjustment between numerical and experimental results were obtained (Figure 25).

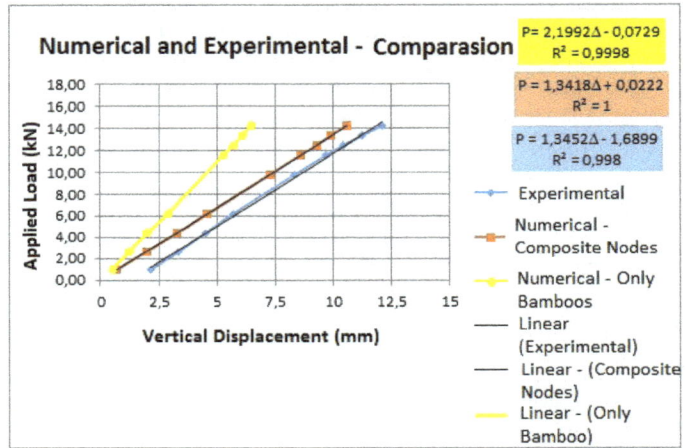

FIGURE 25 - *EXPERIMENTAL AND NUMERICAL ADJUSTMENT – PROTOTYPE II*

Conclusions

The bamboo plane trusses developed in this research participates in the technological rescue of small diameter bamboos - first used by Santos Dumont in 1906, in the structure of 14-bis airplane, but now with 3C connections which depends of PU resins. The 3C connection has a nonlinear elastic behavior in tension and compression, reaching compressive stresses of 8,9 *MPa* with 4% strain in elastic regime and modulus of elasticity of 347 ± 16,5 *MPa*; and tension strength of 12,9 ± 1,6 *MPa*, with modulus of elasticity of 842 *MPa*; in other words, the composite has elastomeric behavior. This explains why Prototype *II*, after removing the 16,9 *kN* limit load that buckled the upper flange (Figure 23), returned to plane form with no apparent damage. It is a very special behavior because any other kind of stiffer connection, using bolts, would most likely have broken the bamboo in the holes. On the other hand, the load of 16,91 *kN* corresponds to 114 times the own weight of Prototype. 3C connections works by gluing the composite in the outer as well as in inner surfaces of the bamboos but probably the gluing of outer surface be sufficient. Although shear resistance stresses reached 4,86 ± 0,96 *MPa* in compression and 5,35 ± 0,78 *MPa* in tension, the aging study of the composite and gluing is an important future research. The rupture of Prototype *I* occurred at the junction of the bars (Figure 22), because in addition to the bamboo be weakened locally by the notch given to the lower flange, the support was in a position that favored the anticipated fracture. Therefore, the support device must distribute the local compressive stresses, what was not the case

Non-Conventional Materials and Technologies – NOCMAT for XXI Century   Materials Research Forum LLC
Materials Research Proceedings 7 (2018) 470-481   doi: http://dx.doi.org/10.21741/9781945291838-45

here.   The *Bambusa tuldoides* bamboos have very high mechanical resistance. For the batch investigated here, it was obtained: - parallel shear strength: $8,7 \pm 0,32$ *MPa*; tension strength: $221 \pm 22,6$ *MPa*; tension modulus of elasticity: $21 \pm 2,5$ *GPa*; compression modulus of elasticity: $16,2 \pm 1,6$ *GPa*.   Moisture content of the samples was $8,9 \pm 0,1$ (%) and the average apparent density was $0,82 \pm 0,1 \ \frac{gf}{cm^3}$.

Acknowledgments

Our thanks to PROPPES (Post-Graduate Program in Structural Engineering) - EEUFMG, for the research funding. We also thank LILD - Laboratory of Free Design Research, PUC-Rio, for the constant partnership in all developments of the LASE - Laboratory of Structural Systems – EEUFMG.

References

[1] Lana, C.A.C. Desenvolvimento de Treliças Planas de Bambu de Pequeno Diâmetro com Bio-conexões Compósitas. Dissertação de Mestrado. Departamento de Engenharia de Estruturas, Escola de Engenharia da Universidade Federal de Minas Gerais, MG, Brasil, 2016.

[2] Ripper, J.L.M and Moreira, L.E. Jogo das Formas – Lógica do Objeto Natural, NAU Editora, Rio, RJ, Brasil, 2014.

[3] Seixas, M.A., Ripper, J.L.M.; Ghavami,K. Construction of mobile bamboo dome covered with textile canvas. In: 16th Non Conventional Building Material and Technologies International Conference, Winnipeg, Canadá, 2015.

Non-Conventional Materials and Technologies – NOCMAT for XXI Century   Materials Research Forum LLC
Materials Research Proceedings 7 (2018) 482-490   doi: http://dx.doi.org/10.21741/9781945291838-46

Influence of Fly Ash on the Fresh and Hardened Properties of a Cement-Based Grout for Mechanized Tunneling

Alfredo Quiroga-Flores[a]*, Flávio de Andrade Silva[b], Romildo Dias Toledo Filho[a]

[a]Universidade Federal do Rio de Janeiro, Rio de Janeiro, Brazil; alquirogaf@coc.ufrj.br, toledo@coc.ufrj.br

[b]Pontificia Universidade Católica do Rio de Janeiro – PUC-Rio, Brazil; fsilva@puc-rio.br

Abstract. This work is focused on studying the influence of fly ash on the fresh and hardened properties of a grout formed by Portland cement and sodium-activated bentonite. Fly ash partially replaced the cementitious material as 10, 30 and 50% of binder weight. This mixture constitutes a part of a two-component grout used for fulfilling the annular gap between the tunnel structure and the surrounding soil which occurs during tunnel excavation by using a tunnel boring machine. The remaining part is a chemical admixture for accelerating the grout hydration and is not the focus of this study. For the fresh state, viscometry tests were carried out imposing increasing and decreasing shear rates. Although fly ash improved the rheology by increasing the partial replacements, bentonite still highly influenced the thixotropy and shear thinning behavior was obtained in all grouts. For the hardened state, hydration and mechanical tests were used. Fly ash delayed the set time proportionally to its content and increased the mechanical compression response in the long term at 92-day age test.

KEYWORDS: Cement, Bentonite, Grout, Fly-Ash ,Tunneling

Introduction

The increasing world population [1] demands infrastructure projects for the massive transportation of people. On this field, tunnels help shorten travels and allow underground transportation in large cities. Tunnels excavated by a tunnel boring machine which advances by cutting and removing a cylindrical segment of soil and placing a tunnel lining structure gather special attention because of its applications in subway projects for populous cities. During construction of this type of tunnel an annular gap is formed between the tunnel structure and the surrounding soil which needs to be filled with a material in order to transmit structure-soil stresses, avoid settlements among many requirements [2]. This material can be a two-component grout formed by component A, a Portland cement-bentonite mixture, and component B an accelerating admixture. Both components are pumped separately and gather each other at the placing point only. Although this kind of grout is widely used in various projects, there is lack of publications studying its properties [3]. For component A, delaying chemical admixtures are often utilized for both improving flowability and retarding the set time. It is important to mention fly ash not only could provide a similar response as the delaying admixture use but also improve the mechanical properties in the long term. Thus the focus of this study is the influence of fly ash on the fresh and hardened properties of component A of this kind of grout. Three fly ash contents were employed and replaced partially Portland cement as follows, 10, 30 and 50%. In the fresh state, rheological behavior was determined by viscometry tests. In the hardened state, Ultrasonic-cement-analysis and mechanical compression tests were employed for describing the grout behavior.

Materials and methods

Materials

The cementitious material produzed in this research is formed by a mix of Portland cement, sodium activated bentonite, fly ash and water. The cement was manufactured by Lafarge, classified as CP II F-32 in the Brazilian standard [5] and formed by 90-94% of clinker+calcium sulphate and 6-10% of carbonatic material. The bentonite comes from local raw deposits in Brazil and was produzed by

Non-Conventional Materials and Technologies – NOCMAT for XXI Century   Materials Research Forum LLC
Materials Research Proceedings 7 (2018) 482-490   doi: http://dx.doi.org/10.21741/9781945291838-46

Bentonit União Nordeste Ind. and Com. Ltda under the product name Permagel. It is a sodium-activated bentonite [4]. On the other hand fly ash was manufactured by Pozo Fly company. Chemical composition of Portland cement (PC) and bentonite (B), in table 1, were acquired by semiquantitative analysis of X-ray fluorescence spectrometer-wavelength dispersive spectroscopy(WDS), model AXIOS(Panalytical) and Sulfur content was obtained by elementary analyzer of Carbone and Sulfur, LECO SC632 at Centro de Tecnologia Mineral, a research center of the Brazilian ministry supporting science, tecnhnology and innovation, Ministério de Ciência, Tecnologia,

Inovações e Comunicações. For the fly ash (FA) composition, see table 1, it was used fluorescence spectroscopy of X-ray dispersive energy with the Shimadzu DX 800 at the Laboratório de Estruturas e Materiais (LABEST) at Instituto Alberto Luiz Coimbra de Pós-graduação e Pesquisa de Engenharia (COPPE) of Universidade Federal do Rio de Janeiro (UFRJ). Granulometry of all materials was determined using laser diffraction in a Malvern Mastersizer 2000 equipment at LABEST and is shown in figure 1.

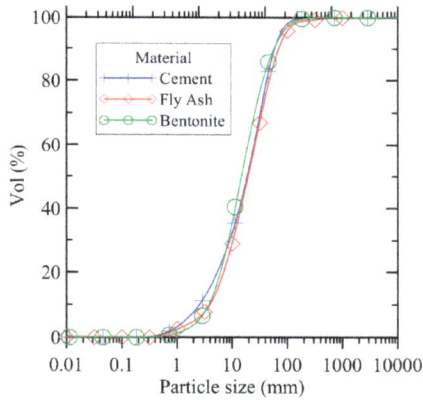

FIGURE 1 – GRANULOMETRY OF MATERIALS

TABLE 1 – CHEMICAL COMPOSITION OF RAW MATERIALS

Basic Oxide	PC	FA	B	Basic Oxide	PC	FA	B
CaO	63	1.9	1.2	ZrO2		0.101	
SiO2	18.3	52.74	59.7	Cr2O3		0.02	
Fe2O3	3.3	4.74	7.7	Rb2O		0.025	
SO3	1.1	1.89	<0.10	Y2O3		0.014	
Al2O3	4.7	33.64	15.8	NbO		0.005	
K2O	0.37	3.46	0.28	V2O5		0.065	
SrO	0.21	0.02	<0.10	Na2O	0.52		2.4
TiO2	0.23	1.31	0.72	MgO	1.8		2.7
MnO		0.04		Cl	<0.10		0.15
ZnO		0.04		LOI	6.1	1.29	9.1

On the other hand, densities were obtained by using a gas pycnometer, Accupyc1340 model of Micromeritics, see table 2.

TABLE 2 – DENSITY

Material	Density(g/cm^3)
Cement	3150
Fly Ash	2350
Bentonite	2491

Mixtures and batching procedure

A reference grout, cement-bentonite only, and three inclusions of fly ash content were studied, 0, 10, 30 and 50% of binder weight replacement, respectively. The quantities of materials are consistent with previous related studies and use of grouts for this kind of use [7], [8] and [9]. A detail of the composing materials is shown in table 3.

TABLE 3 – GROUT COMPOSITIONS

Material/Code: CB40/CB40FAx	CB40	10	30	50
Cement	350	315	245	175
W/b	2.495			
Bentonite	40			
Fly Ash, % replacement	0	10	30	50
Fly Ash weight	0	35	105	175
Total water	873.40			
Cement slurry , w/c	0.5			
Water 1	175			
Bentonite slurry				
Water 2	698.40			

The mixing protocol is detailed in table 4 and a Chandler Engineering Constant Speed mixer with a 4 liter capacity cup was employed. The bentonite and the binders were prepared separately guaranteeing homogenous mixes, in order to do so, total water is split in two as shown in table 3 and will be used as follows. The first three steps allow to prepare a bentonite+water2 mix. In the third one the bentonite mix remains static in a isolated recipient, the objective is to allow water to get deposited among the bentonite flat layers. The remaining part of the water, water 1, is used for producing a cement-fly-ash mix in step 4. The step 5 finally gathers them. High shear mixing was employed in various steps because it guaranteed bentonite dispersion, no floculization, clump avoidance in step 1 and 2 and a homogenous mix when bentonite and binder slurry were joined in step 5 [10,11].

TABLE 4 – MIXING PROCEDURE

Description	Step	Objective	Shearing characteristics	
			Speed (RPM)	Time (min)
Bentonite hydration	1	Dispersion	4000	2
	2	Avoid Floculization	12000	2
	3	Water absortion	24-72 hours	
Grout production	4	Cement+fly ash pre-hydration	4000	2
	5	Binder –bentonite mix	12000	2

Experimental tests

In order to understand the effect of fly ash on the fresh and hardened properties, viscometry was used for establishing its rheological characteristics at the fresh state. Ultrasonic cement analysis (UCA) for the hydration evolution and compression tests of cyilinder specimens for the mechanical characterization.   A viscometer of the Fann Instrument company was used for the rheological experiments imposing increasing and decreasing shear rates corresponding to 3, 6, 30, 60, 100, 200 and 300 rpm. The procedure belongs to the tests methods of the Brazilian Petroleum (PETROBRAS) company [12]. For the hydration tests, UCA tests were carried out in a Chandler Engineering equipment, model 4262 Twin cell – Ultrasonic Cement Analyzer. The grouts were poured in steel cells and a ultrasound wave went through the specimen for 12 days under 23 Celsius degrees and normal pressure. On the other hand, the mechanical compression behavior was obtained by using specimens of 50 diammeter and 100 mm height by following the NBR 5739 Brazilian standard [14]. A Shimadzu AGX-100 kN testing machine, a pair of LVDTs placed at both sides of the specimen and a 0.3 mm/minute displacement rate were used.

Results and discussion
Rheology

All grouts presented a thyxotropic response, high apparent viscosities at low shear shear rates and low viscosities at high shear rates ergo have a shear thinning behavior, see figure 2. Although the Herschel Bulkley (HB) model fitted well the experimental points, the obtained parameters like high yield stresses in table 5, cannot represent the grout responses. Therefore   other rheological parameters are obtained like the apparent yield stress and the experimental shear stress at 3 rpm. The apparent yield stress seems more reliable to represent the stress necessary to start the grout flowing not only compared to the HB curves in figure 2 but also to the experimental shear stress at 3 rpm. By using 10% of fly ash, the hysteresis loop is reduced and less stresses are obtained than the reference. A higher effect on the hysteresis loop is found with a 30% replacement. A 50% fly ash reduces more the shear stresses and flowability is improved [15] however the bentonite viscous effect still influences the grout [16].

Non-Conventional Materials and Technologies – NOCMAT for XXI Century   Materials Research Forum LLC
Materials Research Proceedings 7 (2018) 482-490   doi: http://dx.doi.org/10.21741/9781945291838-46

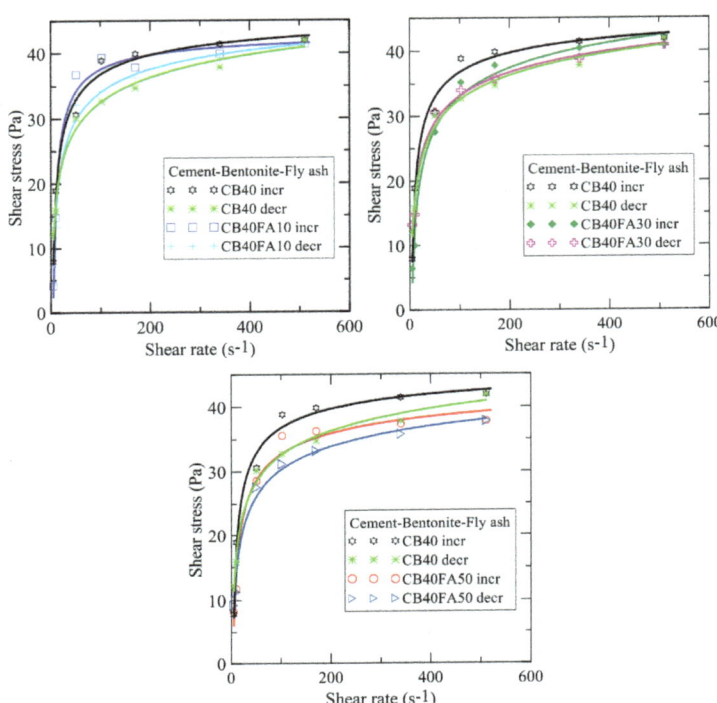

FIGURE 2 – GROUT RHEOLOGICAL BEHAVIOR

TABLE 5 – PARAMETERS OF THE HERSCHEL BULKLEY MODEL AND OTHER RHEOLOGICAL PARAMETERS

Grout	CB40		CB40FA10		CB40FA30		CB40FA50	
Shear rate	Incr	Decr	Incr	Decr	Incr	Decr	Incr	Decr
N	-0.42	-0.14	-0.64	-0.30	-0.22	-0.17	-0.33	-0.20
Yo	48.56	74.21	43.69	52.90	64.71	65.78	48.56	58.23
K	-79.55	/-78.29	-115.08	-73.82	-85.53	-71.37	-72.14	-70.16
Yo app	9.97	11.14	4.29	7.15	4.69	11.14	6.61	7.15
Experimental Shear stress at 3 rpm	7.92	12.26	4.09	8.69	6.39	13.29	8.18	9.20

Hydration and Mechanical compression

Fly ash inclusions produced delay in hydration by comparing the set points of all grouts in figure 3 and table 5. Being CB40 considered as reference, the 10% ,30% and 50% partial replacements produced set point delays of 1 hour and 38 minutes, 2 hours 4 minutes and 4 hours 43 minutes respectively due to the dilution effect caused by fly ash [17],[18].

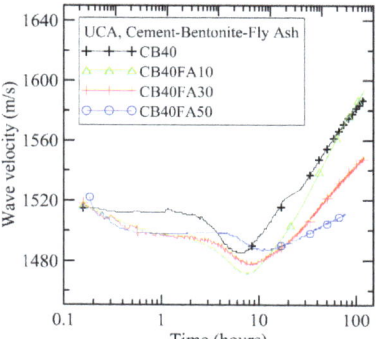

FIGURE 3 – WAVE PASSING THROUGH THE GROUT SPECIMENS IN THE UCA TEST

TABLE 5 *– UCA RESULTS*

Grout	Set point	Minimum wave velocity (m/s)
CB40	5h59'30'	1485.38
CB40FA10	7h37'30"	1470.76
CB40FA30	8h03'30"	1476.74
CB40FA50	10h42'30"	1487.12

The mechanical compression behavior is shown in figures 4, 5 and 6. At 28 days after batching, a 10% fly ash use slightly improves the strength, a gain of 2%, but a reduction is perceived when fly ash content is increased 70% and 45% of reference strength for CB40FA30 and CB40FA50 accordingly , see figures 4 and 5.

The stiffness gain of modulus of elasticity presents as similar response as the comparisons show in figure 5.

Non-Conventional Materials and Technologies – NOCMAT for XXI Century   Materials Research Forum LLC
Materials Research Proceedings 7 (2018) 482-490   doi: http://dx.doi.org/10.21741/9781945291838-46

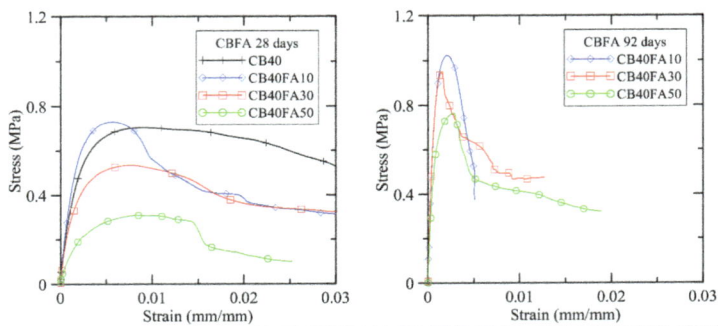

FIGURE 4 – MECHANICAL COMPRESSION OF THE TWO-COMPONENT GROUTS

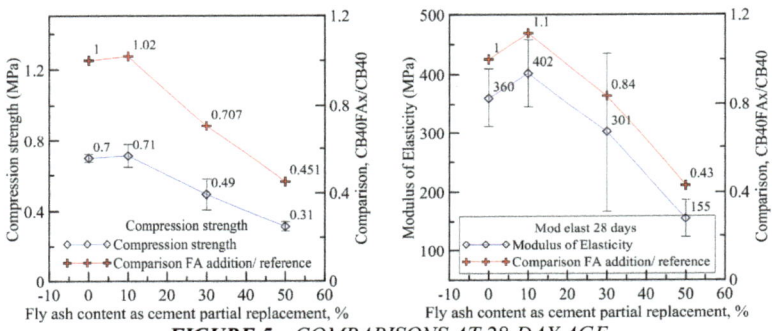

FIGURE 5 – COMPARISONS AT 28-DAY AGE

In addition, fly ash proved to increase the mechanical strength in the long term. Figure 6 indicates all the grouts with fly inclusions improved their strength reaching 150%-240% of their 28-day-age strength. Likewise, modulus of elasticity of every fly ash grouts is stiffer ranging 3- 5 times their values at 28 days. Results are consistent with previous studies on fly ash use [17]

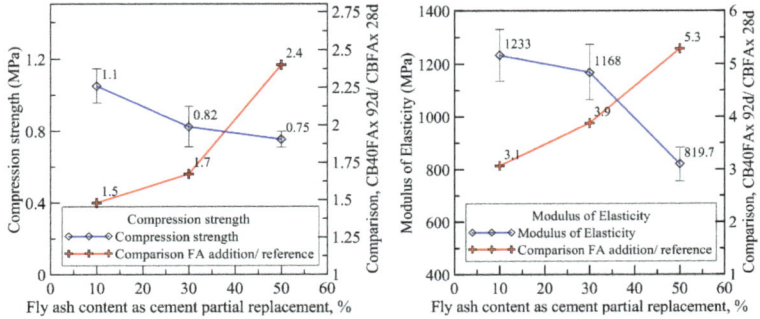

FIGURE 6 – COMPARISON AT 92-DAY AGE

Conclusions

By carrying out this research it is concluded that in the fresh state shear thinning behavior with hysteresis loops is found in all grouts. Although the Herschel Bulkley fits well the experimental points, more care in the parameter analysis is needed by obtaining other parameters like the apparent yield stress. The fly ash addition improves the rheology but bentonite effect is still dominant. In the hardened state, hydration was delayed proportionally to the fly ash increase but mechanical compressive strength was gained in the long term.

References

[1] Worldometers, World Population Clock. Online http://www.worldometers.info/world-population. Acessed on August,2017.

[2] EFNARC, Specification and Guidelines for the use of specialist products for Mechanized Tunneling (TBM) in Soft Ground and Hard Rock. European Federation of Producers and Contractors of Specialist Products for Structures, United Kingdom, 2005.

[3] Jefferis S. Cement-Bentonite Slurry Systems",In Grouting and Deep Mixing, American Society of Civil Engineers. 2012.

[4] da Luz A.B, Oliveira C.H. Argila – Bentonita, In Rochas Minerais Industriais – Usos e especificações. CETEM-MCT 239-253,2008.

[5] ABNT NBR 11578, Cimento Portland Composto – Especificação. Associação Brasileira de Normas Técnicas, Rio de Janeiro, 1991.

[6] ABNT NBR 10908, Aditivos para argamassa e concreto- Ensaios de uniformidade. Associação Brasileira de Normas Técnicas, Rio de Janeiro, 1990.

[7] Peila D, Borio L, PELIZZA S. The behavior of a two-component back-filling grout used in a tunnel boring machine. ACTA Geotechnica Slovenica 2011.

[8] COPPETEC-PEC. Influência da fração volumétrica e tipo de macro fibra polimérica na tenacidade à flexão do concreto pré-moldado a ser utilizado na obra da Linha 4 do Metro do Rio de Janeiro,17928. Relatório preliminar. Parte 1: Análise do Graute e "Slurry". Rio de Janeiro , 2014.

[9] Quiroga-Flores A. Physical and mechanical behavior of a two component cement-based grout for mechanized tunneling application. Master's degree thesis. Rio de Janeiro, Universidade Federal do Rio de Janeiro, 2015.

[10] Jefferis S."Effects of mixing on bentonite slurries and grouts". In Proceeding Conference on Grouting in Geotechnical engineering.

[11] Jefferis S, Private communications, 2014.

[12] Petrobras, Procedimentos e Métodos de Laboratório destinados à Cimentação de Poços Petrolíferos. Rio de Janeiro, Petrobras-Schlumberger-Halliburton-BJ Service. 2005.

[14] ABNT NBR 5739, Concreto - Ensaio de Compressão de corpos de prova cilindricos. Associação Brasileira de Normas Técnicas, Rio de Janeiro, 2007.

[15] Bentz D.P, Ferraris Ch. Rheology and setting of high volume fly ash mixtures. Journal of Cement and Concrete Composites 2010; 32(4) 265-270. https://doi.org/10.1016/j.cemconcomp.2010.01.008

Non-Conventional Materials and Technologies – NOCMAT for XXI Century   Materials Research Forum LLC
Materials Research Proceedings 7 (2018) 482-490   doi: http://dx.doi.org/10.21741/9781945291838-46

[16] Van Olphen H. Forces between suspended bentonite particles. Clays and Clay Minerals. Shell Development Company, Exploration and Production Research Division, Houston, Texas.

[17] Wang X. Effect of fly ash on properties evolution of cement based materials. Journal of Construction and Building materials 2014; 69: 32-40. https://doi.org/10.1016/j.conbuildmat.2014.07.029

[18] Deschner F, Winnefeld F, Lothenbach B, Seufert S, Schewesig P, Dittrich S, Goetz-Neunhoeffer F, Neubauer J. Hydration of Porland cement with high replacement by siliceous fly ash. Journal of Cement and Concrete Research 2012; 42: 1389-1400. https://doi.org/10.1016/j.cemconres.2012.06.009

Non-Conventional Materials and Technologies – NOCMAT for XXI Century   Materials Research Forum LLC
Materials Research Proceedings 7 (2018) 491-504   doi: http://dx.doi.org/10.21741/9781945291838-47

Structural Behavior of an Innovative Bamboo Arch Structure

Ayman S. Mosallam[a]*, Khosrow Ghavami[b], Rahil Shrivastava[a],
Mohamed A. Salama[c]

[a]University of California, Irvine, USA, e-mail: mosallam@uci.edu*, shrivasr@uci.edu

[b]Pontificia Universidade Católica do Rio de Janeiro, Brazil; e-mail: ghavami@puc-rio.br

[c]University of Helwan, Cairo, Egypt; email: masalama@aucegypt.edu

Abstract. Bamboo has a very long history with human kind; it is one of the oldest building material used for construction. Over 1,200 bamboo species have been identified worldwide and it has been used widely for household products and extended to industrial applications by advancement in technology and increased market demand. In this study, a structural evaluation of a full-scale bamboo truss arch structure was performed and its performance was verified by developing a finite element analysis (FEA) using ANSYS® FEA code. The simulated numerical results were compared with experimental results. The time increasing force was applied on the numerical model to simulate deflection and rotations at the joints.   Results of this study indicated that one of the major components that controls the arch structure stiffness is the joint elements that need further studies in order to improve upon the current design. Furthermore, results of the study showed that the proposed design can be used as a cost-effective and eco-friendly building system that can be utilized for rapid construction of emergency shelter and other housing applications.

Keywords: Bamboo, Arches, Sustainability, Large-Scale Tests, Numerical Analysis, Green Construction

Introduction

Bamboo provides many benefits to the environment including (i) it is one of the fastest growing renewable resource known, in addition to its ability to replenish itself rapidly and its growth and harvest have virtually no negative effects on the environment, (ii) bamboo produces greater biomass and 30% more oxygen than a hardwood forest of the same area, (iii) bamboo absorbs two-third more carbon dioxide (CO_2) from the atmosphere than any other plant, and in turn releases two-third more oxygen, producing super-oxygenated, pure air, and (iv) bamboo has higher strength-to-weight ratio as compared to other construction building materials such as steel with higher seismic resilience. Over the years, many researchers evaluated the use of bamboo in construction [1, 2, 3, 4, 5, and 6]. One of the pioneering work on the use of bamboo as internal reinforcement of concrete members was evaluated by the US Naval Civil Engineering Laboratory in 1966 [4].   A comprehensive coverage on the use of bamboo internal reinforcement was reported by Ghavami in 1995 [5].   The benefits of the use of bamboo in lieu of different conventional materials was reported by Bhalla et al. [3].

The use of bamboo in construction applications is common in several Asian and South American countries, however, with the world's green movement, many North American states and European countries started to utilize such sustainable materials in a limited scale. Some of the signature bamboo structures that have been in the past few decades include the Green School in Bali, Indonesia (Fig. 1-A), the Tongji University of China solar-powered bamboo house, that was constructed as the university's official entry to the First European Solar Decathlon that was held in Madrid in 2010 (Fig. 1-B), and the Japanese Noodle Restaurant Jakarta, Indonesia (1-C), among many others worldwide.   In addition, bamboo has been used in constructing light-weight pedestrian bridges with spans up to fifty meters (Fig. 1-D). The applications of bamboo in constructing pedestrian bridges was reported by several researchers [6].

Non-Conventional Materials and Technologies – NOCMAT for XXI Century   Materials Research Forum LLC
Materials Research Proceedings **7** (2018) 491-504   doi: http://dx.doi.org/10.21741/9781945291838-47

FIGURE 1- *EXAMPLES OF BAMBOO CONSTRUCTION APPLICATIONS: (A) GREEN SCHOOL IN BALI, INDONESIA, (B) TONGJI UNIVERSITY OF CHINA SOLAR-POWERED BAMBOO HOUSE, (C) JAPANESE NOODLE RESTAURANT JAKARTA, INDONESIA, (D) BAMBOO PEDESTRIAN BRIDGES*

In this study, structural behavior of an innovative bamboo arch structure is analyzed both experimentally, and numerically using ANSYS® code in order to understand the load path, deformation and stability under sustained loading. Details of experiments, numerical modeling and simulation results are presented in the following sections. The results this study indicated the feasibility of using such bamboo arch for low-cost, temporary housing and mass emergency shelters applications.

Arch structure components & support system
Properties of Bamboo
Due to the fact that the physical and mechanical properties of bamboo vary with the age and the height of the culm [8], five coupon specimens were evaluated in order to determine the average physical dimensions of the arch bamboo components (see Fig. 2). Table (1) presents a summary of the results.

Non-Conventional Materials and Technologies – NOCMAT for XXI Century   Materials Research Forum LLC
Materials Research Proceedings **7** (2018) 491-504   doi: http://dx.doi.org/10.21741/9781945291838-47

FIGURE 2 - BAMBOO COUPON SPECIMENS

TABLE 1 - *VARIATION OF DIMENSIONS IN A FOOT-LONG (25.4 MM) BAMBOO SAMPLES USED IN CONSTRUCTING THE BAMBOO ARCH STRUCTURE*

Specimen	Outside diameter (OD)		Average diameter (d_o)	Inside diameter (ID)		Average diameter (d_i)	Thickness (t_o)
	side1	side2		side1	side2		
	inches (meter)						
1	1.5 (0.46)	1.5 (0.46)	1.5 (0.46)	1.2 (0.37)	1.1 (0.34)	1.2 (0.37)	0.3 (0.09)
2	1.6 (0.49)	1.6 (0.49)	1.6 (0.49)	1.2 (0.37)	1.2 (0.37)	1.2 (0.37)	0.4 (0.09)
3	1.5 (0.46)	1.6 (0.49)	1.5 (0.46)	1.2 (0.37)	1.2 (0.37)	1.2 (0.37)	0.3 (0.09)
4	1.4 (0.43)	1.3 (0.40)	1.4 (0.43)	1.1 (0.34)	1.1 (0.34)	1.1 (0.34)	0.3 (0.09)
5	1.4 (0.43)	1.4 (0.43)	1.4 (0.43)	1.1 (0.34)	1.1 (0.34)	1.1 (0.34)	0.3 (0.09)
Average OD			**1.5 (0.46)**	**Average ID**		**1.2 (0.37)**	

Properties of Nylon Rope (Joining Elements)

Nylon is a synthetic polymer, also known as polyamide. Some of the advantages of using such material for joining different bamboo members is primarily the simplicity of the joint detail and the ability of making the joints by a person with minimum construction skills. In addition, Nylon is durable, has high strength-to-weight ratio, and has excellent abrasion resistance [7]. However, in wet environment, strength degradation of the Nylon may reach up to 20% of its dry strength. As the moisture content increases, modulus of elasticity and other mechanical properties are reduced depending on the type of Nylon. In case of fire, Nylon melts, instead of burning which is makes it very favorable in the joining applications due to the vulnerability of bamboo to fire. However, when it when it breaks down during fire, it produces hazardous smoke that contains hydrogen cyanide. However, because the amount of Nylon materials that are used in the joints are relatively small, this negative effect may have little impact on the overall fire performance of the bamboo structure. It

Non-Conventional Materials and Technologies – NOCMAT for XXI Century   Materials Research Forum LLC
Materials Research Proceedings 7 (2018) 491-504   doi: http://dx.doi.org/10.21741/9781945291838-47

also has good resistant to ultra-violet, insects, fungi, animals, as well as molds, mildew and rot which are also very critical elements that affect any bamboo or wood structure in general. However, Nylon has low resistance to alkalis, iodine and acids. There are several types of nylon ropes that are commercially available.

For this application, modules of elasticity plays a major role in determining the joint behavior as well as ties (see Figure 3) as it will be explain in this paper. The experimental longitudinal modulus of elasticity of the Nylon rope was calculated using experimental results shown in Figure 4 and found to be 12.9 MPa.

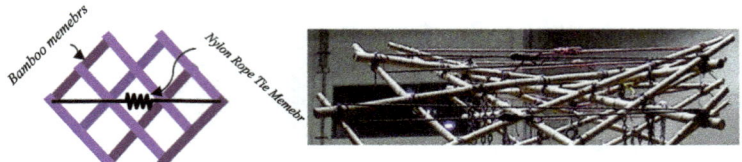

FIGURE 5 - SPRING ACTION OF NYLON ROPE

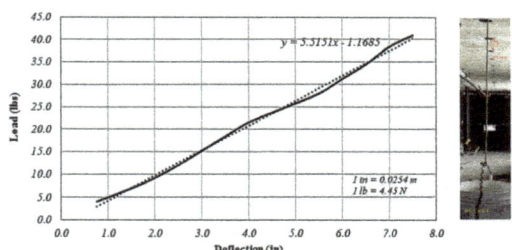

FIGURE 4 - LOAD-DEFLECTION CURVE OF NYLON ROPE

Geometry and Support System

In constructing the arch structure, the bamboo sticks were connected using nylon ropes forming 609.6mm X 609.6mm (2'X 2') rhombus as shown in Figure (5). High-strength prestressed Dwydag® steel rods and nylon ropes were used to fasten the base of the structure. Each two end bamboo sticks were connected together by clove hitch knot/general utility hitch (see Figure 6). This simple joining method provide sufficient stiffness to create the condition of moving members together under moderate loads. Also, this detail makes it easier to tie the knot around the bamboo sticks. It can also be tighten using one end and tied in a bight which provide the resilience and redundancy to the structure.

Non-Conventional Materials and Technologies – NOCMAT for XXI Century   Materials Research Forum LLC
Materials Research Proceedings 7 (2018) 491-504   doi: http://dx.doi.org/10.21741/9781945291838-47

FIGURE 5- RHOMBUS FORMED BY JOINING BAMBOO ELEMENTS

FIGURE 6 - SUPPORT CONDITIONS FOR THE BAMBOO STRUCTURE AND CLOVE HITCH KNOT

Experimental program
Loading Pattern and Joining Details
The loads were applied at four joint points sysmmterically as shown in Figure 7. In each experiment, the load was applied progressively at a constant loading steps of 44.80kN (10.0 lbs.). Figure 7 shows the point load locations and loading fixture.

FIGURE 7 - POINT LOAD LOCATIONS AND LOADING FIXTURE

Joints Designation Codes

In order to avoid errors in collecting experimental measurements, each joint was assigned a distinct number and each rhombus was assigned a different designation number. After assigning each nodes identification number, the coordinates were recorded by assuming X-axis in longitudinal, Y-axis in elevation, and Z-axis in transverse directional change for each nodes to form the geometry. Details of the coding system can be found in Ref. [8].

Both instantaneous and short-term creep and recovery tests were performed on the bamboo arch structure. Prior to conducting the final tests, several simple loading/unloading test runs were performed in order to improve and finalize the design of both the support system and the joining details. Simultaneously, finite element (FE) simulation for each joining and support details were conducted in order to verify the progressive improvements in both the design and arch structure details prior to conducting the final non-destructive full-scale tests. Figure 8 presents the observational nodes and rhombus which were analyzed during the experiment. Using measuring tapes, the observational rhombus change in distance and angles were recorded. In a rhombus, 'α', 'β' are the angles measurements parallel to z- and x- axes, respectively (see Figure 8-b). The change in both direction angle for each rhombus is graphically represented. Initially the rhombus angles, 'α', 'β' were 90° respectively.

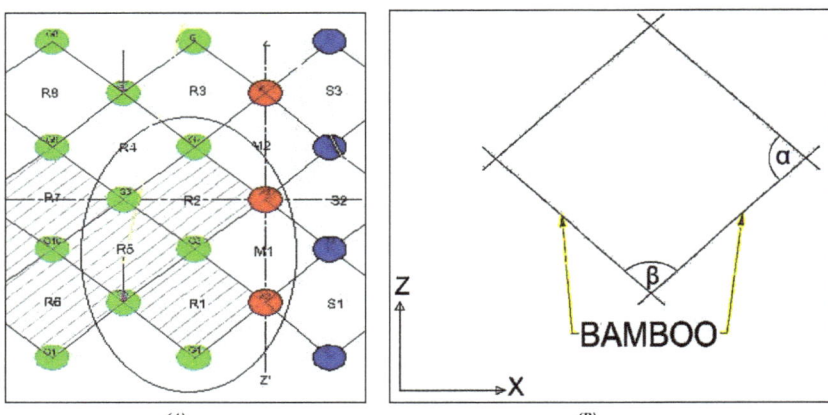

(A)   *(B)*

FIGURE 8 - *(A) OBSERVATIONAL AREA, (B) ANGLE NOTATION IN A RHOMBUS*

Test #1: Joints with Nylon Ropes Only

Case #1: Loading of Structure (Uniform Loading with Load Step of 4.5 kg (10.0lbs.)

In this test, the change in angle data was recorded that is used to capture the alternated expansion and contraction of the loaded arch rhombus. When the value of 'α' increases and the value of 'β' decreases, expansion takes place and the opposite situation results in a contraction. This behavior is referred to as "*scissor action*" at the joint. This was clearly observed in all tests where angle change of the quarter side of bamboo was recorded. Consider the observational the change in angles for rhombuses R1, R2 and R5, shown in Figure 8-A, were recorded and results are presented in Figure 9.

Non-Conventional Materials and Technologies – NOCMAT for XXI Century   Materials Research Forum LLC
Materials Research Proceedings 7 (2018) 491-504   doi: http://dx.doi.org/10.21741/9781945291838-47

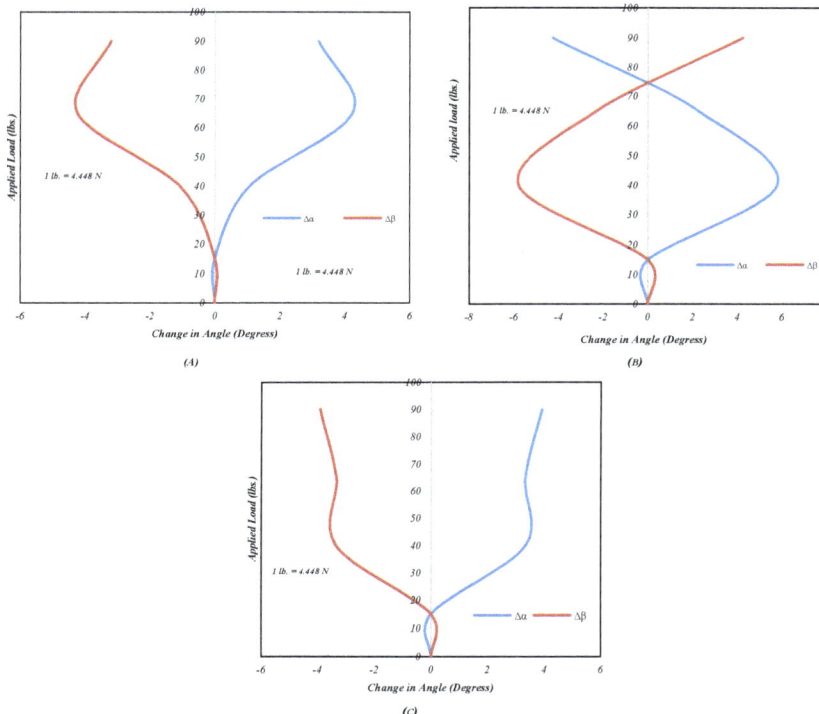

FIGURE 9 - *(A) CHANGE IN ANGLE IN RHOMBUS 'R1' (B) CHANGE IN ANGLE IN RHOMBUS 'R2',*

(C) CHANGE IN ANGLE IN RHOMBUS 'R5'

As shown in Figure 9, as the load increased, rhombuses R1 and R2 expanded, while rhombus R5 contracted. This angular changes verify the arch structure's scissor action for load transfer and distribution.

Now, in order to understand the structural deflection pattern, vertical deflections at three locations, namely; (i) at the quarter, (ii) at half, and (iii) at the third-quarter of the arch span were measured at each increment of load (refer to Figure 10). As mentioned earlier, the tests performed in this study are non-destructive with a main objective is to evaluate the bamboo arch structure stiffness and optimize the design to achieve an acceptable deflection limit states that confirm with different building codes. One can see that in this phase of the multi-phase study, relatively small loads were applied. As shown in this figure, the measured central vertical deflection at a load level of 40.8 N (90.0 lbs.) was 152.0 mm (6.0″).

Non-Conventional Materials and Technologies – NOCMAT for XXI Century   Materials Research Forum LLC
Materials Research Proceedings **7** (2018) 491-504   doi: http://dx.doi.org/10.21741/9781945291838-47

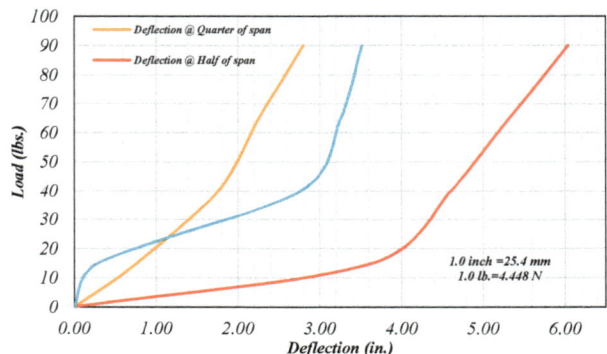

FIGURE 10 - *LOAD VS. VERTICAL DEFLECTION AT DIFFERENT LOCATIONS*

Short-Term Creep & Recovery Test: Bamboo is considered a viscoelastic material, and hence it is sensitive to sustained loading. In order to evaluate the time-dependent behavior of such structures, creep and recovery tests are essential. In this study, several short- and medium-term creep tests were performed in order to assess the bamboo arch creep behavior. In the short-term creep test, two sustained load levels were applied, namely; at 66.8 N (15.0 lbs.) and at 400.34 N (90.0 lbs.) for a period of twelve hours. During the loading period, deflections at different locations were recorded. A summary of the short-term creep tests is presented in Figure 11. In order to observe the recovery behavior of the bamboo arch, unloading was performed by removing the loads at the same rate of 4.50 kg (10.0 lbs.) that was adopted in loading the arch. During the unloading stage, angular changes of each rhombus along with vertical deflection at different locations were recorded. The time-dependent angular results are presented in Figure 12.

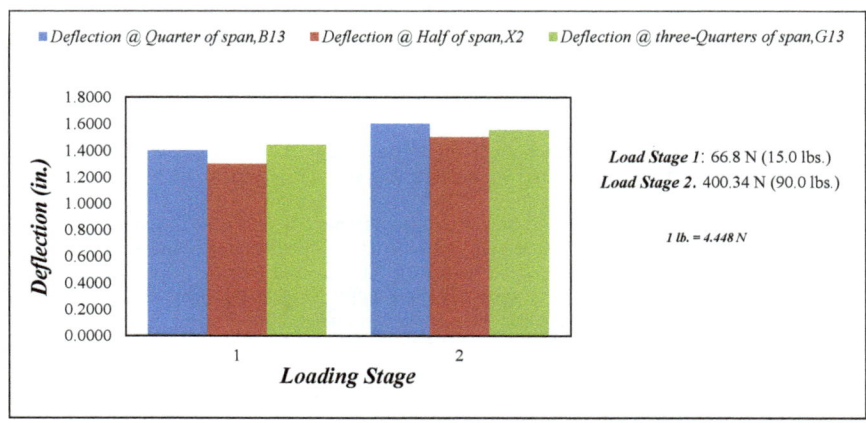

FIGURE 11 - *SHORT-TERM DEFLECTION CREEP AFTER 72 HOURS OF LOADING AT DIFFERENT LOCATIONS*

Non-Conventional Materials and Technologies – NOCMAT for XXI Century   Materials Research Forum LLC
Materials Research Proceedings 7 (2018) 491-504   doi: http://dx.doi.org/10.21741/9781945291838-47

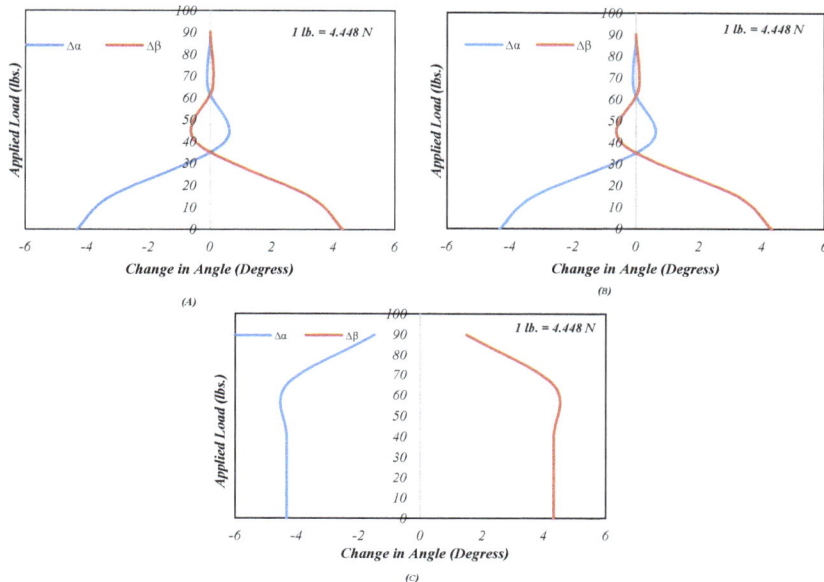

*Figure 12 -Change in Angle During Load Recovery: (a) Rhombus 'R1', (b) Rhombus 'R2' (c)
Rhombus 'R5'*

Test #2: Joints Tied by Nylon Ropes and Cable Ties

Based on the experimental results that indicated that unstiffened joints caused large local joint movement that was translated into large vertical deflection that grew with time. In order to resolve this issue and in order to increase the stiffness of the joints, additional reinforcement to the joints was introduced using commercial electric cable ties with tensile strength of 329.50 N (74.0 lbs.). The cable ties were looped around the two crossing bamboo forming a stiff joint (see Figure 13).

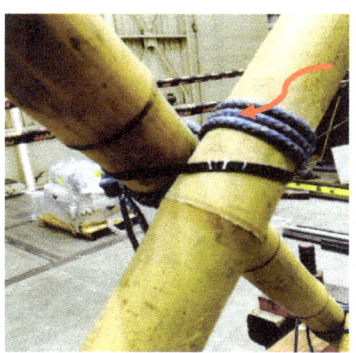

Additional Cable Ties

FIGURE 13 - *CABLE TIES LOOPED AROUND THE JOINT*

Non-Conventional Materials and Technologies – NOCMAT for XXI Century   Materials Research Forum LLC
Materials Research Proceedings 7 (2018) 491-504   doi: http://dx.doi.org/10.21741/9781945291838-47

Case #1: Loading of Structure (Uniform Loading with Load Step of 4.5 kg (10.0lbs.)
Identical procedures similar to those adopted in test #1 described earlier were followed in this experiment. In this case, the angular changes of observational rhombus and vertical central deflection at every load increment were measured. Figure 14 presents the experimental results for rhombus R1, R2, R5 and other observational rhombus. Results obtained from test #1 showed the same behavior which verifies that a pair of three rhombuses and a pair of two rhombuses expanded and contracted, respectively.

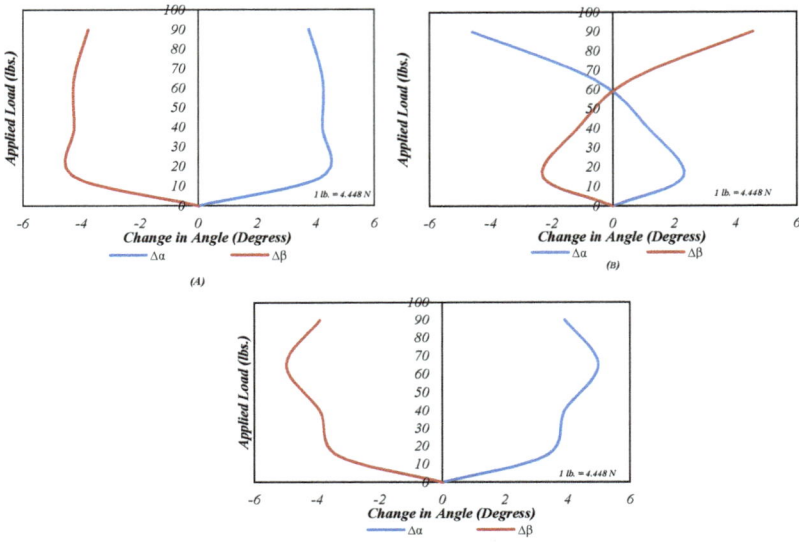

FIGURE 14 - *CHANGE IN ANGLE DURING LOADING: (A) RHOMBUS 'R1', (B) RHOMBUS 'R2' (C) RHOMBUS 'R5'*

Experimental results indicated that increasing joints stiffness by using additional cable ties resulted in substantial reduction of vertical deflection at different locations. For example, the arch mid-span deflection was tremendously reduced from deflection get reduced from 15.2 cm (6.0″) to 6.4 cm (2.5″) which is more than 100% reduction as compared to the deflection of the arch with unstiffened joints. Again, this proves the major influence of the joint stiffness of this type of structure on the overall serviceability of the arch structure.

Creep test
This test was performed to evaluate the viscoelastic behavior of the bamboo arch structure with the modifications described in test #2. The duration of this short-term creep test was forty-eight hours (2,880 minutes) of loading. It is well known that both short- and long-term structural performance of bamboo relies heavily on moisture content and creep [10]. Although creep behavior of bamboo may be susceptible to loading parallel to the fibers, loads perpendicular to the fibers which induce longitudinal splitting behavior are commonly reported as the dominant mode of failure in bamboo members [11]. Based on this fact, electrical strain gauges were poisoned on the bamboo in a

Non-Conventional Materials and Technologies – NOCMAT for XXI Century   Materials Research Forum LLC
Materials Research Proceedings **7** (2018) 491-504   doi: http://dx.doi.org/10.21741/9781945291838-47

direction parallel to the longitudinal fibers. The strain of the central node was observed via two strain gauges that were bonded; one of the top of bamboo stick and another on the bottom. The 100.0lbs (448.0N) sustained point load was applied at the central node as shown in Figure 15. The weight of the steel plates (*each weighing 111.20N/25.0 lbs.*) were added making the total sustained point load to be 448.0N (100.0 lbs.). Over the duration of test, both room temperature and relative humidity were recorded. Figure 16 presents a sample of the creep curve for the bamboo arch structure. As shown in the figure, the primary (instantaneous) creep increased up to 0.0015, after which the rate of change was negligible.

The large value of the primary creep is attributed to the creep component of the low level of the nylon materials. However, since the stiffness of the bamboo is much higher than that for the nylon ropes, the contribution of the bamboo creep was found to be minimal due to the lower value of the imposed sustained load of 448.0 N (100.0 lbs.). The negligible value of the transit (or secondary creep) is attributed to two factors including low strain level and the short duration of the test. In addition, since the strain gages were bonded to the top central bamboo portion of the structure, so as the load was applied, stretching of the nylon robs took place, allowing the top central bamboo members to come to a straight flat position (see Figure 17) where the original arched strain was neutralized resulting in a minimal amount of time dependent strain as observed in this test. In fact the highest deformable portion is towards the support where the arched portion will became more strained as the load is applied and sustained over time.

FIGURE 15 - DETAILS OF LOADING REGIME OF THE BAMBOO ARCH STRUCTURE

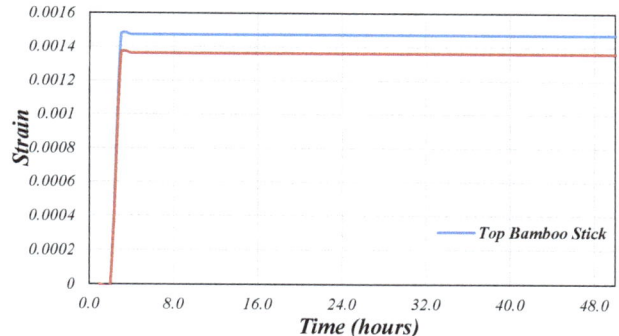

FIGURE 16 - TYPICAL SHORT-TERM CREEP CURVES

Non-Conventional Materials and Technologies – NOCMAT for XXI Century   Materials Research Forum LLC
Materials Research Proceedings **7** (2018) 491-504   doi: http://dx.doi.org/10.21741/9781945291838-47

*FIGURE 17 - AS THE LOAD WAS APPLIED THE ARCHED CENTRAL PORTION OF
BAMBOO MEMBERS FLATTENED*

Numerical modeling and comparison with the experimental results

The FEA numerical model was first created using ANSYS code in order to replicate test #1 results, in order get the accuracy of the numerical model and analysis result. The same linearly-increasing force function was applied by only to reach 400.34kN (90.0lbs) load. The deformed shape and central vertical deflection obtained from both finite element simulation and test #1 results were compared. The numerically simulated deformed shape is presented in Figures 18-a. In this figure, the blue and dashed gray lines are elements of structure after deformation and initially unloaded structure respectively. The applied load vs. vertical central node deflection curves that were obtained from both the FE model and experimental test # 1 is shown in Figure 18-b. One can see form this figure, an excellent correlation between the numerical and experimental results was achieved. Based on this results, it is confirmed that the numerical finite element model can be used for further analysis for design improvement.

Conclusions

In this study, the behavior of an innovative bamboo arch structure was performed. The study was divided in two interrelated parts; experimental and numerical. In the experimental phase, the bamboo arch structure was tested under light-weight loading to examine both the instantaneous and short-term creep behavior. One of the important finding is the major influence of joint detail and stiffness on both short- and long-term behavior of the arch structure. For this reason, more work is needed to improve upon the joint details using non-conventional methods.

Non-Conventional Materials and Technologies – NOCMAT for XXI Century   Materials Research Forum LLC
Materials Research Proceedings **7** (2018) 491-504   doi: http://dx.doi.org/10.21741/9781945291838-47

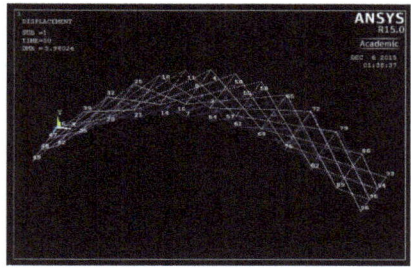
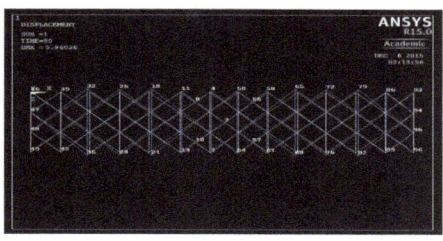

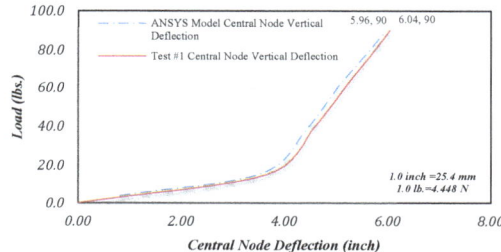

FIGURE 18 - *FE MODELING AND COMPARISON WITH EXPERIMENTAL RESULTS*

References

[1] Henrikson, R. Jorg Stamm: Bamboo designer and master builder, ecobamboo, Colombia. http://www.bamboosun.com/bios/jorgstamm.html, 2008.

[2] Nurdiah, E. A. The potential of bamboo as building material in organic shaped buildings. Procedia-Social and Behavioral Sciences, 216, pp. 30-38, 2016. https://doi.org/10.1016/j.sbspro.2015.12.004

[3] Bhalla, S., Gupta, S., Sudhakar, P., and R. Suresh. Bamboo as green alternative to concrete and steel for modern structures. Journal of Environmental Research and Development, Vol. 3, No. 2, October-December, 2008.

[4] Francis, E. Brink, and J. R. Paul. Bamboo Reinforced concrete Construction. US Naval, Civil Engineering Laboratory, Port Hueneme, California, 1966.

[5] Ghavami, K. Bamboo low cost and energy saving construction materials, Proc. International Conference on Modern Bamboo Structures, p. 28-30, October, Changsha, China, 2007.

[6] Xiao, Y., Zhou, Q., and B. Shan. Design and construction of modern bamboo bridges, ASCE Journal of Bridge Engineering 15, no. 5, p. 533-541, 2009. https://doi.org/10.1061/(ASCE)BE.1943-5592.0000089

[7] Tong, N., Zhu, C., Zhang, C., & Zhang, Y. Study on Raman spectra of aliphatic polyamide fibers. Optik-International Journal for Light and Electron Optics, 127(1), 21-24, 2016. https://doi.org/10.1016/j.ijleo.2015.09.180

Non-Conventional Materials and Technologies – NOCMAT for XXI Century   Materials Research Forum LLC
Materials Research Proceedings 7 (2018) 491-504   doi: http://dx.doi.org/10.21741/9781945291838-47

[8] Shrivastava, R. Structural Evaluation of Bamboo Arch Structure, Master thesis, University of California, Irvine, California, USA, 113ps., 2015.

[9] Chauhan, L., S. Dhawan, and S. Gupta. Effect of age on anatomical and physicomechanical properties of three Indian bamboo species. J. of the T.D.A.. 46, pp.11-17, 2000.

[10] Xu, Q., Harries, K., Li, X., Lui, Q., and J. Gottron. Mechanical properties of structural bamboo following immersion in water. Engineering Structures, Vol. 81, December, pp. 230–239, 2014. https://doi.org/10.1016/j.engstruct.2014.09.044

[11] Gottron, J., Harries, K.A., and Q. Xu. Creep behavior of bamboo, Construction and Building Materials, Vol. 66, September, pp. 79–88, 2014. https://doi.org/10.1016/j.conbuildmat.2014.05.024

Non-Conventional Materials and Technologies – NOCMAT for XXI Century   Materials Research Forum LLC
Materials Research Proceedings 7 (2018) 505-515   doi: http://dx.doi.org/10.21741/9781945291838-48

Effect of the Residual Fine Elements and Contribution on Fresh Properties of Self-Levelling Mortars

Henrique Duarte Sales Carvalho [a], Malik Cheriaf [b] and Janaíde Cavalcante Rocha [c] *

[a] Universidade Federal de Santa Catarina, UFSC, Brasil

hdsc2012@gmail.com [a], malik.cheriaf@gmail.com [b], janaide.rocha@ufsc.br [c],

Abstract. The self-levelling mortar are a fluid product that has as a low viscosity and high fluidity. This building material requires no vibration and levelling, having very low thickness (30 mm). The purpose of this study was to evaluate fresh properties of self-levelling SLU using three by-products: bottom-ashes, fine elements of quarry limestone and fine elements from recycled-concrete-waste. Portland Cement OPC was partially replaced by 0,10, 15 e 20% of fines. The water/binder was maintained constant at 0.50 and mix proportion 1:2 cement and sand (by weight) was used. Fresh properties were examined: plastic shrinkage, curling and and heat measurements using a semi-adiabatic calorimeter. A superplasticizer (polycarboxylate) was used to assure a fluidity. During drying (HR=50%) a mass loss was monitored. To the same cement content fine from bottom-ash act as efficiently to reduce a shrinkage. The curling phenomenon is high (more than 1,70 to mortar based on fine from recycled-concrete-waste. Higher early curling was obtained to mortar with quarry limestone. A good performance could be assured with ternary mixes with bottom ashes (curling less than 8mm). Self-levelling mortar systems with more than 20% of fine produced the mixes with negative effect on segregation and bleeding, so not recommended to self-levelling mortar purposes.

Keywords: Bottom Ashes, Fine From CDW Aggregates, Self-Levelling Mortars, Fresh Properties, Shrinkage

Introduction

Self-leveling mortars are characterized by their ability to spread quickly and self level, without the need for manual or mechanical intervention. The homogeneity of these compounds is necessary to ensure the endurance and durability characteristics of the final product, therefore, it is imperative that the mixture exhibits high stability [14]. The most important properties in a self-leveling mortar system, considering its viability and its final characteristics, are: self leveling, low viscosity, fast hardening, fast setting time, fast strength gain, dimensional stability, high end strength, surface durability and a strong adhesion with the substrate. To attaint all these requirements, self-leveling mortars often contain a wide variety of organic and inorganic compounds [18].

The use of self-leveling mortar system offers environmental and economic advantages. Among these advantages, it is possible to highlight the use of residues from other processes, such as a source of admixtures to replace part of the Portland cement, and to improve the characteristics of the final product, mainly the rheological properties [7]. In Spain, natural coarse aggregates may be substituted by Recycled Concrete Waste (RCW) to concrete production in proportions of 20% to 100% [5,6], however the fine fraction of RCW is not allowed. The greater difficulty in the use of fine RCW is related to their high water absorption [20]. In order to guarantee self-leveling, a greater amount of fines is required to achieve a effective fluidity [13]. Therefore, the use of the fine portion of this residue can be a sustainable solution.

Mortar and concrete are porous materials, which are susceptible in volume variations due to shrinkage during hydration and drying. In particular, the development of cement based self-leveling mortars still limited because of problems such as cracking or warping of corners [10 and 11]. These cracks are caused by shrinkage occurring from drying. The shrinkage in self-leveling systems is influenced by the high content of fine particles required to achieve effective fluidity [9 and 15]. The

Non-Conventional Materials and Technologies – NOCMAT for XXI Century   Materials Research Forum LLC
Materials Research Proceedings 7 (2018) 505-515   doi: http://dx.doi.org/10.21741/9781945291838-48

fine particles demand more water, due to the greater specific surface area, increasing the tendency of greater volumetric variation during drying.

The Coal Bottom Ash (CBA) when used as a substitute for the natural fine aggregate can improve internal curing, and reduce capillary forces, hydration temperature and consequently the shrinkage in concretes [8]. Concrete mixtures which incorporate bottom ash as fine aggregates for partial or total replacement of natural sand exhibit better dimensional stability. The structure of porous particle in bottom ash was beneficial for decreasing shrinkage in concrete samples. At the age of 180 days, concrete mixtures with bottom ashes FBC 2 (50%), FBC 3 (75%) and FBC 4 (100%) presented 21.79%, 34.62% and 37.17%, respectively, demonstrating lower shrinkage values by drying compared to the control mixture [17].

In this context, the present study aims to produce self-leveling mortars using fines of residues (CBA and RCW) as partial substitutes for Portland cement and contribute to sustainable development.

Materials

The materials used in the research were Portland Cement CP V - ARI (Hight initial strength), Limestone Powder (LP), Recycled Concrete Waste (RCW) and Coal Bottom Ash (CBA) fines. As an aggregate, natural sand was used, with fine granulometry (grains less than 2 mm) and a superplasticizer (SP) admixtures with chemical base polycarboxylate.

Method

Material characterization

The material characterization tests were carried out according to the procedures established in Brazilian standards (Table 1).

TABLE 1 - MATERIALS CHARACTERIZATION TESTS.

Granulometric analysis	ABNT NBR NM 248/2003	Sand
Specific gravity	ABNT NBR NM 52/2009	Sand and fines (LP, RCW and CBA)
Specific gravity	ABNT NBR NM 23/2001	Portland Cement
Specific Area	ABNT NBR NM 76/1998	Portland Cement and fines (LP, RCW and CBA)

The laser particle sizer was performed on the fines (LP, RCW and CBA) using the Microtrac S3500 equipment. Chemical analysis of fine materials (CP, LP, RCW and CBA) was carried out using Energy Dispersive X-Ray Spectrometer 700 (EDX 700) Shimadzu.

Mortar sutdy

The fines (LP, RCW and CBA) were used as partial substitutes of the Portland cement CP V ARI, in content of 10%, 15% and 20%. The substitution of the cement by the fines was performed according to the absolute volume of the materials, in relation for the difference between the specific gravity of the fines. The sand and fines were added to the dry.

Initially, experiments were carried out to evaluate the consistency properties of the mortar in the fresh state. For this purpose, the mortar was dosed to determine the optimum admixtures percentage and the content of fines. The mixtures were composed in a ratio by weight of 1: 2 (cement: aggregate) and w/(water cement ratio) of 0.5.

Once tests at the fresh state were conducted and mortars with satisfactory fluidity and cohesion properties for self-leveling systems were reached, tests were developed in the hardened state to evaluate the temperature of hydration, drying shrinkage and curling.

Non-Conventional Materials and Technologies – NOCMAT for XXI Century   Materials Research Forum LLC
Materials Research Proceedings 7 (2018) 505-515   doi: http://dx.doi.org/10.21741/9781945291838-48

Flow test

The flow test had was conducted to evaluate the chemical admixtures content for reaching a flow value between 24 and 27 cm. The tests were performed shortly after preparation of the mixtures, using a mini-cone. After removal of a mini cone, the flow was evaluated over a glass plate marked with 20 cm, 25 cm and 30 cm diameters. The final value of the recorded flow was considered as the mean value between two perpendicular measurements.

The self-leveling mortars chosen were those which behaved cohesive without signs of segregation or bleeding and presented a good visual appearance. The selected mortars were identified according to the material used and the cement replacement content (Table 2).

TABLE 2 - *IDENTIFICATION OF SELF-LEVELING MORTARS.*

Identification	Cement Consumption (m³)	Cement Replacement Content	Fines Content (%)		
			LP	RCW	CBA
FC 10	536,4	10%	10		
FC + RCD 10	534,2		5	5	
FC + CZP 10	530,1		5		5
FC + RCD +CZP 10	529,4		5	2.5	2.5
FC 15	527,9	15%	15		
FC + RCD 15	517,9		7.5	7.5	
FC + CZP 15	528,4		7.5		7.5
FC + RCD + CZP 15	516,3		7.5	3.75	3.75

Evolution of hydration temperature of mortars

A semi-adiabatic calorimeter was used to evaluate the temperature evolution of self-leveling mortars. The mortars were placed, still under the fresh state, in a cylindrical container of polystyrene. Thereafter, the container was placed in a thermal box. Temperature measurements from the samples were collected through thermocouples and processed into a Hewlett-Packard data logger, model 34970A, connected to a computer for data storage, with readings at every 20 seconds.

Drying shrinkage

In order to evaluate displacements of the corners (curling), an experiment on plates was carried out in a manner similar to those developed by other researchers [11, 16] whom have evaluated the drying shrinkage in self-leveling mortar.

Tests to measure Curling and the linear shrinkage of self-leveling mortars were performed simultaneously with the same sample. The displacement values were verified by Linear Variable Differential Transformer (LVTDs) and processed automatically through a data acquisition equipment (Data Logger), which was connected to a computer for data storage. After the mortar started setting, equipments for linear shrinkage and curling data collection was connected for a period of 5 consecutive days. The temperature ($23° \pm 2$) C and relative humidity ($60\% \pm 5$) were monitored throughout the test period.

Results and analysis

Physical and chemical properties

The physical and chemical properties of cement and fines are shown in Table 3.

Non-Conventional Materials and Technologies – NOCMAT for XXI Century   Materials Research Forum LLC
Materials Research Proceedings 7 (2018) 505-515   doi: http://dx.doi.org/10.21741/9781945291838-48

TABLE 3 - CHEMICAL COMPOSITIONS AND PHYSICAL PROPERTIES OF MATERIALS.

Oxides	Cement Portland CP V	Limestone Filer	Fines CBA	Fines RCW
Chemical analysis (%)				
Oxides				
CaO	74.795	54.788	1.733	41.202
SiO_2	12.303	4.321	40.819	29.048
Al_2O_3			37.458	
Fe_2O_3	4.458	0.775	5.713	9.805
K_2O	2.071	0.472	5.197	3.905
SO_3	1.984			
TiO_2				1.447
CO_2	3.243		6.674	13.22
Physical properties				
Specific gravity (Kg/m^3)	3040	2680	1920	2460
Blaine Fineness (m^2/Kg)	505	228	218	502

Only the fines from bottom ash presented aluminum in a significant amount (37%), in contrast the CaO content (1.7%) was much lower than the other materials. The bottom ash showed the lowest specific gravity, probably due to its high internal porosity.

Natural sand 1.2 to 0.075 mm and fineness modulus of 1.39 was used in mortar composition. Approximately 13% of the natural sand showed grains inferior to 0.15 mm.

All fine elements, LP and RCW were less than 0.075 mm and 70% of bottom ash. To LP all particle was less than 50 μm. The addition with the largest grain sizes was CBA, with all grains less than 0.15 mm.

Flow test
The results of flowing tests in relation to SP additive, content for self-leveling mortars with 10% and 15% of substitution are presented in Figure 1 and 2, respectively.

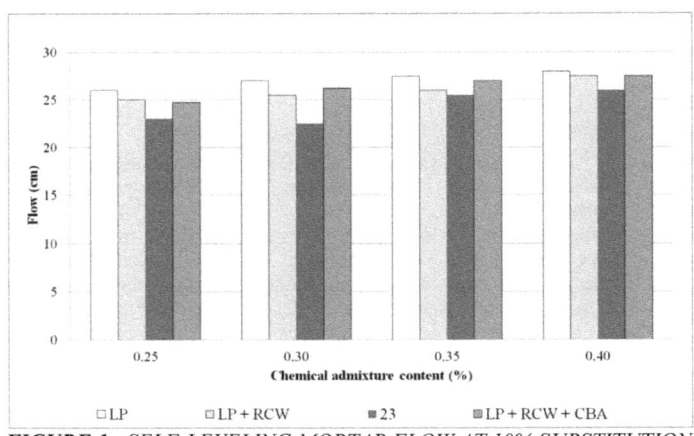

FIGURE 1 - SELF-LEVELING MORTAR FLOW AT 10% SUBSTITUTION.

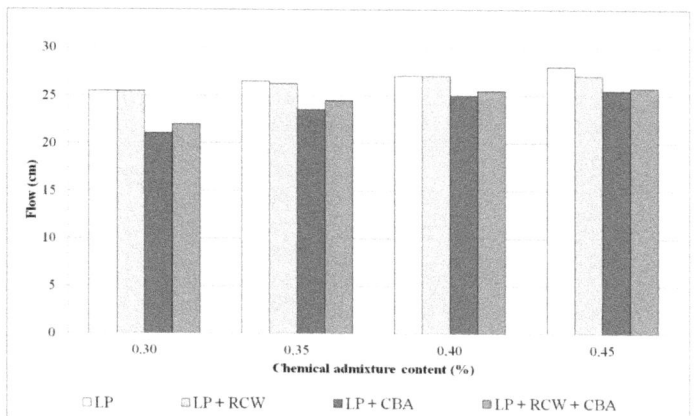

FIGURE 2 - SELF-LEVELING MORTAR FLOW AT 15% SUBSTITUTION.

The maximum content of the SP chemical admixture used was 0.40%, above from this value, the flow of the mortars did not change significantly. In chemical admixtures, most of the mixtures with 0.45% and 0.50% showed signs of bleeding and /or segregation.

It was verified that the flow decreased with the increase of the fines content, in all the mixtures studied. The same results were found by other researchers [9, 12]. The results can be explained by the increase of the fineness and the specific surface area of the fines, because with the increase of the fine content it becomes necessary more water to wet the surface of the particles to maintain the same fluidity.

Mixtures with 20% cement substitution (from cements to the fines) were also performed, in the same way as mixtures with 10% and 15% substitution. However, samples composed with LP + CBA and LP + CBA + RCW presented segregation, in addition to providing lower scattering values. Thus, the scattering results for substitution contents of 20% were not considered satisfactory.

The scattering results of the selected mortars with the respective additives content are summarized in Table 4.

TABLE 4 - CHOSEN MORTARS.

Identification	Additive Content (%)	Flow (cm)	Cement Replacement Content (%)
FC 10	0.25	26.0	
FC + RCD 10	0.35	26.5	
FC + CZP 10	0.40	26.0	10%
FC + RCD + CZP 10	0.30	26.25	
FC 15	0.30	25.5	
FC + RCD 15	0.30	25.5	
FC + CZP 15	0.40	25.0	15%
FC + RCD + CZP 15	0.40	25.5	

Evolution of hydration temperature of mortars

In order to determine the initial setting time of mortar the points of the curve were chosen where the gradual increase in temperature began. To determine the time required to reach the maximum temperature, the points with the highest temperature peak were chosen in the graph. The results are shown in Figures 3 and 4 for 10% and 15% substitution, respectively.

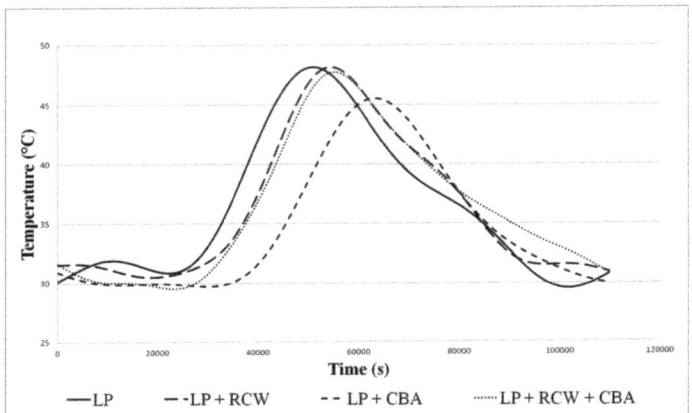

FIGURE 3 *- EVOLUTION OF THE TEMPERATURE VERSUS TIME FOR MORTAR WITH 10% REPLACEMENT.*

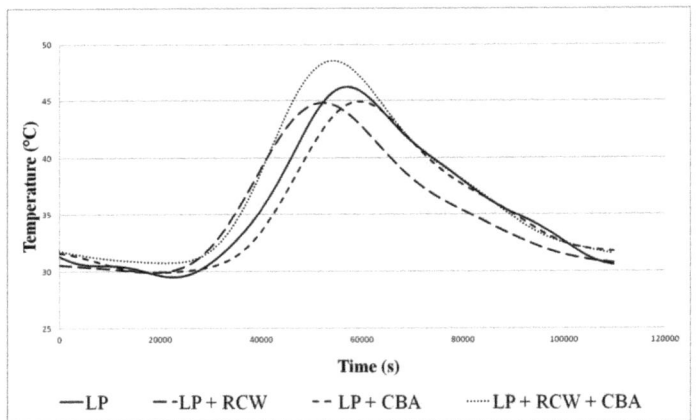

FIGURE 4 *- EVOLUTION OF THE TEMPERATURE VERSUS TIME FOR MORTAR WITH 15% REPLACEMENT.*

For 10% of substitution, the LP + CBA 10 mixture presented the highest period of acceleration of the hydration reactions (approximately 8 hours). This fact occurred due to the higher content of SP chemical admixture used to achieve the required flow. The other mixtures showed initial setting time varying between 5 and 6 hours, approximately.

For mortars with a 15% substitution, all samples showed shorter initial setting time than the reference (FC 15). The mortars LP + CBA 15 and LP + RCW 15 presented the lowest maximum temperatures. Although the maximum temperature difference between all mixtures produced has been small (± 30 ° C), a mixture with the lowest maximum temperature may help reduce shrinkage in self-leveling mortars. Lower temperatures during the hydration process tends to reduce the amount of water that evaporates, and consequently the dimensional variation tends to decrease.

Dimensional variation (Linear shrinkage)
Results for linear shrinkage are shown on Figure 5.

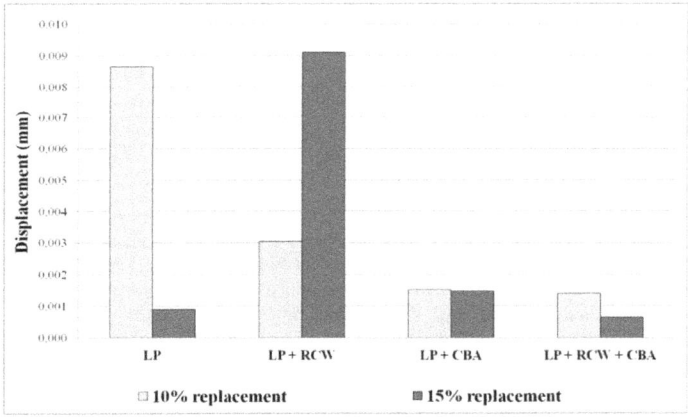

FIGURE 5 - LINEAR SHRINKAGE VALUES.

For the mortars with 10% of substitution, all the mixtures with recycled fines were more efficient than the reference, since they presented smaller displacements. However for 15% of substitution, the mortar LP + RCW 15 presented the worst result of all samples tested. Furthermore, the mortar LP + RCW + CBA 15 presented the smallest linear shrinkage shifts (0.006 mm) of the eight mixtures analyzed. The LP + CBA mixtures did not present great displacement variations with the increase in the content of fines (10% to 15%). Based on the results, it was possible to conclude that the mixtures LP + CBA and LP + RCW + CBA presented good results for both percentages of substitutions studied (10 and 15%).

Dimensional stability (curling)
The average values of the final displacements due to the curling phenomenon generated in self-leveling mortars produced are summarized in the Figure 6.

Non-Conventional Materials and Technologies – NOCMAT for XXI Century   Materials Research Forum LLC
Materials Research Proceedings 7 (2018) 505-515   doi: http://dx.doi.org/10.21741/9781945291838-48

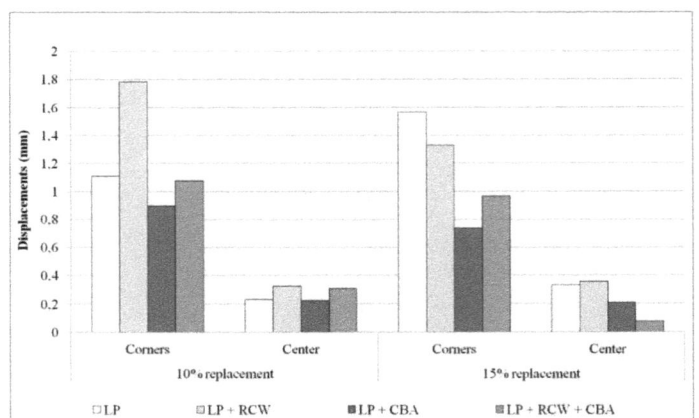

FIGURE 6 - *AVERAGE DISPLACEMENT IN THE CORNERS AND CENTERS OF THE PLATES.*

For 10% of substitution, only the sample LP + RCW 10 presented values of median displacement of the corners, greater than the reference (FC). For 15% of substitution, all the self-leveling mortars were more efficient than the reference and the sample LP + CBA 15 presented the best result of all mortars produced. It was possible to identify a reduction in the displacement of corners results, due to increase the fine content, exception for the reference mortar.

The dimensional variations due to the drying process of the plates generate as results ascending curves, which are predominantly more accentuated over the corners (curling) [11]. The self-leveling mortar FC + CZP 15 achieved 0.74 mm of mean displacement of the corners for 130 hours of testing (Figure 7).

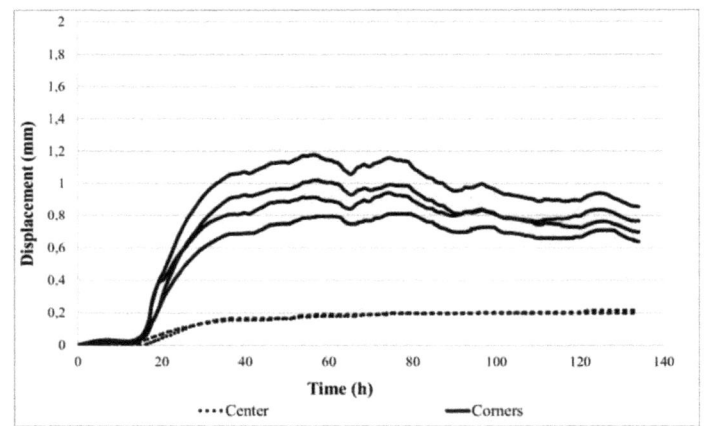

FIGURE 7 - *VALUES OF CURLING OF SELF-LEVELING MORTAR LP + CBA 15.*

The lowest values of shrinkage are related to the effect of internal curing as a function of the addition bottom ash. The physical characteristics of bottom ash (high porosity) confer a higher initial water absorption. As time proceeds the water absorbed initially is gradually released during the drying process of the mortar. Therefore, the internal relative humidity of the mortar undergoes minor variations and the drying shrinkage can be reduced [17, 19]. Furthermore, mixtures with lower maximum hydration temperatures (LP + CBA 15) can reduce water evaporation and thus the shrinkage tends to be lower [8].

It was visually verified that none of the samples produced cracking. Higher shrinkage values can give rise to cracks and consequently contribute with an access for external aggressive agents that generate deterioration in the material, such as, salts, acids, microorganisms, water, among others. The influence of these external agents on the self-leveling mortar tends to reduce the durability of the material.

Conclusions

In the self - leveling mortar mixture it was possible to verify the use of fine elements in form of both recycled fines and bottom ash (RCD and CBA), with 10 and 15% cement replacement and maximum superplasticizer of 0.40 by weight of binder (%wt).

- Mortars based on RCW demonstrated flow values similar to those of the reference mixture (LP), and an adequate consistency and good visual appearance.
- Mortars composed of CBA fine demonstrated more efficiency, no bleeding was observed, despite of reducing the flow in contrast to the reference. Therefore, the use of LP + CBA in binary binder compositions required higher amounts of SP chemical admixtures.
- Mortars formulated with CBA fines show a tendency towards to segregation and a higher superplasticizer consumption ($\geq$ 0.45% wt) of the binder, when higher binder substitution contents was used ($\geq$ 20%);
- The ternary mixtures (LP + RCW + CBA) presented flows similar to the reference (LP).

The main goal of the present research was to evaluate the shrinkage, as it is considered one of the main problems generated in self-leveling mortars to the screeds production. According to the curling effect and maximum hydration temperature it can indicated that the LP + CBA 15 mortar presented the most efficient results. The LP + RCW + CBA 15 sample was the most efficient fine reducing shrinkage. In this way, it was possible to conclude that the use of bottom ash fines (CBA) reduce the drying shrinkage. Finally, we can conclude that the mortars LP + CBA 15 and LP + RCW + CBA 15 were the most efficient mixtures to be used on fluid screeds.

Acknowledgments
The authors grateful to the Brazilian governmental agencies CNPq and CAPES.

References

[1] ABNT - ASSOCIAÇÃO BRASILEIRA DE NORMAS TÉCNICAS, NBR NM 52 Agregado Miúdo – Determinação da massa específica e massa específica aparente (2009).

[2] ABNT - NBR NM 23 Cimento Portland e outros materiais em pó – Determinação da massa específica (2001).

[3] ABNT - NBR NM 248 Agregados – Determinação da composição granulométrica (2003).

[4] ABNT - NBR NM 76 Cimento Portland - Determinação da finura pelo método de permeabilidade ao ar (Método de Blaine) (1998).

Non-Conventional Materials and Technologies – NOCMAT for XXI Century   Materials Research Forum LLC
Materials Research Proceedings 7 (2018) 505-515   doi: http://dx.doi.org/10.21741/9781945291838-48

[5] ACHE, Use of recycled aggregates for production of structural concrete. Asociación Científica del Hormigón Estructural, Madrid, 2005 (in Spanish).

[6] AENOR, EHE-08 regulation of structural concrete, Spain, 2008 (in Spanish).

[7] ALRIFAI, A.; AGGOUN, S.; KADRI, A.; KENAI, S.; KADRI, E.H. Paste and mortar studies on the influence of mix design parameters on autogenous shrinkage of self-compacting concrete. Construction and Building Materials, v. 47, p. 969-976, 2013. https://doi.org/10.1016/j.conbuildmat.2013.05.024

[8] ANDRADE, L.B.; ROCHA, J.C.; CHERIAF, M. Influence of coal bottom ash as fine aggregate on fresh properties of concrete. Construction and Building Materials, v. 23, p. 609-614, 2009. https://doi.org/10.1016/j.conbuildmat.2008.05.003

[9] BENABED, B. KADRI, E. H. AZZOUZ, L. KENAI, S. Properties of self-compacting mortar made with various types of sand. Cement & Concrete Composites, v. 34, p.1167-1173, 2012. https://doi.org/10.1016/j.cemconcomp.2012.07.007

[10] GEORGIN, J.F.; AMBROISE, J.; PÉRA, J.; REYNOUARD, J.M. Development of self-leveling screed based on calcium sulfoaluminate cement: modelling of curling due to drying. Cement & Concrete Composites, v.30, p. 769–778, 2008. https://doi.org/10.1016/j.cemconcomp.2008.06.004

[11] LE-BIHAN, T.; GEORGIN, J.F.; MICHEL, M.; AMBROISE, J.; MORESTIN, F. Measurements and modeling of cement base materials deformation at early age: The cause of sulfo-aluminous cement. Cement and Concrete Research, v.42, p. 1055-1065, 2012. https://doi.org/10.1016/j.cemconres.2012.04.004

[12] LIBRE, N.A.; KHOSHNAZAR, R.; SHEKARCHI, M. Relationship between fluidity and stability of self-consolidating mortar incorporating chemical and mineral admixtures. Construction and Building Materials, v. 24, p. 1262-1271, 2010. https://doi.org/10.1016/j.conbuildmat.2009.12.009

[13] LÓPEZ,D.C.; FONTEBOA B. G.; BRITO J.; ABELLA, F. M.; TABOADA, I.G.; SILVA, S. Study of the rheology of self-compacting concrete with fine recycled concrete aggregates. Construction and Building Materials. v. 96, p. 491-501, 2015. https://doi.org/10.1016/j.conbuildmat.2015.08.091

[14] MEHDIPOUR, I.; RAZZAGUI,M.S.; AMINI, K.; SHEKARCHI, M. Effect of mineral admixtures on fluidity and stability of self-consolidating mortar subjected to prolonged mixing time. Construction and Building Materials, v. 40, p. 1029-1037, 2013. https://doi.org/10.1016/j.conbuildmat.2012.11.108

[15] ONISHI, K.; BIER, T. Investigation into relations among technological properties, hydration kinetics and early age hydration of self-leveling underlayments. Cement and Concrete Research, v.40, p. 1034-1040, 2010. https://doi.org/10.1016/j.cemconres.2010.03.004

[16] PÉRA, J.; AMBROISE, J. New applications of calcium sulfoaluminate cement. Cement and Concrete Research, v.34, p. 671-676, 2004. https://doi.org/10.1016/j.cemconres.2003.10.019

[17] RAFIEIZONOOZ, M.; MIRZA, J.; SALIM, M. R.; HUSSIN, M. W.; KHANKHAJE, E. Investigation of coal bottom ash and fly ash in concrete as replacement for sand and cement.

Non-Conventional Materials and Technologies – NOCMAT for XXI Century   Materials Research Forum LLC
Materials Research Proceedings **7** (2018) 505-515   doi: http://dx.doi.org/10.21741/9781945291838-48

Construction and Building Materials, v. 116, p. 15 - 24, 2016.
https://doi.org/10.1016/j.conbuildmat.2016.04.080

[18] SEIFERT, S.; NEUBAUER, J.; NEUNHOEFFER, F.G. Spatially resolved quantitative in-situ phase analysis of a self-leveling compound. Cement and Concrete Research, v.42, p. 919-927, 2012. https://doi.org/10.1016/j.cemconres.2012.03.012

[19] WYRZYKOWSKI, M.; GHOURCHIAN, S.; SINTHUPINYO, S.; CHITVORANUND, N.; CHINTANA, T.; LURA, P. Internal curing of high performance mortars with bottom ash. Cement & Concrete Composites, v. 71, p. 1 – 9, 2016. https://doi.org/10.1016/j.cemconcomp.2016.04.009

[20] ZHAO, Z,; REMOND, S.; Damidot, D.; Xu, W. Influence of fine recycled concrete aggregates on the properties of mortars. Construction and Building Materials, v.81, p.179-186, 2015. https://doi.org/10.1016/j.conbuildmat.2015.02.037

Non-Conventional Materials and Technologies – NOCMAT for XXI Century   Materials Research Forum LLC
Materials Research Proceedings 7 (2018) 516-522   doi: http://dx.doi.org/10.21741/9781945291838-49

Bamboo Leaves Ashes: A Mineral Addition for Building Construction

Loïc Rodier[1*]; Ernesto Villar Cociña[2]; Holmer Savastano Junior[1]

[1]Department of Biosystems Eng., Faculty of Animal Science and Food Engineering, University of São Paulo, Av. Duque de Caxias Norte, 225, Jardim Elite, 13635-900 Pirassununga SP, Brazil; rodierloic@gmail.com, holmersj@usp.br

[2]Department of Physics, Central University of Las Villas, Santa Clara 54830, Cuba; evillar@uclv.edu.cu

Abstract. In this study the potential use of bamboo leaves ashes as pozzolanic materials have been investigated. Bamboo leaves were calcined at 600°C for 2 h and then were passed through a sieve opening of 325 mesh. Mineralogical, physical-chemical analysis and electrical conductivity measurements were carried out to evaluate the pozzolanic activity of the resulting bamboo ashes. Thermal analysis and mechanical tests were applied to determine the effect of ashes on the hydration kinetics of ordinary Portland cement. The results showed that bamboo leaves ashes present an amorphous phase and high silica content (70%). Electrical conductivity measurements confirmed the high pozzolanic activity of bamboo leaves ashes. Partial replacement of cement by 20 wt.% of bamboo leaves ashes leads to a decrease of heat of hydration of cement pastes. The compressive strength of mortars containing 20 wt.% of bamboo leaves ashes are similar to their counterparts without mineral additions (Control) at 7 days. At 28 days, the use of bamboo leaves ashes leads to the refinement of the pore structure, hence an increase of the compressive strength of mortars. According to the results, bamboo leaves ashes can be used as pozzolanic materials to increase the mechanical performance and the durability of building materials.

Keywords: Bamboo Leaves, Mineral Addition, Pozzolanic Activity, Cement Hydration, Compressive Strength

Introduction

Bamboo is the vernacular or common term of members of a particular taxonomic group of large woody grasses whose height varies between 10 centimeters and 20 meters [1]. Bamboo belongs to the Poaceae family, such as sugarcane and to the subfamily Bambusoideae [1]. More than 1000 species with rapid growth are counted, the maturity is reached after 5 years [2]. Bamboos are found mainly in the tropics, but they also exist in the subtropical and temperate zones of all continents [1].

In the pursuit of eco-friendly and sustainable building, the use of bamboo increased in civil construction. The increase of consumption of bamboo led to an increase of wastes such as leaves. To decrease the amount of these wastes in landfills, the potential use of this residue as partial replacement of cement was investigated [3-5].

Villar-Cociña et al. [3] studied the pozzolanic activity of bamboo leaf ashes and the kinetic parameters characterizing the pozzolanic reaction. The authors showed the bamboo leaf ashes are a high pozzolanic materials and a kinetic-diffusive model describes the pozzolanic reaction in calcium hydroxide solution/bamboo leaf ashes system.

Dhinakaran and Chandana [4] studied the effect of bamboo leaf ashes on compressive strength and durability of concrete. The authors showed that the optimum amount of bamboo leaf ashes in concrete is 15 wt. %.

Singh et al. [5] studied the effect of bamboo leaf ashes on hydration of cement. The authors showed that mortars containing 20 wt. % of bamboo leaf ashes present similar compressive strength than mortars without mineral additions.

Non-Conventional Materials and Technologies – NOCMAT for XXI Century   Materials Research Forum LLC
Materials Research Proceedings 7 (2018) 516-522   doi: http://dx.doi.org/10.21741/9781945291838-49

However, there is a lack of study on the influence of bamboo leaf ashes on kinetic of hydration of cement with combination of other mineral additives. The aim of this study is to study the potential use of bamboo leaf ashes in building construction using several techniques such as pozzolanic activity and compressive strength tests. The results obtained by bamboo leave ashes will be compared to the ones obtained by other mineral additions such as sugar cane bagasse ashes, sugar cane leaves ashes and glass powder. Secondly, investigate the influence of bamboo leaf ashes in binary and ternary binder on hydration of cement pastes by using isothermal calorimetry.

Materials and methods
Materials
The bamboo was collected in the University of São Paulo, Pirassununga Campus, Brazil. Sugar cane bagasse was collected in Abengoa sugar cane factory (Pirassununga, Brazil). The glass powder residue was provided by the recycling company MASSFIX, São Paulo, Brazil.

The bamboo leaves and sugar cane bagasse wastes were calcined at 600°C for 60 min with a heating rate of 10°C/min [6]. Then, a quick cooling was carried out to allow the formation of a vitreous phase. The glass powder was ground in a jar rolling mill model TE-500/2 (TECNAL, Brazil) at 50 rpm during 6 h. All materials were sieved at 325 mesh before testing. The cement used in this study was the cement CPV ARI, correspondent to ASTM C150 Type I.

Methods
The chemical composition of ashes and cement was carried out by the X-ray fluorescence (XRF), using the Philips Venus 100 minilab XRF Spectrometer. Mineralogical characterization was carried out by a Rigaku Miniflex 600 X- Ray Diffractometeer.

Electrical conductivity tests were realized following the method described by Villar-Cociña and al. [3]

The mineral additive (5.25 g) was mixed with a saturated calcium hydroxide solution (250 ml) at 40°C. The conductivity measurements began immediately after the introduction of the mineral additive in the solution.

The effect of mineral additions on hydration of cement pastes was studied using isothermal calorimetry. The binder and water (Table 1) were introduced in a flask and then putted in an isothermal calorimeter TAM Air (TA Instruments) maintained at 25°C during the experiment. The water/binder ratio was keeping equal to 0.4 for all formulations. The monitoring of the heat released by the cement paste was carried out for 48 h.

TABLE 1 – FORMULATION OF CEMENT PASTES

Specimens	Cement	Bamboo leaf ashes	Sugar cane bagasse ashes	Sugar cane leaf ashes	Water/Binder
Control	100	0	0	0	0.5
BAM20	80	20	0	0	0.5
BAM10SCBA10	80	10	10	0	0.5
BAM10SCLA10	80	10	0	10	0.5

The effect of mineral additions on mechanical properties of mortars was studied compressive strength tests. Cylindrical specimens of mortar (diameter = 4 cm, length = 8 cm) were elaborated using cement, sand and mineral additions according to ASTM C109 [7]. After elaboration, mortars were kept in the molds for 24 hours at laboratory temperature. Then, mortars were cured at 60°C and 100% of relative humidity. The compressive strength of mortars was determined using a universal

Non-Conventional Materials and Technologies – NOCMAT for XXI Century   Materials Research Forum LLC
Materials Research Proceedings 7 (2018) 516-522   doi: http://dx.doi.org/10.21741/9781945291838-49

testing machine EMIC DL3000 with load speed of 0.5 mm/min at early days (7 days). Four specimens for each formulation were used to determine the compressive strength early ages.

The volume of permeable voids of mortars was determined in accordance with Standard specifications of ASTM C642 (2006) [8]. This method is based on the measurement of the weight of the specimen at several state such as dry, saturated and saturated/boiled.

Results and dicussion

Table 2 shows the chemical composition of ashes. In bamboo leaf (BLA) and sugar cane leaf (SCLA) ashes the major component is silica following by CaO, K_2O, SO_3, MgO and P_2O_5. In sugar cane bagasse ashes (SCBA), silica (SiO_2) is the major component following by K_2O, Al_2O_3, Fe_2O_3, CaO, P_2O_5, MgO, SO_3 and TiO_2. The main elements present in GPR are silica (SiO_2) followed by CaO, Na_2O and MgO. The presence of alkali in GPR can provokes the alkali-silica reaction in the cementitious material elaborated with GPR.

According to ASTM C618 [9], only BLA, SCLA and GPR respond to the minimum value of 70% of the sum of SiO_2, Al_2O_3, and Fe_2O_3. Moreover, the loss of ignition of all ashes is less than 10 wt. %, which indicates that all mineral additions contain a low organic matter content.

TABLE 2 – CHEMICAL COMPOSITION OF MINERAL ADDITIONS

Oxides (%)	Sugar cane bagasse ashes	Sugar cane leaf ashes	Bamboo leaf ashes	Glass powder
Na_2O	0.200	<0.001	<0.001	11.400
MgO	4.750	1.480	1.840	2.580
Al_2O_3	12.300	0.579	0.632	0.688
SiO_2	36.200	71.600	70.500	70.000
P_2O_5	5.420	1.440	1.670	0.033
SO_3	4.380	2.690	2.870	0.274
Cl	0.171	0.892	0.851	0.060
K_2O	12.800	4.950	5.140	0.226
CaO	7.100	7.470	7.860	14.000
TiO_2	1.980	0.074	0.063	0.059
MnO	0.303	0.315	0.345	0.010
Fe_2O_3	8.760	0.373	0.468	0.195
CuO	0.028	0.002	0.010	0.060
ZnO	0.085	0.010	0.010	0.031
LOI	5.370	8.080	7.790	7.460

Figure 1 (a) and (b) shows the XRD patterns of SCBA, SCLA, BLA and GPR. All mineral additions present a broad band located between $2\theta = 15°$ and $2\theta = 35°$, which is a characteristic of amorphous material [3, 10]. Moreover, Quartz is the main crystalline phase presents in all ashes due to temperature of calcination of the material. The presence of other crystalline compounds such as sylvite and calcite is due to the use of fertilizers in the soil where the plant was collected. In GPR, the only crystalline compound found is the calcite.

Non-Conventional Materials and Technologies – NOCMAT for XXI Century   Materials Research Forum LLC
Materials Research Proceedings 7 (2018) 516-522   doi: http://dx.doi.org/10.21741/9781945291838-49

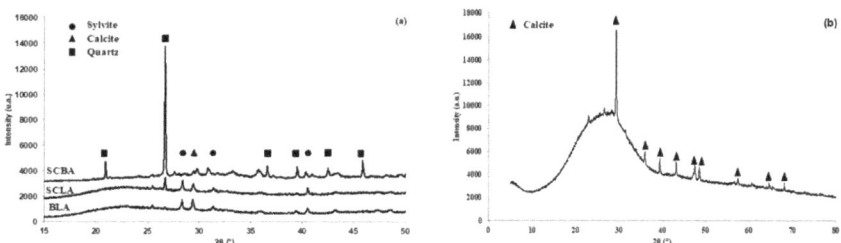

FIGURE 1 – DRX PATTERNS OF (A) SCBA, SCLA AND BLA AND (B) GPR

Figure 2 presents the electrical conductivity versus time of the mineral additions in the calcium hydroxide saturated solution. The electrical conductivity of all mineral additions-lime solution decreases with time. This behavior is attributed to the pozzolanic reaction between amorphous silica and the calcium hydroxide, which produce CSH [3]. The formation of CSH decrease the quantity of ions Ca^{2+} in solution hence a decrease of the electrical conductivity.

The loss of conductivity at 70 h is calculated according the following equation:

$$\xi = (C_0 - C_{70}) * 100 / C_0 \tag{1}$$

With ξ the loss of conductivity (%), C_0 the initial conductivity at t = 0 (mS/cm) and C_{70} the initial conductivity at t = 70 h (mS/cm)

The loss of conductivity (ξ) was calculated for all formulations. The loss of conductivity of the SCBA-lime solution (42.49%) is lower than GPR-lime solution (50%), SCLA-lime solution (60.51%) and BLA-lime solution (79.18%). This behavior is due to the amount of silica presents in the ashes observed in Table 1. This result indicates that BLA has a higher pozzolanic activity than the other mineral additions.

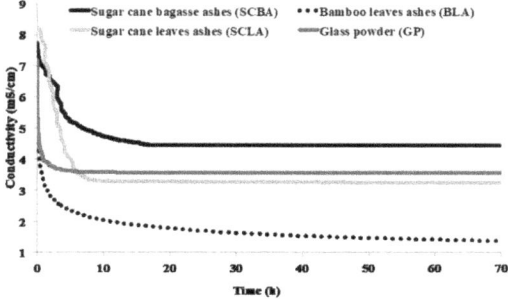

FIGURE 2 – ELECTRICAL CONDUCTIVITY OF MINERALS ADDITIVES IN SATURATED CALCIUM HYDROXIDE SOLUTION

Figure 3 presents the heat of hydration of cement pastes without mineral additions and partially replaced by 20 wt.% of BLA (BAM20), 10 wt.% of BLA and 10 wt.% of SCBA (BAM10SCBA10) or 10 wt.% of SCLA (BAM10SCLA10).

The use of mineral additions modifies the cement hydration. Indeed, the heat of hydration of BAM10SCLA increases of 16% in comparison of control mortars. This behavior is due to the fineness of SCLA [11].

In the case of BAM20 and BAM10SCBA10, the heat of hydration decreases of 25 and 48% respectively. This behavior is due to the dilution effect [12], which causes by the decrease of the amount of cement in the bonder.

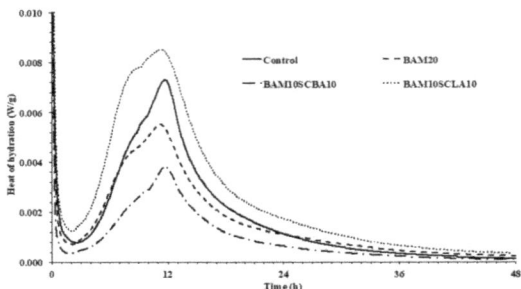

FIGURE 3 – HEAT FO HYDRATION OF CEMENT PASTES WITHOUT MINERAL ADDITIONS (CONTROL), CONTAINING MINERAL ADDITIONS (BAM20, BAM10SCLA10, BAM10SCBA10)

Figure 4 shows the compressive strength of mortars containing bamboo leaf ashes (BAM20) and without mineral additive (Control) at 7 and 28 days.

The results show the effect of bamboo leaf ashes on compressive strength with time. At 7 days, according to standard deviation, the mortars containing 20 wt.% BLA (BAM20) presents similar compressive strength than control mortars.

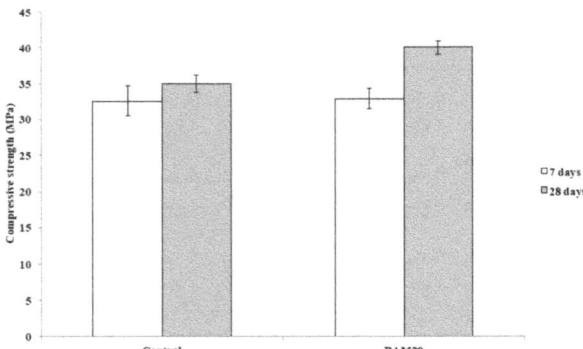

FIGURE 4 – COMPRESSIVE STRENGTH OF MORTARS WITHOUT MINERAL ADDITIONS (CONTROL) AND MORTARS CONTAINING 20 WT.% OF BLA (BAM20) AT 7 AND 28 DAYS

It is known that the partial replacement of cement by mineral additions can decrease the compressive strength of the material at early age. In this case, this phenomenon did not take place. At

Non-Conventional Materials and Technologies – NOCMAT for XXI Century   Materials Research Forum LLC
Materials Research Proceedings 7 (2018) 516-522   doi: http://dx.doi.org/10.21741/9781945291838-49

28 days, an increase of 14% of the compressive strength of BAM20 was observed in comparison to control mortars. This increase is due to the pozzolanic reaction between the ashes and the lime produced during the hydration [13]. The product of this reaction (C-S-H) decreases the volume of permeable pores (VPP) (Table 3) of the mortars by filling the pore solution thus an increase of the mechanical properties.

TABLE 3 – VOLUME OF PERMEABLE PORES OF MORTARS AT 7 AND 28 DAYS

Specimens	VPP at 7 days (%)	VPP at 28 days (%)
Control	19	18
BAM20	15	11

Conclusion

Bamboo leaf ashes are a promising mineral addition for use as partial replacement of cement in binary or ternary binders. Based on the experimental results, the following conclusion can be drawn:

- Bamboo leaves ashes present an amorphous phase and high silica content (70%).
- Electrical conductivity measurements confirmed the high pozzolanic activity of bamboo leaves ashes in comparison of SCBA, SCLA and GPR.
- The partial replacement of cement by 20 wt.% of bamboo leaves ashes leads to a decrease of 25% of the heat of hydration of cement pastes.
- The partial replacement of cement by BLA permit to decrease the volume of permeable pores of mortars and increase the compressive strength in comparison to mortars without mineral additions (Control).

References

[1] Wong KM, Bamboo-the amazing grass: a guide to the diversity and study of bamboos in South East Asia, International Plant Genetic Ressources Institute/ University of Malaysia, Malaysia, 2004.

[2] International Network for Bamboo and Rattan (INBAR). Priority species of bamboo and rattan. New Delhi, India, 1994, 68.

[3] Villar-Cociña E, Valencia-Morales E, Gonzalez-Rodrıguez R, Hernandez-Ruız J. Kinetics of the pozzolanic reaction between lime and sugar cane straw ash by electrical conductivity measurement: A kinetic–diffusive model. Cem Concrete Res 2003;33:517-524. https://doi.org/10.1016/S0008-8846(02)00998-5

[4] Dhinakaran G, Gangava HC. Compressive Strength and Durability of Bamboo Leaf Ash Concrete. Jordan J Civil Eng 2016;10:279-289. https://doi.org/10.14525/JJCE.10.3.3601

[5] Singh NB, Das SS, Singh NP, Dwivedi VN. Hydration of bamboo leaf ash blended Portland cement. Indian J Eng Mater Sci 2007;14:69-76.

[6] Dwivedi VN, Singh NP, Dasa SS, Singh NB. A new pozzolanic material for cement industry: bamboo leaf ash. Int J Phys Sci 2006;1:106-111.

[7] ASTM C109. Standard test method for compressive strength of hydraulic cement mortars (using 2-in. or [50-mm] cube specimens), 2013.

[8] ASTM C642. Standard Test Method for Density, Absorption, and Voids in Hardened Concrete, 1997.

Non-Conventional Materials and Technologies – NOCMAT for XXI Century   Materials Research Forum LLC
Materials Research Proceedings 7 (2018) 516-522   doi: http://dx.doi.org/10.21741/9781945291838-49

[9] ASTM C618. Standard Specification for Coal Fly Ash and Raw or Calcined Natural Pozzolan for Use in Concrete, 2015.

[10] Morales EV, Villar-Cociña E, Frías M, Santos SF, Savastano H. Effects of calcining conditions on the microstructure of sugar cane waste ashes (SCWA): Influence in the pozzolanic activation. Cem Concr Comp 2009;31:22-28. https://doi.org/10.1016/j.cemconcomp.2008.10.004

[11] Li HL, Yang L, Xie YJ. Effect of Fineness on the Properties of Cement Paste", Key Eng Mater 2015;629-630:366-370. https://doi.org/10.1016/j.cemconcomp.2008.10.004

[12] Bahurudeen A, Kanraj D, Dev VG, Santhanam M. Performance evaluation of sugarcane bagasse ash blended cement in concrete. Cem Concr Comp 2015;59:77-88. https://doi.org/10.1016/j.cemconcomp.2015.03.004

[13] Taylor HF. Cement chemistry. Thomas Telford, 1997. https://doi.org/10.1680/cc.25929

Non-Conventional Materials and Technologies – NOCMAT for XXI Century   Materials Research Forum LLC
Materials Research Proceedings 7 (2018) 523-532   doi: http://dx.doi.org/10.21741/9781945291838-50

Mechanical Properties of Adobe Reinforced with Wheat Fibers in Iran (Persia)

Reza Nasiri [a], Khosrow Ghavami [b], Marzieh Kadivar [c]*, Amir Maghami [d]

[a] South Pars Gas Company, Iran; nasiri.reza586@gmail.com

[b] PUC-Rio, Brazil; ghavami@puc-rio.br

[c] Shiraz University of Technology, Shiraz, Iran, ma.kadivar@sutech.ac.ir

[d] Tehran University, Tehran, Iran, amir.maghami.96@gmail.com

Abstract. In adobe constructions, earthen blocks are made of a clayey-sandy-silt mixture with or without straw and/or additives. Based on a survey of traditional earth used in adobe constructions in Iran (Persia), the most commonly reinforcement's fibers used are wheat straw, animal or human hair and some additives such as lime, hydrated salts, natural gum, and animal blood. In this study, the effects of some traditional amendments on the mechanical properties (Compressive, Bending and Tensile Strength) of adobes are investigated. The soil for the test samples is a mixture of sand, silt and clay which was taken from the mountain of the city of Yazd situated in center of Iran. Then the adobes were sun-dried without significant warping or cracking. Main considered variables were the percentages of the additives for establishing the best mixing trace. It has been observed that the strength of the adobe blocks increases significantly with the presence of straw and amendments. For a constant wheat fiber/soil ratio, the increase in additives increases the soil's mechanical strengths including Compressive, Bending and Tension Strength. The optimum level of wheat fiber content was achieved with 50 percent of soil volume addition and the optimal level of lime content was achieved with 10%. However compressive, bending, and tensile strength increases with the increase in the natural gum contents.

Keywords: Adobe, Natural Gum, Lime, Compressive, Bending, Tensile Strength

Introduction

The building and construction industry accounts for up to 40% of the world's energy [1] combined with approximately 40% of its raw material usage [2, 3]. It has been reported that the structural system accounts for 25% of the building's environmental impact [4]. Industrial materials such as steel, aluminum, cement, ceramic, glass, are consuming high energy in their productions and produce high amount of CO_2 and other polluting gases. The centralized industries use all their modern method of publicity selling their polluting construction materials, without considering that our globe has limited non-renewable materials and the profit is not the only aim in our globe. The control of energy resources conduces to concentration of income and increases poverty and social inequalities and generates environmental contamination. In the beginning of the twentieth century, the developing countries adopted the university programs of the centralized industrialized countries without considering their own culture and the local engineering practices. An example is building constructions with earth, which is not being taught at the technical schools, engineering and architectural departments of the universities. Therefore, the soil, which has been used since the beginning of western civilization in Persia, was not taught and studied in the past as an engineering material [5]. The oldest and the most traditional building material is earth and up to 30% of the world's population still is living in earthen constructions [6]. In Iran, the majority of earth buildings are adobe constructions which have been traditionally used in order to construct not only simple dwellings but also large and elaborate monumental structures, such as Tchogha Zanbil, Narin Qal'eh,

Non-Conventional Materials and Technologies – NOCMAT for XXI Century   Materials Research Forum LLC
Materials Research Proceedings 7 (2018) 523-532   doi: http://dx.doi.org/10.21741/9781945291838-50

Great Mosque of Kabir Neyriz, Badr Abad's Tel-castle , Moorchekhort Castle, and Dome of Soltaniyeh which are presented in Figure.1.

(a)   (b)   (c)

(d)   (e)   (f)

FIGURE 1 - LARGE AND ELABORATE MONUMENTAL STRUCTURES IN IRAN: A) GREAT MOSQUE OF KABIR NEYRIZ, B) NARIN QAL'EH, C) TCHOGHA ZANBIL , D) BADR ABAD'S TEL-CASTLE, E) MOORCHEKHORT CASTLE, F) DOME OF SOLTANIYEH

Adobe is the best known and one of the most versatile methods of using earth for construction. Adobe bricks (Khesht in Persian language) have been used for thousands of years and are probably one of the first man-made construction materials. According to the archaeological evidence, the mud brick was invented between 12,000 and 10,000 years ago, while the molded production of adobe developed in Mesopotamia 7,000 years ago [7].

The term "adobe" is of Spanish/Persian origin and is used to describe construction materials, usually mud bricks, which were fabricated from moistened earth and were sun dried [8].

Although earth constructions are known for being unstable with low durability, this is something of a myth that must be refuted. They are remarkably durable and in most cases, the fragility of the material is not structural but merely superficial [9]. As a result of their natural composition, the deterioration in earthen buildings caused due to lack of maintenance as most inhabitant of these houses are in general low income families. The defect in earth construction classified in three categories: Mechanical facor such as cracks (Fig 2-a), Biologic factors such as termite (Fig 2-b), and Physical factor such as moisture (Fig 2-c) [5]. These defects proved that existing soil at a construction site cannot always be totally suitable for making adobes. For this reason cheap, safe and natural materials should be used for soil improvement. Cement, lime, fly ash, bituminous materials, chlorite, natural or industrial resins can be used as additive materials [10].

In this study, the effects of some traditional amendments on the mechanical properties (Compressive, Bending and Tension Strength) of adobes are investigated.

Non-Conventional Materials and Technologies – NOCMAT for XXI Century   Materials Research Forum LLC
Materials Research Proceedings **7** (2018) 523-532   doi: http://dx.doi.org/10.21741/9781945291838-50

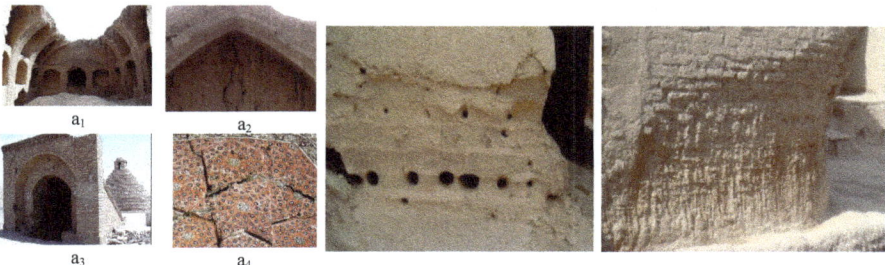

A   B   C

***FIGURE 2** - THE DETERIORATION PROBLEMS OF EARTHEN BUILDINGS: A) CRACKS, A₁) PART OF MOORCHEKHORT CASTLE (ISFAHAN, IRAN), A₂) WALL OF A MOSQUE (YAZD, IRAN), A₃) AB ANBAR (SEMNAN, IRAN), A₄) DOME OF SOLTANIYEH'S ROOF (ZANJAN, IRAN), B) BIOLOGIC FACTORS SUCH AS TERMITE (SEMNAN, IRAN), C) MOISTURE (WALLS OF ORDINARY OLD BUILDING ,YAZD, IRAN).*

Materials and methods
Materials
Based on the state-of-the-art research studies and field applications, one of the key factors for the improved performance of adobe construction is its composition. The constituent of the adobe in Iran consists of: soil (clayey-sandy-silt), fibres and other additives including oils and natural resins. Wheat fiber was selected in this study to increase the performance and durability of the blocks as it is represented by Bouhicha et al [11].

Soil
Earth blocks are to contain mineral soil with aggregate less than 25.4mm (1 in.) diameter and soluble salts less than 2% [12]. The soil, used in this study, was obtained from the town of Yazd in the centre of Iran. Soil grain granulometry curve is shown in figure1 and the chemical composition of the used soil is presented in Table 1.

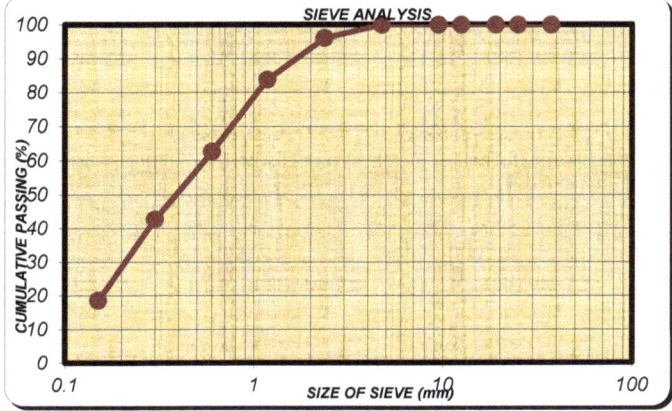

***FIGURE 3** - SOIL GRAIN GRANULOMETRY CURVE*

Non-Conventional Materials and Technologies – NOCMAT for XXI Century   Materials Research Forum LLC
Materials Research Proceedings 7 (2018) 523-532   doi: http://dx.doi.org/10.21741/9781945291838-50

Wheat husk

Wheat husk or wheat fibre which is called in Persian language as Kah, was considered as the reinforcing fibres. Kah is the residue of wheat production. This wheat straw is light in weight with golden yellow colour. The Kah fibres are taken from the 2015 harvest in Iran. It has been sieved through a No.10 sieve before mixing with soil matrix.

TABLE 1 - RESULTS OF XRF TEST FOR THE SOIL

Chemical Composition	Amount %
TiO2	1.024
P2O5	
Cu	
SO3	
MnO	0.096
Na2O	1.18
K2O	2.94
Fe2O3	5.13
MgO	4.13
CaO	9.19
L.O.l	13
Al2O3	8.13
SiO2	51.49

Lime

The quantity of lime plays a major role in stabilization of soil. The chemical reaction of clay soils treated with lime is responsible for its modification of the engineering properties [13]. Lime with specific weight of 548 kg /m3 has been used in testing program. Lime density is 2.35 gr/cm3 according to experimental tests.

Natural gums

Natural gums are hydrophilic polysaccharides composed of monosaccharide units joined by Glycosidic bonds. They are gelatinous when moist but harden on drying [14]. In this research natural gums extracted from the almond trees in the town of Shiraz in Iran were used. The almond gums are soluble and dispersible in water.

Methods

To investigate the effects of lime and natural gum on adobe stabilized with wheat fibre, compressive, bending and tensile strength tests on samples containing 0, 6, 10, 12, 14, and 16% weight of lime, and 0, 3, 4, 5, 6, and 7% weight of natural gum were considered. In addition the effect of curing time on samples after 3, 14, and 28 days of age has been investigated. Three samples were tested for each considered variable. The first set of soil samples without fibres and additives were tested as a reference material. The second set contained different percentages of lime, and the third samples set were soil mixed with different percentages of natural gum. The mixture soil and kah is shown in figure 4. The samples are pressed into a mold to form an adobe brick. After the adobe brick is removed from the mold, it must dry in the open air for a month or more before it can be used as shown in Figure.5.

Experimental procedures
Standard samples were prepared at room temperature (21°C) by compacting soil at the respective optimum moisture contents and also optimum amount of wheat fiber (50%) with different percent of lime, and natural gum for the experimental procedure. The specimens were air dried and the variables chosen were: determining compressive, bending, and tensile behavior of adobe matrix and adobe mixed with additives.

Compressive Strength
Compressive strength is one of the most important properties contributing to adobe durability. It can be used as an indicator of adobe quality. To measure the compressive strength of adobe blocks, the set-up which is shown in Fig.6.a was used. Standard cubes (50 mm in size) were prepared and the samples in this experiment were subjected to a loading of 0.2 MPa per second up to failure as per the standard specified in ASTM C 39.

FIGURE 4 - *THE MIXTURE OF SOIL AND KAH*

Tensile strengths
The test based on ASTM C307-83 "Standard Method for Tensile Strength of Chemical Resistant Mortar, Grouts, and monolithic Surfacing" was used in this investigation. The width and thickness of the briquette molds at the waist line was 25 mm. The tests were realized in a universal testing machine shown in Fig.6.b. The speed of the load application was 0.25 MPa per second up to failure.

Results and discussion
The test results with treated additives compared with those results without any additives and improvements achieved in compressive, bending and tensile strength were evaluated against control measurements.

    Based on these tests, for compressive, bending and tensile strength of Adobe, the following results were obtained.

Bending strengths
The three-point bending test has been used as a simple means to measure bending strength of specimens. Standard cube specimens 40 × 40 × 160 mm in size were prepared and the samples in this experiment were subjected to a loading to failure. See Fig.6.c for compression device.

Non-Conventional Materials and Technologies – NOCMAT for XXI Century   Materials Research Forum LLC
Materials Research Proceedings 7 (2018) 523-532   doi: http://dx.doi.org/10.21741/9781945291838-50

FIGURE 5 - *SPECIMEN PREPARATION AND DRYING IN LABORATORY*

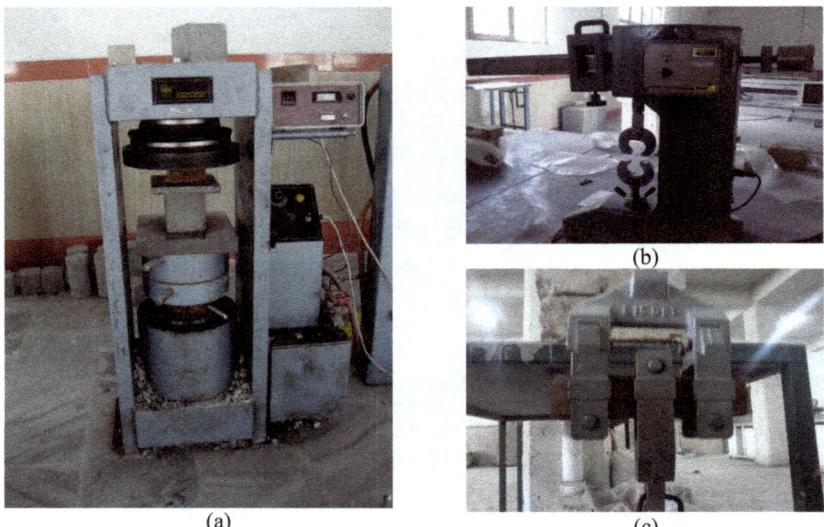

(a)   (c)

FIGURE 6 - *INSTRUMENT, A: COMPRESSIVE STRENGTH SET-UP, B: TENSILE STRENGTH TESTING MACHINE*

Compressive strength

The influence of lime and natural gum on the compressive strength are shown in Fig.7.a There is an increase in compressive strength via adding lime and natural gum, which indicated that the optimum percentage of lime is 10 percent and increase of replacement of soil with natural gum the compressive   strength were found to increase as the percentage of the replacement increases. It was observed that with addition of more than 10 percent lime there was no relative improvement in strength of soil. This might be due to non-availability of reactive silica.

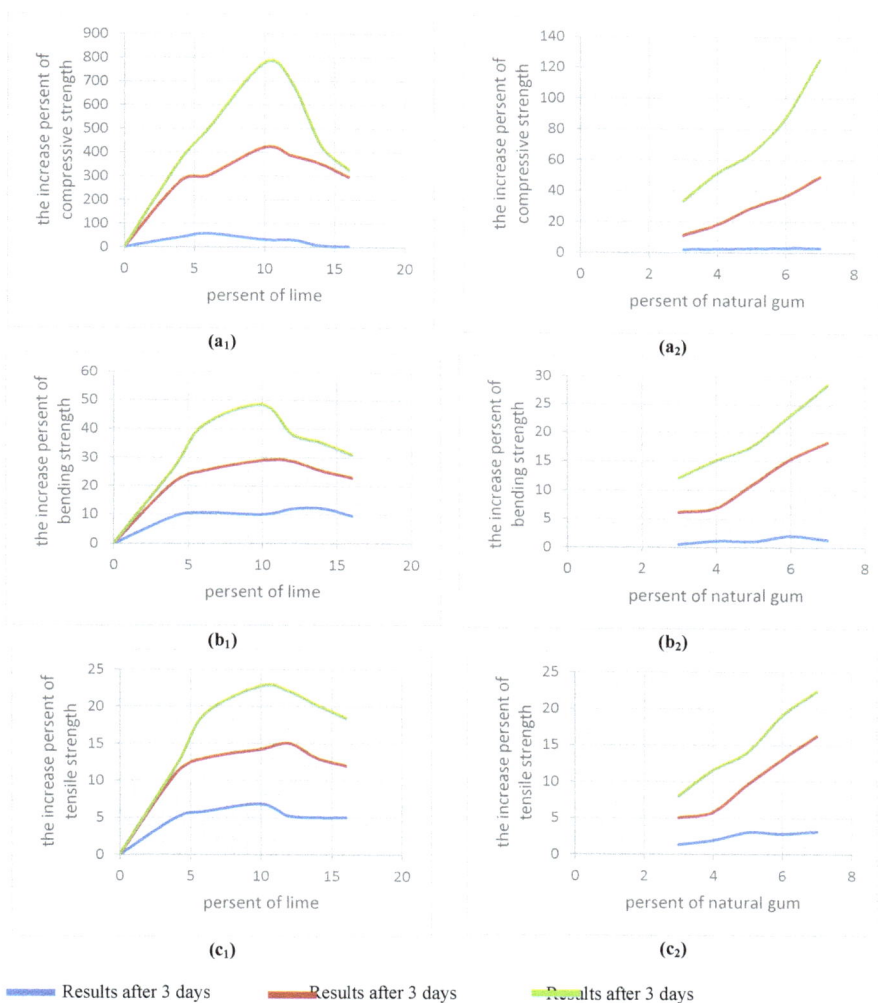

FIGURE 7 - *RELATIVE COMPRESSIVE STRENGTH OF SPECIMENS AT DIFFERENT TIMES COMPARED TO CONTROL SPECIMENS*

Bending strength

From Fig. 7.b, it may be understood that the bending strength of the mixture increases until adding 11% of lime, and then decreases. This phenomenon is due to the saturation state in chemical reactions between soil, lime and water. Adding natural gum also led to improved bending strength in the specimens in all ages (3, 14, and 28 days).

Non-Conventional Materials and Technologies – NOCMAT for XXI Century   Materials Research Forum LLC
Materials Research Proceedings 7 (2018) 523-532   doi: http://dx.doi.org/10.21741/9781945291838-50

Tensile strength

The results show that the lime and natural gum blended adobe had higher tensile strength compared to that of the adobe without additives as can be seen in Fig.7.c It was found that the adobe could be improved by adding lime up to the maximum limit of 10% as the same level of the optimal content of lime that was achieved with replacement in the previous tests.

The influence of time

The effects of curing time in the experiments were investigated regarding the different amount of lime and natural gum in the mixture, and considered 3, 14 and 28 days after molding. The results are shown in figure 8. As illustrated in figure 8, the rate of strength growing in third days with the mixtures containing lime and natural gum is very low which indicates that adobe with these kinds of additives needs more curing time. Figure 8. Shows the influence of time on tested specimens with optimum amount of lime in the mixture of soil. It can be observed that after 3 days the specimens are still wet and more flexible than specimens with more curing time.

a1   a2

b1   b2

c1   c2

FIGURE 8 - *A1,B1, AND C1: THE SPECIMENS THAT ARE BROKEN AFTER 3 DAYS. A2, B2, C2: THE SPECIMENS THAT ARE BROKEN AFTER 28 DAYS*

Conclusions

Adobe is one of the oldest building materials. Although technologically simple, it has the potential to produce durable structures of a sustainable nature, provided that it is protected from the factors causing its deterioration. In this study the application of lime and natural gum in the strength of adobe was investigated. The lime was selected as one active pozzolanic material and the natural gum as an additive to increase the cohesive forces added to the soil-wheat fiber mixture with optimum amount of wheat fiber. Following conclusions can be made based on the obtained data:

The optimum amount of wheat fiber in the mixture of soil- wheat was 50% in which the highest amount of strength was achieved.

Adding lime in the mixture of soil with 50% wheat fiber, increased the compressive, bending and tensile strength of adobe with the optimum amount of 10%. Lime.

The compressive strength of the soil and wheat straw (Kah-gel) mixture (50% Kah) with adding 10% of lime (the optimum content), increased up to 20 times, after 28 days curing. This increase was about 5 and 3.5 times for bending and tensile strength respectively.

For constant Kah/soil ratio, adobe containing natural gum affects the results of compressive, bending, and tensile strength which increases with the increase of the natural gum contents.

The same comparison of curing time made for strength of soil-wheat fiber-natural gum mixture and results showed an increase of about 40 times in compressive strength of specimens for 7% natural gum after 28 days curing. This increase was 21 and 7 times for bending and tensile strength respectively.

References

[1] Lippiatt, B. "Selecting cost-effective green building products: BEES approach." Journal of Construction Engineering and Management, 1999, 125(6), 448. https://doi.org/10.1061/(ASCE)0733-9364(1999)125:6(448)

[2] Meadows, D. "ASTM international and sustainable development keeping pace with a new global market." Standardization News, 2004, 32(4), 30.

[3] Pulselli, R.M. "Emergy analysis of building manufacturing, maintenance and use: Embuilding indices to evaluate housing sustainability." Energy and Buildings, 2007, 39(5), 620. https://doi.org/10.1016/j.enbuild.2006.10.004

[4] Webster, M. D. "The Relevance of Structural Engineers to Green Building Design." Proc, Metropolis & Beyond, ASCE, New York, NY, 2005, 60.

[5] Kadivar. M, Rahnema.H, Ghavami.K, Faghihi. L. "Influence of nano silica and natural gum on permeability of Kah-Gel (mud- straw plaster)." 15th NOCMAT conference. Brazil, 2014, 91.

[6] Binici, H., Aksogan, O., Bodur, M. N., Akca, E., & Kapur, S. "Thermal isolation and mechanical properties of fibre reinforced mud bricks as wall materials." Construction and Building Materials, 2007, 21(4), 901-906. https://doi.org/10.1016/j.conbuildmat.2005.11.004

[7] Morton, T. "Earth Masonry – Design and Construction Guidelines, IHS BRE Press: Watford." 2008.

[8] Brown, P.W. & Clifton, J. R "Adobe I: The Properties of Adobe. Studies in Conservation. ", 1978, 23(4), pp. 139-146.

[9] MariaIsabel G. Beas. Traditional Architectural Renders on Earthen Surfaces, Presented to the faculties of the universityof Pennsylvania in Partial Fulfillment of the Requirements for the degree of master of science, 1991.

Non-Conventional Materials and Technologies – NOCMAT for XXI Century   Materials Research Forum LLC
Materials Research Proceedings 7 (2018) 523-532   doi: http://dx.doi.org/10.21741/9781945291838-50

[10] Mitchell. J. K. "Soil Improvement, (State of the Art Report)," in Proc. X. Soil Mechanics and Foundation Eng. Conference, Stockholm, Sweden, 1981.

[11] Bouhicha, M., Aouissi, F., and Kenai, S. "Performance of composite soil reinforced with barley straw." *Cement and Concrete Composites,* 2005, 27(5), 617-621. https://doi.org/10.1016/j.cemconcomp.2004.09.013

[12] New Mexico Administrative Code. "2006 New Mexico Earthen Building Materials Code." New Mexico Administrative Code, <http://www.nmcpr.state.nm.us/nmac/parts/title14/ 14.007.0004.htm>, 2008.

[13] Bell F.G. "Lime stabilization of clay minerals and soils", Engineering Geology, 1996, Vol 42, pp 223-237. https://doi.org/10.1016/0013-7952(96)00028-2

[14] Hawley's Condensed Chemical Dictionary, p. 581.

Non-Conventional Materials and Technologies – NOCMAT for XXI Century   Materials Research Forum LLC
Materials Research Proceedings 7 (2018) 533-540   doi: http://dx.doi.org/10.21741/9781945291838-51

Glued Laminated Bamboo and Timber Association: Laboratory Testing of Industrialized Products

Gisele Fernandes de Oliveira [a, *], Divino Eterno Teixeira [b], Júlio Eustáquio de Melo [c], Vanda Alice Garcia Zanoni[d], Jaime Gonçalves de Almeida[e], Sérgio Alberto de Oliveira Almeida[f]

[a] Undergraduate at Universidade de Brasília – UnB, Brasil; gisele.gfo@gmail.com

[b] Laboratório de Produtos Florestais – LPF/SFB/MMA, Brasil; divinot@gmail.com

[c] Universidade de Brasília – UnB, Brasil; alej@unb.br

[d] Universidade de Brasília – UnB, Brasil; vandazanoni@unb.br

[e] Universidade de Brasília – UnB, Brasil; jagal@unb.br

[f] Laboratório de Produtos Florestais – LPF/SFB/MMA, Brasil; seralm44@hotmail.com

Abstract. The proposition and development of a flex material employing bamboo and timber laminated and glued (Gluelam+Gluebam), which can be fabricated industrially, is a way of reducing the impacts of conventional construction and, most importantly, enable the promotion of small farmers. The application of this material in furniture and urban equipment inserts rural population in the productive system, reaffirming their cultural identity, generating income and promoting employment. However, in order to introduce a new material such as the one that is been proposed, the performance of resistance and structural behavior tests are necessary so that its characteristics are understood, ensuring a technical certification for its application and, above all, the security of users. Thereby, this article intends to characterize the physical-mechanical behavior of bolted connections in a bamboo and timber flex material, aiming its application in the production of a bus stop to be built in Brasilia. The bamboo species used was *Dendrocalamus asper* and the timber *Eucalyptus urograndis*, commercially called Lyptus®. Methods consisted in carrying out laboratory compression tests with representative samples of a bolted connection from the bus stop structure, both parallel and perpendicular to grain. Each test was performed with six specimens, and the results were interpreted from the maximum load of rupture, as in accordance with the ASTM D 5652 – 95 international standard test method. The average values of resistance found were 1786,4 kgf in compression parallel to grain, and 1063,84 kgf in compression perpendicular to grain, proving that the connection with the flex material can be applied in light structures such as the bus stop.

Keywords: Gluelam, Gluebam, Laboratory Testing, Industrialized Production, Bolted Connection

Introduction

As a tropical plant, bamboo represents great potential of adaptation in Brazil. Despite the vast number of native species in this country, the *Dendrocalamus asper* stands out for being an introduced species of giant bamboo that occurs in many locations, including the Federal District (DF). Its clumps can produce mature culms annually, without the need of replanting. The growth of a bamboo pole is faster than any other plant, reaching the average height of 30 meters, in a 3 to 6 months period [1]. Therefore, bamboo presents great agricultural potential.

Dendrocalamus asper is a highly resistant and durable species. Its culms can reach 20 to 30 meters height, with a diameter of 8 to 20 cm and 11 to 20 mm thickness in internal walls [1]. Despite the flexibility of uses that bamboo presents, the geometry of its poles present certain limitations for its use in construction [2]. The glued laminated bamboo (GlueBam) technique represents a solution

Non-Conventional Materials and Technologies – NOCMAT for XXI Century   Materials Research Forum LLC
Materials Research Proceedings 7 (2018) 533-540   doi: http://dx.doi.org/10.21741/9781945291838-51

to enhance the use of bamboo to an industrial production scale, allowing it to be certified and standardized, as timber is.

Likewise bamboo, timber is an ancient building material. When associated with sustainable forest management and cultivation practices, timber exploitation can provide renewable and quality material for several applications. Lyptus® is a hybrid commercial species (*E. urograndis*), created from the crossbreeding of *Eucaluptus grandis* with *Eucalyptus urophylla*. It is produced through reforestation and considered a high strength wood, ideal for use in construction, and high density, which allows easy machining.

The use of glued laminated timber (GlueLam) provides many advantages over the use of sawn wood, such as the production of industrial components of larger dimensions and variable cross-sections, and architectural freedom [3].

The Forest Products Laboratory of the Brazilian Forest Service (LPF/SFB/MMA) has been contributing with the sustainable management of the Amazon forest through the characterization of wood species that do not have as much commercial visibility yet , as a way of reducing the pressure over the most used species [4]. Similarly, the development of a flex material, associating the use of timber to the use of bamboo, can contribute to the diversification of commercialized species.

The purpose of this paper is to characterize bolted connections using the flex material, aiming its application in the production of a bus stop, to be built in Brasília. This paper is part of the research project called "Glued Laminated Bamboo and Timber: development of a flex material for the production of furniture", which intends to offer a local, unique industrialized option for the furniture and urban equipment production in DF and the Integrated Development Region of the Federal District and Surrounding (RIDE).

Methodological proceedings

Tests were conducted in accordance with the international standard test methods for bolted connections in wood and wood-based products ASTM D 5652 – 95 [5], with six specimens in compression parallel to grain and six specimens in compression perpendicular to grain. Instead of bolts, ½ inch pins were used, one in each test specimen. No nuts or washes were used, in order to prevent the influence of load concentration during the tests. The dimensions of the specimens can be verified in Figures 1 and 2.

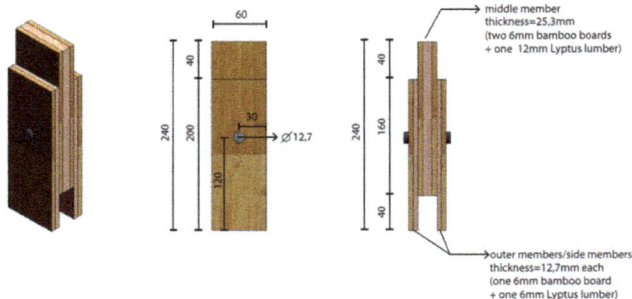

FIGURE 1 - TEST SPECIMENS FOR TESTING GLUELAM+GLUEBAM PIN CONNECTION PARALLEL TO GRAIN IN COMPRESSION. PERSPECTIVE, SIDE VIEW AND FRONT VIEW. DIMENSIONS IN MM.

Non-Conventional Materials and Technologies – NOCMAT for XXI Century   Materials Research Forum LLC
Materials Research Proceedings **7** (2018) 533-540   doi: http://dx.doi.org/10.21741/9781945291838-51

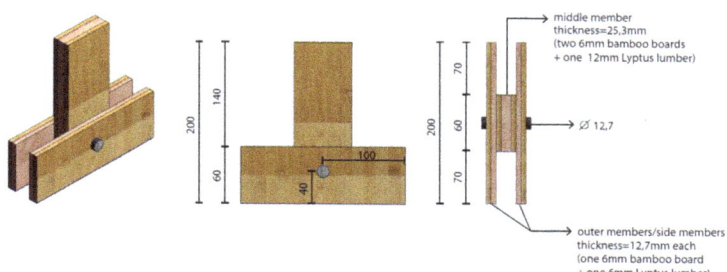

FIGURE 2 - *TEST SPECIMENS FOR TESTING GLUELAM+GLUEBAM PIN CONNECTION PERPENDICULAR TO GRAIN IN COMPRESSION. PERSPECTIVE, FRONT VIEW AND TOP VIEW. DIMENSIONS IN MM.*

Glued laminated *Dendrocalamus asper* bamboo boards and Lyptus® lumber were selected, then conditioned in accordance with Brazilian standard ABNT NBR 7190 [6] – (65±1)% humidity and (20±3)°C temperature – until mass stabilization. Dimensions of the boards and lumber were 700x84x15mm.

The pieces were then flattened, thinned and cut into four 700x70x12mm Lyptus® boards, eight 700x70x06mm Lyptus® boards and sixteen 700x70x06mm bamboo boards. These boards were glued, combining bamboo and wood, with the Tekbond PUR adhesive, in a 200g/m² grammage (10,22g in between boards). Within the 30 minute time frame specified by the adhesive manufacturer for partial cure, the samples were taken to the press, where they stood for 12 hours under a 10kgf/cm² pressure, measured in a torquemeter (101 N.m/bolt, two bolts in the press).

The pieces were then cut into components for the test specimens, in the dimensions 200x60mm, resulting in twelve 25,3mm thick pieces in a bamboo-timber-bamboo combination and twenty-four 12,7mm thick pieces in a timber-bamboo combination. These components were put back into the conditioning chamber, and had their densities measured after 72 hours.

For calculating the density, six pieces of the middle members and six pieces of the outer member of the test specimens were used. Each piece was measured in its width and length in the center, and in thickness using the average thickness of its center and two far ends. Afterwards, they were weighted in a leveled scale of a 0,01g precision.

All the pieces were then drilled using a ½ inch drill and the test specimens were assembled as detailed in Figures 1 and 2. The tests were carried out in an universal machine USM-600. The speed of load application was 1mm/minute, with supports and bearing blocks as in Figure 3.

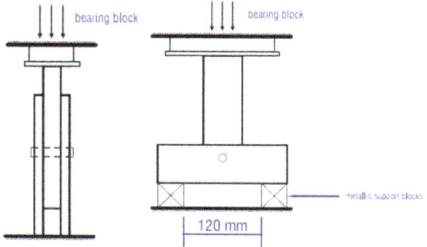

FIGURE 3 - *SUPPORT SCHEME FOR TESTING PIN CONNECTIONS PARALLEL AND PERPENDICULAR TO GRAIN IN COMPRESSION.*

Non-Conventional Materials and Technologies – NOCMAT for XXI Century   Materials Research Forum LLC
Materials Research Proceedings 7 (2018) 533-540   doi: http://dx.doi.org/10.21741/9781945291838-51

Results and discussion

Table 1 shows the values for apparent density of the pieces, at a 12% moisture content.

TABLE 1 - APPARENT DENSITY FOR SIDE AND MIDDLE MEMBERS OF THE TEST SPECIMENS.

Piece	Minimum density (g/cm³)	Maximum density (g/cm³)	Average density (g/cm³)
200x60x12,7mm	0,71	0,82	0,758
200x60x25,3mm	0,70	0,78	0,731
Average density for all the components			0,744

Previous studies have found values for the apparent density of $0,773g/cm^3$ for Lyptus®, $0,720g/cm^3$ for *Dendrocalamus asper*, and $0,722g/cm^3$ for the bamboo- Lyptus®-bamboo panels, classifying the flex material as a light material [7]. It is noticeable that the test specimens present similar apparent density.

Table 2 presents maximum load of rupture for each specimen, parallel and perpendicular to grain in compression, along with the 5% exclusion load and coefficient of variation for each test assembly.

TABLE 2 - MAXIMUM LOAD OF RUPTURE OF CONNECTIONS WITH STRENGHT PARALLEL AND PERPENDICULAR TO GRAIN IN COMPRESSION.

Compression testing	Load of rupture per specimen (kgf)						Average (kgf)	Coefficient of variation
	1	2	3	4	5	6		
Parallel to grain	1451	1635	1709	1830	2050	2046	1786,84	13,2%
Parallel to grain with 5% exclusion	1060	1244	1318	1439	1659	1655	1395,83	-
Perpendicular to grain	962	1076	1090	1211	990	1054	1063,84	8,2%
Perpendicular to grain with 5% exclusion	817	931	945	1066	845	909	918,83	-

ASTM D 5652 – 95 [5] specifies the conduction of ten tests, or a sufficient number of tests to present a coefficient of variation of 15% to 30%. As shown in Table 2, the performance of six tests was enough to reach a coefficient of variation lower than the recommended minimum, which makes it a statistically sufficient number of tests for the analysis. On Graph 1, the variation of the maximum load in each test can be verified, in absolute values and in a 5% exclusion rate.

There has been three different combinations of assembly for the side members of the specimens. Test specimens 1, 2 and 3 were assembled with bamboo on the outer faces, specimen 4 with wood on the outer faces, in both parallel and perpendicular to grain tests. Specimens 5 and 6 were alternated, with bamboo on the outer face in one side, and wood on the other side for tests in compression parallel to grain, and so was specimen 5 for tests in compression perpendicular to grain. Specimen 6 tested perpendicular to grain was similar to specimen 4. These combinations can be verified on Tables 3 and 4.

Non-Conventional Materials and Technologies – NOCMAT for XXI Century   Materials Research Forum LLC
Materials Research Proceedings 7 (2018) 533-540   doi: http://dx.doi.org/10.21741/9781945291838-51

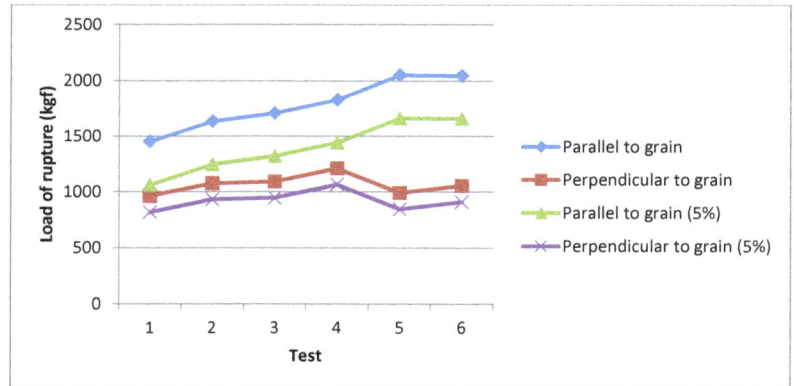

GRAPH 1 - *MAXIMUM LOAD OF RUPTURE AND 5% EXCLUSION LOAD OF RUPTURE IN TESTS CONDUCTED WITH 1 PIN CONNECTIONS FOR COMPRESSION PARALLEL AND PERPENDICULAR TO GRAIN.*

TABLE 3 - *AVERAGE MAXIMUM LOAD IN COMPRESSION TESTS PARALLEL TO GRAIN FOR 1 PIN CONNECTION, IN VARIABLE COMBINATIONS OF TEST SPECIMEN ASSEMBLY*

	Bamboo on the outer face	Wood on the outer face	Alternate outer faces
Average maximum load (kgf)	1598,33	1830,00	2048,00
Specimen representation			

TABLE 4 - *AVERAGE MAXIMUM LOAD IN COMPRESSION TESTS PERPENDICULAR TO GRAIN FOR 1 PIN CONNECTION, IN VARIABLE COMBINATIONS OF TEST SPECIMEN ASSEMBLY*

	Bamboo on the outer face	Wood on the outer face	Alternate outer faces
Average maximum load (kgf)	1042,67	1132,50	990,00
Specimen representation			

Statistics analysis using the Tukey's test proved that there has not been significant variation on the results related to the different combinations of assembly in the test specimens.

On tests parallel to grain the specimens breakage happened along the grain, at the central region of the outer members, aligned with the pin hole, after slip. Specimens 4 and 6 had detachments between bamboo and wood boards. On tests conducted perpendicular to grain, rupture also occurred on the outer members, following the grain, aligned with the pin hole, after the inferior grains had been bent. There was no full detachments, except for the ends of specimens 4 and 6. In all specimens, even those that have suffered detachments, wood and bamboo worked together, breaking at the same point. Figure 4 shows three broken specimens.

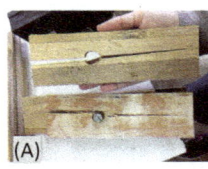

FIGURE 5 - TYPES OF RUPTURE OF TEST SPECIMENS AFTER TESTING PIN CONNECTIONS PARALLEL AND PERPENDICULAR TO GRAIN IN COMPRESSION. (A) DETACHMENT, (B) AND (C) RUPTURE ON THE DIRECTION OF GRAIN.

As a comparison to the results, a paper on two pin fastenings with wood species ipê (*Tabebuia serratifolia*), cumaru (*Dipterix odorata*), maçaranduba (*Manilkara huberi*) e angelim vermelho (*Dinizia excelsa*) was used, considering those are the most used in timber structures, in general [8]. Table 5 presents the results for tests in compression parallel and perpendicular to grain for two ½ inch pin connections for each of these species, and their densities.

TABLE 5 - LOAD OF RUPTURE IN COMPRESSION PARALLEL AND PERPENDICULAR TO GRAIN FOR TWO ½ INCH PIN CONNECTIONS.

Species	Apparent density (g/cm³)	Load of rupture parallel to grain (kgf)	Load of rupture perpendicular to grain (kgf)
Ipê *T. serratifolia*	1,12	6853,61	4639,70
Cumaru *D. odorata*	1,11	6170,40	5258,37
Maçaranduba *M. huberi*	1,08	5433,96	4749,77
Angelim vermelho *D. excelsa*	1,02	5041,47	4275,67

Source: adapted from Siqueira; Melo and Teixeira [8].

Siqueira's, Melo's and Teixeira's [8] results were divided by two to find the load in each pin. Then, the relative load of rupture expected in relation to density was calculated, based on the apparent density of the bamboo- Lyptus®-bamboo panels - 0,744(g/cm³) – using simple proportion, in order to compare them to the maximum loads obtained from the tests on the one pin connections using the flex material. Tables 6 and 7 show these relations.

TABLE 6 - *EXPECTED LOADS OF RUPTURE IN COMPRESSION PARALLEL TO GRAIN FOR ONE ½ INCH PIN FOR THE APPARENT DENSITY OF 0,744 (G/CM³).*

Species	Load of rupture for one pin (kgf)	Expected load of rupture for the apparent density = 0,744(g/cm³)
Ipê T. serratifolia	3426,80	2276,37
Cumaru D. odorata	3085,20	2067,91
Maçaranduba M. huberi	2716,98	1871,69
Angelim vermelho D. excelsa	2520,73	1838,65
Average		2013,65

TABLE 7 - *EXPECTED LOADS OF RUPTURE IN COMPRESSION PERPENDICULAR TO GRAIN FOR ONE ½ INCH PIN FOR THE APPARENT DENSITY OF 0,744 (G/CM³).*

Species	Load of rupture for 1 pin	Expected load of rupture for the apparent density = 0,744(g/cm³)
Ipê T. serratifolia	2319,85	1541,04
Cumaru D. odorata	2629,18	1762,26
Maçaranduba M. huberi	2374,88	1636,02
Angelim vermelho D. excelsa	2637,83	1924,06
Average		1715,845

Considering the average load of rupture for 1 pin connections between panels of the flex material - 1786,84kgf in compression parallel to grain and 1063,84kgf in compression perpendicular to grain – it is proved that the resistance of the panel in this type of connection is relatively close to the resistance of wood that are considered hard and of high density.

Conclusion

The results show that the resistance of the flex material made of *Dendrocalamus asper* and *Eucalyptus urograndis* (Lyptus®) in pin connections is close to the resistance of commercial species of wood generally used in timber structures. Therefore, the flex material panels can be applied in light structures such as the Bus Stop. The material that is being proposed presents many advantages, such as being lightweight and flexible. The character of being unprecedented and the importance of specific tests for bolted or pin connections between wood components or similar materials is also highlighted here, as there is still little bibliography available on the subject.

Aknowledgements

The authors would like to acknowledge the The Forest Products Laboratory of the Brazilian Forest Service, Ministry of Environment (LPF/SFB/MMA) and the Bamboo and Natural Fibers Research and Application Center of the University of Brasília (CPAB/UnB) for providing their installations and team for the preparation and conduction of the tests, as well as the Fund for Research Support of the Federal District (FAPDF) for financing the research project.

References

[1] PEREIRA, M.A.R; BERALDO, A.L. Bambu de Corpo e Alma. Bauru, SP: Canal 6 editora, 2008.

[2] BERALDO, A.L; RIVERO, L.A. Bambu Laminado Colado (BLC). Floresta e Ambiente, V. 10, n.2, p.36 - 46, ago./dez. 2003.

[3] MOODY, R.C.; HERNANDEZ, R. Glued-laminated timber. In: SMULSKI, Stephen. Engineered wood products-A guide for specifiers, designers and users. ISBN-096556736-0-X. Madison, WI: PFS ResearchFoundation: p. 1-1–1-39. Capítulo 1. 1997.

[4] MELO, J. E.; CAMARGOS, J. A. A Madeira e seus Usos. 1. ed. Brasília, DF: Ministério do Meio Ambiente, 2016. v. 1. 204p.

[5] AMERICAN SOCIETY FOR TESTING AND MATERIALS (ASTM). D 5652-95: Standard Test Methods for Bolted Connections in Wood and Wood-Based Products. West Conshohocken, United States, 2000.

[6] ASSOCIAÇÃO BRASILEIRA DE NORMAS TÉCNICAS (ABNT). NBR 7190/97: Projeto de estruturas de madeira. Rio de Janeiro, 1997.

[7] MORO, E.A.; ALMEIDA, J.G. Caracterização de painel de bambu combinado com madeira para confecção de mobiliário. PIBIC/CNPq, 2016.

[8] SIQUEIRA, M.J.; MELO, J.E.; TEIXEIRA, D.E. Characterization of pin fastenings with four amazon tropical timber species. Preceedings: COBEM. Nov. 5-9, 2007.

Non-Conventional Materials and Technologies – NOCMAT for XXI Century   Materials Research Forum LLC
Materials Research Proceedings 7 (2018) 541-551   doi: http://dx.doi.org/10.21741/9781945291838-52

Production and Characterization of Bamboo/Wood-Made Shape to be used in *Longboards*

Divino Eterno Teixeira [a*], Giovana de Almeida Martins Azevêdo [a]

[a] University of Brasilia, Brazil

ABSTRACT. This work aims at the production and characterization of the *amescla* (*Trattinnickia burseraefolia*) and bamboo *(Dendrocalamus asper)* plywood panels as elements of a flex material of low environmental impact. These panels were analyzed based on three different treatments: wood; material made up by 60% of wood and 40% of bamboo; and material made up by 60% of bamboo and 40% of wood. The tests performed to compare the performance of each of those treatments were as follows: apparent specific mass; moisture content; thickness swelling; water absorption; static bending; stress wave timer; and, shear and flaw on the wood on the gluing lines. The findings enabled analyzing which of the three material compositions better fits the production of shapes to be used in longboards. The treatment made up by 60% of bamboo and 40% of wood performed best in such analyses, showing better resourcefulness of the bamboo as element of plywood to be employed in the production of shapes to be used in longboards.

Keywords: Bamboo, Wood, Flex Material, GLB, *Dendrocalamus Asper*, Longboard

Introduction

Bamboo is a material that abounds in tropical and subtropical regions. Since it is renewable, low-cost, easy to grow, of high-strength and great versatility of potential uses, this material has high potential to contribute with the reduction of the uncontrolled cutting of wood in rainforests. This paper focuses on the study and evaluation of the behavior of the bamboo and wood as glued laminate, working together to make up a flex material that enables large-scale production, while minimizing the environmental impacts mainly caused by the civil construction. Brazil houses some bamboo species that fit into this use, such as the *Dendrocalamus asper, which was used in this work together with amescla* (*Trattinnickia burseraefolia)* veneers. This paper aims at the production and characterization of *amescla* (*Trattinnickia burseraefolia)* and bamboo *(Dendrocalamus asper)* plywood panels as elements of a flex material of low environmental impact. These panels were analyzed based on three different treatments: wood; material made up by 60% of wood and 40% of bamboo; and material made up by 60% of bamboo and 40% of wood.

Methodological procedures

Selection and preparation of material

The first stage of the work is devoted to the material preparation and selection. It selected *Dendrocalamus asper stems* in good conditions and with the fewest biotic attacks possible, from the Bamboo and Natural Fibers Research and Application Center, to produce bamboo slats that will make up the plywood.

The slats were produced at the carpentry shop of the Forestry Products Laboratory (LPF/SFB) where these were cut and planed until reach the dimensions of 310 x 25 x 2 mm (length, width and thickness, respectively). The veneers were also cut in the ideal dimensions to make the plywood panels, i.e., 310 x 310 x 2 mm (length, width and thickness, respectively).

Building the panels

The bamboo slats were acclimatized at the temperature of (20±3)°C (68±37°F) and moisture content of (65±1)% at the LPF/SFB until the moisture could be stabilized and then taken to produce the sheets. At this stage, the sheets were sanded with #80 sandpaper, glued on the sides with Cascorez white glue and compressed in a hand-operated press machine (four clips and two clamps) for four hours. This procedure is shown in Figure 1. Each sheet is made up by 12.5 slats with the dimensions of 310 x 310 x 2 mm (length, width and thickness, respectively).

Figure 1(A)   Figure 1(B)   Figure 1(C)

FIGURE 1 - *ASSEMBLING THE BAMBOO SHEETS; 1(A) GLUING THE BAMBOO SLATS; 1(A) BAMBOO SLATS GLUED, MAKING UP SHEETS; 1(C): PRESSING OF THE BAMBOO SHEETS.*

To build the plywood panels, sheets were glued overlaid (one on the other); crossed following the directions of the fibers (one sheet on the horizontal, and the next on the vertical position); and, alternating materials (wood or bamboo). In this procedure we used the polyurethane-base (PUR) glue (brand: Tekbond; density: 1.08 to 1.14 g/cm3 and viscosity of 6000 to 8500 cps). The glue quantification was measured on a digital scale (brand: PGD) to ensure the 150 g/m2 grammage specified by the manufacturer - in this case, 14.41 g of glue on each layer. Soon after gluing, the panels of 310 x 310 x 10 mm (length, width and thickness, respectively) were taken to the hydraulic pressing machine (brand: Indumec) for four hours, under pressure of 10 kgf/cm2 (Figure 2).

Figure 2(A)   Figure 2(B)   Figure 2(C)   Figure 2(D)   Figure 2(E)   Figure 2(F)

FIGURE 2 - *ASSEMBLING THE PLYWOOD PANELS; 2(A): APPLICATION OF GLUE ON THE SHEET; 2(B): SPREADING THE GLUE ON THE SHEET; 2(C): BAMBOO SHEET WITH PUR GLUE; 2(D): ASSEMBLAGE OF THE PANEL LAYERS; 2(E): PANEL PRESSING IN HYDRAULIC PRESSING MACHINE; 2(F): PANEL PRESSING IN HYDRAULIC PRESSING MACHINE.*

This procedure was repeated to build the three different types of 5-layer plywood (treatments) analyzed in this work, which differ for the combination of bamboo and wood, as described in Table 1.

Non-Conventional Materials and Technologies – NOCMAT for XXI Century   Materials Research Forum LLC
Materials Research Proceedings 7 (2018) 541-551   doi: http://dx.doi.org/10.21741/9781945291838-52

TABLE 1 - TREATMENTS AND QUANTIFICATION OF MATERIALS DURING THE ASSEMBLAGE OF THE PLYWOOD PANELS

Treatment	Description	Quantity of Panels	Quantity of sheets (by panel)		Quantity of Bamboo Slats (per panel)
			Wood	Bamboo	
M	100% wood	3	5	0	0
MB	60% wood 40% bamboo	3	3	2	25
BM	60% bamboo 40% wood	3	2	3	37,5

M: Wood; MB: 60% Wood and 40% Bamboo; BM: 60% Bamboo and 40% Wood

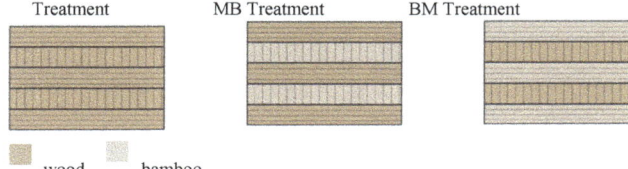

Treatment   MB Treatment   BM Treatment

wood   bamboo

Cutting the bodies of evidence

The bodies of evidence used in the tests were cut from the panels newly-produced to each treatment, as shown in Figure 3. Treatments are divided in three types with the following quantities and dimensions for length, width and thickness, respectively:

- Type-A (4 per panel): 230 x 50 x 2 mm, for the test of density and static bending;
- Type-B (4 per panel): 115 x 25 x 2 mm, for the test of shear on the gluing line;
- Type-C (10 per panel): 25 x 25 x 2 mm, for the test of absorption, swelling and moisture.

Figure 3(A)   Figure 3(B)   Figure 3(C)   Figure 3(D)   Figure 3(E)

FIGURE 3 - *PROCESS OF CUTTING THE BODIES OF EVIDENCE; 3(A): CUTTING OF THE TYPE-A BODY OF EVIDENCE; 3(B): CUTTING OF THE TYPE-B BODY OF EVIDENCE; 3(C): CUTTING OF THE TYPE-C BODY OF EVIDENCE; 3(D): CUT BODIES OF EVIDENCE; 3 (E): BODIES OF EVIDENCE CUT AND SEPARATED BY TYPE.*

Physical and mechanical test
Apparent Specific Mass (ASM)

At this stage, the Type-A bodies of evidence were used to determine their densities and compare these according to the different treatments and based on the EN 310:1993 standard. In each body of evidence, the following were measured: mass (with a digital scale); thickness, checked on five different points (assisted by the dial indicator); width, checked on three different points (assisted by a digital pachymeter); and length, checked on two different points (assisted by the digital pachymeter) - as shown in Figure 4. Based on the arithmetic averages of these measures, the respective volumes

(V) and densities (D) of each body of evidence were calculated, as described in the standard. Therefore, the values of Apparent Specific Mass (ASM) for each treatment were found.

Figure 4(A)   Figure 4(B)

FIGURE 4 - *MEASURING THE BODIES OF EVIDENCE; 4(A): MEASURING THE WIDTH WITH DIGITAL PACHYMETER; 4(B): MEASURING THE THICKNESS WITH A DIAL INDICATOR.*

Moisture Content
To define the moisture content, six Type-C bodies of evidence of each treatment were use. These were weighed with a digital scale and then put in the greenhouse at (103±2) ^{0}C (217±35^{0}F), as shown in Figure 5. After stabilizing the mass, the moisture content was verified according to the ASTM D1037:1999 standard.

FIGURE 5 - *BODIES OF EVIDENCE IN THE GREENHOUSE*

Thickness swelling and water absorption
In order to find the thickness swelling and water absorption indexes, 24 Type-C bodies of evidence were used for each treatment. These were immersed in water, and their thickness and mass were checked on the periods of 2 hours and 24 hours, according to the ASTM D1037:1999 standard, and as shown in Figure 6.

Figure 6(A)   Figure 6(B)   Figure 6(C)

FIGURE 6 - *PROCESSES OF THE MOISTURE CONTENT TEST; 6(A) TYPE-C BODIES OF EVIDENCE; 6(B): BODIES OF EVIDENCE IMMERSED IN WATER; 6(C) WEIGHING THE BODIES OF EVIDENCE.*

Non-Conventional Materials and Technologies – NOCMAT for XXI Century   Materials Research Forum LLC
Materials Research Proceedings 7 (2018) 541-551   doi: http://dx.doi.org/10.21741/9781945291838-52

Static Bending

The Type-A bodies of evidence, with the fibers in the parallel direction, were used to the static bending test (Figure 7). Twelve bodies of evidence were used to each type of treatment. The dimensions and procedures follow the guidance provided for in the EN 310:1993 standard to obtain the analyses of the moduli of elasticity, rupture and the consequent strengths of the materials.

Figure 7(A)   Figure 7(B)   Figure 7(C)

FIGURE 7 - PROCESSES OF THE STATIC BENDING TEST; 7(A): TYPE-A BODY OF EVIDENCE POSITIONED IN THE MACHINE; 7(B): START OF THE STATIC BENDING TEST; 7(C): BODY OF EVIDENCE BROKEN IN THE STATIC BENDING TEST.

Stress Wave Timer

The Stress Wave test was performed to find the Dynamic Modulus of Elasticity (MOEd). The test measures the time the body of evidence takes to propagate the waves caused by a metal pendulum part of the Stress Wave Timer (Figure 8). Twelve Type-A bodies of evidence were used to each treatment.

Figure 8(A)   Figure 8(B)   Figure 8(C)

FIGURE 8 - PROCESSES OF THE STRESS WAVE TEST; 8(A): POSITIONING OF THE TYPE-A BODY OF EVIDENCE IN THE EQUIPMENT; 8(B): START OF THE STRESS WAVE TEST; 8(C): END OF THE STRESS WAVE TEST.

Shear on the gluing line

The shear on the gluing line test employed 12 Type-B bodies of evidence (Figures 9(A); 9 (B)) for each treatment, as instructed by the ABNT NBR ISO 12466-1: 2006 standard, with the support of the Universal Assays Machine USM-600 (Figures 9(C); 9(D)). The values of the loads presented in the test were divided by the respective areas of the bodies of evidence to find the gluing shearing strength.

Figure 9(A)   Figure 9(B)   Figure 9(C)   Figure 9(D)

FIGURE 9 - *PROCESSES OF THE SHEAR ON THE GLUING LINE TEST; 9(A): TYPE-B BODIES OF EVIDENCE BEFORE THE SHEAR TEST; 9(B): TYPE-B BODIES OF EVIDENCE AFTER THE SHEAR TEST; 9(C): SHEAR TEST; 9(D:) SHEAR TEST.*

Statistical Analyses

The statistical analyses were performed using the Excel and SPSS 13.0 softwares. The experimental outline, based on the three types of material analyzed, is a variance analysis with 5% of significance level to make up the results, according to the Tukey's test.

Results and discussion

Physical and Mechanical Tests of the plywood panels.

Apparent Specific Mass (ASM)

An analysis of the results disclosed in Table 2, as regards the Apparent Specific Mass for the different treatments studied, shows that the BM treatment stands out for the Turkey's at 5% significance level, because it presents higher mean. This can be explained by the fact that this was the material with more quantity of bamboo in the composition, although the variation among the three values is not very irregular. For the values of the standard deviation and the coefficient of variation presented, a highlight is the MB treatment with values lower than those for the other treatments.

TABLE 2 - MEAN VALUES OF THE APPARENT SPECIFIC MASS OF THE PLYWOOD PANELS

Treatment	ASM	Standard	# of Samples	Coefficient of Variation (%)
M	0.515 [a]	0.02	12	4.1
MB	0.608 [b]	0.01	12	1.8
BM	0.651 [c]	0.02	12	3.9

M: Wood; MB: 60% Wood and 40% Bamboo; BM: 60% Bamboo and 40% Wood
a, b, c: Equal letters mean treatments equal to 5% of significance

Moisture Content

The Moisture Content mean values are presented in Table 3. Variation is not so sharp among the percentages found to each treatment and the magnitudes pursuant to the standards to perform the other tests, since these are close to 12%.

TABLE 3 - MEAN MOIST CONTENT VALUES FOR THE PLYWOOD PANELS

Material	Moisture Content	Standard	# of samples	Coefficient of
M	10.67	0.60	6	5.62
MB	10.29	0.32	6	3.09
BM	10.52	0.23	6	2.20

M: Wood; MB: 60% Wood and 40% Bamboo; BM: 60% Bamboo and 40% Wood

Thickness swelling and water absorption

Table 4 shows the mean values for Water Absorption after two hours of immersion (AA - 2h) and after twenty-four hours of immersion (AA - 24h) to the different treatments. In both periods, the M treatment presents mean higher than the others, as well as the standard deviation and the coefficient of variation, although the Tukey's test shows that no treatment reached 5% of significance likeliness. It can be observed that the higher the quantity of bamboo in the treatment composition, the lower the Water Absorption mean for the bodies of evidence.

TABLE 4 - WATER ABSORPTION MEAN VALUES OF THE PLYWOOD PANELS

	Treatment	Mean	Standard	# of samples	Coefficient of Variation (%)
AA – 2h	M	30.51 [c]	4.23	24	13.9
	MB	25.33 [b]	2.42	23	9.6
	BM	15.41 [a]	1.86	24	12.1
AA – 24h	M	66.57 [c]	6.97	24	10.5
	MB	50.65 [b]	4.71	23	9.3
	BM	37.26 [a]	3.43	24	9.2

M: Wood; MB: 60% Wood and 40% Bamboo; BM: 60% Bamboo and 40% Wood;
a, b, c: Equal letters mean treatments equal to 5% of significance

Regarding the Thickness Swelling mean values (Table 5), in turn, the M treatment presented the lowest mean in both periods, although it was outstanding in the previous test, with values higher than the others. It would be worth to evaluate how the properties of the different materials influence on the plywood performance.

TABLE 5 - THICKNESS SWELLING MEAN VALUES OF THE PLYWOOD PANELS

	Treatment	Mean (%)	Standard	# of samples	Coefficient of Variation
IE – 2h	M	-0.74 [a]	1.15	24	-156
	MB	0.17 [b]	1.16	23	675
	BM	0.40 [b]	1.28	24	319
IE – 24h	M	0.58 [a]	1.11	24	191
	MB	1.61 [b]	1.41	23	88
	BM	2.10 [b]	1.47	24	67

M: Wood; MB: 60% Wood and 40% Bamboo; BM: 60% Bamboo and 40% Wood;
a, b, c: Equal letters mean treatments equal to 5% of significance

It is worth mentioning that a sample of the MB treatment presented delamination in the first period of immersion in water and was excluded from the test, as shown in Figure 10.

FIGURE 10 - DELAMINATION OF THE BODY OF EVIDENCE.

Flexão Estática
The Moduli of Elasticity (MOE) and Rupture (MOR) achieved in the Static Bending Test are described in Table 6, which highlights the BM treatment for its higher means in both moduli, although the MB treatment is outstanding for its higher values of standard deviation and coefficient of variation. The coefficient of variation among the results of each treatment is high, what is clearly shown in the Turkey's test in which none of them resembles the other regarding the values achieved.

TABLE 6 - MOE AND MOR MEAN VALUES OF THE PLYWOOD PANELS

	Treatment	Mean	Standard	# of samples	Coefficient of Variation (%)
MOE	M	6632 [b]	510.78	12	7.7
	MB	4510 [a]	1974.46	12	43.8
	BM	11712 [c]	1294.78	12	11.1
MOR	M	53,89 [b]	5.58	12	10.4
	MB	38,29 [a]	10.16	12	26.5
	BM	90,32 [c]	8.54	12	9.5

M: Wood; MB: 60% Wood and 40% Bamboo; BM: 60% Bamboo and 40% Wood;
a, b, c: Equal letters mean treatments equal to 5% of significance

Stress Wave Timer
The Stress Wave Timer test was performed to find the values of the dynamic Moduli of Elasticity (MOEd) of each treatment, in a non-destructive way. Table 7 shows that the BM treatment is outstanding in the results with mean and standard deviation values much higher than those of the M and MB treatments that, in turn, present little variation with each other.

TABLE 7 - MOED MEAN VALUES OF THE PLYWOOD PANELS

Treatment	Mean	Standard	# of samples	Coefficient of Variation (%)
M	5525.54	617.51	12	11.2
MB	5245.27	549.44	12	10.5
BM	11112.81	972.49	12	8.8

M: Wood; MB: 60% Wood and 40% Bamboo; BM: 60% Bamboo and 40% Wood

The test was complemented with the correlation between the dynamic Modulus of Elasticity (MOEd), a non-destructive method, and the Moduli of Static Elasticity (MOE) and of Rupture (MOR), with destructive methods. Therefore, the graphs in Figures 11 and 12 are presented with the results evidencing that both for MOE x MOEd and for MOR x MOEd, the correlations with R^2 value higher than 0.8 (r higher than 0.9).

Non-Conventional Materials and Technologies – NOCMAT for XXI Century   Materials Research Forum LLC
Materials Research Proceedings **7** (2018) 541-551   doi: http://dx.doi.org/10.21741/9781945291838-52

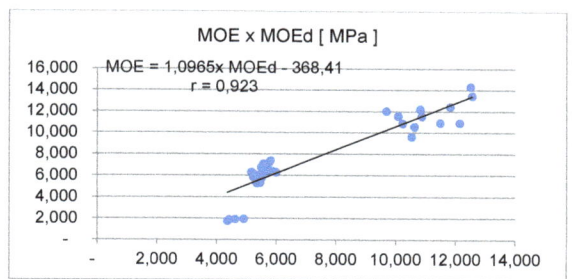

FIGURE 11- GRAPH OF THE CORRELATION BETWEEN MOE AND MOED OF THE PLYWOOD PANELS

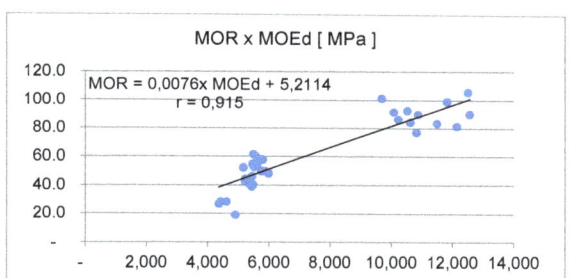

FIGURE 12 - CORRELATION BETWEEN MOR AND MOED OF THE PLYWOOD PANELS.

Shear on the gluing line

To evaluate the quality of the gluing line the shear test is performed for each treatment. The test results are described in Table 8. The difference between their respective strengths is small and, so, the values for the mean, standard deviation and coefficient of variation are similar. This fact has also been evidenced in the Tukey's test, in which the three treatments were defined with variance equal to 5% of significance. Therefore, one could say that the treatment differences do not significantly change the shear strength on the gluing line.

TABLE 8 - SHEAR STRENGTH MEAN VALUES OF THE PLYWOOD PANELS

Treatment	Mean	Standard	# of samples	Coefficient of
M	1.69 [a]	0.67	12	39.4
MB	1.83 [a]	0.53	12	29.2
BM	2.30 [a]	0.78	12	33.9

M: Wood; MB: 60% Wood and 40% Bamboo; BM: 60% Bamboo and 40% Wood;
a, b, c: Equal letters mean treatments equal to 5% of significance

Still in this context, one can also observe the values of flaw on the wood presented by each treatment in the same test (Table 9). Here, the MB treatment was a highlight for its much lower mean and standard deviation and coefficient of variation much higher than the other ones. This is also shown in the Tukey's test, through the values of the M and BM treatments equal to 5% of significance, i.e., these were considered similar values.

TABLE 9 - FLAW ON THE WOOD MEAN VALUES OF THE PLYWOOD PANELS

Treatment	Mean	Standard	# of samples	Coefficient of Variation
M	92 [b]	3	12	3.6
MB	50 [a]	26	12	52.0
BM	90 [b]	5	12	5.2

M: Wood; MB: 60% Wood and 40% Bamboo; BM: 60% Bamboo and 40% Wood;
a, b, c: Equal letters mean treatments equal to 5% of significance

Conclusion

The study of the Dendrocalamus asper bamboo and Trattinnickia burseraefolia wood as elements of a flex material used to produce longboard shapes was important to differentiate and characterize the best-performing treatment.

The analysis of the tests of apparent specific mass, moisture content, thickness swelling and water absorption, static bending, stress wave and shearing strength on the gluing line performed, regarding the strength of each treatment, found higher results for the treatment with 60% of bamboo and 40% of wood in relation to the other types of plywood.

These results are important not only to analyze the composition of the proper flex material to produce longboard shapes, although this is the research focus, but also to characterize a material that could be very helpful in the effort to reduce deforestation, since 60% of its composition is based on an abundant and renewable raw-material: the bamboo.

Acknowledgement

To the research team, Jaime Gonçalves de Almeida, Divino Eterno Teixeira, Vanda Alice Zanoni, Sérgio Alberto de Oliveira Almeida and Jonatas Pereira da Silva, responsible for the project organization and development, and to the technician João Evangelista Anacleto for performing the mechanical assays. To my work fellows Carolina Simões, Gisele Fernandes, Natália Côrtes and Marina Patury. To the Institutional Scientific Initiation Scholarship Program (ProIC/UnB) for the scholarship. To the team of the Bamboo and Natural Fibers Research and Application Center (CPAB/UnB) and to the team of the Forestry Products Laboratory of the Brazilian Forestry Service (LPF/SFB).

References

[1] ABNT - ASSOCIAÇÃO BRASILEIRA DE NORMAS TÉCNICAS. Madeira compensada - Qualidade de colagem. Rio de Janeiro, 2006. (Brazilian Standard NBR ISO 12466-1).

[2] AMERICAN SOCIETY FOR TEST AND MATERIALS. ASTM D - 1037. Standard test methods for evaluating properties of wood-based fiber and particle panel materials. Annual Book of ASTM Standards, Philadelphia, 1999.
BERALDO, A.L.; PEREIRA, M.A.R. Bambu de corpo e alma. Bauru, SP., 2007.

[3] ESPELHO, J. C. C.; BERALDO, A. L. Avaliação físico-mecânica de colmos de bambu tratados. Revista Brasileira de Engenharia Agrícola e Ambiental, Campina Grande, v. 12, n.6, 2008.

[4] EUROPEAN STANDARDS. EN 310: Wood-based panels - determination of modulus of elasticity in bending and of bending strength. United Kingdom. 1993.

Non-Conventional Materials and Technologies – NOCMAT for XXI Century   Materials Research Forum LLC
Materials Research Proceedings 7 (2018) 541-551   doi: http://dx.doi.org/10.21741/9781945291838-52

[5] LIMA, D. M.; AMORIM, M.M.; LIMA JÚNIOR, H.C.; BARBOSA, N.P.;WILRICH, F.L. Avaliação do comportamento de vigas de bambu laminado colado submetidas à flexão. Ambiente Construído, Porto Alegre, v. 14, n. 1, Jan/Mar 2014.

[6] MELO, JÚLIO EUSTÁQUIO DE; CAMARGOS, JOSÉ ARLETE ALVES . A madeiras e seus usos. 1st Edition. Brasilia, 2011.

[7] RIVERO, L. A.; BERALDO, A. L. Laminado colado e contraplacado de bambu. Universidade Estadual de Campinas. Faculdade de Engenharia Agrícola. Campinas - SP. August, 2003.

Non-Conventional Materials and Technologies – NOCMAT for XXI Century   Materials Research Forum LLC
Materials Research Proceedings 7 (2018) 552-559   doi: http://dx.doi.org/10.21741/9781945291838-53

Investigation of Self-Healing Phenomenon in High Performance Fiber Reinforced Microconcrete with Steel and Sisal Fibers

Tamara Nunes da C. Moreira[a], Saulo Rocha Ferreira[b],
Romildo Dias Toledo Filho[c]

[a,c] UFRJ, Brazil; tamaranunes@coc.ufrj.br, toledo@coc.ufrj.br

[b] UFLA, Brazil, ferreira.sr@deg.ufla.br

Abstract. High performance fiber reinforced concretes present advantages when compared with conventional concretes, such as greater tenacity, durability and crack control. Nevertheless, it can have great environmental appeal if part of the cement is replaced with blast furnace slag. Combining the lower environmental impact achieved by the adequate selection of the matrix with the self-healing potential of this type of composite, a greater economic and environmental value can be added to the family of advanced materials. The purpose of this research is to evaluate the self-healing phenomenon of high performance microconcrete reinforcement (0,64% in vol.) with steel and sisal fibers (l_f =13 mm). The addition of vegetable fiber aimed to guarantee a better distribution of moisture in the matrix and to potentiate the phenomenon, but also to act as secondary reinforcement. The samples were cured for 28 days and then, under tensile loads, a single crack (200 µm) was induced. The micro-cracked specimens were subjected to wet and dry cycles for 3 months to allow the healing. After this period, the samples were re-tested under tensile loads until collapse. The evaluation of the healing potential was performed through two mechanical indexes: index of mechanical recovery (ITR_1 and ITR_2), which was the relation between the maximum stresses before and after conditioning. Microscopic analyzes were performed to evaluated the crack closure, as well as thermal analysis of the filler filling material to identify the hydration products present in the crack. The results indicate that partial/total recovery of mechanical behavior occurred and sisal fiber acted as nucleating agent.

Keywords: Self-Healing, Sisal, Steel, Crack Control, High Performance

Introduction

Self-healing of concrete can be defined as the ability of cracks to diminish autogenously in width time (Ozbay et al, 2013). Concrete structures are susceptible to present cracks and remain with it maybe all life service if it size do not overcome a maximum crack opening of 400 µm in non-aggressive environment and 200 µm in an aggressive one (NBR 6118), although the presence of these cracks can represent prejudice to concrete and the shortening of life cycle. In 2013 the US spent 5,2 billion to maintenance of bridges and in Netherlands 1/3 of construction annual budget was spend on maintenance structures and indirect costs related to that (TC RILEM, 2012).

There are a lot of mechanisms to provide cracks diminish and even total sealing. These can be provide through capsular or vascular systems (Blaiszik et al, 2010), bacterial insertions (Snoeck et al, 2014), superabsorbent polymers (Kim and Schlangen, 2010) and intrinsic (Toledo, 1997) (Moreira, 2017).

According to Wu et al (2012) four types of self-healing exists: precipitation of calcium carbonate at the crack, cracks blocks by water dirty and concrete particles, late hydration of cementitious materials and C-S-H expansion. The first and fourth types are more defunded in literature. Yang, 2009 could completely seal cracks of 50 µm of width and partiality seals cracks between 50 µm and 150 µm. Researches ((Ferrara et al 2014), (Snoeck et al, 2014), (Krelani, 2015)) provide the crack to

Non-Conventional Materials and Technologies – NOCMAT for XXI Century   Materials Research Forum LLC
Materials Research Proceedings 7 (2018) 552-559   doi: http://dx.doi.org/10.21741/9781945291838-53

study self-healing through bending tests, resulting in multiple and smaller cracks, which help the self-healing to occur. In order to study the phenomenon in a concentrated away, a single crack was provide through a notch and the crack was induced by direct tensile (Moreira, 2017). This study is based on the same principle of one crack and concentrated phenomenon.

Blast-furnace slag was used due it property of slow hydration, therefore, promoting healing in mature concretes. According to Mehta&Monteiro, 2006, finely ground blast-furnace slag is a nonmetallic self-cementing product consisting essentially of silicates and aluminosilicates of calcium and other bases that contribute to concrete in at least three aspects. The first is the slow reaction; which means that the rates of heat liberation and strength development will also be slow. Second, the reaction is lime-consuming instead of lime-producing. And the third is regarding pore size distribution of this pozzolanic concrete is smaller due to the products formation occurs filling up the capillary spaces, which improves durability and crack control.

Vegetable fibers, also used in this research, have in its majority porous systems that can absorb water and also work as a network to water percolate inside the concrete specimen. In this study the sisal fibers were added to promote a better and faster healing, nucleating the carbonation reaction and sealing the cracks.

This paper aims to evaluate the self-healing capacity of high performance microconcrete reinforced with steel and sisal fibers. In order to accomplish that, tensile tests were performed at age of 28 days to promote a crack induction until 0.2mm, controlled by LVDTs. Then, the specimens were submitted to three months of wet and dry cycles [24h/24h]. Along this time microscopic pictures were made to visualize the crack behavior and at 1+3 months a final tensile test were performed to evaluate the mechanical proprieties. Indexes were created to measure the mechanical recovery of the specimens. The index of mechanical recovery (ITR_1) is the maximum tensile strength after treatment by the maximum tensile strength at crack induction test (28 days) and the second index (ITR_2) is the post treatment tension corresponding to the respective displacement of crack induction maximum tension by crack induction maximum tension at the 28 days age test.

Methods and materials
Materials
The used matrix presented the mix design of Table 1 with Brazilian cement CPV ARI, blast furnace slag, sand, water, and a polycarboxylate superplasticizer (SP) - Glenium 51 - manufactured by BASF (bwoc). The steel fibers (BEKAERT OL13/.16mm) were added in a total of 0,64% and the sisal fibers, also with 13 mm, have a volume fraction of 0,64% by total volume. The water/binder ratio was 0.18 which leads to a supposed high number of un-hydrated grains in concrete. Therefore, this mix was designed to promote the healing capacity, by delayed hydration of cement and further hydration of slag. In order to provide a more sustainable concrete, as showed at mix design, approximately 45% of cement content was replaced by slag which is a waste of steelwork. This material has cementitious characteristics and further delayed hydration then cement which corroborate to enhance of healing properties at later ages. The used matrix is self-compacted concrete (SCC) by compressive strength at 28 days of 86 MPa, according to NBR 7215/96.

Non-Conventional Materials and Technologies – NOCMAT for XXI Century   Materials Research Forum LLC
Materials Research Proceedings 7 (2018) 552-559   doi: http://dx.doi.org/10.21741/9781945291838-53

TABLE 1 - MIX DESIGN OF INVESTIGATED HPFRCC

Constituent	Dosage [kg/m³]
Cement type V	600
Slag	500
Water	200
Superplasticizer	33
Sand 0-2 mm	980
Steel fibers	50
Sisal fibers	7

Mix Procedure

The concrete mixture was produced using a planetary mixer at a room with controlled temperature (21 ± 1°C). In order to produce a self-compacting concrete the cement and slag were mixture during 2 minutes for homogenization. Water and superplasticizer were then poured into the mixture during other 2 minutes and then the paste was mixed for 8 minutes. The sand was then added in and the mortar was mixed for 8 minutes until it became homogenous. The final procedure consisted on the addition of sisal and steel fibers. Finally, the mixture was mixed 5 more minutes to obtain the final product.

The casting was procedure in order to promote an ideal fiber distribution. A slab of 30 x 600 x 900 mm was casting by pouring the planetary mixer barrel in order to asses a continuous flow and thus guiding most of the fibers to a parallel direction to the surface slab. However, the spot which the specimen was withdrawn of the slab has influence on mechanical performance, as showed on [8] and wall effect may also happen.

After 24 hours, the slab was demolded and placed into a fog room (RH ≥95%) to moist cure for 28 days. The slab was, then, cut into smaller size specimens (30 x 60 x 300 mm) and forwarded to crack induction through direct tensile.

Three samples were randomly chosen of the slab for crack induction and 2 other specimens chosen for control. The last ones do not have crack inducted but were submitted to the same wet and dry (W&D) cycles, as control specimen.

Crack induction was performed at age of 28 days through tensile test. In order to concentrate tension and promote one crack opening, a notch of 10 mm was performed on each side in the middle zone of the specimen

Tensile crack formation and morphology

The crack was inducted in the specimen by means of displacement-controlled tensile test at age of 28 days. Crack induction was performed at Shimadzu 300 kN by a velocity of 0,01mm/min until reach a crack opening of 0.2 mm, controlled by two LVDTs placed aside of the specimen, as seen at Figure 1.

Non-Conventional Materials and Technologies – NOCMAT for XXI Century   Materials Research Forum LLC
Materials Research Proceedings **7** (2018) 552-559   doi: http://dx.doi.org/10.21741/9781945291838-53

FIGURE 1 - TENSILE TEST SETUP FOR CRACK INDUCTION WITH TWO LVDTS PLACED ASIDE

Curing conditions
After crack induction, the specimens were submitted to wet and dry cycles [24h/24h] during 3 months in a water reservoir and a room with controlled temperature and humidity ($21\pm1°C$ and $65\pm3\%$). Samples without crack induction (control) were also submitted to the same condition.

Tensile test
After 3 months of wet and dry cycles the specimens submitted to previous crack induction were re-tested under tensile forces until final collapse. The control samples were also tested until collapse, using the same method described at item 2.3.

Index of mechanical recovery
In order to evaluate the mechanical behavior after crack induction and wet and dry treatment two index were created.
   The first index (ITR_1) is calculated dividing the maximum tensile strength after treatment by the maximum tensile strength at crack induction test (28 days).
   The second index (ITR_2) is calculated dividing the post treatment tension corresponding to the respective displacement of crack induction maximum tension by crack induction maximum tension at the 28 days age test.

Thermogravimetric analysis
Thermogravimetric analysis was performed in a thermo-balance TGA/DTG/DSC simultaneous, model SDT Q600 manufactured by TA Instruments ©, using 10 mg of smashed and partially dried material spread uniformly on the aluminum bottom of the thermal analyzer. The pyrolysis process were performed at heating rates of 10°C/min in a nitrogen flow of 100 ml/min and the temperature of the furnace was programmed to rise from room temperature to 1000°C. In the analysis the derivative thermogravimetric curves (DTG) have been used to determine the weight loss and to identify the decomposition of the material at given temperatures.

Results and discussion
Figure 2 present the results obtained for direct tensile of three specimen submitted to wet and dry cycles during 3 months. The curves in gray were from the crack induction stage and the curves in black present the crack inducted specimens' behavior of tensile stress x displacement post wet and dry cycles. Displacement axis is represented by an average of two LVDTs placed aside of the specimen, according to what was exposed on item 2.3. The samples presented a similar crack opening varying between 99 μm and 131 μm. The average of first crack stress was $5,73\pm0,63$ MPa, that is equal to the average of all maximum stress because the softening behavior of these samples. They also presented an average stiffness of 41 GPa.

Non-Conventional Materials and Technologies – NOCMAT for XXI Century   Materials Research Forum LLC
Materials Research Proceedings 7 (2018) 552-559   doi: http://dx.doi.org/10.21741/9781945291838-53

After wet and dry cycles it is possible to observe that all the specimens presented a hardening behavior instead of the softening at crack induction, this may be associated to a densification on the fiber-matrix interface. In order to prove that, CT Scan are being performed. The maximum tension of the re-tested specimens were 5,33±0,35 MPa and 19 GPa of stiffness.

Values of mechanical recovery indexes are presented in Table 2. From that we can conclude that two specimens had 18% and 25% of stress gain in comparison to 28 days crack induction specimen at the same displacement and one specimen lost 24% of its maximum stress. In terms of maximum stress after the first crack (ITR_1) the lost was only about 5% for the last specimen while the other two achieve 23% and 29% more stress than the original ones.

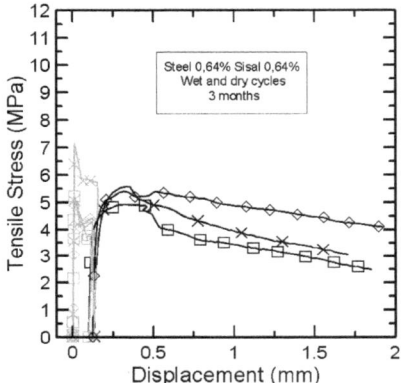

FIGURE 2 - RESULTS OF STRESS X DISPLACEMENT CURVES FOR CRACK INDUCTION (GRAY) AND AFTER 3 MONTHS OF WET AND DRY CYCLES (BLACK)

TABLE 2 - INDEX OF MECHANICAL RECOVERY AFTER 3 MONTHS OF WET AND DRY

Specimen	ITR_1	ITR_2
A	0,95	0,76
B	1,29	1,25
C	1,23	1,18

CYCLES

Figure 3 shows in orange the stress-displacement curves of two control specimens in comparison to the three samples (in grey) submitted to crack induction and healing, already presented at Figure 2. It is possible to observe that all the specimens presented a similar maximum stress. The non-cracked ones had 5,72±0,30 MPa while the healed ones achieved 5,33±0,35 MPa. It is important to highlight that the control specimens presented a softening behavior like the specimen at crack induction on 28 days. After the wet and dry cycles all healed specimen presented a hardening behavior. As already pointed out that may be explained by a change on the fiber-matrix bond. For more information see [6].

Figure 4 shows the crack induction of one of the specimens (a) and after three months of wet and dry cycles (b). It is possible to observe that the phenomenon of self-healing has occurred; sealing almost completely a crack of near 100 μm. Through the observation of the crack sealing phenomenon

Non-Conventional Materials and Technologies – NOCMAT for XXI Century   Materials Research Forum LLC
Materials Research Proceedings 7 (2018) 552-559   doi: http://dx.doi.org/10.21741/9781945291838-53

it was possible to see the presence of a white material closing the crack. Therefore a thermoanalysis was performed in order to identify that product. Figure 5 present the results, showing peaks in correspondence of the dehydroxilation of C-S-H (80-150°C), CAH and C-A-S-H from the aluminum phase (150-300°C) and a high content of the decarbonation (620-700°C). The presence of the aforementioned substances confirms that the products appearing in the cracks are products of the delayed hydration and carbonation reactions, which are the cause of the healing produced during the wet and dry stages.

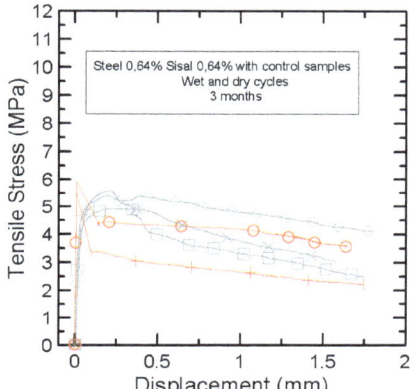

FIGURE 3 - *RESULTS OF STRESS X DISPLACEMENT CURVES FOR CONTROL SPECIMEN (ORANGE) AND AFTER 3 MONTHS OF WET AND DRY CYCLES (GRAY)*

(a)

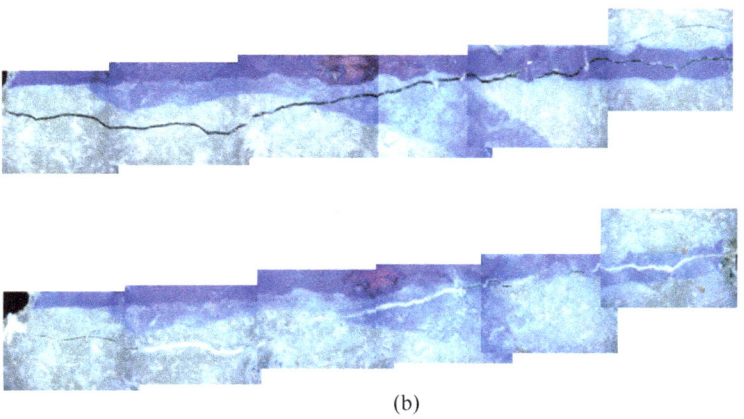

(b)

FIGURE 4 - *SELF-HEALING OF ONE SPECIMEN OF HPFRC: (A) CRACK INDUCTION AT 28 DAYS (B) CRACK ALMOST CLOSED AFTER THREE MONTHS OF WET AND DRY CYCLES*

Non-Conventional Materials and Technologies – NOCMAT for XXI Century   Materials Research Forum LLC
Materials Research Proceedings 7 (2018) 552-559   doi: http://dx.doi.org/10.21741/9781945291838-53

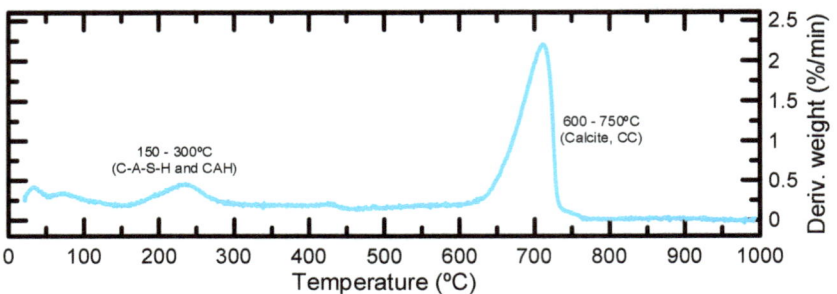

FIGURE 5 -DIFFERENTIAL THERMOGRAVIMETRIC CURVES FOR THE FILLING MATERIAL OF THE HEALED CRACK.

Figure 6 shows the influence of sisal fibers on self-healing. It can be seen that the hydration products grow on the fiber, in other words, the fiber develop a nucleation roll providing a faster sealing of the crack with calcium carbonate as the natural fibers work bridging both sides.

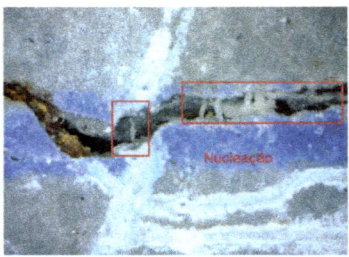

FIGURE 6 - SISAL WORKING AS A NUCLEATION AGENT OF CALCIUM CARBONATE

Conclusion

This paper shows the results of the healing capacity of HPFRC specimens submitted to tensile loads. The follow conclusions can be made from the present study:

- Two from three specimens healed with time, presenting greater tensile stress than 28 days at crack induction.
- Control specimens showed a softening behavior while the samples with crack induction presented a pseudo-hardening behavior.
- The filling material presented high amount of calcium carbonate, CAH and C-A-S-H proving that self-healing is a combination of delayed hydration and carbonation reactions;
- Cracks with width below 100 µm close almost completely.

Non-Conventional Materials and Technologies – NOCMAT for XXI Century   Materials Research Forum LLC
Materials Research Proceedings 7 (2018) 552-559   doi: http://dx.doi.org/10.21741/9781945291838-53

References

[1] NBR 6118: Concrete structures projects – Proceedings

[2] BLAISZIK, B., KRAMER, S., OLUGEBEFOLA, S., et al. "Self-healing polymers and composites", Annual Review of Materials Research, v. 40, pp. 179 – 211, 2010. https://doi.org/10.1146/annurev-matsci-070909-104532

[3] SNOECK, D., VAN TITTELBOOM, K., STEUPERAERT, S., et al. "Self-healing cementitious materials by the combination of microfibres and superabsorbent polymers", Journal of Intelligent Material Systems and Structures, v. 25, n. 1, pp. 13–24, 2014. [4]Kim and Schlangen, 2010.

[5]Toledo, R. "Composite materials reinforced with natural fibers: experimental characterization." 1997. 483 p. PhD thesis (Civil Engineering). PUC-Rio, Rio de Janeiro, 1997.

[6] MOREIRA, T. "Investigation of self-healing phenomenon in high performance fiber reinforced concrete with steel and sisal ". 2017. 158 p. Master thesis (Structure and Material in Civil Engineering) – COPPE, Federal University of Rio de Janeiro, Rio de Janeiro. 2017.

[7] WU, M., JOHANNESSON, B., GEIKER, M. "A review: Self-healing in cementitious materials and engineered cementitious composite as a self-healing material", Construction and Building Materials, v. 28, n. 1, pp. 571–583, 2012. https://doi.org/10.1016/j.conbuildmat.2011.08.086

[8] FERRARA, L., KRELANI, V., CARSANA, M. "A "fracture testing"based approach to assess crack healing of concrete with and without crystalline admixtures", Construction and Building Materials, v. 68, pp. 535–551, 2014. https://doi.org/10.1016/j.conbuildmat.2014.07.008

[9] KRELANI, V. "SELF-HEALING CAPACITY OF CEMENTITIOUS COMPOSITES". 2015. 256 p. PhD thesis (Civil Engineering) - POLITECNICO DI MILANO, Milano, 2015.

[10] Mehta, P. K., Monteiro, P. J., & Carmona Filho, A. (1994). Concrete: structure, properties materials. Pini.

Non-Conventional Materials and Technologies – NOCMAT for XXI Century   Materials Research Forum LLC
Materials Research Proceedings 7 (2018) 560-568   doi: http://dx.doi.org/10.21741/9781945291838-54

Binary Cements with High Coal Waste Contents: Properties and Behaviour Against CO_2

Moisés Frías [a,*], Lidia García [a], Laura Caneda-Martínez [a], M. Isabel Sánchez de Rojas [a], Rosario García [b], Raquel Vigil [b], Íñigo Vegas [c], Sagrario Martínez-Ramírez [d].

[a] Eduardo Torroja Institute for Construction Science (IETcc-CSIC), Madrid, Spain; mfrias@ietcc.csic.es, lidia.gm94@hotmail.com, laura.caneda@ietcc.csic.es, srojas@ietcc.csic.es

[b] Universidad Autónoma de Madrid, Geomateriales - Unidad Asociada (CSIC-UAM), Madrid, Spain ; rosario.garcia@uam.es, raquel.vigil@uam.es

[c] Tecnalia, Parque Tecnológico de Bizcaia, 28160 Derio, Spain; inigo.vegas@tecnalia.com

[d] Institute for the Structure of Matter (IEM-CSIC), Madrid, Spain; sagrario@iem.cfmac.csic.es

Abstract. It is well known that there are several scientific, technical and environmental advantages of incorporating active additions to the cement, due to its ability to react chemically with the portlandite generated during the hydration reaction of the cement particles, to give more dense and compact matrices. The coal wastes are an alternative source of obtaining future ecological pozzolans, fundamental pillar of the main strategy of the Circular Economy. As a result, an improvement in the blended cement performances was obtained. However, some aspects regarding the durability due to CO_2 reaction have not been solved. A wide range of pozzolans (silica fume, fly ash, natural pozzolan, natural metakaolinite) are found in kaolinite-based industrial wastes (paper waste, potable water treatment), which are an excellent alternative for the socio-economic development of a country. The current work presents the scientific-technical advances of coal wastes as supplementary cementing material for the manufacture of low clinker cements and their influence under the action of CO_2. The obtained results show that, after thermal activation, the products obtained have a high pozzolanic activity in the range of 550 and 650°C and the kinetics reaction is similar to other pozzolans of silico-aluminous nature. After exposure to CO_2 environment, the behaviour of the pozzolanic material is similar to cement, used as reference, when the percentage of substitution is low; while the 50% blended cements showed a rapid carbonation process.

Keywords: Coal Wastes, Blended Cement, Carbonation, Durability

Introduction

It is well established that Portland cement manufacture is an important contributor to CO_2 emissions, which accounts for approximately 5-7 % of the global carbon releases.[1] In order to increase the environmental sustainability of cement industry, several promising approaches have been, and are being studied. Nonetheless, partial clinker substitution with alternative cementitious materials seems to be the measure that shows more potential for the reduction of CO_2 emissions, since it simply reduces the clinker content on cement, significantly decreasing the amounts of both fuel and raw materials required per ton of cement produced. It must be highlighted that the cementitious materials chosen for this purpose (also known as pozzolanic materials) are usually by-products of other industries, such as fly ash or blast furnace slag, which would otherwise be treated as waste, therefore contributing to the Circular Economy and achieving an additional environmental purpose.[2-4]

The first condition for a material to be considered as a pozzolanic addition is that it must contain a high concentration of amorphous SiO_2 and Al_2O_3. Consequently, clays (especially Metakaolin) have

Non-Conventional Materials and Technologies – NOCMAT for XXI Century   Materials Research Forum LLC
Materials Research Proceedings 7 (2018) 560-568   doi: http://dx.doi.org/10.21741/9781945291838-54

been widely studied as supplementary cementitious materials. As pozzolanic activity requires a low degree of crystallinity, a heat treatment is frequently necessary, but even so, the process emits much less CO_2 than the corresponding production of clinker.[5]

Coal mining wastes can be regarded as an ecological source of pozzolanic materials, since they are mainly composed of silico-aluminous materials, among which kaolinite can be found and subsequently transformed into metakaolinite. Nonetheless, until recently, research on this topic has been scarce.[6] Some of the co-authors of this work have proven over the last years that, after thermal activation in the range of 550-650 °C, carbon wastes present aptitudes as a pozzolanic addition [7, 8] and that blended pastes and mortars elaborated with this by-product exhibit suitable physical and mechanical properties.[9, 10]

Even though a good physical-mechanical behavior is clearly a priority when it comes to cement performance, durability is also a high relevant aspect and, therefore, it should not be disregarded. Resistance to carbonation is one of the most desirable attributes for cement-based materials in terms of durability, since carbonation is considered a major cause of steel bar deterioration in reinforced concrete. During the process carbon dioxide from the atmosphere reacts with the portlandite resulting from cement hydration, hence decreasing pH and making steel vulnerable to corrosion.

Pozzolanic products also react with portlandite in cement, and therefore, addition of supplementary cementitious materials to Portland cement have a substantial impact on its behavior against CO_2. Additionally, microstructure of the blended pastes is seriously affected by pozzolanic additions, which influences carbon dioxide permeability. Consequently, it is not easy to initially predict how blended pastes behave under CO_2 exposure and experimental studies are required.[11] Hence, in this work, blended pastes containing 20 and 50% of activated carbon waste have been subjected to an accelerated carbonation test in order to assess their behavior against CO_2. For comparison purposes, ordinary Portland cement specimens have been additionally studied. The properties of the specimens, such as weight, pH and compressive strength have been surveyed throughout the study and carbonation depth has been measured by the phenolphthalein spray method, from which the rate of the process has been estimated. In addition, mineralogical and morphological studies have been conducted on the samples by means of X-ray diffraction and scanning electron microscopy.

Materials and methods

Materials

The raw coal mining waste (CW) was supplied by a Spanish Coal Group (Sociedad Anónima Hullera Vasco-Leonesa), located in the León province. The carbon waste was ground in order to achieve a particle size below 90 µm and was thermally activated at 600 °C in an electric laboratory furnace for 2 hours, since these have been previously identified as the optimal activation conditions from an economic, environmental and technical standpoint.[12] The activated products (ACW) were additionally sifted through a 90 µm sieve.

A commercial ordinary Portland cement (OPC) of type CEM I 52.5 R (according to the current European standard) was used in this work. The chemical composition of the cement and carbon waste (both activated and raw) is listed in Table 1.

TABLE 1 – Chemical composition (%) of OPC and activated (ACW) and raw coal mining waste (CW) used in this work

	OPC	CW	ACW
SiO_2	20.80	49.79	56.63
Al_2O_3	5.70	21.77	25.29
Fe_2O_3	2.89	4.07	4.64
MnO	0.03	0.08	0.08
MgO	1.89	0.64	0.77
CaO	58.99	3.84	4.20
Na_2O	0.93	0.13	0.17
SO_3	4.11	0.27	0.27
K_2O	1.36	2.74	3.09
TiO_2	0.15	1.07	1.17
P_2O_5	0.26	0.13	0.14

Regarding the mineralogy, according to earlier research studies,[13] activated coal wastes are primarily composed of quartz, calcite, mica and hematite. On the other hand, the kaolinite present in the raw carbon waste is transformed into metakaolinite.

Additionally, the water-reducing admixture Sikament®-FF has been employed in the preparation of the blended pastes.

Specimen Preparation
Paste specimens were prepared at a water/binder ratio (w/b) of 0.5 and, in the case of blended pastes, OPC was partially replaced by 20 and 50% of activated coal waste. In the case of 20 and 50 % blended pastes, 8 and 15 g of water-reducing admixture were added, respectively. The mixtures were molded into 1 cm x 1 cm x 6 cm prisms (30 per formulation) and, after being demolded, they were cured by immersion in water for 28 days. Once the curing process was finished, the specimens were dried at 40 °C for 24 hours.

Accelerated Carbonation Test
The dried specimens were stored at a relative humidity of 65 % for two weeks, in order to homogenize and stabilize the internal humidity within the cement matrix. Exposure time begins at this point, in which the specimens were exposed twice a day to a 100 % CO_2 atmosphere for 15 minutes.

Instrumental Techniques
Chemical characterization was studied by X-ray Fluorescence (XRF) using a Philips PW 1404 spectrometer instrument. pH was measured with an Mettler Toledo SevenMulti pH-meter, provided with a Mettler Toledo InLab Expert Pro pH electrode. Mineralogical analyses were carried out by X-ray diffraction (XRD) in a Bruker AXS D8 Advanced diffractometer equipped with a Cu anode. Morphological observations were obtained using a Hitachi S4800 scanning electron microscope.

Non-Conventional Materials and Technologies – NOCMAT for XXI Century   Materials Research Forum LLC
Materials Research Proceedings 7 (2018) 560-568   doi: http://dx.doi.org/10.21741/9781945291838-54

Results and discussion
Effect of carbonation on the properties of the blended pastes
In order to preliminary assess the grade of alteration of the specimens under study due to CO_2 exposure, their weight and pH were periodically measured. Figure 1 shows the evolution of these parameters over time, from which it was observed that both OPC and 20% blended pastes remain virtually unaltered during the course of the experiment. Nevertheless, pastes containing 50% ACW exhibit a substantial increase in weight, especially at early ages, which can be attributed to the capture of CO_2 and the subsequent formation of $CaCO_3$ through reaction with portlandite. This consumption of portlandite leads to lower pH values and, thus, an analogous decreasing trend was detected in this parameter for 50% substituted pastes. It must be noted that addition of pozzolanic materials in cement matrices imply a higher degree of porosity at early ages in the resulting materials, as a consequence of the dilution effect of cement clinker.[14] Therefore, particularly low carbonation resistances are usually expected during the first stages of exposure, which is in line with the behavior shown in Figure 1.

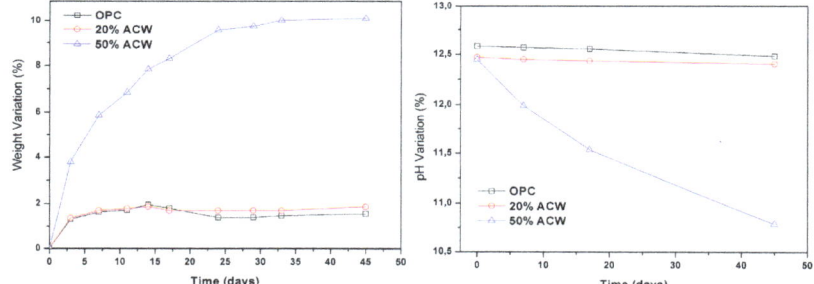

figure 1 – *variation over time of weight (left) and pH (right) measured for specimens of the different formulations under study.*

Moreover, considering that one of the most critical negative impacts of carbonation is the loss of protection of steel bars from corrosion in reinforced concrete, it is of value to identify the carbonation depth in the specimens. Given that this loss of protection arises as a consequence of the reduction of pH values, the carbonation front has been estimated through the treatment of cross sections of the specimens with a phenolphthalein pH indicator solution. According to the results of the test (Figure 2), nearly the entire cross section of the OPC specimens remains uncarbonated throughout the study, while for 20% blended pastes, only a minor carbonation front is detected by the end of the testing period (0.8 mm at 45 days). Conversely, for pastes prepared with 50% of ACW, complete carbonation of the section is perceived at 45 days.

Non-Conventional Materials and Technologies – NOCMAT for XXI Century   Materials Research Forum LLC
Materials Research Proceedings 7 (2018) 560-568   doi: http://dx.doi.org/10.21741/9781945291838-54

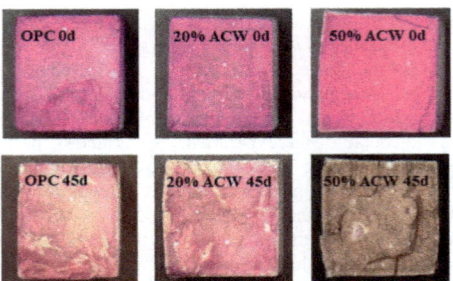

figure 2 – color change experimented by the different specimens after spraying their cross sections with phenolphthalein indicator solution.

Assuming that the diffusion rate of CO_2 is constant,[14] the relationship between carbonation depth (X_C) and time (t) can be expressed by Eqn. 1:

$$X_C = A\sqrt{t} \qquad (1)$$

Where A is an empirical constant that reflects the rate of the process. Accordingly, as it is shown in Figure 3, a linear correlation can be observed between the measured carbonation depths and the square root of the time of exposure, providing the rate of carbonation through the slope of the fitted regression line. Thus, a rate of 0.761 mm/√d has been found for 50% blended pastes, while only a value of 0.126 mm/√d was obtained in pastes containing 20% of addition. OPC samples showed negligible rates of carbonation (0.043 mm/√d).

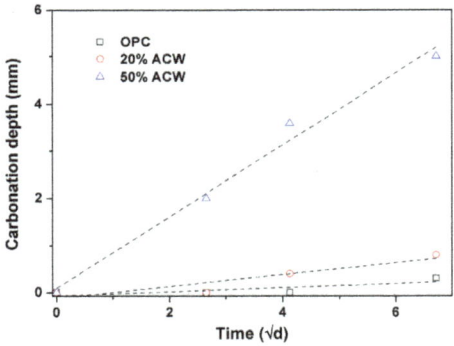

figure 3 – evolution of carbonation depth over time for the three types of formulations studied.

Effect of carbonation on the mechanical behavior
The effect of carbonation on the mechanical properties of the cement-based materials it is also an aspect of utmost importance. It is known that the $CaCO_3$ that precipitate as a result of carbonation of portlandite presents higher mechanical properties than the reactants. This, together with the reduction

of porosity related to the deposition of carbonates in the interstices of the paste, tend to improve the compressive strength of carbonated Portland cement materials.[11,15] Nonetheless, as reflected in Figure 4, the carbonated blended samples contemplated in this study evolved toward lower compressive strength values as exposure time progressed.

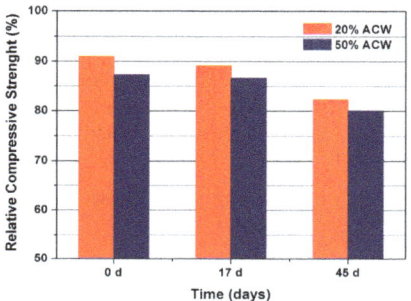

figure 4 *– development of relative compressive strength with respect to opc specimens.*

Mineralogical characterization

XRD studies have been performed on the specimens under study at 0 and 45 days making it possible to monitor the carbonation reaction by following the changes in intensity of the most important crystalline phases implicated in the process: portlandite (18.01 2θ) and calcite (29.4 2θ). The minerals detected in each sample by DRX are shown in Table 2. In both OPC and 20% blended pastes a slight reduction in the portlandite peak intensity is detected as a consequence of its reaction with CO_2. Similarly, an increase in the calcite peaks can be distinguished. The remaining crystalline phases identified in the diffractograms – ettringite (9.1 2θ) and belite (32-34 2θ) – show no significant changes.

TABLE 2 *– Minerals present in the specimens under study measured at 0 and 45 days of co_2 exposure.*

	OPC		20% ACW		50% ACW	
	0 d	45 d	0 d	45 d	0 d	45 d
Belite	x	x	x	x	x	x
Calcite	x	x	x	x	x	x
Ettringite	x	x	x	x	x	
Portlandite	x	x	x	x	x	
Quartz			x	x	x	x

On the other hand, for samples containing 50% ACW the mineralogical modifications resulting from carbonation are more than evident. Portlandite peaks are no longer detected at 45 days and a large increase in the calcite main signal is observed. Furthermore, ettringite seems to be also affected by carbonation, since there is no evidence of its presence in the sample, which indicates the high degree of carbonation reached for these samples under the aggressive environment in which they were tested.[16] In addition, an intense peak of quartz (26.6 2θ) has been identified in the blended pastes, deriving from the carbon wastes used in the preparation of the specimens.

Morphological characterization by SEM
Scanning Electron Microscopy (SEM) has been performed on the different types of specimens at the experimental starting date as well as at termination of exposure to CO_2 (45 days). For OPC samples, a dense surface is initially observed (Figure 5), in which substantial amounts of portlandite and anhydrous phases of cement are detected (primarily dicalcium silicate, being the phase that shows the slowest rate of reaction). Similar results are found once the exposure is concluded, locating portlandite only at 140 μm from the surface. Moreover, anhydrous phases are also distinguished only a few micrometers from the edge of the sample, thus bearing a great similarity to the unexposed material. Equally, 20% blended pastes present a similar aspect, also showing no signs of carbonation.

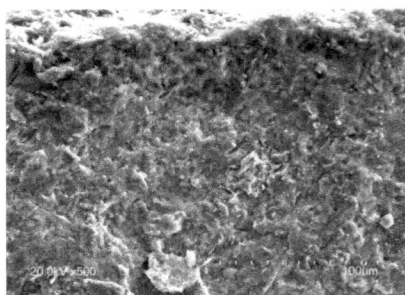

figure 5 *– morphological observation of opc pastes at 0 days (left) and 45 days of co_2 exposure (right).*

Conversely, according to SEM images, a replacement level of 50 % in the pastes leads to an uneven and highly porous surface at 0 days of exposure, presumably due to the dilution effect of cement clinker. However, at 45 days a more regular morphology is observed (Figure 6) and porosity is substantially reduced as a consequence of the deposition of calcite in the pores during carbonation. Accordingly, calcite crystals are easily found in the SEM images and no trace of portlandite is detected.

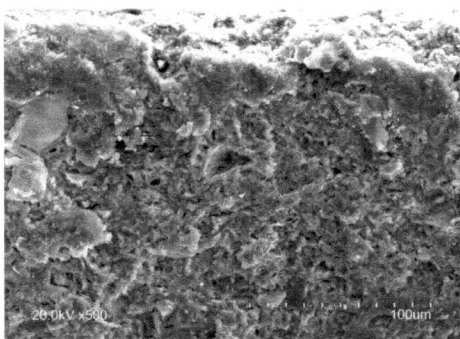

figure 6 *– sem image of 50% blended pastes after 45 days of co_2 exposure.*

Conclusions

The main conclusions that can be drawn from the results presented in this study are as follows:

1. After thermal activation, coal mining wastes are capable of reacting with portlandite in cement matrices, thus presenting pozzolanic properties. Consequently, this activation process allows for the reutilization of this type of industrial wastes and, at the same time, it can be exploit for the manufacture of eco-efficient cements, known for their lower CO_2 emissions.

2. According to the physical and mechanical characterization performed in this work, pastes containing high percentages of activated carbon waste (50%) are far more sensitive to carbonation than those prepared with lower ACW content (20%). Rates of carbonation for 50% blended pastes are one order of magnitude higher than those calculated for 20 % ACW pastes, and two orders higher compared to OPC specimens.

3. 20% blended pastes behave similarly to OPC pastes against CO_2, although minor signs of carbonation are detected after 45 days of aggressive CO_2 exposure.

4. With respect to mechanical properties, all the blended pastes analyzed in this work tend to display lower compressive strength values after CO_2 exposure.

5. This study paves the way for further studies to determine the behavior of carbon waste blended cements against CO_2 after exposure to natural environment and under the influence of atmospheric conditions.

Acknowledgements

This research has been supported by the Spanish Ministry of Economy and Competitiveness and the European Regional Development Fund (Project Ref. BIA-2015-65558-C3-1,2,3-R) (MINECO/FEDER) and as well as the Spanish Training Program, co-financed by the European Social Fund (MINECO/FSE)(ref. number: BES-2016-078454). The authors are also grateful to the Sociedad Anónima Hullera Vasco-Leonesa, SIKA (Madrid, Spain) and to the Spanish Cement Institute (IECA) for their assistance with this research.

References

[1] Benhelal E, Zahedi G, Shamsaei E, Bahadori A. Global strategies and potentials to curb CO_2 emissions in cement industry. Journal of Cleaner Production 2013;51:142-61. https://doi.org/10.1016/j.jclepro.2012.10.049

[2] Kline J, Kline C. Cement and CO_2: what is happening. IEEE Transactions on Industry Applications 2015;51:1289-94. https://doi.org/10.1109/TIA.2014.2339396

[3] Gartner E, Hirao H. A review of alternative approaches to the aeduction of CO_2 emissions associated with the manufacture of the binder phase in concrete. Cement and Concrete Research 2015;78:126-42. https://doi.org/10.1016/j.cemconres.2015.04.012

[4] Salas DA, Ramirez AD, Rodríguez CR, Petroche DM, Boero AJ, Duque-Rivera J. Environmental impacts, life cycle assessment and potential improvement measures for cement production: a literature review. Journal of Cleaner Production 2016;113:114-22. https://doi.org/10.1016/j.jclepro.2015.11.078

[5] Mohammed S. Processing, effect and reactivity assessment of artificial pozzolans obtained from clays and clay wastes: a review. Construction and Building Materials 2017;140:10-9. https://doi.org/10.1016/j.conbuildmat.2017.02.078

[6] Beltramini B, Suarez ML, Guilarducci A, Carrasco MF, Grether RO. aprovechamiento de residuos de la depuración del carbón mineral: obtención de adiciones puzolánicas para el cemento Portland. Rev. Tecnol. y Ciencia 2010;3:7-18.

[7] García Giménez R, Vigil de la Villa R, Frías M. From coal-mining waste to construction material: a study of its mineral phases. Environ.Earth Sci. 2016;75:478. https://doi.org/10.1007/s12665-016-5494-8

[8] García-Giménez R, Mencía RVV, Rubio V, Frías M. The transformation of coal-mining waste minerals in the pozzolanic reactions of cements. Minerals 2016;6(64).

[9] Vegas I, Cano M, Arribas I, Frías M, Rodríguez O. Physical-mechanical behavior of binary cements blended with thermally activated coal mining waste. Constr.Build.Mater. 2015;99:169-74. https://doi.org/10.1016/j.conbuildmat.2015.07.189

[10] Frías M, Sanchez De Rojas MI, García R, Juan Valdés A, Medina C. Effect of activated coal mining wastes on the properties of blended cement. Cem.Concr.Compos. 2012;34(5):678-83. https://doi.org/10.1016/j.cemconcomp.2012.02.006

[11] Šavija B, Luković M. Carbonation of cement paste: understanding, challenges, and opportunities. Constr.Build.Mater. 2016 8/1;117:285-301.

[12] Vigil de la Villa R, Frías M, García-Giménez R, Martínez-Ramirez S, Fernández-Carrasco L. Chemical and mineral transformations that occur in mine waste and washery rejects during pre-utilization calcination. International Journal of Coal Geology 2014;132:123-30. https://doi.org/10.1016/j.coal.2014.07.014

[13] Frías M, De La Villa RV, De Rojas MS, Medina C, Juan Valdés A. Scientific aspects of kaolinite based coal mining wastes in pozzolan/ Ca(OH)$_2$ system. J Am Ceram Soc 2012;95(1):386-91.

[14] Ashraf W. Carbonation of Cement-Based Materials: Challenges and Opportunities. Constr.Build.Mater. 2016 9/1;120:558-70.

[15] Fabbri A, Corvisier J, Schubnel A, Brunet F, Goffé B, Rimmele G, et al. Effect of carbonation on the hydro-mechanical properties of Portland cements. Cement and Concrete Research 2009;39:1156-63. https://doi.org/10.1016/j.cemconres.2009.07.028

[16] Goñi S, Gaztañaga M, Guerrero A. Role of cement type on carbonation attack. J. Mater. Res. 2002;17:1834-42. https://doi.org/10.1557/JMR.2002.0271

[17] Black L, Breen C, Yarwood J, Garbev K, Stemmermann P, Gasharova B. Structural features of C–S–H(I) and its carbonation in air—A Raman spectroscopic study. Part II: Carbonated phases. J Am Ceram Soc 2007;90(3):908-17. https://doi.org/10.1111/j.1551-2916.2006.01429.x

Non-Conventional Materials and Technologies – NOCMAT for XXI Century   Materials Research Forum LLC
Materials Research Proceedings 7 (2018) 569-581   doi: http://dx.doi.org/10.21741/9781945291838-55

The Behavior of Heat Treated *Dendrocalamus Giganteus* Bamboo Subjected to Water and Humidity

A. Azadeh[a,*], K. Ghavami[a]

[a]Pontifícia Universidade Católica do Rio de Janeiro, (PUC-Rio) Brasil

[a]arashazadeh@yahoo.com*, ghavami@puc-rio.br

Abstract. The objective of present study is investigation about the effect of heat treatment in different temperatures and time exposures on hygroscopic property of *Dendrocalamus giganteus bamboo*. The required time to stabilizing of samples from dry condition to 75% RH Relative Humidity) with different geometrical shapes and treated in 3 different temperatures studied at the first experiment. Next, the shrinkage and water absorption of bamboo segments in three Longitudinal, radial and tangential directions, due to heat treatment in 7 different temperatures from ambient temperature to 225°C during 3hrs and 24hrs is investigated. The water absorption test result shows that by increasing the heat treatment temperature and time exposure the hygroscopic property of bamboo is reduced. The heat treatment at higher temperatures can slow down the rate of the short term water absorption.

Keywords: Bamboo, Heat Treatment, Humidity, Water Absorption, *Dendrocalamus giganteus*

Introduction

Due to energy crisis started in the world since the seventies of the twentieth century, some researchers decided to change their research orientation from investigation about physical, chemical and mechanical properties of conventional materials to the nonconventional materials such as bamboo and vegetal fibers at macro, meso, micro and nano levels [1]. Among the herbaceous materials, Bamboo is an all-purpose, popular and useful substance as a non-timber forest product. The advantages of bamboo as a construction and building material are rapid growth, being renewable, affordability, workability and high mechanical resistance especially in tensile stress comparable with steel. Bamboo as an optimized and ready tubular element can be used as scaffolding, bridges, trusses, beams, columns and many other types of building materials. Bamboo as other organic, arboreal, woody and herbaceous materials, by exposing in moisture, humidity or water, reacts by absorbing water and swelling. Despite of insufficient study about bamboo [2], there are various investigations about physical behavior of different types of hard and soft woods due to water absorption with various methods of heat treatment [3][4][5][6][7][8][9][10]. The result of these studies shows the heat treatment can improve the dimensional stability and resistance to biological deterioration for different types of woods. The same as wood, by exposing the bamboo in different temperatures with different time exposure, not only the appearance and mechanical property but also the physical and hygroscopic property of bamboo such as capacity of water absorption and swelling is changed [11]. After heat treatment, wood becomes rather hydrophobic [12].

Bamboo can be used as reinforcement in different types of mortars and concrete then the destructive interaction between bamboo strips and water base mortars and concrete can be determinative. While the concrete starts to shrinkage during the curing period, bamboo strips start to swelling due to water absorption which may cause some longitudinal cracks in concrete [13][14]. Constraining the dimension changes and creating more compatibility between construction elements, makes more constructive interaction between materials. In general, more immutability and consistency and fewer changes in dimension due to water and humidity absorption is more

Non-Conventional Materials and Technologies – NOCMAT for XXI Century   Materials Research Forum LLC
Materials Research Proceedings 7 (2018) 569-581   doi: http://dx.doi.org/10.21741/9781945291838-55

acceptable for all kind of materials which are used in construction as structural or non-structural elements.

The present research tries to measure the variation of hygroscopicity, shrinkage and swelling due to heat treatment in different temperatures and time exposures by doing some tests. The first one explains the Humidity Stabilization Time (HST) test. This test is carried out to establish the required time for each different geometrical shape to reach from dry condition to the 75% RH, because all the specimens which are used for mechanical tests, before execution of the test program, should have been stabilized at the temperature and humidity 27°C ± 2°C and 70% ± 5% RH respectively [19][20][7][3]. Four distinct geometrical shapes with three different heat treatments have been considered for doing this test. The purpose of second part is to investigate about the shrinkage and swelling of bamboo specimens in three Longitudinal, radial and tangential directions, due to heat treatment in 7 different temperatures from ambient temperature to 225°C during 3hrs and 24hrs. The specimens in this investigation have been made from the middle part of five years old *Dendrocalamus giganteus bamboo (D.G. bamboo)* culm taken from Sao Paulo state. The culms had been dried on the climatic conditions for one year.

Bamboo structure
Bamboo is a natural composite with two principal components; sclerenchyma (fibers) and parenchyma (matrix) as can be seen in Figure 1a. Sclerenchyma is considered as the high resistance parallel fibers along the entire length of the bamboo and parenchyma acts as the matrix of the composite. The image prepared by stereomicroscope in Figure 1a shows the parenchyma, sclerenchyma and veins. The two images with two different scales prepared by SEM at PUC-Rio show a general view of the porous texture for parenchyma and dense texture for sclerenchyma or fibers in Figure 1b and the porosity of parenchyma in Figure 1c.

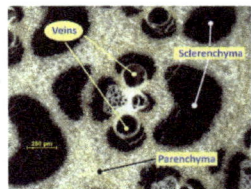 

    (a) bamboo structure   (b) general view of the sclerechyma   (c) porosity of parenchyma
                                 and parenchyma texture
FIGURE 1 - IMAGES PREPARED FROM STEREOMICROSCOPE AND SEM

Experimental procedure
The effect of heat on bamboo structure
The chemical composition of bamboo material constitutes of hemicellulose, cellulose, lignin and extractives [21]. The temperature range of decomposition of hemicellulose and cellulose is 200–380°C and 250–380°C respectively, while lignin decomposition has a wide range from 180°C up to 900°C [22][23]. TGA (thermal gravimetric analysis) is the preliminary analysis to get knowledge about the behavior of a material when heated. The TGA result for a *D.G. bamboo* with 10°C/min heating rate [24] has been done by authors in PUC-Rio shows that the weight change and degradation of bamboo starts about 150°C as can be seen in Figure 2.

The continuous red line is the weight loss percentage regarding to temperature and the blue dotted line shows the mass loss derivation. The test heating range starts from ambient temperature up to the torrefaction temperature. Torrefaction is a thermal treatment at temperatures of 200–300 °C for the

Non-Conventional Materials and Technologies – NOCMAT for XXI Century   Materials Research Forum LLC
Materials Research Proceedings 7 (2018) 569-581   doi: http://dx.doi.org/10.21741/9781945291838-55

purpose of upgrading solid biomass fuel [25]. At the thermal torrefaction range, the property of material and the mechanical resistance changes dramatically. the minimum temperature accepted for heat treatment according to different author researches is 100°C but based on wood species the minimum temperature for wood degradation begins at 100°C for pine but 130-150°C for oak [26]. The upper bound temperature due to initial tests on bamboo samples and TGA is considered 225°C and the interval between the selected temperatures considered 25°C started from 100°C. Then, the test temperatures defined 25°C (ambient temperature), 100°C, 125°C, 150°C, 175°C, 200°C and 225°C.

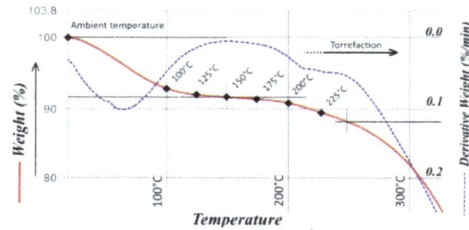

FIGURE 2 - *TGA TEST FOR D.G. BAMBOO HAS BEEN DONE IN PUC-RIO*

Humidity Stabilization Time (HST)
Each selected bamboo segment has six faces with different rate of water and humidity absorption. There are two cross sections (transversal faces), two radial faces, two tangential faces, one external and one internal as shown in Figure 3. Transversal sections contain the veins which are the surface with the highest water and moisture absorption. The longitudinal cuts which are called radial faces go through parenchyma, sclerenchyma or veins. The radial faces are not as permeable as the transversal ones. The permeability of the internal and the external faces of bamboo are less than the radial faces.

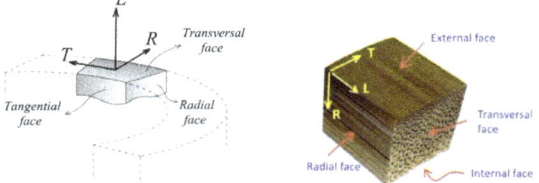

FIGURE 3 - *DIFFERENT DIRECTIONS AND FACES IN A CUBIC BLOCK OF BAMBOO*

Four different geometrical shapes as given in Table 1, selected to investigate the effect of heat and the geometrical shapes, on the moisture absorption.

TABLE 1 - *SAMPLE DIMENSIONS FOR DIFFERENT TEST TYPES*

Test type	Designation	Shape
Tensile test	T	*Bamboo thickness* — 2 mm — 100 mm
Longitudinal Shear and bending test	LS	*Bamboo thickness* — 100 mm — 15 mm
Transversal Tensile test	TT	5 mm — *Bamboo thickness* — 40 mm
Water Absorption	WA	*Bamboo thickness* — 15 mm — 15 mm

The approximate size of each shape resembles nearly the size of each mechanical test specimen. Tensile test specimen has a narrow width about 2 mm but longitudinal shear test and bending test specimen shapes have two equal sides and transversal tensile test specimens are transversally cut segments. For each test type mentioned in Table 1, three specimens with three different heat treatment, ambient temperature (25°C), 150°C and 200°C during 3 hours, have been considered.

At first, the specimens dried in 60°C for one week and then moved inside a desiccator with saturated salt water solution for HST. By utilizing saturated salt water solution, the humidity can be controlled and remains constant about 75% Rh [27][28]. By measuring the weight changes until stabilization, the HST for each shape can be achieved.

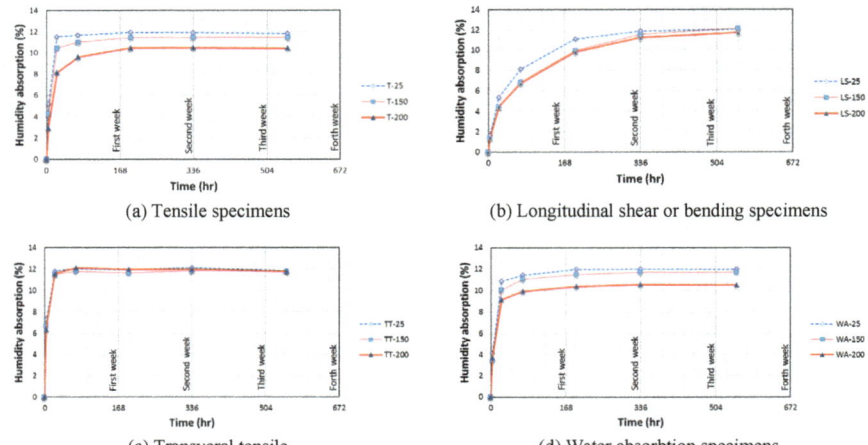

(a) Tensile specimens   (b) Longitudinal shear or bending specimens

(c) Transveral tensile   (d) Water absorbtion specimens

FIGURE 4 - *RELATION BETWEEN TIME AND HUMIDITY ABSORPTION FOR FOUR DIFFERENT TYPES OF SPECIMENS TREATED IN 25°C, 150°C AND 200°C DURING 3HRS*

A general observation to the Figure 4c and Figure 4d reveals that humidity stabilization time for transversal tensile and water absorption specimens is about two days. The humidity absorption of tensile specimens for various heat treatments is different. The T-25 and T-150 specimens approach to the asymptote line of stabilization, quicker than the T-200 also approaching to the stabilization for T-

25 is quicker than T-150 but both of them are equilibrated after three days as shown in Figure 4. In addition, the Figure 4 shows that the humidity absorption ability in heated specimen in 200°C is reduced about 20% and T-200 is equilibrated after about one week but there is an important inconsistency between transversal tensile samples with the three other shapes (WA, T and LS). The humidity absorption of all three transversal tensile specimens (TT-25, TT-150, TT-200) are the same and about 12% and there are not any significant differences between them as can be seen in Figure 4c. But in the three other groups, the specimens treated in 200°C, the hygroscopic property is decreased regarding to other samples. Proposed stabilization times for Transversal tensile (TT), water absorption (WA), tensile (T) and longitudinal shear (LS) specimens are 2 days, 3 days, one week and 3 weeks respectively.

Bamboo shrinkage and water absorption
Sample preparation for shrinkage and water absorption
Two different time exposure 3h and 24h with seven different temperatures 25°C, 100°C, 125°C, 150°C, 175°C, 200°C and 225°C are considered in the test program. The time and temperature exposure of the tests in form of matrix of samples are shown in Figure 5a. In this matrix there are 6 columns with 7 rows.

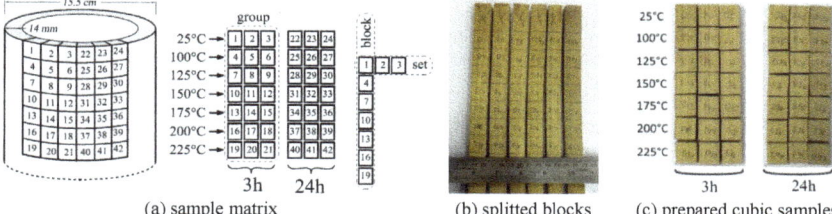

| (a) sample matrix | (b) splitted blocks | (c) prepared cubic samples |

FIGURE 5 - SAMPLE SELECTION FOR SHRINKAGE AND WATER ABSORPTION TESTS

Each group has three blocks and seven sets of samples, and each set has three specimens. This type of sample selection can guarantee the same condition in terms of property, thickness and same fiber amount and distribution in each block.

The specimens prepared from the internode of *Dendrocalamus giganteus bamboo* (*D.G.*) at the middle part of bamboo culm with about 15.5 cm diameter and 14 mm wall thickness as can be seen in Figure 5b and c. Following the thickness of block which is about 14 mm the two other dimensions in tangential and longitudinal directions considered close to 14 mm to make a cubic shape samples. By using a metal grinding polishing machine and water, the veins are cleaned and all the opposite faces are finished parallel together and left for 2 weeks to dry and adaptation to the ambient humidity (average Temp.=24°C & RH=60%). The samples put in the oven with 60°C for one week and put in desiccator with humidity absorber for 3 days then the dried weight and the dimensions of each sample is registered.

Influence of temperature on the shrinkage of samples
After measuring the dried weight and dimensions, each set of samples was put in the oven at 6 different temperatures during 3 and 24 hours. The samples after the heating process shown in Figure 6 and is not observed any color changes up to 150°C for specimens treated for 3hrs and 24hrs.

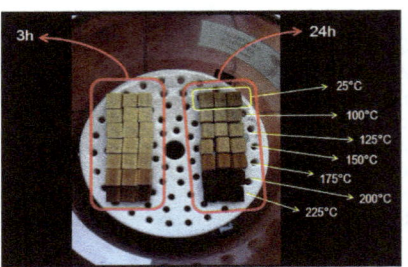

FIGURE 6 - THE COLOR CHANGES OF THE TWO GROUPS OF SPECIMENS TREATED DURING 3 AND 24 HOURS AFTER REMOVING FROM OVEN

The visual observation of the samples in Figure 6 shows that the set of samples (3 specimens) which are heated for 24 hours in 225°C became brittle. The color of the specimens changed to dark, close to charcoal with remarkable shrinkage which can be observed visually. After removing from oven and stabilization in desiccator, the weight and dimensions are registered. The weight loss due to increment of temperature for 3 hours and 24 hours has been shown in Figure 7. As it can be seen in Figure 7, the fast rate of weight changing started from 175°C for the samples treated in 24 hours and 200°C for the samples treated in 3 hours. Also after 150°C, the weight loss of samples treated in 175°C, 200°C and 225°C for 3 hours are nearly equal to the samples treated in 150°C, 175°C and 200°C respectively but treated for 24 hours. It can prove that the effect of time exposure for heat treatment after about 150°C is considerable.

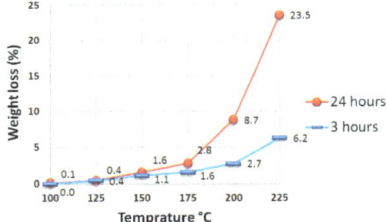

FIGURE 7 - THE WEIGHT LOSS FOR TWO DIFFERENT TIME EXPOSURE AND SIX DIFFERENT TEMPERATURE

The dimension changes in three longitudinal, tangential and radial directions are presented in the Figure 8. As it can be seen for temperatures more than 200°C, the shrinkage of both groups of specimens in oven for 3h and 24h, increases remarkably. The dimension measurement has been done by a caliper rule (precision: ±0.02 mm) at the middle of the specimens. The shrinkage for both group of samples treated in 3 and 24 hours up to 150°C is less than one percent but the fast rate of dimension variations for radial and tangential directions started after 150°C. The longitudinal shrinkage for both groups of samples treated in 3 and 24 hours up to 225°C is less than 0.3% and is negligible regarding to dimension changes in tangential and radial directions.

Non-Conventional Materials and Technologies – NOCMAT for XXI Century   Materials Research Forum LLC
Materials Research Proceedings 7 (2018) 569-581   doi: http://dx.doi.org/10.21741/9781945291838-55

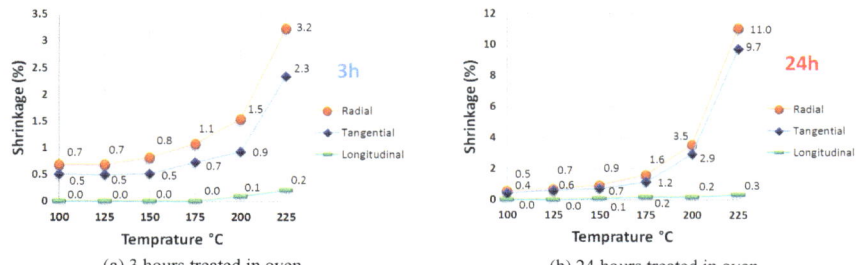

(a) 3 hours treated in oven   (b) 24 hours treated in oven

FIGURE 8- *SAMPLE SHRINKAGE IN THREE RADIAL, TANGENTIAL AND LONGITUDINAL DIRECTIONS*

Short term and long term water absorption test

The specimens used for the shrinkage test in section 2.3.2 are used to establish water absorption of bamboo. The dimension and weight of dried samples before putting in the distilled water with ambient temperature are registered. The water absorption at the start of the procedure is high, and then the registration time intervals are selected short and increased progressively. The selected registration time intervals are 6 hours, one day, three days, one week, two and half weeks and finally four weeks. At the first 6 hours or short-term water absorption for the specimens treated in 3h and 24h, by increasing the heat exposure time and temperature, the rate of water absorption is decreased as can be seen in Figure 9. For the specimens treated in 3 hours the water absorption decreases about 20% and for the specimens treated within 24 hours this value can reduce from 45% at 25°C to about 10% at 225°C.

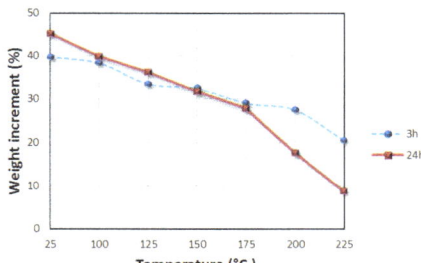

FIGURE 9 - *SHORT TERM WEIGHT INCREMENT AFTER 6 HOURS WATER ABSORPTION FOR THE TWO GROUPS OF SPECIMENTS SUBJECTED TO HEAT FOR 3 AND 24 HOURS*

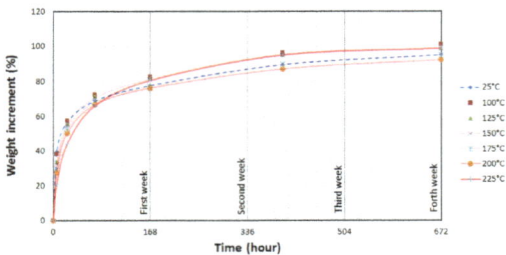

FIGURE 10 – LONG-TERM WEIGHT INCREMENT OF HEAT TREATED BAMBOO SAMPLES DUE TO WATER ABSROPTION - 3 HOURS SUBJECTED TO HEAT

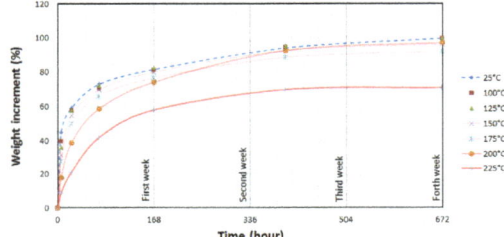

FIGURE 11 - LONG TERM WEIGHT INCREMENT OF BAMBOO SAMPLES DUE TO WATER ABSROPTION - 24 HOURS SUBJECTED TO HEAT

The results of the natural water absorption for the heat-treated samples during 4 weeks (28 days) which is considered as long-term water absorption have been presented in Figure 10 and Figure 11. The set of samples treated in 225°C for 24 hours are close to charcoal and their water absorption behavior are not the same as other sets as can be seen in Figure 11. It means the hygroscopic property of samples has been changed remarkably. It can be seen at the short term period, there are differences between the water absorption of the samples but in long term all of the samples absorb more or less the same amount of water from 90% to 100% of the their dry initial weight. The only exception is the set of samples treated in 225°C for 24 which absorb about 70% of the dry weight of samples. It shows at the beginning of torrefaction temperature, the hygroscopic property of bamboo changes dramatically.

After four weeks, the weight and dimension of specimens are registered due to natural water absorption (long term water absorption in ambient pressure). By using a vacuum chamber (forced water absorption), the air pressure is lowered to 0.3 bar to remove any possible air bubble trapped inside the samples. Then the immersed samples released in atmosphere pressure to absorb more water. This procedure is repeated to stabilizing the weight of the samples. The weight and dimensions are measured after this procedure.

Non-Conventional Materials and Technologies – NOCMAT for XXI Century   Materials Research Forum LLC
Materials Research Proceedings 7 (2018) 569-581   doi: http://dx.doi.org/10.21741/9781945291838-55

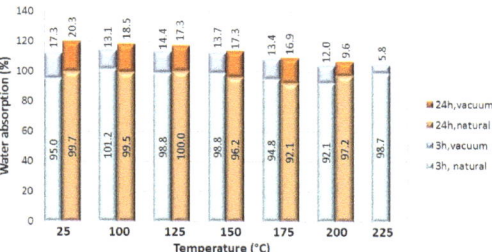

FIGURE 12 - WATER ABSORPTION FOR THE SPECIMENS SUBJECTED TO HEAT FOR 3 HRS AND 24 HOURS AFTER FOUR WEEKS NATURAL WATER ABSOPTION AND AFTER USING VACUUM

Water absorption of the immersed specimens subjected to heat in 3hrs and 24hrs, for natural water absorption and after using vacuum presented in Figure 12. The specimens subjected to heat at 225°C for 24 hours removed from the results because of the excessive shrinkage and lack of hygroscopic property regarding to the other samples. In average, the capacity of water absorption for both heat treatment in 3 and 24 hours by increasing the temperature is reduced up to 15%. In addition, the effectiveness of using vacuum for the samples treated in higher temperatures is reduced.

Water absorption model
Bamboo is a material with two rapid and slow water soaking phases [29]. The test results in previous section show that the rate of water absorption for all of the samples at the early hours is considerable. After a few hours up to about one day the specimens absorb 40% to 60% of the total water absorption named rapid phase of water absorption, then little by little this rate is reduced and nearly 2 or 3 days later, the samples absorb water with a slow rate up to saturation. Then the water absorption has three steps: a) rapid sorption b) transition sorption c) slow sorption. The steps have been shown in Figure 13.

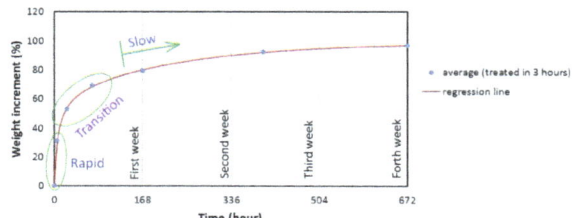

FIGURE 13 - THREE DIFFERENT PHASES OF WATER ABSOPTION: RAPID, TANSITION AND SLOW SORPTIONS

There are some different methods to model the water absorption. The simplest model is a non-exponential empirical modal defined by Peleg [30] which used for one phase water absorption [31]. The two phases water absorption equation defined by Mohsenin [32] is not converge to an asymptote by time increment due to have a linear component in the equation. Also the results in Figure 10 and Figure 11 cannot be fitted well by using the Peleg and Mohsenin models as fitting curve equations. The suggested model in this investigation presented in Eqn.1 has four coefficients, which can be achieved by curve fitting.

$$W_a(t) = R_1 \left(1 - e^{-t/R_2}\right) + S_1 \left(1 - e^{-t/S_2}\right) \tag{1}$$

In equation.1, $W_a(t)$ is the water absorption percentage consists of two rapid and slow water absorption components. Each component has two coefficients; R_1 and R_2 for rapid and S_1 and S_2 for slow water absorption, which are presented in Table 3 and Table 4.

TABLE 2 - *THE COEFFICIENTS OF WATER ABSORPTION CURVE FITTING EQUATION FOR SPECIMENS TREATED IN 3 HOURS*

Coefficient	Temperature (°C)						
	25	100	125	150	175	200	225
R_1	53.50	53.44	51.60	51.18	52.46	49.48	43.80
R_2	4.801	5.250	6.433	6.693	8.385	8.528	12.30
S_1	41.92	48.01	47.35	47.92	43.07	42.40	55.31
S_2	188.7	170.6	153.6	163.0	191.0	164.14	150.6

TABLE 3 - *THE COEFFICIENTS OF WATER ABSORPTION CURVE FITTING EQUATION FOR SPECIMENS TREATED IN 24 HOURS*

Coefficient	Temperature (°C)						
	25	100	125	150	175	200	225
R_1	55.34	52.90	53.65	51.35	46.47	37.32	14.05
R_2	3.872	4.715	5.891	7.001	7.581	11.97	15.27
S_1	44.55	47.22	46.51	45.16	45.58	61.50	56.96
S_2	179.1	178.0	165.9	159.4	147.2	179.4	110.9

The two coefficients of R_2 and S_2 present the approximate time of the start and the end of transition period respectively. Summation of R_1 and S_1 shows the total capacity of water absorption. The coefficient of determination for the samples treated in 3 and 24 hours are between 0.9982 up to 0.9992 and shows the high dependency between the experiments with curve fitting equation. It should be considered it is valid for natural water absorption without using the vacuum to forced water absorption.

Dimension changes due to water absorption
The rate of dimension changes due to water absorption in three L-R-T directions is different. Bamboo expansion in longitudinal direction caused due water absorption is less than 0.5% but in tangential direction varies from 5% to 7% and in radial direction has the maximum expansion within 8% to 12%. Dimension changes for the specimen subjected to heat for 3 hours and 24 hours shown in Figure 14a and Figure 14b. By increasing the heat and time exposure, the bamboo expansion due to water absorption is reduced.

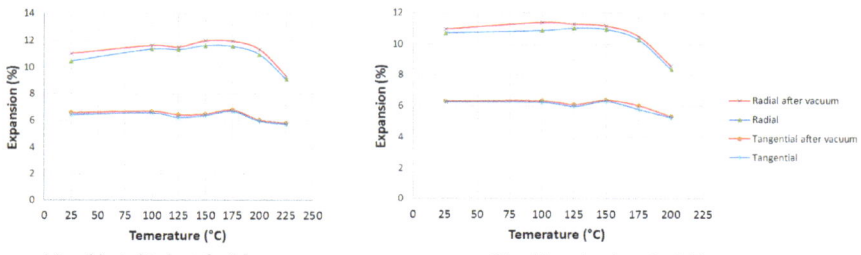

(a) subjected to heat for 3 hours   (b) subjected to heat for 24 hours

FIGURE 14 - *DIMENSIONAL CHANGES DUE TO WATER ABSORPTION IN RADIAL AND TANGENTIAL DIRECTIONS*

As a result, by increasing the heat and time exposure, the capacity of water absorption as well as the bamboo expansion is reduced. For the samples treated within 3 hours, up to 175°C is not observed any remarkable changes in expansion due to water absorption. The same happens for the sample treated within 24 hours but the reduction of expansion starts about 25°C less in 150°C.

Conclusion
Bamboo, the same as the other woody products has reaction with humidity and water absorption. The results of humidity stabilization time shows that the rate of humidity absorption is mainly related to geometrical shape and element size then to heat treatment. In general, heat treatment of bamboo slows down the rate of water and humidity absorption. The result of short term water absorption shows that the rate of water absorption reduces considerably from 7.5% per hour for not treated specimens to about 3% per hour for the specimens subjected to 200°C for 24 hours. The weight increment due to natural water soaking for all the specimens treated in different temperatures is around 100%. After natural water soaking, the vacuum chamber is used in order to absorb the complementary water and fully saturation; the not treated specimens absorb more water up to 20% but for the samples treated in more temperatures, the ability of water absorption in vacuum is limited to 6%. In general, heat treatment up to 200°C can reduce the bamboo water absorption capacity up to 15% relative to non-treated samples. The effect of heat treatment in 175°C for 24 hours is the same as heat treatment in 200°C for 3 hours and both can reduce significantly the expansion in radial and tangential directions caused by water absorption. Bamboo water absorption follows a three phases pattern; rapid, transition and slow for all treated and not treated samples (for cubic size about 15 mm each side).

Acknowledgement
The authors would like to thank all those colleagues who contributed to present investigation and special thanks to Prof. Holmer from University of Sao Paulo (FZEA-USP) and Alexandr Zhemchuzhnikov for his assistance in laboratory. Also, the financial support given by Brazilian financing agency CNPq and CAPES is appreciated.

References
[1] K. Ghavami, "Introduction to nonconventional materials and an historic retrospective of the field," in *Nonconventional and Vernacular Construction Materials: Characterisation, Properties and Applications*, Woodhead Publishing, 2016, pp. 37–61. https://doi.org/10.1016/B978-0-08-100038-0.00002-0

Non-Conventional Materials and Technologies – NOCMAT for XXI Century   Materials Research Forum LLC
Materials Research Proceedings **7** (2018) 569-581   doi: http://dx.doi.org/10.21741/9781945291838-55

[2] C. Eng, J. E. Waning, and Y. T. Mekonnen, "An Experimental Investigation of the Effects of Moisture Content on the Mechanical Properties of Bamboo and Cane."

[3] ISO 13061-3, "Physical and mechanical properties of wood — Test methods for small clear wood specimens." 2014.

[4] S. Metsä-Kortelainen, T. Antikainen, and P. Viitaniemi, "The water absorption of sapwood and heartwood of Scots pine and Norway spruce heat-treated at 170 C, 190 C, 210 C and 230 C," *Eur. J. Wood Wood Prod.*, vol. 64, no. 3, pp. 192–197, 2006. https://doi.org/10.1007/s00107-005-0063-y

[5] J. Chrastil, "Quantitation of water absorption, swelling, and drying of biological materials. Soaking of rice and soaking and drying of wood," *J. Agric. Food Chem.*, vol. 37, no. 4, pp. 965–968, 1989. https://doi.org/10.1021/jf00088a030

[6] S. Malasri, A. Pourhashemi, P. Aung, M. Harvey, and R. Moats, "Water Absorption of Wooden Pallets," *Int. J. Appl.*, vol. 2, no. 9, 2012.

[7] C. C. Gerhards, "Effect of moisture content and temperature on the mechanical properties of wood: an analysis of immediate effects," *Wood Fiber Sci.*, vol. 14, no. 1, pp. 4–36, 2007.

[8] O. Unsal, S. Korkut, and C. Atik, "The effect of heat treatment on some properties and colour in eucalyptus (Eucalyptus camaldulensis Dehn.) wood," *Maderas. Cienc. y Tecnol.*, vol. 5, no. 2, pp. 145–152, 2003. https://doi.org/10.4067/S0718-221X2003000200006

[9] D. S. Korkut and S. Hiziroglu, "Experimental test of heat treatment effect on physical properties of red oak (Quercus falcate Michx.) and southern pine (Pinus taeda L.)," *Materials (Basel).*, vol. 7, no. 11, pp. 7314–7323, 2014. https://doi.org/10.3390/ma7117314

[10] W. Scheiding, M. Direske, and M. Zauer, "Water absorption of untreated and thermally modified sapwood and heartwood of Pinus sylvestris," *Eur. J. Wood Wood Prod.*, vol. 74, no. 4, pp. 585–589, 2016. https://doi.org/10.1007/s00107-016-1044-z

[11] Y. M. Zhang, Y. L. Yu, and W. J. Yu, "Effect of thermal treatment on the physical and mechanical properties of phyllostachys pubescen bamboo," *Eur. J. Wood Wood Prod.*, vol. 71, no. 1, pp. 61–67, 2012. https://doi.org/10.1007/s00107-012-0643-6

[12] W. Shangguan, Y. Gong, R. Zhao, and H. Ren, "Effects of heat treatment on the properties of bamboo scrimber," *J. Wood Sci.*, vol. 62, no. 5, pp. 383–391, 2016. https://doi.org/10.1007/s10086-016-1574-3

[13] K. Ghavami, "Bamboo as reinforcement in structural concrete elements," *Cem. Concr. Compos.*, 2005. https://doi.org/10.1016/j.cemconcomp.2004.06.002

[14] A. Azadeh and H. H. Kazemi, "New Approaches to Bond between Bamboo and Concrete," *Key Eng. Mater.*, vol. 600, pp. 69–77, 2014. https://doi.org/10.4028/www.scientific.net/KEM.600.69

[15] S. Amada, T. Munekata, Y. Nagase, Y. Ichikawa, A. Kirigai, and Y. Zhifei, "The Mechanical Structures of Bamboos in Viewpoint of Functionally Gradient and Composite Materials," *J. Compos. Mater.*, vol. 30, no. 7, pp. 800–819, May 1996. https://doi.org/10.1177/002199839603000703

[16] T. Tan, N. Rahbar, S. M. Allameh, S. Kwofie, D. Dissmore, K. Ghavami, and W. O. Soboyejo, "Mechanical properties of functionally graded hierarchical bamboo structures," *Acta Biomater.*, vol. 7, no. 10, pp. 3796–3803, 2011. https://doi.org/10.1016/j.actbio.2011.06.008

Non-Conventional Materials and Technologies – NOCMAT for XXI Century   Materials Research Forum LLC
Materials Research Proceedings **7** (2018) 569-581   doi: http://dx.doi.org/10.21741/9781945291838-55

[17] K. Ghavami, C. de S. Rodrigues, and S. Paciornik, "Bamboo: functionally graded composite material," *Asian J. Civ. Eng*, vol. 4, no. 1, pp. 1–10, 2003.

[18] W. Liese, "The anatomy of bamboo culms. International Network for Bamboo and Rattan (INBAR)," 1998.

[19] R. M. Rowell and R. L. Youngs, "Dimensional Stabilization of Wood in Use.," DTIC Document, 1981. https://doi.org/10.2737/FPL-RN-243

[20] A. C. ICBO, "162: Acceptance criteria for structural bamboo," *ICBO Eval. Serv. Ltd., California, USA*, 2000.

[21] H. P. S. Abdul Khalil, I. U. H. Bhat, M. Jawaid, A. Zaidon, D. Hermawan, and Y. S. Hadi, "Bamboo fibre reinforced biocomposites: A review," *Mater. Des.*, vol. 42, pp. 353–368, Dec. 2012. https://doi.org/10.1016/j.matdes.2012.06.015

[22] W. Jin, K. Singh, and J. Zondlo, "Pyrolysis Kinetics of Physical Components of Wood and Wood-Polymers Using Isoconversion Method," *Agriculture*, 2013. .

[23] L. Gašparovič, Z. Koreňová, and Ľ. Jelemenský, "Kinetic study of wood chips decomposition by TGA," *Chem. Pap.*, vol. 64, no. 2, pp. 174–181, Jan. 2010.

[24] Z. Jiang, Z. Liu, B. Fei, Z. Cai, and Y. Yu, "The pyrolysis characteristics of moso bamboo," *J. Anal. Appl. Pyrolysis*, vol. 94, pp. 48–52, 2012. https://doi.org/10.1016/j.jaap.2011.10.010

[25] W.-H. Chen, J. Peng, and X. T. Bi, "A state-of-the-art review of biomass torrefaction, densification and applications," *Renew. Sustain. Energy Rev.*, vol. 44, pp. 847–866, 2015. https://doi.org/10.1016/j.rser.2014.12.039

[26] B. M. Esteves and H. M. Pereira, "Wood modification by heat treatment: A review," *BioResources*, vol. 4, no. 1, pp. 370–404, 2009.

[27] D. E. Goldberg, *Fundamentals of chemistry*. McGraw-Hill, 2006.

[28] M. J. Wheeler, S. Russi, M. G. Bowler, and M. W. Bowler, "Measurement of the equilibrium relative humidity for common precipitant concentrations: facilitating controlled dehydration experiments," *Acta Crystallogr. Sect. F Struct. Biol. Cryst. Commun.*, vol. 68, no. 1, pp. 111–114, 2012. https://doi.org/10.1107/S1744309111054029

[29] J. Khazaei, "Water absorption characteristics of three wood varieties," *Cercet. Agron. în Mold.*, 2008.

[30] M. PELEG, "An empirical model for the description of moisture sorption curves," *J. Food Sci.*, vol. 53, no. 4, pp. 1216–1217, 1988. https://doi.org/10.1111/j.1365-2621.1988.tb13565.x

[31] N. Abu-Ghannam and B. McKenna, "The application of Peleg's equation to model water absorption during the soaking of red kidney beans (Phaseolus vulgaris L.)," *J. Food Eng.*, vol. 32, no. 4, pp. 391–401, 1997. https://doi.org/10.1016/S0260-8774(97)00034-4

[32] N. N. Mohsenin, "Physical properties of plant and animial materials. Vol. 1. Structure, physical characterisitics and mechanical properties.," *Phys. Prop. plant animial Mater. Vol. 1. Struct. Phys. characterisitics Mech. Prop.*, vol. 1, 1970.

Non-Conventional Materials and Technologies – NOCMAT for XXI Century   Materials Research Forum LLC
Materials Research Proceedings 7 (2018) 582-591   doi: http://dx.doi.org/10.21741/9781945291838-56

Mechanical Behavior of Natural Composites Subjected to Condensation Process

A.C. Bolaños *, M. L. Sánchez, J. D. Caicedo

* Universidad Militar Nueva Granada

u1101968@unimilitar.edu.co

Abstract. The use of composite materials made with natural materials has increased in several areas of engineering, mainly in the elaboration of structural elements, in which the material can be subjected to the action of atmospheric agents for long periods, can generate changes in the microstructure of the material and consequently affect its mechanical behavior. Among environmental phenomena that can cause a rapid degradation of natural composite, the exposure of the material to humidity conditions is emphasized. This work presents the determination of the physical properties of composites made with fibers and vegetal resin after being exposed to cycles of wetting under conditions of variable temperature. For the experimental tests, composite panels with Guadua and Arundo Dónax fibers immersed in a vegetal-based polyurethane were elaborated. For the modification of the surface of the fibers, an alkaline treatment was established. The degradation of the material was obtained using an accelerated weathering tester. To simulate the effect of moisture a condensation process was employed. The results were compared with experimental results obtained for panels that have not undergone degradation process.

Keywords: Natural Composites, Moisture Content, Mechanical Properties, Condensation Process

Introduction

Nowadays the use of natural fibers to replace the use of synthetic fibers constantly employed as reinforcement of elaborated materials with polymeric resins has grown in multiple industry sectors [1]. Due to high costs of typical raw material used on engineering field, the future sustainability of natural reservoir and the thread to the environment currently is essential to use natural materials, which are *environment friendly* and at the same time useful to development, production and fabrication of compounds [2-4].

The bio-polymeric matrix compounds and the natural fibers reinforcements are highly useful due to their environmental advantages [5-8]. Bamboo fibers have mechanical properties favorable to some uses; nevertheless, they can be more fragile than other natural fibers because of the extra content of lignin that covers its surface [9]. The Cane (Arundo Donax) is considered an invasive species, it grows at the river and it is estimated that this species is between the hundred (100) species of the grasses family (Poaceae) more invasive of the world [10-12]. For that reason it is important to consider its exploitation because at the same time the application of the bamboo contributes to the environment preservation.

This research job is based on the analysis of the humidity effect on the physical properties of boards (panels) made out of vegetal origin materials. Two agglomerate panels were made in order to develop this work, the first one with *Guadua Angustifolia Kunth* Fiber and the other with Cane *Arundo Donax* Fiber.; using vegetal resin based on *higuerilla*, used as binder. To simulate board's materials degradation, they were exposed to an accelerated aging process

Materials and methods

The panels were made with randomly distributed fibers and with Higuerilla-based vegetal resin as matrix of the compound. The fibers used as reinforcement were preserved following the specifications of Colombian Technical Standard NTC 5301-07 [14]. Next, the methodology used for

Non-Conventional Materials and Technologies – NOCMAT for XXI Century   Materials Research Forum LLC
Materials Research Proceedings **7** (2018) 1-6   doi: http://dx.doi.org/10.21741/9781945291838-1

the selection of the material, the extraction of the fibers, the elaboration of the slabs and the process of aging is presented.

Material selection

The fibers used were selected from the overcrowding of bamboo culms of the species Guadua Angustifolia Kunth aged between three (3) and five (5) and from the upper culms of Arundo Dónax cane with an average of eighteen (18) months. The physical properties of the fibers: moisture content, absorbability and relative density were determined according to the recommendations of ASTM D 4442-07 [15]. The results are presented in Table 1.

TABLE 1 - PHYSICAL PROPERTIES

	Physical feature	Average values
Guadua Angustifolia Kunth	Density (g/cm^3)	0.70 ± 0.20
	Absorption percentage (%)	45.00 ± 2.40
	Humidity content (%)	9.00 ± 4.00
Cane Arundo Dónax	Density (g/cm^3)	0.61 ± 0.20
	Absorption percentage (%)	57.99 ± 2.40
	Humidity content (%)	8.37 ± 4.00

Extraction and fibers treatment

This process was done through a chemical-mechanical combined method. In order to break the bonds of the lignin molecules joining the fibers, an alkaline treatment using a chemical solution based on sodium hydroxide in a concentration of 5% was used. The selected culms were immersed in the solution for 48 hours at a controlled temperature, after which time they were washed with plenty of distilled water and oven dried at the point to achieve constant mass.

A machine composed by a pair of rollers with teeth coupled to a motor, which rotate crushes and separate the fibers in order to make easier the manual extraction of the fibers, this step allows us to extract the fibers one by one and then elaborate the material compound. The crusher used and the obtained fibers are shown in figures A and B respectively.

(A)   (B)

FIGURE 1 – EXTRACTION OF THE FIBERS (A) CRUSHER, (B) VEGETAL FIBERS

Non-Conventional Materials and Technologies – NOCMAT for XXI Century   Materials Research Forum LLC
Materials Research Proceedings 7 (2018) 1-6   doi: http://dx.doi.org/10.21741/9781945291838-1

Inside the panels fabrication process was used a 300 x 300 mm mold, with adjusted thickness of 10 mm as shown in the figure 2. The materials involved were mixed in a 70% fiber and 30% resin ratio.

FIGURE 2 – MOLD.

The resin used presents the physical features exposed on the Table 2. This data is provided by the supplier: color, density, status, boiling point and the shape of each one of the two components that conforms it.

TABLE 2 – RESINA FEATURES (DATA PROVIDED BY SUPPLIER)

Feature	Component A	Component B
Physical status	Liquid	Liquid
Color	Brown	Amber
Consistency	Viscous	Viscous
Boiling point	190°C	313°C
Relative density	1.25gr/cm3	0.98gr/cm3

The volumes formula provided by the supplier was used in order to calculate the resin volume, by the equations 1 and 2.

$$V_R = V_A + V_B \quad (1)$$
$$V_A = 1.5V_B \quad (2)$$

Where:

- V_R= Resin total volume
- V_A= Volume of component A
- V_B= Volume of componente B

In the Table 3 is presented the mass distribution for the two components used for the elaboration of the material compound to guarantee the appropriated ratio on components percentage.

TABLE 3 – *DISTRIBUTION OF THE RESIN*

Material	Density (gr/cm3)	Volume (cm3)	Mass (gr)
"A"	1.25	86.4	108
"B"	0.98	129.6	127

The fabrication of the panels was carried through the method of manual molding and compaction by compression. This process consists on applying fibers layers randomly on a 30 x 30 x 5 cm Steel mold as shown in Figure 3. At the same time, the impregnation of the resin to the fibers was made, and all the process was done manually. Then, a previous compaction was made to erase lumps and to fill vacuum (little empty spaces) and so, to guarantee an uniform distribution.

FIGURE 3 – *MOLD WITH FIBER-RESIN MIX.*

For the compaction, the hydraulic press shown in Figure 4 was used, this in order to obtain the desired and uniform thickness; this press has a test frame of high rigidity conformed by four columns with capacity of 100 tons, a digital indicator, a pressure transducer of 10,000 psi and a manual pump. The mold is placed and the compaction load is applied, this load was between 10 and 15 tons. For adequate compaction and uniform, a time of 18 hours was required for each composite material.

FIGURE 4 – *HIDRAULYC PRESS*

Non-Conventional Materials and Technologies – NOCMAT for XXI Century   Materials Research Forum LLC
Materials Research Proceedings **7** (2018) 1-6   doi: http://dx.doi.org/10.21741/9781945291838-1

Once after the 18 hours the panel is unmolded and is left to be cured during 24 hours under controlled temperature, searching the resin to reach it maximum resistance. After the curing process, the two compound panels shown in figures 5 and 6 were obtained.

FIGURE 5 – GUADUA CURED BOARD

FIGURE 6 – CANE CURED BOARD

Once the panels already cured, they were cutting in order to have test tubes of each one of the compound materials shown in figure 7, their dimension were 30 x 5 cm.

FIGURE 7 – TEST TUBES

To assess the degradation of the material due to its exposure to moisture over prolonged periods of time, its aging was simulated. This procedure was performed on the Accelerated Weathering Tester machine shown in Figure 8, according to ASTM standard G154-16 with cycle B, which has a programmed irradiation intensity of 0.71 W / m2 for 8 hours, water spray for 15 minutes and a condensation phase for 3 hours and 45 minutes. In total, the composite was exposed to a total of 48 hours in cycles of 12 hours for each of the phases [16].

Non-Conventional Materials and Technologies – NOCMAT for XXI Century   Materials Research Forum LLC
Materials Research Proceedings 7 (2018) 1-6   doi: http://dx.doi.org/10.21741/9781945291838-1

FIGURE 8 – ACCELERATED AGING CHAMBER

After the two 12 hours cycles were finished, the two slabs were taken out and a comparative visual analysis was carried out in order to verify the same physical properties initially given to the fibers.

Physical properties
Humidity (moisture) content
The humidity (moisture) content is performed according to ASTM D 4442-07 [15], this standard suggests recommendations and presents a method of oven drying. The requirements for performing the test are shown below:

1- Two (2) samples were taken for each material, test material (Guadua, Arundo Donax).
2- Controlled temperature oven at 105 ± 2 °C.
3- Precision balance of at least 0.01 gr resolution.
4- Relative humidity of the samples has to be less than 70 %.
5- To guarantee a high quality of the data, verification for non-moisture accumulation of the oven is needed, it has to be checked before the test, same for the ventilation.

The initial mass of the sample is determined. Then it is put in the oven for 24 hours at a constant temperature of 105 ° C. After 24 hours, the sample is removed and its dry weight is determined. Having these two weights, and using the Eqn.3, the value of the moisture content can be calculated:

$$\%CH = \frac{A-B}{B} * 100 \qquad (3)$$

Where:
- %CH is the moisture content percentage
- A is the mass sample at natural humidity condition
- B is the mass sample at oven drying condition

Absorption capability and effective absorption
The effective absorption capacity was performed following the recommendations of ASTM D 570-14 [17]. First, the initial mass is determined in its natural state and then immersed in distilled water for 24 hours. After that time, the sample is removed from the water and dried with an absorbent towel of superficial form; and is weighed and the mass is obtained in a dry saturated condition. Finally, the sample is carried to the oven until it has a constant mass, and the dry mass is determined in the oven. The effective absorption capability was determined using equation 4.

Non-Conventional Materials and Technologies – NOCMAT for XXI Century   Materials Research Forum LLC
Materials Research Proceedings 7 (2018) 1-6   doi: http://dx.doi.org/10.21741/9781945291838-1

The material is then baked in the oven for 24 hours and its mass is determined in dry condition and the absorption capability is determined with Eqn.5.

$$\%AE = \frac{P_i - P_{sss}}{P_i} * 100 \qquad (4)$$

$$\%A = \frac{P_s - P_{sss}}{P_s} * 100 \qquad (5)$$

Where:

- %A: Absorption capability
- %AE: Effective absorption capability
- Pi: Mass of the fibers beam at natural humidity condition
- Ps: Mass at oven drying condition
- Psss: Mass if the fibers beam at dry satured condition

Results and analysis
Qualitative analysis
Visual analysis for test tubes before and after being exposed to the accelerated aging chamber in a 48 hours cycle is presented in the following sections.

Guadua Angustifolia Kunth
Before: In figure 9, it is shown that the material is uniform without any irregularity besides it has a brown color because of the natural color of Guadua and resin.

FIGURE 9 – TEST TUBES OF GUADUA COMPOUND MATERIAL BEFORE BEING IN THE AGING CHAMBER.

After: Later the test tube to be exposed on the aging chamber, it seems that the tube had a loss of agglomeration and uniformity, besides of color change as well as irregularities at its edges, as shown in figure 10.

Non-Conventional Materials and Technologies – NOCMAT for XXI Century   Materials Research Forum LLC
Materials Research Proceedings **7** (2018) 1-6   doi: http://dx.doi.org/10.21741/9781945291838-1

FIGURE 10 – TEST TUBES OF GUADUA COMPOUND MATERIAL AFTER BEING IN THE AGING CHAMBER.

Caña Arundo Donax

Before: In Figure 11 it is observed that the material is uniform and with no irregularities, in addition it has a soft but strong color.

FIGURE 11 – CANE COMPOUND MATERIAL TEST TUBES BEFORE THE AGING ACCELERATED MACHINE EXPOSITION.

After: In Figure 12 it is observed that the material after being exposed presented loss of agglomeration and uniformity, likewise it has a more radiant color and irregularities on its edges.

FIGURE 12 – CANE COMPOUND MATERIAL TEST TUBES AFTER THE AGING ACCELERATED MACHINE EXPOSITION.

Physical features

The results obtained in relation to the physical properties of each of the compound materials, after the accelerated aging chamber exposition process as follows: moisture (humidity) content, in the saturated state at two (2) hours and in the saturated state at twenty-four (24) hours.

TABLE 4 – PHYSICAL FEATURES

Physical features	GUADUA	ARUNDO DONAX
Humidity content (%CH)	6.85 ± 0,51	4.22 ± 0,51
Effective absorption (%A) 2 Hours	2.56 ± 1,41	2.70 ± 1,41
Effective absorption (%A) 24 Hours	8.93 ± 2,20	10.53 ± 2,20
Absorption capability (%CA) 2 Hours	9.59 ± 1,46	7.04 ± 1,46
Absorption capability (%CA) 24 Hours	19.61 ± 2,61	21.15 ± 2,61
Swelling (%H) 2 Hours	2.66 ± 0,01	0.18 ± 0,01
Swelling (%H) 24 Hours	12.98 ± 0,02	8.69 ± 0,02

It was realized that the compound material with Guadua natural fiber and Common Cane compound presented a moisture content of 6.85% and 4.22% respectively after two hours, Guadua Angustifolia Kunth content being higher. After 24 hours, moisture contents of 9.8% and 9.62%, for Guadua and Common Cane respectively were observed, being Guadua a bit higher difference. This allows us to say that Guadua Angustifolia Kunth had a higher water content.

Regarding the percentage of effective absorption after 2 hours the results were similar for the two panels, considering the Arundo Donax cane slightly larger. The percentage of effective Absorption at 24 hours of Arundo Donax cane board had the highest effective absorption of 10.53% while that of Guadua Angustifolia Kunth obtained 8.93%.

Conclusions

It was observed that for both of compounded materials (Guadua and Arundo Donax Cane), their capabilities of absorption and effective absorption increased considerably from 2 hours to 24 hours, e.g., there was an increase of double or more than its initial value.

The swelling represent the change that the sample presents in terms of thickness variations, as it can be viewed on Table 4, specimens demonstrate a considerable increase of the dimensions from 2 to 24 hours, since the 2 hours have a swelling of 2.66% and of the 24 hours of 12.98% for the Guadua and a swelling of 0.18% at 2 hours and 24 hours an 8.69% for the Arundo Donax. Between both of the compounded materials, Guadua Angustifolia Kunth is the one which has higher swelling, this can be explained because it presents a moisture content higher than Arundo Donax's one.

Acknowlegment

This paper is a derivative product of the project (INV-ING-2392) financed by the Vice-rectory of Research of Universidad Militar Nueva Granada-validity (2017).

Referencias

[1]   Chand N, Fahim M. Tribol Nat Fibre Polym Compos. Hard-cover ed.; 2008

[2]   Jawaid M, Abdul Khalil HPS. Cellulosic/synthetic fibre reinforced polymer hybrid composites: a review. Carbohyd Polym 2011; 86:1–18. https://doi.org/10.1016/j.carbpol.2011.04.043

[3]   Puglia D, Biagiotti J, Kenny JM. A review on natural fibre-based composites – part II. J Nat Fibres 2005; 1:23 – 65. https://doi.org/10.1300/J395v01n03_03

[4]   J.D. Badia, T. Kittikorn, E. Strömberg, L. Santonja-Blasco, A. Martínez-Felipe, A. Ribes-Greus, M. Ek, S. Karlsson, Water absorption and hydrothermal performance of PHBV/sisal

Non-Conventional Materials and Technologies – NOCMAT for XXI Century   Materials Research Forum LLC
Materials Research Proceedings 7 (2018) 1-6   doi: http://dx.doi.org/10.21741/9781945291838-1

biocomposites, elservier, Polymer Degradation and Stability 108 (2014) 166 – 174. https://doi.org/10.1016/j.polymdegradstab.2014.04.012

[5] H. Ventura, J. Claramunt, M.A. Rodríguez-Perez, M. Ardanuy, Effects of hydrothermal aging on the water uptake and tensile properties of PHB/flax fabric biocomposites, elservier, Polymer Degradation and Stability (2017) 129 – 138. https://doi.org/10.1016/j.polymdegradstab.2017.06.003

[6] O. Faruk, A. Bledzki, H. Fink, M. Sain, Biocomposites reinforced with natural fibers: 2000 - 2010, Prog. Polym. Sci. 37 (2012) 1552 – 1596. https://doi.org/10.1016/j.progpolymsci.2012.04.003

[7] N.M. Barkoula, S.K. Garkhail, T. Peijs, Biodegradable composites based on flax/ polyhydroxybutyrate and its copolymer with hydroxyvalerate, Ind. Crops Prod. 31 (2010) 34 – 42. https://doi.org/10.1016/j.indcrop.2009.08.005

[8] Bledzki, Composites reinforced with cellulose based fibres, Prog. Polym. Sci. 24 (2002) 221 – 274. https://doi.org/10.1016/S0079-6700(98)00018-5

[9] H.P.S. Abdul Khalil, I.U.H. Bhat, M. Jawaid, A. Zaidon, D. Hermawan, Y.S. Hadi, Bamboo fibre reinforced biocomposites: A review, elservier, Materials and Design 42 (2012) 353 – 368. https://doi.org/10.1016/j.matdes.2012.06.015

[10] Deltoro Torró, V., Jiménez Ruiz, J. & Vilán Fragueiro X. M. Bases para el manejo y control de Arundo donax L. (Caña común). Colección Manuales Técnicos de Biodiversidad, 4. Conselleria d'Infraestructures, Territori i Medi Ambient. Generalitat Valenciana. (2012) Valencia.

[11] Zedler, J. B. Causes and consequences of invasive plants in wetlands: opportunities, opportunists, and outcomes. Critical Reviews in Plant Sciences (2004) 23(5): 431-452. https://doi.org/10.1080/07352680490514673

[12] Lowe, S., Browne, M., Boudjelas, S. & M. De Poorter. 100 of the World's Worst Invasive Alien Species: A selection from the Global Invasive Species Database. Published by The Invasive Species Specialist Group (ISSG) a specialist group of the Species Survival Commission (SSC) of the World Conservation Union (IUCN). (2000) 9 -13.

[13] Norma Técnica Colombiana NTC 5301 – 07 "Preservación y secado del culmo de guadua angustifolia Kunth".

[14] NORMA ASTM D 4442 – 07 Standard Test Methods for Direct Moisture Content Measurement of Wood and Wood-Base Materials.

[15] NORMA ASTM G154 – 16 Standard Practice for Operating Fluorescent Ultraviolet (UV) Lamp Apparatus for Exposure of Nonmetallic Materials.

[16] ASTM D570 – 98 Standard Test Method for Water Absorption of Plastics.

Non-Conventional Materials and Technologies – NOCMAT for XXI Century   Materials Research Forum LLC
Materials Research Proceedings **7** (2018) 592-603   doi: http://dx.doi.org/10.21741/9781945291838-57

Electrochemical Differences on the Passivity State of Reinforced Concrete for Two Concrete Design Methods

J.A. Briceño-Mena[a], M. Balancán-Zapata[a], P. Castro-Borges[a,*]

[a] Centro de Investigación y de Estudios Avanzados del IPN, Unidad Mérida, km. 6 Ant. Carr. a Progreso, 97310 Mérida, Yucatán, México

mena1703@hotmail.com, mercedes.balancan@cinvestav.mx, pcastro@cinvestav.mx*

Abstract. This work aims to detect, through electrochemical information, differences in the passivity state of reinforced concrete specimens manufactured with two different concrete design methods: the method one (M1) considers the ultimate resistance of the element and the method two (M2) considers the accommodation of the aggregates in the final element. There is a diversity of methods in codes and standards that provide information about the proportion of aggregates and cement content, based on the materials conditions and mechanical resistance of the final element. However, there is a few information about durability issues of reinforced concrete structures (RC) made in accordance with those methods and their differences during the passivation state. Small 150x150x300 mm beams with 6 rods embedded at 15, 20 and 30 mm and two water/cement (w/c) ratios of 0.65 and 0.45 for each method were made with Composed Portland Cement (CPC 30R). They were exposed in a tropical marine environment at 50 m from the sea in the north of Yucatan Peninsula, during a period of 700 days (passivity state). Corrosion rate, corrosion potential, resistivity, and internal conditions (relative humidity and temperature) measurements were performed periodically. In general, the beams designed with M2 behaved similar during the passivity state regardless the depth of cover, the w/c ratio and also the M1 method perse.

Keywords: Concrete design method, corrosion, electrochemical monitoring, concrete internal conditions, marine environment.

Introduction

Concrete structures must be designed, constructed, and supervised based on codes, manuals, and standards to ensure their durability and safety and, therefore, to comply with the service life and conditions for which they were designed. However, the increasing demands of the loading conditions of reinforced concrete structures, as well as the aggressive actions of nature, increasingly require the development of new technologies for the design of concrete elements [1]. In addition, they must be more resistant and durable, and contribute to reducing pollutants in the manufacture of building materials (CO_2 emission, 5% of global emissions) [2]. For these reasons, it is increasingly common to find new studies and proposals that generate elements with a greater performance by using additives, new design techniques, new materials, among others.

Concrete is mainly composed of three elements, cement, aggregates, and water. The fine (sand) and coarse (gravel) aggregate make up between 60 and 75% of the volume of a concrete mix. Therefore, it is not surprising that selecting them is an important step in designing durable and high-performance concretes. Because of their size, distribution, and physical properties, to mention a few, aggregates, play an important role for the strength, durability, and economy of concrete elements [3]. The aggregates come from different sources: natural as rivers or lagoons, or manufactured as crushed stone, and can be identified or related according to their resistive characteristic [4]. However, even among aggregates of the same source, there are differences that generate very different concretes [5]. This work is focused to compare two design methods in terms of the effect of concrete cover, water/cement ratio, and internal conditions of temperature and relative humidity that could affect the passivity state over the tested period of reinforced concrete specimens exposed to the natural marine environment of Yucatan Peninsula.

Experimental procedure

Specimen design

Concrete beams (24) were designed using two methods: Method 1 (M1), based on the ultimate strength of the elements [6] and Method 2 (M2) considering the arrangement of the aggregates in concrete elements [7]. Of the total beams, 12 were designed through M1 and 12 through M2. Of the specimens of each design method, six were constructed for the electrochemical monitoring and six for the chemical monitoring.

The dosages used in each design can be seen in Table 1. Among coincidences, both methods consider the internal characteristics of the aggregates (sand and gravel) as the moisture content and the absorption. In addition to the above, the M2 considers as part of the design process an estimate of the volume of the aggregates to obtain an appropriate relation between them that provides the least amount of the percentage of voids. If the quantities of the materials used are analyzed, it is possible to observe that in M2 the cement content is lower than M1 and the content of aggregates (sand and gravel) is higher. Specifically, to have the least content of voids between the coarse particles, the amount of sand used is 7% higher in the M2 for both w/c ratios.

TABLE 1 - DOSAGE OF MATERIALS ACCORDING TO DESIGN[1]

Material	Method 1		Method 2	
	w/c 0.45	w/c 0.65	w/c 0.45	w/c 0.65
Cement	408	277	404	274
Sand	681	739	728	790
Gravel	800	868	758	823
Water	187	180	185	177

Portland Cement (CPC) Type 1 and crushed aggregates of limestone from the region were used. The two w/c ratios used, 0.45 (low) and 0.65 (high), are considered to produce concrete of good and poor quality, respectively. The geometry of the specimens consisted of a cross-section of 150 mm by 150 mm and length of 300 mm. As reinforcement, six bars of carbon steel of 9.525 mm ($^3/_8$") diameter and 500 mm length were used. Prior to the placement of the reinforcements in the molds, the study area was delimited with an epoxy and insulation tape leaving the center section (150 mm) of the reinforcement uncoated. Reinforcement's ends were completely isolated with the application of an epoxy and subsequently placed in a plastic tube. For the electrical contact between the reinforcing steel and the measuring equipment, the ends of each reinforcement were drilled, and a copper wire was placed (figure 1).

---

[1] The amounts provided are kg/m^3 and corrected for moisture content of the aggregates

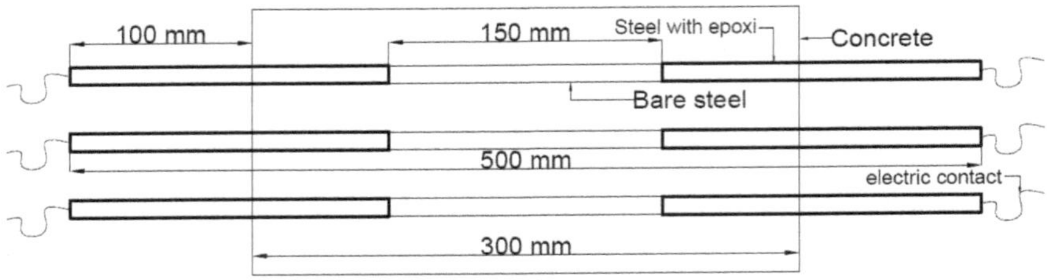

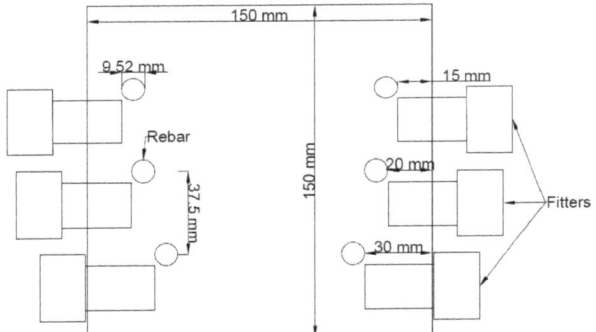

FIGURE 1 - SPECIMEN SCHEME

The steel was located along the specimens at three different depths, 15, 20 and 30 mm. Six fitters, to follow the internal conditions of relative humidity and temperature of the concrete, were installed at the depth of the reinforcements (figure 2).

FIGURE 2 - REINFORCEMENT AND FITTERS POSITION.

The curing process was performed for a period of 28 days in a controlled humidity chamber at 100% RH.

Environmental exposure
The specimens (beams) were exposed at the marine station of the Research and Advanced Studies Center of the National Polytechnic Institute, Merida Unit, (CINVESTAV-Merida) located in the municipality of Telchac Puerto, at the north of Yucatan, Mexico, at 50 m from the seashore and at 0.95 m from the ground height. The reinforcement located at 30 mm was positioned at the bottom of the beam. The orientation of the beams had the influence of the prevailing winds, VP, from the northeast (figure 3).

Non-Conventional Materials and Technologies – NOCMAT for XXI Century   Materials Research Forum LLC
Materials Research Proceedings 7 (2018) 592-603   doi: http://dx.doi.org/10.21741/9781945291838-57

FIGURE 3- SPECIMENS EXPOSURE.

Electrochemical analysis

Of the 12 specimens, six had reinforcement for electrochemical measurements (corrosion rate, i_{corr}, corrosion potential, E_{corr}, and electrical resistivity, ρ) and six were left without reinforcement for chemical analysis (ion chloride, Cl⁻, and carbonation). However, in this work, we will focus only on the electrochemical aspect. The electrochemical measurements were carried out for a period of approximately 700 days in approximate 90-day periods with a portable corrosion meter (figure 4) which, through a confinement system, can provide E_{corr} in mV; the electric resistance of the concrete (R_s), from which the ρ is obtained in kΩ-cm, through equation (1).

$$\rho = 2\,R_S D \qquad (1)$$

Where: D is the counter electrode diameter of the corrosimeter. The i_{corr} is obtained in $\mu A/cm^2$ [8], by using the linear polarization resistance technique, R_p, through equations (2) and (3).

$$R_P = \frac{\Delta E}{\Delta I} \qquad (2)$$

$$i_{corr} = \frac{B}{R_P} \qquad (3)$$

The literature establishes two criteria for the constant B, 26 mV in presence of corrosion and 52 mV in absence of corrosion. The corrosimeter uses the value of 26 mV, which has been shown to be an acceptable value for evaluations under site conditions [9].

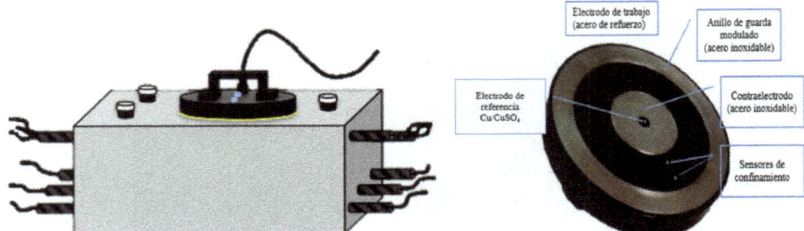

FIGURE 4 - EQUIPMENT FOR ELECTROCHEMICAL MONITORING.

Internal measurements of relative humidity (RH) and Temperature (T).
With the use of a portable probe (figure 5) the relative humidity and internal temperature of the concrete were measured to identify the influence of them on the depassivation of the reinforcing steel or its impact at early ages on the electrochemical behavior of reinforcements according to different design methods.

FIGURE 5 - RELATIVE HUMIDITY AND TEMPERATURE MONITORING.

Results and discussion
The evaluation criteria of the electrochemical parameters obtained in this work can be consulted in Mexican standards [10]–[12].

The results of E_{corr}, figures 6a and 6b, show a tendency towards the uncertainty of having or not corrosion in a few reinforcements. Figure 6a shows a different behavior than expected since the reinforcement located at greater depth, that is, to which the aggressive agents would take longer to arrive, is presenting more negative values in the potential which would indicate a depassivation of the reinforcement, as well as the concrete with low w/c ratio in which is embedded. This w/c ratio corresponds to dense concretes with lower moisture content which leads to a slow mobility of Cl⁻ and therefore, to a reduction in the accumulation of Cl⁻ on the surface of reinforcing steel [13]. If the reinforcement located in the concrete with high w/c ratio and at a depth of 15 mm is observed, there is a slight tendency to the depassivation as expected. On the other hand, in the specimens designed with M2, figure 6b, it can be observed that most of the reinforcements present a tendency to have corrosion uncertainty at the end of the tested period. It is observed that the potentials exhibit a difference between 0.45 and 0.65 w/c ratios. The reinforcements from the M2 present the most negative potential at the end of the tested period. However, once in the zone of uncertainty, this behavior tends to disappear and the reinforcements at 15 and 20 mm have similar electrochemical activity.

Non-Conventional Materials and Technologies – NOCMAT for XXI Century   Materials Research Forum LLC
Materials Research Proceedings **7** (2018) 592-603   doi: http://dx.doi.org/10.21741/9781945291838-57

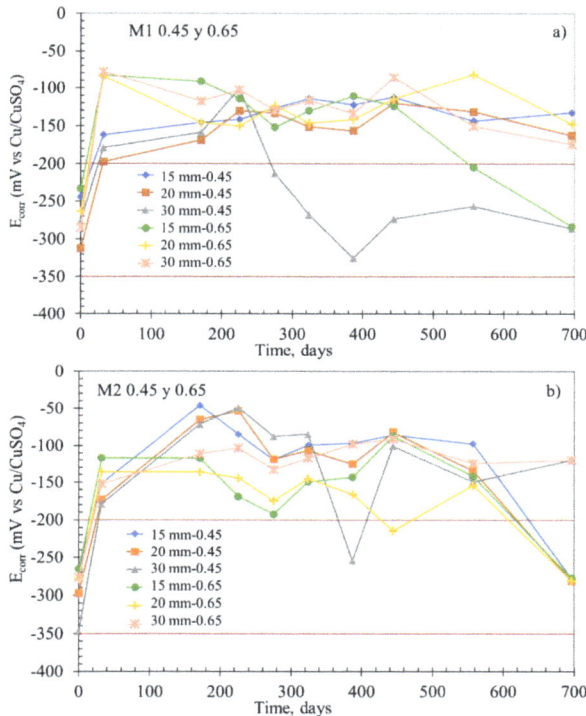

FIGURE 6 - *ECORR VS TIME.*

Figures 7a and 7b present the results obtained from the i_{corr} analysis. As in E_{corr}, the results obtained from the specimens of M1 show dispersion in the data, unlike what happens in concrete with M2, which, in the first 300 days, presents a similar behavior in all the reinforcements. The two previous behaviors can be attributed to the differences in method design since, as mentioned above, M2 has a lower content of cement and more aggregates, which causes more tortuosity and less penetration of the aggressive agents towards the reinforcement. However, the behavior of the reinforcement at 20 mm in the concrete of M2, for a 0.65 w/c ratio, would imply that, despite having a good behavior at initial periods, it undergoes a possible depassivation after a certain time, and this may be due to the characteristics of the aggregates with which the mixture was made. In the region of the Yucatan Peninsula, aggregates come from limestone and could influence in the durability of the concrete due to its high porosity [14].

Non-Conventional Materials and Technologies – NOCMAT for XXI Century   Materials Research Forum LLC
Materials Research Proceedings 7 (2018) 592-603   doi: http://dx.doi.org/10.21741/9781945291838-57

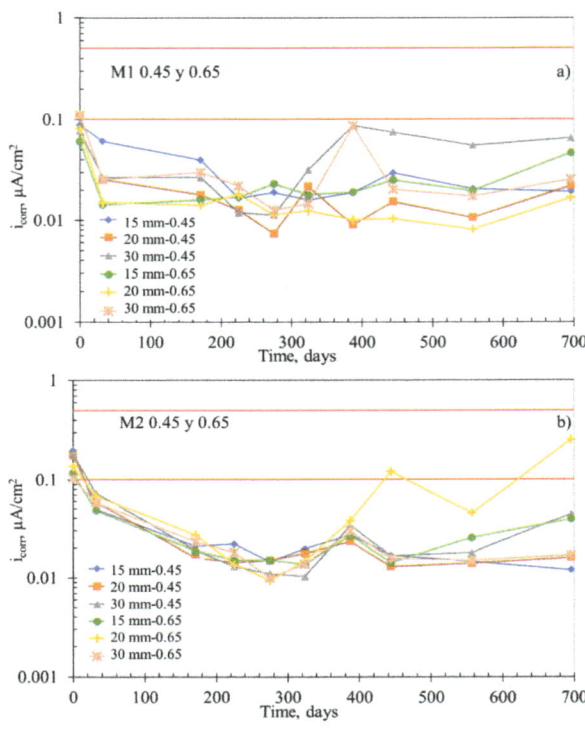

FIGURE 7 - ICORR VS TIME.

On the other hand, figures 8a and 8b present the behavior of the apparent resistivity of the concrete design through the two methods where, at least at 700 days of exposure, there is no influence of the resistivity of the concrete in the electrochemical behavior of reinforcements regardless of the design method.

Non-Conventional Materials and Technologies – NOCMAT for XXI Century   Materials Research Forum LLC
Materials Research Proceedings **7** (2018) 592-603   doi: http://dx.doi.org/10.21741/9781945291838-57

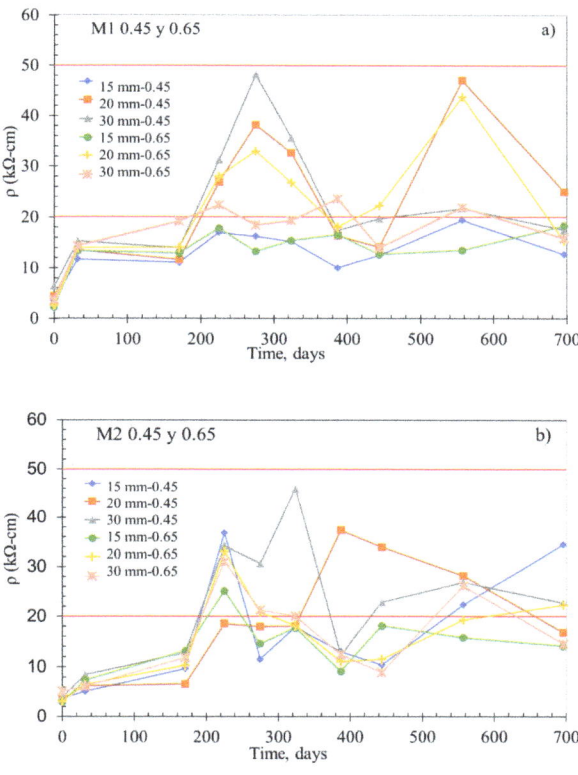

FIGURE 8 - *P VS TIME.*

Figures 9 and 10 show the fluctuations of temperature and relative humidity with respect to time, which can be attributed to the different internal microstructure that present the aggregates where, although they come from the same place, this would not be a restriction to have a different porosity or fragility between them. The above could be manifested in their coefficients of thermal expansion which could cause microcracks [5]. However, it can be observed that, regardless the depths of the reinforcements, or the design method used (figures 9a and 9b), there is no clear difference between 15, 20 or 30 mm.

Non-Conventional Materials and Technologies – NOCMAT for XXI Century   Materials Research Forum LLC
Materials Research Proceedings 7 (2018) 592-603   doi: http://dx.doi.org/10.21741/9781945291838-57

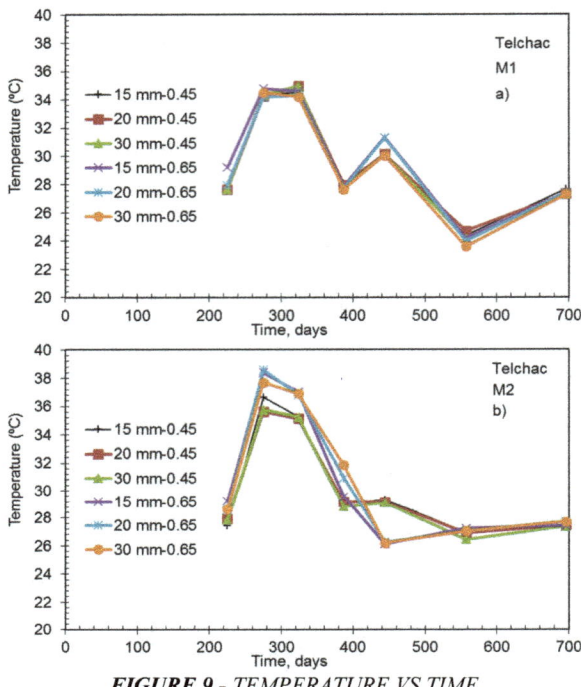

FIGURE 9 - *TEMPERATURE VS TIME.*

Figure 10 shows the average temperatures for each design method and w/c ratio and, as in the previous figure, there is no marked difference between both methods that could justify [15], [16] the observed electrochemical evolution in figures 7-9.

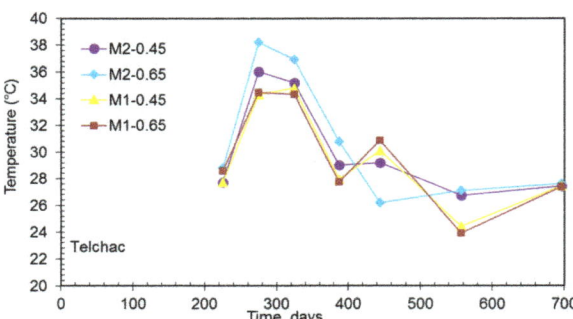

FIGURE 10 - *AVERAGE TEMPERATURE VS TIME.*

Non-Conventional Materials and Technologies – NOCMAT for XXI Century   Materials Research Forum LLC
Materials Research Proceedings **7** (2018) 592-603   doi: http://dx.doi.org/10.21741/9781945291838-57

In general, the behavior of RH follows an identical pattern for both methods, except for some specific cases, like beams with 0.45 w/c ratio with M1 at 315 days of exposure, where a high RH was observed (figure 11a). In the same way, beams with 0.45 w/c ratio from M2 showed a low RH at 550 days of exposure (figure 11b). Both cases could be considered as isolated and collected information may not be enough to find relations among them. This is confirmed by the information provided in Figure 12.

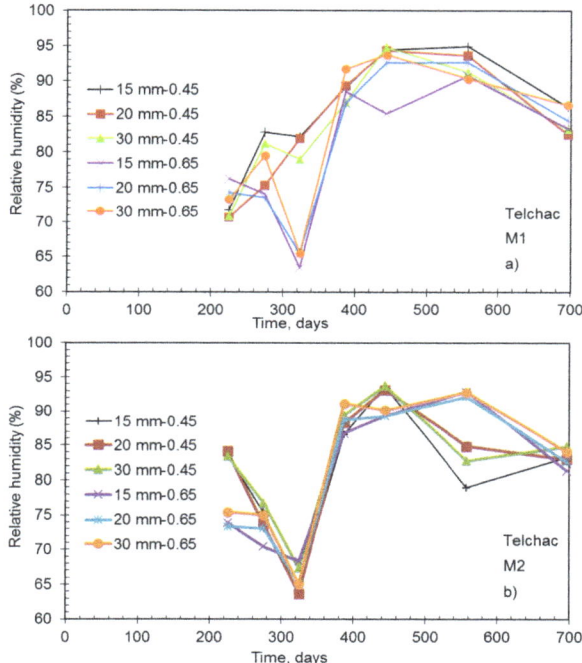

FIGURE 11 - *RELATIVE HUMIDITY VS TIME.*

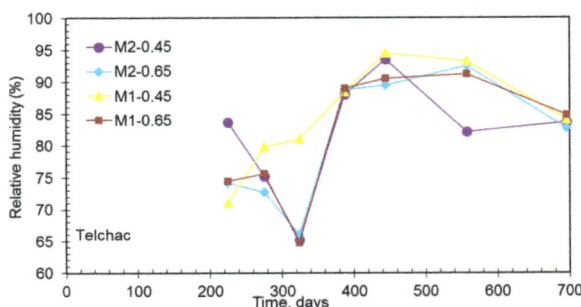

FIGURE 12 - *AVERAGE RELATIVE HUMIDITY VS TIME.*

Conclusions

During the passivity state corresponding to the tested period (two years) the electrochemical techniques (i_{corr} and E_{corr}) were able to show slightly different tendencies between both concrete design methods regardless resistivity and internal measurements of relative humidity and temperature. A different treatment of the electrochemical information could provide clearer tendencies. Under these conditions, and in general, the beams designed with M2 behaved similar during both years of exposure regardless the depth of cover, the w/c ratio and also the M1 method perse. The most unfavorable condition for M2 (0.65 w/c ratio, cover depth of 2 cm) was the 2-cm case that could be attributed to several reasons like non-visual detected microcracking or double penetration effect of the aggressive agents.

References

[1] R. D. J. M. Steenbergen, C. P. W. Geurts, and C. A. Van Bentum, "Climate change and its impact on structural safety," *HERON*, vol. 54, no. 1, pp. 3–36, 2009.

[2] E. Worrell, L. Price, N. Martin, C. Hendriks, and L. O. Meida, "Carbon Dioxide Emission from the Global Cement Industry," *Annu. Rev. Energy Environ.*, vol. 26, pp. 303–329, 2001. https://doi.org/10.1146/annurev.energy.26.1.303

[3] Portland Cement Association, "http://www.cement.org/cement-concrete-applications/concrete-materials/aggregates," 2017. .

[4] U. V. Tilak and A. N. Reddy, "Effect of Different Percentage Replacement of Weathered Aggregate in Place of Normal Aggregate on Young's Modulus of Concrete to Produce High strength and Flexible/Ductile Concrete for use in Railway Concrete Sleepers," *SSRG Int. J. Civ. Eng.*, vol. 2, no. 11, pp. 24–29, 2015. https://doi.org/10.14445/23488352/IJCE-V2I11P105

[5] G. E. Troxell, H. E. Davis, and J. W. Kelly, *Composition and Properties of Concrete*. 1968.

[6] ACI211-91, "Standard Practice for Selecting Proportions for Normal, Heavyweight, and Mass Concrete (Standard Practice for Selecting Proportions for Normal, Heavyweight, and Mass Concrete)," *Am. Concr. Inst. committe 211*, no. Reapproved, pp. 1–38, 2002.

[7] V. A. O'Reilly, *Métodos para Dosificar Mezclas de Hormigón*. Cuba: Científico-Técnico, 1990.

[8] C. Andrade and C. Alonso, "Corrosion rate monitoring in the laboratory and on-site," *Constr. Build. Mater.*, vol. 10, no. 5, pp. 315–328, 1996. https://doi.org/10.1016/0950-0618(95)00044-5

[9] I. Troconis, O., Romero, A., Andrade, C., Helene, P., & Díaz, Manual de inspección, evaluación y diagnóstico de corrosión en estructuras de hormigón armado. Red Durar, 1998.

[10]NMX-C-495-ONNCCE, "Industria de la construcción–Durabilidad de estructuras de concreto reforzado–Medición de potenciales de corrosión del acero de refuerzo sin revestir, embebido en concreto-Especificiones y método de ensayo," 2015.

[11]NMX-C-514-ONNCCE, "Industria de la construcción-Resistividad eléctrica del concreto hidráulico-Especificaciones y método de ensayo," 2015.

[12]NMX-C-501-ONNCCE, "Industria de la construcción-Durabilidad de estructuras de concreto reforzado-Medición de velocidad de corrosión en campo-Especificaciones y método de ensayo," 2015.

[13]K. Petterson, "Corrosion threshold value and corrosion rate in reinforced concrete," Stockholm, 1992.

[14]J. L. Y. Chan, R. C. Solís, and É. I. Moreno, "Influencia de los agregados pétreos en las características del concreto," *Ingeniería*, vol. 7, no. 2, pp. 39–46, 2003.

[15]A. a Sagues, E. I. Moreno, W. Morris, and C. Andrade, "Carbonation in Concrete and Effect on Steel Corrosion," 1997.

[16]L. Bertolini, B. Elsener, P. Pedeferri, E. Redaelli, and R. Polder, *Corrosion of Steel in Concrete - Prevention, Diagnosis, Repair.* 2013. https://doi.org/10.1002/9783527651696

Non-Conventional Materials and Technologies – NOCMAT for XXI Century   Materials Research Forum LLC
Materials Research Proceedings 7 (2018) 604-621   doi: http://dx.doi.org/10.21741/9781945291838-58

Durability Analysis of a Bamboo Community Center in Cambury, Brazil

Sven Mouton[1], Karen Allacker[1], Khosrow Ghavami[2], Han Verschure[1]

[1] Faculty of Engineering Sciences, KU Leuven, Heverlee 3001, Belgium

[2]Faculdade de Enghenaria, Pontifícia Universidade Católica do Rio de Janeiro (PUC), Rio de Janeiro 38097, Brazil

Abstract. Bamboo is a non-conventional material often used in constructions. This paper discusses the durability of the material by analyzing the construction of a Community Center in the town of Cambury in Brazil, built more than a decade ago. The building is constructed using various locally available non-conventional materials and technologies (NOCMAT), i.e. bamboo, adobe blocks and rammed earth. The Community Center has been constructed by the Belgian non-profit organization Bamboostic in collaboration with KU Leuven and PUC-Rio. Beside the analysis of the durability, defects in the original design and material performance considering local weather conditions are discussed in this paper. The analysis is based on visual inspection and interpretation.

Keywords: Adobe, Bamboo, Durability, Cracking, Rammed Earth

Introduction / problem statement

Non-conventional materials, such as bamboo, rammed earth and adobe are local materials in Brazil, which are however not widely accepted. These materials clearly have important advantages compared to conventional materials, such as a low environmental impact, low cost and better mechanical performance (e.g. in earthquakes). [1] In order to broaden the general acceptance of these non-conventional materials, it is important that the construction quality is guaranteed, in the first place through extensive material research. [2] Most studies on material durability however are carried out in controlled laboratory settings. Such controlled conditions do not correspond to realistic outdoor environments where many factors play a role on the durability such as wind, temperature, humidity, salinity, solar radiation, orientation and wear and tear of usage. Little research has been done on how structures react to such realistic impact elements of climate. This paper aims at contributing to this knowledge gap by analyzing the behavior of these materials over time. More specifically, the goal is to analyze the performance of a building, i.e. the Community Centre in Cambury Brazil, consisting of bamboo, rammed earth and adobe which was constructed more than a decade ago.

The Community Centre consists of three different constructions built in different time periods, (2005-2011). The main bamboo construction of the Community Center was built in 2005-2006 and hence this case study is seen as an opportunity to analyzing the durability and consequences of the building design over time. The Community Centre was built by the non-profit organization Bamboostic in collaboration with KU Leuven and PUC-Rio in order to offer local inhabitants new means of revenue (in construction) and to improve the quality of local construction techniques with respect for their traditional ways of building (wattle and daub). A tech-transfer occurred by organizing workshops and constructing community buildings such as the one in Cambury. During these workshops and construction works, the methods of bamboo treatment, cutting and jointing were explained to the members of the bamboo cooperative in Cambury. The supporting bamboo structure is made with Dendrocalamus Giganteus (columns) and Phyllostachys Pubescens (trusses and wind braces). Rammed earth and adobe bricks are used as infill for the walls.

Non-Conventional Materials and Technologies – NOCMAT for XXI Century   Materials Research Forum LLC
Materials Research Proceedings 7 (2018) 604-621   doi: http://dx.doi.org/10.21741/9781945291838-58

Methodology

The Community Center in Cambury is analyzed in detail, focusing on the quality of the structure and its durability. Firstly, it is analyzed how and which bamboo building techniques were applied. Secondly, the degradation of the bamboo after a period of one decade is investigated. Conducting a destructive research on the bamboo canes wasn´t possible because the Community Center is still in use. Therefore, the degradation is analyzed based on visual inspection and interpretation of these observations by linking them to the knowledge at the research institutes involved and the expertise of the authors whereof the corresponding author assisted building the Community Center.

Material

Dendrocalamus Giganteus, diameter 18-20 cm, is used for all the columns and *Phyllostachys Pubescens* (also known as Moso) is used for the roof trusses and the wind braces. Their diameter is smaller, between 10-12 cm. Guadua is worldwide the preferred bamboo species to build with because it has a good strength and straightness, it has been studied extensively and gained significant appraisal due to its use in famous constructions such as the buildings by S. Velez. [3] However, it is locally not available and was hence not applied in the Community Center (to avoid additional building costs importing the bamboo). The Dendrocalamus Giganteus canes were cut free of costs at a nearby plantation owned by the state park Serra do Mar in Ubatuba. With their permission over 20 sufficiently aged bamboos were cut early in the morning (4 till 7 o`clock) at a waning moon. They were left to dry in the bush for 15 days maintaining their vertical position but lifted from the ground onto a stone. This common method of curing is to dry the bamboo through evaporation. Subsequently these canes were cut in 4m lengths and treated by immersion in a borax acid solution (10% Octoborato in a bath of 5.000 liters water). To facilitate submersion each internode was perforated on the top and bottom applying a 6mm drill. Fig 1 shows the immersion pool where the canes were submerged for three days. Afterwards the treated culms were stacked on top of the immersion pool leaking the fluids left in the internodes back into the pool. The Phyllostachys Pubescens canes were bought in large quantity at a bamboo plantation in the interior of São Paulo and given the same preservation treatment. The bamboos were then set up in a tripod drying construction for 3 months (Fig. 2) and stored in a covered but ventilated construction, lifted from the ground (Fig. 3).

FIGURE 1- IMMERSION POOL © SVEN MOUTON

FIGURE 2- DRYING OF BAMBOO © SVEN MOUTON

FIGURE 3 - BAMBOO STORAGE @ SVEN MOUTON

The strength of these bamboos was analyzed and the validity of the structural design verified by testing several canes from the construction in the laboratory at the PUC University in 2005. These tests were conducted by an engineer and Bamboostic project member under supervision of Prof. K. Ghavami, co-author of this article. The research was not completed but its preliminary results showed similar outcomes to other lab testings' on Dendrocalamus Giganteus and Phyllostachys Pubescens. [4] The tests confirmed that the quality of the canes of the community center were sufficient to carry the various loads, as calculated in the design phase. The procedure for the tests on tensile, bending and shear strength was in line with ISO 22157. [5] As stated in this ISO code the specimen was taken at several points of the bamboo culm (cut as a 'bowtie'): top, middle and bottom. Due to insufficient tested samples of both bamboos species to obtain representative values, the results obtained only indicate that the samples correspond more or less to the before mentioned available data on mechanical properties of both types of bamboo.

For the rammed earthen walls, a formwork was made with naval plates 18 mm of 220 mm x 110 mm screwed together onto a 35cm cement foundation using M14 bolts put on the outside of the formwork thus defining the wall size. The mixture applied was 9 wheelbarrows red earth, 4,5 wheelbarrows beach sand, two 20 kg sacks of lime (10%) and 2 cups of a Baba de Cupim solution, an oil-based stabilizer. Then 4 or 5 buckets of this mixture (about 19 cm of earth) are put into the formwork and rammed to one firm layer of about 10 cm thick. This is repeated until the formwork is filled entirely. The final layer is best protected by a cement layer or a wooden board. The latter was used in the Community Center. For the adobe bricks the same earth mixture was used, applying a brick making machine where three bricks at the same time were manually pressed. Afterwards the bricks were piled up in the sun to dry.

Community center project
Location
The community center is built in Cambury, a coastal town 50 km from Ubatuba (latitude 23°22′03.9"S, 44°47′37.8"W). It is situated in the Atlantic Rainforest, more specifically in the State Park ´Serra Do Mar-Nucleo Picinguaba` in São Paulo. The Atlantic Rainforest is a South American forest which extends along the Atlantic coast of Brazil largely between the megacities São Paulo and Rio de Janeiro. Over 85% of this principal biome has been deforested due to farming, ranching and charcoal making. [6] In this coastal region, the construction of the BR101 highway in the early seventies attracted land grabbers, real estate companies and speculators looking to develop the region using not always legal methods to oust original inhabitants. Soon 80% of the land in Cambury was appropriated by two landowners (which up to today remains indifferent), expelling original inhabitants, forcing them to move to less accessible areas or other cities at the coast. [7]

To protect the remaining forest the government created various parks such as the state park ´Serra do Mar` which was founded in 1977, thereby however also significantly restricting the activities of its residents. For 150 years Quilombos (people of African origin, mainly descendants from escaped coffee plantation slaves) and Caiçaras (descendants from indigenous people mixed with Europeans and Africans), inter-familiarly mixed into a homogeneous community, lived their traditional way of life based on agriculture, hunting and fishing. The formation of the state park however imposed constraints on fruit collection, agriculture and logging wood. Fishermen were no longer allowed to harvest specific trees to make their canoes, plantations where original forest needed to be cleared to grow crops were prohibited. These environmental restrictions, but also its secluded geographical location, a rough sea complicating fishing, illiteracy of the people made income generation very limited. [8] Today, approximately 50 families, both Quilombos and Caiçaras, live in Cambury, with primitive fishing, manioc growth and tourism as their main activities. The Bamboostic project advocates other means of revenue in the ecological construction sector and tries to consolidate the unity of the community by communal buildings.

The climate of Brazil varies considerably, from the tropical north to more temperate zones south of the Tropic of Capricorn (23°26' S latitude). The climate in Ubatuba, due to its geographical location at the sea surrounded by the mountainous Atlantic rainforest, is warm and humid. According to Köppen and Geiger, Ubatuba´s climate is classified as Af, meaning a tropical rainforest climate, albeit with noticeably cooler and warmer periods of the year. Regions with this climate typically feature tropical rainforests. A tropical climate is influenced by the low-pressure area of the calm zones with little air movement resulting in cloudy and humid conditions throughout the year. [9] The relative humidity in Ubatuba is between 84-88%,

Non-Conventional Materials and Technologies – NOCMAT for XXI Century   Materials Research Forum LLC
Materials Research Proceedings 7 (2018) 604-621   doi: http://dx.doi.org/10.21741/9781945291838-58

favoring molds and fungi. The average annual temperature remains constantly high with slight variations. Ubatuba´s average is 25,2°C in the warmest month February and 19°C in the coldest month July. The rain season extends several months with high level of levels of precipitation, even in the driest month. In a year, the average rainfall in Ubatuba is 2552 mm. The driest month is June, with 84 mm of rain. The greatest amount of precipitation occurs in January, with an average of 343 mm. Buildings in this context hence need to deal with heat, strong solar radiation, high levels of air humidity and torrential rainfall and architects are challenged to design buildings that offer comfortable spaces without requiring mechanical cooling systems. [10]

Architectural Design
Clients´ requirements
For the design 3 main requirements were put forward by the local association of Cambury:

1. To provide a communal space to hold meetings, school activities or other events and several separate rooms to host classes and to store material.
2. To form a perceived geographical center of the town.
3. To integrate the building within the surrounding landscape and the existing school located on the same terrain.

Organization
Fig. 4 illustrates in what period each part of the construction was built. The main building of 106 m² plus annex of 53 m² was finished in June 2006 and contains a multifunctional space with a small stage and kitchen, a classroom for the pre-school and an open work atelier for the bamboo workers. The highest point of the main roof is 5,95 m high, whilst a free span inside the main building of 4,95 m is reached. In August 2009 a separate office for the association was concluded and in September 2011 two rooms to store tools & sport materials (surfboards, soccer balls, etc.) for extra-curricular activities and a library/computer room. In August 2017 a new bamboo project, a communal bakery, is being built.

The multifunctional space, being the most public zone, forms the core of the Community Center with the semi-public spaces designed around it. Circulation is foreseen around this multifunctional space connecting all other rooms and with its low earthen walls (1,1m - 1,5m) facilitating contact. To link the multifunctional area with the existing school and the other public zones on the terrain, the Community Center is carefully positioned.

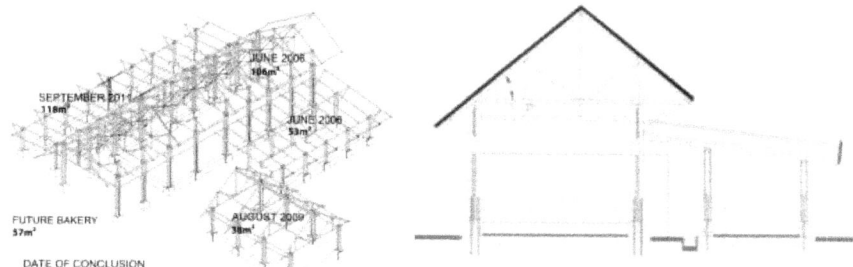

FIGURE 4 - DATES OF CONSTRUCTION
© SVEN MOUTON

FIGURE 5 - CROSS SECTION OF MAIN BUILDING (2006) © SVEN MOUTON

The cross section in Fig 5 shows that the truss is not completely symmetric, the far-right corner has been cut away and the columns are placed underneath the last vertical subdivision. This effect is not provided out of structural need, but to accomplish a visual effect for visitors, opening up the structure from the front side and giving the whole a lighter, aired and grand look. The two lower roofs were added for the class rooms and storage rooms.

The roof cladding of the main roof exists of ceramic tiles. The dead load of this roof equals 38,5 kg/m² (with a total weight of 4100 kg). Originally, the lower roof seen in Fig 5 was covered with a green roof (55 kg/m², with a total weight of 2926 kg), but due to the heavy rain fall and fierce solar radiation, the grass burned. The original green roof was therefore replaced by the undulated plates (20kg/m²). Wind loads (in July) were estimated at a low 2,7 m/s = 10 km/h, temporary extreme wind gusts can clearly be much higher (mainly summertime).

Ventilation
The terrain is situated 50 meters land inward from the beach. The center is oriented in the direction of the sea to catch the main wind for ventilation. By raising the roof sufficiently high and by avoiding perpendicular walls blocking airflow inside the building, the ventilation flow is optimal. Under warm and humid conditions higher wind velocities have a positive effect on the physiological as well as psychological wellbeing. The height of the building aids the buoyancy or stack effect; air will flow in when the warmer indoor air rises up through the building and escapes at the top, therefore the design foresees both lateral sides open. The rising warm air reduces the pressure at the base of the building, drawing colder air in when there is a lack of natural airflow and stagnant air. [11]

Additionally, the sheer force of the wind is a key factor in the design. The impact of this force is larger when a construction gains in height (needed for the ventilation). As studied at PUC-Rio in push-over (bamboo) constructions [12], in order to have adequate wind-bracing, the triangulation of the construction needs to be well studied and executed in good order and detail. Elevating a building with wind-bracing only at the end can have detrimental consequences during (frequent) storms. The use of four columns, with the cross-bracing of both lateral trusses proved to be sufficient to act as wind-bracing. As studied by F. Demets [13], placing lateral structures in between four columns provide a wind bracing in itself, much like the fixed connections of the spatial trusses by Prof. Eng. Arthur Vierendeel. Such trusses do not have the usual triangular voids as seen in typical wind-bracing, but rather employ rectangular openings and rigid connections in the elements, which (unlike a conventional truss) must also resist substantial bending forces. This rigid connection is also formed by the four culms that are laterally cross-connected such as shown in Fig 6.

FIGURE 6 - *WIND-BRACING AND COLUMNS © SVEN MOUTON*

Weathering protection
Another main factor in bamboo architecture is protection against sun radiation, precipitation and moist soil. This is a necessity for bamboo, but also for many other non-conventional materials such

Non-Conventional Materials and Technologies – NOCMAT for XXI Century   Materials Research Forum LLC
Materials Research Proceedings 7 (2018) 604-621   doi: http://dx.doi.org/10.21741/9781945291838-58

as timber, earth, straw bale, etc. [14] Only a limited amount of rainfall to a directly affected area is tolerable for these materials. [15] First, the base of the construction needs to be lifted at least 30 cm but preferably more and executed in stone, concrete or another inert/water-resistant material. In this case a cement foundation with an outer layer of small stones of 50 cm high was built. Second, roof crossings need to be foreseen large enough to ensure a weathering protection. The length of a roof crossing depends on the design and height of the building as described by Simon Velez. [16] In this design, the roof has a 90 cm (small roof) and 145 cm (large roof) eave sideways (Fig.5). In the front and back of the building a timber sheathing was designed to protect the bamboo trusses. The office has an eave of 150 cm sideways and the storage rooms 80 cm in the back.

Structural Design

General

No specific building codes on bamboo are available in Brazil. The most appropriate building codes to ensure proper built bamboo constructions and to minimize the risk of collapse or other damage available are (amongst others) the Colombian NSR-10 regulations or ISO-DC625 (in markup). [17] [18] [19] In July 2017, actions are being undertaken by PUC-Rio (Prof Ghavami) and other partners to develop a specific Brazilian building code (ABNT/NBR).

It is a challenge to use as little bamboo as needed for a safe stability without over-using the material, not merely for the cost-factor but also from an environmental perspective. Equally important is to carefully design the joints to ensure cost-effectiveness. The German-Chinese House (Fig 7) at the Expo 2010 in Shanghai with its detailed steel joints for example led to high costs due to the high requirements regarding the appearance of the bamboo structure [20], which for this specific case and requirements were allowable, but for development projects is clearly undesirable.

FIGURE 7 - *GERMAN-CHINESE HOUSE © ARCH DAILY*

Columns

As mentioned above every column exists of 4 culms put onto a 50 cm high base where four wire rods extend for about 60 cm. The first 3 or 4 internodes are punctured and the bamboo canes put on top of these wire rods, as shown in Fig 6. A hole is drilled near to where the wire-rod ends and filled with grout to ensure good stability. The distance between two columns (and the rafters) is 2,5 m. Not only because commercially available distances are often kept at 2,95 m, but also because longer lengths seem to buckle more, and therefore, results in deformation of the lateral wind braces. This perceived deformation is also found in Wellington 2007 [21]. In the other community center constructions only one or two bamboos were used to form the columns but the general system is comparable to the above-mentioned system.

Non-Conventional Materials and Technologies – NOCMAT for XXI Century   Materials Research Forum LLC
Materials Research Proceedings **7** (2018) 604-621   doi: http://dx.doi.org/10.21741/9781945291838-58

Trusses

As the trusses are triangulated, a large span of 4,95 m could be made. The bamboo trusses are all modular and built on the ground before lifted onto its place. Technical knowhow of the local workers about stability issues is limited and therefore the decision was made to work with comprehensible modules that afterwards could be easily repeated in other similar constructions. Fig. 8 shows the asymmetric rafter module in detail. The main construction consists of 9 trusses supported by 18 columns. In the trusses plain fish mouth connections are used. When perfectly executed, this forms a considerably strong connection, but attention needs to be given curving it to the exact diameter at the end of the bamboo that arrives at the 'receiving' bamboo. If not, uneven forces at different places of the wall will crack/split it open where the wall does not touch the 'receiving bamboo'. The system of the fish mouth is very straightforward and well-known; a 'bolt' is twisted into a hook and placed in the 'receiving' bamboo, while a cross-connection is made in the bamboo arriving at it. When the hook is pulled tight on this lateral connector, the joint becomes fixed and is afterwards fixated with concrete/grout that is poured over it (Fig 9). In order to do so, a hole is drilled applying a hole saw of 40 mm above this connection and grout, without holding too much water, is poured into the internode using the mouth of a plastic bottle. When slightly 'hammered' on the sides, the concrete sinks in tight so that after drying a fixed connection is made. Subsequently the drilled holes are clogged using the bamboo cap that was drilled out to tamp it again. A truss roof system is also applied in the association's office similar to the trusses as described above but with a smaller span and symmetrical. The other two constructions used a counter batten roof system.

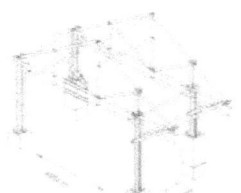

FIGURE 8 - MODULE OF BAMBOO   *FIGURE 9 - FISH-MOUTH*
TRUSS   *CONNECTION ©SVEN MOUTON*

Connections

By applying 4 culms with an interspace of one bamboo, cross-connections could be made that in itself serve as wind-brace (making a fixed joint such as the above mentioned Vierendeel-framework). Rafters could be attached in one direction between the top of the four culms, while two lateral wind-bracing bamboos were put in the other direction (Fig 11). The bamboos are connected by wire rods at every intersecting bamboo. This was done with a large drill (30 cm) puncturing 3 bamboos at once. It is nearly impossible to drill holes with a drill bit that only drills one bamboo, because the same direction can never be found when drilled from other sides. Whenever transverse bolted connections in the bamboos were made (as is the case in these columns) 5/8-inch wire rods with bolts and washers were used to keep similar distances between the bamboos. If no bolts and washers would be placed also in between the bamboos, they would be pushed together making the column skewed. (Fig 10 and 11). The hollow internodes of these connections are filled with specie to limit displacement and prevent cracking.

Non-Conventional Materials and Technologies – NOCMAT for XXI Century   Materials Research Forum LLC
Materials Research Proceedings **7** (2018) 604-621   doi: http://dx.doi.org/10.21741/9781945291838-58

FIGURE 10 - FOUR CULMS PART OF WIND-BRACING FIGURE 11 - LARGE DRILL AND USE OF SPACERS © SVEN MOUTON

Process

During the building process, several adaptations and modifications were made. Because of the extreme sun in summertime the bamboos showed signs of cracking (loud cracking noises). In order to protect the bamboos from warming up they were wrapped in yellow plastic. This however made no difference and was soon removed again. It is advisable to avoid building in summertime, especially when there is no possibility of creating shade, as was the case here. Once the structure was covered the bamboo cracking stopped. Afterwards every crack was filled with a mixture of araldite glue and sawdust. It is necessary to close up every opening otherwise this will become hiding places for insects. Moreover, when rain can enter the bamboo through the cracks, it will deteriorate from the inside out.

Detrimental weathering of the lower part of several bamboo culms was perceived due to solar radiation, even being lifted 50 cm of the ground. Therefore, several columns were encapsulated using adobe bricks as an outside layer filled with dry and salty sand, up to 1,5 meters high measured from the stone base (2 m high from the ground). As can be seen in the original visualizations of the project in Fig 4, these adobe bricks weren't originally foreseen. To have an indication of the weathering impact not all bamboo columns were encapsulated.

For the durability of the construction fixed connections are used providing these are more reliable than lashings. Lashings will slowly get looser, especially through friction caused by gusts of wind-loads. Lashings also greatly depend on the skills of the person placing them and a high moisture level in the air facilitates rotting below the lashing. During the construction of the Community Centre lashings with sisal rope were tried as an ornamental aspect on one of the trusses, but removed after one month because the bamboo was starting to rot where moisture was trapped beneath the lashing. There exists research that by anointing the sisal with a polyester resin a longer durability is possible. [22] Fixed connections that capture few forces were only connected with wire rods, bolts and washers but those that have to withstand stability forces (load & wind) are filled with mortar, especially when there are lateral forces.

The roof of the preschool was firstly built with a green roof. However, the grass could not withstand the heavy precipitation and sun radiation and shortly after died. Possibly using heat-tolerant plants and an irrigation system would have worked but the community cooperative replaced it with recycled plastic roof plates one year after placement.

Durability analysis

The durability analysis is based on four visual inspections between July 2016 and August 2017. The following aspects have been inspected and elaborated on in the subsequent paragraphs: molds and fungi; cracking; attacked bamboo by insects; degradation of earthen walls; corrosion of bolt connections; others. All structural culms were closely observed and are photographically

Non-Conventional Materials and Technologies – NOCMAT for XXI Century   Materials Research Forum LLC
Materials Research Proceedings 7 (2018) 604-621   doi: http://dx.doi.org/10.21741/9781945291838-58

documented. Fig 12 indicates the state of degradation of all bamboo culms. Where severe degradation was found, the culms were annotated in red, average to low degradation from orange to yellow, good condition in green. Overall, it can be perceived that the bamboo structure is in a good condition, apart from some problem zones. Bamboo in less ventilated spaces and exposed bamboos (to weather) show more degradation, as well as horizontal culms suffered more than vertical ones.

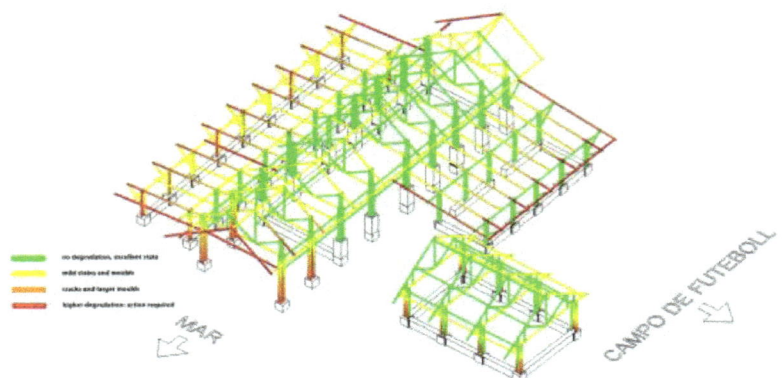

FIGURE 12 - *OVERVIEW OF DEGRADATION OF BAMBOO CULMS (VISUAL INSPECTION MAY 2017)*

Molds and fungi

Although protection against exposure to rain, sun, moist, dry air, etc. was carefully considered in the design process it appeared insufficient on several places. Tropical climates with high temperatures and a high level of humidity favor molds easily; molds occur at the cross-ends and on the surface of culms. The borax treatment prevents largely the proliferation of molds, but washing out by rain reduces their protective efficacy. [23, 24] Also the varnish layer slowly disappears through sun and rain. Possibly due to the lubricity of the outer bamboo layer, which is lower when the bamboo culm is unprotected in comparison to when varnish is applied, mold can attach itself more easily to bamboo [25]. The external surface of the first main truss nearest to the sea, and in lesser severe condition the second and third truss as well, show green and white spots, indicating the presence of wet moisture (Fig.13). In the design, a timber sheathing was intended to be placed at the end of the roof protecting these particular trusses, however due to the depart of the Bamboostic project members this was only done on one side. The bamboos of this first truss need to be replaced, the second and third bamboo truss suffered considerably less damage and by washing and re-varnishing can be maintained. In addition, the timber sheathing or another permanent protection has to be placed. On the other end of the main roof, where the timber sheathing was placed, also deterioration of the bamboo culms is recorded but in milder deteriorated condition. One culm was more deteriorated which indicates that even the timber sheathing is insufficient as protection against the sun and rain (Fig 13) and another solution needs to be sought.

Figure 14 shows an example of insufficient roof eaves facilitating deterioration. These bamboo ends show signs of decay/rot; they turned black and can easily be salvaged. The problem only arises on the end-parts of the bamboo, on the outside of the building. The rest of the bamboo is still of good quality. To solve this, the last counter batten should be inserted more from the end of the roof panel

Non-Conventional Materials and Technologies – NOCMAT for XXI Century   Materials Research Forum LLC
Materials Research Proceedings 7 (2018) 604-621   doi: http://dx.doi.org/10.21741/9781945291838-58

and the affected tops sawn off. Afterwards the new tops should be filled with cement or another filler to protect them against insects entering or nesting in the open part of the culm.

FIGURE 12 - AFFECTED BAMBOO TRUSS NEAR THE SEA ©SVEN MOUTON   *FIGURE 13 - AFFECTED BAMBOO TRUSS NEAR THE SCHOOL. @SVEN MOUTON*   *FIGURE 14 - ROTTING ROOF ENDS BECAUSE OF CONTINUOUS EXPOSURE @SVEN MOUTON*

Bamboos also suffer from exceeding moist levels in the air due to little ventilation (Fig 15). In contrast to the daily usage of the main building, the storage rooms are mostly closed in order to protect the materials stored inside. Moreover, the height of the roof of the storage rooms needed to be lower than the outer left side of the main roof being at its highest point 2,7 m. This low height and the closeness of the construction promotes the humidity in the air resulting in molding and darkening of the bamboo. However, no cracks are noticed such as in the bamboo trusses exposed to the sun and rain. Also, the adobe bricks in these rooms show signs of mold and fungi which isn´t encountered elsewhere in the construction stipulating again the importance of ventilation. To remediate this lack of ventilation, arisen due to a design flaw, is difficult to tackle but can possibly be done by opening the walls more or by installing mechanical ventilation.

Noteworthy is nonetheless that not all bamboos which are relatively exposed to the weather conditions are in decay. The bamboo columns of the association´s office and the main building are still in relative good condition despite of this exposure (Fig 18, 19). It is noticed that horizontal members suffered more easily than vertical ones, probably because of the possibilities of run-off in vertical members that horizontal members do not have and therefore poses more stagnant moist.

*ABOVE: **FIGURE** 15 - MOIST AND MOLD ON BAMBOO IN POORLY VENTILATED SPACE © SVEN MOUTON*

Non-Conventional Materials and Technologies – NOCMAT for XXI Century   Materials Research Forum LLC
Materials Research Proceedings 7 (2018) 604-621   doi: http://dx.doi.org/10.21741/9781945291838-58

*BELOW: **FIGURE 16** - VERTICAL EXPOSED BAMBOO IN GOOD CONDITION © SVEN MOUTON*

Cracking

Cracking is caused by three main reasons namely exceedingly low humidity level, heating of air inside the nodes or cracking due to poor execution. First, as bamboo behaves best in an average of 60-80% air-humidity dry air will cause cracking (hair-cracks), as was perceived empirically by the corresponding author while working with bamboo during wintertime in Europe (with a relative air-humidity around 40%). Second, when bamboo is exposed to direct sun radiation during the warmest period of the day (mid-day/afternoon), the air inside the bamboo nodes heats up, expands and finds its way out by cracking. This is often accompanied by a loud 'explosive' sound when occurring. As mentioned, this happened during the building process of the main construction. It could be remedied by puncturing the internodes so that the air can expand and escape through the entire culm. Nonetheless this might complicate the connection of joints where mortar is required and the internode forms a natural boundary. These both types of cracks occur in lateral direction and are mostly small. In general, these cracks don´t cause problems (Fig. 17), but can become problematic when rain water enters the bamboo culm and starts to rot. Moreover, small insects such as the powder post beetle, borer, etc. can also infiltrate the bamboo through the cracks and attack it from the inside (where the fibers are weaker than the lubricated outside). Traces of small and bigger linear cracking can be found throughout the entire construction but none of them were identified as being a risk for the stability.

FIGURE 17 - *CRACKS THAT POSE NO STRUCTURAL PROBLEM BUT NEED TO BE ADJOURNED © SVEN MOUTON*

It is recommended to tamp these cracks with a mixture of white glue or a two-component glue such as Araldite© mixed with bamboo or wood sawdust. It is found that this woodwork-technique of using filler (glue) and sawdust also works perfectly to fill bigger cracks and other small holes in the culms. Whenever an exceeding load force or direct exposure to sun remains, the crack comes back

Non-Conventional Materials and Technologies – NOCMAT for XXI Century   Materials Research Forum LLC
Materials Research Proceedings 7 (2018) 604-621   doi: http://dx.doi.org/10.21741/9781945291838-58

and worsens. This solution can therefore only be seen as filler, not as a permanent adhesion that cures design or execution flaws. Cracks wider than 1cm and/or running over several nodes in a load-bearing and crucial position can be dangerous and require replacement of the bamboo. In such case, filler should not be applied as a remediation solution.

Fig 18 shows a crack caused by improper execution. Leaving a small opening in a bamboo joint between the higher and the lower bamboo which had to support the upper bamboo has put an excessive lateral force on the upper bamboo causing it to crack. When the upper bamboo, as is the case in the other similar joints, touches the lower one, the forces are directly passed onto the structure that was intended for carrying this load. As the upper structure does not touch the lower, but instead is carried by merely a lateral wired rod, this unforeseen (lateral) load distribution in time succumbed to the weight. Of the 18 similar connections in the main building this occurred only in one of them so therefore it can be concluded that this was not due to a design error but an execution error. Bamboo fibers run from internode to internode and therefore have a very high tension and compression strength in longitudinal direction but not in lateral direction. [26 ref] A proper solution is to ´splint` this bamboo by putting a support on the top and bottom of the cracked bamboo. Replacing this particular bamboo seems arduous. Noteworthy is the flexibility of bamboo in construction, even though there was a structural problem that caused a severe crack, this joint never gave a structural problem to the whole of the construction. No other severe cracks were recorded.

FIGURE 18 - CRACKS THAT POSE A STRUCTURAL PROBLEM AND NEED TO BE RESOLVED   © SVEN MOUTON

Bamboo attacked by insects

During the visual inspection no signs of termites, powder post beetles or other type of small insects, deriving from possible inadequate treatment, were found. However, this visual inspection cannot give a 100% certainty of no attacks by borers, as these do not eat the outer hard high silica skin but the soft inner layers of the bamboo. [24] The danger of the infiltration of insects is that they harm the material from the inside and their presence is invisible, except from leaving small traces outside (powder dust), only to be noticed when the damage has been done. No traces, on none of the bamboos, were found during the examination, and this after 12 years harvesting time (2005). It can preliminary be concluded that the borax treatment against termites and other insects is effective.

In the bamboo rafters from the storage rooms small drilling holes of the borax treatment are still open. A sign of small insects (species unknown to author) nesting inside one of these bamboo holes was encountered. These small drilling holes need to be sealed with a mixture of bamboo saw dust and white glue, similar as described above. Especially open endings and bigger holes offer an ideal place for hiding for insects, birds, rodents and other animals. They nest inside the hollow internode and can spoil its soft inner part.

In the roof structure of the main construction several nests of wasps were found hanging on the bamboos, see Fig. 19. These need to be removed before becoming a danger, even though they are hard to reach. They may become hazardous to people, but do not pose a structural danger to the construction.

FIGURE 19 - WASPS NESTING ON THE BAMBOOS

FIGURE 20 - VANDALIZED RAMMED EARTH © SVEN MOUTON

Degradation of earthen walls

The rammed earth and adobe brick walls, built following the indications of CRATERRE [27] have withstood time and weather conditions well regardless their light exposure to sun and rain. The color and the quality is similar to when it was recently built. Obviously, children have tried to write on the rammed earth walls and have been vandalizing them as shown in Fig. 20. The walls show marks hereof but in general children could not break the walls more than just the corners and smaller bits and pieces. The compaction of the walls provided a smooth and strong lateral surface, which is hard to penetrate. No larger gravel was used in the mixture, and therefore no loose particles could be taken out.

Some of the adobe brick walls were too narrow to imbed the bricks firm enough and are standing loose, see Fig. 21, most probably due to children pushing and pulling at these bricks. It is recommended to remove these isolated walls and redo them connecting them better onto the sides.

Two of the 14 adobe brick walls built around the bamboo columns to protect them have been vandalized as well. The bricks have been placed with an earth/sand/lime mortar which possibly was easy to break. Some of the bricks were taken away, demonstrating clearly the difference in appearance of bamboo in- and outside the sand (Fig. 22). The bamboos that had been inside the sand show, besides lacking the varnished layer, little difference. Because beach sand was used it offers an additional protection against fungi, beetles and termites. Similar adobe bricks need to be used to rebuild the savaged walls. The perfect state of the bamboo culms inside the (salty) sand-filled column is remarkable, and might offer solutions to future constructions where base protection is needed.

FIGURE 21 - *LOOSE NARROW WALLS*   *FIGURE 22* - *BAMBOO INSIDE AND OUTSIDE AN ADOBE BRICK/SAND WALL © SVEN MOUTON*

FIGURE 23 - *MOLD ON EARTHEN WALLS © SVEN MOUTON*

Almost none of the earthen walls manifests mold or fungi, merely the ones at the back of the storage rooms indicate mold at the bottom of the wall, near the stone/cement bases, which stems from lack of ventilation (Fig. 23). These walls stand very close to the edge between two terrains and several trees and plants are growing nearby blocking air to pass. These plants need to be removed as much as possible to let air pass and the mold needs to be washed off with water and chlorine.

In general, if details such as corners, narrow pieces of walls, connections to other walls, etc. were better executed and protected against vandalism, it would improve the quality of the earthen walls substantially. The disadvantage of repairing these walls afterwards is the difficulty to get exactly the same color of the original wall, which makes repairs very visible. Overall, besides these defects, the condition of all adobe brick and rammed earth-walls is robust.

Corrosion of the wire rods and bolts

The corrosion of the bolts and washers is in an advanced state, as can be seen in Fig. 24. Although stainless steel wire rods, bolts and washers were acquired, every single one of them is now corroded due to the saline air from the ocean. The condition of the wire rods inside the bamboo columns cannot be examined but is most likely not as corroded as moist salty air cannot penetrate (constantly) inside the bamboo. It may be assumed that the wire rods inside the bamboo culms filled with grout are even better protected because of the mortar, as is the case in armed concrete degradation where only exposed steel corrodes. It might be a solution to fill every joint where there are wire rods inside with concrete or investigate alternative types of wire rods, bolts and washers, for example in composite materials or fire-treated galvanized steel. The latter is currently being used in the new bakery (2017). These fire-treated galvanized steel rods are mostly used in the naval industry and are half the price of a stainless-steel rebar. As the project is situated in the coastal region where fishery is predominant, accessibility in specialized stores is no problem and hence seen as the best solution for

Non-Conventional Materials and Technologies – NOCMAT for XXI Century   Materials Research Forum LLC
Materials Research Proceedings 7 (2018) 604-621   doi: http://dx.doi.org/10.21741/9781945291838-58

this project. Also, treating them with rust-resistance paint (such as Hammarite©-paint) or daubing with rust-resistance oil (WD-40©) as a regular maintenance procedure could alleviate this problem, even after corrosion has occurred.

FIGURE 24 - CORROSION OF WIRE RODS AND BOLTS © SVEN MOUTON

Others

An additional element observed (Fig. 25) is the darkened color of the bamboos in the storage rooms compared to the bamboos in the other two constructions. The cause of this darkened color is unknown. It might be caused by a difference in bamboo type or quality, treatment, humidity, etc. Most likely it can be ascribed to the higher level of humidity inside the storage rooms (see section 5.1). At first sight, this does not pose any structural problem. However, the only way to examine if this color difference has any impact on the inner wall is to perform destructive testing, which for obvious reasons was not performed.

Another problem that has arisen in the construction are the undulated plates which proved to be of very poor quality and are currently no longer being sold. After a certain time, they lose their strength and deform, as can be seen in Fig. 26. If future budget allows, it is advisable to replace these roof panels with more durable roof panels. It will not only benefit the look of the building but also better protect the bamboo structure underneath.

FIGURE 25 - DARKENING OF THE BAMBOO © SVEN MOUTON   FIG 26: POOR QUALITY OF ROOF PANELS ©SVEN MOUTON

Discussion

It is noticed that the columns (Dendrocalamus) and the roof battens (Phyllo/Moso) show different forms of decay. The Dendrocalamus (cut, treated and dried carefully on site) was fairly untouched by decay, while some of the Moso suffered more even though these were less exposed. It is unclear if the differences in decay are due to differences in bamboo quality, or rather due to differences in the way these have been applied (columns vs. roof battens, horizontal vs vertical position). Further investigation is needed to clarify these variations in level of decay.

The analysis furthermore revealed the importance of the connection material. Even though galvanized wire rods were used, this proved insufficient to prevent rusting. Destructive investigation should be done in order to see whether the wire rods covered in mortar inside the bamboo are also deteriorating, which could pose a structural problem. Also tests need to be executed on the results of the fire-treated galvanized steel as is used in the new bakery project (2017).

When the bamboos were no longer covered with a finishing layer of varnish (due to peeling off in time without maintenance), a larger degree of deterioration could be noticed. This could however also be due to the fact that the varnish only peeled off in places where weathering could take place (insufficiently protected places through design), making the role of varnish more unclear when it comes to protecting the bamboo. More research needs to be done as to the effective role of varnish in bamboo-constructions, reality however seems to concur the need for this protection. Also, it has been perceived that not all varnish types are as efficient. The brand that has given the best result so far is OsmoColor ©.

Conclusion

The main factors in deterioration of bamboo are known to be radiation, precipitation, humidity, moist soil and insects. This paper demonstrates that none of these factors alone can cause bamboo decay. There has to be a multiple of these factors present. The outer bamboo trusses of the main building and the bamboo ends in the pre-school all exposed to sun, rain and air humidity demonstrate clearly signs of deterioration. Nevertheless, other bamboo culms such as in the associations´ office and the main building, likewise exposed to radiation, rain and humidity, show no signs of decay (no molds, no cracks).

It is moreover often assumed that the age of the bamboo construction determines the decay of the bamboo and that the decay is accelerated when a construction is poorly designed or executed. This is partially contradicted in this paper: the visual inspection revealed that the stage of decay of the bamboo canes used in the storage rooms, which is poorly ventilated and covered by a bad quality roof panel, is beyond the deterioration of most bamboos in the main building built five years earlier. The association´s office (2009) is in the best condition showing little signs of decay. Maintenance of the construction is crucial to avoid or at least diminish bamboo culms are put at risk and endanger the structural capacity of the building. Structural bamboo culms must therefore have adequate spacing for cleaning and removal of mold and should always be covered with a finishing coat of varnish and to be repeated at least every five years. Also, cracks need to be closed with a glue mixed with bamboo saw dust to avoid attacks from insects and rain/humidity from entering. Proper maintenance will augment the durability of any bamboo structure but cannot resolve problems caused by bad designing. Overall, besides small failures, the community building can be considered a well-preserved success. After a decade, the construction shows no severe defects and is stilled used on a daily base by the local community. [28]

References

[1] Moraes, L. A. F., De La Ohayon, P., Ghavami, K. (2011). Application of non-conventional materials: Evaluation criteria for environmental conservation in Brazil. Key Engineering Materials, vol.517, p.20-26. https://doi.org/10.4028/www.scientific.net/KEM.517.20

[2] Achila, H., Ansell, P., Walker, P. (2012). Low Carbon Construction Using Guadua Bamboo in Colombia, Key Engineering Materials, vol. 517, p. 127-134. https://doi.org/10.4028/www.scientific.net/KEM.517.127

[3] Hebel, D., Heisel, F., Javadian, A., Wielopolski, M., Schlesier, K., Griebel, D. (2013). Local alternatives: essay replacing steel with bamboo. The Economy of Sustainable Construction.

Ruby Press, Berlin, p. 44-55.

[4] Minke, G. (2012). Building with bamboo. Design and technology of a sustainable architecture. Birkhauser, Basel, 158p.

[5] ISO Norm. Retrieved from https://www.iso.org/standard/36150.html.

[6] Prado, G., Esteves, R., Ramires, M., Begossi, A. (2015). The Caiçaras and the Atlantic Forest coast: Insights on their resilience, UNISANTA Bioscience, vol. 4 (3): p. 189-196.

[7] Comunidades Quilombolas no Brasil - Litoral Norte - Camburi., CPISP, Commission Pro-Indio of São Paulo. Retrieved from www.cpisp.org.br/comunidades/html/i_brasil_sp.html.

[8] Santos, E.L. (2013). Estação Memoria Cambury, São Paulo. ECA-USP, Dissertation Master, 2004-2006, 191p.

[9] Lauber, W. (2005). Tropical architecture. Sustainable and humane building in Africa, Latin America and South-East Asia. Prestel, Munich, 199 p., p85 'fundamentals of climatically appropriate building and relevant design principals'.

[10] Tablada, A., Blocken, B., Carmeliet, J., De Troyer, F., Verschure, H. (2009) Airflow conditions and thermal comfort in naturally-ventilated courtyard buildings in a tropical-humid climate, Building and Environment 44, p. 1943-1958. https://doi.org/10.1016/j.buildenv.2009.01.008

[11] Krautheim, M., Pasel, R., Pfeiffer, S., Schultz-Granberg, J. (2014). City and wind. Climate as an architectural instrument. Dom Publishers, Berlin, 201 p., p. 120

[12] Bhavna, S., Mitch, D., Harries, K.A., Ghavami, K. (2011). Pushover behavior of bamboo portal frame structure, International Wood Products Journal, vol. 2(1), p. 20-29. https://doi.org/10.1179/2042645311Y.0000000003

[13] Demedts, F. (2006). Naar een nieuwe houtskeletbouwwijze: van wereldbeeld tot bouwmethode; bijdrage tot humaan-ecologisch bouwen, Doctoral degree at the Department of Architecture, Building and Planning for the University of Eindhoven; Supervisors: P. Schmid and G. De Roeck.

[14] Ghavami, K (2009). Non-conventional materials and technologies; applications and future tendencies, Proceedings of the 11th International Conference NOCMAT 6-9 Sept 2009), Bath, UK.

[15] Quoc-Bao, B., Morel, J., Hans, F., Walker, P. (2014). Effect of moisture content on the mechanical characteristics of rammed earth, Construction and Building Materials 54, p163-169. https://doi.org/10.1016/j.conbuildmat.2013.12.067

[16] Velez, S., Von Vegesack, A. (2013). Simon Velez and Bamboo Architecture: Growing Your Own House, Vitra Design Museum, 253p.

[17] Krause, J. Q., De Andrade De Silva, F., Ghavami, K., Da Fonseca, O., Dias Tolhedo Filho, R. (2016). On the influence of Dendrocalamus Giganteus bamboo microstructure on its mechanical behavior. Construction and building materials, vol. 127, p.199-209. https://doi.org/10.1016/j.conbuildmat.2016.09.104

[18] Ghavami, K., Arnal, I.P., Sandretti, F., O'Brian, M., Brajovic, M.S. (2005). Engineering properties of entire bamboo column of species Guadua: contribution to a sustainable development. Bamboo Lab. Barcelona, p. 33-50.

[19] Janssen, J.J.A. (2000). Designing and building with bamboo. INBAR, Beijing, 187p.

[20] Heinsdorff, M. (2011). The bamboo architecture. Design with nature. Hirmer Verlag,

Non-Conventional Materials and Technologies – NOCMAT for XXI Century   Materials Research Forum LLC
Materials Research Proceedings 7 (2018) 604-621   doi: http://dx.doi.org/10.21741/9781945291838-58

Munich, 208p.

[21] Wellington, M, Kenmochi, S., Cometti, N, Leal, P. (2007). Avaliação de estrutura de bamboo como element constructive para casa de vegetação, Engenharia Agricola, Jaboticabal, vol.27(1), p100-109.

[22] Kruse, J.Q., Ghavami, K. (2009). Transversal reinforcement in Bamboo Columns, Proceedings of the 11th International Conference on Non-Conventional Materials and technologies (NOCMAT 6-9 September 2009), Bath, UK.

[23] Tang, T.K.H., Schmidt, O., Liese, W. (2009). Environment-friendly short-term protection of bamboo against moulding, Timber Dev. Assoc. of India, vol. 55, p.8-17.

[24] Liese, W., Kumar, S. (2003). Bamboo preservation compendium, Centre for Indian Bamboo Resource and Technology; First edition, 231p.

[25] Jianchao, D., Fuming, C., Wang, Ge., Daochun, Q., Xiaoke, Z., Xuquan, F. (2014). Hydrothermal aging properties, moulding and abrasion resistance of bamboo keyboard, Holz als Roh- und Werkstoff vol. 72(5), p. 659-667. https://doi.org/10.1007/s00107-014-0828-2

[26] Houben, H., Guillaud, H. (1989). Craterre. Traité de construction en terre. Editions Parenthèses, Marseille, 349 p.– p.341 'Pathologie, Symptômes'.

[27] Hidalgo López, O. (2003). The Gift of the Gods. published by the author, Bogotá, 553p.

Non-Conventional Materials and Technologies – NOCMAT for XXI Century   Materials Research Forum LLC
Materials Research Proceedings 7 (2018) 622-631   doi: http://dx.doi.org/10.21741/9781945291838-59

Mechanical Properties of Guadua Elements Subject to Variations in Moisture Content

M.L. Sánchez *[a], L.Y. Morales[a], Y. Lara[a], Y. Díaz[a], C.D. Rincón[a]

[a] Universidad Militar Nueva Granada, Colombia

martha.sanchez@unimilitar.edu.co, luz.morales@unimilitar.edu.co,
u1101884@unimilitar.edu.co, u1101862@unimilitar.edu.co, u1101909@unimilitar.edu

Abstract. Due to the housing crisis presented by some Latin American countries, which have abundant natural resources, the development of non-conventional materials has become a necessity today. Colombia is one of the countries that identifies itself as a regional reference for its development in the study and application of alternative materials for its use in construction. Colombia is recognized fundamentally to the wide availability of natural resources, among which the Bamboo of the species Guadua Angustifolia Kunth. Despite the fact that the use for structural purposes of Guadua elements is normalized in Title G-12 of the Colombian Regulation of Resistant Earthquake Construction (NSR-10), it is necessary to consider that being a natural material, volumetric variations can occur. Especially when the material is subjected to changes in moisture between the saturation point of the fibers and the equilibrium moisture. The objective of this paper is to evaluate the influence of the moisture content on the mechanical properties of elements extracted from the base, middle and top of bamboo culms aged 4 to 6 years, after immunization with boric acid solution. The experimental characterization is focused on the determination of compression, bending and shear strength according to the recommendations of the Colombian Technical Standards. The results are compared with those proposed in the Colombian Regulation of Resistant Earthquake Construction (NSR-10).

Keywords: Bamboo, Moisture Content, Compression, Shear, Bending

Introduction

Currently the use of non-conventional materials has gained importance in several areas of civil engineering. The strength of the bamboo culms, their low weight as well as their hollow cylindrical structure make this type of material a viable option for the elaboration of structural elements that are subjected to the action of axial and flexural loads during its service life [1]. However, it has been found, that as a lignocellulosic material both the physical and mechanical properties of the material can be noticeably affected when the material is subjected to wet conditions for prolonged periods [2, 3].

In recent years, the study of the factors that affect the mechanical performance of structures made from the use of lignocellulosic materials has been reported in the specialized literature [4-6]. Mvondo et al. (2017) analyzed the effect of moisture content on tensile and flexural strength of three species of tropical wood: Milicia excelsa, Nauclea diderrichii and Erythrophleum suaveolens, varying the moisture content between 10% and 30%. This work not only demonstrates a loss of mechanical strength with increasing moisture content, but also establishes the effect of moisture on the main functional groups present in the wood [7].

According to Okhio et al. (2011), the analysis of the influence of moisture content on bamboo elements may become more complex than wood elements [1]. This is because in the culms the humidity can vary not only throughout the culm length, but also throughout its cross section [5]. Research results on the physical and mechanical properties of bamboo culms have focused on evaluating the effect of moisture content on the specific weight, dimensional stability and

Non-Conventional Materials and Technologies – NOCMAT for XXI Century   Materials Research Forum LLC
Materials Research Proceedings 7 (2018) 622-631   doi: http://dx.doi.org/10.21741/9781945291838-59

compressive strength of bamboo culms of Dendrocalamus giganteus species , analyzing the age of harvest and the conditions of growth (height above sea level) [8].

It has been shown that it is possible to degrade the mechanical properties of bamboo elements when subjected to wet conditions for extended periods of time [9, 10]. Xu et al. (2014) analyzed the behavior of bamboo culms after immersion in water for 1 and 7 days, thus simulating the behavior of the material when it is subjected to the effect of rainy periods. These studies showed that with the exposure to heavy rains, bamboo elements not only lose mechanical strength, but also go from having a fragile behavior to a very ductile behavior [10]. On the other hand, Jakovljevic et al. analyzed the influence of the moisture content on the mechanical properties (tensile, compression and static bending) of bamboos of species Pseudosasa amabilis y Pleioblastus amarus. The study was based on analyzing the performance of the material after being for three weeks in the wet chamber. The results confirmed that regardless of the type of stress, mechanical strength is significantly reduced when the material has a moisture content of about 60% [11].

In Colombia, experimental studies have been carried out in the last decade. In these works it has been verified the influence of moisture content on the strength of elements of Guadua culms [12-14]. According to Gutiérrez and Takeuchi (2014) bamboo elements of Guadua Angustifolia Kunth do not present a significant reduction of the tensile strength parallel to the fiber in the range of humidities in which the material is used for structural purposes [12]. However, results presented by Dumar (2014) demonstrate that both the flexural strength and the modulus of elasticity decreases linearly by every 0.01% that the moisture content in the Guadua angustifolia Kunth increases, when the material is below its proportionality limit [13].

Based on the literature consulted, this paper presents an experimental methodology for the evaluation of the effect of moisture content on the mechanical properties of specimens extracted from Guadua Angustifolia Kunth bamboo culms, analyzing its influence in the determination of the maximum stresses to be used in the resistant earthquake design of bamboo structures.

Materials and methods

For the development of this work , bamboo culms of the Guadua Angustifolia Kunth species were selected, healthy, free of defects, with average age of 4 years immunized by injection of solution of pentaborate and boric acid (according to the recommendation of the Colombian technical norm NTC 5301) . The material used comes from the municipality of Calarcá, Quindío. From each of the culm regions, base (B), middle (M and, top (T) were cut and labeled test pieces according to specifications of the Colombian technical standard NTC 5525 [15].

The determination of the moisture content was performed immediately after completion of the mechanical characterization tests, establishing for each specimen the mass loss, expressed as a percentage of the oven dry mass, using Eqn. 1.

$$CH = \frac{w_h - w_s}{w_s} 100 \tag{1}$$

Where

CH   is the moisture content, in %
w_h is the mass of the wet specimen, in g
w_s is the mass of the oven-dried specimen, in g

For the determination of the basic density, prismatic specimens were prepared, with an average width of 25 mm, height 25 mm and thickness equal to the thickness of the culm wall. The density (mass, oven dried, per unit volume of green) of each sample was obtained by Eqn. 2.

$$DB = \frac{w_s}{V_v} \qquad (2)$$

Where

DB   is the basic density, in g/cm^3
w_s is the mass of the oven-dried specimen, in g
V_v is the wet (green) volume of the specimen, in cm^3.

The volumetric shrinkage from the initial wet condition to the dry final condition was calculated according to Eqn. 3.

$$ICV = \frac{V_v - V_s}{V_v} \, 100 \qquad (3)$$

Where

V_v is the wet (green) volume of the specimen, in cm3.
V_s is the dry volume of the specimen, in cm^3.
ICV   is the rate of volumetric shrinkage, in %

For the realization of axial compression tests, specimens of equal length in diameter were prepared. The geometric properties are presented in Table 1.

Table 1 - Geometric properties of specimens used in compression test

Specimens	External diameter (mm)	Thicknes (mm)	Height (mm)	Net Area (mm^2)
Top without node	93.60 ±0.28	13.78 ± 0.72	101.08 ± 0.66	3453.40 ± 140.82
Top with node	97.12 ± 3.7	9.73 ± 2.09	101.09 ± 2.82	2631 ± 395.03
Middle without node	106.59 ± 2.43	12.12 ± 0.9	104.53 ± 8.81	35993 ± 20,85
Middle with node	110.87 ± 0.73	13.10 ± 0.83	111.86 ± 1.21|	4023.89 ± 246.03
Base without node	134.35 ± 1.13	14.49 ± 0.27	128.74 ± 1.57	5456.76 ± 142.38
Base with node	138.04 ± 0.04	19.95 ± 1.05	136.89 ± 2.71	7397.07 ± 324.07

The test was performed in a universal test machine by placing the specimens in such a way that the center of the moving head remained vertically above the center of the cross-section thereof and initially applying a preload of 1 kN to accommodate the specimen in the testing machine. The load was continuously applied at a speed of 0.01 mm / s, until failure occurred. The maximum compression stress was determined according to Eqn. 4:

$$\sigma = \frac{F}{A} \qquad (4)$$

Where

σ is the compression stress, in MPa
F   is the compression load, in N.

Non-Conventional Materials and Technologies – NOCMAT for XXI Century   Materials Research Forum LLC
Materials Research Proceedings **7** (2018) 622-631   doi: http://dx.doi.org/10.21741/9781945291838-59

A   is the net cross-sectional area, in mm^2.

For the realization of the shear tests, specimens of equal length in diameter (L=D) were prepared. The geometric properties are presented in Table 2.

Table 2 - *Geometric properties of specimens used in shear test*

Specimens	Thickness (mm)	Height (mm)
Top without node	9.92±0.33	87.94±0.11
Top with node	10.22±0.26	100.84±0.15
Middle without node	13.56±0.41	105.83±0.17
Middle with node	13.45±0.28	105.76±0.13
Base without node	23.19±0.39	136.40±0.09
Base with node	23.39±0.24	136.21±0.11

For the determination of the maximum strength, a universal test machine was used, in which the specimen was supported at the lower end, on two quarters of its surface, applying the load at the upper end, on the two quarters parts that were not supported. This way of supporting and applying the load to the specimen produces four cutting areas (see Figure 1). This test allowed the determination of the maximum load at which the specimen failed, as well as the number of areas that failed. The maximum parallel shear strength was determined according to Eqn. 5.

$$\tau = \frac{F}{\sum(t*L)} \tag{5}$$

Where

τ is the shear strength, in MPa
F   is the applied load, in N
t   is the thickness of specimen , in mm
L   is the height of the specimen, in mm

FIGURE 1 - *SHEAR TESTS*

Non-Conventional Materials and Technologies – NOCMAT for XXI Century   Materials Research Forum LLC
Materials Research Proceedings 7 (2018) 622-631   doi: http://dx.doi.org/10.21741/9781945291838-59

For the determination of the modulus of rupture (MR) a three-point bending test was performed (See Figure 2). For the test were cut rectangular beams of 500 mm in length and square section according to the thickness of the culm region. A distance supports of 300 mm was established. The physical properties are presented in Table 3. The rupture modulus (MR) was calculated using Eqn. 6.

FIGURE 2 - *STATIC BENDING TESTS*

$$MR = \frac{3FL}{2b\,d^2} \qquad (6)$$

Where

F   is the applied load, in N
d   is the thickness of specimen , in mm
L   is the distance between supports, in mm
b   is the width of the beam, in mm

Table 3 - *Geometric properties of specimens used in bending test*

Specimens	thickness (mm)	Width (mm)
Top	9.85±0.45	9.62±0.63
Middle	10.38±0.25	10.49±0.39
Base	20.95±0.37	21.30±0.09

Non-Conventional Materials and Technologies – NOCMAT for XXI Century   Materials Research Forum LLC
Materials Research Proceedings 7 (2018) 622-631   doi: http://dx.doi.org/10.21741/9781945291838-59

Results ans discussions

The volumetric contraction index and the basic density of specimens extracted from the Guadua culms were determined according to the specifications of NTC 5525. From the results, it is possible to estimate the saturation point of the fibers (FSP) of the material. The results are presented in Table 4.

Table 4 - *Saturation points of fibers*

Region	4ICV (%)	DB (g/cm^3)	FSP (%)
Top	10.43±1.36	0.57±0.05	21.26±1.09
Middle	11.74±0.78	0.58±0.02	22.42±1.14
Base	13.59±0.83	0.61±0.08	24.78±0.83

For the static bending test, a center loading across a beams with 300 mm of span was used. Results are shown in Figure 3.

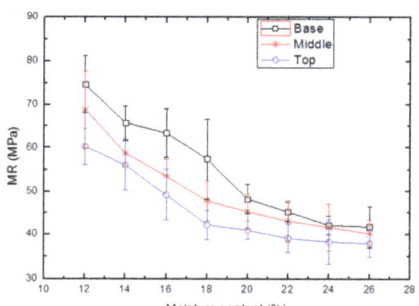

FIGURE 3 - *RESULTS OF STATIC BENDING TESTS*

Figure 3 shows the variation of the static flexural strength for specimens extracted from the top, middle and base of the Guadua culms. The results presented indicate a non-linear decrease of the average stress values. Depending on the region of the culms, the reduction in the value of the rupture modulus can be about 45%, when the material has a moisture content close to the point of fibers.

In order to analyze the influence of the moisture content on the strength to parallel compression of Guadua culms, graphs of maximum stress were performed as a function of the moisture content. The results are presented in Figure 4.

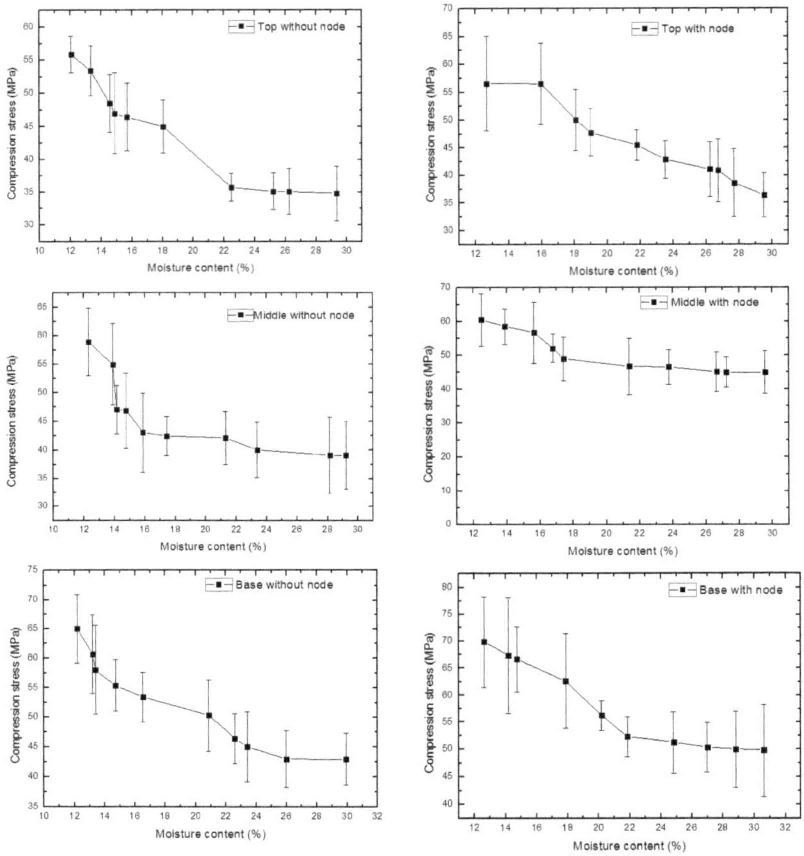

Non-Conventional Materials and Technologies – NOCMAT for XXI Century   Materials Research Forum LLC
Materials Research Proceedings **7** (2018) 622-631   doi: http://dx.doi.org/10.21741/9781945291838-59

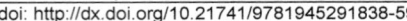

FIGURE 4 - *Results of compression tests*

From the results presented in Figure 4 it is possible to perceive that independently of the region of the culm, the compression stress undergoes a reduction that can be up to 30% when the material has a moisture content between 12% and the saturation point of the fibers. The reduction in the value of the basic compression stress can affect the determination of the maximum allowable stresses that must be considered in the structural design, mainly when the material has a moisture content between the saturation point of the fibers and the equilibrium moisture of the region.

To analyze the influence of moisture content on the shear behavior of bamboo cylindrical specimens were tested. The results are presented in Figure 5.

Non-Conventional Materials and Technologies – NOCMAT for XXI Century   Materials Research Forum LLC
Materials Research Proceedings **7** (2018) 622-631   doi: http://dx.doi.org/10.21741/9781945291838-59

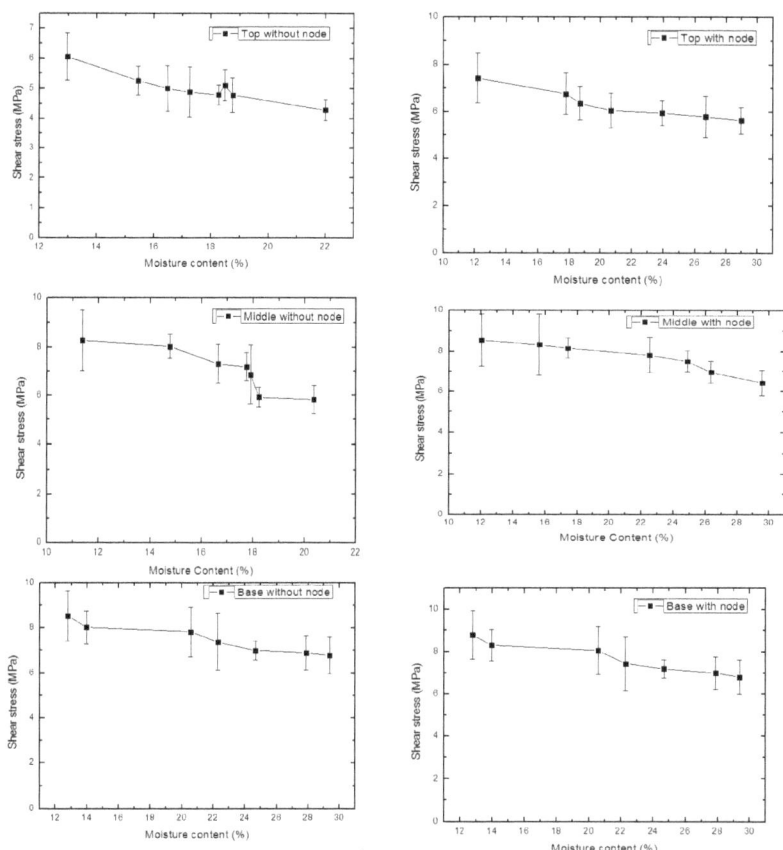

FIGURE 5 - *RESULTS OF shear TESTS*

From the results presented in Figure 2, it is possible to notice a decrease in the shear strength as the moisture content increases. This reduction is more significant for specimens extracted from the top and middle of the culms (25% - 30%). Specimens extracted from the base of *Guadua* culms had a stress reduction of less than 23%.

From the results obtained it is possible to determine the variation of the maximum permissible stresses for each of the culm regions. The results can be compared with results presented in the Colombian Regulation of Resistant Earthquake Construction (NSR-10) [16] (See Table 5).

Non-Conventional Materials and Technologies – NOCMAT for XXI Century   Materials Research Forum LLC
Materials Research Proceedings 7 (2018) 622-631   doi: http://dx.doi.org/10.21741/9781945291838-59

TABLE 5 *- maximum permissible stresses*

Moisture content (%)	Top without node (MPa)	Top with node (MPa)	Middle without node (MPa)	Middle with node (MPa)	Base without node (MPa)	Base with node (MPa)	NSR-10 [16]
12	0.97	1.19	1.33	1.37	1.37	1.41	1.20
14	0.84	1.09	1.29	1.34	1.29	1.33	1.13
16	0.80	1.02	1.17	1.31	1.25	1.29	1.07
18	0.78	0.97	1.15	1.25	1.18	1.19	1.00
20	0.77	0.95	1.10	1.20	1.12	1.15	0.96
22	0.82	0.93	0.95	1.11	1.11	1.12	0.96

Conclusions

In this paper, the influence of the moisture content on the mechanical properties of elements extracted from the base, middle and top of bamboo culms was evaluated. The experimental characterization was focused on the determination of compression, bending and shear strength.

From the results it is possible to verify that the increase of the moisture content reduces the mechanical strength of the material. The decrease of the values of maximum stress is more significant for moisture values between 12% and saturation point of the fibers and is accompanied by important dimensional changes that impair the mechanical performance of the material. The results show that above the point of saturation of the fibers does not modify the mechanical behavior of the culms.

Acknowledgment

This paper is a derivative product of the project (INV-ING-2389) financed by the Vice-rectory of Research of Universidad Militar Nueva Granada-validity (2017).

References

[1] Okhio C.B, Waning J.E, Mekonnen Y.T. An Experimental Investigation of the Effects of Moisture Content on the Mechanical Properties of Bamboo and Cane. JSAB 2011:7-14.

[2] Chen H., Miao M., and Ding X. Influence of moisture absorption on the interfacial strength of Bamboo and cane and Cane/vinyl ester composites. Composites Part A 2009: 40,(12): 2013-2019. https://doi.org/10.1016/j.compositesa.2009.09.003

[3] Gerhards, C.C. Effect of Moisture Content and Temperature on the Mechanical Properties of Wood: An Analysis of Immediate Effects. Wood Fiber Sci 1981:14(1): 4-36.

[4] Jiang Z., Wang H., Tian G., Liu X., Yu W. Moisture and bamboo properties. Bioresources 2012; 7(4):5048–58.

[5] Chen T.Y, Shih C.H, Chen H.C. Determination of the moisture content at fiber saturation point of bamboo by nondestructive testing method with stress wave timer. In: Proceedings of the fifth world conference on timber engineering, Montreaux, Switzerland 1998:2: 820–821.

[6] Wakchaure M., Kute S. Effect of moisture content on physical and mechanical properties of bamboo. Asian J Civ Eng (Build Housing) 2012: 13(6):753–763.

[7] Mvondo R., Meukam M., Jeong J., Meneses D. Nkeng E. Influence of water content on the mechanical and chemical properties of tropical wood species, Results in Physics 2017:7: 2096–2103. https://doi.org/10.1016/j.rinp.2017.06.025

[8] Xiaobo L. Physical, chemical, and mechanical properties of Bamboo and cane and Cane and its utilization potential for fiberboard manufacturing. Thesis 2004: 27-30.

[9] Pérez-Peña N., Valenzuela L., Diaz-Vaz J.E., Ananías R.A. Prediction of equilibrium moisture content in wood in relation to the specific gravity of the cell wall and environmental variables. Maderas, Cienc. Tecnol 2011:13(3):253-266.

[10] Xu Q., Harries K., Li X., Liu Q., Gottron J. Mechanical properties of structural bamboo following immersion in water, Eng Struct 2014:81: 230–239. https://doi.org/10.1016/j.engstruct.2014.09.044

[11] Jakovljevic S., Lisjak D., Alar Z., Penava F. The influence of humidity on mechanical properties of bamboo for bicycles. Constr Build Mater 2017:150:35–48. https://doi.org/10.1016/j.conbuildmat.2017.05.189

[12] Gutiérrez M., Takeuchi C. Efecto del contenido de humedad en la resistencia a tensión paralela a la fibra del bambú Guadua Angustifolia Kunth. Scientia et Technica 2014: 19(3):245-250.

[13] Dumar J. Determinación de la variación de la resistencia a flexión y módulo de elasticidad longitudinal de la guadua angustifolia kunth, con el contenido de humedad. Thesis 2015:1-118.

[14] Giraldo G., Wbeimar G. Resistencia de la sección transversal de la guadua sometida a compresión. Reporte técnico 2004:1-28.

[15] Instituto Colombiano de Normas Técnicas. Métodos de ensayo para determinar las propiedades físicas y mecánicas de la Guadua angustifolia Kunth (NTC5525). ICONTEC, Bogotá, 2007.

[16] Reglamento Colombiano de Construcción Sismo Resistente NSR-10. Asociación colombiana de Ingeniería Sísmica, 2010.

Non-Conventional Materials and Technologies – NOCMAT for XXI Century   Materials Research Forum LLC
Materials Research Proceedings 7 (2018) 632-639   doi: http://dx.doi.org/10.21741/9781945291838-60

Shear Behavior of the Lamined Bamboo Guadua (LBG)

Caori Takeuchi [a,*], Dorian Linero [b]

[a] Universidad Nacional de Colombia, Bogotá (Colombia); cptakeuchit@unal.edu.co

[b] Universidad Nacional de Colombia, Bogotá (Colombia); dllineros@unal.edu.co

* Universidad Nacional de Colombia, Facultad de Ingeniería, 406-301; cptakeuchit@unal.edu.co

Abstract. The shear behavior parallel and perpendicular to the fiber of laminated bamboo Guadua, LBG, was studied by testing six different types of samples. The definition of the types was made taking into account the direction of the fibers, the orientation of the slats and the shear plane. In all configurations, the shear between the parenchyma matrix and fibers was studied.
The samples whose fibers and boards were parallel to the direction of the load presented a crack parallel to the fiber whose surface mostly coincides with the expected cutting plane. The average value of shear strength of LBG was 6.0MPa, when the fibers are parallel to the loading direction and the contact surface between slats is perpendicular to the shear plane. In the samples where the fibers are perpendicular to the loading direction and to the contact surface between slats, the crack propagated approximately 45° from the vertical and the shear strength was 2.9MPa. In the samples in which the direction of the load was perpendicular to the fiber and parallel to the slats, the crack was parallel to the expected shear plane. In some of these samples, the crack path did not coincide with the expected shear plane, occurring in other parallel planes with lower fiber density. This case presented the smallest shear strength with average value of 2.3MPa.

Keywords: Laminated Bamboo Guadua, Shear Behavior, Strength, Cracks

Introduction

Bamboo *Guadua* is a fastest-growing plant, located at tropical and subtropical zones, mainly in Central and South America. The culm of this plant has hollow circular cross-section which external diameter and wall thickness greater than the other bamboo types. The internal structure of the *Guadua* presents 40 percent of cellulose large fibers embedded in 60 percent of a parenchyma matrix, approximately. The fibers are oriented parallel to the culm and are concentered near the outer wall. This kind of bamboo has been used to build houses in some Colombian regions. Currently, there are buildings and bridges made with this material in natural and laminated form.

In the laminated process, first slats of rectangular cross-section are extracted from *Guadua* culms, then these slats are joined and glued with an adhesive and finally the blocks are conformed and named laminated bamboo Guadua (LBG).

The behavior of the bamboo *Guadua* has been studied since some years [1-5], obtained elasticity modulus, strength and other mechanical properties. Particularly, some authors have determined the shear strength of LBG although of experimental tests as blocks subjected to direct shear and notched beams subjected to bending and shear [6-8]. Only one fiber direction of the LBG with respect to shear plane has been studied in these researches.

This work describes the shear behaviour of LBG by means of experimental tests of direct shear [9]. The fibers direction and the orientation and disposition of the slats with respect to the expected shear plane determine six different configurations of the direct shear test. This work studies the influence of these configurations on the shear strength and the crack pattern of LBG.

Laminated bamboo guadua

The laminated bamboo Guadua (LBG) is a composite material, which is formed by long slats of rectangular cross-section glued and pressed together. Each slat is derived from a *Guadua* culm, preserving the fibers direction s parallel to its longitudinal axis, as shown in **FIGURE 2**.

Non-Conventional Materials and Technologies – NOCMAT for XXI Century   Materials Research Forum LLC
Materials Research Proceedings **7** (2018) 632-639   doi: http://dx.doi.org/10.21741/9781945291838-60

The plane associated to the long side of the rectangular cross-section in the contact of two slats, is defined as contact surface between slats. The unit vector **r** is normal to this surface, as shown in **FIGURE 2**.

Formaldehyde melanin urea is used as adhesive, whose strength is greater that the parenchyma matrix. Consequently, the failure is associated to the cracking of the parenchyma matrix, without debonding between the slats.

Experimental test

The standard test ASTM 143 – 94 allows evaluating the shear strength in the timber, where the shear plane is parallel to the grain orientation [10]. The same test protocol was used on LBG specimens, considering different directions of fibers and of slats with respect to the expected shear plane.

The geometry, the loads and the boundary conditions of the LBG block are shown in **FIGURE 1**. The force vector $\mathbf{t} = -p\,\mathbf{j}$ is applied on a portion of the top face of the block and in the negative direction of the y-axis. The surface normal to **i**, located at $x = 30\text{mm}$ corresponds to the expected shear plane. LBG specimen is subjected to a mean shear stress τ at the loading direction **j**, acting on the expected shear plane of vector normal **i**. The mean shear stress τ corresponds to the component σ_{xy} of the stress state and it is equal to the applied load p divided by the area of the expected shear plane A_s.

In configurations A, B and C, the fibers direction **s** of LBG block is parallel to the loading direction **j** and perpendicular to **i**, as shown in **FIGURE 2**. Particularly, the normal to long side of the cross-section of the slats **r** is parallel to z-axis in the configuration A (**FIGURE 2(A)**), and **r** is parallel to x-axis in the configurations B y C. The expected shear plane matches the contact surface between slats in configuration C (**FIGURE 2(C)**), and it does not match in the configuration B (**FIGURE 2(B)**).

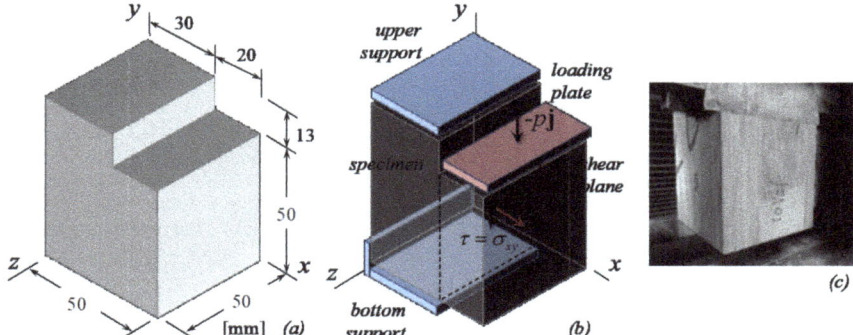

FIGURE 1 – *Experimental Test: (a) geometry of specimen, (b) applied loads and boundary conditions, and (c) photo.*

Non-Conventional Materials and Technologies – NOCMAT for XXI Century   Materials Research Forum LLC
Materials Research Proceedings 7 (2018) 632-639   doi: http://dx.doi.org/10.21741/9781945291838-60

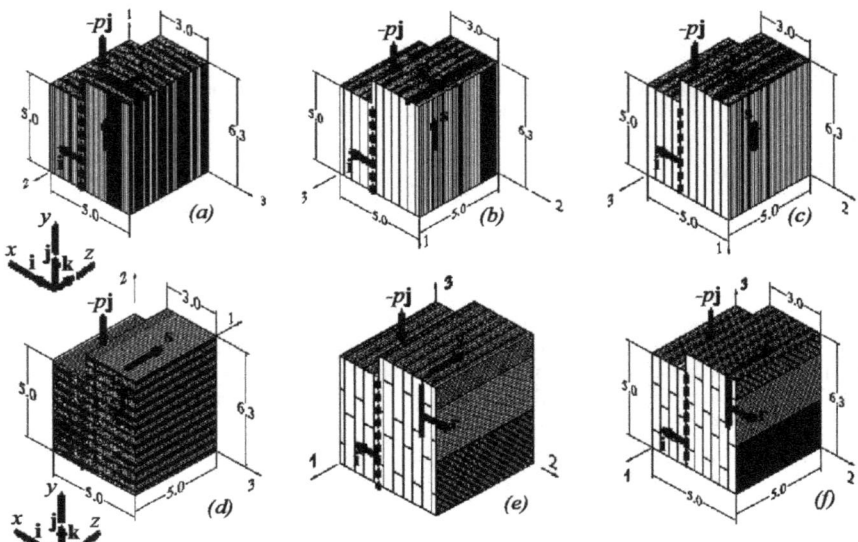

FIGURE 2 – *Configurations of experimental test: (a) Configuration A, (b) Configuration B, (c) Configuration C, (d) Configuration D, (e) Configuration E, and (f) Configuration F.*

In configurations D, E and F, the fiber directions of the LBG block are perpendicular to the loading direction **j** and are parallel to **i**. Particularly, the normal to the long side of the cross-section of the slats **r** is parallel to y-axis in the configuration D (**FIGURE 2(D)**), and **r** is parallel to x-axis in the configurations E y F. The expected shear plane matches the contact surface between slats in configuration F (**FIGURE 2(F)**), and it does not match in the configuration E (**FIGURE 2(E)**).

The fibers do not cross the expected shear plane in the six configurations, in other words, the fiber direction **s** is perpendicular to the normal of the expected shear plane **i**. This allows observing the shear capacity of the parenchyma matrix. If the fibers are perpendicular to the shear plane, the high shear strength of the fibers produces bending of the specimen and cracks of perpendicular tension [8].

Analysis of results
The mechanical response of a specimen is represented by means of the relationship between the mean shear stress τ and the plate displacement δ divided by the height l of the shear plane at the loading direction.

LBG specimen photos after the test show the fiber direction **s**, the normal to the long side of the slats **r** and the crack path which unit normal vector is **n**.

Shear test of LBG specimen with fibers parallel to the loading direction
Ten LBG specimens with Configuration A were tested. In all the samples, the crack pattern corresponds closely to the expected shear plane, as shown the normal vector **n** in **FIGURE 3(A)** and **FIGURE 3(B)**. Particularly, **FIGURE 3(C)** indicates that the material fracture occurs in the parenchyma matrix and its path avoids the fibers, but tends a straight line parallel to z-axis.

Non-Conventional Materials and Technologies – NOCMAT for XXI Century   Materials Research Forum LLC
Materials Research Proceedings 7 (2018) 632-639   doi: http://dx.doi.org/10.21741/9781945291838-60

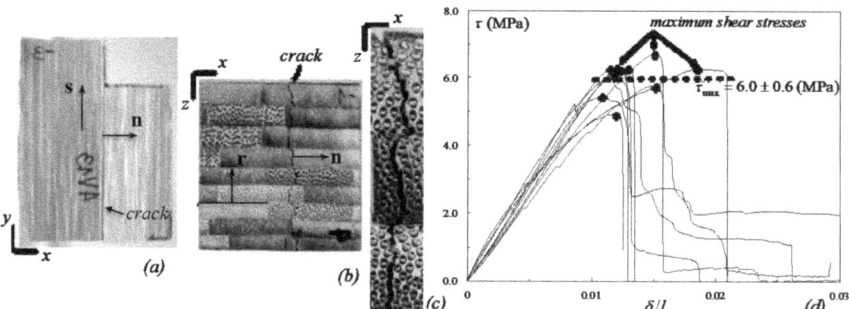

FIGURE 3 – *Results of configuration A: crack pattern: (a) in xy-plane, (b) in xz-plane bottom side, (c) close up in xz-plane, and (d) mechanical response.*

The mechanical response of the samples indicates linear behavior and low dispersion in the initial stage (**FIGURE 3(D)**). Likewise, shows behavior brittle and moderate dispersion in the post-peak stage. The maximum mean shear stress exhibits an average value of 6.0MPa and standard deviation of 0.6MPa.

In Configuration B, six of the ten samples indicate a crack without the contact surfaces between slats, as shown in **FIGURE 4(B)**. Like in previous configuration, the material behavior is brittle and the crack occurs in the parenchyma matrix. The mechanical responses of four samples are considered outliers (**FIGURE 4(D)**), and these are excluded because exhibit crack patterns different to the expected shear plane. The average value of maximum shear stress is 5.7MPa and its standard deviation is 0.5MPa.

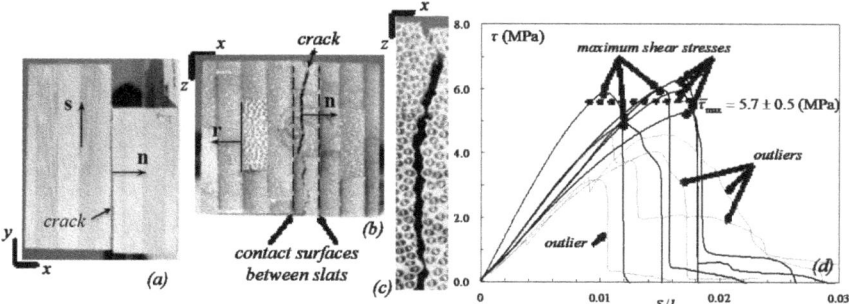

FIGURE 4 – *Results of configuration B: crack pattern: (a) in xy-plane, (b) in xz-plane bottom side, (c) close up in xz-plane, and (d) mechanical response.*

Ten specimens with Configuration C were tested, where one sample presented outlier values. **FIGURE 5(B)** and **FIGURE 5(C)** show that the crack pattern matches approximately with the contact surface between slats. The average value of maximum shear stress is 4.7MPa and its standard deviation is 0.6MPa.

Non-Conventional Materials and Technologies – NOCMAT for XXI Century   Materials Research Forum LLC
Materials Research Proceedings **7** (2018) 632-639   doi: http://dx.doi.org/10.21741/9781945291838-60

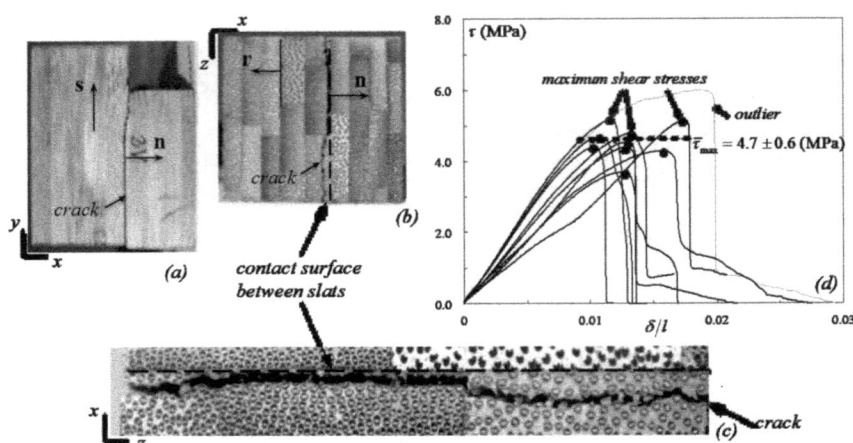

FIGURE 5 – *Results of configuration C: crack pattern: (a) in xy-plane, (b) in xz-plane bottom side, (c) close up in xz-plane, and (d) mechanical response.*

Shear test of LBG specimen with fibers perpendicular to the loading direction
The crack path of the ten samples of the Configuration D is oriented 45 degrees with respect to the expected shear plane, such as, the angle between the vectors **i** and **n** is 45 degrees, as shown in **FIGURE 6(A)**. The close up of the **FIGURE 6(B)** indicates that the crack occurs in the parenchyma matrix and crosses the contact surfaces between the slats. The mechanical response curves in **FIGURE 6(D)** show that the dispersion of the initial slope is low and the dispersion of the maximum shear stress is moderate. The latter has an average value of 2.9MPa and a standard deviation of 0.3MPa. Likewise, the curve indicates more ductile of the material than the mechanical response of the previous configurations.

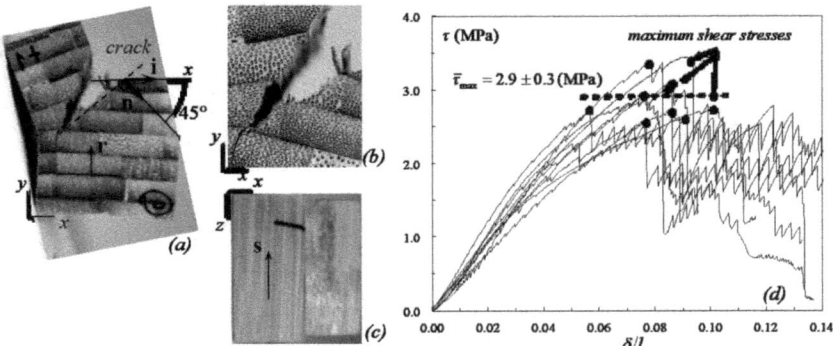

FIGURE 6 – *Results of configuration D: crack pattern: (a) in xy-plane, (b) close up in xy-plane, (c) in xz-plane upper side, and (d) mechanical response.*

Non-Conventional Materials and Technologies – NOCMAT for XXI Century   Materials Research Forum LLC
Materials Research Proceedings 7 (2018) 632-639   doi: http://dx.doi.org/10.21741/9781945291838-60

Small differences of the crack pattern in the ten samples of Configuration E were observed. Initially the crack is orientated 45 degrees with respect to **i**, then this is aligned near a contact surface between slats (SBS). In some samples, the crack is near the first surface between slats, as shown in **FIGURE 7(A), FIGURE 7(B)** and **FIGURE 7(F)**. In others samples, the crack is near the second (**FIGURE 7(C)**) or the thirty contact surface between slats (**FIGURE 7(D)**). The mechanical response curves have low dispersion of the initial slope and the maximum shear stress. The latter has an average value of 2.2MPa and a standard deviation of 0.3MPa.

Ten samples of Configuration F were tested, where four shown outlier results. The other five samples presented two cracks parallel to the contact surface between slats. In some samples, the cracks is located near the first surface between slats, one grown from top to down and the another grown from down to top, as shown in **FIGURE 8(A)**. In other cases, the top – down crack is located near the first surface between slats, but the down – top crack begins with 45 degrees of orientation and then is aligned near the second surface between slats, as shown in **FIGURE 8(B)** and **FIGURE 8(C)**. High dispersion of the initial slope and the maximum shear stress is observed in the mechanical response curves of **FIGURE 8(D)**. The maximum shear stress has an average value of 2.3MPa and a standard deviation of 0.3MPa.

In configurations D, E and F, the saw-tooth form of the mechanical response curves is produced by the sudden change of the crack path in the parenchyma matrix, required to avoid the fibers. Generally, the maximum shear stress of the mechanical response corresponds to the shear strength of LBG.

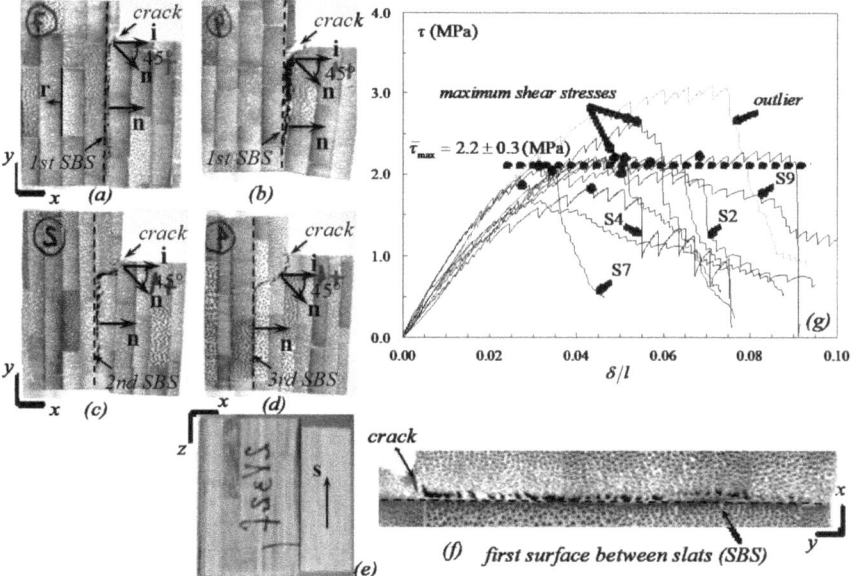

FIGURE 7 – *Results of configuration E: crack pattern in xy-plane: (a) sample 7, (b) sample 9, (c) sample 2, (d) sample 4, (e) crack pattern in xz-plane, (f) close up in xy-plane of sample 7, and (g) mechanical response.*

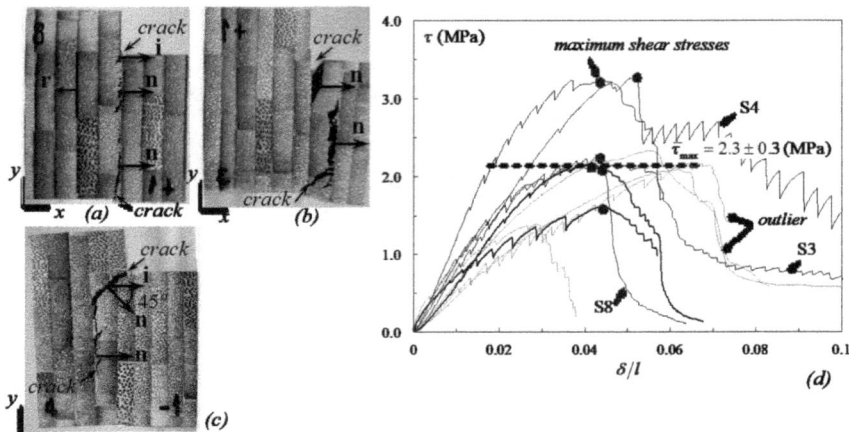

FIGURE 8 – *Results of configuration F: crack pattern in xy-plane: (a) sample 8, (b) sample 3, (c) sample 4, and (d) mechanical response.*

Conclusions

The shear strength and the crack pattern of the laminated bamboo Guadua are studied in this work, considering that the fibers do not cross the expected shear plane. Two fiber orientations are tested: fibers parallel to the loading direction (Configurations A, B and C) and fibers perpendicular to the one (Configurations D, E and F). The disposition of the slats is also studied, when the contact surface between slats is parallel to the expected shear plane (Configurations B, C, E and F) or when it is perpendicular (Configuration A and D). Likewise, the cases where the expected shear plane is within a slat (Configurations B and E) and the expected shear plane matches with the contact surface between slats (Configurations C and F) are analyzed.

The shear strength of LBG tests with the fibers parallel to the loading direction is significantly greater than the tests with fibers perpendicular to the one. In the samples where the contact surface between slats is perpendicular to the expected shear plane, the shear strength is 6.0MPa for fibers parallel to the loading direction and 2.9MPa for fibers perpendicular to the load.

The slats direction modifies the shear strength by about 5 percent for fibers parallel to the loading direction and 24 percent for fibers perpendicular to the load. Instead, the position of the contact surface between slats with respect to shear plane changes the strength by about 17 percent for fibers parallel to the loading direction and 4 percent for fibers perpendicular to the load.

In the configurations where the contact surface between slats (Configuration B, C, E and F) is parallel to the shear plane, the crack path tends to align with a contact surface between slats. Close to this surface, the parenchyma matrix is broken while the adhesive zone remains glued. Instead, when the fibers are perpendicular to the loading direction and to the contact surface between slats (Configuration D), the crack is propagated approximately 45 degrees with respect to expected shear plane.

The smallest strength shear the LBG is obtained when the fibers are perpendicular to the loading direction and the contact surface between slats is parallel to the shear plane.

Non-Conventional Materials and Technologies – NOCMAT for XXI Century   Materials Research Forum LLC
Materials Research Proceedings 7 (2018) 632-639   doi: http://dx.doi.org/10.21741/9781945291838-60

References

[1] Dixon, P.G., et al., *Comparison of the structure and flexural properties of Moso, Guadua and Tre Gai bamboo.* Construction and Building Materials, 2015. **90**: p. 11-17. https://doi.org/10.1016/j.conbuildmat.2015.04.042

[2] García, J.J., C. Rangel, and K. Ghavami, *Experiments with rings to determine the anisotropic elastic constants of bamboo.* Construction and Building Materials, 2012. **31**: p. 52-57. https://doi.org/10.1016/j.conbuildmat.2011.12.089

[3] Pacheco, C. and C. Takeuchi. Tension strength perpendicular to the fiber in Guadua angustifolia (Bamboo). in Brazilian Conference on Non-Conventional Materials and Technologies in Ecological and Sustainable Construction. NOCMAT 2006. 2006. Salvador de Bahia (Brazil).

[4] Takeuchi, C., et al., Experimental Determination of Allowable Stresses for Bamboo Guadua Angustifolia Kunth Structures. Key Engineering Materials, 2012. **517**: p. 76-80. https://doi.org/10.4028/www.scientific.net/KEM.517.76

[5] Trujillo, D. and L.F. López, 13 - Bamboo material characterisation, in Nonconventional and Vernacular Construction Materials. 2016, Woodhead Publishing. p. 365-392. https://doi.org/10.1016/B978-0-08-100038-0.00013-5

[6] Archila, H., M. Ansell, and P. Wallker. Measurement of the in-plane shear moduli of bamboo-guadua using the iosipescu shear test method. in 10th World Bamboo Congress. 2015. Korea.

[7] Correal, J., et al., Experimental evaluation of physical and mechanical properties of Glued Laminated Guadua angustifolia Kunth. Construction and Building Materials, 2014. **73**: p. 105-112. https://doi.org/10.1016/j.conbuildmat.2014.09.056

[8] Cortes, J.C., et al. Assesment of the influence of glue type in the mechanical behavior of glued laminated guadua (bamboo) in International Conference on Non-conventional Materials and Technologies (NOCMAT 2010). 2010.

[9] Takeuchi, C., Caracterización mecánica del bambú guadua laminada para uso estructural, in Facultad de Ingeniería - Doctorado en Ciencia y Tecnología de Materiales. 2014, Universidad Nacional de Colombia: Bogotá.

[10] ASTM, *Standard Test Methods for Small Clear Specimens of Timber.* 2000, ASTM International: United States of America.

Non-Conventional Materials and Technologies – NOCMAT for XXI Century   Materials Research Forum LLC
Materials Research Proceedings 7 (2018) 640-649   doi: http://dx.doi.org/10.21741/9781945291838-61

Concrete with Recycled Aggregates: Physical Properties and External Sulfates Attack

Eduardo da Cruz Teixeira[a*], Camila Macêdo Medeiros[a], M.A. Padilha Jr [a],
Normando Perazzo Barbosa [b], J. A. Rossignolo [c]

[a]IF Sertão PE – campus Salgueiro, Brazil; educrtx@hotmail.com,
camilamedeirosm@gmail.com, jr_padilhamarcos@hotmail.com

[b]UFPB, Brazil; nperazzob@yahoo.com.br

[c]USP, Brazil; e-mail: j.a.rossignolo@gmail.com

Abstract. The search for new materials that can bring more sustainability and reduce the costs of construction has proven target of several studies. When it comes to concrete, building waste and demolition material have been extensively studied, however, it is necessary to advance the durability studies recycled concrete. Thus, this research studied concrete with recycled aggregates, the absorption properties, consistency, strength and durability aspects, such as external sulfates attack. The research was developed in the IFSPE, campus Salgueiro, Brazil, with aggregates generated by construction and demolition waste in the city. The results show the direct relationship between the physical properties of concrete and behavior of the material to attack by sulfates, 02 percentage of substitution of recycled aggregate were consistent with the properties of conventional concrete, completed the compatibility of the use of recycled concrete without structural.

Keywords: Concrete, Recycled, Sulfate Attack

Introduction

The industry of civil construction is recognized as one of the most important activities to the economic and social development, however it still behaves as a great generator of environmental impacts, being by the consumption of natural resources, or by landscape modification or residue generation [1].

The sector is one of the largest consumers of natural resources, either as construction materials, or as raw material of these. It is estimated that the civil construction uses between 20 and 50% of natural resources consumed by the society. In the case of wood, about 50% of the amount extracted is consumed in the world as a building material [2]. The consumption of natural aggregates varies between 1 and 8 tons / habitants Research Group Unconventional Materials, Salgueiro, Brazil year, in 2014 the aggregate industry for construction presented a demand of about 740 million tons of crushed rock and sand (ANEPAC, 2014). It is because of this high consumption, that around the big cities, sand and natural aggregates start to become scarce. In São Paulo the natural sand, mostly traveling distances over 100 km,

Bringing the cost to values around R\$25/m³. In the state of Pernambuco, there is only one licensed deposit, to run, the other make the exploitation of irregular manner.

In generating waste, it is estimated that for every ton of municipal solid waste collected, 2 tons of waste are collected from the construction activity, or demolition [4]. The main problem of this type of waste, from the environmental point of view, is its irregular deposition, causing the creation of points of waste, contaminating the soil. Moreover, from a financial point of view, this irregular disposal burdens the municipal administrations, which end up having to be responsible for removal and disposal of such accumulate waste.

Given this context, taken by the premise of the economic development of the city of Salgueiro and the Central Hinterland of Pernambuco, in northeastern Brazil, IF Hinterland PE, campus of Salgueiro, the research groups ITCE (Innovation and Technology in Civil Engineeringand

Non-Conventional Materials and Technologies – NOCMAT for XXI Century   Materials Research Forum LLC
Materials Research Proceedings 7 (2018) 640-649   doi: http://dx.doi.org/10.21741/9781945291838-61

development of unconventional materials, both based in Salgueiro campus, have been studying the technical and economic viability of sustainable concrete using aggregates derived from waste from the construction industry (RCD).

The characteristics of the used waste influence the properties of concrete with recycled aggregates. This has been proven by several studies [5], [6], [7], [8], [9], Studies have consolidated the viability of the use of recycled aggregates in concrete for non-structural purposes, however, studies of mechanical behavior, durability and microstructure are being developed in order to use recycled concrete for structural purposes and ensure its technical and financial viability.

It is essential that the concrete structure continues to perform its intended function. Several analyses have been conducted to better understand the durability of concrete made with recycled aggregate. This property of concrete is defined as its ability to resist the action of the weather, attacks by sulphates, abrasion or other deterioration process that is, the durable concrete will retain its original form, quality and usability when exposed to a particular environment.

According to [10], this is an area of research that needs to be further explored, especially in Brazil, since the use of concrete with recycled aggregates will only gain market share when we get to know very well its durability in comparison with conventional concrete. Therefore, this research studied concrete with recycled aggregates, the absorption properties, and consistency, strength and durability issues, such as attacks by sulphates.

Methodology
Study area
The study area of this research is the municipality of Salgueiro, Figure 1. Salgueiro is a Brazilian city in the state of Pernambuco, northeastern Brazil. It belongs to the Mesoregion of the Hinterland of Pernambuco and Microregion of Salgueiro, being located west of the state capital, being far from it 513 km. It has a territorial extension of 1 733.7 km², with 6.75 sq km in urban area, having its estimated population in 2014 to 59 409 inhabitants. The municipal headquarters has an average temperature of 26.0 ° C, having Caatinga as its original and predominant vegetation. Approximately 80.7% of the population living in the urban area its Human Development Index (HDI) is 0.669, considered average compared to the state value. The provision of services and industry stands out as the main income generators for the municipality (Salgueiro Hall, 2013).

FIGURE 1 - *SALGUEIRO LOCATION, BRAZIL.*

Known as the "Crossroads of the Northeast" as it is in the most central part of the Northeast - It can be considered equidistant from almost all Northeastern capitals - Salgueiro It is the main city in the hinterland's central region of Pernambuco,holding at the regional level, a diversified trade. It is located in the municipality the central site of operation of the Transnordestina, railroad that connects the Suape Port, Southcoast of Pernambuco, to the Brazilian cerrado of Piaui and to the Pecem Port, Ceara. Salgueiro is still crossed by the Transposition canals of São Francisco River, worksthat promise to take the river water to Ceara, to the highland of Paraiba and to the highland of Rio Grande do Norte.

In Salgueiro – PE it is seen the civil construction growth in the city, once it hosts nowadays two large construction Works of Federal Government: the Transnordestina and the São Francisco River Transposition. It also has received the Federal Institute of Technologic Education, as source of personal qualification, the construction of the city Mall, besides the government programs, as Minha Casa, Minha vida. These aspects contribute to the population, Market and, consequentlyhousing growth in the city. The Zum Magazine (2013) said that the small Salgueiro (PE) became the new center of irradiation of important axes of the country's infrastructure. A strong new evidence of that growth is that the city, called as "Development Desk", had a 88% of growth in the number of Individual microentrepreneursin one year, from 2012 to 2013 (Salgueiro City Hall).

Materials

The recycled small aggregate originated by the process of crushing of RCD used in this work is from the city of Salgueiro-PE, and it was the material used as the research basis performed by the Highland IF PE, Campus of Salgueiro. It is a material which will be used in substitution of the natural small aggregate, the washed sand.

The washed sand is from river beds and has a maximum diameter of 2,40mm, fineness modulusof 1,41, according to [12], and the specific mass 2,36g/cm³ according to [13]. The recycled small aggregatehas maximum diameter of 2,26mm, fineness modulus of 2,40, according to [12] and specific mass of 2,28g/cm³, according to [13].

The big aggregate used in this work was the conventional aggregate from granitic rock, with maximum diameter of 25,4mm according to [12], shape index of 2,1, according to NBR [14], wear and abrasion of 37% according to [15] and powdery material level of 9,7%, according to [16].

The cement used was the CPIII, 40 Mpa, the high-oven Portland Cement, has addition of waste of 35% to70% in mass, which provides properties as: low heat of hydration, greatest impermeability and durability, being recommended to large and highly aggressive construction works, for its resistance to sulfates.

The trace used either in the reference concrete as in the concrete with 25%, 50% and 75% of RCD recycled aggregate was 1:2:2:0,5, according to Table 1, the mixture and densification were mechanically performed. 12 proof samples were molded for each type of concrete, being 6 of them for the mechanical resistance test, 3 for absorption by capillarity and 3 for the sulfates attack.

TABLE 1 - PROPORTION OF THE CONCRETE MIXTURE

	Mix	w/c	recycled small aggregate replacement content
Family 1	1:2:2	0,5	0%
Family 2	1:2:2	0,5*	25%
Family 3	1:2:2	0,5*	50%
Family 4	1:2:2	0,5*	75%

Methods

A The samples and cure were made according to [17]: Concrete - Procedure for molding and maturing bodies of the test pieces, the 6 test samples with the reference concrete and 6 with concrete made with recycled material for each replacement fraction were broken based on [18]: Concrete - compression testing of cylindrical test samples, in a hydraulic press, where it was obtained the compressive strength of each one.

Initially the bottom edges were coated with silicone to a height of 20 mm, so that water does not penetrate the sides of specimens. Then, the sample was placed on a plate on supports so that the

Non-Conventional Materials and Technologies – NOCMAT for XXI Century   Materials Research Forum LLC
Materials Research Proceedings 7 (2018) 640-649   doi: http://dx.doi.org/10.21741/9781945291838-61

bottom of the specimen stayed in contact with water. So, the water was put to a height of 5 mm above the bottom of specimens. There were 4 weighing for each sample. Measurements were made at the times 3 h, 6 h, 24 h and 72 h from the moment that the specimen were placed in contact with water on the plate. The absorption coefficient (k) of each period was calculated by dividing the change in mass of the specimen by the area of the base, according to Equation 1 [19].

$$k = (A-B)/S$$

Where:
k = capillary absorption coefficient (g/cm2);
A =sample mass that remains with one of the faces in contact with the water during a specific period of time, in g;
B = dry sample mass, as soon as it reaches the temperature of $(23 \pm 2)°$ C, in g;
S= base area of the sample (cm2).

The height of the maximum internal capillary rise should be expressed in centimeters.
    In order to use a fast experimental method, the analysis of attacks by sulfates, he samples were immersed in an anhydrous potassium sulfate solution, at a concentration of 10% in distilled water, the temperature of the environment was kept constant at 40 ° C (+/- 2 ° C), as shown in Figure 2.

FIGURE 2 - CYLINDRICAL SAMPLE IMMERSED IN SULFATE SOLUTION.

Resultsand discussions
The concrete prepared in the Construction Laboratory of the Federal Institute of Education, Science and Technology, campus Salgueiro were subjected to the characterization tests described in the methodology using Brazilian technical standards, presenting the results of absorption by capillarity, mechanical resistance to compression and the behavior of concrete immersed in sulfate solution was evaluated.

Absorption of the concretes
The absorption of concrete is an important property because it reveals the concrete capacity to suffer attacks by external agents, besides being related to the porosity of it.
    Concrete behavior was observed before the determination of the water absorption through capillarity rise, shown in Table 02.

TABLE 2 - ABSORPTION COEFFICIENTS [g/cm²].

Non-Conventional Materials and Technologies – NOCMAT for XXI Century   Materials Research Forum LLC
Materials Research Proceedings 7 (2018) 640-649   doi: http://dx.doi.org/10.21741/9781945291838-61

	3 h	6 h	12 h	24 h	48 h	72 h
Family 1	0,1273	0,1910	0,3184	0,3821	0,3821	0,4458
Family 2	0,1273	0,1910	0,3821	0,4458	0,4458	0,5732
Family 3	0,1910	0,3184	0,5732	0,6369	0,6369	0,8280
Family 4	0,3184	0,4458	0,7643	0,8917	1,0191	1,0828

The four prepared concrete in this paper presented the following capillary absorption coefficient values over time, according to Figure 3 and 4.

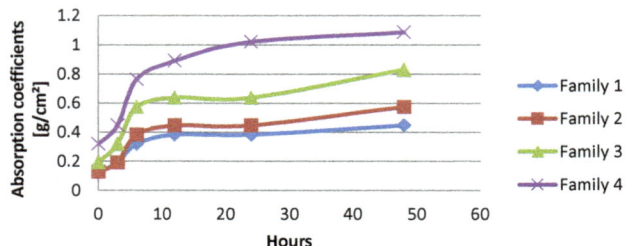

FIGURE 3 - *ABSORPTION COEFFICIENTS [g/cm²] X TIME*

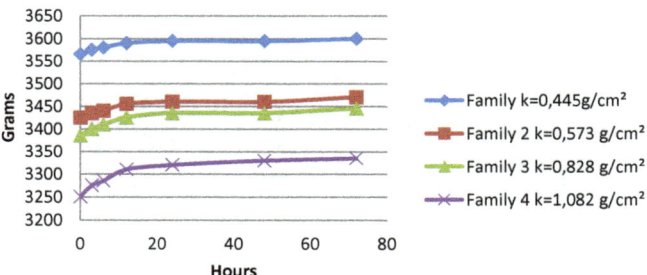

FIGURE 4 - *ABSORPTION (grams) X TIME (hours)*

The concrete with partial replacement of recycled have shown absorption coefficients for higher capillarity than the reference concrete (family 1), it is clear that the larger the replacement content, the higher the concrete's ability to absorb water, consequently more porous. The concrete with 75% recycled aggregate replacement recorded absorption of 1,082 g/cm², differently from the reference concrete which recorded valueof 0,455 g/cm².

After 72 hours of the start of the test, the samples were broken to the diametrical compression, being able to identify the maximum height that the water has reached into the concrete, according to Figure 5.

Non-Conventional Materials and Technologies – NOCMAT for XXI Century   Materials Research Forum LLC
Materials Research Proceedings 7 (2018) 640-649   doi: http://dx.doi.org/10.21741/9781945291838-61

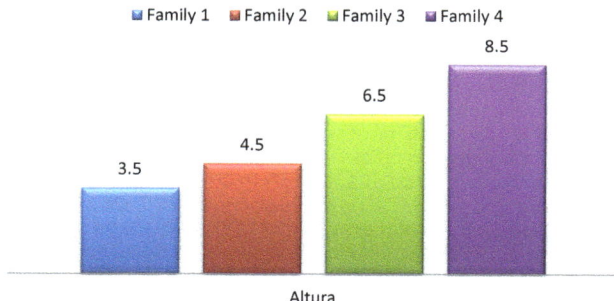

FIGURE 5 - *HEIGHT OF THE INTERNAL CAPILLARY RISE IN THE SAMPLES.*

Resistance to compression of the concretes

The concretes (families 1, 2, 3 e 4) have shown the following results of resistance to compression according to Table 3.

TABLE 3 - RESISTÊNCIA À COMPRESSÃO DOS CONCRETOS.

	Family 1	Family 2	Family 3	Family 4
sample 1	30,97 MPa	29,00 MPa	16,75 MPa	17,19 MPa
sample 2	26,47 MPa	28,73 MPa	13,21 MPa	13,88 MPa
sample 3	25,21 MPa	27,10 MPa	13,07 MPa	17,53 MPa
sample 4	28,43 MPa	26,63 MPa	16,16 MPa	12,60 MPa
sample 5	23,64 MPa	29,30 MPa	20,10 MPa	14,81 MPa
sample 6	27,52 MPa	27,15 MPa	16,71 MPa	16,38 MPa
Standard Deviation	2,56	1,15	2,62	1,96
Arithmetic Mean	27,04 MPa	27,99	16,00 MPa	15,40 MPa

The Figure 6 illustrates the mean resistance to compression of the prepared concretes, with the broken samples at 28 days old.

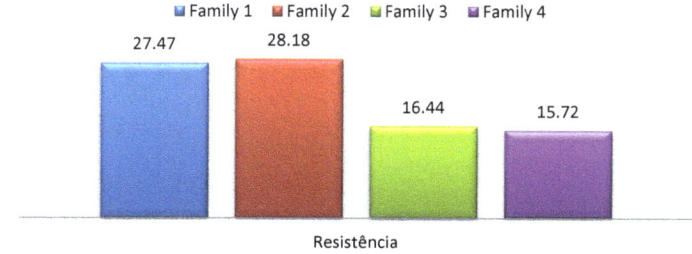

Non-Conventional Materials and Technologies – NOCMAT for XXI Century   Materials Research Forum LLC
Materials Research Proceedings 7 (2018) 640-649   doi: http://dx.doi.org/10.21741/9781945291838-61

FIGURE 6 - RESISTANCE TO THE COMPRESSION OF THE CONCRETES.

It is noted that the concrete with 25% of partial replacement of the natural small aggregate for the recycled small aggregate (Family 2) recorded mechanical resistance of 28,18Mpa, slightly larger than the resistance recorded by the reference concrete (family 1). The concrete of Family 2, although has recorded a higher level of mechanical resistance to compression in comparison to the reference concrete, has shown larger absorption coefficient by capillarity ($k = 0,573$ g/cm²).

The concretes with replacement levels of 50% and 75% of recycled aggregate (family 3 e 4) has shown low values, 16,44 e 15,72 MPa respectively, in which occurs also larger absorption coefficients than the reference, $k = 0,828$ g/cm² (family 3) e $k=1,082$ g/cm² (family 4).

Sulfate attack

The concretes were submitted to the procedure described in this work's methodology, in which the samples, immersed in anhydrous potassium sulfate solution, 10% of concentration in distilled water. The dimensions (diameter of the faces and height) of the samples were measured As were measured every 24 hours with a digital caliper rule, allowing identification of the sensitive volume change of these, the results are described in the Table 4.

TABLE 4 - THE RESULTS OF TABLE 4 ARE ILLUSTRATED IN FIGURE 7:

	Family 1	Family 2	Family 3	Family 4
day 1	1.570,80 cm³	1.570,80 cm³	1.570,80 cm³	1.570,80 cm³
day 2	1.570,80 cm³	1.570,80 cm³	1.570,80 cm³	1.570,80 cm³
day 3	1.570,80 cm³	1.570,80 cm³	1.570,80 cm³	1.570,80 cm³
day 4	1.570,80 cm³	1.570,80 cm³	1.575,90 cm³	1.574,48 cm³
day 5	1.570,80 cm³	1.570,80 cm³	1.579,59 cm³	1.574,48 cm³
day 6	1.570,80 cm³	1.572,21 cm³	1.583,28 cm³	1.574,48 cm³
day 7	1.572,33 cm³	1.572,21 cm³	1.583,28 cm³	1.578,16 cm³
day 8	1.572,33 cm³	1.572,21 cm³	1.586,99 cm³	1.578,16 cm³
day 9	1.581,23 cm³	1.572,21 cm³	1.590,70 cm³	1.581,85 cm³
day 10	1.583,45 cm³	1.574,00 cm³	1.594,42 cm³	1.585,26 cm³
day 11	1.584,22 cm³	1.575,00 cm³	1.598,15 cm³	1.586,05 cm³
day 12	1.591,50 cm³	1.577,88 cm³	1.601,89 cm³	1.586,05 cm³
day 13	1.591,50 cm³	1.586,95 cm ³	1.605,64 cm³	1.629,31 cm³
day 14	1.592,00 cm³	1.589,12 cm³	1.613,17 cm³	1.632,82 cm³
day 15	1.592,00 cm³	1.590,38 cm³	1.616,194cm³	1.633,63 cm³
Vol. variation %	1,331%	1,231%	2,853%	3,846%

Non-Conventional Materials and Technologies – NOCMAT for XXI Century   Materials Research Forum LLC
Materials Research Proceedings **7** (2018) 640-649   doi: http://dx.doi.org/10.21741/9781945291838-61

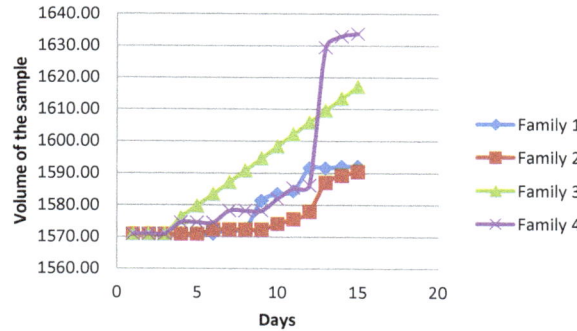

FIGURE 7 - VARIATION OF THE VOLUME OF THE SAMPLES.

The concrete with 75% of replacement by the RCD aggregate (Family 4) had a variation of 3,846% of the volume when immersed in sulfate solution for 15 days, the concrete of Family 3, with 50% of substitution and the reference (Family 2 and 1) have shown volume variation values of 2,853%, and the concrete with 25% replacement rate and reference (family 2 e 1) have shown volume variation approximate values, being 1,331% the reference concrete, with 0% of recycled aggregate and 1,231% the concrete of family 2, with 25% of recycled aggregate in its mixture, according to Figure 8.

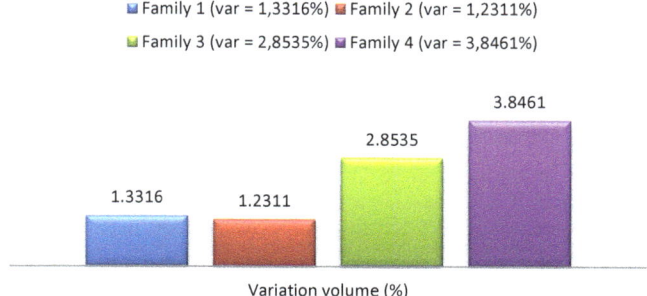

FIGURE 8 - *VOLUME VARIATION OF THE CONCRETE (%)*

It is noticed also, that in the first five days, the concrete remain with volume variation equal to zero, even with solution at high sulfate concentration and environment with elevated temperature.

Conclusion

For use of construction waste as fine aggregate in concrete, according to limits of Brazilian Standard, it is necessary the washing to reduce powders content rate, especially the recycled aggregate with significant fractions of ceramic material.

Non-Conventional Materials and Technologies – NOCMAT for XXI Century   Materials Research Forum LLC
Materials Research Proceedings 7 (2018) 640-649   doi: http://dx.doi.org/10.21741/9781945291838-61

The substitution rate of 25% of the recycled aggregate by the conventional presented a good performance in comparison to the reference concrete, results of resistance to compression, absorption coefficient by capillarity and volume variation in potassium sulfate solution, demonstrates the feasibility of using construction waste in concrete as aggregate.

The high concentration of potassium sulfate solution and the short period of the test is not recommended by the Brazilian standard, which advises a concentration of 5%, however, in this work, it was made an experimental test for behavior analysis of concrete immersed in the solution with temperature of 40 ° C.

Concretes with replacement content ≥ 50% of recycled aggregate have higher coefficients of absorption by capillarity, allowing more easily the entering of external agents, however the concrete with 25% of replacement had a similar behavior to the reference concrete.

This paper reports and recommends the reinforcement of concrete durability studies with recycled aggregate, including in an analysis with longer period to attacks by sulfate.

References

[1] Silva, Alex Fabiane Fares da.Gerenciamento de resíduos da construção civil de acordo com a resolução Conama n° 307/02: Estudo de caso para um conjunto de obras de pequeno porte / Alex Fabiane Fares de Silva. — 2007. Dissertação (mestrado) – Universidade Federal de Minas Gerais, Escola de Engenharia.

[2] John, V. M. Reciclagem de resíduos na construção civil: Contribuição à metodologia de pesquisa e desenvolvimento., 2000. 102f. Tese de livre docência – Escola Politécnica, Universidade de São Paulo, São Paulo.

[3] ANEPAC- Associação Nacional das Entidades de Produtores de Agregados para Construção. Revista Areia e Brita, 2014, Rio de Janeiro.

[4] Bidone, F.R.A. (2001) Resíduos sólidos provenientes de coletas especiais: eliminação e valorização. PROSAB ABES/RJ. 240 p.

[5] Leite, M. B. Avaliação de Propriedades Mecânicas de Concretos Produzidos Com Agregados Reciclados de Resíduos de Construção e Demolição. Porto Alegre, 2001. Tese (Doutorado em Engenharia Civil) – Escola de Engenharia, Universidade Federal do Rio Grande do Sul, Porto Alegre, 2001.

[6] AwahstI, A.; Chauhan, S. S. Using AHP and Dempster e Shafer Theory For Evaluating Sustainable Transport. Environmental Modeling and Software, v. 26, n. 6, p. 787-796, jun. 2011. https://doi.org/10.1016/j.envsoft.2010.11.010

[7] Vidal, L. A. et al. Applying AHP to Select Drugs To Be Produced by Anticipation in A Chemotherapy Compounding Unit. Expert Systems with Applications, v. 37, n. 2, p. 1528-1534, mar. 2010. https://doi.org/10.1016/j.eswa.2009.06.067

[8] Issai, M. T. et al. Intelligent Timetable Evaluation Using Fuzzy AHP. Expert Systems withApplications, v. 38, n. 4, p. 3718-3723, abr. 2011.

[9] Vieira, G. L.; DAL MOLIN, D. C. C.; LIMA, F. B. Resistência e Durabilidade de Concretos Produzidos Com Agregados Reciclados Provenientes de Resíduos de Construção e Demolição. Revista Engenharia Civil da Universidade do Minho, v. 19, p. 5-18, 2004.

Non-Conventional Materials and Technologies – NOCMAT for XXI Century   Materials Research Forum LLC
Materials Research Proceedings 7 (2018) 640-649   doi: http://dx.doi.org/10.21741/9781945291838-61

[10] Pereira, E.; MEDEIROS, M. H. F; LEVY, S. M.; Durabilidade de concretos com agregados reciclados: uma aplicação de análise hierárquica. Ambiente Construído, Porto Alegre, v. 12, n. 3, p. 125-134, jul./set. 2012. https://doi.org/10.1590/S1678-86212012000300009

[11] Prefeitura de Salgueiro, disponivel em: <<http://salgueiro.pe.gov.br/ >>.

[12] ABNT NBR NM 248:2003 - Agregados - Determinação da composição granulométrica, Rio de Janeiro.

[13] ABNT NBR NM 52:2009 - Agregado miúdo - Determinação da massa específica e massa específica aparente, Rio de Janeiro.

[14] ABNT NBR 7809:2006 - Agregado graúdo - Determinação do índice de forma pelo método do paquímetro - Método de ensaio, Rio de Janeiro.

[15] ABNT NBR NM 51:2001 - Agregado graúdo - Ensaio de abrasão "Los Ángeles", Rio de Janeiro.

[16] ABNT NBR NM 46:2003 - Agregados - Determinação do material fino que passa através da peneira 75 um, por lavagem, Rio de Janeiro.

[17] ABNT NBR 5736/2003-: Concreto – Procedimento para moldagem e cura de corpos-de-prova, Rio de Janeiro.

[18] ABNT NBR 5739, 2007 - Concreto - Ensaios de compressão de corpos-de-prova cilíndricos, Rio de Janeiro.

[19] ABNTNBR 9779, 1995- Argamassa e concreto endurecidos -Determinação da absorção de água por capilaridade, Rio de Janeiro.

Non-Conventional Materials and Technologies – NOCMAT for XXI Century   Materials Research Forum LLC
Materials Research Proceedings 7 (2018) 650-660   doi: http://dx.doi.org/10.21741/9781945291838-62

Applicability of Artemis Views Software in Public Works and Planning

Luana Maris Pedrosa Cruz [a, *], Carmen Couto Ribeiro[a], Danielle Oliveira Meireles[a], Sidnea Eliane Campos Ribeiro[a], Tadeu Starling[b]

[a] Escola de Engenharia da Universidade Federal de Minas Gerais – UFMG, Brasil;
luanampc@hotmail.com, carmencouto@oi.com.br, daniellemdo@gmail.com,
sidneaecr@gmail.com

[b] Architecture School - Universidade Federal de Minas Gerais – UFMG, Brasil;
tadeustarlingbh@oi.com.br

Abstract. This study deals with the applicability of the Artemis Views management software in planning and controlling public works, focusing mainly on the importance of the compatibility with the project so as to minimize risks and reach a good construction performance, interfaced by the Request for Proposals Act 8666:93. A research was conducted on the requirements needed for planning and controlling public works based on the directives established by the ABNT NBR 15575:2013 Performance Norm, concerning the useful life expectancy, the efficiency of materials, and the sustainability of buildings. This study was conducted based on the analysis of the application of preventive and corrective actions provided by the Artemis Views management software aiming at meeting the legal directives established by the Request for Proposals Act 8666:93. The results obtained demonstrate feasibility in adopting the Artemis Views software with the purpose of diminishing losses of materials and increasing efficiency in the utilization of public funds, consequently improving building health. The analysis of the relation between the Request for Proposals Act and the Performance Norm provides evidences that an effective integration of project designs, resources and planning processes increases the performance and the efficiency in the employment of materials, thus promoting sustainability in public works.

Keywords: Request for Proposals Act 8666:93, Planning, Performance in the execution of public works

Introduction

This study analyzes the applicability of construction management software programs in compliance with norms established for bidding processes as provisioned by the RFP (Request For Proposal) Law 8666:93 [1], so as to establish specific requirements for public works planning according to ABNT norm NBR 15575:2013 [2].

The work focuses mainly on the importance of a project compatibility level, so as to minimize risks and achieve a good construction performance, in interaction with ABNT NBR 15575:2013 Performance Norm that lists guidelines for the development of residential construction projects, providing performance, efficiency and sustainability directives for construction projects.

In Brazil, public works are ruled by Law 8666:93, which regulates article 37 of subsection XXI of the Federal Constitution and institutes standards for bidding processes and contracts to be made by Public Administration offices. The general standards and norms regarding the bidding processes and administrative contracts are provisioned by RFP (Request For Proposal) Law 8666:93, just as it was listed in the Law Decree 2.300:86, permitting that all companies interested in taking part in bidding processes concerning public works paid for with public funds, while theoretically preventing benefits given by governing rulers to corporations to the detriment of others [3].

So, as to identify any incongruence and polemical points in the realm of bidding processes, an analysis based on the concepts implemented by the PDCA Cycle was made through a verification

Non-Conventional Materials and Technologies – NOCMAT for XXI Century   Materials Research Forum LLC
Materials Research Proceedings 7 (2018) 650-660   doi: http://dx.doi.org/10.21741/9781945291838-62

carried out with the Artemis Views software. The purpose of the analysis is to reach full compliance with public works planning process improvement method so, as to fulfill all bidding process requirements accordingly to building quality and maintenance indicators.

In order to contribute for the consolidation of a theoretical basis and the establishment of development alternatives and for the implementation of production planning and control systems, it was necessary to evaluate the Artemis Views management software, which is applicable to Engineering services and public works, to assure the efficacy of its implementation.

The evaluation of compliance with continuous improvement tools in public works planning processes to ensure licitation process requirements are met through analyses of quality, efficiency and sustainability indicators of buildings is based on the strategy to produce studies on public works planning and execution so, as to guarantee the projects will have an improved quality performance.

The conception and project phases play a remarkable role and are directly associated with final quality and efficiency delivered, being then of utmost importance to assure quality and sustainability of products and efficiency in projects [4].

The project design process could be defined as an activity or service that is an integral part of the construction process, responsible for the development, organization, registration and transmission of physical and technical characterization specifically determined for those construction works, which are to be considered during the project execution phase [5].

Planning is defined as a managerial process based on a specific demand, involving the establishment of goals and of procedures that are to be adopted to accomplish goals and explain that planning will only be efficient when produced in association with control and monitoring [6].

In the second half of the 20th Century, the technological complexity, the increased volume of investments and the need of safety played a remarkable role in the expansion of quality control. It became fundamental to first assure the quality of products, services, installations and equipment, which then originated the Total Quality Control principles. At the time of the Industrial Revolution the main concern was with the product, while later in the globalization era, quality control focused mainly on the production process [7], according to the figure 1.

FIGURE 1 - *THE ADVANCEMENT OF QUALITY IN BRAZIL – FEUDAL SYSTEM /*
INDUSTRIAL REVOLUTION / WORLDWAR1 / WORLDWAR2 / COLD WAR
GLOBALIZATION - EMPHASIS ON PRODUCT / EMPHASIS ON PROCESS

Source: Fernandes, 2001

Non-Conventional Materials and Technologies – NOCMAT for XXI Century   Materials Research Forum LLC
Materials Research Proceedings **7** (2018) 650-660   doi: http://dx.doi.org/10.21741/9781945291838-62

Therefore, still according to [7], the emphasis on the quality concept gradually transferred from product to process, since the project will only be a quality product if a qualified, reliable, safe and efficient process is behind it.

Ever since the first years of last century, industrial organizations already knew about the three mass-production processes: specification, production and inspection. Taylor (considered the father of scientific administration) recommended plan-do-see as a reference for planning the basic stages of a productive process [8]. These processes are enchained in an open, linear and simple sequence and represent the operating structure of industries at that time (figure 2).

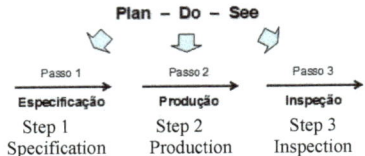

FIGURE 2 - *TAYLOR'S CONTROL CONCEPT AND THE THREE MASS PRODUCTION PROCESSES, Source: Moen and Norman (2007)*

The PDCA methodology was developed by Walter A. Shewhart in the 1930s, and was championed by William Edward Deming, and sequentially was successfully employed by Japanese companies as of 1950s to increase quality in processes (PACHECO et al, 2010). Also known as the Deming Cycle, PDCA is the acronym for each stage of the cycle: Plan; Do; Check; and corrective Action [9].

The PDCA cycle is a quality tool that facilitates decision-making and guarantees the accomplishment of those goals necessary for company survival [10]. Even though it is simple, the cycle, which represents clear advancements to reach efficacy in planning, is composed of 4 phases, according to the figure 3:

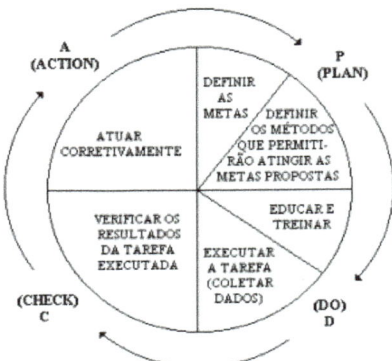

FIGURE 3 - *DEMING CYCLE OR PDCA CYCLE – ACTING CORRECTLY / DEFINING GOALS / DEFINING METHODS THAT ENABLE THE ACCOMPLISHMENT OF GOALS / EDUCATE AND TRAIN / PERFORM THE TASK (GATHERING DATA) / VERIFYING RESULTS OF TASK ACCOMPLISHED. Source: SEBRAE, 2016*

As stated by [11], one of the most important stages in this cycle is planning, as it contributes by creating structural and strategic conditions to organizations that are facing challenges. The planning stage is primordial since it provides for knowledge expansion concerning the existing problems and prepares organizations to deal with them.

The Improvement Method (of PDCA Cycle) is a system or process management tool that provides a path to reach goals established for the products of corporate systems [12]. The implementation of PDCA in organizations requires the understanding of some key elements that interfere with the activities, such as: (i) what a process is; (ii) what the variability of processes means; (iii) what control items and verification items are; and (iv) which are the quality tools that can aid in the implementation of the cycle.

Among several different management methods, the PDCA program provides support for the execution of public works, by enhancing project compatibility process improvements and the conformity with bidding process rules.

The bidding process can be understood as a management procedure through which the legally responsible individual makes the selection of interested parties that are in conformity with the call by adopting previously established criteria and objectives and decides which is the most advantageous proposal for the contract [13].

Law 8666 was enacted on June 21, in 1993. It governs article 37, subsection XXI of the Federal Constitution that provisions Norms for bidding processes and administrative contracts regarding constructions, services, there including publicity, procurements, alienation and leasing within the Powers of the Federal Government, of States and the Federal District, and of Municipalities [1].

Article 37, XXI of Federal Constitution, which caused the enacting of Law 8.666:93 (known as Contracts and RFP-Request for Proposals Law), features the following text:

XXI – pointing out cases listed in the legislation, public works, services, procurements and alienations shall be realized through public bidding licitation processes that ensure equality in conditions to all bidding participants, with clauses that establish payment bound to the effective realization of the conditions established in the proposal, according to the law, which will only allow provide for requirements of technical and economic qualification that are indispensable to guarantee the contractual obligations are fulfilled [1].

By establishing a margin of preference it is possible for the Administration to acquire products and services at a higher price than the lowest bid proposed in the licitation process. In international bidding processes, prices may be quoted in foreign currencies [1].

Basic and Executive Projects are compulsory documents to be produced in licitation processes regarding public works and engineering services in modalities such as competition, price quotation and request, but not for procurement of goods. The law does not impose as a requirement for the Administration the effective availability of finances; however, it demands that the funds must have been listed in the Budget Law [1].

The Basic Project is specifically determined by the Law as a set of required elements that are sufficient and feature adequate precision levels to characterize the works or services, the set of works or services, while being produced in conformity with orientations provided by preliminary technical studies that ensure technical viability and the adequate treatment of environmental impacts caused by the project, enabling the assessment of construction costs and the definition of construction methods and deadlines. This project must be produced previously to publishing the request for proposals, presenting a management plan for the public works and clearly identifying all construction elements.

The Executive Project should guide the operations. However, it is not compulsorily required to be presented previously to the bidding process, and may be produced concurrently to drawing up the contract, if endorsed by the Administration.

Data concerning modality, competition, price quotation and request to produce proposals for the Licitation process are made available in the table 1:

TABLE 1 – DATA TO PARTICIPATE IN THE LICITATION PROCESS – RFP LAW 8666:93

Modality	COMPETITORS	BUDGETED COSTS	REQUEST
Participants	Open to any bidder	- Registered bidder - Bidder that complies with conditions and makes registration in up to 3 days	- Requested bidders (registered or not), minimum of three. - Registered bidders that previously (up to 24 hours before) manifest interest.
Capacitating	Capacitating Stage	In advance (Registration Data)	In advance (Registration Data)
Object	- Public works, services, and procurements of any value. - Procurement and alienation of real estates. - Concession of severable use rights - Concession of services - International Licitation process - Price registering	- Engineering services and construction up to R\$ 1.5 million. - Procurement of goods and services up to R\$650,000.	- Engineering services and construction up to R\$ 150,000. - Procurement of goods and services up to R\$80,000.
Commission	Minimum of 3 members, at least 2 effective servants	Minimum of 3 members, at least 2 effective servants	Minimum of ONE single servant (small units, small staff)

Source: Pereira Júnior (2003)

Public works, services and suppliers procured by the Administration must be divided into different requests for proposals whenever such initiative brings advantages, so as to increase competitiveness and attract bidders able to supply the desired object [3].

Every single product must feature characteristics in conformity with the objectives and functions for which it was designed, based on performance criteria. In the 1990s, Brazil took a remarkable measure to promote the rationalization in production processes. The initiative originated from the increased number of production systems on the market, which initially generated conflicts in requirements and performance standards consequently hindering building behaviors [14].

Within such context, in 2008 the first version of the Performance Norm NBR 15575 was published determining procedures to constructors and designers, and to the construction materials industry that must be fully followed, some of which were unprecedented. In 2013, the CAU (Architecture and Urbanism Council), the Norm 15575 was updated and published. Taking into account the conditions of implementation and users' requirements, the Norm established and assessment method and the requisites (qualitative characteristics) and criteria (quantitative measures) to be fulfilled [2].

As opposed to traditional norms that list the characteristics of products based on their use, performance norms determine those properties that are necessary for different construction elements, regardless of the matter constituting the products. In the first case, the product must be used in conformity with its characteristics. In the second, the product must be developed and employed in conformity with the needs of the construction [15].

The NBR 15575:13 determines the responsibility of each player – developer, designer, constructor, supplier and user. This plays an important role nowadays since the constructor is responsible for the integration process [16]. The table 2 displays some of the responsibilities listed by ND.

TABLE 2 – RESPONSIBILITIES LISTED BY ND

Developer	Assess the conditions of the site, identify potential risks, specify the standards of the building (minimum, intermediate, maximum) and provide the required technical reports and studies.
Designer	Develop the project and specify the products fulfilling all performance requirements established specify in memorials and drawings the Project Life Span of each system that is part of the project.
Manufacturer/ Supplier	Specify the life cycle of products and supply results that are in conformity with the performance standards
Constructor	Guarantee that the performance of the system (and not of the product) features the required performance; produce Manuals regarding Utilization, Operation and Maintenance
User	Maintain the building as specified in the maintenance plan

Source: ASBEA, 2012

Users must comply with requirements established to ensure safety, usability and sustainability. For each of these topics, specific requests are listed based on the following factors [2].

NBR 15575: 2013 is divided into six parts. The first part of the Norm works as a reference index, specifying, whenever possible, items in the building parts (structure, flooring, vertical sealing, roofs and hydro sanitation systems). The second part of the Performance Norm deals with the requirements for the structure of residential buildings. The third part of NBR 15.575 was the one that was most modified in the revision process, and now regulates not only internal flooring systems – as it did before – but also the external ones. Parts 4, 5, and 6 refer to internal and external vertical sealing systems, roofs and hydro sanitation systems, respectively. [2]

The joint action undertaken by all these players brings benefits for all. The figure 4 illustrates how the continuous maintenance increases durability in buildings.

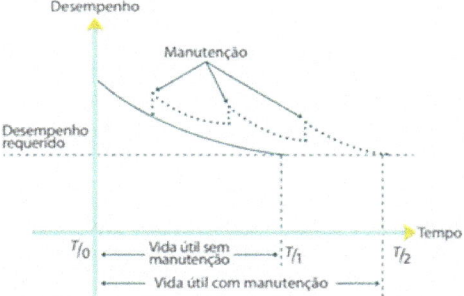

FIGURE 4- CONTINUOUS MAINTENANCE INCREASES THE DURABILITY OF THE BUILDING -PERFORMANCE/ MAINTENANCE/ REQUIRED PERFORMANCE/TIME/LIFE-SPAN WITH NO MAINTENANCE/ LIFE SPAN WITH MAINTENANCE

Source: NBR 15575/CBIC

In order to promote improvements in the performance of buildings, several different software programs are used to provide a more efficient management process, among them, Artemis Views.

Non-Conventional Materials and Technologies – NOCMAT for XXI Century   Materials Research Forum LLC
Materials Research Proceedings 7 (2018) 650-660   doi: http://dx.doi.org/10.21741/9781945291838-62

This software traces resources through a system that includes portfolio management, cost control and planning reports.

Portfolio management is a dynamic process by means of which a list with projects is constantly updated and reviewed. With this process new projects are assessed, selected and priorities are established; existing projects can be accelerated, interrupted or have their priority reduced; and resources are allocated to projects in action [17].

The American management software program, Artemis Views (Artemis Management System) carries out the verification of project processes approaching the importance of the production of a quality project, clearly indicating likely future problems when the project is badly structured [18]. Principles linked to the company's strategic management are used in the effective management of risks and in the analysis of added values [19].

In this context, the applicability of Artemis Views software was analyzed in order to contribute for improvements in the compatibility of public projects, with the employment of the concepts stated by the PDCA cycle in public works and planning and with the objective of fostering good performance of buildings.

Methodology

The applicability of Artemis Views software in the public works planning process aiming at providing improved performances of the constructions based on Law 8666:93 was analyzed through the following stages:

i.   Analysis of the directives established by Law 8666:93 in order to evaluate the criteria required in the bidding process that determine the company winner of the licitation;

ii.   Definition of requirements to define the conception, development, and compatibility of Projects based on the criteria provisioned by the Performance Norm;

iii. Assessment of efficacy in the implementation of the PDCA cycle for continuous improvements of the project's development process;

iv.   Analysis of the applicability of Artemis Views software (Artemis Management System) in project planning in conformity with Law 8666:93.

Results and discussion

The results of the analysis of the implementation of Artemis Views software in the full compliance process of public works and planning, which is done through the traceability of resources with a portfolio management system, cost control, time and planning reports. The objective is to assess the applicability of tools adopted to plan and execute public works so as to enhance good performance of buildings, accordingly to Law 8666:93.

Analysis of the directives established by Law 8666:93 in order to evaluate the criteria required in the bidding process that determine the company winner of the licitation

As of the analysis of the RFP (Request For Proposals) Law 8666:93, it was verified that the adoption of a single criteria – lowest price – to decide the winner in the bidding process, should consider other important factors, such as the analysis of the company's operational structure and financial conditions, their experience in similar projects, the level of quality of the project and construction, a guarantee that ABNT standards will complied with, and so on, which, overall, would enable a better decision considering the best and most advantageous proposal as ruled by the Law.

Currently, there are a number of companies on default, which have interrupted or did not complete the construction contracted, or deliver works with low quality levels, and show durability that does not meet desirable levels, non-compliance with specifications etc. Therefore, bringing to low levels the respectability that Construction Engineering could afford.

Non-Conventional Materials and Technologies – NOCMAT for XXI Century   Materials Research Forum LLC
Materials Research Proceedings 7 (2018) 650-660   doi: http://dx.doi.org/10.21741/9781945291838-62

Consequently, the main evaluations of the directives of the RFP (Request For Proposals) Law demonstrated that, in general, in the request for proposals and bidding processes the norms and standards that rule the processes are not fully complied with as established by the law.

In the processes that involve contracts, agreements and associations, which are also ruled by Law 8.666:93, the main goal of inspections aimed at determining whether the legal norms were complied with concerning control, service rendering and conformity of services.

Definition of requirements to define the conception, development, and compatibility of Projects based on the criteria provisioned by the Performance Norm 15576:13 and on the implementation of the PDCA cycle for continuous improvements in the project development process

Analysis was made on the directives established by Law 8666:93. The analysis observed which criteria are to be adopted for the development of projects, making it clear that by improving the planning stage, it is possible the project will comply with a continuous improvement process management in conformity with the execution of the project.

Problems concerning detailing assignments were diagnosed through the use of proposal management tools. The purpose was to demonstrate the importance of and the need to implement the tool in a company.

Regarding the development of the project, topics to be approached are: legal aspects, surrounding, intentions of project contracting party, construction type, materials to be employed, constructive techniques, technologies that will be employed during the construction, functionality of space, environmental comfort, final user's needs in the building, costs, and so on. All aiming at meeting compliance with Performance Norm 15575:13.

Therefore, based on the continuous improvement method, stated by the PDCA cycle, evaluation criteria and requirements were established to provide for an efficient planning of the project execution process, delivering satisfactory results with its implementation.

Analysis of the applicability of Artemis Views software (Artemis Management System) in project planning in conformity with Law 8666:93

The management software Artemis Views (Artemis Management System) supports the verification of project processes. It manages and controls all requirements listed by the project through utilizing cost planning tools through a graphic application based on Windows. It includes online early warning indicators and comprehensive reports that feature the visibility of the performance of projects in order to ensure preventive and corrective initiatives are carried out.

The analysis sought to find possible flaws in the RFP (Request for Proposals) Law regarding the lack of excellence in the execution of public work projects. Consequently, it was verified how a company wins a public works bidding process without being demanded to produce a finely detailed technical report on interface with the PDCA method, demonstrating the efficacy of management software Artemis Views in the verification of the process.

The software combines in one single environment all the data on projects and resources providing for display, modification, processing and development of reports, equally dealing with structured data (deadlines, costs, etc) and non-structured data (documents, risks, emails, notes, among others), as shown in figure.

Non-Conventional Materials and Technologies – NOCMAT for XXI Century   Materials Research Forum LLC
Materials Research Proceedings **7** (2018) 650-660   doi: http://dx.doi.org/10.21741/9781945291838-62

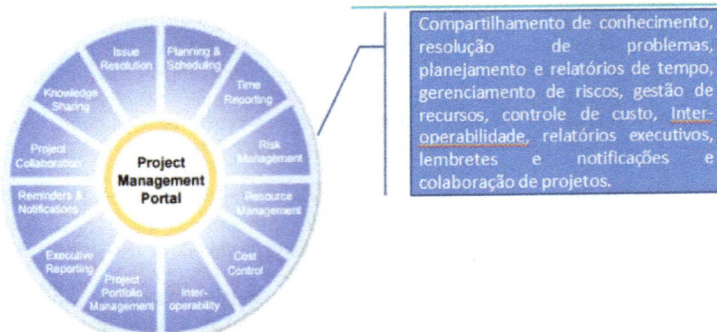

FIGURE 5 - *ARTEMIS VIEWS PROJECT MANAGEMENT PORTAL (2017)*

Artemis Views provides different sets of Corporate Project Management tools. Each of them supports specific user functions in the organization. The tools are integrated permitting that several different users simultaneously access updated information. The system is based on a client/server mode and uses SQL databases that are commercially available for their repository. The software supports relational databanks, such as Oracle, Microsoft SQL Server and Sybase working on operational systems like Windows NT, Sun Solaris or HP-UX.

The set of Artemis products hold up an open interface environment to support the integration with other corporate tools. It also features integration with systems like Microsoft Project and Enterprise Resource Planning (ERP), which are available for the project's cost accounts.

The software creates and updates the project's timetables and schedules through the Gantt chart and worksheets. It employs cost planning tools using interactive performance indicators. It also includes online early warnings, indicators and comprehensive reports that provide access to performances in order to ensure adequate and timely management initiatives are undertaken.

Artemis Views offers quality project management reports, including panels, charts, and interactive performance indicators, displaying detailed charts and enabling facility in analysis of causes and tendencies, all provided in PDF or Excel format files.

Consequently, the problems regarding the detailed configuration were diagnosed through the proposed management tool with the purpose of demonstrating its importance and the necessity that it be implemented in a company. As a result, with the continuous improvement method, evaluation methods and criteria were established to make project planning more efficient, as show in table 3.

The applicability of Artemis Views software is made clear with the integration of management tools that make it viable to monitor the execution of the project, ensuring the control of risks and promoting continuous improvements in the process to guarantee efficiency in the performance of the building.

Non-Conventional Materials and Technologies – NOCMAT for XXI Century   Materials Research Forum LLC
Materials Research Proceedings 7 (2018) 650-660   doi: http://dx.doi.org/10.21741/9781945291838-62

TABLE 3- ARTEMIS VIEWS TOOLS

Artemis Views
Cost control
Produces graphic reports
Plans processes graphically
Generates budget reports
Graphically compares a selected project version with baseline version
Manages / compares several project versions simultaneously, using different instances of Gantt Chart
Establishes and maintains Inter project relations
Constructs and updates project chronograms using electronic worksheets
Analyzes scenarios of different projects without duplication of data

Source: Author(2017)

Conclusion

RFP (Request for Proposals) Law 8666:93, evaluated regarding the standardization of public works bidding processes, proved to be inefficient in regards to its main goals concerning the construction sector, specifically in the enforcement of regulations. The level of quality of the construction and the guarantee of compliance with Performance Norm NBR 15575:13, are some primordial items that must be taken into account, in addition to the single criteria as the law stands, to decide which company will be the winner in the request for proposals process.

The performance of the public works achieved based on bidding processes must be considered from the early conception of the project. Therefore, hopefully changes and clear improvements will be made to process from its conception to the execution planning stage of a building to enhance qualified production based on the considerations about Law 8666:93 integrated to the Performance Norm 15575:13.

The suitability and feasibility of Artemis Views management software were clearly proven in public works execution planning processes based on Law 8666:93. It was verified that the program is able to unite existing management tools as it integrates methods and provides greater efficiency in public works planning processes.

References

[1] Law No. 8.666 of June 21, 1993. Regulates art. 37, item XXI, of the Federal Constitution, establishes rules for bids and contracts of the Public Administration and provides other measures. Presidency of the Republic. Available at www.planalto.gov.br Access on May 10, 2016.

[2] BRAZILIAN ASSOCIATION OF TECHNICAL STANDARDS. NBR 15575: Housing Buildings - Elaboration. Rio de Janeiro, 2013.

[3] PEREIRA JUNIOR, Jessé Torres. Comments on the law of public procurement bids and contracts. 6. ed., Rev., Current. E ampl. Rio de Janeiro - RJ. 2003.

[4] FABRÍCIO, Márcio Minto. The Simultaneous Project in the Construction of Buildings. Thesis (PhD in Engineering) - Department of Civil Engineering, Polytechnic School of the University of São Paulo, São Paulo, 2002.

[5] MELHADO, S.B. Management and Coordination of Building Projects, Introduction to the Theme. SP, 2004.

[6] FORMOSO, Carlos Torres; LIEDTKE, Renata; JOBIM, Margaret Souza Schmidt. Quality Management in the Design Process. Federal University of Rio Grande do Sul, Porto Alegre, 1998.

[7] BERTEZINI, A. L. Methods of Evaluation of the Process of Architectural Design in the Construction of Buildings Under the View of Quality Management. 2006. 193 f. Dissertation (Master in Civil Engineering) - Polytechnic School, University of São Paulo, São Paulo, 2006.

[8] ISHIKAWA, Kaoru. TQC - total quality control: strategy and quality management that ensure the prosperity of the company. São Paulo, IMC-International Educational Systems, 1986.

[9] PACHECO, A.P.R. et al. The PDCA cycle in knowledge management: a systemic approach. Available at: Accessed on: 14 Oct. 2010.

[10] SEBRAE. The PDCA cycle. Available at: <http://www.biblioteca.sebrae.com.br/bds/bds.nsf/0f5e363a16336c5e03256c67006799a/49b285 ddc24d11ef83257625007892d4/$FILE/NT00041F72.pdf>. Accessed on: 09 Oct. 2016.

[11] - AGUIAR, Silvio. Integration of Quality Tools into PDCA and the Six Sigma Program. INDG, 2006.

[12] CAMPOS, V. F. Managing the routine of everyday life. Belo Horizonte: INDG Tecnologia e Serviços Ltda., 2004.

[13] - GASPARINI, DIOGENS. Administrative law. SP, 8th edition, Saraiva, 2003.

[14] NEELY, A. Measuring business performance. London: The Economist Newspaper and Profile Books, 1998.

[15] BRAZILIAN CHAMBER OF CONSTRUCTION INDUSTRY (CBIC). Performance of residential buildings: guidance guide for compliance with the standard ABNT NBR 15575: 2013. 300f. Brasilia DF.

[16] ASBEA: Architecture and Urbanism Projects and Services Scope Handbook: 2012, Real Estate Industry

[17] COOPER, R. G .; EDGETT, S. J .; KLEINSCHMIDT, E, J. New product portfolio management: practices and performance. Journal of Product Innovation Management. V. 16, p. 331-351, 1999. https://doi.org/10.1016/S0737-6782(99)00005-3

[18] Views, ARTEMIS. Www.konsultex.com.br/artemis/arquivose.../views7datasheet.pdf, accessed on 07/20/17.

[19] Views, ARTEMIS. Http://www.aisc.com/solutions/views/, accessed 7/20/17.

Non-Conventional Materials and Technologies – NOCMAT for XXI Century   Materials Research Forum LLC
Materials Research Proceedings 7 (2018) 661-670   doi: http://dx.doi.org/10.21741/9781945291838-63

Thin Steel Rings as a Feasible Alternative to Connect Bamboo Culms

J.J García, C. Benítez, L. Villegas, R. Morán

Escuela de Ingeniería Civil y Geomática, Universidad del Valle, Cali, Colombia

jose.garcia@gmail.com, caritobenitez29@hotmail.com, laura.villegas@correounivalle.edu.co, richard.moran@correounivalle.edu.co

Abstract. Bamboo is highlighted by its high axial strength, lightness, tubular shape and high sustainability. Bamboo connections cannot be easily built due to the hollow cylindrical shape and the high dimensional variation and transverse weakness of the culms, which tend to fail by longitudinal splitting. Joints recommended by the Colombian construction code use bolts, curved cuts and mortar injection to increase the transverse strength. These joints require high labor intervention and cannot be used in massive projects using prefabricated processes.

We explore the use of thin steel semi-rings to connect bamboo culms. As the rings (formed with two semi-rings) are thin enough they can conform to the irregularities of the culm after being tighten. Thus, the ring applies an external pressure to the culm and creates a compressive circumferential stress distribution, which counteracts the creation of longitudinal splits. Moreover, even after initial fissures the ring avoids the separation of the parts, which generates ductile modes of failure.

We performed experiments of *Guadua angustifolia* culms wrapped with steel rings under axial and transverse loading. Axial experiments showed maximum loads in the range 7500-18000 N with signs of high plastic deformations that usually begin at around 5000 N. Experiments under transverse loading showed maximum loads in the range 12000-20000 N with ductile types of failure.

Keywords: Guadua Angustifolia, Bamboo Joints, Steel Rings

Introduction

Various challenges are posed to connect bamboo culms [1]. First, there is a wide dispersion in the diameter and thickness of the elements, which is inherent to natural materials [2, 3]. Secondly, the hollow cylindrical shape of the culms avoids using standard connections developed for other materials, e.g. connector developed for solid wood members [4, 5]. Finally, the fiber reinforcement along the axial direction facilitates the creation of longitudinal splits that usually cause complete failure under different type of loading [2, 6, 7]. Hence, the high strength of the material under axial loading cannot be fully used.

Many bamboo culm connections have been proposed [8, 9, 10, 11]. Vernacular connections use lashing and curved cuts to fit bamboo culms [12]. Bolted connection include fish-mouth cuts and screw bars to tight the culms, which must be drilled beforehand. More sophisticated connections use mechanized metal parts, adhesives or FRP reinforcements [13, 8, 10]. Some of them are very expensive and unpractical to use in massive projects. Hence, grouted and bolted connections were adopted in the Colombian code [14]. They work reasonably well, but the construction process is relatively inefficient. Since each connection must be customized to the size of the culms, construction process is lengthy and increase hand-labor cost. In addition, holes in the culms tend to generate splitting failures [15, 16].

To overcome the aforementioned difficulties, a thin steel clamp system is proposed. This pilot study was aimed at characterizing the mechanical performance of a clamp-culm system under axial and transverse load.

Non-Conventional Materials and Technologies – NOCMAT for XXI Century   Materials Research Forum LLC
Materials Research Proceedings 7 (2018) 661-670   doi: http://dx.doi.org/10.21741/9781945291838-63

Description of the system

Clamps formed by two semi-rings were analyzed in this study. Other configurations may be devised, such as clamps formed by three one-third-rings, which may be more versatile for other type of joints. A pilot study showed that thick clamps, such as those used to connect electrical lines to concrete poles, were too heavy and stiff to follow the irregularities of the culm during fastening. Hence, all semi-rings in this study were cold formed from thin plates 25.4 mm wide and 3.2 mm thick made of SAE 1020 steel with a yield strength of 372 MPa and 15% failure deformation.

Each semi-ring (Figure 1) poses a curved portion that becomes in contact with the culm, followed at each side by two sharp curves in the transition to the flats ends, lugs or "ears". In the ears are drilled the holes that allow to fasten the clamp around the culm using bolts of 9.5 mm (3/8") diameter.

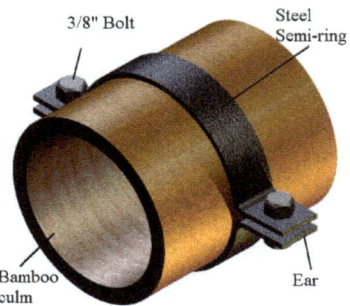

FIGURE 1 - SCHEMATIC REPRESENTATION OF A CLAMP-CULM PAIR

When the bolts are tightened, the ears of one semi-ring get in contact with the corresponding ears of the other semi-ring. An external pressure is generated over the culm on the area of contact while the semi-rings are under circumferential tension. Then, the circumference of the culm decreases and the length of the clamp increases. At the final stage of adjustment, there is a small gap between the ears in which the culm is not in contact with the semi-rings. We chose the dimensions of the semi-rings to keep this small gap to a minimum as it may be the focus of initial splitting.

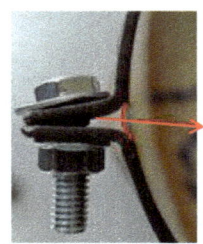

FIGURE 2 - GAP LEFT BETWEEN THE SEMI-RINGS AFTER TIGHTENING THE BOLTS

The contact pressure generated between the culm and the clamp depends on the degree of adjustment. If the ears become in contact after fastening the bolts, the pressure is a function of the difference between the initial (before tightening) external circumference of the culm and the inner

Non-Conventional Materials and Technologies – NOCMAT for XXI Century   Materials Research Forum LLC
Materials Research Proceedings 7 (2018) 661-670   doi: http://dx.doi.org/10.21741/9781945291838-63

curved central length of the semi-rings. Therefore, this difference or "interference" was used in this study to indirectly asses the contact pressure and to define the geometry of the clamps that may be suitable for a range of culm diameters. The word "interference" is used due to the similarity of this problem with that of a ring adjusted with interference around a tube. The generated pressure is dependent on this interference. The maximum bolt adjustment was attained when the ears reached maximum contact. Another alternative measure of adjustment is the torque applied to the bolts, however, due to the high number of factors that affect friction, this measure was discarded.

The culm-ring system can be used to transmit load between culms through the clamps. For instance, two culms with their axes aligned can be joined with short steel plates connected to the ears. In this case, a tensile axial load tries to extract the clamp from the culm. Transverse load can also be transmitted between the culms through the clamps. For a first assessment of the feasibility of the system to transmit loads, the following simple experimental methods were devised.

Methods

Guadua angustifolia (GA) culms were used in the tests. All culms were borax treated and oven dried with practices well stablished [8]. Moisture content was measured following the provisions of the NTC5525 [9]. Average moisture in the tested culms was 10.6% (COV = 0.14).

In this study, the interference I_c between the clamp and the culm was measured as,

$$I_c = P_c - 2\,L_a - 2\,x \tag{1}$$

where P_c is the perimeter of the culm, L_a is the interior curved length of one semi-ring and x is the final gap between the ears of the semi-rings, as shown in Figure 2.

To determine the force needed to draw the clamp for the culm, the set-up described in Figure 3 was used. First, a clamp was tightened around a small piece of culm of about 400 mm. Next, the ears of the clamp-culm system were supported on the upper flat end of a steel pipe, while the bottom face of the pipe was supported on the testing machine. The load was then applied to the upper portion of the culm trying to draw the clamp from the culm. Seven specimens were tested, average diameters and thicknesses of the culms were 113 mm (COV = 0.06) and 12 mm (COV = 0.11), respectively.

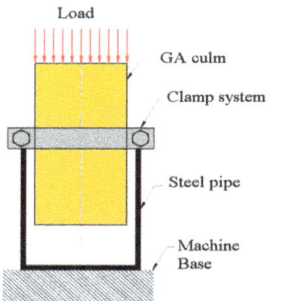

FIGURE 3 - EXPERIMENTAL SET-UP DEVISED TO ASSESS THE CAPACITY OF THE CLAMP-CULM SYSTEM TO SUPPORT AXIAL LOADING

To assess the capacity of the system to sustain transverse loading, the arrangement of Figure 4 was used. Two loading modes were considered, in the first (mode *a*), the load was transmitted perpendicular to the ears and in the second (mode *b*), the load was transmitted parallel to the ears. To avoid failures by bending, short culms of 600 mm were used. The culms were supported at the ends with clamps. The test was performed using a WPM (former Democratic Republic of Germany) testing machine. A pre-compression load of 196 N was applied to accommodate the supports that were not perfectly aligned due to the irregularities of the culms. Seven specimens were tested for each loading mode. For tests in mode a, average diameters and thicknesses of the culms were 122 mm (COV = 0.04) and 12 mm (COV = 0.07), respectively. For mode b, average diameters and thicknesses of the culms were 112 mm (COV = 0.03) and 15 mm (COV = 0.23), respectively. A summary of the mechanical interference I_c for each test, calculated with Eqn. (1), is presented in Table 1.

TABLE 1 – SUMMARY OF THE MECHANICAL INTERFERENCE IN TESTS

Specimen Number	Experimental interference I_c (mm)		
	Axial Tests	**Transverse Tests (Mode a)**	**Transverse Tests (Mode b)**
1	9.3	9.5	9.7
2	11.5	7.9	6.4
3	13.1	6.6	13.8
4	17.9	9.5	5.7
5	14.2	14.6	9.7
6	15.7	6.2	15.9
7	10.8	4.6	11.4
Mean	13.2	8.4	10.4
COV	0.23	0.39	0.36

The initial stiffness was determined as the secant slope of the force versus displacement curves at 10% and 40 % of the maximum load [10].

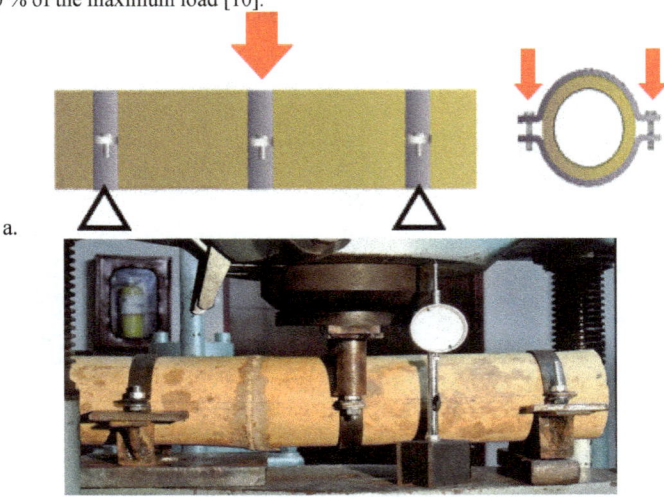

a.

Non-Conventional Materials and Technologies – NOCMAT for XXI Century   Materials Research Forum LLC
Materials Research Proceedings 7 (2018) 661-670   doi: http://dx.doi.org/10.21741/9781945291838-63

b.

FIGURE 4 - EXPERIMENTAL SET-UP DEVISED TO ASSESS THE CAPACITY OF THE CLAMP-CULM SYSTEM TO SUPPORT TRANSVERSE LOADING, A. PERPENDICULAR TO THE EARS AND B. PARALLEL TO THE EARS

Results

Axial load versus displacement curves (Figure 5) show a short initial portion approximately linear followed by a curve with decreasing slope. Only the specimen 1, with interference of 9.3 mm slipped at a maximum load of 7848 N. Significant plastic deformations were developed in the clamps (Figure 6.a) and the tests were sttoped at displacements of around 20 mm, when the system was still sustaining load. No sign of failure was detected in the culms on the contact zone with the clamps, however, some splitting was observed near the opposite end of the load application (Figure 6.b). The initial stiffness was 2165.4 N/mm (COV = 0.61).

The Pearson correlations (R^2) between maximum load and culm diameter, culm thickness and interference, were, respectively, equal to 0.77, 0.62 and 0.67.

Non-Conventional Materials and Technologies – NOCMAT for XXI Century   Materials Research Forum LLC

Materials Research Proceedings 7 (2018) 661-670   doi: http://dx.doi.org/10.21741/9781945291838-63

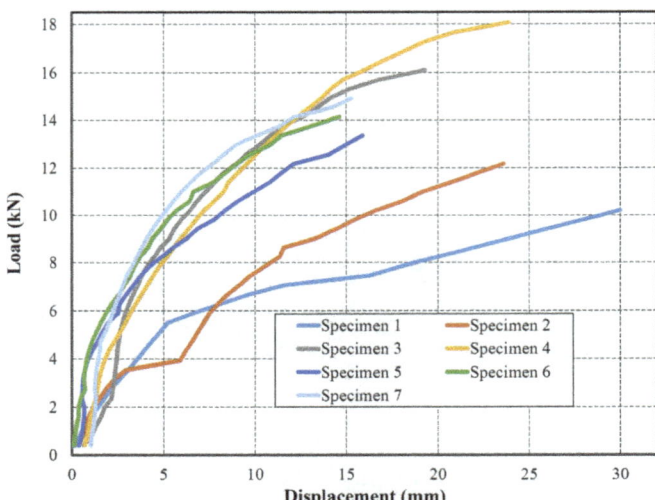

FIGURE 5 - *LOAD-DISPLACEMENT CURVE OBTAINED IN THE AXIAL LOADING TESTS*

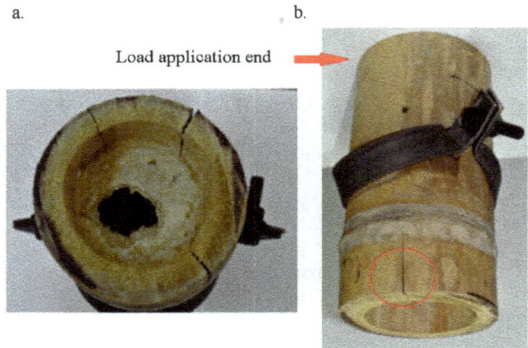

FIGURE 6 - *PLASTIC DEFORMATION OF THE CLAMP AND SPLITTING IN THE CULM IN THE OPPOSITE END OF THE LOAD APPLICATION*

For the first mode of transverse loading, differences in the form of the load-displacement curves can be observed among the seven specimens (Figure 7.a). For instance, the curve for the specimen 1 shows a monotonic reducion of the slope while the curves for the other specimens do not depict a clear trend. This uneven behavior may be attributed to variations in the settlement of the specimens at the supports, due to the geometrical irregularities of the culms. The same dispersion in the form of the load-displacement curves is also present for the transverse loading b (Figure 7.b). The initial stiffnes was 1561 N/mm (COV = 0.34) and 1658 N/mm (COV = 0.28) for loading cases a an b, respectively.

Non-Conventional Materials and Technologies – NOCMAT for XXI Century   Materials Research Forum LLC
Materials Research Proceedings 7 (2018) 661-670   doi: http://dx.doi.org/10.21741/9781945291838-63

a.

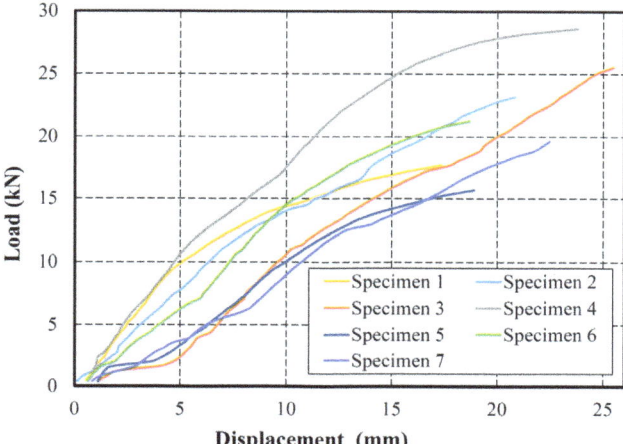

b.

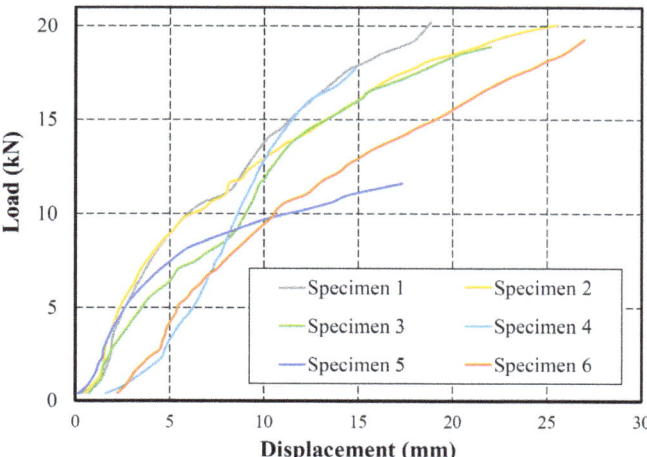

FIGURE 7 - LOAD-DISPLACEMENT CURVES FOR TRANSVERSE LOADING CASES A (AT THE TOP) AND B (AT THE BOTTOM)

For loading case *a*, there was a separation from the culm of the bottom semi-ring at loads of about 10-12 kN (Figure 8.a). After this separation, the upper semi-rings locally deforms and indents the culm. However, the system continues to support the load, showing a ductile behavior. For loading case *b*, the upper ears indent the culm with no sign of sppliting, and the system exhibits a ductile behavior (Figure 8.b).

Non-Conventional Materials and Technologies – NOCMAT for XXI Century   Materials Research Forum LLC
Materials Research Proceedings 7 (2018) 661-670   doi: http://dx.doi.org/10.21741/9781945291838-63

For loading case *a*, the Pearson correlation coefficients between load and culm diameter, culm thickness, and distance to the nearest natural node, respectively, were equal to 0.72, 0.18, and 0.07. The corresponding coefficients for loading case *b* were -0.27, 0.55, and -0.84. This suggest that the load is proportional to the diameter for loading case *a*, while a shorter distance to the natural node increases the maximum load for case *b*. Results also suggest that the thickness of the culm does not greatly affects the maximum load.

FIGURE 8 - DEFORMATION OF THE CULM-CLAMP SYSTEM FOR TRANSVERSE LOADING MODES A AND B

Discussion

A clamp system was proposed that is a feasible option to develop joints for bamboo culms. With this system, there is no need to drill holes in the culms, which avoids the generation of stress concentrators that cause longitudinal splitting and early failure at the joints. Thus, the system exhibits a ductile behavior.

Under axial loading, the interference appears to be a good predictor for maximum loading. Based on this interference, it is possible to design four or five sets of clamps of different sizes that may work well for a range of culm diameters. Hence, during the assembly of a structure, the worker may choose the appropriate set based on the circumference of the culm, which can be easily measured. The range (discarding the result for specimen with the lower interference that slipped in the culm) of maximum axial loads (13000 – 18442 N) are of the order of magnitude of those than can be transmitted with more elaborated joints using injection of concrete (NSR-2010) [11].

This is a pilot study that shows the feasibility of the system to develop joints for bamboo culms. However, there are important issues that have to be addressed, such as the behavior of the system after days or weeks of adjustment, or the capacity when multiple clamps are devised to transmit the

Non-Conventional Materials and Technologies – NOCMAT for XXI Century   Materials Research Forum LLC
Materials Research Proceedings 7 (2018) 661-670   doi: http://dx.doi.org/10.21741/9781945291838-63

load. In addition, theoretical models using finite element analysis should be developed to fully characterize the mechanical behavior of the system. In turn, results of validated finite element models may be used to develop simple formulas for assessing important mechanical parameters such as the initial stiffness, which are needed to design the joint and to assess the deflections of a bamboo structure.

References

[1] M. Vahanvati, «The Challenge of connecting bamboo,» de *10th World Bamboo Congress*, Damyang, 2015.

[2] O. Arce-Villalobos, Fundamentals of the design of bamboo structures. Doctoral dissertation, Eindhoven: Eindhoven University of Technology, 1993.

[3] B. Sharma, Seismic Performance of Bamboo Structures. Doctoral dissertation, Pittsburgh, USA: University of Pittsburgh, 2010.

[4] L. Villegas, R. Moran y J. García, «A new joint to assemble light structures of bamboo slats,» *Construction and Building Materials*, pp. 61-68, 2015. https://doi.org/10.1016/j.conbuildmat.2015.08.113

[5] R. Moran, C. Benitez, H. Silva y J. García, «Desing of steel connector for structural bamboo members,» de *AMDM 2016 Terce Congreso Internacional Sobre Tecnologías Avanzadas de Mecatrónica, Diseño y Manufactura*, Cali, 2016.

[6] B. Sharma, K. Harries y K. Ghavami, «Methods of determining transverse mechanical properties of full-culm bamboo,» *Construction and Building Materials*, pp. 627-637, 2013. https://doi.org/10.1016/j.conbuildmat.2012.07.116

[7] R. Moran, K. Weeb, K. Harries y J. García, «Edge bearing tests to assess the influence of radial gradation on the transverse behavior of bamboo,» *Construction and Building Materials*, pp. 574-584, 2017. https://doi.org/10.1016/j.conbuildmat.2016.11.106

[8] J. García, C. Rangel y K. Ghavami, «Experiments with rings to determine the anisotropoc elastic constants of bamboo,» *Construction and building of materials*, pp. 52-57, 2012. https://doi.org/10.1016/j.conbuildmat.2011.12.089

[9] Instituto Colombiano de Normas Técnicas y Certificación , NTC 5525 - Métodos de ensayo para determinar las propiedades físicas y mecánicas de la Guadua angustifolia Kunth, Bogotá: ICONTEC, 2007.

[10] F. Bruhl y U. Kuhlmann, «CONNECTION DUCTILITY IN TIMBER STRUCTURES,» de *World Conference on Timber Engineering*, Auckland, 2012.

[11] AIS, Reglamento colombiano de Construcción Sismo Resistente NSR-10, Bogotá: AIS, 2010.

[12] A. Widyowijatnoko, R. Fitranto y R. Intan, «Proposing Joints for Bamboo Tensegrity,» de *10th Wolrd Bamboo Congress*, Damyang, 2015.

[13] J. Janssen, «Designing and building with bamboo. Inbar, Technical report No. 20,» INBAR, Eindhoven, 2000.

[14] D. Jayanetti y P. Follet, «Bamboo in construction: an introduction. Technical report INBAR,» Inbar, Beiging, 1998.

[15] G. Minke, Building with bamboo, Birkhauser Verlag AG, 2012.

[16] O. Hidalgo, Manual de construcción con bambú, Bogotá: Estudios técnicos colombianos - Construcción rural, 1981.

Non-Conventional Materials and Technologies – NOCMAT for XXI Century   Materials Research Forum LLC
Materials Research Proceedings 7 (2018) 661-670   doi: http://dx.doi.org/10.21741/9781945291838-63

[17] P. Laroque, Design of a low cost bamboo footbridge. Master Thesis, Massachusets: Massachusets Institute of Technology, 2007.

[18] K. Ghavami y L. Moreira, «Development of a new joint for bamboo space structures,» *Transactions on the Built Environment,* pp. 3-12, 1996.

[19] L. Moreira y K. Ghavami, «Limits states analysis for bamboo pin connections,» *Key Engineering Materials,* pp. 3-12, 2012.
https://doi.org/10.4028/www.scientific.net/KEM.517.3

Non-Conventional Materials and Technologies – NOCMAT for XXI Century   Materials Research Forum LLC
Materials Research Proceedings 7 (2018) 671-675   doi: http://dx.doi.org/10.21741/9781945291838-64

Experimental Analysis of the Variation of Electrical Resistivity of Concrete after Fracture

Fabiano Viana Oliveira da Cunha Médice [a, b, *]; Pedro Henrique Alves Martins [a];
Eduardo Brandão Diniz Lage [a]; Maria Teresa Paulino Aguilar [a]

[a] Universidade Federal de Minas Gerais, Brasil; fmedice@hotmail.com,
pedromartins.eng@hotmail.com, brandaolage@yahoo.com.br, teresa@ufmg.br

[b] Centro de Estudos em Engenharia Eletrônica e Automação, Brasil

Abstract. Techniques for concrete quality control are usually associated with the investigation of the mechanical properties and durability of this material. Usually the preparation processes and conditions of the samples are slow and labor intensive, and are mostly destroyed during analysis, preventing further experiments. So no alternative destructive methods become necessary for effective monitoring and inspection to assess the condition of the structure and when maintenance or repair is needed. These techniques should be able to identify any problems of durability before these become serious. The durability of concrete largely depends on the properties of their microstructure, such as the distribution of pore size and shape of the interconnections between them. A network of finer pores with less connectivity leads to a lower permeability. A structure may undergo changes caused by fractures or defects generated during the manufacturing process, handling or working conditions generating a porous microstructure with a higher degree of interconnections. On the other hand, it results in increased permeability and reduced durability in general. The main idea behind most of the electrical resistivity techniques is to seek alternative ways to quantify the conductive properties of the concrete microstructure. In general, the electrical resistivity of the concrete can be described as the ability of the concrete supporting the transfer of ions subjected to an electric potential difference. This paper aims to evaluate the electrical resistivity of the concrete after the fracture.

Keywords: Electrical Resistivity Concrete, Non-Destructive Testing, Fracture in Concrete Structures

Introduction

According to Malhotra & Carino [1] unlike pipes and welded metal structures, non destructive testing for concrete are relatively new techniques. This is due to the inhomogeneity of the concrete, which is a composite, which may be made of different materials, creating new properties. In addition, the authors report that all concrete properties depends on the handling during the curing time, which there is no hand very specialized work [1].

For Malhotra &Carino [1] there are two classes of non-destructive testing methods for concrete. The first consists of methods to estimate the mechanical resistance and the second includes methods to measure other characteristics such as moisture content, density, thickness, permeability, and electrical resistivity. In the second class, are included methods for identifying defects such as delaminations, voids and cracks in the concrete. In the second class, electrical methods are widely used to locate and measure the reinforcement thickness, moisture, and corrosion potential through the electrical inductance [1]. This is because when making a cement mixture with water, a series of hydration and hydrolysis reactions occur to form a viscous slurry of cement, which perform the curing, becomes a porous solid structure [2,3,4]. As these reactions occur, ions disappear, causing the structure to become less conductive. To rehydrate the concrete, water will enter these pores, releasing ions the large majority of K^+, Na^+, Ca^{++} and $OH^- SO^-$; allowing ionic conduction between the electrodes [2,3,4,5].

According Layssi et al [2] and Santos [4] there are two electrical resistivity measurement processes that are widely used. The uniaxial method, which is measured via two electrodes, usually metal plates; and four points method, which is measured at four points on the surface of the concrete. The method of four points is most advisable for measurements in the field, but it is more sensitive to surface changes to measurement. Being non-destructive tests, and can be used in any structure without significant damage.

This electrical resistivity of concrete is given by the impedance of the same [2]. The value represents the concrete ability to allow the transit of ions into its structure when applied to one electrical potential difference. Thus, this value can be associated with the characteristics of its structure, ie, the distribution of pore size and shape of the interconnections between them. A network of finer pores with less connectivity leads to a lower permeability. The extent to which the permeability increases, the electrical resistivity decrease [2,3,4,5,6]. For the permeability increase, it is necessary for most pores are accessible. This access is the deterioration of the concrete, which may be chemical or physical processes [7]. This occurs because all deterioration results in cracks, which interconnect the microstructures [7].

This study aims to evaluate the feasibility of the electrical resistivity test for the detection of cracks.

Materials and methods

According Layssi et al [2], Santos [4], Elkey & Sellevold [5] and Junior Medeiros et al [8] water-cement ratio, type and amount of aggregate, its mineral and/or chemical additives, the degree of hydration and the time and curing temperature, alter the electrical properties of the concrete. In order to test feasibility, the curing time between the samples were changed, to submit different properties.

18 concrete samples using Portland cement (CPIV) that passed through the submerged curing to be performed measurements were made. Medium sand was used and crushed zero mass with trace of 1: 2: 2 and water/cement ratio of 0.6. After curing, the measurements were carried out in the normative standards. In all measurements, the specimens were kept submerged in a container to supply the local water company (COPASA) for 48 hours according to ABNT NBR 9204: 2012 [9] but rather than humid chamber, submersion was used.

The induced fracture was performed by the block compression test. Using the uniaxial electrical resistivity method (Figure 1) and the four-point method (Figure 2), we measured the resistivity of the cylindrical specimens of 10 x 20 cm before and after an induced fracture. By Wenner probe Resipod manufacturer has worked with the frequency of 40Hz for the four-point method, we also used the same value in the frequency generator for the uniaxial method. For security reasons, he worked with an 8V supply. Due to this value, Chant resistance was necessary to measure the current and thus measure the resistivity of the sample.

Non-Conventional Materials and Technologies – NOCMAT for XXI Century   Materials Research Forum LLC
Materials Research Proceedings **7** (2018) 671-675   doi: http://dx.doi.org/10.21741/9781945291838-64

FIGURE 1 - MEASUREMENT OF ELECTRICAL RESISTANCE WITH UNIAXIAL METHOD

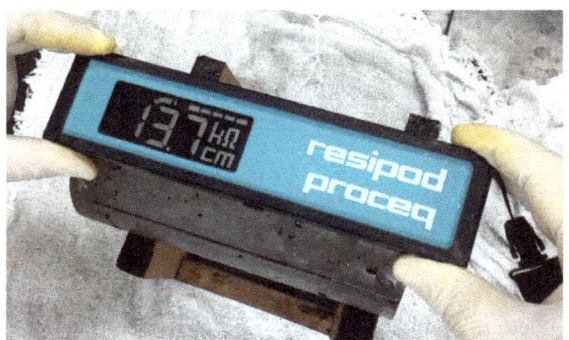

*FIGURE 2 - MEASUREMENT OF ELECTRICAL RESISTANCE WITH FOUR POINTS
METHOD*

Four measurements were made using the probe Wenner spaced 90 degrees each measurement along the surface of the samples. The uniaxial measurement is made by the average of five measurements by varying the resistances of five different Chant.

Results and discussions

Table 1 shows the results of measurements performed by uniaxial and four points methods. For the preparation of Table 1 was measured electrical resistivity of concrete eighteen test bodies by the methods of uniaxial and four points before and after performing a compressive strength test. Therefore, in the table were added two columns of test and divided into three columns, in which the measurements are resistivity prior to compression and after compression the percentage difference between them. The last column shows the change in electrical resistivity between the measuring methods after the fracture.

Non-Conventional Materials and Technologies – NOCMAT for XXI Century   Materials Research Forum LLC
Materials Research Proceedings 7 (2018) 671-675   doi: http://dx.doi.org/10.21741/9781945291838-64

TABLE 1 - RESULTS OF MEASUREMENTS

SAMPLE NAME	TWO-POINT UNIAXIAL METHOD			FOUR-POINT METHOD			VARIATION BETWEEN METHODS
	Before (Ωm)	After (Ωm)	% REDUCED	Before (Ωm)	After (Ωm)	% REDUCED	
CP4 -1	94.78	80.10	15.49%	207.50	170.50	17.83%	2.34%
CP4 -2	93.98	68.85	26.75%	212.50	142.50	32.94%	6.19%
CPIV -3	101.37	66.67	34.23%	207.00	135.75	34.42%	0.19%
CPIV -4	90.48	71.06	21.47%	205.25	137.75	32.89%	11.42%
CPIV -5	99.30	77.07	22.38%	221.00	167.75	24.10%	1.71%
CPIV -6	95.62	74.70	21.88%	216.00	155.75	27.89%	6.02%
CPIV -7	102.28	74.09	27.56%	227.75	184.00	19.21%	8.35%
CPIV -8	115.83	74.93	35.31%	251.75	163.25	35.15%	0.16%
CPIV -9	107.38	71.39	33.51%	236.75	151.50	36.01%	2.50%
CPIV -10	96.31	67.63	29.78%	209.75	138.00	34.21%	4.43%
CPIV -11	92.45	69.45	24.88%	197.50	148.50	24.81%	0.07%
CPIV -12	93.97	71.27	24.16%	210.25	144.25	31.39%	7.24%
CPIV -13	97.22	77.07	20.72%	195.50	158.50	18.93%	1.80%
CPIV -14	97.01	77.45	20.16%	219.25	165.75	24.40%	4.24%
CPIV -15	90.54	62.47	31.00%	175.25	114.75	34.52%	3.52%
CPIV -16	63.94	53.59	16.20%	135.50	101.25	25.28%	9.08%
CPIV -17	61.51	51.68	15.97%	126.50	103.75	17.98%	2.01%
CPIV -18	66.32	52.98	20.12%	135.50	110.25	18.63%	1.49%

The CP's 16, 17 and 18 were made with the same trait, but they were not conducted curing. This were designed in order to demonstrate if any changes occur in the measured values to the curing done in work was done under different conditions. Observing thus the influence of curing would make a difference in the analysis of subsequent electrical resistivity to crack. But it can be observed by the values obtained experimentally that, for this case, the same curing is carried out in different ways, the percentage reduction of the electrical resistivity values is approximate in both cases.

According to Table 1, both the electrical resistivity measurement methods, reduction in the value after the fracture. These values decrease from 15 to 35% after the fracture. Even if the values of resistivity are not the same in both methods, the percentage variation between processes possessed an error of up to 11.42%. However, the values are maintained from the range of 15 to 35%. This error can be approximately 10% due to variation of angle 90° by measuring the resistivity of the block. It happens because in the four-point method, some cracks may not cause variation in measurement, causing the error. All resistivity measured by the four-point method are greater than the uniaxial, which can exemplify that some cracks may not participate in the measurement of resistivity.

During the experiments, some values of uniaxial resistivity were higher than before being fractured. This may be due to lack of contact between the plates of electrodes and concrete. To be broken in some measurements, there was the intimate contact between the concrete and the plate, requiring the exchange of the contactor in order to reduce this variation.

Conclusion

For in both methods of measurement values were reduced, electrical resistivity can report the existence of cracks in the concrete. By varying the electrical resistivity of 10% over the samples, we recommend this method as supplementary order to check cracks and / or defects in the structure. This is desirable, because for values close to or above 15%, the structure might be collapsed.

Aiming to have a control of cracks and cracks inside the concrete and important to keep measuring the electrical resistance during the execution of the work and the end of it. When you get the linearity of the electrical resistivity advised to register this value range and used as a standard for the detection of a possible fracture.

References

[1] Malhotra, V. M.; Carino, N. J. Handbook on Nondestructive Testing of Concrete Second Edition. [S.l.]: CRC press, 2004.

[2] Layssi, H. et al. Electrical Resistivity of Concrete. Concrete International, v. 37, n. 5, p. 41-46, 2015.

[3] Mccarter, W. J. et al.. Two-point concrete resistivity measurements: interfacial phenomena at the electrode–concrete contact zone. Measurement Science and Technology, v. 26, n. 8, p. 085007, 2015. https://doi.org/10.1088/0957-0233/26/8/085007

[4] Santos, L. Avaliação da resistividade elétrica do concreto como parâmetro para a previsão da iniciação da corrosão induzida por cloretos em estruturas de concreto. 2006. 161 f. Dissertação (Mestrado em Estruturas e Construção Civil). Departamento de Engenharia Civil e Ambiental, Universidade de Brasília, Brasília/DF, 2006.

[5] Elkey, W.; Sellevold, E. J. Electrical resistivity of concrete, 1995.

[6] Badilla, V.; Zamora I Mestre, J.-L. Correlación de diferentes métodos no destructivos de detección superficial de anomalías en el hormigón armado. Actas de CONPAT 2015, Instituto Superior Tecnico, p. 3.3-9309., 2015.

[7] Wang, K. et al. Permeability study of cracked concrete. Cement and Concrete Research, v. 27, n. 3, p. 381-393, 1997. https://doi.org/10.1016/S0008-8846(97)00031-8

[8] Medeiros-Junior, R. A. et al. Investigação da resistência à compressão e da resistividade elétrica de concretos com diferentes tipos de cimento. Revista Alconpat, v. 4, n. 2, p. 113-128, 2014. https://doi.org/10.21041/ra.v4i2.21

[9] ABNT. NBR 9204. Concreto endurecido — Determinação da resistividade elétrico-volumétrica — Método de ensaio, Procedimento, Rio de Janeiro, 2012.

Non-Conventional Materials and Technologies – NOCMAT for XXI Century   Materials Research Forum LLC
Materials Research Proceedings 7 (2018) 676-684   doi: http://dx.doi.org/10.21741/9781945291838-65

Evaluation of Pre-Early Age Strength of Super Fine Ggbs Mortars

Korde Chaaruchandra[a*], Matthew Cruickshank[a], Roger P. West[a], John Reddy[b]

[a]Trinity College Dublin, Dublin, Ireland; kordec@tcd.ie, mcruiks@tcd.ie, rwest@tcd.ie,

[b]Ecocem Ltd, East Point, Dublin, Ireland; john.reddy@ecocem.ie

Abstract. This paper reports on a study exploring the use of a super-fine ground granulated blast-furnace slag (SF GGBS) for the precast industry wherein pre-early age strength becomes prominent for de-shuttering of formwork and optimizing productivity. The results will be discussed for research undertaken to study the effect of temperature curing methods, accelerating admixtures and SFGGBS on the compressive and flexural strength of standard mortars to establish the efficacy of the cement fineness and the new accelerator in very early age strength attainment. In this study a rapid hardening cement (RHC) is compared to a 50% SFGGBS to examine the effect of curing temperature (at 20 and 35 °C) and two new types of accelerating admixture on strengths at 6 different age of testing (16hr, 1, 2, 3 ,7 , 28 days). In comparison to pure RHC at 16 hrs, the SFGGBS mortars without admixture achieved only 35 % of flexural strength and about 30% of compressive strength at low temperature; whereas nearly 96 % of flexural strength and 100% of compressive strength at high temperature. With the addition of admixture 1, a new more effective admixture developed to activate SFGGBS, mortars achieved over 70 % of flexural strength and 60% of compressive strength at low temperature whereas over 128 % of flexural strength and 120% of compressive strength at high temperature. At 24 hrs the results are nearly similar to 100% RHC and with further increase in the age of the testing the results are found to broadly equate; thus demonstrating the effective performance of admixture 1 and higher temperature in activating SFGGBS for achieving better pre-early age strength.

Keywords: Accelerating Admixture, Mortar, Pre-Early Age, Strength, Super-Fine GGBS

Introduction

Ground granulated blast-furnace slag (GGBS) has gained wide acceptance as a supplementary cementitious material for on-site concrete work for both technical and environmental reasons: its use reduces the likelihood of thermal cracking in deep pours; it has particular advantages for aspects of durability, especially chloride and sulphate attack resistance; and, increasingly importantly, it has a much lower carbon footprint than Portland cement (NRMCA, 2012). However, the virtue of having slower hydration and strength development (Roy and Idorn,1982, Escalante and Sharp, 2001, Escalante et. al., 2001), leading to lower core temperatures in its early life, becomes a significant deterrent to GGBS concrete's use in pre-cast works where very early de-moulding and lifting is essential to ensure the productivity necessary to be competitive in the construction industry.

Pre-cast companies in temperate climates use several means to ensure that the concrete has sufficient early age strength to lift pre-tensioned pre-stressed concrete slabs out of the shutters as soon as 16 hours after pouring. These techniques include the use of Rapid Hardening Cement (RHC), an accelerating admixture and, most effectively, thermal activation (Soutos, 2014), usually by heating the beds to about 35°C. Hence in the present study the samples are cured under temperatures of 20°C and 35°C.

In terms of early age strength development, the pozzolanic reactions of GGBS and fly ash do not really start until after a day or so after mixing, and so their contribution to concrete strength at one day is not significant. At one day, for 30 % fly ash, 50 % GGBS and 70 % GGBS the strength is around 70 %, 50 % and 30 % respectively as compared to CEM I (Clear, 2011). A standard mortar is normally employed to observe the cement's behaviour early age in the absence of coarse aggregate (IS EN 196-1, 2005). Thus, similarly, mortars with CEM I at 7 days have been compared with 30% and 60 % GGBS at 20°C and 40°C (Cakir and Akoz, 2008) in which compressive strength at 20°C with respect to CEM I (at 33.7 MPa) is 92 % and 78% and at 40°C (at 36.9 MPa) is 92 % and 88%. Similarly, early age flexural strength gain at 20°C in comparison to CEM I (at 6.8 MPa) is 96 % and 88% and at 40°C (at 7.0 MPa) is 90 % and 87%, respectively.

This paper describes research conducted on mortars to characterise the effects of employing two types of admixtures, one novel accelerating admixture and the second a proprietary accelerator which are designed to activate the GGBS to overcome the significant strength losses which exist at 16 hours when up to 70% cement substitution with GGBS and super-fine GGBS (SFGGBS) are used, though with no adverse long term effect on strength. The results are compared with a control sample i.e. 100% RHC (CEM I 42.5R). Comprising over 200 individual test results, carried out in accordance with European standard IS EN196 - 1: 2005, the flexural (72 tests) and compressive strength (144 tests) development with age, up to 28 days, with GGBS inclusion at rates of 50 % will be described in the paper. Furthermore, the effectiveness of these chemical and geometrical activators will be examined for concrete under enhanced thermal activation, namely an elevated (35°C) curing temperatures compared to the normal 20°C. By examining the load deflection plots for mortar prisms in flexure, estimates of early age Young's modulus of elasticity are also made, which are essential for calculating anticipated elastic shortening of the slabs at transfer and longer term shrinkage losses. At each stage, duplicate tests ensure a low coefficient of variation for compressive strength results is achieved, increasing confidence in the effects being observed.

Methodology of research
Materials
This experimental study on standard mortars is undertaken with RHPC (CEM I 42.5R) as the base cement, a siliceous sand as per standard guidelines (IS EN196-1, 2005) with various substitutions of GGBS and potable water. Two types of admixtures are used which are termed as Admixture 1(Ad-1: ECOCEM Admixture) and Admixture 2 (Ad-2: Centrament Rapid FastKick 105). The super-fine GGBS used has a particle size of 7 microns compared to normal particle size of GGBS with a fineness of 11 micron.

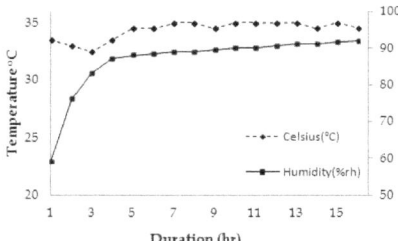

FIGURE 1 - TEMPERATURE AND HUMIDITY PROFILE / CYCLE

Non-Conventional Materials and Technologies – NOCMAT for XXI Century   Materials Research Forum LLC
Materials Research Proceedings 7 (2018) 676-684   doi: http://dx.doi.org/10.21741/9781945291838-65

TABLE 1 - MIX PROPORTION / MOULD

	Mix 1 (NONL/H)	Mix 2 (U51L/H)	Mix 3 (U52L/H)
Fine (gm)	1350	1350	1350
CEM I (gm) (42.5)R	450	225	225
SFGGBS (gm)	0	225	225
Water (ml)	225	225	225
Admix. (ml)	0	4.5(Ad 1)	4.5(Ad 2)

Mix Proportion

Three different mixes are prepared with a combination of CEM I R, SFGGBS and admixtures. The proportions of these mixes are listed in Table 1. Each mix proportion listed is for one mould. A mortar batch comprises enough for 4 moulds with 20% extra for wastage. The mortar has cement to sand ratio of 1:3 and water to cement ratio of 0.5. The nomenclature of the samples can be understood from the two examples: U51L – Superfine with 50% GGBS and Admix-1 at low temperature and HU51H Superfine with 50% GGBS and Admix-1 at high temperature.

Curing Methods

Since this study is to primarily simulate the environment in the precast industry; two types of curing conditions are adopted. These moulds filled with mixes are a) kept in a box which is wrapped with wet jute cloth and further covered with plastic sheet; it is left in atmospheric conditions, which is approximately 20°C and b) kept in a tray which in turn is warped with jute cloth and placed in a temperature and humidity control chamber where the temperature is kept at approximately 35 °C and relative humidity of 90%. Typical temperature and humidity condition for one cycle of testing is as shown in Figure 1. Samples are de-moulded after 16hrs and placed in the curing chamber which is maintained at approximately 20 °C. Samples are kept for up to 28 days and the respective samples are only removed as required for the age-based tests.

Specimen and Testing

The prismatic specimens of size 160 x 40 x 40 mm are de-moulded at 16 hrs, numbered and placed in a curing tank at 20 °C. Two types of tests are performed, namely flexural and compressive tests at 6 different ages (that is, 16hr, 1, 2, 3, 7 and 28 days). For every category there are two specimens, thus 2 x 6 ages gives, in all, 12 samples for each case. These prism pairs are first tested under flexure with a 3-point bending test set-up under deflection controlled loading, After flexural failure by rupture near the middle of the prism, the resulting 4 sample halves are tested under 40 x 40 mm uniform axial compression with the same deflection controlled test equipment. The broad matrix of samples for flexural test is 1 (50 % SFGGBS) x 2 (temperatures conditions 20 °C and 35 °C) x 3 (Admixtures none, Ad 1 and Ad 2) x 6 (ages 16 hrs, 1, 2, 3, 7, 28 days) x 2 sample pairs for flexural and 4 each for compression. Thus there are a total of 72 samples for flexural testing and 144 samples for compression testing. These tests are only part of a more comprehensive study in which varied proportions of GGBS, both fines and super-fines at 30%, 50% and 70% GGBS substitutions are tested. The present paper reports only on the study of 50% fine and SFGGBS.

Non-Conventional Materials and Technologies – NOCMAT for XXI Century   Materials Research Forum LLC
Materials Research Proceedings **7** (2018) 676-684   doi: http://dx.doi.org/10.21741/9781945291838-65

(a)   (b)

FIGURE 2 - *TEST STEP (A) FLEXURE TEST (B) COMPRESSION TEST*

Results and discussion

Flexural Strength

The two characteristics derived from the flexural test are the modulus of rupture (MoR) and flexural modulus of elasticity (MoE). The former is synonymous with the maximum bending stress and the latter is derived from the deflection under the 3-point bending experiment. These parameters are studied to check the performance of the admixtures to activate the early strength gain in the cement matrix. It is a clearly know fact that the addition of GGBS reduces early age strength (Roy and Idorn, 1982, Escalante and Sharp, 2001, Escalante et. al., 2001), however the admixtures used in this study are employed to identify their relative performance in activating the very early age strength development process in GGBS concrete. Thus, it can be observed from the MoR and MoE graphs in Figure 2 and 3, 4 and 5 respectively that admixture 1 is able to activate the strength gain in GGBS at an early age, that is, up to 3 days. Admixture 2 is also able to activate, however, it is not consistent for all the age groups and most of the strength gains are not as high as admixture 1.

At 16 hrs, for MOR (Figure 2), when compared to the control sample at low temperature, 50% GGBS without admixtures develops 35 % of the baseline strength whereas with admix-1 it develops 70% of such strength. With increase in curing temperature there is a 95% rise in strength of control samples. Compared to control samples at high temperature, GGBS samples at high temperature register 45%, 60% and 40% of MOR without admix, with admix-1 and admix 2 respectively. Overall the GGBS samples with admix-1 at high temperature achieve a maximum rise in strength as compared to the control sample at high temperature.   At 1 day, 2 days and 3 days admixture 1 gives better performance, and when combined with high temperature curing, helps in achieving nearly 75%, 90 % and 100 % of the strengths respectively in comparison to control samples at the high temperature regime.

Similarly, at 16 hrs, for the MoE (Figure 4), when compared to the control sample at low temperature, 50% GGBS without admixtures develops 30 % of the baseline modulus whereas with admix-1 it develops 45%. With increase in curing temperature there is 13% rise in modulus of the control samples. Compared to the control sample at high temperature, GGBS samples register 70%, 100% and 90% of MOE for samples without admix, admix-1 and admix 2 respectively. Overall the GGBS samples with admix-1 at high temperature achieve an equivalent strength as compared to the control sample at high temperature. At 1 day, 2 days and 3 days, again admixture 1 combined with high temperature curing helps in achieving nearly 90%, 105 % and 110 % of the modulus respectively in comparison to control sample at high temperature regime.

At 7 and 28 days for MOR and MOE, the GGBS samples along with both admixtures, at high temperatures are able to achieve 90% strength whereas admixture 1 achieves more than 100% strength. The complete MOR and MOE curves from 16 hr to 28 days can be seen in Fig. 3 & 5.

Compressive Strength

The compressive strength is the most important parameter for the precast industry as it is the one test which is conducted in the factory to confirm that it is safe to transfer the pre-stress to the concrete at the earliest possible age. It can be observed (Figure 6) for the strengths at various ages that the temperature curing has proved to be effective for the early age strength gain. Also in Figure 6 it is evident that admixture 1 is able activate the strength gain process to deliver a higher strength in almost all the cases, but not as high as the baseline strength at either temperature.

At 16 hrs (Figure 6), when compared to the control sample at low temperature, 50% GGBS without admixtures develops 30 % of the baseline strength whereas with admix-1 it develops 60% strength. With increase in curing temperature there is a 120% rise in strength of the control samples. Compared to control sample at high temperature GGBS samples register 45%, 55% and 60% of strengths for samples without admix, admix-1 and admix 2 respectively. Overall the GGBS samples, with admix-1 at high temperature achieve a bit less strength as compared to admix 2 but improves at later stages. At 1 day, 2 days and 3 days, again admixture 1 combined with high temperature curing helps in achieving nearly 70%, 80 % and 90 % of the strengths respectively in comparison to the control sample at the high temperature regime.

At 7 days, GGBS samples at low temperature achieve nearly 80% strength while admix -1 achieves 100% strength and, at high temperature, samples achieve nearly 90% of the baseline strength while admix -1 achieves 100% strength. Similarly at 28 days, GGBS samples at low temperature achieve more than 110% strength and at high temperature samples achieve nearly 100% strength. The complete compressive strength curves from 16 hr to 28 days can be seen in Fig. 7.

Conclusions

This experimental research has thus investigated flexural and compressive strengths of mortars with and without 50 % SFGGBS and at low and high temperature curing along with two different types of admixtures. The following conclusions are drawn from this study:

a. The early age flexural and compressive strengths of samples with 50% SFGGBS are lower than the control samples, however the temperature curing and admixtures helps to achieve a higher strength. At 16 hrs, a 50% SFGGBS sample combined with high temperature and admixture 1 is able to achieve a higher strength than the control sample at low temperature for MOR, MOE and compressive strength by 128%, 113% and 120% respectively.

b. With an increase in the age of mortars from 1 to 7 days, 50% SFGGBS samples combined with high temperature and admixture 1, consistently achieve an equivalent strength. At 28 days the samples with high temperature and admixture 1 combined; MOR, MOE and compressive strength surpass the strength of the control sample at low temperature.

c. The use of admixture 1 is proving to be very effective in increasing the MOR, MOE and compressive strength of mortars in both low and high temperature as compared with admixture 2.

d. Also, the combined effect of admixture 1 and high temperature up to 35 °C have a positive effect on the 28 day strength resulting in increase in strength under MOR, MOE and compression.

e. It still remains a challenging aspect to achieve a better strength against control samples at high temperature. Nevertheless the same can be enhanced by a change of water cement ratio, reduction of GGBS substitution, a further increase in temperature of curing and/or new admixtures, which need to be researched.

Non-Conventional Materials and Technologies – NOCMAT for XXI Century   Materials Research Forum LLC
Materials Research Proceedings 7 (2018) 676-684   doi: http://dx.doi.org/10.21741/9781945291838-65

The environmental advantages of precast concrete production will inevitably lead to higher proportions of pre-cast concrete being used world-wide. In motivating designers to specify precast concrete in preference to site-poured concrete not only must the technical requirements be met but the cost must also be minimised, which means that factory productivity must be at least be maintained. The use of GGBS will mitigate against this unless the pre-early strength of concrete can be guaranteed through changes to mix constituents and curing processes without extra cost above current. This is a significant challenge and this paper in particular brings out the possibility of mitigating it to a certain extent with the use of admixtures for activating super-fine GGBS.

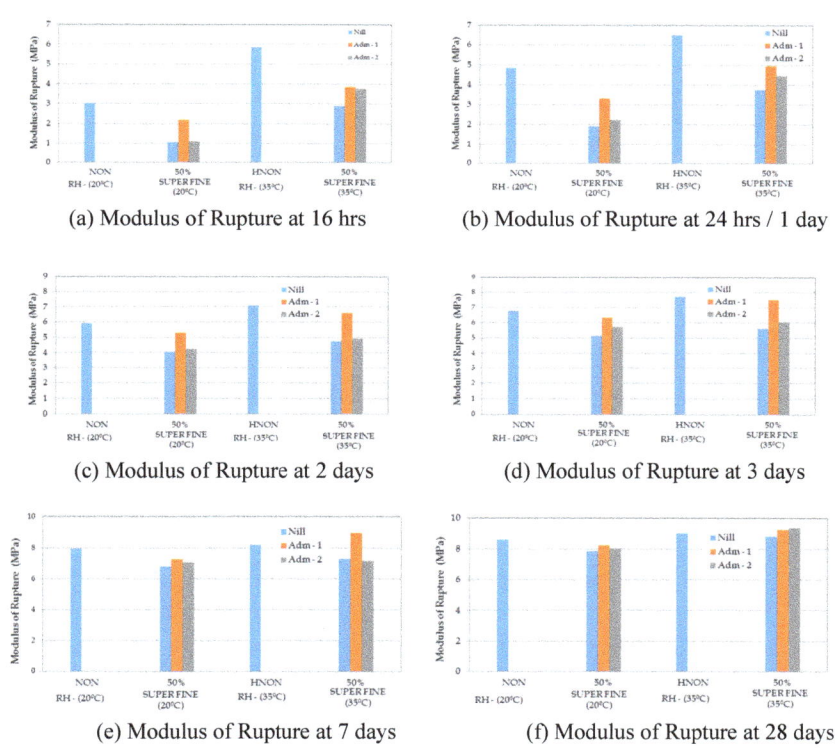

(a) Modulus of Rupture at 16 hrs   (b) Modulus of Rupture at 24 hrs / 1 day

(c) Modulus of Rupture at 2 days   (d) Modulus of Rupture at 3 days

(e) Modulus of Rupture at 7 days   (f) Modulus of Rupture at 28 days

FIGURE 2 - MODULUS OF RUPTURE FOR FINE AND SUPER-FINE GGBS

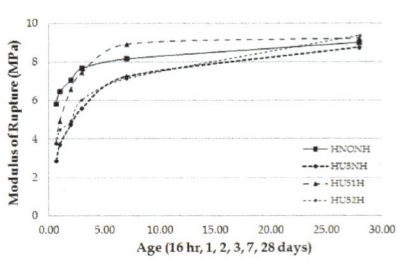

(a) Low temperature (20°C)

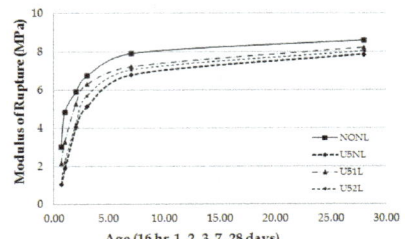

(b) High temperature (35°C)

FIGURE 3 - *MODULUS OF RUPTURE OF FINES FOR DIFFERENT ADMIXTURES AND*

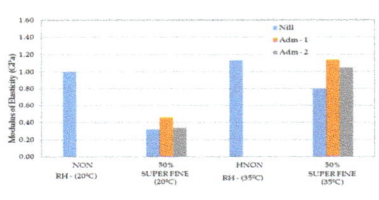

(a) Modulus of Elasticity at 16 hrs

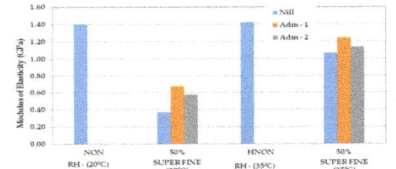

(b) Modulus of Elasticity at 24 hrs / 1 day

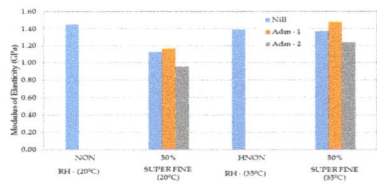

(c) Modulus of Elasticity at 2 days

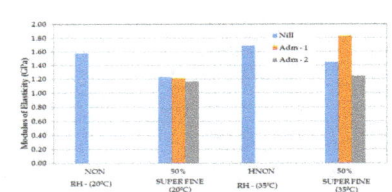

(d) Modulus of Elasticity at 3 days

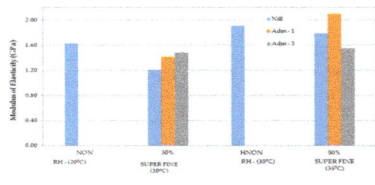

(e) Modulus of Elasticity at 7 days

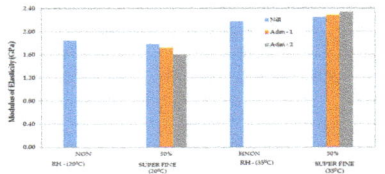

(f) Modulus of Elasticity at 28 days

FIGURE 4 - *MODULUS OF ELASTICITY FOR SUPER-FINE GGBS*

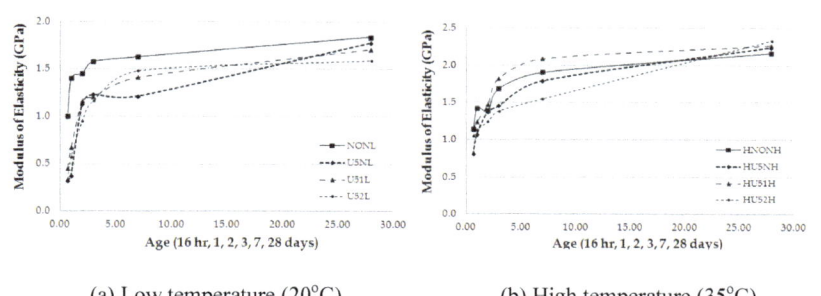

(a) Low temperature (20°C)   (b) High temperature (35°C)

FIGURE 5 *- MODULUS OF ELASTICITY OF FINES FOR DIFFERENT ADMIXTURES AND*
TEMPERATURES

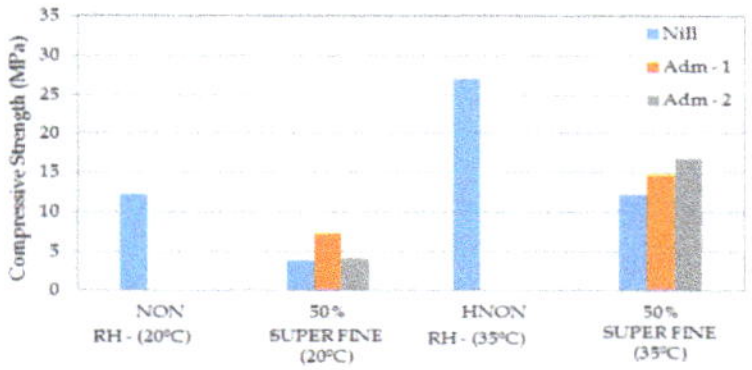

(a) Compressive strength at 16 hrs 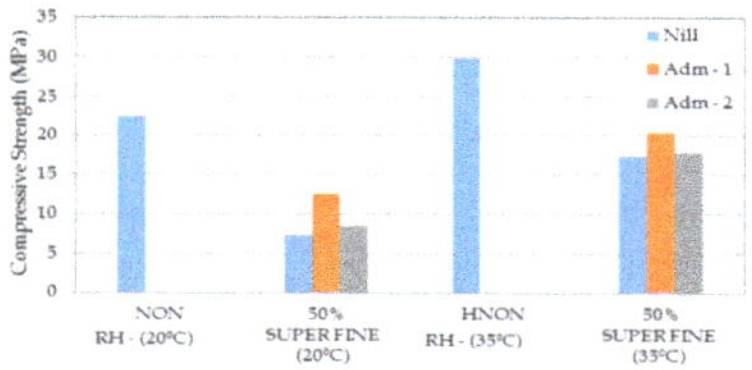   (b) Compressive strength at 24 hrs / 1 day

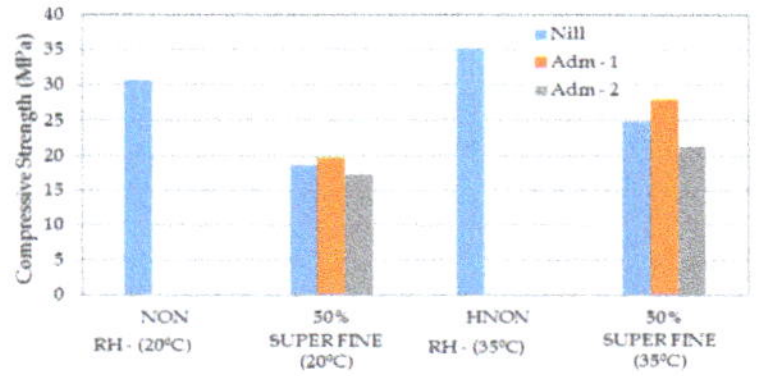

(c) Compressive strength at 2 days 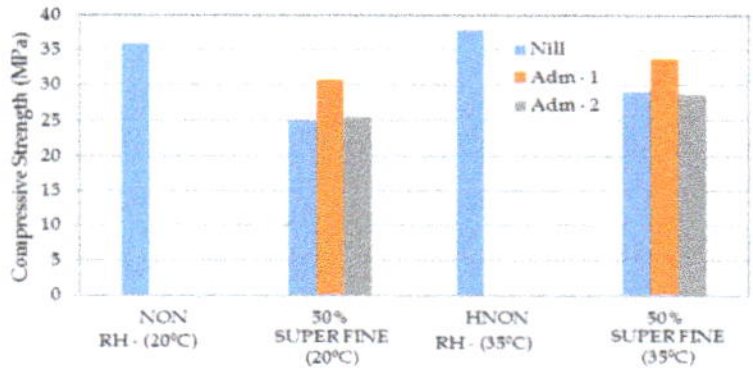   (d) Compressive strength at 3 days

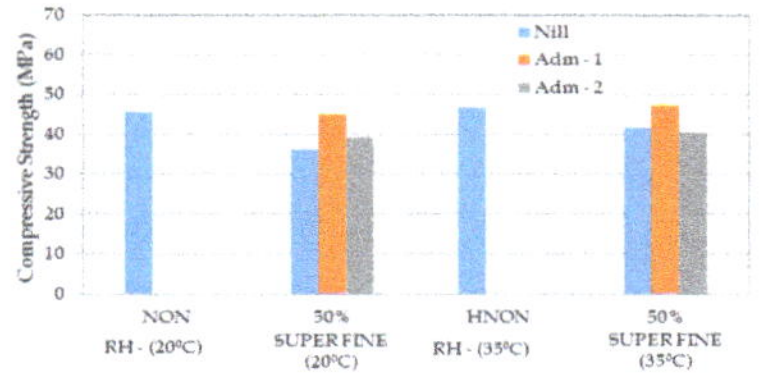

(e) Compressive strength at 7 days 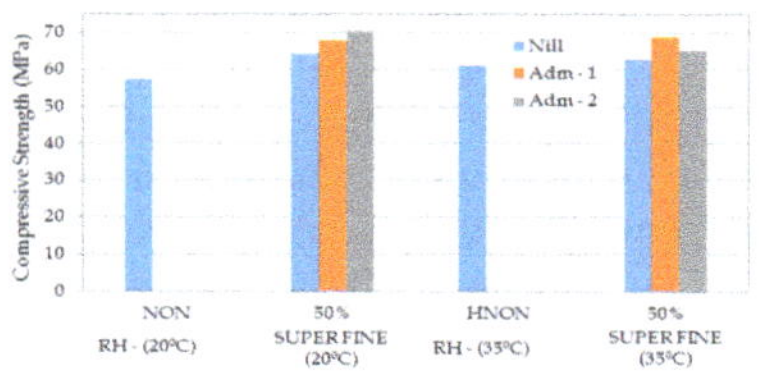   (f) Compressive strength at 28 days

FIGURE 6 *- COMPARISON OF COMPRESSIVE STRENGTH GAIN FOR FINES*

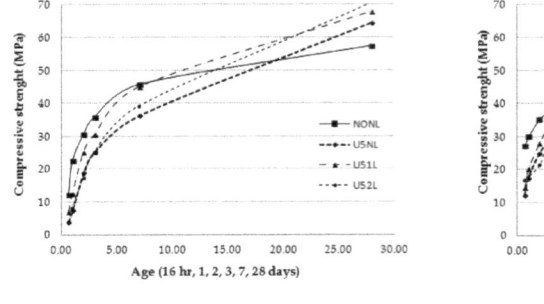 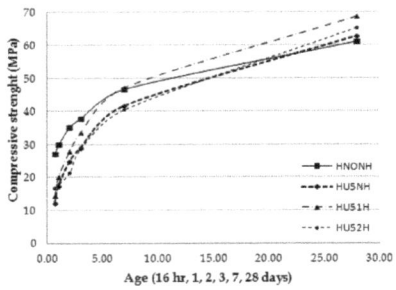

(a) Low temperature (20°C)   (b) High temperature (35°C)

FIGURE 7 - COMPRESSIVE STRENGTH GAIN OF FINES FOR DIFFERENT ADMIXTURES AND TEMPERATURES

References

[1] Clear C. A., 2011, "Cement type/ early age properties", Concrete Today, BRMCA British Ready-Mixed Concrete Association brmca.org.uk, pg. 12-14.

[2] Cakir O. & Akoz F., 2008, "Effect of curing conditions on the mortars with and without GGBFS", Construction and Building Materials, 22, 308-314. https://doi.org/10.1016/j.conbuildmat.2006.08.013

[3] NSAI, IS EN 196-1:2005, "Methods of testing cement – Part 1: Determination of strength", Dublin.

[4] Soutsos M., Hatzitheodorou A., Kanavaris F. and Kwasny J., 2014, "Effect of temperature on the strength development of mortar mixes with GGBS and fly ash", Magazine of Concrete Research; Dec2014, Vol. 66 Issue 24, p1277-1285, 9p.

[5] National Ready Mixed Concrete Association (NRMCA), 2012, "Concrete CO_2 fact sheet", NRMCA Publication Number 2PCO2.

[6] J.I. Escalante-García, J.H. Sharp, 2001, "The microstructure and mechanical properties of blended cements hydrated at various temperatures", Cement Concrete Research, 31, 695– 702. https://doi.org/10.1016/S0008-8846(01)00471-9

[7] J.I. Escalante, L.Y. Gómez, K.K. Johal, G. Mendoza, H. Mancha, J. Méndez, 2001, "Reactivity of blast-furnace slag in portland cement blends hydrated under different conditions", Cement Concrete Research ,31, 1403– 1409. https://doi.org/10.1016/S0008-8846(01)00587-7

[8] D.M. Roy, G.M. Idorn, 1982, "Hydration, structure and properties of blast furnace slag cements, mortars and concrete", ACI Journal, 79 (6), 444– 457.

Non-Conventional Materials and Technologies – NOCMAT for XXI Century   Materials Research Forum LLC
Materials Research Proceedings 7 (2018) 685-694   doi: http://dx.doi.org/10.21741/9781945291838-66

New Cellular Geopolymer Concretes (CGC) Based on Blast Furnace Slag and Spent FCC Catalyst

A. Font, J. Monzó, L. Soriano, M.V. Borrachero, J. Payá*

Grupo de Investigación en Química de los Materiales (GIQUIMA), Instituto de Ciencia y Tecnología del Hormigón ICITECH, Universitat Politècnica de València, Spain.

fonpeal@gmail.comfonpeal, jmmonzo@cst.upv.es, lousomar@upvnet.upv.es,
vborrachero@cst.upv.es, jjpaya@cst.upv.es

Abstract. In current society, the study and development of new construction materials with the aim of improve its sustainability and clean production represents an important goal to be achieved by scientific communities worldwide. This work presents the preliminary studies of new cellular geopolymer concretes (CGC), which combines energy and economic savings achievable through the use of cellular concrete, with the reduction of greenhouse gas emissions well-known linked with the substitution of traditional Portland cement binder by new geopolymer binder. Blast furnace slag (BFS) and spent FCC catalyst (FCC) were used as mineral precursors in CGC preparation, and two methods of pore formation were evaluated in each matrix: (i) mechanical, by addition of commercial grade synthetic surfactant (sodium lauryl sulfate); and (ii) a combined procedure, by mixing of this surfactant and aluminum powder. Pore distribution analyses were also performed. The results confirm the viability of producing cellular geopolymer concretes based on FCC as well as BFS by combined both foaming and aerating techniques. The obtained CGCs were mechanically characterized (5-8 MPa), yielding densities 30% lower respect to OPC systems, showing a compacted matrix and a homogeneous pore structure.

Keywords: Cellular Concrete, Geopolymer, Aluminum Powder, Sodium Lauryl Sulfate

Introduction

Traditional cellular concrete (TCC) is a lightweight material consisting either Portland cement (OPC) paste or mortar with an homogeneous pore structure, created by the addition of suitable reagents. Depending on the method of air entrapped into the matrix, cellular concrete is conventionally classified in three groups [1]: (i) by chemical reaction of metallic powders in high alkaline medium, generating a gas (aerated concrete or gas concrete); (ii) by mechanical procedures with adding bubble stabilizer or surfactant (foamed concrete); and (iii) by combination of chemical and mechanical procedures. The air-bubbles are entrapped into the fresh cement matrix and they are stabilized before setting. As a result a low density composite with moderate strength and good performance in acoustic and thermal insulations is achieved. It combines great properties of constructive and insulation materials [2-5].

The production of TCC requires a large amount of Portland cement; about 70% by weight is conformed by this material [6]. Since, it is well known that the production of Portland cement requires a high-energy demand, the consumption of non-renewable raw materials [3] and the emission of greenhouse gases (around 5-7% of global CO_2 emissions) [7], it is evident that alternative ways to produce cellular concretes are needed. Thus, the use of new alkali-activated cement, commonly referred to as "Geopolymer foam concrete (GFC)", are currently on study [8-11], in order to contribute to the current phenomenon of global warming and the corresponding environmental impact.

Geopolymers (or alkali activated materials) are inorganic polymeric materials which manufacture involves a chemical reaction between amorphous alumino-silicate raw materials and alkali silicate-hydroxide solutions (in high concentration) yielding stable three-dimensional polymeric structures.

Non-Conventional Materials and Technologies – NOCMAT for XXI Century   Materials Research Forum LLC
Materials Research Proceedings 7 (2018) 685-694   doi: http://dx.doi.org/10.21741/9781945291838-66

As a result the production of "green concrete" with a lower both energy requirement and carbon footprint can be achieved [12]. Several types of by-products and/or wastes are generally reported as suitable precursors for preparing geopolymeric binders such as fly ash, blast furnace slag or calcined clay. Besides the viability of using spent FCC catalyst, a waste from the petroleum industry, as a precursor in geopolymers by means alkali activation was demonstrated by some authors [13].

In cellular concrete, the chosen method of pore formation, (by gas release, by foaming or by combined method) influences the microstructure, the resultant pore structure, and thus its properties. For foamed concretes, in order to develop an optimal and stable void system into the matrix, the controlled range of paste consistency is vital therefore a controlled water/binder is required [9,14]. In case of aerated concrete, in order to improve the reaction of aerating agent and the development of a material with great air-void system, the alkalinity of the medium, a controlled consistency and fast setting time are the primordial factors [1]. The improvement of structural and functional properties is directly affected by a stable structure and spherical cell structure. Besides, to obtain composites with uniform density the pores have to be uniformly distributed. The materials with a void system where macropores are dominant are reported to reduce the density significantly and it involves an important strength drop. The compressive strength of cellular concretes decreases exponentially with a reduction in density [8,9,15].

In this research the use of geopolymeric binder is proposed for the preparation of alternative cellular concrete: cellular geopolymer concrete (CGC). To this end, two precursors were tested: blast furnace slag (BFS) and spent FCC catalyst (FCC), also to compare and value the potential of the geopolymeric binders to cellular systems production, two methods of air entrapping were carried out: (i) foaming by means the addition of commercial grade synthetic surfactant (sodium lauryl sulfate, SLS); and (ii) by combining mechanical (foaming) and chemical procedures, with the mixing of this surfactant (SLS) and metallic aluminium powder (A). The materials obtained were compared to that obtained by using Portland cement (OPC) (obtained by the same void-forming methods), meaning a traditional cellular concrete (TCC) system. Density, compressive strength development and air-void distributions for TCC and CGC were analyzed.

Materials, description and methods
Materials
Two precursors were used to prepare the CGC mixes object of the present study. On the one hand, FCC, that was supplied by BP Oil Company (Grao de Castellón, Spain). FCC is a waste from the petrochemical industries rich in SiO_2 (47.76%) and Al_2O_3 (49.26%), and is characterized by its mixed structure which is amorphous and contains some crystalline phases of zeolites type faujasite ($Na_2Al_2Si_4O_{12} \cdot 8(H_2O)$), as well as mullite ($Al_6Si_2O_{13}$). On the other hand, BFS that was supplied by Cementval S.A (Puerto de Sagunto, Spain). This precursor is an amorphous industrial by-product rich in CaO (40.15%), SiO_2 (30.53%) and Al_2O_3 (10.55%), which contains some crystalline phases such as calcite ($CaCO_3$) and merwinite $Ca_3Mg(SiO_4)_2$. A previous milling treatment of the FCC as well as BFS was necessary in order to improve their behaviour as a geopolymer precursors, obtaining: d_{50}= 17.1µm for the FCC; and d_{50}= 26.6µm for the BFS. A ball mill (Gabrielli Mill-2) was used for grinding the raw materials and the particle sizes were measured by means Malvern Instruments Mastersizer 2000. Ordinary Portland cement (OPC), used in TCC mixes, was supplied by Lafarge (Puerto de Sagunto, Spain).

Alkali activating solution used for the synthesis of geopolymers was made from a mix of sodium hydroxide pellets (Panreac, 98% purity) and a waterglass solution (Merck, 8% Na_2O, 28% SiO_2 and 64% H_2O) The alkaline activator stoichiometry is defined by X/Y factor, where X is the Na^+ molality and Y represents the SiO_2/Na_2O molar ratio.

Sodium lauryl sulfate (SLS) is a synthetic foaming agent (bubble stabilizer agent), an alkyl sulfate from (Panreac); and commercial aluminum powder (A) for the gas production in the paste (by its

Non-Conventional Materials and Technologies – NOCMAT for XXI Century   Materials Research Forum LLC
Materials Research Proceedings 7 (2018) 685-694   doi: http://dx.doi.org/10.21741/9781945291838-66

reaction in alkali medium, Eqn. 1.) which was supplied by Schlenk Metallic Pigments GmbH (98% of purity with mean diameter size of 30 μm).

$$Al\ (s) + 3H_2O + OH^-_{\ (ac)} \quad Al(OH)_4^-{\ (ac)} + 3/2\ H_2\ (g) \tag{1}$$

Experimental procedure
The first step to prepare the CGCs samples (as by the use of FCC as BFS) was to ready the alkali solution and wait for 2 or 3 hours until it reached room temperature. Then this solution is putting in the mechanical mixer and shaken for 30 seconds. The solid precursor was dry pre-mixed with SLS or A/SLS (in each case) and it was added during the follow 60 seconds into the mixer. The CGC was plus mixed for 90 seconds. For the TCC reference material the OPC dry pre-mixed with SLS or A/SLS (in each case) and it was mechanically mixed with water for 90 seconds. Due to the expansion process of the pastes, no compacting treatment was carried out, in order to avoid the gas escaping from the cementing matrix during the setting process. For each resulting paste, six cube specimens of dimensions 4x4x4 cm^3 were molded and cured in a wet chamber (23°C and 100% R.H) for 24 hours. After this, the free surface of the cubes had to be cut out with a saw blade and the specimens were de-molded. Finally the samples were kept in wet chamber until testing at 7 days. An overview of samples and its composition can be found in Table 1. C-S, F-S and B-S samples were prepared with SLS, and C-AS, F-AS and B-AS were prepared with a mixture of aluminum powder and SLS. The water/binder ratio (w/b, binder was the amount of precursor) was varied for each mixture, in order to have an appropriate viscosity for good air-void development in the matrix.

TABLE 1 - OVERVIEW OF SAMPLES AND ITS COMPOSITION

	Precursor	A (% by wt)	SLS (% by wt)	Liquid phase	
				w/b	Alkali solution (X/Y)
C-S	OPC	-	2	0.45	-
C-AS		0.05	0.2		
F-S	FCC	-	2	0.5	7.5/1.7
F-AS		0.05	0.2		
B-S	BFS	-	2	0.35	
B-AS		0.05	0.2		

Considering the natural density as the volumetric mass density (mass per unit volume), it was determined by means of the weight of the cubic samples before compressive strength testing. The compression test was carried out by means of an INSTRON 3282 universal testing machine. The natural density assessment and compression test were performed for six specimens of each cellular concrete dosage, and averages and standard deviation values were calculated.

The pore system characterization of the CGC was investigated by a combination of optical microscopy (OM) and field emission scanning electron microscopy (FESEM). A cube of each mixture was crushed in a porcelain mortar. A small piece (7–10 mm) from the inner part of the cube was selected and immersed in acetone for 30 minutes and dried at 65°C for 40 minutes. FESEM micrographs of these samples covered with carbon were taken by an ULTRA 55-ZEISS electron microscope (at 50x, 100x and 200x magnifications), and the pore diameters were measured. On the other hand, a 4x4x4 cm^3 cube of each mixture was cut into slices 2 cm thick, perpendicular to the cast face, using a diamond rotary saw. The samples were observed by a Leica S8 APO optical microscope. Pictures were taken by a Leica DFC 420 digital camera and the images were processed

Non-Conventional Materials and Technologies – NOCMAT for XXI Century   Materials Research Forum LLC
Materials Research Proceedings **7** (2018) 685-694   doi: http://dx.doi.org/10.21741/9781945291838-66

using Leica LAS image analysis software. Magnifications from 8x to 80x were selected with a pixel representing 12 microns.

Results and discussion

Figure 1 shows the results of natural densities and compressive strengths comparing both: (i) the method of air entrapped into the matrix for each raw material; and (ii) the development for TCC and CGC (depending of the raw material base).

Comparing the method of bubbles generation in the matrix, in both systems (traditional and geopolymeric), the combination of agents (SLS plus A), which involves a pore generation by physical and chemical action (AS systems), yield lower natural densities than when merely S was used (pore generation by physical method), as can be seen in Figure 1a. Considering a lower natural density as the optimal result, the improvement observed in each material was: 18% for samples with OPC, 10% for CGC with FCC and 22% for CGC with BFS. The chemical contribution of A in pore-system generation with the mechanical effect of SLS involves an excellent combination of gas-bubbles from H_2 release and air-burbles from foaming agent into the fresh matrix of composites (C-AS, F-AS, B-AS). The reaction of metallic aluminum has a direct influence in binder rheology: it involves a significant rise in matrix temperature, and subsequently decreases the setting time. This fact allows a higher percentage of combined gas and air bubbles get caught into the matrix. As a result, both for the TCC and CGC, a better pore-system development is achieved that was successfully stabilized before hardening, than those obtained for the foamed samples (C-S, F-S, B-S).

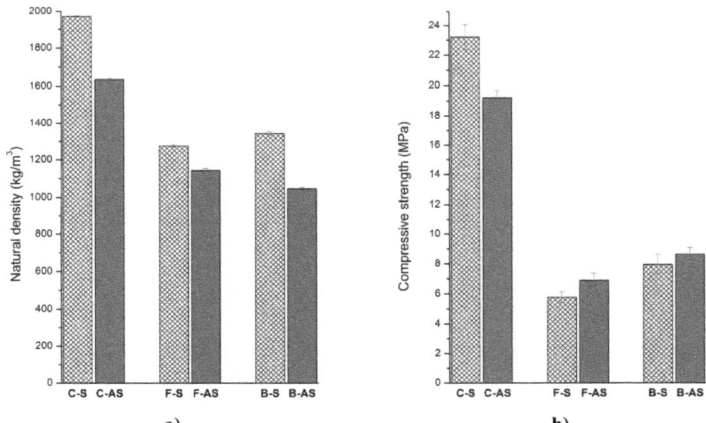

a)   b)

FIGURE 1 - RESULTS FROM PHYSICAL TESTS: A) NATURAL DENSITIES; B) COMPRESSIVE STRENGTHS.

In general, the value of natural density obtained for CGC mixes with FCC (samples F-S and F-AS) either with BFS (samples B-S and B-AS) was below than those resulting for traditional mixes with OPC (samples C-S and C-AS). When the SLS was used as unique foaming agent, the values of natural densities reached were: 1972 kg/m³ for C-S, 1276 kg/cm³ for F-S (35% lower than C-S density) and 1345 kg/cm³ for B-S (30% lower than C-S density). This means that the effectiveness of air entrapping process is better for CGC. On the other hand, when A is added in combination of SLS, the values of natural densities were: 1636kg/m³ for C-AS, 1143 kg/cm³ for F-AS (30% lower than C-AS density) and 1046 kg/cm³ for B-AS(36% lower than C-AS density). These results highlight the fact that the high alkalinity medium of geopolymeric systems compared to the traditional OPC

Non-Conventional Materials and Technologies – NOCMAT for XXI Century   Materials Research Forum LLC
Materials Research Proceedings 7 (2018) 685-694   doi: http://dx.doi.org/10.21741/9781945291838-66

systems, allows a faster and more aggressive transformation rate of aluminum. Thus, a larger part of the hydrogen evolved before setting, trapping gas bubbles into the matrix and preventing the foam breakdown. In the case of TCC, at 20°C curing temperature, the hydrogen evolution is slow and the setting of the system is produced before the complete oxidation of the aluminum: the evolved gas during the paste setting escaped without producing expansion and, therefore part of the air from the foaming action is either escaped.

Focused on the assess for CGC samples, for the samples with SLS, 5% of natural density improvement is reached by using FCC as the precursor, but in contrast, for the samples with SLS and A, 8% of natural density improvement is reached by using BFS as the precursor. The greatest difference is observed between CGC based on BFS (sample B-AS respect sample B-S). From these results two discussions can be presented:

1) In order to get a stable void-system into the matrix by foaming effect from a surfactant addition, the paste must have a proper consistency. Both CGC samples have the same alkalinity rate but differ in water/binder ratio. Due to the higher mean diameter and specific surface area of FCC particles, a large water/binder ratio (w/b = 0.5) is required compared to BFS system (w/b = 0.35). Consequently the resulting CGCs have a different rheology: samples with BFS were more fluid than samples with FCC precursor. Thus, the consistency of F-S samples results better for the inclusion of foam into the matrix before setting compared to B-S samples.

2) When chemical method is used by the generation of gas within the mixture and complement the SLS mechanical effect, in order to prevent the escape of the H_2 and assurance a great foam generation, the paste must have a proper consistency and fast setting time. The reactivity of BFS is faster than the reactivity of FCC, and the CGC based on BFS have a greatest polymerization at mildly low temperature conditions than CGC based on FCC. Furthermore, as stated above, the reaction of metallic aluminum involves a decrease in setting time. Thus, for B-AS sample a complete aluminum reaction was developed and, due to the increase temperature, the consistence of the paste is enough to entrap a great volume of both air and gas resulting an optimal void-system.

Regarding de mechanical characteristics, for CGC samples the compressive strength reached was below than it for TCC in all of the experiences (Fig. 1b.).

As it was introduced, authors argue that the compressive strength decreases exponentially with a reduction in density of cellular concrete [15]. According to it, OPC samples have better mechanical behavior when its natural density is higher (C-S sample) than when its natural density decreases (C-AS sample). Concretely, for C-S it is yielded 23.22 MPa and for C-AS 19.20 MPa, representing a 17.3 % loss. However, for CGC samples there is an increase of compressive strength when natural density is lower due to the use of combined method (foaming by SLS and gas forming by A). For F-S samples it is yield 5.76 MPa and for F-AS 6.88 MPa, representing 16.3 % gain. And for B-S samples it is yield 7.93 MPa and for B-AS 8.62 MPa, representing 8 % gain.

It should be necessary remember that, in current studies, where merely foamed agent or aerating agent is used to generate void matrix, it is discussed that despite the advantages of cellular concrete over traditional concrete, the lower mechanical properties of the former are a drawback [11]. In this respect, these results showed an advantageous behavior of CGC respect TCC. To sum up, when combined method of pore-formation was used, the CGC (by the use of both proposed precursors), it is yield lower natural densities with higher compressive strength.

The above studied properties of cellular concrete, such natural density and compressive strength, are directly consequence by controlling the nature, size and distribution of voids in production of the material. Furthermore, the compressive strength is also influenced by the void/paste ratio as well as spacing and interconnecting among the bubbles.

The pore-system evaluation was carried out for the CGC samples (as based on FCC as well as based on BFS) from the previous physical and mechanical characterization, which represents the optimal behavior comparing to its reference material (TCC based on OPC) in terms of natural densities. In Figures 2-5, OM images and FESEM micrographs of the CGCs are shown, in order to assess the influence of void formation method in pore-system development for the samples of each precursor (FCC and BFS).

The macropores are formed due to the expansion of paste and micropores appear in the walls between the macropores. In this investigation, the following criteria to distinguishing the range of pores was considered: Micropores are pores under or equal to 100 μm and macropores are pores greater than 100 μm.

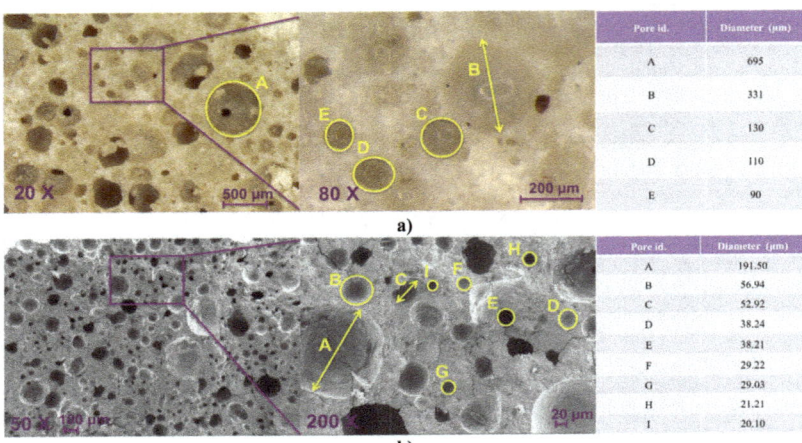

Pore id.	Diameter (μm)
A	695
B	331
C	130
D	110
E	90

a)

Pore id.	Diameter (μm)
A	191.50
B	56.94
C	52.92
D	38.24
E	38.21
F	29.22
G	29.03
H	21.21
I	20.10

b)

FIGURE 2 - PORE SYSTEM EVALUATION FOR F-S SAMPLES: A) OM IMAGES WITH MEASURED VOIDS; B) FESEM IMAGES WITH MEASURED VOIDS.

When foamed method is used by the SLS addition, the resultant void-system was:
1) By using FCC as a precursor (F-S samples) macropores from 191.5 μm to 695 μm and micropores from 20.10 μm and 90 μm could be identified (Fig. 2). There were many bubbles per unit area usually with spherical shape and there were not interconnected each other.
2) With the use of BFS as a precursor (B-S samples) there was identified a macropores distribution with sizes from 470 μm to 160 μm and micropores from 96.44 μm to 15.63 μm (Fig. 3). Differently from FCC case pores were irregular in shape and with interconnections among them. At micro scale (Fig. 3b.), it can be note that the biggest pores had the presence of smaller ones inside their internal walls.

Non-Conventional Materials and Technologies – NOCMAT for XXI Century   Materials Research Forum LLC
Materials Research Proceedings **7** (2018) 685-694   doi: http://dx.doi.org/10.21741/9781945291838-66

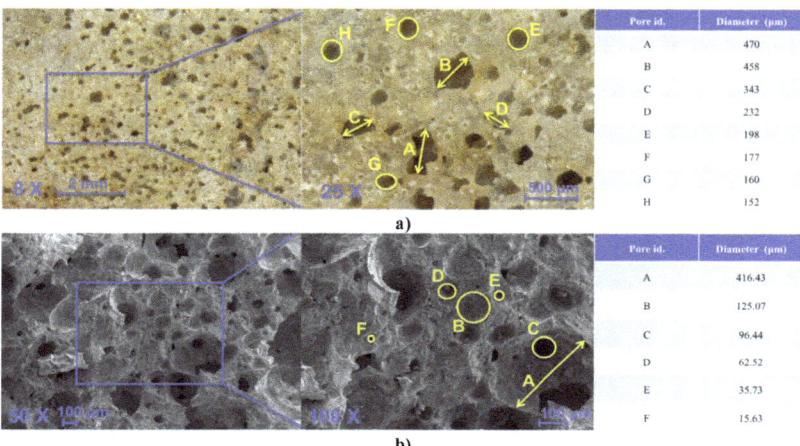

Pore id.	Diameter (μm)
A	470
B	458
C	343
D	232
E	198
F	177
G	160
H	152

a)

Pore id.	Diameter (μm)
A	416.43
B	125.07
C	96.44
D	62.52
E	35.73
F	15.63

b)

FIGURE 3 - *PORE SYSTEM EVALUATION FOR B-S SAMPLES: A) OM IMAGES WITH MEASURED VOIDS; B) FESEM IMAGES WITH MEASURED VOIDS.*

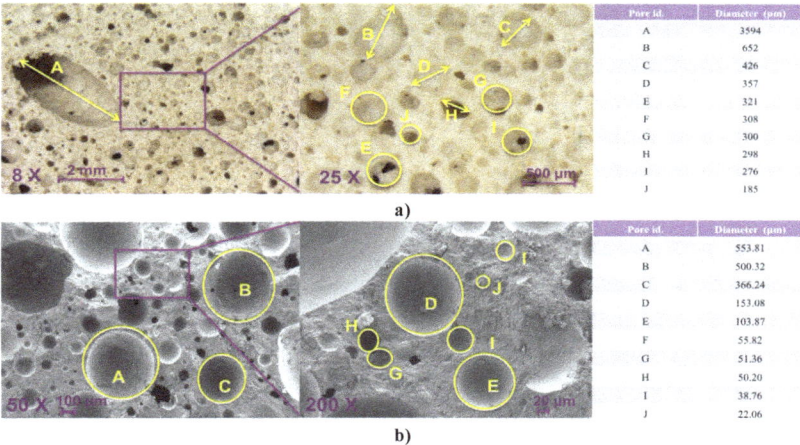

Pore id.	Diameter (μm)
A	3594
B	652
C	426
D	357
E	321
F	308
G	300
H	298
I	276
J	185

a)

Pore id.	Diameter (μm)
A	553.81
B	500.32
C	366.24
D	153.08
E	103.87
F	55.82
G	51.36
H	50.20
I	38.76
J	22.06

b)

FIGURE 4. *PORE SYSTEM EVALUATION FOR F-AS SAMPLES: A) OM IMAGES WITH MEASURED VOIDS; B) FESEM IMAGES WITH MEASURED VOIDS.*

On the other hand, when A powder was added to complement the foaming method in the samples, the resultant void-system was:
1) By using FCC as a precursor (F-AS samples), macropores from 3594 μm to 511 μm and micropores from 22.06 μm and 89 μm could be identified (Fig. 4). Besides, a higher size of macropores was resulted the micro scale range was lower that when merely SLS was added. A

Non-Conventional Materials and Technologies – NOCMAT for XXI Century   Materials Research Forum LLC
Materials Research Proceedings 7 (2018) 685-694   doi: http://dx.doi.org/10.21741/9781945291838-66

similar pore structure to that the obtained in the F-S samples has been achieved. Homogeneous bubble distribution with spherical shape and no interconnection each other, have been observed.

2) Samples with BFS as a precursor (B-AS) presents a predominance of 2611-260 µm macropores and of 91.55-24.56 µm micropores (Fig. 5). The topography of the matrix is dense presenting pores with spherical shape in macro scale, but irregular forms and some interconnection among them in micro scale. Comparing with the B-S samples, a higher size of macropores was resulted and the micro scale range was lower, the same relation than the above discussed for FCC samples (F-S respect F-AS).

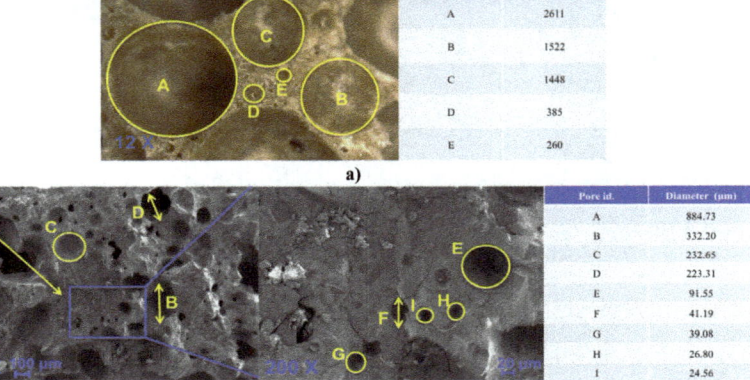

a)

b)

FIGURE 5. *PORE SYSTEM EVALUATION FOR B-AS SAMPLES: A) OM IMAGES WITH MEASURED VOIDS; B) FESEM IMAGES WITH MEASURED VOIDS.*

Conclusions

Despite of the most controllable pore-forming process, by foaming method using SLS, as there are no chemical reactions involved, this investigation reveals that the combination of mechanical and chemical methods involves a greatest effectiveness in both TCC and CGC production. With the addition of 0.2 % of SLS and 0.05% of A (AS mixtures) might be possible to reduce the natural density 18% in TCC, 10% in CGC with FCC and 22% in CGC with BFS, than when merely SLS (2% by weight) was used. Furthermore, for the TCC systems with AS addition a direct relationship between natural density and compressive strength was found, however, for the CGCs (by the use of both proposed precursors), lower natural densities with relative higher compressive strengths were obtained.

In this research, the AS combined method, on the one hand CGCs made from FCC as a precursor with 1143 kg/cm^3 natural density and 6.88 MPa; and on the other hand CGCs made from BFS with 1046 kg/cm^3 natural density and 8.62 MPa. These values of density constitute an improvement (reduction in natural density) of 30% and 36% respectively, respect of TCC systems.

The resultant pore system of both geopolymeric systems (FCC and BFS) by using SLS and A was higher in macro-pore scale and lower in micro-pore scale respect of the resultant with a foaming effect.

Non-Conventional Materials and Technologies – NOCMAT for XXI Century   Materials Research Forum LLC
Materials Research Proceedings 7 (2018) 685-694   doi: http://dx.doi.org/10.21741/9781945291838-66

This paper reveals that the geopolymeric systems based on the reuse of industrial by-products allows obtain cellular concretes with lower densities by using the same percentage of foaming addition, comparing to traditional systems based on OPC.

Acknowledgments

The authors acknowledge the financial support from Universitat Politècnica de València (UPV) through internal project GEOCELPLUS. The authors would like also to express special grateful to the Electronic Microscopy Service of the UPV. The first author thanks Esther Cuellar for her help with the English language.

References

[1] N. Narayanan, K. Ramamurthy, Structure and properties of aerated concrete: a review. Cement and Concrete Composites, 22 (2000) 321-329. http://dx.doi.org/10.1016/S0958-9465(00)00016-0.

[2] Md. Azree Othuman, Y.C. Wang, Elevated-temperature thermal properties of lightweight-foamed concrete. Construction and Building Materials 25 (2011) 705-716. https://doi.org/10.1016/j.conbuildmat.2010.07.016.

[3] W. Wongkeo, P. Thongsanitgarn, K. Pimraska, A. Chaipanich, Compressive strength, flexural strength and thermal conductivity of autoclaved concrete block made using bottom ash as cement replacement materials. Materials and Design, 35 (2005) 434-439, http://dx.doi.org/10.1016/j.matdes.2011.08.046.

[4] A.J. Hamad, Materials, production, properties and application of aerated lightweight concrete: Review, International Journal of Materials Science and Engineering, 2 (2014) 152-157, http://dx.doi.org/10.12720/ijmse.2.2.152-157.

[5] E. Namsone, G. Sahmenko, A. Korjakins, Durability Properties of High Performance Foamed Concrete, Procedia Engineering, 172 (2017) 760-767, https://doi.org/10.1016/j.proeng.2017.02.120.

[6] K.H. Mo, U.J. Alengaram, M.Z. Jumaat, S.P. Yap, S.C. Lee, Green concrete partially comprised of farming waste residues: a review, Journal of Cleaner Production 117 (2016) 122-138, http://dx.doi.org/10.1016/j.jclepro.2016.01.022.

[7] D.N. Huntzinger, T.D. Eatmon, A life-cycle assessment of Portland cement manufacturing: comparing the traditional process with alternative technologies, Journal of Cleaner Production 17 (2009) 668-675, https://doi.org/10.1016/j.jclepro.2008.04.007.

[8] J.G. Sanjayan, A. Nazari, L. Chen, G.H. Nguyen, Physical and mechanical properties of lightweight aerated geopolymer, Construction and Building Materials 79 (2015) 236-244, http://dx.doi.org/10.1016/j.conbuildmat.2015.01.043.

[9] Z. Zhang, J.L. Provis, A. Reid, H. Wang, Geopolymer foam concrete: an emerging material for sustainable construction, Construction and Building Materials, 56 (2014) 113-127, http://dx.doi.org/10.1016/j.conbuildmat.2014.01.081.

[10] H. Esmaily, H. Nuranian, Non-autoclaved high strength cellular concrete from alkali activated slag, Construction and Building Materials 26 (2012) 200-206, http://dx.doi.org/10.1016/j.conbuildmat.2011.06.010.

[11] R.A. Aguilar, O.B. Díaz, J.I.E García, Lightweight concretes of activated metacaolín-fly ash binders with blast furnace slag aggregates, Construction and Building Materials 24 (2009) 1166-1175, http://dx.doi.org/10.1016/j.conbuildmat.2009.12.024.

[12] P. Duxon, J.L Provis, G.C Lukey, J.S.J van Deventer, The role of inorganic polymer technology in the development of 'green concrete', Cement and Concrete Research 37 (2007) 1590-1597, https://doi.org/10.1016/j.cemconres.2007.08.018.

[13] M.M. Tashima, J.L. Akasaki, V.N. Castaldelli, L. Soriano, J. Monzó, J. Payá, M.V. Borrachero, New geopolymeric binder based on fluid catalytic cracking catalyst residue (FCC), Materials Letters 80 (2012) 50–52, http://dx.doi.org/10.1016/j.matlet.2012.04.051.

[14] E.K.K. Nambiar, K. Ramamurthy, Air-void characterization of foam concrete, Cement and Concrete Research 37 (2007) 221-230, https://doi.org/10.1016/j.cemconres.2006.10.009.

[15] K. Ramamurthy, E.K.K. Nambiar, G.I.S. Ranjani, A classification of studies on properties of foam concrete, Cement and Concrete Composites 2009; 31 (2009) 388-396, http://dx.doi.org/10.1016/j.cemconcomp.2009.04.006.

Non-Conventional Materials and Technologies – NOCMAT for XXI Century   Materials Research Forum LLC
Materials Research Proceedings 7 (2018) 695-705   doi: http://dx.doi.org/10.21741/9781945291838-67

Reusing Composite Materials from Decommissioned Wind Turbine Blades

Lawrence C. Bank[a]*, Franco R. Arias[b], T. Russell Gentry[c], Tristan Al-Haddad[d], Jian-Fei Chen[e], Ruth Morrow[f]

[a] City College of New York, New York, USA, lbank2@ccny.ccny.edu

[b] City College of New York, New York, USA, farias01@citymail.cuny.edu

[c] Georgia Institute of Technology, Atlanta, GA, USA, russell.gentry@coa.gatech.edu

[d] Georgia Institute of Technology, Atlanta, GA, USA, tristan.al-haddad@coa.gatech.edu

[e] Queen's University Belfast, Belfast, UK, j.chen@qub.ac.uk

[f] Queen's University Belfast, Belfast, UK, ruth.morrow@qub.ac.uk

Abstract. The very rapid growth in wind energy technology in the last 15 years has led to a rapid growth in the amount of non–biodegradable, thermosetting FRP composite materials used in wind turbine blades that will need to be managed of in the near future. A typical 2.0 MW turbine with three 50 m blades has approximately 20 tonnes of FRP material and an 8 MW turbine has approximately 80 tonnes of FRP material (1 MW ≈ 10 tonnes of FRP). Calculations show that 4.2 million tonnes will need to be managed globally by 2035 and 16.3 million tonnes by 2055 if wind turbine construction continues at current levels and with current technology. Three major categories of end-of-life (EOL) options are possible – disposal, recovery and reuse. Reuse options are the primary focus of this paper since landfilling and incineration are environmentally harmful and recovery recycling methods are not economical. The current work reports on different architectural and structural options for reusing parts of wind turbine blades in new or retrofitted housing projects. Large-sized FRP pieces that can be salvaged from the turbine blades and potentially useful in infrastructure projects where harsh environmental conditions (water and high humidity) exist. Their non-corrosive properties make them durable construction materials. The approach presented is to cut the decommissioned wind turbine blades into segments that can be repurposed for structural and architectural applications for affordable housing projects. The geographical focus of the designs presented in this paper is in the coastal region of the Yucatan on the Gulf of Mexico where low-quality masonry block informal housing is vulnerable to severe hurricanes and flooding. In what follows, a prototype 100m long wind blade model provided by Sandia National Laboratories is used as a demonstration to show how a wind blade can be broken down into parts, thus making it possible to envision architectural applications for the different wind blade segments.

KEYWORDS: Architecture, Composite Materials, FRP, Housing, Recycling, Reuse, Waste, Wind Turbines

Introduction

This paper presents a potential approach to tackling the looming issue of waste glass fiber reinforced polymer (GFRP) composite material to come from decommissioned wind turbine blades. The approach presented is to cut the decommissioned wind turbine blades into segments that can be repurposed for structural and architectural applications for affordable housing projects. The geographical focus of the designs presented in this paper is in the coastal region of the Yucatan on the Gulf of Mexico where low-quality masonry block informal housing is vulnerable to severe hurricanes and flooding (Bank, 2017). Global, U.S. and Mexico wind industry statistics are shown to give a sense of the large quantity of out-of-service wind blades that will need to be recycled in the

Non-Conventional Materials and Technologies – NOCMAT for XXI Century   Materials Research Forum LLC
Materials Research Proceedings 7 (2018) 695-705   doi: http://dx.doi.org/10.21741/9781945291838-67

near future. In what follows, a prototype 100 m long wind blade model provided by Sandia National Laboratories (Griffith, 2013) is used as a demonstration to show how a wind blade can be broken down into parts, thus making it possible to envision architectural applications for the different wind blade segments.

Wind industry

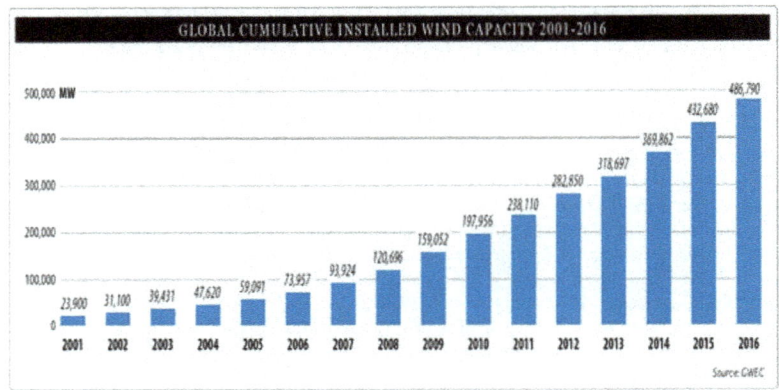

FIGURE 1. GLOBAL CUMULATIVE CAPACITY (GWEC, 2017)

The very rapid growth in wind energy technology in the last 15 years has led to a dramatic increase in the amount of non–biodegradable, thermosetting FRP composite materials used in wind turbine blades that will need to be managed in the near future. A typical 2.0 MW turbine with three 50 m blades has approximately 20 tonnes of FRP material and an 8 MW turbine has approximately 80 tonnes of FRP material (1 MW ≈ 10 tonnes of FRP, see Arias (2016). Calculations show that a total of 4.2 million tonnes of wind blade FRP waste will need to be managed globally by 2035 and 16.3 million tonnes by 2055. As of December, 2016, the global cumulative installed wind capacity was 486,790 MegaWatts (MW) (Fig. 1). What is more striking in this figure is the fact that the global cumulative capacity has been increasing on a year-to-year basis over the last 5 years. However beneficial, this sharp growth in the wind industry also leads to substantial amounts of composite material to be dealt with in the future.

Wind Industry – Mexico
The Mexican wind industry has been installing wind turbines over the past nine years. Wind industry giants such as Gamesa, Acciona and Vestas have played a crucial role in this development. Fig. 2 shows a forecast of the installed capacity (12,823 MW total) for each of Mexico's provinces in 2020. It is worth noting that the two provinces with the largest anticipated wind energy capacities in 2020 are Oaxaca and Yucatan. Large amounts of wind turbine blades facing inevitable decommissioning close to vulnerable communities on the Gulf of Mexico coast would make it feasible to transport blade segments to these communities for reuse.

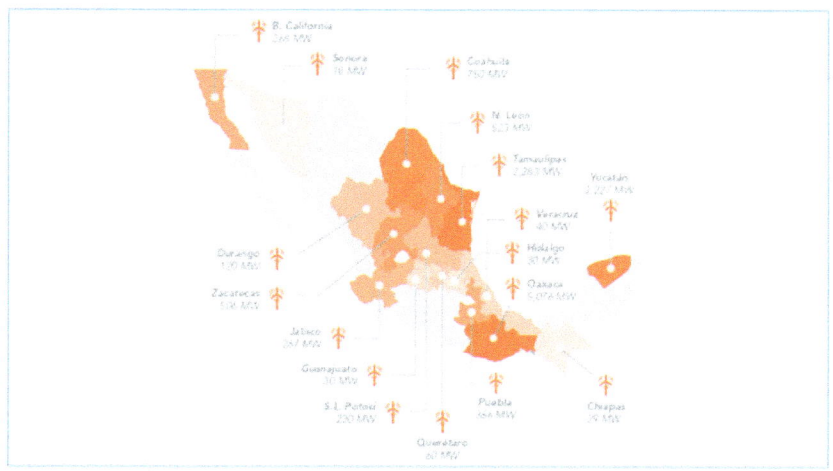

FIGURE 2. MEXICO 2020 FORECAST (MWEA, 2017)

End-of-life options

Three end-of-life (EOL) options for FRP wind blades are currently possible – disposal, recovery and reuse. The two options to dispose of FRP composites at the present involve landfilling or incineration (with or without energy recovery and/or silica ash recovery). Recovery options consist of reclamation of the constituent fibers or the resins by thermo–chemical methods or recovering of small pieces of granular FRP material for use as filler material in concrete or other composites by cutting, shredding or grinding. Reuse options consist of reusing the entire FRP blade or large parts of it in new structural applications. Reuse options are the primary focus of this paper since landfilling and incineration are environmentally harmful and recovery methods are not currently economical for GFRP, of which the vast majority of current blades are made.

The 100-meter long SNL-100-01 prototype wind blade

SNL-100-01 is a publically-available 100 meters long prototype wind turbine blade model that was designed by Sandia National Laboratories (Griffith, 2013). It has GFRP in most of the shell structure, and a smaller amount of carbon fiber reinforced polymer (CFRP) composite material in the shear web and spar caps. The geometry is defined by 25 different airfoil shapes. A total of 393 solid and sandwich composite material lay-ups are used in the blade. The software tool NuMAD (2015) was used build three-dimensional models of the wind blade. Model parameters are airfoil type, station parameters, division points, material models, composite materials, and shear web division points. Fig. 3 shows a finalized version of the SNL-100-m blade. The different colors in the figure represent the 393different material lay-ups in the blade. Most current blades in the 40-50 meter length range are exclusively made of GFRP. Future longer blades (80 m and above) will also use CFRP. This study used the glass and carbon blade model as an example. Rhino 3D (2017) was used to render the blade. Rendering the blade with its actual material thicknesses is required to extract realistic segments for architectural applications. Fig. 4 shows an isometric view of the SNL-100-01 blade.

Non-Conventional Materials and Technologies – NOCMAT for XXI Century   Materials Research Forum LLC
Materials Research Proceedings 7 (2018) 695-705   doi: http://dx.doi.org/10.21741/9781945291838-67

Architectural concepts for decommissioned 100 m long wind blade segments
Root-Foundation System

The root segment of the SNL-100 blade has circular and elliptical cross sections. The top and bottom halves of the blade shell consist of a variety of FRP materials arranged in different layers: gelcoat, resin, triaxial fiber fabrics, and unidirectional fiber (Griffith, 2013). At its thickest point, the root has an FRP thickness of 110.6 mm. The starting chord length for the root is 5.7 meters; it eventually reaches a maximum of 7.5 meters. A typical 2 MW 2000s-era wind blade of about 50-60 m in length would have a root chord ranges from 3 to 4 m in length.

Based on Yucatan 2014 surveys (Matta, 2016), flooding is the second most damaging environmental occurrence to informal houses (wind and rain being the first). Elevating homes is proposed to avoid flooding damage. By cutting the wind blade root section (closest to the turbine hub) into short segments, platforms suitable for home elevations can be obtained. The resulting platforms have cylindrical or elliptical cross sections. One meter high platforms of different root sections are shown alongside a rendering of a typical rectangular masonry house with dimensions of 7 m long, 5 m wide and 2.7 m high are shown in Fig. 5. The platforms would have to be driven into the ground. If a higher elevation for a house is desired, larger segmented platforms can be extracted from the root to provide adequate embedment. Additionally, the inside of the platform may be filled with an aggregate. Fig. 6 shows houses being elevated off the ground by such platforms (scaled down for the 100 m blade to fit the size of the house).

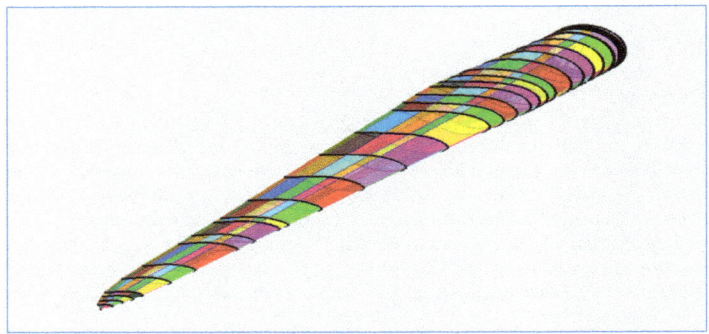

FIGURE 3. THE SNL-100-01 NuMAD BLADE MODEL (NuMAD, 2015)

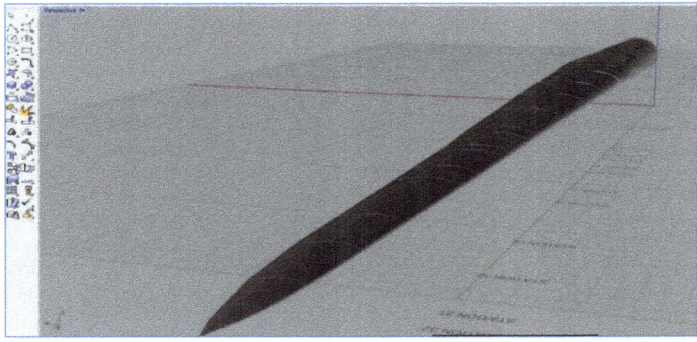

FIGURE 4. THE SNL-100-01 BLADE RENDERED IN RHINO 3D

Non-Conventional Materials and Technologies – NOCMAT for XXI Century   Materials Research Forum LLC
Materials Research Proceedings **7** (2018) 695-705   doi: http://dx.doi.org/10.21741/9781945291838-67

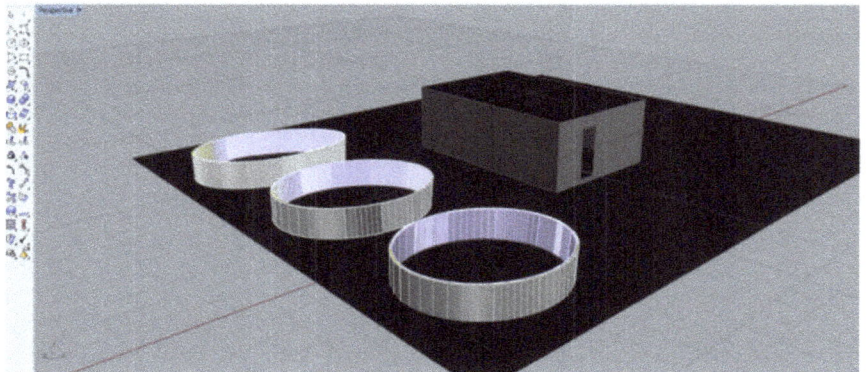

FIGURE 5. ROOT SECTIONS ALONGSIDE HOUSE MODEL

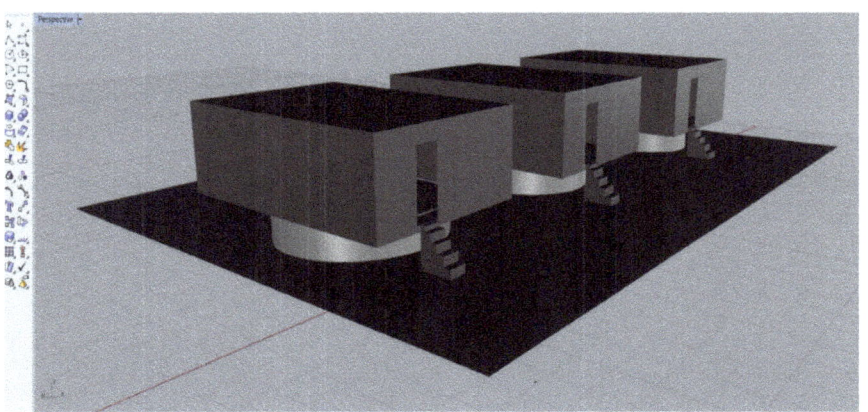

FIGURE 6. ELEVATED HOUSES

Door/Window Applications

Three shear webs connect the top and bottom halves of the blade shell and run along the length of the blade. The shear webs are made of a sandwich composite that has a 60-millimeter-thick foam core (shown in yellow in Fig. 7) in the center and 3-millimeter carbon fiber skins (show in black) on either side of the core. Two of the shear webs are 80 m long and are connected to the carbon fiber spar caps (which are solid carbon fiber and provide the flexural stiffness to the blade). The third shear web is 40 m long at the end of the blade and provides rigidity to the trailing edge (the flatter edge) near the tip of the wind blade. These straight, slender pieces of FRP material are excellent for

Non-Conventional Materials and Technologies – NOCMAT for XXI Century   Materials Research Forum LLC
Materials Research Proceedings **7** (2018) 695-705   doi: http://dx.doi.org/10.21741/9781945291838-67

applications involving doors, window shutters, flooding barriers, structural insulated panels (SIPs) and facades.

Fig. 7 shows the three shear webs extracted from the SNL-100-01 wind blade model. Two rectangular virtual cutting planes, one meter apart, are superimposed onto the shear webs. These two planes show the longitudinal cuts that would be needed to produce solid straight one meter high sandwich panels. Fig. 8 and Fig. 9 show cut-out segments of the shear webs being used as doors and windows covers for the model house, respectively.

FIGURE 7. SHEAR WEBS

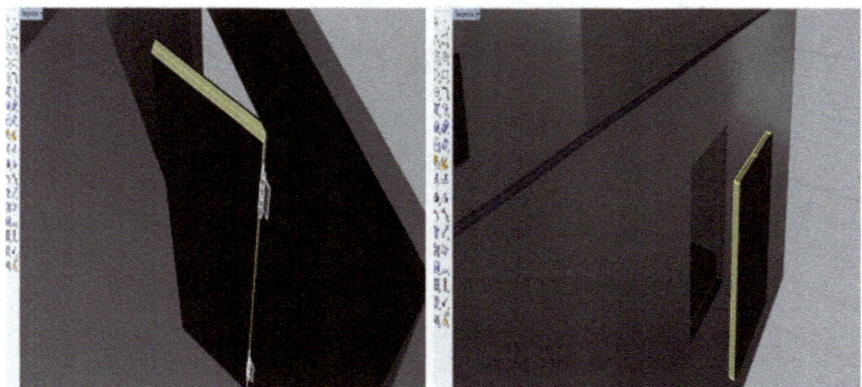

FIGURE 8. SHEAR WEB – DOOR   FIGURE 9. SHEAR WEB – WINDOW COVER

Roof Frames

In Fig. 10, the leading-edge (the rounded panels at the front of the blade shell) has been removed from a blade segment to leave the three shear webs and the trailing-edge panels. The leading top and bottom panels have then been separated to make two roof frames. The shear webs (discussed above) run up to 80 m down the blade length, are over 2 m high at their highest point, most importantly, they

Non-Conventional Materials and Technologies – NOCMAT for XXI Century   Materials Research Forum LLC
Materials Research Proceedings **7** (2018) 695-705   doi: http://dx.doi.org/10.21741/9781945291838-67

are straight, making them easy to line up geometrically with other structural elements found in housing construction. To extract the roof frames, the panels must be sliced at an angle so that the cut bisects the shear webs and passes through the joint at the tip of the trailing-edge as seen in Fig. 10. Bisecting the blade segment in this manner results in one roof frame from the bottom of the blade and another from the top. These two roof frames are similar, but not identical. Both have the same cross-sectional length, height, and material construction. However, the top roof frame has a concave-down roof curvature, while the bottom roof frame has an inflection point midway along its roof curvature. This geometric difference can be observed in Fig. 11, where both half-trusses are displayed next to one another.

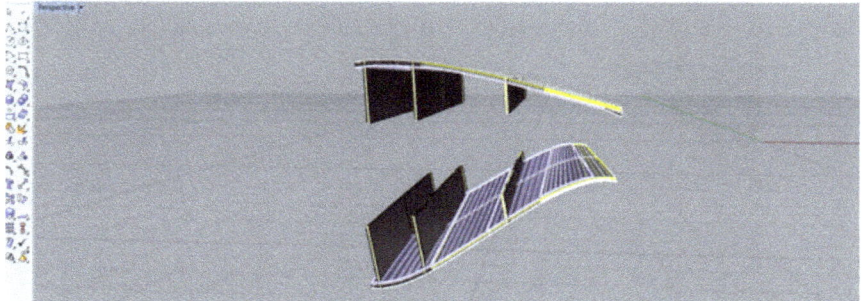

FIGURE 10. BLADE SEGMENT CUT

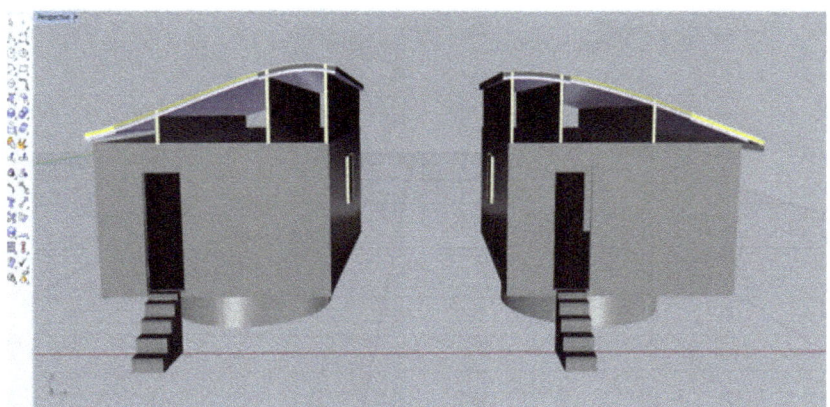

FIGURE 11. ROOF FRAMES

Interlocking Roof System

The FRP material found in the panels of the leading-edge of the blade comprises up to a third of the entire wind blade structure. A substantial amount of the FRP material can be salvaged from the leading-edge (rounded front) portion of a decommissioned blade. It is important to understand the geometric arrangement and material composition of the panels if large-sized FRP blade segment are

Non-Conventional Materials and Technologies – NOCMAT for XXI Century   Materials Research Forum LLC
Materials Research Proceedings **7** (2018) 695-705   doi: http://dx.doi.org/10.21741/9781945291838-67

to be extracted and used for architectural/structural applications. Fig. 12 shows two blade segments extracted from 30 and 47 meters from the root of the blade. Also in Fig. 12, the leading edge FRP panels are cut and extracted from the rest of the blade segments. These panels are the reduced down into smaller shells, as shown in Fig. 13, where the virtual cuts are represented by the black planes.

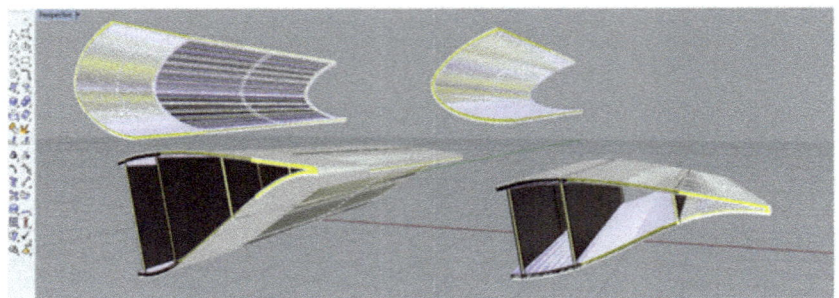

FIGURE 12. SEPARATED LEADING EDGE SEGMENTS (TOP)

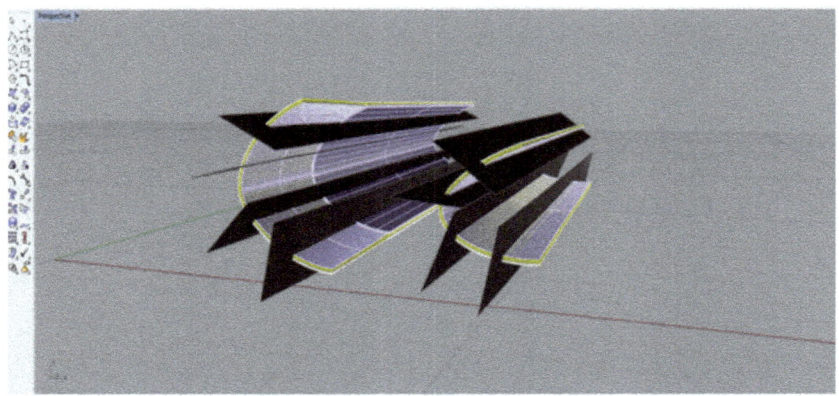

FIGURE 13. LEADING EDGE VIRTUAL CUTTING PLANES

Arranging the cut-out segments in a configuration as seen in Fig. 14 yields a possible roofing system for affordable housing.

Non-Conventional Materials and Technologies – NOCMAT for XXI Century   Materials Research Forum LLC
Materials Research Proceedings 7 (2018) 695-705   doi: http://dx.doi.org/10.21741/9781945291838-67

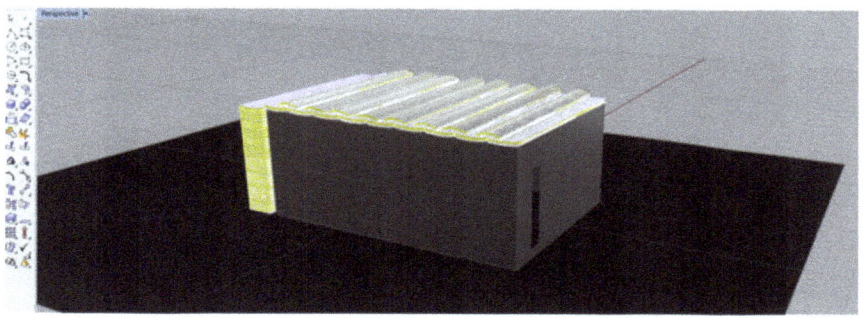

FIGURE 14. INTERLOCKING ROOF PANEL CONFIGURATION

There are two important issues to consider when extracting the leading edge panels and repurposing them for use in a roofing system. First, there is a slight warping in blade geometry that is due to the twist along the length of the wind blade. The most noticeable effect of this can be observed at the transition between the root and wing segment of the blade. As the blade tapers off, the angle of twist decreases and the warping is less significant. Nonetheless, the extracted leading edge shells will not be perfectly straight, making the interlocking roof configuration difficult to arrange. The gaps formed in between the shell segments may prove to be an issue when providing a leak-proof roofing system. This may be addressed by cutting a small amount of materials off to produce tapered edges (in thickness direction) to make them perfectly fit. An alternative is to seal the gaps due to misfit with other materials.

Second, as the wind blade tapers off after reaching its maximum chord length, the material thickness for the foam decreases. This means that at some points along the blade closer to the root the FRP pieces will be thicker than the pieces found further down the blade. This will raise some concerns when selecting curved segments for the interlocking-roof system. Having segments that are too thin or thick with respect to their adjacent segments will create additional gaps that need filling. However, this would not be an issue if all the concave segments are tapered in one direction and all the convex segments are tapered in the opposite direction.

A Possible Vision
Building or retrofitting an affordable housing community with salvaged wind blade parts might resemble something like that depicted in Fig. 15. In this representation the root-foundation system has been used in all the houses, elevating them off the ground. All doors and windows have been modelled using the shear webs panels. Lastly, the roof frame and the interlocking roof configuration are shown in different scenarios.

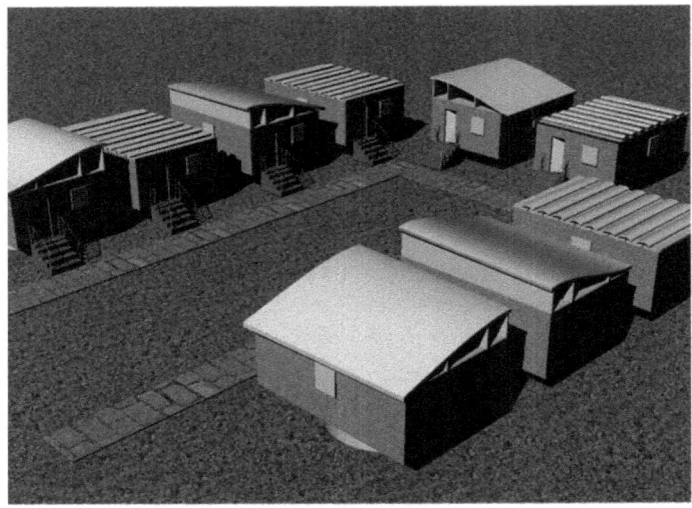

FIGURE 15. RETROFITTED AFFORDABLE HOUSING CONCEPT

Conclusions

The purpose of this paper was to present a feasible solution to the impending issue of recycling decommissioned wind turbine blades. As discussed, a wind turbine blade, in this case, the SNL-100-01, can be presented as a wireframe model. From this rudimentary wireframe model, a blade can be reassembled as a 3D model for better architectural/mechanical analysis. Computer software like NuMAD and Rhino 3D made it possible to extract segments from a decommissioned SNL-100-01 wind blade and find real-life structural applications for affordable housing communities. However, this paper is a first step towards making the disposal of wind blades efficient and environmentally friendly. In time, this process of repurposing decommissioned wind blade parts must be further researched with regards to mechanical systems (MEP), structural analysis, logistics, and detailing, other architectural and infrastructural applications; cost and ease of dis- and reassembly; social accessibility and acceptability.

Aknowledgements

This work was supported by NSF grant numbers 1321464 and 1701413, and a grant from the Department for the Economy Northern Ireland under the US-Ireland R&D Partnership Programme (Project No. USI 116).

Non-Conventional Materials and Technologies – NOCMAT for XXI Century   Materials Research Forum LLC
Materials Research Proceedings 7 (2018) 695-705   doi: http://dx.doi.org/10.21741/9781945291838-67

References

[1] Arias, F. (2016). "Assessment of Present/Future Decommissioned Wind Blade Fiber-Reinforced Composite Material in the United States." Independent Study, Dept. of Civil Engineering, City College of New York, United States.

[2] Bank, L.C (2017). EAGER/Collaborative Research: Overcoming Barriers to Diffusion and Adoption of Sustainable and Resilient Building Materials in Coastal Areas of Southern Mexico. Project Outcomes Reports, 1321464. Available at www.research.gov Research Spending & Results.

[3] Griffith, T. (2013). The SNL100-01 Blade: Carbon Design Studies for the Sandia 100-meter Blade. SAND2013-1178. Sandia National Laboratories, Albuquerque, New Mexico. http://prod.sandia.gov/techlib/access-control.cgi/2013/131178.pdf

[4] Matta, F., (2016). EAGER/Collaborative Research: Overcoming Barriers to Diffusion and Adoption of Sustainable and Resilient Building Materials in Coastal Areas of Southern Mexico. Project Outcomes Report 1321489. Available at www.research.gov Research Spending & Results.

[5] NuMAD (2015). Numerical Manufacturing and Design for Wind Turbine Blades, Sandia National Laboratories. http://energy.sandia.gov/energy/renewable-energy/wind-power/rotor-innovation/numerical-manufacturing-and-design-tool-numad/

[6] Rhino 3D. (2017). Rhinoceros 3D, Robert McNeel & Associates, Seattle, USA. www.rhino3d.com

Non-Conventional Materials and Technologies – NOCMAT for XXI Century   Materials Research Forum LLC
Materials Research Proceedings 7 (2018) 706-715   doi: http://dx.doi.org/10.21741/9781945291838-68

An Evaluation of Electric Arc Furnaces Dust as a Replacement for Cement in Mortars

Margareth da Silva Magalhães[a,*], Flora Faleschini[b], Carlo Pellegrino[b], Katya Brunelli[c]

[a] Dept. of Civil Construction and Transport, Universidade do Estado do Rio de Janeiro, Brazil; margarethsm@yahoo.com.br

[b]Dept. of Civil, Environmental and Architectural Engineering, University of Padua, Italy

[c]Dept. of Industrial Engineering, University of Padua, Italy

Abstract. The dust resulting from the steel producing by electric arc furnaces is categorized as a hazardous material by world environmental protection agency and, therefore, needs to be considered for recycling. The incorporation of electric arc furnaces dust (*EAFD*) in mortar is considered a good solution for this problem. In this research *EAFD* in as-received condition from the dust collection system of a carbon steelmaking factory was characterized and evaluated for their performance in mortar. *EAFD* characterization included the determination of mineralogical and chemical composition and physical characteristics. The performance of *EAFD* in mortar was evaluated through determination of the setting times with *EAFD* content up to 20%. Compressive and flexural strength of mortars were also studied by using compressive and flexural tests. The tests results indicated that the *EAFD* studied is shown to be suitable for use in mortar. However, the incorporation of elevated content of *EAFD* in the mixture can retard the setting times because of high zinc oxide content. Results also indicated that mortar containing 5% and 10% of *EAFD* presented similar compressive strength to the reference mortar at the age of 28 days and the mortar with 20% of *EAFD*, in the same age, presented around of 94% of the reference mortar. Flexural strength presented a similar behavior of compressive strength.

Keywords: Setting Time, Mechanical Properties, Electric Arc Furnace Dust

Introduction

Steel produced by electric arc furnaces uses recycled material as prime input and the product is totally recyclable. However, the industry does not produce only steel products, by-products, co-products and wastes are also generated and the big challenge is to find valuable uses for them through the combined application of technology and economics.

The electric arc furnace dust (*EAFD*) is one of the solid wastes formed in the steelmaking process. *EAFD* is a high density material composed of very fine and heterogeneous distribution particle size and generally contains up to 40% zinc [1] and 50% iron [2, 3]. It is also accompanied with a variety of harmful heavy metals, such as, cadmium, lead, and chromium [4-6], and thereby frequently categorized as a hazardous material by world environmental protection agency, and needs to be considered for recycling or immobilization process. According to Havlik et al. [7] and Buzin et al. [8], the composition of *EAFD* is variable and depends on input materials used during steelmaking and processing conditions.

In recent years, the use of *EAFD* in materials has become an attractive alternative to disposal and has attracted the attention of many researchers. Some researchers are being carried out on the utilization of *EAFD* in different methods, such as, ceramic materials [9, 10] and cement based materials [6, 9, 11-13]. The aim of these initiatives is not only to utilize the dust as raw material, but also inhibit the action of harmful elements to the environment by their confinement [8].

Non-Conventional Materials and Technologies – NOCMAT for XXI Century   Materials Research Forum LLC
Materials Research Proceedings 7 (2018) 706-715   doi: http://dx.doi.org/10.21741/9781945291838-68

Xuefeng and Yuhong [14] reported that the quality of cement produced with *EAFD* meets the Chinese specifications for cement. Further, it was reported that the use of *EAFD* in cement is more economical that the use of iron ore. Al-Zaid *et al.* [6] reported that the addition of 3% *EAFD* to concrete specimens improved the compressive and splitting strength. A 15-30% increase in the 7 and 28 days compressive strength of concrete was also reported by Macray [15]. Nevertheless, Vargas *et al.* [11] verify that cement pastes containing 5% of *EAFD* presented similar compressive strength to the reference (0% *EAFD*) at the age of 28 days and the cement pastes with 15% and 25%, in the same age, presented 80% of the reference strength. Several other works reported some benefits on durability [4, 6, 12] and slump retention [4] when *EAFD* was used in cement based materials.

Despite these researchers have found excellent benefits in the use of limited *EAFD* addition, the presence of zinc oxide in the dust acts as a retardant in cement setting time and this has been led to the proposal of using *EAFD* as a set retarder in concrete in hot climates to adjust the hydration rate [16-23]. According to Al-Zaid *et al.* [6] and Fares *et al.* [16], the extension of setting time may exceed 33 h with just an addition of 3% of dust. Hamilton and Sammes [24] verified that specimens with 10% of EAFD had the same 28-days compressive strength to that of the reference specimens after 56 days of ageing, displaying a significant delayed strength evolution. However, Balderas *et al.* [23] demonstrated that the addition of 8 and 10% of treated *EAFD* with a H_2SO_4 solution decreases early-age compressive strength, but increased it significantly after seven days. This behavior was related to the delayed hydration process often observed in blended cements with *EAFD*.

The retarding effect due to zinc oxide presence in *EAFD* has hindered, until now, its application in blended mixtures at elevated dosage (more than 3%) and little literature exists about using *EAFD* addition on mortar, particularly at high dosages. Thus, the purpose of this work is to provide additional information about the effects of elevated *EAFD* addition on the mortar behavior. In this research, a simple set test was used to analyze changes in set of blended cements containing up to 20% *EAFD*. The flexural and compression tests were also used to characterize the mechanical properties of *EAFD* mortars.

Experimental program
Materials
Portland cement, electric arc furnace dust (*EAFD*), sand and tap water are the materials used to prepare all the mortar mixes. The cement used is a rapid setting ordinary Portland cement type I 52.5R, as defined in the BS-EN 197-1 [25], with a density of 3080 kg/m³. *EAFD* is used in as-received condition from the dust collection system of a carbon steelmaking factory in Italy, and has a density of 3488 kg/m³. The natural sand has a 2 mm maximum particles size, apparent density of 2.64 g/cm³ and saturated surface-dried (s.s.d.) density of 2.76 g/cm³.

The chemical compositions of Portland cement and *EAFD* were obtained with X-Ray Fluorescence (*XRF*) analyses, and are shown in Table 1. As it is possible to see, the composition of this dust is rich in *Zn*, *Fe* and *Ca* oxides, with few quantity of *Si*, *Mg*, *Mn*, *Pb* and *Cl* oxides. Also limited amount of harmful heavy-metals is present, such as *Cr*, *Cd* and *Ni*. Figure 1 shows instead particles size distribution of cement and *EAFD*, as they are obtained using laser diffraction technique. *EAFD* has a wide range of particle sizes, with coarser particles having a diameter up to 478 µm. The characteristics of the distribution are: d_{10} equal to 2.57 µm; median particle size d_{50} of 10 µm and d_{90} is 67.58 µm. Cement is characterized instead by d_{10} = 1.35 µm, d_{50} = 10.68 µm and d_{90} = 32.75 µm, being finer than the *EAFD* used in this work.

Non-Conventional Materials and Technologies – NOCMAT for XXI Century   Materials Research Forum LLC
Materials Research Proceedings 7 (2018) 706-715   doi: http://dx.doi.org/10.21741/9781945291838-68

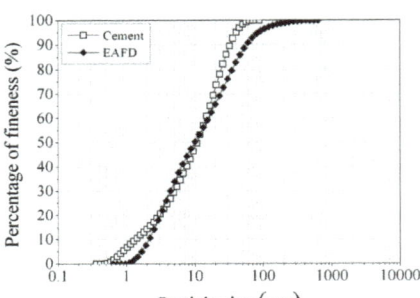

Figure 1 - *Particle size distribution of cement and EAFD.*

Table 1. *Major chemical compounds of ordinary Portland cement and EAFD.*

Compound	SiO$_2$	Al$_2$O$_3$	Fe$_2$O$_3$	CaO	SO$_3$	MgO	K$_2$O	MnO	Na$_2$O
Cement (wt %)	21.61	5.77	2.67	61.39	6.11	2.58	0.98	0.064	2.00
EAFD (wt %)	3.89	-	35.92	13.32	1.07	2.52	-	3.92	-

Compound	ZnO	PbO$_2$	Cr$_2$O$_3$	NiO	CdO	CuO	Cl
EAFD (wt %)	31.34	1.72	0.85	0.44	0.09	0.26	2.06

Microstructure and morphology of dust were studied with Scanning electron microscope (*SEM*) images. For this scope, a Cambridge Stereoscan 440 SEM, equipped with a Philips PV 9800 energy-dispersive X-ray spectrometer (*EDS*), was used. Images were taken in the backscattered electrons (*BSE*) mode, using an accelerating voltage of 25 keV; the composition of the particles was analyzed through the *EDS* spectra. Figure 2 shows that *EAFD* appears as composed of ultrafine and spherically shaped particles, which composition is mainly formed by zinc and iron, in addition to manganese, chromium, calcium and silicon.

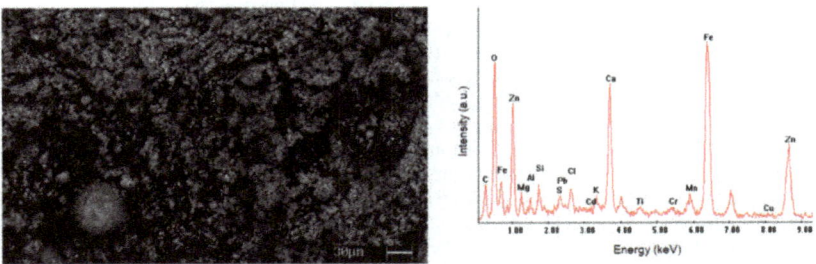

Figure 2 - *SEM image and respective EDS spectrum of EAFD sample.*

X-ray diffraction (*XRD*) analysis of *EAFD* sample was performed using a Siemens D500 diffractometer, with a stepped and continuous scanning device, with *CuKα* radiation (λ=1.5405 Å), at operating conditions of 40 kV and 30 mA. Five main phases were identified, which are magnetite (*Fe$_3$O$_4$*), franklinite (*ZnFe$_2$O$_4$*), zincite (*ZnO*), calcium ferrite (*CaFe$_2$O$_4$*) and lime (*CaO*). A quantitative analysis was made via the Rietveld method, taking into account only the above main phases, with the aim of understanding the amount of minerals containing iron and zinc. Results of this analysis indicate that magnetite and franklinite represent together about 61% of the phases in weight, whereas zincite, calcium ferrite and lime constitute respectively 18.7%, 11.7% and 8.5%.

Mixture details

Two series of experimental programs were designed, which involved tests on 12 different mixtures. The first series consisted of 8 mixtures, used to study the effect of *EAFD* content on setting time of blended cements. Here, *EAFD* content used varies from 0% to 20%, partially substituting the cement. The second series was considered to assess the effect of *EAFD* amount on fresh properties and mechanical strength of mortars, prepared with w/b ratio of 0.50. In this case, four mixtures with different amounts of *EAFD* were used, being 0% (reference), 5%, 10% and 20% expressed on cement weight. Mortars mixture proportions are listed in Table 2. The binder to sand ratio (*b:s*) was maintained constant in all the mixtures and is equal to 1:3 (by mass). The sum of cement and *EAFD* was considered as binder (*b*) to calculate the water content.

Table 2 - *Mortars mixture proportions (kg/m³).*

Mix	Cement (%)	EAFD (%)	w/b	Quantities (kg/m³)			
				Cement	EAFD	Sand	Water
Ref	100	-	0.50	523.11	0.00	1569.34	261.56
M01	95	5	0.50	497.45	26.18	1570.90	261.82
M02	90	10	0.50	471.74	52.42	1572.46	262.08
M03	80	20	0.50	420.16	105.04	1575.60	262.60

EAFD: electric arc furnace dust; w/b: water/binder ratio.

Concerning specimens manufacture, the preparation of the mixtures was performed in a mechanical mixer with 5l capacity, and, after mixing, they were cast in steel molds. Specimens with 0% and 5% *EAFD* were demolded after 24 h, whereas specimens with 10% and 20% *EAFD* after 48h, due to delayed hardening. Specimens were cured until time of testing in 100% *RH* conditions.

Experimental methods

Initial and final setting times were determined using an automatic setting time machine based in the Vicat apparatus [26]. This equipment measures paste resistance to the penetration of a needle under a load of 300 g. The time elapsed between zero and the instant at which the distance between the needle and the base-plate is 6 ± 3 mm is taken as the initial set time. In this test, final setting time was considerate as the time elapsed between zero and the instant at which the needle penetrates the paste to a maximum depth of 3 mm. Instant zero is considered from the moment when mixing water is added to the mixture. Set tests were carried out in a room with relative humidity and temperature of $54 \pm 2.0\%$ and $19 \pm 1.0°C$, respectively.

The fresh properties of mortars were determined through consistency and density tests. A small flow cone for conventional flow table test was used to quantify the deformability of fresh mixtures according to EN 1015-3 [27], and the bulk density of fresh mortar was measured according to EN 1015-6 [28]. Lastly, flexural and compressive strength of hardened mortar specimens were determined in accordance with EN 1015-11 [29]. For the former, a *3PBT* (three point bending test) was conducted on $160 \times 40 \times 40$ mm prismatic specimens; for the latter, compressive strength was determined on each half of the prism specimens, obtained after the *3PBT*. Compressive and flexural tests were performed at the ages of 1, 3, 7 and 28 days.

Experimental results

Setting time dependency on EAFD content

Changes in initial and final setting times were assessed for mixtures containing up to 20% of *EAFD* and *w/b* ratio of 0.35. In this study, the terms initial and final setting are used to describe an arbitrarily chosen stage of setting. Here initial setting time, which indicates the time when the paste has become unworkable, was defined at a penetration depth of 34 mm. Final setting time was fixed instead when the needle penetrates the paste to a maximum depth of 3 mm. Figure 3 shows the results: as expected, both initial and final setting time displays a strong dependency on dust addition, which is directly dependent on the amount of the addition.

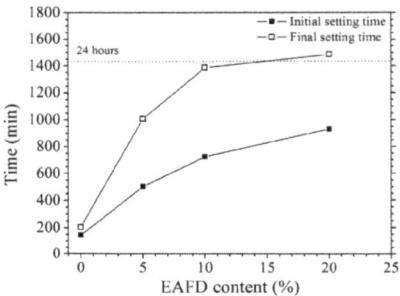

Figure 3 - *Setting times of cement (reference) and pastes with 5%-20% EAFD.*

Initial setting time increased from 146 min in the reference paste to 503 min, 725 min and 930 min for the mixtures containing 5%, 10% and 20% of *EAFD*, respectively. The same occurred for the final setting time, which increased from 206 min in the reference paste to 1007 min, 1385 min and 1485 min for the mixtures containing 5%, 10% and 20% of *EAFD*, respectively. The relative increase in initial setting time was of 340% (5% *EAFD*), 500% (10% *EAFD*) and 640% (20% *EAFD*); in the final setting time it was of 490% (5% *EAFD*), 670% (10% *EAFD*) and 720% (20% *EAFD*), if compared to the ordinary cement paste.

It is worth noting that the observed retardation can be directly assigned to the effect of Zn^{2+} introduction into the paste, which is considered to be an inhibitor of C_3S or cement hydration [17, 18]. Indeed, according to Chen *et al*. [18], Zn^{2+} delays the early hydration of C_3S, which could arise from the precipitation of calcium zincate ($CaZn_2(OH)_6 \cdot 2H_2O$). Calcium zincate coats C_3S grains, thus reducing material transport necessary for C_3S hydration, causing the delayed set observed in this work. According to Vargas *et al*. [11], the "dormant period", during which the hydration rate is very low and the paste is workable, is longer in *EAFD*-cement paste than in cement paste and depends on *EAFD* content. Chen *et al*. [30] reported that the *pH* will arise in time due to the hydration of C_3S, leading to calcium zincate dissolution, and as a result, the degree of C_3S hydration would increase. Furthermore, it was reported that other heavy metals, such as Cu^{2+}, Pb^{2+} and Cr^{3+} may promote C_3S hydration.

Consistency and bulk density of fresh mortars

Consistency (flow diameter) and bulk density of fresh mortar depending on the amount of cement substitution are presented in Figure 4. All mortars achieved a plastic consistency with flow diameter between 204–237 mm. The addition of *EAFD* improves the consistency of the mixtures. This result is in a good agreement with other literature findings [4, 16, 31], which reported that workability and slump retention of cement-based materials with *EAFD* were enhanced. A slight decrease of up to 4.2

Non-Conventional Materials and Technologies – NOCMAT for XXI Century   Materials Research Forum LLC
Materials Research Proceedings 7 (2018) 706-715   doi: http://dx.doi.org/10.21741/9781945291838-68

% was observed in bulk dry density values, although the dust has higher density than cement. This may be due to an increase in mortar porosity due to *EAFD* addition, as also observed in [32].

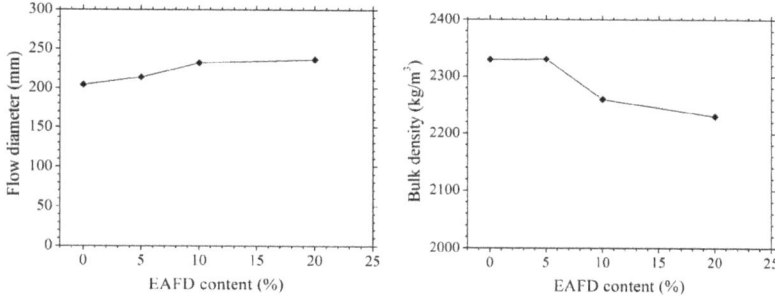

Figure 4 - Consistency and bulk density of fresh mortars.

Compressive and flexural strength of EAFD mortars

Compressive and flexural strength of mortars with 5%, 10% and 20% *EAFD* were experimentally evaluated after 1, 3, 7 and 28 days of curing. Results are then compared to the respective strengths of a reference mixture. Figure 5a presents the compressive strength development depending on the amount of cement substitution. It can be observed that at 1 day of curing, the specimens without *EAFD* had superior resistance, when compared to the mortars containing *EAFD*, independent of the amount added. Between 3 and 7 days, mortars with 5% *EAFD* present a relatively high rate of compressive strength development, similar to that of reference mortar. The rapid development of compressive strength of 5% *EAFD* mortar at 3 and 7 days indicates a rapid hydration during this period. However, as the *EAFD* content increases, the rate of compressive strength development in 10% and 20% *EAFD* mortars decreases significantly. After three days of curing, the compressive strength value of the mortar with 10% of *EAFD* is 69% of its 28 days-compressive strength; instead, the 3-days compressive strength of the 20% *EAFD* mortar is 3% of its 28 days- strength. It is significant to note that this gap is recovered during time; indeed, at 7 days, compressive strength of 10% and 20% *EAFD* mortar are respectively 81% and 71% of its 28 days-compressive strength.

When compared to the strengths of a reference mixture (Figure 5b), it is visible that specimens with 5% *EAFD* presented an enhancement of compressive strength for maturation age higher than 3 days, of about + 2.8%, + 5.15% and +2.84% at 3, 7 and 28 days respectively. These results confirm ones observed in previous studies [11, 22]. At 10% of substitution, a strength reduction is observed only at early age (-3.3% at 3 days), but at the following ages it is not affected. A remarked strength decay is instead displayed by 20% *EAFD* mortar, particularly at the early age (-96% at 3 days); then, this gap is reduced as maturation time increases (-19% ay 7 days and - 6% at 28 days).

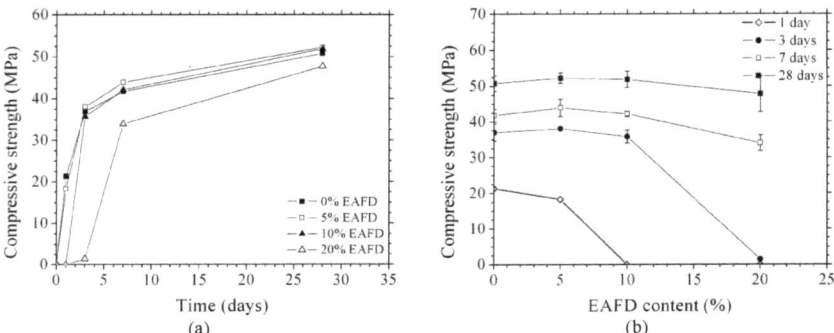

Figure 5 - *Compressive strength development and b) (b) Effect of EAFD replacement on compressive strength of mortars at 1, 3, 7 and 28 days.*

A similar evolution of compressive strength can be obtained analyzing flexural strength results at Figure 6; also in this case, at low substitution ratio (5% *EAFD*) the differences between the reference and *EAFD* mortars are limited, whereas as the *EAFD* content increases, a lower strength is observed at early age. Furthermore, *EAFD* did not exhibit significantly influence on flexural to compressive strength ratio. According to the results, flexural strength of all mixtures is around 15% of compressive strength.

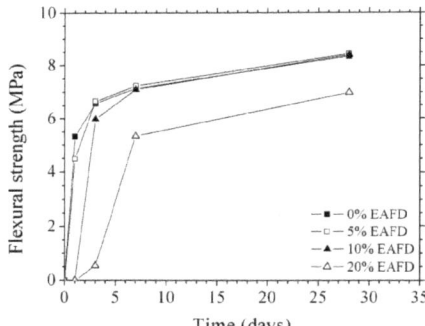

Figure 6 - *flexural strength development of mortars at 1, 3, 7 and 28 days.*

Conclusions

The experimental campaign carried out in this work provides additional information about the effects of high *EAFD* dosage without any preliminary treatment on setting times of blended cements and mechanical behavior of mortar. The setting process of cement is delayed as the *EAFD* content is increased; nevertheless, the hydration reactions seem to take place after 3 days. An enhancement in the consistency of mortars with *EAFD* was obtained in all the analyzed cases. Instead, fresh densities were little influenced by *EAFD* content.

Non-Conventional Materials and Technologies – NOCMAT for XXI Century   Materials Research Forum LLC
Materials Research Proceedings 7 (2018) 706-715   doi: http://dx.doi.org/10.21741/9781945291838-68

Results allow to identify an optimum dosage of *EAFD*, which has been estimated in 5% on cement weight. It was observed that cement blended with 5% *EAFD* had, since the beginning of the hydration process, greater strength than the cement mortar. At 28 days of maturation, 10% *EAFD* mortars displayed almost the same strength of the reference mortar. In case of 20% of addition, the 28-days strength difference with the conventional is also highly reduced if compared to the early-age strength.

Acknowledgments

The authors acknowledge the Quality Improvement Program for Universities of the Ministry of Education and Culture (*CAPES*) for the financial support and Acciaierie Venete SpA for supplying the *EAFD*.

References

[1]   Sobrinho, P.J.N., Espinosa, D.C.R., Tenorio, J.A.S. (2003). Characterization of dusts and sludges generated during stainless steel production in Brazilian industries. Ironmaking & Steelmaking 30(1): 11–17. https://doi.org/10.1179/030192303225009506

[2]   Orhan, G. (2005). Leaching and cementation of heavy metals from electric arc furnace dust in alkaline medium. Hydrometallurgy 78: 236–245. https://doi.org/10.1016/j.hydromet.2005.03.002

[3]   Salihoglu, G., Pinarli, V. (2008). Steel foundry electric arc furnace dust management: stabilization by using lime and Portland cement. J Hazard Mater 153: 1110-1116. https://doi.org/10.1016/j.jhazmat.2007.09.066

[4]   Maslehuddin, M., Awan, F.R., Shameem, M., Ibrahim, M., Ali, M.R. (2011). Effect of arc furnace dust on the properties of OPC and blended cement concretes. Constr Build Mater 25: 308–312. https://doi.org/10.1016/j.conbuildmat.2010.06.024

[5]   Ruiz, O., Clemente, C., Alonso, M., Alguacil, F.J. (2007). Recycling of an electric arc furnace flue dust to obtain high grade ZnO. J Hazard Mater 141(1): 33–36. https://doi.org/10.1016/j.jhazmat.2006.06.079

[6]   Al-Zaid, R.Z., Al-Sugair, F.H., Al-Negheimish, A.I. (1997). Investigation of potential uses of electric-arc furnace dust (EAFD) in concrete. Cem Concr Res 27(2): 267–278. https://doi.org/10.1016/S0008-8846(96)00204-9

[7]   Havlik, T., Turzakova, M., Stopic, S., Friedrich, B. (2005). Atmospheric leaching of EAF dust with diluted sulphuric acid. Hydrometallurgy 77(1-2): 41–50. https://doi.org/10.1016/j.hydromet.2004.10.008

[8]   Buzin, P.J.W.K., Heck, N.C., Vilela, A.C.F. (2016). EAF dust: An overview on the influences of physical, chemical and mineral features in its recycling and waste incorporation routes. J Mater Res Technol doi:10.1016/j.jmrt.2016.10.002. https://doi.org/10.1016/j.jmrt.2016.10.002

[9]   Sikadilis, C., Mitrakas, M. (2007). Utilization of electric arc furnace dust as raw material for the production of ceramic and concrete building products. J Environ Sci Health Part A 41(9): 1943–1954. https://doi.org/10.1080/10934520600779240

[10]   Stathopoulos, V.N., Papandreou, A., Kanellopoulou, D., Stournaras, C.J. (2013). Structural ceramics containing electric arc furnace dust. J Hazard Mater 262: 91–99. https://doi.org/10.1016/j.jhazmat.2013.08.028

[11]   Vargas, A., Masuero, A.B., Vilela, A.C.F. (2006). Investigations on the use of electric arc furnace dust (EAFD) in pozzolan-modified Portland cement I (MP) pastes. Cem Concr Res 36(10): 1833–1841. https://doi.org/10.1016/j.cemconres.2006.06.003

[12]   Al Mutlaq, F.M., Page, C.L. (2013). Effect of electric arc furnace dust on susceptibility of steel to corrosion in chloride-contaminated concrete. Constr Build Mater 39: 60–64. https://doi.org/10.1016/j.conbuildmat.2012.05.008

[13]   Alqam, M. (2012). Potential reuse of electric arc furnace dust (EAFD) in concrete. Jordan J Civ Eng 6: 174–185.

[14]   Xuefeng, X., Yuhong T (1998). Application of electric arc furnace dust in cement production. Iron Steel (Peking) 33(6): 61–4.

[15]   Macray, D.R. (1985). Electric arc furnace dust: disposal, recycle and recovery. Technical report Center for Metal Production. CMP 85-2, Pittsburgh, USA.

[16]   Fares, G., Al-Zaid, R.Z., Fauzi, A., Alhozaimy, A.M., Al-Negheimish, A.I., Khan, M.I. (2016). Performance of optimized electric arc furnace dust-based cementitious matrix compared to conventional supplementary cementitious materials. Const Build Mater 112(1): 210-221. https://doi.org/10.1016/j.conbuildmat.2016.02.068

[17]   Taylor, H.F.W. (1997). Cement Chemistry, second ed. Thomas Telford Press, London. https://doi.org/10.1680/cc.25929

[18]   Chen, Q.Y., Tyrer, M., Hills, C.D., Yang, X.M., Carey, P. (2009). Immobilisation of heavy metal in cement-based solidification/stabilisation: A review. Waste Manage 29(1): 390-403. https://doi.org/10.1016/j.wasman.2008.01.019

[19]   Arliguie, G., Grandet, J. (1990). Influence de la composition d'un ciment Portland sur son hydratation em presence de zinc. Cem Concr Res 20(3): 517–524. https://doi.org/10.1016/0008-8846(90)90096-G

[20]   Murat, M., Sorrentino, F. (1996). Effect of large additions of Cd, Pb, Cr, Zn, to cement raw meal on the composition and the properties of the clinker and the cement. Cem Concr Res 26: 377–385. https://doi.org/10.1016/S0008-8846(96)85025-3

[21]   Brehm F.A., Vargas A.S., Moraes C.A., Masuero A.B, et al. (2001). Characterization and use of EAF dust in construction. Japan–Brazil Symposium on Dust Processing-Energy-Environment. In Metallurgical Industries, 3, Proceedings, EPUSP, São Paulo, 2001. 173–181.

[22]   Balderas, A., Navarro, H., Flores-Velez, L.M., Dominuez, O. (2001). Properties of Portland cement pastes incorporating nanometer-sized franklinite particles obtained from electric arc furnace dust. J Am Ceram Soc 84(12): 2909-2913. https://doi.org/10.1111/j.1151-2916.2001.tb01114.x

[23]   Olmo, I.F., Chacon, E., Irabien, A. (2001). Influence of lead, zinc, iron (III) and chromium (III) oxides on the setting time and strength development of Portland cement. Cem Concr Res 31: 1213–1219. https://doi.org/10.1016/S0008-8846(01)00545-2

[24]   Hamilton, I.W., Sammes, N.M. (1999). Encapsulation of steel foundry bag house dusts in cement mortal. Cem Concr Res 29: 55 - 61. https://doi.org/10.1016/S0008-8846(98)00169-0

[25]   EN 197-1 (2011). Cement. Composition, specifications and conformity criteria for common cements, Brussels, Belgium.

[26]   EN 196-3 (2016). Methods of testing cement. Determination of setting times and soundness. Comité Européen de Normalisation, Brussels, Belgium.

[27]   EN 1015-3. (2006). Methods of test for mortar for masonry: Part 3: Determination of consistency of fresh mortar (by flow table). Brussels: Comité Européen de Normalisation.

[28]   EN 1015-6 (2007). Methods of test for mortar for masonry: part 6: determination of bulk density of fresh mortar. Comité Européen de Normalisation, Brussels, Belgium.

[29]   EN 1015-11 (2007). Methods of test for mortar for masonry: part 11: determination of flexure and compressive strength of hardened mortar. Comité Européen de Normalisation, Brussels, Belgium.

[30]   Chen, Q.Y., Hills, C.D., Tyrer, M., Slipper, I., Shen, H.G., Brough, A. (2007). Characterisation of products of tricalcium silicate hydration in the presence of heavy metals. J Hazard Mater 147(3): 817–825. https://doi.org/10.1016/j.jhazmat.2007.01.136

[31]   Magalhães, M.S., Faleschini, F., Pellegrino, C., Brunelli, K. (2017). Effects of Electric Arc Furnace dust (EAFD) addition on setting and strength evolutions of cement pastes and mortars. Eur J Civ Environ Eng: 1-13.

[32]   Obaidat, Y.T. (2014). Assessment of the potential for using EAFD in cement paste in Jordan. Eur J Civ Environ Eng 20(3): 332-343. https://doi.org/10.1080/19648189.2015.1036127

Non-Conventional Materials and Technologies – NOCMAT for XXI Century   Materials Research Forum LLC
Materials Research Proceedings 7 (2018) 716-724   doi: http://dx.doi.org/10.21741/9781945291838-69

Study of Mix Proportion of Structural Concrete with Recycled Aggregate of CCW for Application in Precast Slabs

Enildo Tales Ferreira[1a]; Aluísio Braz de Melo[1b]; Viviana Maria Zanta[1c*];
Ubiratan Henrique Oliveira Pimentel[1d]

[a] enildotales@gmail.com; [b] aluisiobmelo@hotmail.com; [c] zanta@ufba.br; [d]
ubiratan.hop@gmail.com

Abstract. Technologies for recycling municipal solid waste are being created to minimize the extraction of raw materials and to improve the sanitation system of cities and their inhabitants' quality of life. In this context, this work presents results of a mix proportions study for Portland cement concrete, using recycled aggregates, collected in a CCW beneficiation plant. The objective was to evaluate the potential of these aggregates in structural concrete (fck = 25 MPa), applied in precast concrete slabs. In the experimental program of the research, based on the specific norms, the material and concrete characterization stages were carried out with natural and recycled aggregates. It complements the large dosage study, considering the proportions (25%, 50%, 75% and 100%) of replacing the natural aggregates with the recycled aggregates, an evaluation of the physical-mechanical properties of these concretes. What this study sought was a compatibility between the higher substitution content of the natural aggregates for the recycled aggregates and an adaptation to the physical-mechanical requirements of the structural concrete, applied to the precast slabs. Upon this analysis, the dosage chosen was that with 75% substitution of natural aggregate for recycled aggregate, to be used in the experiments with this structural element. The results of the flexing tests on the precast slabs allowed to claim that the recycled aggregate used meets the expectations of resistance and deformation inherent to the precast element.

Keywords: Civil Construction Waste (CCW), Recycling, Recycled Aggregates, Concrete and Precast Slab

Introduction

The 21st century will be a time of challenges to humankind in almost every aspect of human life on the planet. Catastrophes motivated by the influence of man on the environment start to make the news all over the world, many times due to the lack of skills of public officials as well as the unreasonable interests of private companies involved. In addition to causing environmental losses beyond calculation, these catastrophes also generate social, economic and cultural damage. In this context, apart from the need to consume great quantities of natural resources, the civil construction industry also modifies the landscape of the cities, and generates solid waste (CCW) intrinsic to such activity. Encouraging the recycling of this waste by using it (the recycled aggregates) in structural concrete is a way of broadening the possibilities of applying such material and reducing the consumption of raw materials, thus producing environmental and economic advantages.

However, to warrant the quality of the concretes containing recycled aggregates (CCW) it is important to conduct some assessment based on the recommendations of Ângulo & John (2006), as they state that the density of the recycled coarse aggregate influences compression strength results for the concrete with such aggregates, as well as this concrete's modulus of elasticity and water absorption. Still, according to these two authors, when the density of the recycled aggregate (CCW) is greater than 2.2 g/cm^3, it can be sent to the market of conventional structural concrete, whereas less dense aggregates must be sent to less demanding markets. Therefore, this is an essential requisite when intending to determine the application potential of recycled aggregates (CCW) in structural concrete. Another testing procedure proposed by Leite (2001) concerns the assessment of the water absorption ratio (for 24 hours) of the recycled aggregates (CCW), due to their usual greater porosity (direct relation with the mortar adherence to the surfaces of the recycled coarse aggregate), which may greatly alter the water/cement relation of the concrete containing this type of recycled aggregate and, consequently, reduce its compression strength. Thus, in the study of dosage for structural concretes with recycled aggregates (CCW) such assessment must be strictly done and make part of a set of tests recommended by the norms.

This paper covers a broad study on recycled aggregates in a municipal CCW processing plant, with the aim to assess its potential use in structural concrete (f_{ck} = 25 MPa), applied in precast slabs. However, before providing a definition of structural concrete with recycled aggregates (CCW) to be tested in precast slabs, three goals must be reached in search of a stricter assessment of the potential use of the material processed in this plant, namely:

1. Meet the specific standard requirements for recycled aggregates, NBR 15116 (ABNT, 2004);
2. Verify if the procedures used in the plant resulted in coarse recycled aggregates with appropriate density (> 2.2 g/cm^3);
3. Determine the potential of the recycled aggregates to meet the requirements of structural concretes.

Materials

To assess the performance of the recycled concrete (CCW) applied to precast slabs, this study used the natural materials available from the city of João Pessoa, Brazil, and the recycled materials obtained from a recycling plant (USIBEN) built and managed by the city administration. Material was collected from three lots of recycled coarse aggregates (lots 1, 2 and 3) after 30+ days to assess the variability of the material resulting from the processing in question. After being classified as a recycled coarse aggregate, gravel (Dmáx = 19 mm) was found to have a greater density than 2.2 g/cm^3 and, for this reason, it was chosen to produce recycled concretes to replace the natural aggregate of similar classification. The natural aggregates used were sand (quartz; Dmáx = 2.36 mm) and gravel (granite; Dmáx = 19 mm). To knead the reference concrete and the recycled concrete, deionized water was used, since it is free from mineral salts and is highly pure as far as chemical properties. This water was also used to pre-humidify the recycled coarse aggregates. The experiment used binder, Portland cement, CPV-ARI, used to produce both natural and recycled concretes, due to its greater pureness, that is, due to its chemical composition with a lower percentage of lime-based filler. Another reason for choosing this type of cement was its characteristic of having high early strength and because it is vastly used in the precast industries in the area. We have chosen to use an additive for the concrete to reach the adequate consistency for application in precast slabs (Slump test = 60 mm +/- 10 mm) without changing the original water/cement ratios of the concrete mix. As a function of the level of recycled CCW aggregate in the tested dosages, a superplasticizer was used as additive -- Adiment Premium, which is polycarboxylate-based, in percentages ranging from 0.2 % and 0.6% of the cement mass.

Methods

To characterize the natural aggregates used, the tests were conducted according to NBR 7211 standard (ABNT, 2005). The recycled aggregates used to prepare the recycled concrete was gray in color, A class, according to NBR 15116 standard (ABNT, 2004). The experiment program used comprises four steps, whose checking takes place in a classifying or excluding sequence:

Aggregates characterization – Conducting assessment to determine which recycled and natural aggregates would be selected, according to the ABNT standards, to be used in structural concrete.

Non-Conventional Materials and Technologies – NOCMAT for XXI Century   Materials Research Forum LLC
Materials Research Proceedings 7 (2018) 716-724   doi: http://dx.doi.org/10.21741/9781945291838-69

Concrete dosage study – Assessing the performance of the concrete with better potential of compression strength, by using the method of Helene and Terzian (1992), by constructing dosage diagrams for the reference (natural) concrete) and recycled concretes, as a function of the mean compression strength of 34 MPa (fck = 25 MPa). The study considered the proportion variation of Portland cement (rich, normal and lean concrete), the percentage of natural aggregate replaced with the recycled aggregate (0%, 25%, 50%, 75% and 100%) and the different water/cement ratios in the concrete dosages, named "Ref" (reference), with no recycled coarse aggregates; "Rec25" with 25% of such aggregates to substitute natural gravel; and so on: "Rec50" with 50%; "Rec75" with 75% and "Rec100" with 100%.

Concrete physical-mechanical properties – Assessing the following physical-mechanical properties: compression, flexural strength, modulus of elasticity, water absorption and void ratio. In this case, based on the dosage diagrams, we only tested the recipes of the reference concretes and recycled concretes which reached the predicted compression strength (34 MPa, corresponding to the minimum characteristic strength for the study of precast slabs; fck = 25 MPa). These new concrete dosages were named "CeRef" with 0% replacement; "CeRec25" with 25%; "CeRec50" with 50%; "CeRec75" with 75% and "CeRec100" with 100%. This procedure allowed selecting a single dosage of recycled concrete, with a greater level of natural aggregate replaced with recycled aggregate, to be compared with the reference concrete for the flexing test performance of the precast slab, assessed in the following section.

Precast slab experiments – Assessing deformations and flexural strength of the compared precast slabs. The precast slabs (Figure 1) measuring 1.35 m x 3,00 m x 0.12 m (width x length x thickness) were produced under the same lab conditions, with the same amount of reinforcement, by using precast beams and ceramic blocks. The reference concretes (CeRef) and recycled concretes (CeRec75) were applied to the slab beams and concrete topping, following the same procedures of compacting and water curing (surface protected with wet burlap). To run the test, the same type of strength was applied (to the thirds relative to the slab length), under the same support conditions and deformation assessment. To this, five linear variable differential transformers (LVDT) were installed on strategic spots (Figure 2) to register the slab's deformation values. Three of them (2, 3 and 5) were distributed in line, in transverse position, next to the central area of the slab; spot no. 3 coincided with the longitudinal axis of the slab, while the other two (1 and 4) were placed near the line of the slab's length thirds. The description of the items considered to calculate the slab's deformations can be seen in Chart 1.

Non-Conventional Materials and Technologies – NOCMAT for XXI Century   Materials Research Forum LLC
Materials Research Proceedings **7** (2018) 716-724   doi: http://dx.doi.org/10.21741/9781945291838-69

a) Location of the five LVDTs

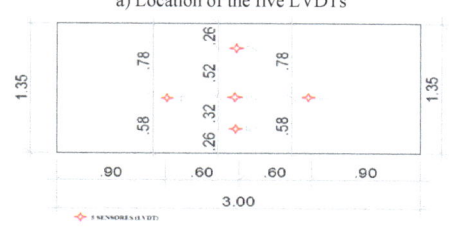

b) Slab dimensions and location of beams: upper view

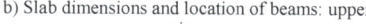
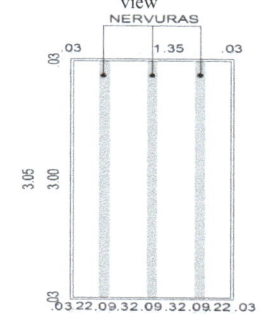

c) Slab cross section

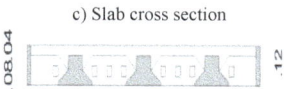

d) Slab topping reinforcing mesh setup and distribution

e) Compacting and leveling

FIGURE 1 – SLAB DIMENSIONS AND PRODUCTION PROCEDURES. Source: Ferreira (2013)

FIGURE 2 – EVALUATING SLAB DEFORMATIONS AND STRENGTH AT THE FLEXING TEST. Source: Ferreira (2013)

Non-Conventional Materials and Technologies – NOCMAT for XXI Century   Materials Research Forum LLC
Materials Research Proceedings 7 (2018) 716-724   doi: http://dx.doi.org/10.21741/9781945291838-69

CHART 1 – DESCRIPTION OF ITEMS CONSIDERED TO CALCULATE SLAB DEFORMATIONS. Source: Ferreira (2013)

CONSIDERED ITEMS	DESCRIPTION
Concretes – CeRec75 CeRef	fck = 25 MPa and Ecs = 25.6 GPa (experimental value set out by Ferreira (2013)
Steel (CA 60)	Es = 210.000 MPa - (NBR 6118:2007)
Infill element	Ceramic blocks 8cm high, 32cm wide and 20cm long
Strength element	Beams 8 cm high (h_o), 9 cm wide (b_w) and 300 cm long
Reinforcement of each joist	2 ⌀ 5.0 mm (steel cross section corresponding to 0.392 cm^2)
Additional reinforcement of each joist	1 ⌀ 5.0 mm (upper edge)
Slab topping reinforcing mesh	⌀ 5.0 mm rebar mesh; 20 cm spacing - (NBR14859-1:2002)
Slab cross section	12 cm high ($h_c + h_0$) and 41 cm effective width (b_f)
Load considered	Permanent load (g_1) due to its own weight - (1.5 kN/m^2) Predicted accidental load (q_1) for roof slab - (0.5 kN/m^2) Load from roof weight (g_2) - (0.7 kN/m^2) Total load – p = $g_1 + g_2 + 0.3q$ = 2.74 KN

Results and Discussion

The tests to characterize the aggregates confirmed that the recycled gravel (Dmáx = 19 mm) provided density over 2.2 g/m^3, which meets the requirements for use in conventional structural concretes (ÂNGULO & JOHN; 2006). In this case, it must be remembered that only this recycled gravel was used to replace natural gravel (Dmáx = 19 mm). At this initial phase of assessment, the compression strength results of all concretes that used these aggregates reached values between 24.9 MPa and 55.6 MPa (Table 1), which also shows good quality for this type of recycled aggregate. Other tests set out at NBR 9917 (ABNT, 2009) and NBR 15116 (ABNT, 2004) also qualified the recycled aggregate, especially when it comes to being within the contaminant thresholds. For each dosage of recycled aggregate used, the variation of additive content used to maintain the water/concrete ratio and the Slump (60 ± 10 mm) is considered low. Constructing dosage diagrams for each level of natural aggregate replaced with recycled aggregate was possible due to the data shown in Table 1.

Based on the materials and procedures used in this study, the results of Table 1 show that:

•   The behavior of concretes of greater values of compression strength is coherent with the greater amount of Portland cement combined with the lower water/cement ratio;
•   The increase of natural coarse aggregates replaced with recycled aggregates, as expected, resulted in reductions in the concretes' compression strength.

Setting up dosage diagrams for the various concretes studied in the initial phase was important to determine the mix of structural concretes (CeRef; CeRec25; CeRec50; CeRec75 and CeRec100) that reached the predicted strength of 34 MPa, which are considered – in this study – adequate to meet the desired strength (25 MPa). Figure 3 shows the dosage diagram for the recycled concrete (CeRec75),

Non-Conventional Materials and Technologies – NOCMAT for XXI Century   Materials Research Forum LLC
Materials Research Proceedings **7** (2018) 716-724   doi: http://dx.doi.org/10.21741/9781945291838-69

used to exemplify the identification of the mix corresponding to the dosage resistance, with which physical-mechanical property tests were run.

TABLE 1 – *RESULTS OF INITIAL TESTS CONDUCTED WITH THE SLAB DOSAGES FOR SLUMP (60 ± 10 mm). Source: Ferreira (2013)*

		Additive (%)	Cement consumption C (kg/m^3)	Density (kg/m^3)	f_{cm} (28 days) MPa
1:6.5 (1: 2.83: 3.67) Lean concrete a/c = 0.63	Ref	0.25	282	2293	28.5
	Rec25	0.29	282	2293	31.1
	Rec50	0.42	278	2264	27.3
	Rec75	0.49	277	2250	26.5
	Rec100	0.55	276	2243	**24.9**
1:5.0 (1:2.06: 2.94) Normal concrete a/c = 0.50	Ref	0.20	359	2335	33.0
	Rec25	0.27	360	2333	39.5
	Rec50	0.42	349	2271	34.1
	Rec75	0.51	345	2242	33.7
	Rec100	0.57	350	2239	31.2
1:3.5 (1: 1.30: 2.2) Rich concrete a/c = 0.37	Ref	0.25	487	2371	50.1
	Rec25	0.28	484	2357	**55.6**
	Rec50	0.46	482	2350	49.7
	Rec75	0.57	479	2335	49.2
	Rec100	0.60	479	2335	44.0

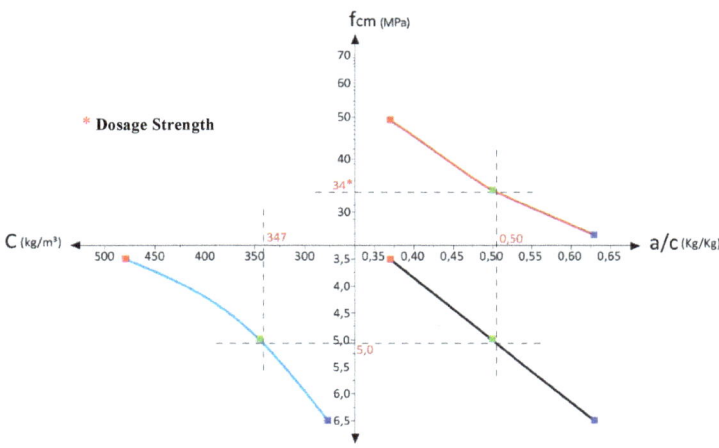

FIGURE 3 –*DOSAGE DIAGRAM FOR RECYCLED CONCRETE – CeRec75. Source: Ferreira (2013)*

Not ● lean concrete ● normal concrete ● rich concrete

Non-Conventional Materials and Technologies – NOCMAT for XXI Century   Materials Research Forum LLC
Materials Research Proceedings 7 (2018) 716-724   doi: http://dx.doi.org/10.21741/9781945291838-69

The curves generated in the 3 quadrants of the dosage diagram for recycled concrete (CeRec75) and the projection of the mean compression strength value curves (f_{cm}) over such curves, in this case 34 MPa, determined — on the abscissa and ordinate axes — the water/cement ratio values (w/c), the aggregates total mass values (m) and the cement consumption values (C). Repeating this process over the dosage diagrams for the other concretes (CeRef; CeRec25; CeRec50; and CeRec100) allowed determining the corresponding mixes, as shown in Table 2. All structural concretes studied, prepared with different percentages of recycled coarse aggregate, need to use different amounts (ranging from 0.2% to 0.6%, depending on the greater presence of recycled aggregate in the mixes) of superplasticizer, to keep constant the Slump of 60 ± 10 mm, which is adequate for the application in precast slabs.

TABLE 2 – *RESULTS OF PHYSICAL-MECHANICAL PROPERTIES FOR RECYCLED AND REFERENCE CONCRETES. Source: Ferreira (2013)*

Note: fck = 25 MPa (For all concretes)

Concretes	CeRef.	CeRec25	CeRec50	CeRec75	CeRec100
1 : m	1 : 4.9	1 : 5.9	1 : 5.0	1 : 5.0	1 : 4.6
Single mix (1: a: p_n : p_r)	1:2.01:2.89:0.0	1:2.52:2.54:0.85	1:2.06:1.47:1.47	1:2.06:0.74:2.21	1:1.86:0.0:2.74
w/c	0.50	0.58	0.50	0.50	0.47
C (kg/m³)	365	312	347	347	374
Additive (%)	0.20	0.47	0.50	0.56	0.55
Compression strength fc- (MPa)	33.7	32.4	34.6	33.6	35.5
Flexural strength (MPa)	4.20	3.40	3.10	3.43	3.52
Modulus of elasticity (GPa)	25.8	23.9	27.1	25.6	25.0
Water absorption (%)	3.51	4.70	5.28	6.05	6.79
Void ratios (%)	7.8	10.1	10.9	12.6	14.2
Specific mass (g/m³)	2.20	2.17	2.17	2.08	2.07

Regarding the results of the physical-mechanical properties of the concretes CeRef, CeRec25, CeRec50, CeRec75 and CeRec100 shown in Table 2, CeRec75 is the optimal recycled structural concrete, due to the best fit of a greater amount of natural aggregate replaced with recycled aggregate, and the one that least compromises such properties.

It must be highlighted that these physical-mechanical properties of the structural concretes are important to set the dimension of precast slabs and to assess their displacements to be measured in experiments following the assessment. Thus, this aspect reinforces the argument for choosing the CeRec75 concrete over the others (CeRef, CeRec50 e CeRec75). In fact, CeRec75 proved to be optimal concrete and the best fit of a greater amount of natural aggregate replaced with recycled aggregate and the one that least compromises the physical-mechanical properties, since their results were the closest to those obtained with the reference concrete. Therefore, among the various structural concretes produced with recycled coarse aggregates only the CeRec75's behavior is compared to the CeRef reference concrete, when applied in the production of precast slabs. Based on the results shown in Figure 4, obtained in the flexing tests conducted with the respective precast slabs, the following can be stated:

Non-Conventional Materials and Technologies – NOCMAT for XXI Century   Materials Research Forum LLC
Materials Research Proceedings 7 (2018) 716-724   doi: http://dx.doi.org/10.21741/9781945291838-69

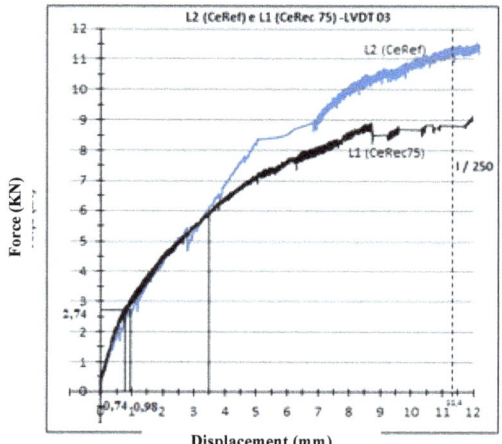

FIGURE 4 –STRENGTH X DISPLACEMENT CHART FOR THE SLAB FLEXING TEST
(CeRef) AND (CeRec75) – RECORDS FOR THE LVDT 03. Source: Ferreira (2013)

• Both slabs, produced with the reference concrete (CeRef.) and the recycled concrete (CeRec75), respectively, behaved similarly for displacements up to 3.5mm, corresponding to the force of 6.0 kN. In this interval, the slab with recycled concrete (CeRec75) proved to be a little less prone to deformation;

• Considering the displacements between 3.5mm and 11.4mm, equivalent to the threshold values (L/250) recommended by NBR6118 (ABNT, 2007), the slab with recycled concrete (CeRec75) proved to be more prone to deformation than the reference concrete (CeRef.). It should be noted that for the maximum displacement of 11.4mm of the recycled concrete slab (CeRec75), it was necessary to apply a 9kN force, that is, 18% less force than that found (11 kN) in the reference concrete slab (CeRef). However, this force (9kN) is nearly three times as big as the value of 2.74 kN, which was considered to calculate the slab displacements to meet the combination of almost permanent loads in action;

• For the applied force of 2.74kN, both the theoretical displacements (with recycled concrete = 2.19 mm and reference concrete = 1.74 mm) and the experimental displacements (with recycled concrete = 0.74 mm and reference concrete = 0.98 mm), found for two slabs, were much inferior to the threshold value of 11.4mm, recommended by the norm. Therefore, this finding confirms the good performance of the slab produced from recycled concrete (CeRec75).

Conclusions
The results obtained in the flexural tests using produced slabs, considering the materials used in the experiments, allow us to state that the slab produced from recycled concrete (CeRec75) meets the expectations of resistance and deformation intrinsic to the precast element (slab). This suggests a great potential for application of structural concretes with recycled aggregates in precast slabs, which are usual in small constructions and building repairs, just the type of work which, on average, as a whole, contributes the greatest percentage of all CCW generated in cities.

Non-Conventional Materials and Technologies – NOCMAT for XXI Century   Materials Research Forum LLC
Materials Research Proceedings **7** (2018) 716-724   doi: http://dx.doi.org/10.21741/9781945291838-69

From the results obtained, it can be concluded that the coarse aggregates processed in the plant have technical viability to be used in the production of precast slabs on a large scale. It may contribute, for example, to establishing a market for recycled aggregates and, in a certain way, to collaborating with the responsible destination of part of the CCW generated in the city.

References

[1] Associação Brasileira de Normas Técnicas. NBR 15116: Agregados reciclados de resíduos sólidos da construção civil: Utilização em pavimentos e preparo de concreto sem função estrutural – Requisitos. Rio de Janeiro, 2004.

[2] NBR 7211: Agregados para concreto - Especificação. Rio de Janeiro, abril 2005.

[3] NBR 6118: Projeto de Estrutura de Concreto: Procedimento. Rio de Janeiro, junho 2007.

[4] NBR 9917: Agregados para concreto – Determinação de sais, cloretos e sulfatos solúveis

[5] NBR 14859-1: Requisitos, Parte 1: Lajes unidirecionais. Rio de Janeiro, maio 2002.

[6] Ferreira, E.T. Contribuição ao estudo do potencial de aproveitamento de agregados reciclados de RCC produzidos na USIBEN - João Pessoa – em concreto estrutural aplicado em lajes pré-moldadas. Salvador, 2013. 276p. Tese (Doutorado) – Faculdade de Arquitetura da Universidade Federal da Bahia

[7] Helene, P.R.L.; Terzian, P. Manual de Dosagem e Controle do Concreto. São Paulo: PINI, 1992.

[8] John, V. M; Ângulo, S. C; Kahn. H. Controle da qualidade dos agregados de resíduos de construção e demolição reciclados para concretos a partir de uma ferramenta de caracterização. Construção e Meio Ambiente – Editores Miguel Aloysio Sattler e Fernando Oscar Pereira – Porto Alegre: ANTAC, 2006 – Coleção Habitare, vol. 7, 260 p.

[9] Leite, M.B. Avaliação de propriedades mecânicas de concretos produzidos com agregados reciclados de resíduos de construção e demolição. Porto Alegre, 2001. 270p. Tese (Doutorado) - Escola de Engenharia Civil da Universidade Federal do Rio Grande do Sul, Porto Alegre- RS, 2001.

Non-Conventional Materials and Technologies – NOCMAT for XXI Century   Materials Research Forum LLC
Materials Research Proceedings 7 (2018) 725-732   doi: http://dx.doi.org/10.21741/9781945291838-70

Preliminar Assessments for Decreasing Cement Content on Concretes Made with Recycled Aggregates

Bruno Luis Damineli [a *], Javier Mazariegos Pablos [b]

[a] Institute of Architecture and Urbanism, Brazil: bruno.damineli@usp.br

[b] Institute of Architecture and Urbanism, Brazil: pablos@sc.usp.br

Abstract. Nowadays, Construction and Demolition Waste (CDW) are produced in large volumes every day. Thinking in decreasing landfill disposal, the inclusion of these wastes as recycled aggregates in concrete is an effective strategy since the volume of aggregates required are high. So, this is one of the most studied ways for developing more sustainable concretes.

However, the replacement of natural aggregates by recycled aggregates (RA) in concrete often signify an increase in cement content. Considering that cement production is the heaviest environmental burden of concrete, the real sustainability of using RA can be put under check. Increasing cement content is more prejudicial than the disposal of construction waste, at least in places where space for landfills are not a problem. Besides of environmental impact, costs increase too.

This study presents preliminary a simple strategy – packing and dispersion of aggregates – that could allow decreasing cement content even using RA in concrete mixtures. Some concretes using current dosage method were designed, with natural aggregates and replacing them by recycled ones. In a third step, mixture with RA were optimized by packing and dispersion of particles. There were achieved concretes with very similar ratio "cement content / compressive strength" (here called Binder Intensity), which signifies same cement use efficiency. A better use of RA depends, rather than their quality, also from better concrete design methods.

Keywords: Concrete, Binder Index, CO_2, Recycled Aggregates

Introduction

Building construction is the most impacting chain in terms of resources consumption and CO_2 emissions by manufacturing new materials. Due to that, recycling has been increased interest since it can avoid landfill disposal of Construction and Demolition Waste (CDW); use of new resources to produce mew materials; and CO_2 emissions from production [1].

However, considering the concrete chain, the use of CDW as Recycled Aggregates (RA) replacing natural ones seems to often increase cement content for a same performance. Cement production process released around 842 kg CO2 / ton clinker by the year 2014 average, value that is decreasing very slowly since 2010 despite industrial efforts [2]. This is by far the most important environmental load from concrete. Other strategies are needed, such as the optimization of cement use in concrete design. "Sustainable" concretes must to consider the cement content variation, which is commonly forgotten or neglected in literature – works that does not consider the impact of RA on cement consumption of resulting concretes. RA from CDW generally have specific characteristics that can impact on that, such as the high level of source and porosity variability [3].

Once the apparent density is known, (there are many works trying to develop faster and more accurate ways for this task [4]), it is possible to use some tools for increasing sustainability of concretes made with RA. The design using particles packing and dispersion technology, which showed to be effective for decreasing cement content in concretes, reaching mixtures with 50 to 75% lower cement content compared to market benchmark [5]. The increase in cement efficiency use can be measured by either, the Binder Index (BI), or the "cement content (kg/m3)/compressive strength (MPa) ratio". BI can compare objectively information about the efficiency of cement use [6].

The objective of this paper is to preliminary investigate the potential of using packing and dispersion of particles for decreasing cement content in concretes made from 100% RA. This could compensate the loss of cement efficiency generated by higher porosity and heterogeneity of RA. BI is used to assess and compare the dosage efficiency of mixtures, allowing a more direct way to understand the effect of RA on the sustainability of concretes taking into account the cement content x compressive strength performance.

Methodology

Materials

It was acquired RA from a Brazilian supplier. Aggregates were produced from CDW with no treatments (mix of mortar, ceramic, concrete, clay and even soil). They were not separated by porosity, only by 3 distinct particle size distributions in the plant.

Particle size distribution of 3 different CDW RA were measured by sieving test (Figure 1). It was used the complete sieve series of ASTM E11 [7]. The more accurate particle size determination is for measuring and optimizing the packing by the chosen method, item 2.2.

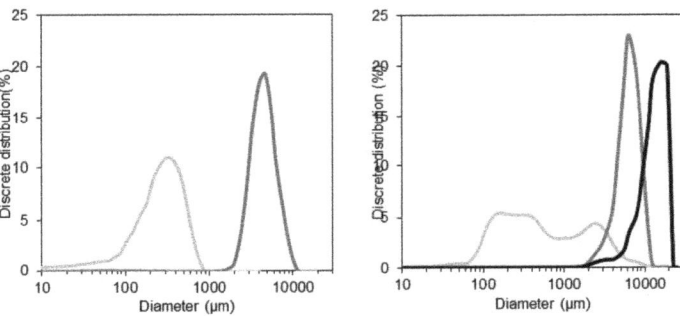

Figure 1 – *Particle size distributions of: a) Natural aggregates (left to right, lighter grey to darker grey: Sand 1 Natural and Gravel 1 Natural); b) CDW recycled aggregates (left to right, lighter grey to black: Sand 2 recycled, Gravel 2 recycled, Gravel 3 recycled).*

Table 1 summarizes water absorption, real density and apparent density measured by [8-10]. Aggregates are very low quality since they have high water absorption and low apparent density compared to real one. Apparent density is useful for designing mixtures by volume.

Table 1 – *Water absorption, real density and apparent density of Natural and CDW RA.*

	Water absorption (%)	Real density (Kg/dm^3)	Apparent density (Kg/dm^3)
Sand 1 (nat)	0.3	2.82	2.81
Sand 2 (RA)	19.5	2.63	2.06
Gravel 1 (nat)	0.4	2.71	2.71
Gravel 2 (RA)	19.0	2.62	2.05
Gravel 3 (RA)	19.2	2.61	2.06

It was used a typical CP V-ARI Brazilian cement strength class 40 MPa, high initial strength. Specific surface area = 1.72 m^2/g by BET [11], Gemini 2375, Micromeritics; real density = 3.05

Non-Conventional Materials and Technologies – NOCMAT for XXI Century   Materials Research Forum LLC
Materials Research Proceedings 7 (2018) 725-732   doi: http://dx.doi.org/10.21741/9781945291838-70

kg/dm^3, Helium Picnometer Quantachrome MVP 5DC. It is a free mineral admixtures cement, so variations sources were decreased, allowing analysis only in RA influence.

It was used a commercial policarboxilate dispersant.

Concrete design optimization – principles

Mixtures were designed for assessing the efficiency of cement use in concretes with RA compared to same mixtures with natural aggregates, using a method of design that does not care about packing [12]; and, in a second step, optimizing packing in concrete with RA, allowing: 1) determine the loss of efficiency in cement use for low-quality RA; and 2) assess how much efficiency is possible to recover optimizing packing using same low-quality RA.

The theory of packing of particles is based on the fact that it is possible to produce a system with lower void volume using different sizes of particle fractions if smaller ones fill the voids between larger ones. The lower the voids of aggregates skeleton, the lower the paste volume required for filling that and enabling flowing – which starts when paste surpass voids volume. As compressive strength in an ordinary simple cement-water paste depends mostly on water/cement ratio, a better packing project makes possible to use of lower paste content for a same flow and strength, resulting in a concrete with same performance and lower cement content – higher cement use efficiency. The lack of concrete design methods is not to care about measuring voids for that optimization. In optimized packing, flow is not dependent of paste increase, but of voids volume decrease. Several particles packing theories are known (as example [13]), based on calculus of the volume of voids. In this paper, [14] was used.

For an efficient system, fine particles need to be fully dispersed since they tend to agglomerate due to their low mass and high surface area (higher attraction forces). Agglomerates increases viscosity since they block the mobility of flow lines, generate voids increasing water consumption, and decrease surface area available for hydration. To obtain dispersion with economic criteria, it is necessary to use dispersants in optimized content (minimum required for highest flow [15]). In this work, dispersant content was determined experimentally for cement with three different water/cement (w/c) ratio, by two flowing tests: mini-slump test (flow in low shear rate, yield stress, Figure 2a; and Marsh funnel time flow (indirect measurement of viscosity, flow in high shear rates, Figure 2b). For mini slump, the higher the spreading, the lower the yield stress and the higher the flow; for Marsh funnel, the lower the time the paste surpass the hole, the lower the viscosity and the higher the flow. The optimum content would be that one where flow reach the highest potential with lowest dispersant. Dispersant content varies for different w/c; 1% of dispersant (of cement mass) iwas the minimum required for a good dispersion (it can provide flow to pastes with w/c>0.25, the case of the concretes of this study). The use of low dispersant content also complies to economic aspects.

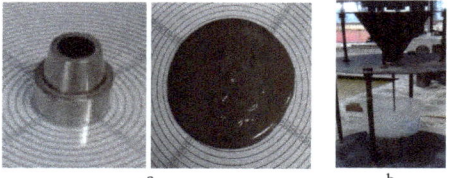

a   b

Figure 2 *– dispersant content determination by: a) mini slump test (yield stress); b) Marsh funnel time flow (viscosity, high shear rate).*

Non-Conventional Materials and Technologies – NOCMAT for XXI Century   Materials Research Forum LLC
Materials Research Proceedings 7 (2018) 725-732   doi: http://dx.doi.org/10.21741/9781945291838-70

Experimental planning
Table 2 shows concrete mixtures, in volume (packing optimization consider space occupation of voids). Mass converting was made by apparent density. There were mixed two reference concrete (Ref) with natural granite aggregates and two concretes using 100% RA (Rec).

Table 2 *– Concretes designed. Legend: Cem = cement CPV; S1n = Sand 1 (Natural); G1n = Gravel 1 (Natural); S2r = Sand 2 (Recycled); G2r = Gravel 2 (Recycled); G3r = Gravel 3 (Recycled).*

Mix	Composition (% vol)								Voids aggr	Voids total
	Cem	S1n	G1n	S2r	G2r	G3r	Water	Disp (% cem mass)		
Ref-01	16.1	38.1	45.7	-	-	-	23.4	0	26.6	22.9
Ref-02	16.1	38.1	45.7	-	-	-	16.0	1	26.6	22.9
Rec-01	12.2	-	-	35.1	-	52.7	13.1	1	24.7	21.9
Rec-02	12.2	-	-	35.1	24.0	28.7	12.0	1	15.7	11.4

Concretes were designed with the aim of comparing performance of concretes made from natural aggregates and a non-packing cared dosage methodology [12] against: a) concrete with 100% RA using the same non-packing methodology (Rec-01); and b) concrete made from 100% RA using [14] packing method (Rec-02). As comparison, Ref-01 and Ref-02 are 1:5 mass proportion (parameters from [12]), mortar 50% (same as Rec-01 and Rec-02). Rec-02 is resulting from packing optimization on Rec-01. Voids between aggregates were 26.6% for Ref-01 and Ref-02, following 24.7 for non-packing concrete with RA (Rec-01). Rec-02, however, achieved 15.7% voids between aggregates. For the overall voids calculation, including paste (in this paper paste is a fixed 100% cement), Ref-01 and Ref-02 had 22.9% voids; Rec-01, 21.9%; and Rec-02 had 11.4%. As the voids calculations are made on dry materials data, % volume of materials is presented in the basis of 100% total dry materials.

Experimental procedures
Before concrete mixing, RA were pre-saturated in water for 24 hours, aiming to avoid the absorption of water during mixing process, a problem related in literature [16] that can highly decrease flowability and affect process control. A 24h pre-saturation guarantees that most of pores will be filled in the mixing moment.

Concretes were mixed by 10 minutes in a simple 120-liter concrete mixer, a low-knowledge process. Water and dispersant were placed, cement was added for 1 minute, followed by 2 minutes-mixing. Aggregates were placed, from coarsest to finest, 1 minute between each one.

Rheological behavior was measured by the simple slump test [17]. Six 10x20 cm cylindrical specimens were cast for measuring compressive strength at 7 and 28-day ages (three specimens each age) [18].

After 1 day, specimens were demolded and taken to the humidity chamber (temperature $23 \pm 2°C$, relative humidity ~ 100 %,) [18] and kept under moist cure up to the compressive strength test (7 and 28 days).

Surface regularization of top and bottom was done by grit. Compressive strength was performed at 7 and 28 days by the [19] procedures, with specimens in wet state.

Non-Conventional Materials and Technologies – NOCMAT for XXI Century   Materials Research Forum LLC
Materials Research Proceedings 7 (2018) 725-732   doi: http://dx.doi.org/10.21741/9781945291838-70

Binder Intensity analysis
In [6], it was presented and discussed a useful tool for assessing the efficiency of cement (and binder in general) use: the Binder Intensity (BI). As definition, BI is the ratio between the total binder content (in this case cement) content, in $kg \cdot m^{-3}$, and the performance (in this case, the 28-day compressive strength, in MPa), Equation 1:

$$BI = binder\ content\ /\ compressive\ strength\ \ (1),\ where$$

This index is expressed in $kg \cdot m^{-3} \cdot MPa^{-1}$ and allows the analysis of the relationship between the total amount of binders (environmental impact, most of CO_2 is released in clinker production) and a performance measurement (the most commonly used for concrete is 28-day compressive strength). The lower the BI, the lower the binder content required for same performance and therefore the higher the eco-efficiency.

Results and discussion
Comparing concretes using 1% of dispersant content, in mass of cement, Ref-02 required 160 l/m^3 of water for reaching a 210mm slump (Table 3). Using same w/c ratio, Rec-01, using the same non-packed design method, reached a 30mm slump level. This can show that flow is reduced when the same mass dosage method and quantity (1 cement: 5 aggregates) is used for concretes with natural and RA. This is due to the lower apparent density of RA aggregates, which implies in a lower paste volume concrete (lower cement and lower water content, in kg/m3) than Ref-02. Flow is given by paste detaching coarse aggregates. This already shows that the same mass design method is not suitable for concretes made with RA.

Table 3 – Rheological behavior (slump flow), hardened parameters (compressive strength) and cement efficiency index (BI) of studied concretes.

Mix	Voids aggr	Voids total	Cement (kg/m^3)	Water (l/m^3)	w/c	Slump	CS7	CS28	BI7	BI28
Ref-01	26.6	22.9	367.3	234	0.65	120	29.1	32.0	12.6	11.5
Ref-02	26.6	22.9	402.7	160	0.41	210	40.3	43.4	10.0	9.3
Rec-01	24.7	21.9	314.4	131	0.42	30	23.9	26.3	13.2	12.0
Rec-02	15.7	11.4	318.3	120	0.38	80	27.5	31.3	11.6	10.2

Rec-01, using 100% recycled aggregates and designed without packing, achieved a very high BI: 12.0 $kg.m^{-3}.MPa^{-1}$. It is a higher value even than Ref-01 (no dispersant), which demonstrates that the indiscriminate use of recycled aggregates increases cement content for a similar performance. The loss of efficiency is around 30% when compared to Ref-02 – a non-controlled concrete made from RA can release 1.5 times CO_2 compared to a concrete made from natural aggregates, in similar conditions of design and dispersant content. This is a significant increase on environmental and economic impact. The landfill disposal saving cannot justify the cement increase.

On the other hand, using packing for re-designing the concrete made from RA, it is reached an increase in cement efficiency. Rec-02 was designed starting from Rec-01 but inserting a new sized-aggregate (G2r). The volume of G2r was determined by the calculation of packing of aggregates by [14] – a concrete with 10% lower voids content. Rec-02 achieved a BI of 10.2 $kg.m^{-3}.MPa^{-1}$. This is a good value compared to Ref-02 (BI = 9.3 $kg.m^{-3}.MPa^{-1}$). Packing allowed Rec-02 to achieve slump of 80mm, a higher value than Rec-01 even using lower water content and same dispersant content. Difference was achieved due to the care about decreasing voids (increase packing of particles). As a result, in terms of efficiency, the lower water content increased compressive strength, decreasing BI to 10.2 kg.m^{-3}.MPa^{-1}, almost 20% less than Rec-01. Compared to Ref-02, the BI was less than 10% higher (Ref-02 BI = 9.3 kg.m^{-3}.MPa^{-1}; Rec-02 BI = 10.2 kg.m^{-3}.MPa-1, a 0.9 kg.m^{-3}.MPa^{-1} difference). This means that, using packing, it can be compensated almost all loss brought by the low-quality CDW RA, even replacing 100% and using RA as they come from market, without any processing, such as particle size separation, fine washing, density separation or even changing surface characteristics [20]. Rec-02 presented lower BI (higher efficiency in the use of cement) than a natural concrete that did not use packing or dispersion (Ref-01, 0% dispersant).

If dispersant content used was the same 1% of cement mass for Ref-02 and Rec-02, for other hand the total amount of dispersant was lower in Rec-02 since it had lower cement content (~90 kg/m^3) – more economic. If same dispersant content would use, in total mass, this probably would help to disperse the higher amount of fines present in Recycled Sand when compared to Natural Sand (Figure 1), neglected on dispersant dosage (based on cement mass). Using same dispersant content, slump and BI could be even closer for a same cost.

Conclusion

The use of CDW as RA in concretes is an usual way found in literature for decreasing landfill disposals and increase building construction sustainability since there are a huge amount of CDW generated every day. However, despite all efforts, researches usually do not consider the increase in cement content that this usually produces in concrete if same performance is required. This is a gap of recycling technology if it has the aim of being sustainable.

This paper explored the potential of increasing the efficiency of concretes made from RA using packing and dispersion of particles. If packing is increased, voids between aggregates decrease and lower paste content is required for same rheological performance. Results showed that the use of packing of aggregates in concretes made from RA is useful for compensating almost all the loss of efficiency of cement use brought by the lower quality of these materials compared to natural ones, using same dispersant content and even replacing 100% of aggregates by very low-quality recycled ones – water absorption ~20%, no processing. Compared to concretes with natural aggregates without dispersant, recycled ones achieved higher efficiency.

A BI practiced on the market (~10 kg.m^{-3}.MPa^{-1}) was reached by concretes with RA using cement with no replacement; low dispersant amount; very low-quality recycled aggregates, replacing 100% of natural aggregates; and a low-quality mixing method (concrete mixer). So iit is feasible in real practice of market, which can contribute for developing the increase of use of RA in concrete considering the efficiency of cement use (not allowing cement content increase for compensating strength loss). This depends, rather than quality or processing of RA, also from better designing concrete methodologies, making it more sustainable.

Acknowledgements

Authors would like to thanks to technicians Paulo Pratavieira and Sergio Trevelin for helping on mixing processes and compressive strength tests.

References

[1] Tam VWY, Tam CM. A review on the viable technology for construction waste recycling. Res Cons and Recycling 2006; 47:209-221. https://doi.org/10.1016/j.resconrec.2005.12.002

[2] CSI WBCSD, "Getting the Numbers Right" (GNR), Cem. Sustain. Initiat. (2013). http://www.wbcsdcement.org/index.php/key-issues/climate-protection/gnr-database.

[3] Angulo SC, Carrijo PM, Figueiredo AD, John VM, Chaves AP. On the classification of mixed construction & demolition waste aggregate by porosity and its impact on the mechanical performance of concrete. Materials and Structures 2010;43:519-528. https://doi.org/10.1617/s11527-009-9508-9

[4] Damineli BL et al. Rapid method for measuring the water absorption of recycled aggregates. Materials and Structures

[5] Damineli BL, Pileggi RG, John VM, Low binder intensity eco-efficient concretes. In: Pacheco-Torgal F et al (editor), Eco-Efficient Concrete (2013), chapter 02, 26-44, Woodhead Publishing Limited, Cambridge, UK / Philadelphia, USA. https://doi.org/10.1533/9780857098993.1.26

[6] Damineli BL, Kemeid F, Silva PA, John VM, Measuring the eco-efficiency of cement use. Cement and Concrete Composites 32 (2010) 555-562.

[7] American Society of Testing Materials. E11: Standard Specification for Woven Wire Test Sieve Cloth and Test Sieves. Washington: 2017.

[8] American Society of Testing Materials. C 127: standard test method for density, relative density (specific gravity) and absorption of coarse aggregate. Washington: 2007. 6p.

[9] ABNT (Brazilian Standard). NM 53: Coarse aggregate - Determination of the bulk specific gravity, apparent specific gravity and water absorption. Rio de Janeiro: 2009 (in Portuguese).

[10] ABNT (Brazilian Standard). NM 52: Fine aggregate - Determination of the bulk specific gravity and apparent specific gravity. Rio de Janeiro: 2009 (in Portuguese).

[11] Brunauer S, Emmett PH, Teller E. Adsorption of gases in multimolecular layers. Journal of American Chemistry Society, v. 60, n. 2, pp. 309-319, 1938. https://doi.org/10.1021/ja01269a023

[12] Helene PRL, Terzian PR. Manual de dosagem e controle do concreto. São Paulo: PINI, 1992. 349p (in Portuguese).

[13] De Larrard F. Concrete mixture proportioning: a scientific approach. Modern Concrete Technology Series, vol. 9. London: E&FN SPON, 1999. 421 p.

[14] Funk JE, Dinger DR, Predictive process control of crowded particulate suspensions applied to ceramic manufacturing. Boston/Dordrecht/London: Kluwer Academic Publishers, 1994. 765p. https://doi.org/10.1007/978-1-4615-3118-0

[15] Damineli BL, John VM, Lagerblad B, Pileggi RG, Viscosity prediction of cement-filler suspensions using interference model: a route for binder efficiency enhancement. Cement and Concrete Research 84 (2016) 8-19. https://doi.org/10.1016/j.cemconres.2016.02.012

[16] Poon CS, Shui ZH, Lam L, Fok H, Kou SC. Influence of moisture states of natural and recycled aggregates on the slump and compressive strength of concrete. Cement and Concrete Research 2004;34:31-36. https://doi.org/10.1016/S0008-8846(03)00186-8

[17] ABNT (Brazilian Standard). NBR NM 67: Concrete – Slump test for determination of the consistency, 1998. 8p (in Portuguese).

Non-Conventional Materials and Technologies – NOCMAT for XXI Century   Materials Research Forum LLC
Materials Research Proceedings **7** (2018) 725-732   doi: http://dx.doi.org/10.21741/9781945291838-70

[18] ABNT (Brazilian Standard). NBR 5738: Concrete – Procedure of molding and curing on concrete test specimens, 2008. 9p (in Portuguese).

[19] ABNT (Brazilian Standard). NBR 5739: Concrete – compression test of cylindrix specimens – Method of test, 2007. 9p (in Portuguese).

[20] Tsujino M, Noguchi T, Tamura M, Kanematsu M, Maruyama I. Application of Conventionally Recycled Coarse Aggregate to Concrete Structure by Surface Modification Treatment. Journal of Advanced Concrete Technology 2007;5:13-25. https://doi.org/10.3151/jact.5.13

Non-Conventional Materials and Technologies – NOCMAT for XXI Century   Materials Research Forum LLC
Materials Research Proceedings 7 (2018) 733-739   doi: http://dx.doi.org/10.21741/9781945291838-71

Biomass-Derived from Bamboo Leaf Ash: Pozzolanic Reactivity

M.J.B. Moraes [a *], J.C.B. Moraes [a], L. Soriano [b], J. Payá [b], J.L.P. Melges [a], M.M. Tashima [a], J.L. Akasaki [a]

[a] Universidade Estadual Paulista (UNESP), Faculdade de Engenharia, Campus de Ilha Solteira, Brasil; maju.bamoraes@gmail.com, jotabassan@gmail.com, jlmelges@dec.feis.unesp.br; maumitta@hotmail.com, jorge.akasaki@gmail.com

[b] ICITECH - Instituto de Ciencia y Tecnología del Hormigón. Universitat Politècnica de València, Valencia, Spain, lousomar@gmail.com, jjpaya@cst.upv.es

Abstract. Bamboo leaf ash (BLA) was assessed as a pozzolanic material in pastes and mortars. Differently from most of research performed, the BLA studied in this paper was obtained through an auto-combustion process. This material was characterized by means of X-ray fluorescence, laser granulometry and powder X-ray diffraction. Pastes of calcium hydroxide/BLA in proportions of 1:1, 1:2 and 2:1 were assessed through thermogravimetric analysis after 7 days of curing. The compressive strength development of Portland cement mortars containing 20% of BLA cured during 3, 7 and 28 days at room temperature were compared with a control mortar (only Portland cement). The obtained results showed that BLA is an alternative pozzolanic material presenting high reactivity and can be used in the production of blended Portland cement mortars.

Keywords: Sustainable Material, Pozzolanic Reactivity, Bamboo Leaf Ash

Introduction

Civil construction development has been increasing along the years. Currently the main material utilized for building construction is the concrete, which ordinary Portland cement (OPC) is the binder component. Although the hydration of Portland cement reaches interesting mechanical properties, the cement production is responsible for greats amounts of CO_2 emission, the manufacture of ordinary Portland cement (OPC) contributes about 5% to global anthropogenic carbon dioxide emissions, besides the cement is manufactured using non-renewable raw material (clay, limestone) and demands a lot of energy to achieve the high temperatures required [1-3].

Researchers found in pozzolanic materials a great potential to mitigate those environmental impacts caused by the cement industry. Pozzolan is an amorphous siliceous or silicoaluminous material that presents low or no cementitious reaction by itself, but, in finely divided form and in the presence of water, it reacts with the calcium hydroxide and forms compounds with cementitious properties [4].

Advantages by using pozzolans replacing partially the Portland cement can be technical and environmental. Regarding to technical advantages, they can be divided in two main effects: chemical and physical. Chemically, the amorphous silica presents in the pozzolan reacts with the portlandite, which is a crystalline formation of calcium hydroxide generated during the hydration of the cement, producing more hydrated products and increasing the resistance of the matrix [5-7]. Physically, the advantage is due the high fineness of the pozzolan, where it causes the nucleation and filler effect. The former physical advantage, nucleation, accelerates the hydration of the particles. In the case of filler effect, it is the process that the pozzolan particles fill the porous of the blended matrix, increasing the cohesive and resistant of final material [8,9].

Related to the environmental advantages, the use of pozzolans reduces the Portland cement consumption and valorizes the residues from industries. In the case of residues valorization, recently the use of wastes from agroindustries as construction material increased. It can be highlighted the

Non-Conventional Materials and Technologies – NOCMAT for XXI Century   Materials Research Forum LLC
Materials Research Proceedings 7 (2018) 733-739   doi: http://dx.doi.org/10.21741/9781945291838-71

residues from the production of sugar cane, rice, olive, palm and corn [10]. In these industries, it is common to obtain a by-product that is usually burned to generate energy, called biomass, and this burning process generates an ash. This ash has no appropriate discard, and employing as construction material can be a fine destination.

Therefore, this paper presents another biomass residue: the bamboo leaf. The bamboo is a faster growing plant, develops in degraded environments and stands out for its high level of carbon sequestration, much more promising than the carbon storage of many other tree species [11]. The bamboo is a natural resource with varied purposes, from food to architecture. In addition, bamboo fiber is a sustainable and low-cost alternative as reinforcement in concrete elements, presenting satisfactory results [12-14]. However, its cultivation generates a by-product: the bamboo leaf, which is burned in landfills, generating the bamboo leaf ash (BLA) which does not have a suitable purpose, becoming a source of pollution. In this way, few researchers studied the pozzolanic behavior of BLA and its potential as a supplementary cementitious material.

Roselló et al. [15] established a relation with the content of silica in the bamboo leaf before and after the calcination. It was possible to observe that the studied bamboo leaf presented a high silica-content phytoliths (silica cells) and the spodogram for BLA was maintained after the calcination at 850°C. These two factors reflected that in the BLA chemical composition, silica is the main compound.

Singh et al. [16] and Dwivedi et al. [17] assessed an Indian bamboo leaf ash produced in a muffle furnace under controlled temperature. Their studies suggested that the BLA reacted with the calcium hydroxide and formed siliceous hydrated products. Mortars with 20% OPC replacement by BLA and 1:3 cement/sand ratio were assessed, at 28 curing days the compressive strength of the samples and it obtained a similar result of the mortars with no cement replacement.

Villar-Cociña et al. [18] studied a bamboo leaf ash obtained from a calcination under controlled temperature at 600°C in an electric furnace for 2 hours. The researches utilized an electrical conductivity and the application of mathematical models for the determination of the kinetic parameters to study the ash reactivity. They concluded that the BLA presented a high pozzolanic reactivity.

Frías et al. [19] produced an ash by controlled burning in an electric furnace at 600°C. They studied the chemical and physical characterization of the bamboo leaf ash, and assessed its reactivity by thermogravimetric analysis and the compressive strength in mortars with 10% and 20% of OPC replacement. The BLA presented an amorphous nature and high reactivity in early ages. When used in blended mortars, the admixtures with OPC replacement presents similar results to the control at 28 and 90 curing days.

Therefore, the aim of this research is to contribute to the study of bamboo leaf ash (BLA) as a pozzolanic, alternative and sustainable material for OPC replacement. Differently from the mentioned researches [15-18], the BLA studied in this paper was obtained by an auto-combustion process. This material was characterized chemically, mineralogically and physically. The pozzolanic behavior was assessed in pastes of calcium hydroxide/BLA by means of thermogravimetric analysis after 7 days of curing. The compressive strength development of Portland cement mortars containing 20% of BLA were compared with a control mortar, which it is composed by only Portland cement.

Experimental program
Materials
The biomass assessed in this research, the bamboo leaf, leaves were collected from a bamboo plantation in the surroundings of the city of Ilha Solteira (São Paulo, Brazil). Then, these leaves were passed by an autocombustion process. The maximum temperature reached in this process was 750°C. After this burning procedure, the bamboo leaf ash (BLA) is obtained. Then, the ash was sieved (300 μm opening) to remove unburned materials, and the passed ash was milled for 50 minutes.

Non-Conventional Materials and Technologies – NOCMAT for XXI Century   Materials Research Forum LLC
Materials Research Proceedings 7 (2018) 733-739   doi: http://dx.doi.org/10.21741/9781945291838-71

The calcium hydroxide ($Ca(OH)_2$) used to evaluate the pozzolanic reactivity of BLA had purity of 95%. The ordinary Portland cement (OPC) used was a Brazilian commercial type, CPV – ARI. This cement was chosen due to its composition, which contains no pozzolan, not interfering in the results. Siliceous sand with a specific gravity of 2.580 g/cm^3 and fineness modulus of 2.12 was used in the mortars preparation.

Methods
Bamboo leaf ash (BLA) characterization
The BLA was characterized chemically, mineralogically and physically by means of X-ray fluorescence (XRF), X-ray diffraction (XRD) and laser granulometry. The XFR was performed in the XRF Philips Magix Pro equipment to study the chemical composition of BLA. Mineralogical characterization was obtained by means of XRD, using the equipment RX Diffractometer Seifert TT 3003 with Cu-Kα radiation and a Ni filter, the X-ray generator intensity and voltage were 20 mA and 40 kV, respectively, where samples have diffractogram analyses for the interval of 2θ between 5° and 60°, with step of 0.02 and accumulated time of 2 seconds. A Mastersizer 2000 instrument of Malvern Instruments was used to obtain the granulometric distribution, mean diameter (D_{med}) and median particle diameter (D_{50}).

Microstructural studies
Microstructural studies were carried out in pastes of calcium hydroxide/BLA (CH/BLA) in proportions of 1:1, 1:2 and 2:1 and water/binder ratio of 0.8. They were assessed by thermogravimetric analysis (TGA) after 7 curing days at 25°C. This test was performed by a TGA 850 Mettler-Toledo equipment, where samples were heated at range from 35 to 600°C, a heating rate of 10°C/min, in an N_2 atmosphere with continuous 75 ml.min^{-1} gas flow, the test used an aluminum crucible with a pinhole lid and then sealed.

Mortars studies
Mortars studies were assessed by their compressive strength of specimens with Portland cement replacement by BLA in percentages of 0 (control) and 20%. The compressive strength was assessed by a EMIC Universal Machine with a 2000 kN load limit, and the mortars were cast in 5 cm x 5 cm x 5 cm molds. Samples were tested after 3, 7 and 28 curing days at 25°C, and they were stored in full immersion in a hydrated lime saturated water to avoid the sample calcium hydroxide solubilization.

Results and discussion
Bamboo leaf ash (BLA) characterization
The chemical characterization of the BLA is shown in Table 1. The ash presented silica as major component (83.56%). Both Indian [16-17] and other Brazilians [15,18-19] ashes also presented this oxide as main compound. It was observed a very low loss on ignition (LOI) of only 0.19%, which is due lower content of organic matter in the ash

TABLE 1 – *Bla chemical composition in percentage*

SiO$_2$	Al$_2$O$_3$	Fe$_2$O$_3$	CaO	MgO	K$_2$O	SO$_3$	P$_2$O$_5$	Cl	TiO$_2$	MnO	Others	LOI*
83.56	2.56	2.63	3.71	1.64	2.38	0.95	1.15	0.44	0.52	0.17	0.10	0.19

*Loss of ignition

Non-Conventional Materials and Technologies – NOCMAT for XXI Century   Materials Research Forum LLC
Materials Research Proceedings **7** (2018) 733-739   doi: http://dx.doi.org/10.21741/9781945291838-71

The XRD diffractogram of BLA is presented in Fig 1. The ash pattern presented a broad band localized between 2θ of 17° and 34°, which is a typical behavior of amorphous materials. Related to crystalline phases, it was possible to identify quartz (SiO_2, PDF Card #0000789) and calcite ($CaCO_3$, PDF Card #0000098). The presence of both minerals may be due some soil contamination, since the leaves were collect from the ground.

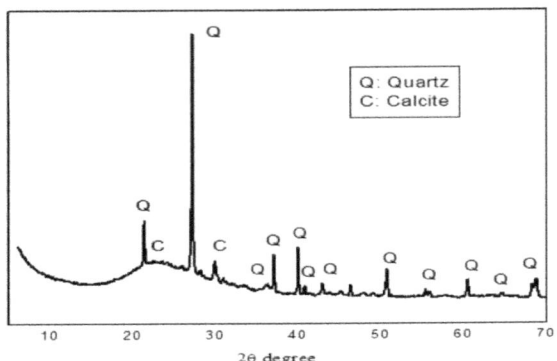

FIGURE 1 – XRD PATTERN OF BLA

The granulometric distribution of BLA is shown in Fig. 2. The calculated mean diameter (D_{med}) and median particle diameter (D_{50}) obtained were 19.86 μm and 14.23 μm, respectively.

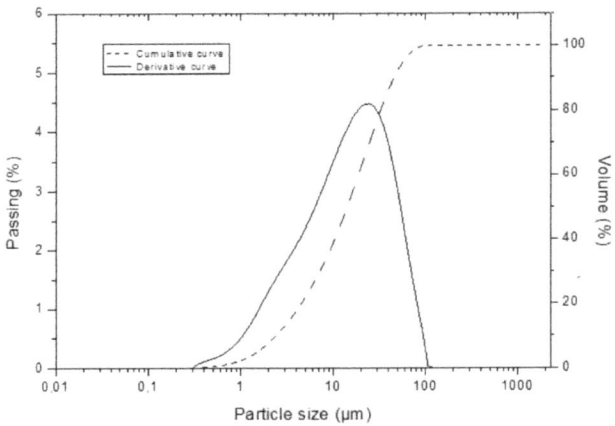

FIGURE 2 - BLA GRANULOMETRIC DISTRIBUTION

Microstructural studies of pastes
Derivative thermogravimetric (DTG) curves from TGA are presented in Fig. 3. The pastes of all ratios showed a dehydration in the range of 100-160°C, which represents the dehydration of C-S-H

Non-Conventional Materials and Technologies – NOCMAT for XXI Century   Materials Research Forum LLC
Materials Research Proceedings 7 (2018) 733-739   doi: http://dx.doi.org/10.21741/9781945291838-71

gel, a product of the pozzolanic reaction. The portlandite $(Ca(OH)_2)$ dehydration occurs at 500 – 550°C. Table 2 shows the percentage of lime fixation [20]. Firstly, it is observed that the paste with ratio 1:2 presented no mass loss at this range of temperature, meaning that the calcium hydroxide was all consumed. Even in a CH/BLA proportion of 2:1 the ash consumed more than half of the calcium hydroxide, showing that is a very reactive pozzolan, in the proportion with the same amount of BLA and calcium hydroxide (1:1), the hydroxide was 80.56% consumed.

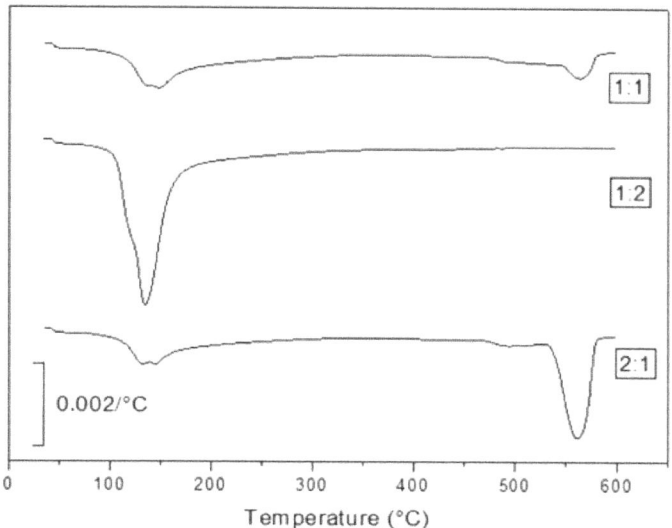

FIGURE 3 - *dtg curves for ch:bla pastes with the following proportions: 1:1, 1:2 and 2:1*

TABLE 2 - *Lime fixation and fixed lime/pozzolan of ch/bla pastes*

Ratio of CH/BLA	Lime Fixation (%)
1:1	80.56
1:2	100.00
2:1	55.20

Compressive strength

The results of compressive strength are shown in Table 3, where Control represents the mortar with no pozzolan and BLA 20 represents the blended admixture with 20% replacement-content.

Non-Conventional Materials and Technologies – NOCMAT for XXI Century   Materials Research Forum LLC
Materials Research Proceedings 7 (2018) 733-739   doi: http://dx.doi.org/10.21741/9781945291838-71

TABLE 3 - *Compressive strength of mortars (MPa) and standard deviation.*

Specimens/Curing days	3 days	7 days	28 days
Control	32.5 ± 1.3	37.2 ± 1.3	45.3 ± 1.1
BLA 20	26.8 ± 1.1	35.7 ± 0.6	44.4 ± 0.7

The mortars presented a continuous compressive strength increase. After 3 days of curing, the mortar with BLA presented lower compressive strength than the control (26.8 and 32.5 MPa, respectively). However, the high pozzolanic activity from BLA was observed after 7 days of curing. In this curing time, the mortar with BLA presented similar results to the control (35.7 and 37.2 MPa, respectively). This behavior can be justified by the high lime fixation observed in the TGA studies for this time of curing. In the 28-day curing time they also presented similar results: Control mortar and BLA mortars reached 45.3 and 44.4 MPa, respectively.

Conclusions

The bamboo leaf ash (BLA) was assessed as pozzolanic material. Characterization studies showed that the BLA presents silica as main compound (83.56%) and an amorphous phase, both requirements to be a pozzolan. In TGA studies of CH/BLA pastes, the ash had high lime fixation after 7 days of curing. This behavior is also observed in mortars studies, where the ash presented similar compressive strength than the control (a mortar without the presence of BLA) after 28 days of curing. These results allow concluding that autocombustion of bamboo leaves was an appropriate procedure for getting a reactive resulting ash and the BLA is a sustainable alternative which has potential to be used as pozzolanic material.

Acknowledgments

The authors acknowledge the Brazilian agency FAPESP – Brazil (process n° 2016/16403-5) for the financial support.

References

[1] Worrell E, Price L, Martin N, Hendriks C, Meida LO. Carbon dioxide emissions from the global cement industry 1, Annu. Rev. Energy Env. 26 (1) (2001) 303–329. https://doi.org/10.1146/annurev.energy.26.1.303

[2] Mo KH, Alengaram UJ, Jumaat MZ, Yap SP, Lee S C. Green concrete partially comprised of farming waste residues: a review. Journal of Cleaner Production 2016, v.117, p. 120 – 138. https://doi.org/10.1016/j.jclepro.2016.01.022

[3] Habert G, Billard C, Rossi P, Chen C, Roussel N. Cement production technology improvement compared to factor 4 objectives. Cement and Concrete Research 2010, v. 40 p. 820–826. https://doi.org/10.1016/j.cemconres.2009.09.031

[4] ASTM C-125-16. Standard Terminology Relating to Concrete and Concrete Aggregates. Current edition approved December 15, 2016. Annual Book of ASTM Standards.5.

[5] Paiva H , Silva A S, Velosa A , Cachim P, Ferreira V M. Microstructure and hardened state properties on pozzolan-containing concrete. Construction and Building Materials 2017, v. 140 p.374–384. https://doi.org/10.1016/j.conbuildmat.2017.02.120

[6] Senhadji y, Escadeillas G, Mouli M, Khelafi H, Benosman. Influence of natural pozzolan, silica fume and limestone fine on strength, acid resistance and microstructure of mortar. Powder Technology 2014, v. 254 p. 314–323. https://doi.org/10.1016/j.powtec.2014.01.046

Non-Conventional Materials and Technologies – NOCMAT for XXI Century   Materials Research Forum LLC
Materials Research Proceedings 7 (2018) 733-739   doi: http://dx.doi.org/10.21741/9781945291838-71

[7] Gutiérrez R M, Effect of supplementary cementing materials on the concrete corrosion control, Rev. Metal Madrid 2003,Vol. Extr. p. 250–255.

[8] Wild S, Khatib J M, Jones A. Relative Strength, Pozzolanic Activity and Cement Hydration in Superplasticised Metakaolin Concrete. Cement and Concrete Research 1996, vol. 26, No. 10, p. 1537-1544. https://doi.org/10.1016/0008-8846(96)00148-2

[9] El-Diadamony H, Amer A A, Sokkary T M, El-Hoseny S. Hydration and characteristics of metakaolin pozzolanic cement pastes. Housing and Building National Research Center 2016, press article.

[10] Aprianti E, Shafigh P, Bahri S, Farahani JN. Supplementary cementitious materials origin from agricultural wastes – A review. Construction and Building Materials 2015, v. 74 p.176–187. https://doi.org/10.1016/j.conbuildmat.2014.10.010

[11] Sohel, M. S. I.; Alamgir, M.; Akhter, S.; Rahman, M. Carbon storage in a bamboo (Bambusa vulgaris) plantation in thedegraded tropical forests: Implications for policy development. Land Use Policy 2015, v. 49, p.142 - 151. https://doi.org/10.1016/j.landusepol.2015.07.011

[12] Agarwal A, Nanda B, Maity D. Experimental investigation on chemically treated bamboo reinforced concrete beams and columns. Construction and Building Materials 2014, vol. 71 p. 610–617. https://doi.org/10.1016/j.conbuildmat.2014.09.011

[13] Ghavami K. Bamboo as reinforcement in structural concrete elements. Cement & Concrete Composites 2005, vol. 27 p. 637–649. https://doi.org/10.1016/j.cemconcomp.2004.06.002

[14] Akeju T A I, Falade F. Utilization of bamboo as reinforcement in concrete for low-cost housing. Structural Engineering, Mechanics and Computation 2001, vol. 2 p. 1463 -1470.

[15] Roselló J, Soriano L, Santamarina M P, Akasaki J L, Melges J L P, Payá J. Microscopy Characterization of Silica-Rich Agrowastes to be used in Cement Binders: Bamboo and Sugarcane Leaves. Microscopy and Microanalysis 2015, v. 21 p.1314–1326. https://doi.org/10.1017/S1431927615015019

[16] Singh N B, Dasa S S, Singh N P, Dwivedi V N. Hydration of bamboo leaf ash blended Portland cement. Indian J Eng Mater Sci 2007, v.14 p. 69–76.

[17] Dwivedi VN, Singh N P, Das S S, Singh NB. A new pozzolanic material for cement industry: Bamboo leaf ash. Int J Phys Sci 2006, vol.1 pg. 106–111.

[18] Villar-Cociña E, Morales E V, Santos S F, Savastano H Jr, Frías M. Pozzolanic behavior of bamboo leaf ash: Characterization and determination of the kinetic parameters. Cem Concr Comp 2011, vol. 33 p. 68–73. https://doi.org/10.1016/j.cemconcomp.2010.09.003

[19] Frías M, Villar-Cociña E, Valencia-Morales E. Characterization of sugar cane straw waste as pozzolanic material for construction: Calcining temperature and kinetic parameters. Waste Manag 2007, vol. 27 p. 533–538. https://doi.org/10.1016/j.wasman.2006.02.017

[20] Payá J, Monzó J, Borrachero MV, Velázquez S, Bonilla M. Determination of the pozzolanic activity of fluid catalytic cracking residue. Thermogravimetric analysis studies on FC3R–lime pastes. Cem Concr Res 2003, v. 33(7) p. 1085-1091. https://doi.org/10.1016/S0008-8846(03)00014-0

Non-Conventional Materials and Technologies – NOCMAT for XXI Century   Materials Research Forum LLC
Materials Research Proceedings 7 (2018) 740-749   doi: http://dx.doi.org/10.21741/9781945291838-72

Effect of RHA Addition on Bond Strength in Steel Fiber Reinforced Cement-Based Composites

Rodrigues Conrado de Souza[a], Jorge Felipe Sérgio Bastos[b*]

[a]CEFET-MG, Minas Gerais, Brazil; crodrigues@civil.cefetmg.br

[b]CEFET-MG, Minas Gerais, Brazil; felipebjorge@hotmail.com

Abstract. Steel fibers are being widely used in concretes and other cement-based composites in order to improve their toughness properties. It's known that concrete has high compressive strength and durability, but its tensile and flexural strength is low, especially when compared to steel. Studies indicate that in cement-based composites the Interfacial Transition Zone (ITZ) is where first cracks appear and their propagation lead to material failure. Mineral additions have been incorporated into cement matrixes with two main objectives: to give a sustainable destination to agroindustrial wastes such as rice husk ash (RHA) and to improve cement-based composites properties. RHA is being produced by burning for thermal energy generation, has high amorphous silica content, which is of high potential for pozzolanic activity. Therefore, RHA is considered as a new hydration product, containing mainly the C-S-H, producing higher density to cement matrix, reducing pores and voids, besides being sustainable material, since part of cement which is highly polluting material is replaced by RHA. The main objective of this paper is to present the results of a research considering the improvement of bond between the cementitious matrix and steel fiber by partial replacing cement with RHA. For that, mortar samples were produced with RHA addition, in which steel fiber was introduced. Afterwards, pull-out tests were performed to assess adherence. As conclusion, it can be stated that RHA incorporation promoted ITZ densification, which caused an improvement in bond strength between the matrix and the steel fiber, indicating the use of that material can contribute to product more resistant steel fiber reinforced cement-based composites.

Keywords: RHA, Bond Strength, Steel Fiber, ITZ

Introduction

Concrete is known for its great capacity to withstand compression efforts and its durability. Although it has some resistance to tensile and bending efforts, its use for this purpose is not very common, given the presence of other more suitable materials, such as steel, for example. This phenomenon is explained by low tenacity of the concrete, which is the ability to absorb stresses, preventing the propagation of small cracks and cracks. In order to improve the performance of the concrete, with respect to the tenacity, studies have been developed through the incorporation of fibers to the matrix.

The use of fibers as reinforcement in fragile building materials has been known for a long time. Clay-reinforced mud huts and mortar for masonry with addition of animal hairs are two examples. The first building materials to use the fibers as a calculated and controlled reinforcement were the cement / asbestos composites, produced in smooth or corrugated thin sections. The use of the fibers allowed the production of very fine elements with good mechanical properties, given the large volume of fibers incorporated. Cement composites reinforced with fibers are more robust materials, which use staple fibers as integral elements in the conventional method of concrete mixing. They are used in low percentages, typically no greater than 1.5% of the volume. Another characteristic of this type of fiber is the length; they are longer, varying between 15 and 65 mm (JOHNSTON [1]).

The first studies of concrete reinforced with steel fibers date back to the 1960s, and addressed the behavior of this composite. Since then reinforcement with steel fibers has been increasingly used. The steel fibers greatly increase the toughness of cement-based composites and were initially used for cracking control. Currently, although they are still used for this purpose, steel fibers have also been used in structural applications, replacing conventional armor, or acting in conjunction with it (BENTUR and MINDESS [2]).

Generally, steel fiber reinforced concretes can be produced using the same production practices as conventional concrete. However, it requires greater attention regarding the dispersion of the fibers in the matrix. The main problem in introducing a given volume of fibers to the blend is the uniform dispersion of this component in order to achieve performance improvements in the mechanical properties of the concrete, also considering the workability required for proper mixing, release and finishing of this composite. Even with the use of superplasticizers, this percentage of incorporation remains around 2% when using conventional concrete mixing practices (BENTUR and MINDESS [2]).

Conventional concrete is a composite made from cement, aggregates, water, additives and additions. During Portland cement production, much of the raw material is spent either as a resource or as a fuel for blast furnaces. In order to reduce the consumption of cement, and consequently its constituent materials, many studies have been directed at the use of mineral additives in the production of concrete. Generally, these additions are inorganic materials, replacing or adding by mass, in the percentages of 20 to 70%, the cementitious material used (MEHTA and MONTEIRO [3]).

According to Sokolovicz et al. [4], the mineral additives usually used are blast furnace slag, fly ash, active silica and ash from agroindustrial residues (RHA, sugarcane ash, eucalyptus ash, among others). These additions may act to promote chemical reactions in the mixture and the chemical reactions originating from amorphous silica, whose structures are formed by short bonds with $Ca(OH)_2$ to form hydrated calcium silicates (C-S-H), are called pozzolanic reactions.

Rice husk, when burned, acquires physical and chemical characteristics that the mineral additions must have to be incorporated to the concrete. Its pozzolanic activity depends on the silica content, crystallinity content and the surface area of the particles. In addition, it should have a low percentage of carbon in its composition. As the rice husk is burned in a controlled manner, the gray particles acquire an amorphous structure with a large surface area. The method of grinding and burning the rice bark directly influences the quality of the ash produced (Kulkarni et al. [5]).

The RHA presents high percentages of SiO2, which qualifies it to be used with mineral addition in cementitious composites in partial replacement of the cement (BATISTA [6]). Table 1 presents the chemical composition of the RHA according to several authors.

TABLE 1 - RHA COMPOSITION

Chemical properties									
Constituent	SiO_2	Al_2O_3	Fe_2O_3	CaO	MgO	SO_3	Na_2O	K_2O	Loss on ignition
Mehta (1992)	87.2	0.15	016	0.55	0.35	0.24	1.12	3.68	8.55
Zhang et al. (1996)	87.3	0.15	0.16	0.55	0.35	0.24	1.12	3.68	8.55
Bui et al. (2005)	86.98	0.84	0.73	1.40	0.57	0.11	2.46	-----	5.14

SOURCE: GIVI et al. [7]

RHA has been used in cementitious composites as pozzolanic material, due to the increase in the properties of these composites, such as strength, durability and their contribution to sustainability in building materials. In the fresh state, the addition of RHA increases the plasticity of the concrete, which allows for easy launching and finishing. It also decreases the specific mass of the composite (Kulkarni et al. [5]).

Non-Conventional Materials and Technologies – NOCMAT for XXI Century   Materials Research Forum LLC
Materials Research Proceedings **7** (2018) 740-749   doi: http://dx.doi.org/10.21741/9781945291838-72

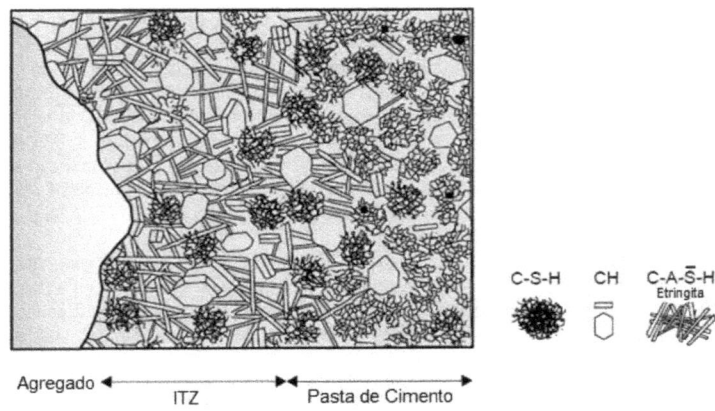

FIGURE 1 –ITZ

SOURCE: METHA and MONTEIRO [3]

ITZ is a phase of the cementitious composites located between the cement paste and the aggregate, either large or small. According to Liao et al. [8], it is composed of two weak layers, the first is a thicker film on the surface of the aggregate composed of CH crystals and C-S-H gels. The second layer is porous, consisting of more CH crystals, some C-S-H gels and little ettringite. Both are weak and through realized studies it was verified that the process of crack propagation occurs in the "crack" existing between the materials of the composite, possibly provoked by the retraction of the cement matrix. The faults started in this slit, they propagated through the interface with the aggregate until they advanced through the matrix, causing the composite fracture. Figure 1 schematizes this region.

According to Ollivier et al. [9], it is possible to densify this region through the use of fine and dispersed mineral additions, promoting the filling of voids and the creation of new hydration products by the pozzolanic activity of these additions.

The objective of this work is to evaluate the adhesion gain between the cementitious matrix and steel fiber by replacing the cement with RHA in the percentages of 10, 20 and 30%.

Methodology
Materials
The materials used in this research were cement, sand, steel fiber, water and RHA.

Cement
The cement used in this research was the Portland High Strength Initial Cement CPV ARI from Holcim. Because it is a purer cement, it is the ideal to evaluate the efficiency of RHA incorporation and its pozzolanic activity in the composite.

Sand
The sand used is the middle river sand, supplied by the company Conprem, located in Rio de Janeiro - RJ, with granulometry according to the grain size curve presented in Figure 2, fineness modulus of 1.65 and maximum characteristic dimension of 1.2mm.

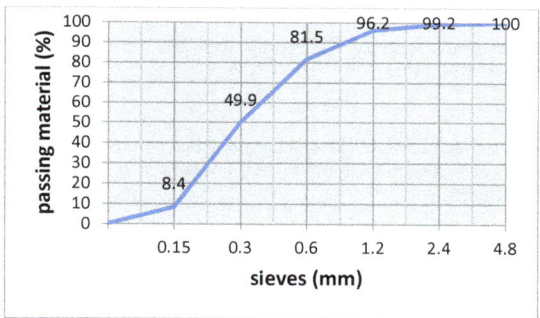

FIGURE 2 – SAND GRAIN CURVE

SOURCE: **OWN AUTHOR**

Steel Fiber

Steel fibers adopted in the research were the DRAMIX type 80 / 60BG fibers (Figure 3), manufactured by Belgo Bekaert according to ASTM A820, located in the city of Belo Horizonte in the state of Minas Gerais.

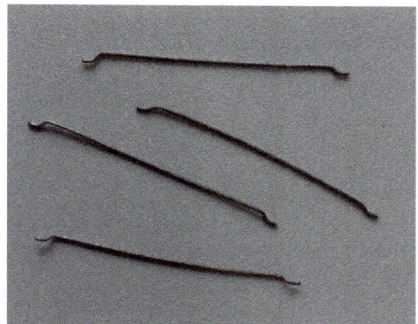

FIGURE 3 – STEEL FIBER

SOURCE: OWN AUTHOR

The fiber characterization is described in Table 2 below.

TABLE 2 – STEEL FIBER PROPERTIES

Fiber	Diameter (mm)	Length (mm)	Tensile strength (N/mm²)	Max. load (N)	Aspect ratio (l/d)
SF01	0,75	60,00	1.100	485,00	80,00

SOURCE: OWN AUTHOR

Non-Conventional Materials and Technologies – NOCMAT for XXI Century   Materials Research Forum LLC
Materials Research Proceedings 7 (2018) 740-749   doi: http://dx.doi.org/10.21741/9781945291838-72

Water
The water used in the tests was supplied via artesian well and stored in the CEFET-MG reservoir, Curvelo Campus, MG. The water Ph is 7,92.

Rice Husk Ash
It will be used the industrialized RHA of the company Pilecco Nobre, called Silcca Nobre. The RHA is produced by uniform and controlled burning and contains a minimum silica content of 95%, according to the manufacturer.

Method
Pullout Tests
Cylindrical test specimens with dimensions of 10cm high by 5cm in diameter were made in different formulations aiming at establishing the optimal percentage of cement replacement by the RHA in order to obtain the best adhesion between the matrix and the steel fiber. Three formulations with percentages of 10, 20 and 30% RHA substitution in mass of cement were used, as in Table 3. The formulations followed the 1: 1 trait with water cement factor of 0.45. The tests were performed at 28 days of curing by immersion in saturated water by calcium hydroxide in order to protect the steel fiber against corrosion.

TABLE 3 – FORMULATIONS

Name	Cement (g)	Sand (g)	Water (g)	RHA (g)	Anchorage (cm)
CPA_PO_2_00	504,9	504,9	227,2	0,00	2,00
CPA_PO_2_10	454,4	504,9	227,2	50,5	2,00
CPA_PO_2_10	403,9	504,9	227,2	101	2,00
CPA_PO_2_30	353,4	504,9	227,2	151,5	2,00

SOURCE: OWN AUTHOR

For the pullout test, a device was coupled to the electromechanical test machine Instron - EMIC 23-300 located at CEFET - MG, Curvelo according to Figure 4 below.

Non-Conventional Materials and Technologies – NOCMAT for XXI Century   Materials Research Forum LLC
Materials Research Proceedings **7** (2018) 740-749   doi: http://dx.doi.org/10.21741/9781945291838-72

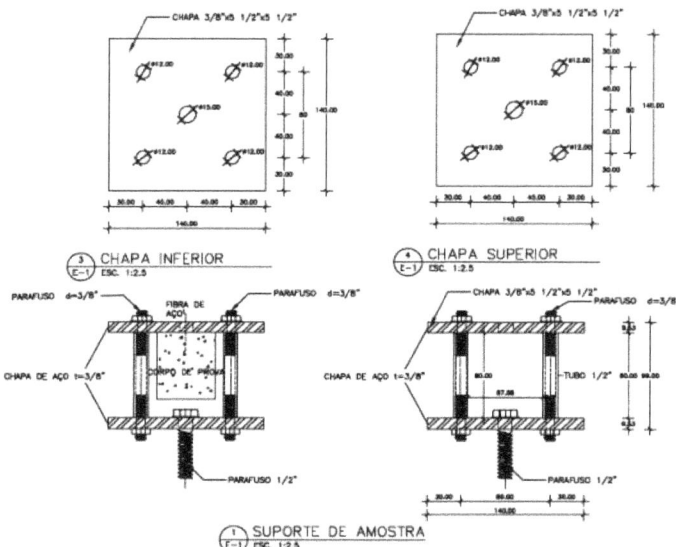

FIGURE 4 – PULLOUT TEST DEVICE

SOURCE: GONZÁLEZ [10]

That test is an adaptation of the Rilem-Rc-6 [11] standard and consists of a procedure in which the clamp coupled to the test machine promotes a vertical effort by pulling the inserted fiber in the test piece that is fixed to the base of the test Machine, according to Figure 4. For this research the length of anchorage (which is the length of the fiber that is inserted inside the matrix) of 20 mm was used. In this test, the steel fiber had the hooks removed so that there was no mechanical influence of the same on the result. For the results to be satisfactory, it is necessary that the claw secure the fiber firmly without "slippage" and that stresses are not promoted to the CP other than those from fiber tearing. The results of this test will depict the bond strength between the cementitious matrix and the steel fiber. Bond strength for this type of test is shown in Figure 5.

Non-Conventional Materials and Technologies – NOCMAT for XXI Century   Materials Research Forum LLC
Materials Research Proceedings 7 (2018) 740-749   doi: http://dx.doi.org/10.21741/9781945291838-72

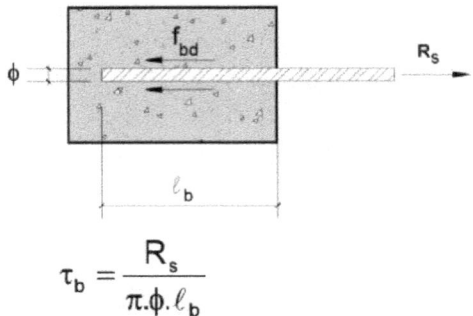

$$\tau_b = \frac{R_s}{\pi.\phi.\ell_b}$$

R$_s$ é a força atuante na barra;
φ   é o diâmetro da barra;
ℓ_b é o comprimento de ancoragem.

FIGURE 5 – BOND STRENGTH

SOURCE: PINHEIRO and MUZARDO [12]

Compressive Strength Tests
For the compressive strength tests, three control CPs were made, consisting of cement and sand only, in the 1: 1 trait and water / cement factor (A / C) 0.45. CPs with RHA incorporation were produced to compare the initial densification results of the cement matrix, in percentages of 10, 20 and 30% replacing cement by volume, with three CPs for each percentage. The specimens were assayed at 28 days of curing, which was by immersion. The tests followed the prescriptions of Nm-101 [13] and were performed in the electromechanical test machine Instron - EMIC 23-300 located at CEFET-MG, Curvelo.

Results
The results of the pullout tests are shown in Table 4, and compiled according to Figure 6.

TABLE 4 – PULLOUT TEST RESULTS

Proof bodies	Max. load (N)	Bond strength (Mpa)
CPA_PO_28_00	29,51	0,6262
CPA_PO_28_10	16,79	0,3563
CPA_PO_28_20	22,25	0,4722
CPA_PO_28_30	23,15	0,4913

SOURCE: OWN AUTHOR

Non-Conventional Materials and Technologies – NOCMAT for XXI Century   Materials Research Forum LLC
Materials Research Proceedings **7** (2018) 740-749   doi: http://dx.doi.org/10.21741/9781945291838-72

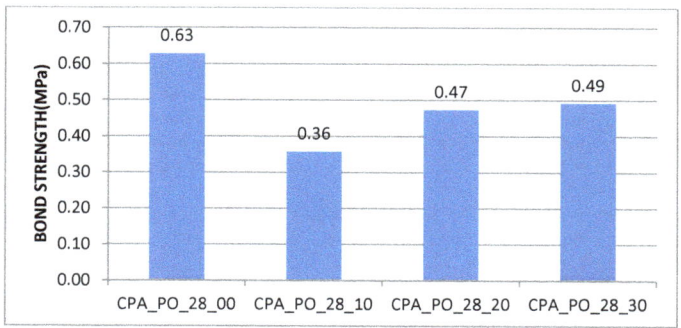

FIGURE 6 – PULLOUT TEST RESULTS

SOURCE: OWN AUTHOR

The results of the compressive strength tests are found in Table 5 and compiled according to Figure 7.

TABLE 5 – COMPRESSIVE STRENGTH RESULTS

Proof Bodies	Max. load (KN)	Max. stress (Mpa)
CPA_COMP_28_00	119,26	60,74
CPA_COMP_28_10	126,07	64,21
CPA_COMP_28_20	118,30	60,25
CPA_COMP_28_30	91,14	46,42

SOURCE: OWN AUTHOR

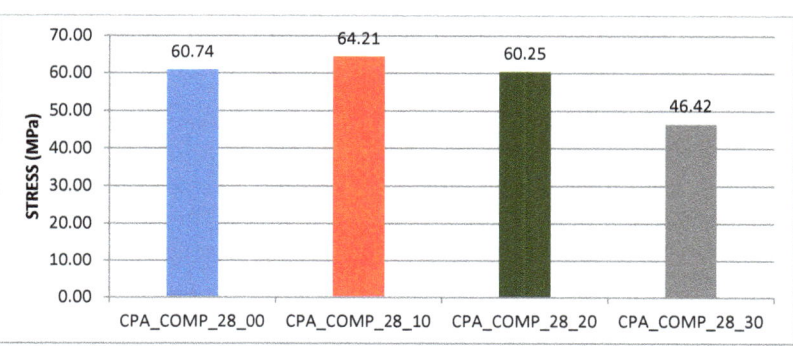

FIGURE 7 – COMPRESSIVE STRENGTH RESULTS

SOURCE: OWN AUTHOR

Non-Conventional Materials and Technologies – NOCMAT for XXI Century   Materials Research Forum LLC
Materials Research Proceedings 7 (2018) 740-749   doi: http://dx.doi.org/10.21741/9781945291838-72

Conclusions

The results for the pullout test were not satisfactory, since the values found for the resistance force were very low when compared to the literature, in which the reference values for CP without addition of RHA are in the order of 355N (GONZÁLEZ [10]) and 295N (ABBAS and KHAN [14]). The molding method must be modified in order to obtain results more suitable to those already known. It is possible that the method of molding and preparing the CPs has not been performed to the satisfaction. It is also possible that the non-use of superplasticizers may have limited the hydration capacity of cement and RHA.

For the axial compression test, the results indicate that the incorporation of the RHA is able to improve the ITZ by densifying the cement matrix increasing the compressive strength of the composite, in percentages between 10 and 20%. It is believed that the ideal value of cement substitution per RHA is close to 15%. It is suggested that new formulations be elaborated in order to establish the ideal combination between the elements of this composite. It is also possible that the use of superplasticizers can contribute to achieving better results.

The results of these tests will serve as reference for future work regarding the incorporation of agroindustrial residues in improving the mechanical properties of cement-based composites.

References

[1] JOHNSTON, C. D. Fiber-reinforced cements and concretes. New York: Taylor & Francis Group, 2001. ISBN 90-5699-694-0.

[2] BENTUR, A.; MINDESS, S. Fiber reinforced cementitious composites. 2. New York: Taylor & Francis Group, 2007. ISBN 978-0-415-25048-1.

[3] MEHTA, P. K.; MONTEIRO, P. J. M. Concrete - Microstructure, Properties and Materials. 3. United States of America: The McGraw-Hill Company, 2006. ISBN 978-0-07-146289-1.

[4] SOKOLOVICZ, B.; ISAIA, G.; GASTALDINI, A. Concrete with ash from natural rice husk: study of the penetration of chlorides in concrete prototypes. X Latin American Congress of Pathology and XII Congress of Quality in Construction. CONPAT 2009, 2009, Valparaiso-Chile.

[5] KULKARNI, M. S. et al. Effect of rice husk ash on properties of concrete. Journal of Civil Engineering and Environmental Technology, v. 1, n. 1, p. 26-29, 2014. ISSN 2349-8404.

[6] BATISTA, A.D. B. Effects of agroindustrial residues on mechanical strength, porosity and permeability of mortars and concretes. 2016. (Master degree). Department of Civil Engineering, CEFET-MG, Belo Horizonte / Brazil.

[7] GIVI, A. N. et al. Contribution of rice husk ash to the properties of mortar and concrete: a review. Journal of American Science, v. 6 (3), p. 157-165, 2010.

[8] LIAO, K. Y. et al. A study on the characteristics of interfacial transition zone in concrete. Cement and Concrete Research, v. 34, p. 977-989, 2004. ISSN 0008-8846.

[9] OLLIVIER, J. P.; MASO, J. C.; BOURDETTE, B. Interfacial transition zone in concrete. Advanced Cement Based Materials, v. 2, n. 1, p. 30-38, 1995. https://doi.org/10.1016/1065-7355(95)90037-3

[10] GONZÁLEZ, R. I. M. Influence of the use of rice hull ash as a mineral addition on portland cement composites reinforced by steel fibers. 2016. (Master degree). Department of Civil Engineering, CEFET-MG, Belo Horizonte / Brazil.

Non-Conventional Materials and Technologies – NOCMAT for XXI Century   Materials Research Forum LLC
Materials Research Proceedings **7** (2018) 740-749   doi: http://dx.doi.org/10.21741/9781945291838-72

[11] RILEM-RC-6. Bond test for reinforcement steel / pull-out test: International Union of laboratories and experts in construction materials, systems and structures: 4 p. 1983.

[12] PINHEIRO, L. M.; MUZARDO, C. D. Adherence and anchorage. In: (Ed.). University of Sao Paulo. São Paulo: USP, 2003. cap. 10.

[13] NM-101. Concrete - Compression test of cylindrical specimens. Mercosur: Brazilian Association of Technical Norms: 5 p. 1996.

[14] ABBAS, M. Y.; KHAN, M. I. Fiber-Matrix Interfacial Behavior of Hooked-End Steel Fiber Reinforced Concrete. Journal of Materials in Civil Engineering, v. 28, n. 11, 2016. https://doi.org/10.1061/(ASCE)MT.1943-5533.0001626

Non-Conventional Materials and Technologies – NOCMAT for XXI Century   Materials Research Forum LLC
Materials Research Proceedings 7 (2018) 750-755   doi: http://dx.doi.org/10.21741/9781945291838-73

Non-Conventional Materials in Civil Construction: A Case Study in Curvelo - Mg

Carvalho, Maria Cristina Ramos[a]; Estevão, Ana Cecília [a, *];Fialho, Patrícia Bhering[a];Alves, Lucas[a]; Cordeiro, Luana Emanuele Cruz[a];Cardoso, Eva Priscila[a]; Figueiredo, Thiago Antunes[a].

[a] Centro Federal de Educação Tecnológica – CEFET-MG

mariacristinaramosdecarvalho@gmail.com, anacestevao@yahoo.com.br, patriciabhering@gmail.com, lucas-alves296@hotmail.com, luanacordeiro@outlook.com, pry_dtna@hotmail.com, thiago-fig@hotmail.com

Abstract. The search for a sustainable civil construction encourages research on the use of alternative products, which are a result of renewable materials and industrial and domestic waste, known as non-conventional materials (NOCMAT). This paper shows a research in progress in Centro Federal de Educação Tecnológica de Minas Gerais (CEFET-MG), whose objectives are to check the use of NOCMAT in existing constructions in Curvelo/Brazil, and to analyze difficulties found using NOCMAT more frequently. Lastly, this research intends to divulge knowledge about NOCMAT among constructors, students and teachers of courses offered by CEFET-MG. The following materials were selected for this research: PET bottles, bamboo, soil and low environmental impact cementitious materials. The first step of the methodology is the theoretical research about the materials. The second step is to identify constructors in the city and carry out interviews to verify how much these materials are used and the difficulties in adopting them. The third step will be to catalog, organize and illustrate the research in the format of a digital booklet. The fourth step will be the digital booklet distribution for the academic community and constructors in the city. In addition, an online research will be done to identify aspects which could be improved. The last step will be a final revision of the digital booklet and its distribution. The digital booklet will allow easy access to the results of the research, which might encourage the use of NOCMAT in a greater scale in the city of Curvelo.

Keywords: PET Bottles, Bamboo, Soil Construction, Cementitious Materials, Booklet

Introduction

The construction is one of the most important activities for the development of society, however, it is considered the sector that causes more environmental impacts, since it uses many natural resources, modifying the environment and generating a large volume of waste (TESSARO et.al, 2012).

All kind of construction causes impacts, does not matter its size or function, interfering on environmental, social or economic sector of the cities. One of the ways to mitigate these impacts is using new or reused materials that produces less waste and those from other constructions, domestic or industrial process. (SPADOTTO, 2011).

The current model of construction systems presents, in general, large generation of waste, such as solids, liquids or gaseous, during the production process, transport and application. Aiming at reducing this impact, it is necessary to create constructive options with low environmental impact that enable more harmony between nature and the built environment.(SCHERER, 2016).

According to Brosler (2011), the Non-Conventional Materials (NOCMAT) can be seen as materials made using resources that can be found on the environment, like natural materials, derivatives of recyclable products or objects that had other functions before joining the construction.

According to Brazilian Association of Non-Conventional Materials (2013), new models of development and industrialization characterized by the harmony of technological and sustainable

Non-Conventional Materials and Technologies – NOCMAT for XXI Century   Materials Research Forum LLC
Materials Research Proceedings 7 (2018) 750-755   doi: http://dx.doi.org/10.21741/9781945291838-73

development are necessary. Innovative solutions of NOCMAT are going to proportionate economic progress and social stability, offering a durable, efficient and effective system of infrastructure with affordable prices.

Although these materials have been shown to be efficient and profitable through a variety of researches, it is noteworthy, that in Brazil, they are still used in an incipient way, facing the resistance of an industry traditionally not very accessible to innovations.

This article is a report of one researching progress in Centro Federal de Educação Tecnológica de Minas Gerais (CEFET-MG), in Curvelo, that has the objective to verify the using of NOCMAT in constructions of the municipality of Curvelo – Minas Gerais and analyze the difficulty to using them in large scale. It intends to provide greater dissemination of NOCMA Tamong builders, students and teachers of the courses offered by the Institution. Four types of NOCMAT were selected for this research: PET bottles, bamboo, earth and cementitious materials with lower environmental impact.

However, the city of Curvelo-MG is located in a region with large production of eucalyptus wood; the most buildings in the city are built with conventional masonry. In relation to the adequate destination of the Construction and Demolition Waste (CDW), it's observed that, for the most part, they are discarded without taking good reuse practices into account.

In addition, this research seeks to give greater visibility to the use of NOCMAT in construction, highlighting mainly the environmental, social and/or economic gains.

Theoretical review
Cementitious materials with lower environmental impact
The application of nature fibers as matrix reinforcement based of cementitious material is awakening a large interest in developing countries, like Brazil, due to disponibility, savings in production/extraction and the environmental benefits (PLESSIS, 2001, *apud* SAFASTANO JUNIOR, 2002). Many materials are used: bamboo, açaizeiro seed fiber, field bush sisal fiber, among others. Fibers originating from residues such as tires, PET bottle and sawmill residues are also used. The researches point out an improvement in the mechanical properties of fiber reinforced composites.

The roof is an indispensable item in construction. The tiles, the main element of the roof, are found in various models and made with many kinds of materials, for example, ceramic, aluminum, zinc-coated steel and asbestos cement. Bamboo's fibers, with the right chemical treatment are adequate for the production of corrugated tiles made with cement and sand mortar (BERALDO *et.al.*, 2003).

FIGURE 1 - TILE MADE WITH BAMBOO'S FIBER
RESEARCH: SAVASTANO JÚNIOR, 2000.

Non-Conventional Materials and Technologies – NOCMAT for XXI Century   Materials Research Forum LLC
Materials Research Proceedings 7 (2018) 750-755   doi: http://dx.doi.org/10.21741/9781945291838-73

Earth constructions

The raw earth is a sustainable material that was used in Brazil since the colonization period. There are two basic ways to build using earth: the taipa, a method in which the earth is beaten using casts to create walls and adobe, the method in which the earth is pressed in blocks and dried before being used. Starting with these two main techniques, the other variations of constructive techniques that base on earth appeared (BERGE, 2009).

According to Minke (2005), taipa has been widely used for centuries as a traditional means of building walls. In the techniques of manufacturing taipa stack, land is dampened in layers up to 15 cm between so-called formwork and then compacted by its own method. The formwork generally consists of two walls parallel and connected by spacers.

When done in the correct way, with the clay well-chosen and compacted, a house built in the technique of taipa stack, will be solid and comfortable from thermal and acoustic point of view. From the aesthetic point of view, these houses are quite pleasant, and even the use of plaster can be dispensed, because the walls are totally smooth, so they allow the application of the paint directly on the walls, after they are ready (SILVA, 2000 *apud*, XAXA, 2013).

FIGURE 2 - MODERN RESIDENCE BUILT IN TAIPA.

RESEARCH: EARTH ARCHITECTURE.

Polyethyleneterephthalate (PET) Bottles

The PET bottles area material that has many uses for the construction of a building. It is used for the manufacture of tiles, green roof, lamps and walls. According to a research carried out by Santos (2015) the project house's foundation and walls are built with PET bottles lying overlapping. They are filled with washed sand or soil-cement, and are joined with a material that is similar to the one used to lay bricks.

Lohamann (2006) developed a constructive system composed of prefabricated plates of bottles of mortar and PET, for the construction of housing of social interest. In this system the plates are prefabricated in wooden cast on which a layer of mortar cement and sand with additive is placed. After the mortar, reinforcing reinforcements are placed on all sides of the shape and the PET bottles are positioned longitudinally. The bottoms of the bottles must be cut, allowing them to be fitted together, and forming columns. Finally the plate is coated with mortar. The electrical and hydraulic installations are placed inside the bottles before the mortar. The fastenings between the panels are made with sheet metal. The plates, in comparison with conventional masonry, have lower weight. Besides, the manufacture and assembly become simpler and faster.

Bamboo constructions

The bamboo is ecologically sustainable. It's found in large in nature and has various possibilities of applicability and simple planting techniques. It is a lightweight and flexible material and still stands out as the largest consumer of carbon dioxide in the plant kingdom, thus being seen as a plant of one of the planet's highest degrees of sustainability (CARBONARI, 2017).

Non-Conventional Materials and Technologies – NOCMAT for XXI Century   Materials Research Forum LLC
Materials Research Proceedings 7 (2018) 750-755   doi: http://dx.doi.org/10.21741/9781945291838-73

It's possible use the bamboo in composition of construction's elements, as a way of generating income and diversification of the economy, as well as applying it in self-construction of dwellings, under a joint effort regime (BARROS, 2004).

According to Ghavami and Marinho (2002) *apud* Barros and Souza (2014) when testing the tensile strength of bamboo, high values are obtained in the longitudinal sense of the fibers, when compared to wood. In this way, it can be concluded that it's possible to replace the wood by the bamboo in pieces submitted to the traction and to obtain a greater capacity of load.

Carbonari (2017) explains that, although the steel's traction resistance is bigger, the relation between traction resistance and the specifics mass results in much higher values than the steel's, certain 4 to 5 times greater, showing that in some cases the steel can be replaced for bamboo.

Barros (2014) points out that, in the case of parts subject to compression, a report analysis, only without limit of resistance of the material is not enough, it is necessary to take into account a slenderness of the structural element, verifying a probability of even coming to fail by buckling. Thus, although there have been results found in relation to tensile strength, bamboo can be used as structural element replacing wood and concrete, and may provide better results, in some cases.

FIGURE 3 - PANEL WITH THE HYDRAULIC AND ELECTRICAL INSTALLATIONS

RESEARCH: LOHMANN, 2006.

FIGURE 4 – BAMBOO BUILDING

FONT: CASA CLAUDIA

Methodology

The methodology used in this research is the combination of two procedures: the bibliographic and field research. In the first step, for each thematic proposal, a theoretical review was done that described the type of NOCMAT including characteristics of mechanical resistance for the intended

Non-Conventional Materials and Technologies – NOCMAT for XXI Century   Materials Research Forum LLC
Materials Research Proceedings 7 (2018) 750-755   doi: http://dx.doi.org/10.21741/9781945291838-73

function, the environmental, social and/or economic gain and involved a comparison of costs in relation to the material with the same function, traditionally used in civil construction.

In the second step, the main builders of the city were identified and conducted interviews, to check the use of NOCMAT and what are the difficulties for its adoption. In addition achievement of the interviews looking for to identify the knowledge of students and teachers about the subject covered here.

The third step will consist of cataloguing, organizing and searching for illustration in digital booklet format. In the fourth step will be held the distribution the primer in the academic community, construction of the city and an online survey to identify points that can be optimized for best result. In the last step, final review of the primer and your distribution.

Results and discussion

A script was drafted a for the semi-structured interviews after the completion of the theoretical review. The script has six questions that were designed considering the recommendations of Marconi and Lakatos (2009) to conducting interviews in qualitative research.

The first question seeks to obtain the opinion of the interviewees about civil engineering as agent that moves the economy, but degrades the environment. Most of the interviewees agreed whit the statement, for various reasons. However, it was highlighted that engineering can help move the economy as well as help the environment, it is enough to unite it to what we call eco development, and professionals still show resistance. Only one of the people interviewed considers a very wrong idea, affirming that, actually, the construction industry seeks to move the economy without harming the environment or attacking as little as possible.

The second question sought to verify if sustainability was present in the professional formation/performance. Most of the professionals and Students interviewed pointed out that sustainability was present in professional or academic life. However, Part of the students of the civil engineering course of CEFET-MG, unit Curvelo, pointed that, although the course has an emphasis on sustainability, the subject isn't adequately approached. One of the students interviewed stressed that the subject is studied largely in theory and little in practice.

One of the engineers pointed out the difficulty in convincing customers to use sustainable materials, because their cost is, in most cases, high. He highlighted the need for research that seeks the development of lower cost products.

The next three questions seek to gauge the knowledge about NOCMAT in a general way and the opinion of the interviewees about them. 50% of the respondents reported knowing and having used NOCMAT. The interviewees believed in this type of material and stressed the need for more research to increase their reliability. Engineers reported that they have already used eucalyptus in one of his works and that the PET bottle and bamboo are good materials, but rely on specialized labor.

Most respondents believe there is a tendency for NOCMAT to grow, because society and government nowadays show greater concern with environmental preservation. Others understand that their large-scale use will be very difficult, due to the resistance of construction professionals

Finally, a question about the possibility to join sustainability, technology and economy was asked. 100% of the interviewed answered positively, however, they said this will take time and very much study about the subject.

Conclusion

Theoretical revision enabled to recognize that the NOCMAT have been used for a long time in construction. They are an environmentally sustainable and low-cost choice.

Through the interviews done until now it was possible to perceive that, although interviewees report knowing these materials, they have an unsatisfactory knowledge, because they believe that they are high-cost materials or of inferior quality to those traditionally used in civil construction, and that need highly skilled labor, contrary to what was verified in the theoretical review.

It is possible to affirm the relevance of this research for the civil engineering course of CEFET-MG, Curvelo unit, and for the municipality because the interviews indicate interest in the theme and suggest that studies should be developed to enable wide use of NOCMAT in buildings of the municipality. The resistance of engineers to using these materials can be minimized by preparation of dissemination materials that are easy to read, prepared according to reliable references, developed in several universities in the country and the world.

The digital booklet will be a quick-access way to divulge the results of this research and can become an incentive material for a larger scale use of NOCMAT in the municipality of Curvelo.

References

[1] BERGE, B. The Ecology of Building Materials, 2° Edition, Architectural Presse, ISBN 978-1-85617-537-1, Elsevier Science, 2009. https://doi.org/10.4324/9780080949741

[2] BRASIL. ASSOCIAÇÃO BRASILEIRA DE MATERIAIS NÃO CONVENCIONAIS. International Conference on Non-conventional Materials and Technologies. 2013. Disponível em: <http://www.abmtenc.civ.puc-rio.br/eventos/>. Acesso em: 28 maio 2017

[3] B, T. M. Materiais não convencionais na construção civil: presente, passado e futuro no processo de conhecimento dos assentados de Mogi Mirim-SP. p.181, 2011. Dissertação de Mestrado - Universidade Estadual de Campinas, Faculdade de Engenharia Agrícola, São Paulo. 2011.

[4] CARBONARI, Gilberto et al. BAMBU – O AÇO VEGETAL. Mix Sustentável, [s.l.], v. 3, n. 25, p.17-25, jan. 2017.

[5] GHAVAMI, Khosrow. Materiais e Tecnologias não Convencionais para o Século XXI. Rio de Janeiro: Puc Rio, 2014.

[6] PROVENZANO, Thaís Lohmann. Desenvolvimento de sistema construtivo em painéis pré-fabricados de argamassa e garrafas plásticas para habitação de interesse social. 2016. 170 f. Dissertação (Mestrado) - Curso de Arquitetura, Ufsc, Florianópolis, 2006.

[7] SANTOS, Igor Almeida de et al. Reciclagem de garrafas pet para fabricação de telhas. Cadernos de Graduação: Ciências Exatas e Tecnológicas, Sergipe, v. 1, n. 17, p.83-90, out. 2013.

[8] SAVASTANO JUNIOR, Holmer; PIMENTEL, Lia Lorena. Viabilidade do aproveitamento de resíduos de fibras vegetais para fins de obtenção de material de construção. Revista Brasileira de Engenharia Agrícola e Ambiental, [s.l.], v. 4, n. 1, p.103-110, abr. 2000. FapUNIFESP (SciELO). http://dx.doi.org/10.1590/s1415-43662000000100019. Disponível em: <http://www.scielo.br/scielo.php?script=sci_arttext&pid=S1415-43662000000100019>. Acesso em: 06 jul. 2017

[9] SPADOTTO, Aryaneet al. Impactos ambientais causados pela construção civil. Joaçaba: Unoesc, 2011. Disponível em: <https://editora.unoesc.edu.br/index.php/acsa/article/viewFile/745/pdf_232>. Acesso em: 28 maio 2017

[10] I CONFERÊNCIA LATINO-AMERICANA DE CONSTRUÇÃO SUSTENTÁVEL X ENCONTRO NACIONAL DE TECNOLOGIA DO AMBIENTE CONSTRUÍDO, 1., 2004, São Paulo. Bambu: alternativa construtiva de baixo impacto ambiental. São Paulo: I Conferência Latino-americana de Construção Sustentável X Encontro Nacional de Tecnologia do Ambiente Construído, 2004. 12 p.

Non-Conventional Materials and Technologies – NOCMAT for XXI Century   Materials Research Forum LLC
Materials Research Proceedings 7 (2018) 756-763   doi: http://dx.doi.org/10.21741/9781945291838-74

Chemical Characterization for the Comparative Study of Peruvian Natural Fibers

[1]Carlos Tenazoa, [2]Samuel Charca*, [3]María Quintana, [3]Elena Flores

[1]Department of Chemical Engineering;

[2]Department of Mechanical Enginnering; samuel.charca@upr.edu

[3]Department of Chemical Enginnering; cfloresb@utec.edu.pe, mquintana@utec.edu.pe

Universidad de Ingenieria y Tecnologia – UTEC, Perú

Abstract. In recent years, natural fibers have acquired a fundamental role in the industry because, besides being available in great abundance, they are bio-renewable. These are mainly composed of lignin, cellulose and hemicellulose, which are natural polymers that shape the cellular wall of plants. In Peru, there is a great variety of natural resources with great potential where fibers can be extracted from or used like such, one of them is the "Ichu", a type of straw that grows abundantly in highland areas, it was used as food for camelids for many years and now, it is simply burned to recover areas. For industrial applications, especially for composite materials, it is important to know the mechanical resistance of these fibers, which is directly related to the cellulose content, because of this it is necessary to characterize them chemically. Therefore, an efficient methodology, based on technical standards TAPPI, has been applied and validated to determine the percentage content of lignin, cellulose and hemicellulose, as well as moisture, ash and extractives of the fibers in order to determine how their performance would be for different applications. For that purpose, the species Jarava ichu from Cusco (3400 masl) and Huancavelica (4100 masl) have been analyzed and compared. The results show that geographic and climatic variations influence the chemical composition, hence higher altitude means the plant had a higher percentage content of lignin and lower content of cellulose.

Keywords: Natural Fiber, Jarava Ichu, TAPPI, Chemical Characterization, Cellulose, Hemicellulose, Lignin, Moisture, Ash, Extractives

Introduction

Nowadays, due to the increase of the environmental pollution, there have arisen innovative ideas that are based on the substitution of synthetic materials for natural materials in order to reduce pollution, such as, the replacement of the synthetic fibers like those of glass and carbon for natural fibers extracted from plants. The advantages of the natural fibers on the synthetic ones are their low cost, low density and biodegradability, associated with high specific properties, which turns them into an eco-friendly material with great potential [1].

In Peru, there is a great variety of natural raw materials with excellent potentialities, which are not currently used. One of them is the Ichu, a type of straw that grows abundantly in the high Andean zones and that for many years has been used as construction material; besides being used as animal food (llamas, alpacas, vicuñas and others). With the development of the technology of construction and the modernization in the upbringing of animals, the Ichu entered a process of disuse, causing the accumulation of this fiber (highly flammable) in the field, increasing the incidence of forest fires [2] [3].

The knowledge of the chemical composition of natural fibers is fundamental to predict their potential use in the industry for which they are to be applied. This consists in determining the content of extractives, which may be formed by salts, sugars, polysaccharides, fats, waxes or resins adhered to the plant; ash, which includes all the minerals present in it; cellulose, which provides plants their strength, rigidity and stability; hemicellulose, which is made up of several sugars and acts as an interface between lignin and cellulose. Finally, lignin, which is an aromatic compound that provides impermeability and protects the plant from any microbial attack [4]. In addition, it is known that geographic and climatic variations influence the chemical composition of these, which for example may be soil type and altitude [5]. For that, the goal of this research is to determine the chemical composition of ichu from two localities, Cusco and Huancavelica, which are at different altitudes and have slightly different environments. This was done in order to see how much influence these geographic and climatic factors in the chemical composition and according to this determine where is more convenient to extract the plant to later be used as natural fiber.

Materials and methodology
Materials
Ichu in mature stage extracted from the root, from Canchis, Cusco (located at 3400 masl) and Huaytará, Huancavelica (located at 4100 masl).

Methodology
Taxonomic characterization
The characterization of two ichu samples from two localities, Cusco and Huancavelica, was carried out in the laboratory of applied botany of the Cayetano Heredia University.

Sample preparation
The plant was crushed up to obtaining a length of about 5 mm, this was done in order to have a more homogeneous sample and to facilitate the contact of this one with the chemical reagents to be used for the determinations.

Chemical composition
The following characteristics of the plant were determined according to the standardized methods in the TAPPI standards (*Technical association of the pulp and paper industry*): Percent content of moisture, ash, extractives, cellulose, hemicellulose and lignin [4].

Moisture
This methodology, based on the standard TAPPI. T 421 om-02 [6], consisted in the use of an electric oven with air circulation for the drying of the vegetal sample.

Ash
Procedure based on the standard TAPPI. T 211 om-02 [7], which consisted in the calcination of the sample in a muffle at 600 °C with a ramp of 9.6 °C / min.

Extractives
The determination was made according to the standard TAPPI. T 204 cm-97 [8] by Soxhlet extraction using ethanol as an organic solvent.

Holocellulose
Holocellulose is the combination of cellulose and hemicellulose. Its determination is based on the technique used by Herbst called the preparation of sodium chlorite [9]. This technique consisted in the oxidation of the lignin of the vegetable sample in an acid medium to obtain only holocellulose.

Non-Conventional Materials and Technologies – NOCMAT for XXI Century   Materials Research Forum LLC
Materials Research Proceedings 7 (2018) 756-763   doi: http://dx.doi.org/10.21741/9781945291838-74

Cellulose
Methodology based on the standard TAPPI, T 203 cm-99 [10], which consisted in determining the fraction of holocellulose that does not dissolve in a solution of 17.5% (m/v) sodium hydroxide. The undissolved fraction represents the cellulose and the dissolved fraction is the hemicellulose.

Lignin
Technique known as the determination of Klason lignin, based on the standard TAPPI. T 222 om-02 [11]. It consisted in using a diluted solution of sulfuric acid to dissolve all the carbohydrates of the sample, remaining only the insoluble lignin that was then filtered. However, a small portion of the lignin is solubilized in the acid solution, so an aliquot of the acidic filtrate was collected and then analyzed in a UV-visible spectrophotometer to determine the percentage of soluble lignin (Goldschimid method) [4].

Analysis by FTIR spectroscopy
FTIR spectroscopy was used to verify the presence of cellulose, hemicellulose and lignin in the sample by analyzing the peaks of the bonds present in the functional groups characteristic of each of the three compounds mentioned above [15].

Methodology validation
The validation was carried out with 100% pure sample of cellulose and lignin with a purity of 96%, both of the brand SIGMA-ALDRICH. This was done in order to verify that the methodologies followed for each case were the ideal ones.

Results and discussion
Taxonomic characterization
The results of the taxonomic study are shown in Table 1, where it can be seen that the two samples of ichu analyzed are of the same species and family, but the place and altitude from which they were extracted are different, so it will be interesting to determine how these factors influence the chemical composition.

TABLE 1 - RESULTS OF THE TAXONOMIC STUDY OF ICHU

Common name	Scientific name	Family	Location
Kiska ichu	*Jarava ichu* Ruíz and Pavón	POACEAE	Huancavelica-Huaytará-4100 masl
Ichu	*Jarava ichu* Ruíz and Pavón	POACEAE	Cusco-Canchis-Tinta- 3400 masl

Comparative study
This study was performed for the leaves of both ichu samples, since they represent the most part of a bush of ichu (Fig. 1), as shown in the results of quantification of a bush (Table 2). Therefore, if this raw material were chosen in the industry, the leaves would have to use in order to make the process economically feasible.

Non-Conventional Materials and Technologies – NOCMAT for XXI Century   Materials Research Forum LLC
Materials Research Proceedings 7 (2018) 756-763   doi: http://dx.doi.org/10.21741/9781945291838-74

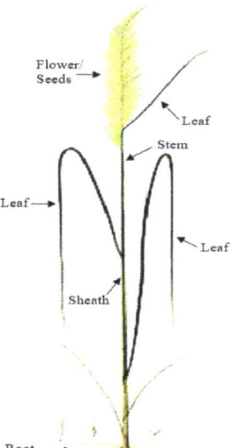

FIGURE 1 - JARAVA ICHU, DRAWING MODIFIED FROM RUIZ, H., PAVÓN, J.,
DRAWINGS OF THE ROYAL BOTANICAL EXPEDITION TO THE VICEROYALTY OF PERU

TABLE 2 - QUANTIFICATION OF A BUSH OF JARAVA ICHU

Part	Jarava ichu
Leaves (g)	138.37
Sheat (g)	100.16
Stem (g)	65.78
Useless (g) (very dry parts or garbage)	13.85

The comparative study of the leaves of both ichu samples was carried out based on the results of the chemical characterization for each one. The results on dry basis are presented in Table 3.

Likewise, the comparison is shown in Figure 2, which shows the following: The ichu of Cusco has slightly lower moisture. This may be due to the fact that the climate in this area is very variable, since it can change from a radiant sun to a torrential rain, in contrast the climate of Huancavelica is very cold and humid almost all year. With respect to the ash, there is not much difference, since this depends on the minerals that has the soil from which the plant was extracted. Also, it can be observed that the ichu of Cusco has slightly more percentage of extractives, this depends on the environment to which the plants are exposed, since these could capture many substances that are present in the environment, which could be very clean or perhaps slightly contaminated [12]. Then it could be concluded that Huancavelica has a less polluted environment than Cusco and that influences the amount of extractives. If the percentage of hemicellulose in both cases is observed, these do not differ much since the amount of sugars present in this type of plant depends on the process of photosynthesis that they perform and in both cases, this process is similar.

Non-Conventional Materials and Technologies – NOCMAT for XXI Century   Materials Research Forum LLC
Materials Research Proceedings 7 (2018) 756-763   doi: http://dx.doi.org/10.21741/9781945291838-74

TABLE 3. *RESULTS OF THE CHEMICAL COMPOSITION OF THE LEAVES OF ICHU OF CUSCO AND HUANCAVELICA ON DRY BASIS*

Determination	Huancavelica (%)	Cusco (%)
Cellulose	25.40 ± 0.77	42.81 ± 0.51
Hemicellulose	23.89 ± 1.92	28.71 ± 1.38
Lignin	37.61 ± 1.15	12.99 ± 0.26
Extractives	13.95 ± 0.36	16.48 ± 0.14
Total	**100.85 ± 4.2**	**100.99 ± 2.29**
Moisture	9.43 ± 0.32	8.60 ± 0.58
Ash	4.34 ± 0.25	4.83 ± 0.18

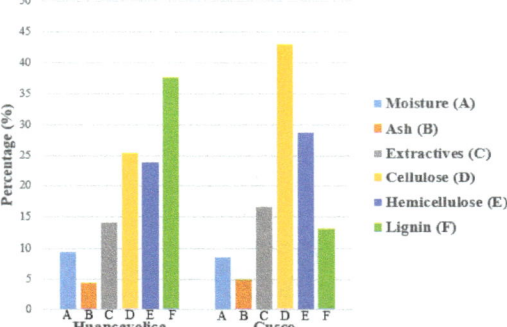

FIGURE 2. *COMPARISON BETWEEN THE CHEMICAL COMPOSITIONS OBTAINED FROM LEAVES OF ICHU OF HUANCAVELICA AND CUSCO*

On the other hand, it can be seen that the ichu of Cusco has almost twice the percentage of cellulose compared to the ichu of Huancavelica and, besides, its percentage of lignin is considerably lower. Therefore, it could be deduced that at a higher altitude the percentage of cellulose is lower and that of lignin is higher, this could be because in these conditions of high altitude, the climatic variations are extreme and therefore the plant will need greater protection against any attack related to this type of climate. Since lignin is responsible for providing this protection, then its content will be greater compared to the percentage amount of cellulose in the plant. Besides, Vaziri in [13] states that the increase in the percentage of lignin at higher altitudes is due to excessive cell growth and cell proliferation, which naturally increases lignin macromolecules with shorter chains that simultaneously increase the molecular weight of the lignin. This matches with what Kiaei states [14], where the percentage content of lignin in Persian ironwood obtained was greater when increasing the altitude from 100 to 700 masl and the percentage content of cellulose was smaller to greater altitude.

Analysis by FTIR spectroscopy
In Figure 3, the spectra obtained from the leaves of the ichu of Huancavelica without treatment (initial), of holocellulose and cellulose obtained from its leaves are compared. In all cases, a band is observed between 3200 to 3320 cm^{-1} corresponding to the O-H (hydroxyl) bonds present in the 3 compounds [15].

Non-Conventional Materials and Technologies – NOCMAT for XXI Century   Materials Research Forum LLC
Materials Research Proceedings 7 (2018) 756-763   doi: http://dx.doi.org/10.21741/9781945291838-74

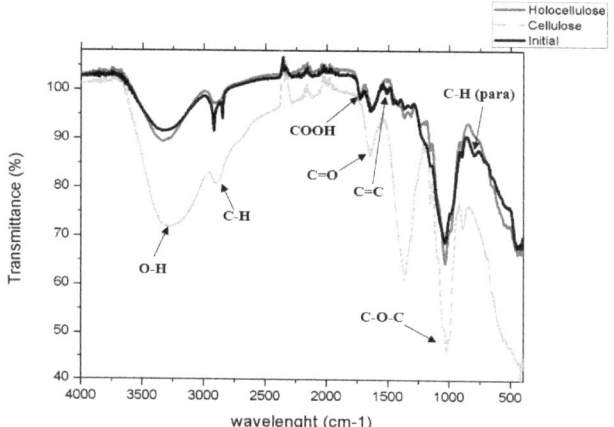

FIGURE 3. *COMPARISON BETWEEN THE SPECTRA OBTAINED FROM THE ICHU LEAF FROM HUANCAVELICA: WITHOUT TREATMENT, HOLOCELLULOSE AND CELLULOSE SAMPLE*

On the other hand, it is observed that the two peaks between 2800 and 2950 cm^{-1} in the sample of the untreated leaf corresponding to the C-H bond present in the 3 compounds has become a single peak in the case of holocellulose and cellulose. This can be due to the fact that some of the C-H bonds present in the 3 compounds disappeared, presuming that some of these 3 compounds was removed, which could be deduced that it was lignin. In addition, the high intensity peak at 1035 cm-1 corresponding to the C-O-C bond (ether) is repeated for the three spectra but for the case of cellulose, the intensity is higher, so it could be deduced that the percentage of cellulose has increased for that case. On the other hand, in the case of the initial sample and holocellulose, the two peaks located between 1650 and 1800 cm^{-1} are maintained, which correspond to the C=O bond present in the hemicellulose and cellulose and the COOH bond present only in the hemicellulose (at 1750 cm^{-1}) [15]. However, for the case of cellulose the peak of the COOH bond has disappeared. Therefore, it could be concluded that the hemicellulose has been removed. With respect to the presence of lignin, the peaks of the C=C bonds of aromatic rings located at 1550 cm^{-1} and the C-H (para) bonds between 800 and 850 cm^{-1} are not distinguished for holocellulose and cellulose, so it is deduced that there is no lignin present for these cases. It is noteworthy that similar spectra were obtained for the ichu of Cusco.

Methodology validation
Finally, the results obtained with samples standard of cellulose and lignin are shown in Table 4, from which it can be observed that with the methodologies based on TAPPI standards, it was possible to recover almost the total of standard products. Therefore, these methodologies are reliable and are already validated.

Non-Conventional Materials and Technologies – NOCMAT for XXI Century   Materials Research Forum LLC
Materials Research Proceedings 7 (2018) 756-763   doi: http://dx.doi.org/10.21741/9781945291838-74

TABLE 4. *VALIDATION OF METHODS BASED ON RECOVERY OF CELLULOSE AND LIGNIN STANDARD*

Product	Brand	Recovery (%)
α-cellulose powder	SIGMA- ALDRICH	99.4 ± 2.2
Lignin, alkali	SIGMA- ALDRICH	94.82 ± 1.8

Conclusions

It was obtained that, at a greater altitude, the percentage content of lignin is greater than that of the cellulose since the plant needs greater protection because the climatic variations are extreme to these conditions of great altitude. Therefore, if the ichu is destined to be used as natural fiber, this should be extracted from a place that does not have much altitude, because of it Cusco would be the best option. On the other hand, although a bush of ichu also contains sheath and stem, the leaves are found in greater proportion, then if it would consider to use ichu in the industry of natural fibers, the leaves would have to be processed, otherwise, the process would not be economically feasible. Finally, it is concluded that the TAPPI standards are optimal for the chemical characterization of any plant or fiber since it was demonstrated that in the validation almost all the standard products of cellulose and lignin were recovered.

References

[1] M. Bajardo et al., "Surface Chemical Modification of Natural Cellulose Fibers," Journal of Applied Polymer Science, Dep. of Chemistry, Univ. of Bologna, Bologna, Italy, 2002.

[2] F. Alegría. "Inventario y uso sostenible de pastizales en la zona colindante a los depositos de relavera de ocroyoc-comunidad san antonio de rancas – pasco," M.S. thesis, Dept. Ambiental, Pontificia Universidad Católica del Perú, Lima, Perú, 2013.

[3] B. Cardelús and T. Guijarro. "Cápac Ñan, el gran camino inca,". Lima, Perú: Santillana S. A., 2012, pp. 98-100.

[4] J. P Saraiva, "Procedimentos para Análise Lignocelulósica". 1st ed. Brasil: Campina grande: Embrapa Algodão, 2010, pp. 11-42.

[5] A. Komuraiah et al. "Chemical composition of natural fibers and its influence on their mechanical properties", in Mechanics of Composite Materials, Vol. 50, No. 3, New York, July, 2014.

[6] TAPPI. T 421 om-02. Moisture in pulp, paper and paperboard. 2002a, 3 p.

[7] TAPPI. T 413 om-93. Ash in wood, pulp, paper and paperboard: combustion at 900°C. 1993, 4 p.

[8] TAPPI. T 204 cm-97. Solvent extractives of wood and pulp. 1997, 4 p.

[9] J. Herbst. The preparation of chlorite holocellulose. Canadian Journal of Chemistry, v. 30, n. 9, p. 668-678, 1952. https://doi.org/10.1139/v52-079

[10] TAPPI. T 203 cm-99. Alpha-, beta- and gamma-cellulose in pulp. 2009, 7 p.

[11] TAPPI. T 222 om-02. Acid-insoluble lignin in wood and pulp. 2002c, 5 p.

Non-Conventional Materials and Technologies – NOCMAT for XXI Century   Materials Research Forum LLC
Materials Research Proceedings 7 (2018) 756-763   doi: http://dx.doi.org/10.21741/9781945291838-74

[12] B. Kord et al. "Comparison of lead content absorption in different parts of Eldar pine (Pinus eldarica Medw.) in Tehran city". Iranian Journal of Forest and Poplar Research, 2015, 18 (2): 265-277.

[13] V. Vaziri *et al.* "Effect of Altitude of Fiber Characteristics, Chemical Composition and Kraft Yield Pulp of Brutian Pine (Pinus brutia)". Journal of Wood and Forest Science and Technology, 2009, 6(1): 1-14.

[14] M. Kiaei *et al.* "Mineral content in relation to radial position, altitude, chemical properties and density of persian ironwood". In Woods, Science and Technology, University of Bio bio, 2015.

[15] M. Bin, E. Jayamani. "Comparative Study of Functional Groups in Natural Fibers: Fourier Transform Infrared Analysis (FTIR)". International conference on Futuristic Trends in Engineering, Science, Humanities and Technology. Gwalior, Indi, 2016.

Non-Conventional Materials and Technologies – NOCMAT for XXI Century   Materials Research Forum LLC
Materials Research Proceedings 7 (2018) 764-774   doi: http://dx.doi.org/10.21741/9781945291838-75

Ichu: New Natural Fibers for Composites and its Extraction Methodology

Sandra Mori[1], Elena Flores[2], Samuel Charca[3,*]

[1]Department of Industrial Engineering; sandra.mori.19@gmil.com

[2]Department of Chemical Enginnering; cfloresb@utec.edu.pe

[3]Department of Mechanical Enginnering; samuel.charca@upr.edu

Universidad de Ingenieria y Tecnologia – UTEC, Perú

Abstract. Natural fibers obtained from plants (vegetables) have been gaining great impulse in the last years, especially in the composites materials industry; besides, studies shows that the annual growth rate will be 10% in the coming years, which will trigger a gap with respect to the current offer. Due to the new legislations, that encourages the use of biodegradable materials, the preference of more users for ecological products and the tendency towards new business economic models focused on zero waste. From these, a great opportunity arises for research and development method and process of extraction of fibers from new natural resources, such as Ichu (grass). This Andean endemic grass is plentiful and its inside contains technical fibers, with potential to be extracted and used as reinforcement in composites materials. Therefore, in this research a method of extraction was designed to obtain technical fiber from Ichu (*Festuca distichovaginata*). Those technical fibers were used as reinforcement of polyester resin and its flexural mechanical properties were measured. In order to make comparable the properties, material index (MI) were used for strength, stiffness and density; results were compared to the literature Yute fiber composites. Finally, the process parameters were established in such a way to minimize the reference cost of production until ~1.34 USD/kg (production rate of 600 kg of fibers per day).

Keywords: Natural Fibers, Composite Material Industry, Ichu, Technical Fibers, Material Index

Introduction
In the recent years, the use of natural fibers (NF) as reinforcement of polymer has grown at a rate of 10-20% per year, as result of the pressure of new legislations that encourage the use of materials and processes more friendly with the environment, especially in the automotive sector [1]-[5]. In addition, 66% of people are willing to pay more, for products that generate a positive social and environmental impact [6]. This promotes the development of NF from new natural sources to take advantage of its characteristics such as low density, good mechanical strength, low cost and its processing is much simpler compared to synthetic fibers [7]-[12].

An unexplored natural resource, from which can extract NF is Ichu. This natural grass abounds in the Andean region and only in Peru is estimated an availability of 76 thousand tons per year, approximately [13]. Another advantage of this grass is the diameter of its leaves, which are small in comparison to other natural resources (bamboo, jute, sisal and others), this small diameter gives an easy processing during the extraction of NF, which can reduces the production cost [14]-[15]. As this grass is endemic, it is not necessary to implement irrigation system or fertilizers, as it grows naturally and invasively in inhospitable territories for other crops, which does not affect the competition of land for food crops [16]. Therefore, the production of Ichu fibers could not only be less costly and also more environmentally friendly, since they do not carry the negative environmental aspects that are criticized in the life cycle assessment of natural fibers from sisal, jute, hemp, flax, others [1], [17].

Natural fibers can be extracted from raw materials using different methods; those methods can be classified in biological (water retting, enzymes), physical (steam explosion, plasma, ozone), mechanical (crushing, grinding or rolling mill) and chemical (alkaline and acids). Depending on the method, some of them are effective to extract NF and others to improve the compatibility of NF to the matrix or resin. Treatment like retting, steam explosion, alkali and acids are effective to reduce the content of non-cellulosic components (hemicellulose, lignin and others), which are presented in the raw materials of processed fibers [18]-[21]. Treatment efficiency depends on the characteristics of the raw material and its chemical composition (species, place of harvest, part of the plant, among others) [22]. The simplest method to assess the performance and quality of the fibers for composites application is the flexural test, with this test properties like flexural strength and stiffness can be determined [23]. One way to ensure the functionality of the new material (for certain application) is combining the main properties, these is called Material Index (MI), in this case properties like strength and stiffness are related with the density for a simple flexure application [24].

Alkali treatment to process raw material to obtain NF is a complex task; due to, it is necessary evaluated different parameters, which can affect the final quality of the fibers. In the experimental area, there are different tools, which can be used to screen and quantify the effect of the different parameters. One of the experimentation methodologies to find the optimal parameters of the processes is the Design of Experiments (DOE) [25]-[28]. This technique provides tools to plan, execute the tests and analyze statistically the data obtained to show objective evidence in the design of processes and/or products [29].

In view of the opportunity to develop new natural fibers for composites materials application, the principal aim of this research is to develop (implement) a proper methodology to extract natural fibers from raw Ichu plants, using a robust DOE methodology, considering the technical pre-feasibility in addition to the analysis of the reference cost of production.

Materials and methodology:

Materials

Raw materials (Ichu plants, Festuca distichovaginata Pilg, Gramineae family) were collected in the province of Canas at 3800 meters over sea level, Cusco (Peru). Alkali treatment was selected to perform the extraction processes, with Sodium Hydroxide (NaOH flakes, 98% purity) diluted in distilled water. Unidirectional laminated were fabricated using unsaturated polyester resin as a binding material.

Experimental Design

The methodology and the experimental procedure were developed in four consecutive stages, based on the DOE methodology using MINITAB 17 (Figure 1). The first stage: initial proposal of the process; it is to define the extraction process that fit to the characteristics of the raw material (RM) previously defined. In the second stage: screening, in this part the proposed method (process) is evaluated in order to identify significant variables that reduce non-cellulosic components (Severity Level or SL); for this purpose, fractional factorial design $2^{(5-1)}$ was used with the following variables to be studied: time of water retted or WR (2 and 192 hours), mechanical defibrillation time or DM (8 and 30 seconds), alkali treatment (AT) time (0.5 and 1 hour), AT temperature (30 and 60 °C) and AT concentration (0,1 and 1 molar NaOH). Third stage: technical optimization, in this part performance metric of the process was established (minimum Severity Level, SL) which allows to reach the technical specifications for its industrial application; for this part completely random design was used in the following SL: 43.52, 49.68, 53.29, 56.89 and 59.45%. Finally, in the fourth stage: cost optimization, it determined the process parameters that comply with the performance metric (SL) and at the same time minimize the reference costs of production, the concentration (0.5 and 1 molar) and temperature (40 and 60 °C) were analyzed through a central composite design.

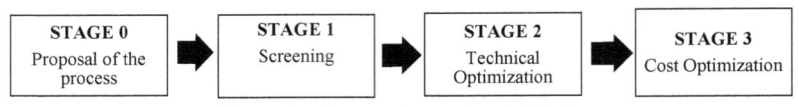

Figure 1 - Phases of experimental design

Chemical Composition of Raw Materials
In order to define the characteristic of the raw materials chemical composition were characterized according to the TAPPI standards T203 cm-99 (alpha-cellulose), T222 om-02 (lignin), T204 cm 97 (extractives) and T421 om-02 (content of moisture).

Severity Level
It is well know that fibers with less content of non-cellulosic components possess better mechanical properties. The SL (also, performance metric) was defined as the relative mass loss (%) in the extraction process and it is determined by gravimetry, according to Eqn. 1 (where h: moisture and M_f: final mass) [19]-[22].

$$LS = \frac{8(1-h) - M_f}{8(1-h)} \times 100\% \qquad (1)$$

Material Index
The materials indexes (MIs) of stiffness and flexural strength are a metric that guarantees the proper functioning of the material; and they were determined through $E^{1/2}/\rho$ and $S^{2/3}/\rho$, respectively (where E is the modulus of elasticity, S is the flexural strength and ρ is the density of the laminate) [24]. S and E were determined by the three-point bending test, based on ISO 14125 standard [30]. Unidirectional laminates were manufactured using the Vacuum Assisted Resin Transfer Molding or VARTM method. The equipment used to perform the flexural test was a universal testing machine (Servo-hydraulic MTS Landmark) with a load cell of 100 kN. All tests were performed under displacement control (1,4 mm / min). Prior to flexural test, density of each laminated mere measured using the buoyancy principle according to the ASTM D 792-08 [31].

Reference Cost of Production
The reference cost of production (CP) to produce one kilogram of Ichu fibers was determined based on an hypothetical plant located in Juliaca, Puno (industrial zone and near the raw materials), considering a production rate of 600 kg of Ichu fibers in a working day for ten years [13]. This cost includes the raw material cost, direct labor (DL) cost and indirect manufacturing costs (IMC) [32]-[33]. The price of raw materials was estimated based on a harvesting process (it does not include plantation for its natural growth) considering the DL (1.8 USD / kg), cost of transport (0.06 USD / kg) and community tax (20%). The number of operators, machinery (balance, reactor, centrifuge, combing and heating system), energy and water consumption were estimated from the line balance (supplier productivity), also maintenance cost (corrective type) is estimated 10% of the value of the corresponding equipment. For the digester (equipment), its price (9376 USD USD/m²), NaOH consumption (0.31 USD USD /kg NaOH) and energy cost was estimated as a function of alkaline treatment parameters [34]-[35]. Table 1 shows the reference costs for one kilogram of Ichu fiber.

Table 1 - *Estimated costs of production elements*

Budgeting		Value (USD/kg of IF)	REF
Raw Materials	Ichu (leaves of Festuca distichovaginata Pilg)	0.33	-
DL	Employees* (5 people)	0.17	-
IMC	Staff* (8 people)	0.35	-
	Other equipments**	0.04	[36]-[37]
	Maintenance	0.00	
	Electric Energy (other equipments)	0.04	[34]
	Water and drainage services	0.02	[38]
	Terrain and infrastructure (1200 m²)	0.06	-

*ESSALUD (9%), CTS (1 salary), Gratifications (2 salaries), Peruvian law

Results and analysis

Stage 1: Initial proposal of the process

Chemical compositions of the selected raw materials (leaves part of the Festuca distichovaginata Pilg) were determined, obtaining the following results: Cellulose 38.07%, hemicellulose 26.52%, lignin 19.92% and other extractives 14.57. The moisture content for the raw materials was 9.33%.

Based on the literature review overall treatment were proposed (Figure 2). The processes consists in submerging the raw material in water (sometimes called retting under environmental conditions), with this treatment some non-cellulosic components can degrade and separate from the cellulose by biodegradation, non external accelerators were used [39]-[40]. After that, retted raw materials were defibrillated mechanically by mechanical tearing (Oster blender), with these raw materials diameter can reduce and therefore facilitate the subsequent sub process [41]. Once the raw materials were defibrillated, processed materials were subjected to alkaline treatment, due to effectiveness and their cost efficiency to remove non-cellulosic components [42]-[43]. Finally, treated materials were rinsed with water, after that dried at 40° C, for 12 hours.

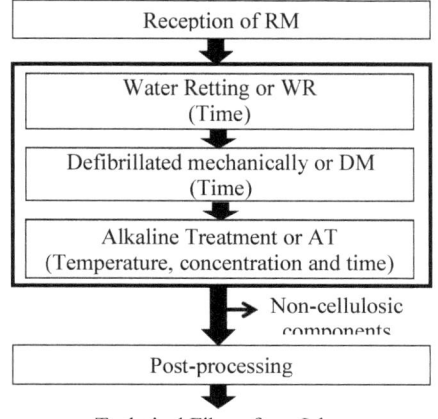

Figure 2 - *Proposed process to obtain technical fiber from Ichu*

Stage 2: Screening

Table 2 shows the results of the treatment described in stage 1. Statistical tests (ANOVA and linear regression) identify that the concentration and temperature of the AT and the time of WR were the parameters that have significantly influence in the Severity Level (p-Value < 0.05) and with a linear adjustment of 98.79% (adjusted coefficient of determination, R^2 adj). Furthermore, in the Pareto chart (Figure 3), it can be seen that the effect of the concentration contributes in great magnitude with respect to the other variables, since at high concentrations of NaOH the non-cellulosic components are solubilized. The second position in the Pareto of the effects is the temperature, due to at high temperatures the kinetics of the reaction (alkaline treatment) increases, with this allowing higher solubilization of non-cellulosic components [43]. The effect of WR time is also identified as significant for the Severity Level, however it is lower than the effect of alkaline treatment, similar results were obtained by Angelini et.al (for Ramie fibers) [44]. On the other hand, the statistical analysis also showed that mechanical defibrillation is not significant for SL, due to the fineness of the RM (reduced diameter in its natural state) and so is not necessary a mechanical process (reduce its size) unlike other natural fibers [17], [19]. Therefore, the evaluation of the extraction process selected and filtered the alkaline treatment to reduce non-cellulosic components.

Table 2 *- Statistical Analysis – Phase 2*

Input variables	DF	Adj. SS	F-Value	p-Value
A: AT concentration	1	1122.69	57.72	0.00
B: AT temperature	1	509.21	26.18	0.00
C: WR time	1	420.83	21.63	0.00
D: AT time	1	79.02	4.06	0.07
E: MD time	1	79.02	4.06	0.07

Note: R^2 = 91.91 %; R^2 adj = 87.87% y DF: Degree freedom

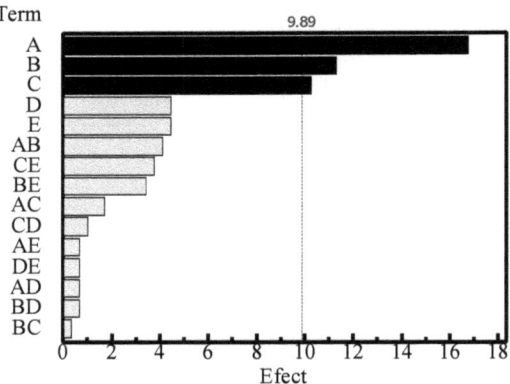

Figure 3 *- Pareto chart of the Efect – Phase 2*

Stage 3: Technical Optimization

Table 3 shows the summarized technical specifications of MIs to be reached, established according to the literature review, taking as reference composites laminated reinforced with Jute fibers, which is one of the most NF used in the industry of composite materials, [45]-[47]. Results of MI for strength

Non-Conventional Materials and Technologies – NOCMAT for XXI Century   Materials Research Forum LLC
Materials Research Proceedings 7 (2018) 764-774   doi: http://dx.doi.org/10.21741/9781945291838-75

and stiffness (longitudinal and transversal direction) for different SL (alkali treatment 1.5 molar 70°C) fiber treatments are showing in the Figure 4. From the results the MI for Longitudinal Strength (L-strength) reach the technical specifications (MI_{Yute}) for all SL; however, MI Longitudinal Stiffness (L-stiffness) only meet for SL of 53,29%. Furthermore, the MIs in transverse direction reach specifications over 53,29% of SL. Therefore, the SL at 53,29% is the target value (minimum) to process fibers that comply with the technical specifications for their industrial application.

Table 3 - *Material Index – Yute composites [Own elaboration]*

Material Index(MI)	Unit	Lower bound	Target
Longitudinal Strength ($S_L^{1/2}/\rho$)	MPa/(Mg/m3)	18.33	20.45
Transversal Strength($S_T^{1/2}/\rho$)	MPa/(Mg/m3)	6.36	6.75
Longitudinal Stiffness ($E_L^{2/3}/\rho$)	GPa/(Mg/m3)	2.25	2.42
Transversal Stiffness ($E_T^{2/3}/\rho$)	GPa/(Mg/m3)	1.21	1.28

S: Flexural Strength (L: longitudinal o T: transversal), E: Modulus of Elasticity y ρ: density; S_L=145 ±21.94 MPa, E_L=10.64 ±1.41 GPa y ρ =1.35 [46]; S_T=25.7 ±2.17 MPa, E_T=2.73 ±0.28 GPa y ρ =1.29 [47]

Figure 4 (a) and (b) show that MI_{Ichu}-L. variability is lower than that for MI_{Yute}-L and its minimum values are above the average (MI_{Yute}-L, dotted lines). This may due to the technique used to manufacture of Ichu-reinforced laminates (VRTM), which provide higher quality of laminates [22]. On the other hand, the behavior of these MI_{Ichu}-L increases, until 53,29% of SL, where they reach their maximum value and after begin to decrease, similar results were obtained by Mamunur et. al. this reduction may due to the fiber damaged during the treatment [48].

In Figure 4 (c) and (d), it is observed that MI_{Ichu}-T shows a progresive increment with respect to the SL; however, from 53.29% the MI_{Ichu}-T Stiffness presents some reduction for some SL values. It is important note that these indexes (MI_{Ichu} T) for Ichu composites using polyester resin are similar to those of Jute composites using epoxy, although this last resin (epoxy) is considered more compatible to NF [22]. Therefore, it is very likely that composites with Ichu fibers will achieve better results than jute composites (using epoxy resin). Moreover, the Ichu fibers obtained can be better than the Jute fibers, because the fiber volume fraction (V_f) of Ichu composites manufactured in this study are between 35-40%; however, the MI-T for jute composites, their volume fractions are higher V_f (52%) [49].

According to the statistical tests (ANOVA and linear regression) as shown Table 4, SL is significant for MI_{Ichu} in the transverse direction (for both, strength and stiffness). Thus may be explained due to the interface suffers direct loading when the load is applied in the transversal direction, consequently transverse loading testing measures the compatibility of the fiber/matrix interface (in other works, quality of the interface). On the other hand, during the longitudinal testing, loads are transfer from matrix to fiber by shear stress along the length of the fiber, as the fibers are large, the interface shear stresses are enough to transfer the total loads; in other works, the longitudinal direction testing can determine the quality of the fiber, as displayed by the result on the Figure 4 (a), where the MI decreases at certain value of SL.

Table 4 - *ANOVA of the MI_{Ichu} regarding the SL– Phase 2*

Response V.	SC Adj.	Valor F	Valor p	R^2 adj(%)
MI L-strength	0.08	0.11	0.746	0.00
MI T-strength	10.07	31.87	0.000	56.26
MI L-stiffness	0.68	1.17	0.068	10.64
MI T-stiffness	0.10	24.72	0.000	49.70

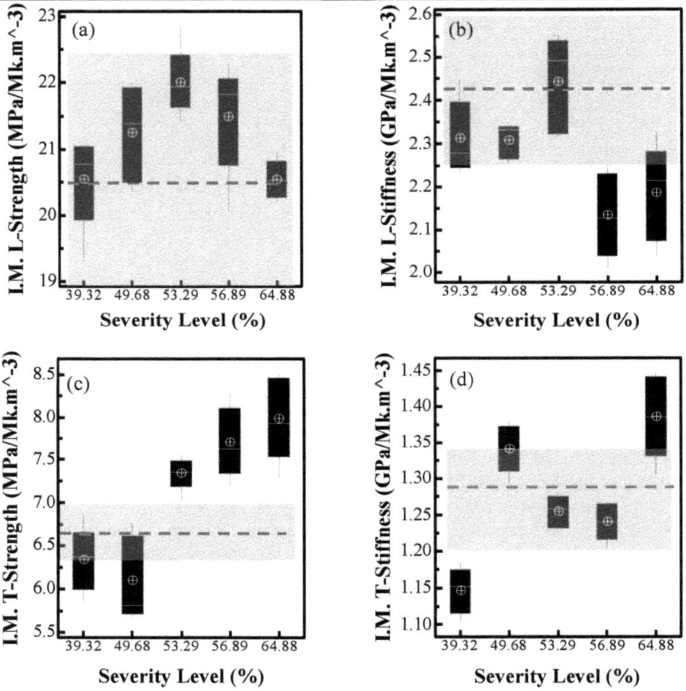

Figure 4 - *Boxplot of IM_{Ichu} (a) L-Strength, (b) L-Stiffness, (c) T-Strength and (d) T-Stiffness to different SL (plum area and red line corresponds to the IM_{jute} and its average, respectively)*

Step 4: Economic Optimization
The analysis of variance (ANOVA) identified the effect of temperature, concentration, temperature^2 and concentration × temperature, as the significant variables which reduce the reference production cost (Value p <0.05), as shown in Table 5. These variables were adjusted to nonlinear-regression model with second order terms with an excellent coefficient of determination-adjusted equal to 96.63%.

Non-Conventional Materials and Technologies – NOCMAT for XXI Century   Materials Research Forum LLC
Materials Research Proceedings 7 (2018) 764-774   doi: http://dx.doi.org/10.21741/9781945291838-75

Table 5 - *ANOVA to the reference cost of production*

Term	DF	SS Adj.	F-Value	P-Value
A: Temperature	1	1.18611	232.60	0.000
B: Concentration	1	0.19892	39.01	0.000
A*A	1	0.66521	130.45	0.000
A*B	1	0.23622	46.33	0.000

Note: R^2 = 97.13 %; R^2 adjust = 96.63% y DF= Degree Fredom

Figure 5 shows the contour plot in order to obtain SL of 53.29%, It is evident that for low temperature the cost is considerable higher, this is due to at low temperature, the treatment time is higher to 10 hours which increases the cost of energy and digester equipment. In addition, temperature behavior is quadratic due to the balance between time and temperature of the treatment that affects to these costs. On the other hand, the optimum point (minimum cost) is found at high temperature (60-65 °C) and at concentrations lower than 0.5 molar. The optimizer tool (MINITAB 17) predicts a minimum cost of 1.34 USD/kg, at concentration and temperature of 0.39 molar and 64 ° C, respectively.

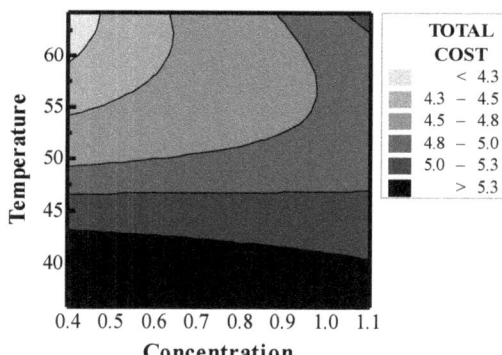

Figure 5 - *Contour plot for referencial cost analysis*

Conclusions

In this work, the effect of various treatments and parameters were investigated in order to reduce the non-cellulosic components present in the leaves of Ichu (Festuca distichovaginata, RM) and obtain fine technical fibers as reinfocemten inmanufacturing of laminated composites. From the screening processes, alkali treatment has significant effects especially, temperature and concentration, less effect was observed in the retting time and mechanical treatment. On the other hand, alkaline treatment at level of severity of 53.29% (relative mass lost in the process), shows excellent results to obtain the MI that guarantee the proper functioning of the fibers as a reinforcement in polyester resin composites. Even more, these composites had better compatibility than Jute fiber with epoxy by MI-transverse (strength). The extraction process designed at 0.39 molar NAOH, 64 °C and 3.5 hours, and then rinsed with water and dried at 40 °C for 12 hours, can reach reference cost of production up to ~1.34 USD/kg of Ichu fibers, through the optimization of concentration and temperature.

Non-Conventional Materials and Technologies – NOCMAT for XXI Century   Materials Research Forum LLC
Materials Research Proceedings 7 (2018) 764-774   doi: http://dx.doi.org/10.21741/9781945291838-75

References

[1]   Adekomaya, Oludaisi, et al. "Negative impact from the application of natural fibers". Journal of Cleaner Production 143 (2017): 843-846. https://doi.org/10.1016/j.jclepro.2016.12.037

[2]   Lucintel (2011) Opportunities in Natural Fiber Composites [Online]. Available in: http://www.lucintel.com/lucintelbrief/potentialofnaturalfibercomposites-final.pdf

[3]   APC. Plastics in automobile markets, vision and technology roadmap. American Plastics Council; 2001.

[4]   European_Commission. Directive 2000/53/EC of the European parliament and of the council de 18 September 2000 on end of life vehicles. Union OJotE; 2000.

[5]   Ferris L. (2017). China Auto Regulatory Trends 2016: New-Energy, Materials Restrictions, Recalls and Emissions Figure Prominently. Publicado: The National Law review. Available in: http://www.natlawreview.com/article/china-auto-regulatory-trends-2016-new-energy-materials-restrictions-recalls-and

[6]   The Nielsen Company (2015) Consumer brands that demonstrate commitment to sustainability outperform those that don't. The Sustainability Imperative. Available in: http://richesses-immaterielles.com/wp-content/uploads/2015/10/9053_Global_Sustainability_Report_Site-Web-RRI.pdf

[7]   Kozlowski, Ryszard M., et al. "Quo vadis natural fibres in 21st century?." Molecular Crystals and Liquid Crystals 627.1 (2016): 198-209. https://doi.org/10.1080/15421406.2015.1137675

[8]   Fiore, V., T. Scalici, and A. Valenza. "Characterization of a new natural fiber from Arundo donax L. as potential reinforcement of polymer composites." Carbohydrate polymers 106 (2014): 77-83. https://doi.org/10.1016/j.carbpol.2014.02.016

[9]   Hyness, N. Rajesh Jesudoss, et al. "Characterization of New Natural Cellulosic Fiber from Heteropogon Contortus Plant." Journal of Natural Fibers (2017): 1-8.

[10]   Indran, S., R. Edwin Raj, and V. S. Sreenivasan. "Characterization of new natural cellulosic fiber from Cissus quadrangularis root." Carbohydrate polymers 110 (2014): 423-429. https://doi.org/10.1016/j.carbpol.2014.04.051

[11]   Seki, Yoldas, et al. "Extraction and properties of Ferula communis (chakshir) fibers as novel reinforcement for composites materials." Composites Part B: Engineering 44.1 (2013): 517-523. https://doi.org/10.1016/j.compositesb.2012.03.013

[12]   Sarikanat, Mehmet, et al. "Determination of properties of Althaea officinalis L.(Marshmallow) fibres as a potential plant fibre in polymeric composite materials." Composites Part B: Engineering 57 (2014): 180-186. https://doi.org/10.1016/j.compositesb.2013.09.041

[13]   Candiotti Sergio, Estudio de la disponibilidad de materia prima Stipa Ichu, Festuca Orthophylla ("Iru Ichu") y Agave Americana, ("Maguey"). Lima, Perú: CONCYTEC (2017).

[14]   Maita L., Mamani N., Medina Y., Perez C., Soto S. Características de la comunidad vegetal "pajonal" en la zona de amortiguamiento (Simbral) de la reserva nacional salinas y aguda blanca, comprendido en junio del 2014.

[15]   Zimniewska, Malgorzata, and Maria Wladyka-Przybylak. "Natural fibers for composite applications." Fibrous and Textile Materials for Composite Applications. Springer Singapore, 2016. 171-204.

[16]   Charca, S., Noel, J., Andia, D., Flores, J., Guzman, A., Renteros, C., & Tumialan, J. "Assessment of Ichu fibers as non-expensive thermal insulation system for the Andean regions". Energy and Buildings, vol 108, pp 55-60, 2015. https://doi.org/10.1016/j.enbuild.2015.08.053

[17]   Dissanayake, Nilmini PJ, et al. "Energy use in the production of flax fiber for the reinforcement of composites". Journal of Natural Fibers, vol 6.4, pp 331-346, 2009.

[18]   Dunne, R., et al. "A review of natural fibres, their sustainability and automotive applications." Journal of Reinforced Plastics and Composites 35.13 (2016): 1041-1050. https://doi.org/10.1177/0731684416633898

[19]   Zakikhani, Parnia, et al. "Extraction and preparation of bamboo fibre-reinforced composites". Materials & Design 63 (2014): 820-828. https://doi.org/10.1016/j.matdes.2014.06.058

Non-Conventional Materials and Technologies – NOCMAT for XXI Century   Materials Research Forum LLC
Materials Research Proceedings 7 (2018) 764-774   doi: http://dx.doi.org/10.21741/9781945291838-75

[20] Kumar, K. Palani, and A. Shadrach Jeya Sekaran. "Some natural fibers used in polymer composites and their extraction processes: A review". Journal of Reinforced Plastics and Composites33.20 (2014): 1879-1892. https://doi.org/10.1177/0731684414548612

[21] Pickering, Kim L., MG Aruan Efendy, and Tan Minh Le. "A review of recent developments in natural fibre composites and their mechanical performance". Composites Part A: Applied Science and Manufacturing 83 (2016): 98-112. https://doi.org/10.1016/j.compositesa.2015.08.038

[22] Faruk, Omar, et al. "Progress report on natural fiber reinforced composites." Macromolecular Materials and Engineering 299.1 (2014): 9-26. https://doi.org/10.1002/mame.201300008

[23] Al-Oqla, Faris M., et al. "A decision-making model for selecting the most appropriate natural fiber–Polypropylene-based composites for automotive applications." Journal of Composite Materials 50.4 (2016): 543-556. https://doi.org/10.1177/0021998315577233

[24] Ashby, Michael F., Hugh Shercliff, and David Cebon. Materials: engineering, science, processing and design. Butterworth-Heinemann, 2013.

[25] Anderson, M. J., Trimming the fat out of experimental Methods. OE Magazine, 2005. 5(6): p. 32-33

[26] Jiju Antony (2003), Design of Experiments for engineers and scientists. Elsevier Science & Technology Books.

[27] Wu C., y Hamada M. (2009). Experiments: planning. analysis. and optimization, Atlanta, USA: Jhon Willey.

[28] Montgomery, D.C., Design and Analysis of Experiments. 2005, New York: Wiley

[29] Gutierrez H y De la Vara R. (2° edición) (2008). Análisis y diseños de experimentos. Mexico McGraw-Hill

[30] ISO, I. 14125: 1998 (E). Fibre reinforced plastic composites–determination of flexural properties (1998).

[31] Standard, A. S. T. M. D792-08, 2008, Standard Test Methods for Density and Specific Gravity (Relative Density) of Plastics by Displacement." ASTM International, West Conshohocken, PA, 2008, DOI: 10.1520/D0792-08." https://doi.org/10.1520/D0792-08

[32] Cashin, James, and Ralph Polimeni. "Contabilidad de costos". McGraw-Hill, 1993.

[33] Smolje, Alejandro. "La gestión de los costos de investigación y desarrollo". Universidad de Buenos Aires, Argentina.

[34] (2017) Alibaba.com [Online]. Available in: https://www.alibaba.com/product-detail/Industry-grade-BV-Approved-99-Caustic_60645483220.html?spm=a2700.7724838.0.0.QbnTsG&s=p

[35] Ministerio de Energía y Minas. Tarifas eléctricas en el Perú. (En. 2016) Available in: http://www2.congreso.gob.pe/sicr/comisiones/2015/com2015enemin.nsf//pubweb/389AD906020DCA44 05257F960071F3B8/$FILE/PPT-MEM2016.PDF

[36] Janjai, Serm. "A greenhouse type solar dryer for small-scale dried food industries: development and dissemination." International journal of energy and environment 3.3 (2012): 383-398.

[37] (2017) Alibaba.com [Online]. Available in: https://www.alibaba.com/product-detail/cotton-baling-machine-waste-paper-packer_60617295839.html?spm=a2700.7724838.0.0.nYqf8Y

[38] EMPSSAPAL   S.A.   "Localidad   Sicuani"   [Online].   Available   in: http://www.sunass.gob.pe/doc/tarifas/empssapal_tarifas_42014.pdf

[39] Liu, Ming, et al. "Effect of harvest time and field retting duration on the chemical composition, morphology and mechanical properties of hemp fibers." Industrial Crops and Products 69 (2015): 29-39. https://doi.org/10.1016/j.indcrop.2015.02.010

[40] Sisti, Laura, et al. "Evaluation of the retting process as a pre-treatment of vegetable fibers for the preparation of high-performance polymer biocomposites". Industrial Crops and Products 81 (2016): 56-65. https://doi.org/10.1016/j.indcrop.2015.11.045

[41] Senwitz, Christian, et al. "Almost Forgotten Resources–Biomechanical Properties of Traditionally Used Bast Fibers from Northern Angola." BioResources 11.3 (2016): 7595-7607. https://doi.org/10.15376/biores.11.3.7595-7607

[42] Amel, B. Ahmed, et al. "Effect of fiber extraction methods on some properties of kenaf bast fiber." Industrial Crops and Products 46 (2013): 117-123. https://doi.org/10.1016/j.indcrop.2012.12.015

Non-Conventional Materials and Technologies – NOCMAT for XXI Century   Materials Research Forum LLC
Materials Research Proceedings 7 (2018) 764-774   doi: http://dx.doi.org/10.21741/9781945291838-75

[43] Manalo, Allan C., et al. "Effects of alkali treatment and elevated temperature on the mechanical properties of bamboo fibre–polyester composites."Composites Part B: Engineering" 80 (2015): 73-83. https://doi.org/10.1016/j.compositesb.2015.05.033

[44] Angelini, Luciana G., et al. "Ramie fibers in a comparison between chemical and microbiological retting proposed for application in biocomposites."Industrial Crops and Products" 75 (2015): 178-184. https://doi.org/10.1016/j.indcrop.2015.05.004

[45] Stevens, Christian. Industrial applications of natural fibres: structure, properties and technical applications. Vol. 10. John Wiley & Sons, 2010.

[46] Das, S., and M. Bhowmick. "Mechanical Properties of Unidirectional Jute-Polyester Composite". Journal of Textile Science & Engineering 5.4 (2015): 1.

[47] Biswas, S., Shahinur, S., Hasan, M., & Ahsan, Q. (2015). Physical, mechanical and thermal properties of jute and bamboo fiber reinforced unidirectional epoxy composites. Procedia Engineering, 105, 933-939. https://doi.org/10.1016/j.proeng.2015.05.118

[48] Rashid, Md Mamunur, et al. "Study of Different Chemical Treatments for the Suitability of Banana (Musa oranta) Fiber in Composite Materials".

[49] Malhotra, Navdeep, Khalid Sheikh, and Sona Rani. "A review on mechanical characterization of natural fiber reinforced polymer composites." Journal of Engineering Research and Studies 3.1 (2012): 75-80.

Non-Conventional Materials and Technologies – NOCMAT for XXI Century   Materials Research Forum LLC
Materials Research Proceedings 7 (2018) 775-784   doi: http://dx.doi.org/10.21741/9781945291838-76

The Effect of using Recycled Materials in Earth-Based Blocks

Dick, K.J[a*], Pienuta, J.[a] Arnold, K.[a]

[a]Biosystems Engineering, University of Manitoba, Winnipeg, Canada

*Kristopher.Dick@umanitoba.ca, Jennifer.Pieniuta@umanitoba.ca , arnoldk@myumanitoba.ca

Abstract. Polystyrene and plastic use is ubiquitous worldwide and has a detrimental impact on the environment due to their inability to decompose. Incorporating waste into earth-based building materials provides an innovative approach to utilize expanded polystyrene (EPS) and polyethylene terephthalate (PET) waste, along with other recyclable materials including paper, plastic and cardboard. The research presented in this paper discusses the material properties of earth combined with various recyclables, including paper, cardboard, with varying percentages of EPS and plastic. The effect on density, moisture content, load-deformation behaviour, and compressive strength of earth test cylinders was evaluated for non-heat treated and heat-treated specimens. A series of tests were conducted on cylinders with various mix designs. The results indicate that an increase in the EPS content results in a decrease in the density, stiffness, and compressive strength, whereas the axial deformation, and Poisson's ratio increased. Subjecting the cylinders to heat prior to testing resulted in an increased compressive strength, but decreased the average density, moisture content, axial deformation and Poisson's ratio for the mixtures. Unheated specimens with plastic exhibited higher compression values when compared to heated cylinders. Compressive strengths ranged from 1.12 to 2.25 MPa for EPS specimens while the PET specimens had a range of 1.23 to 2.23 MPa.

Keywords: Expanded Polystyrene (EPS), Polyethylene Terephthalate (PET), Recycled Material, Earth Building, Compressive Strength

Introduction

The disposal of expanded polystyrene (EPS) is a great concern today as most EPS is ending up in landfills instead of being utilized in sustainable ways. Most EPS, consisting essentially of 98% air, is finding its way to landfills because it is extremely expensive to transport and thus, not economically viable to do so (Babu et al. 2004). This landfill-bound material poses numerous hazards including the potential to harm wildlife if ingested, it can deteriorate soil fertility if mixed in the soil, and it can remain in landfills for centuries due to its inability to decompose (Saikia and de Brito 2012). By utilizing this waste in innovative ways, landfills can expect to receive less waste annually and several environmental concerns will be addressed.

Many researchers have been investigating different means of using different forms of waste EPS to produce sustainable building materials as EPS can undergo large compressive deformations and absorb energy (Krundaeva et al. 2016). EPS earth is also relatively easy to fabricate on a construction site compared to other types of earth, whose fabrication process is complex.

According to Babu et al. (2004), in addition to using earth containing EPS aggregate for building applications, similar mixes have been used for specialized applications including railway track beds, energy absorbing material for the protection of military structures, and in floating marine structures.

Along with EPS, recyclable materials such as paper and cardboard can be utilized in construction materials as they demonstrate favorable binding properties in mixes (Demirbaş 1999). Although these materials do not pose the same challenges presented by EPS, the materials are still produced in vast amounts and can also be utilized in sustainable ways. In some developing countries, informal salvaging of recyclable materials is practiced so one can obtain a small income. If these individuals are aware of sustainable ways to utilize these materials, the recyclable material can also be used to improve their way of living in more ways than one (Demirbaş 1999).

Non-Conventional Materials and Technologies – NOCMAT for XXI Century   Materials Research Forum LLC
Materials Research Proceedings 7 (2018) 775-784   doi: http://dx.doi.org/10.21741/9781945291838-76

The main objective of this research was to investigate the feasibility of using waste EPS, paper, polyethylene terephthalate (PET) and cardboard as aggregates in the production of earth cylinders. Many authors (Babu et al. 2006; Kan 2009; Sayadi 2016; Ramamurthy et al. 2009) have investigated the properties of EPS earth, but no research on EPS earth with paper and cardboard aggregate have been reported.

Since EPS poses more of an environmental concern over paper and cardboard due to its inability to decompose, this research focuses on varying the percentage of EPS aggregate with recycled aggregates. By incorporating the polystyrene aggregate at various volumes in the earth, while keeping the paper and cardboard aggregate volumes constant, a wide range of earth densities can be produced. As EPS is known to have a low density of 10-50 kg/m3, which causes segregation in mixing, and is also hydrophobic (Babu et al. 2004), it is expected that as the percentage of EPS is increased in the mix, the compressive strength of the cylinder will decrease. Although the goal is to produce cylinders that contain recycled unconventional building materials of at least 50% of the total volume, determining the percentage of EPS that results in properties suitable for building materials is the main goal. The effect of subjecting the specimens to heat before conducting compression testing will also be investigated. It is expected that specimens subjected to heat will demonstrate higher compressive strengths and stiffness values over the specimens that are not exposed to heat.

Materials
Cement
To improve long-term strength and workability, general use hydraulic Type 10 Portland Cement/Quikrete (Target Products Ltd., Canada) was used. This Type 10 general use cement satisfies the Canadian CSA A3001 and ASTM C-150 Specification and requires the addition of clean water and aggregate to set.

Aggregate
Coarse aggregate used in the mixture, classified as a loam soil textural class, was collected from Anola, MB. Sieve analysis was previously performed on the soil and was found to contain on average 45.11 % sand, 35.33 % silt and 19.56 % clay (Griffith, 2007). The density of the loose clay was 1.827 g/cm^3.

The recyclable cardboard and paper was collected from the University of Manitoba recycling depot. To prepare the paper and cardboard aggregate, a Powershred® H-8Cd cross-cut shredder was used (Fellowes, USA). This mechanical shredder cut each full piece of paper or cardboard into 4 x 35 mm cross-cut particles. Given that variations from the specified size were observed, an average collected from randomly selecting 10 pieces of shredded material, was recorded. The average size of the paper and cardboard aggregate was 4 x 31 and 4 x 27 mm, respectively.

Densities for the loose material was 0.045 and 0.087 g/cm^3 for the paper and cardboard, respectively.

The waste EPS material, obtained from different sources and collected in 1 x 12 x 12 inch sheets, was broken into bead form using one of two methods. First, a drywall knife was used across the grain to shred the EPS. This method was efficient in breaking the sheet of polystyrene into a uniform size but was noted to be quite time consuming and unlikely to be used in a large-scale operation. The second method involved attaching a 3-in wire cup brush to a 12V Compact Cordless Drill. By starting at the center of each sheet and moving outwards with the wire cup brush, broken down EPS mainly collected in the center hole that enlarged as the circular motion moved towards the outer edges. This method was quicker and resulted in a final aggregate size suitable for mixing with a loose density of 0.011 g/cm^3. The final appearance of the aggregates, prior to mixing with water and cement, is shown in Figure 1.

Non-Conventional Materials and Technologies – NOCMAT for XXI Century   Materials Research Forum LLC
Materials Research Proceedings 7 (2018) 775-784   doi: http://dx.doi.org/10.21741/9781945291838-76

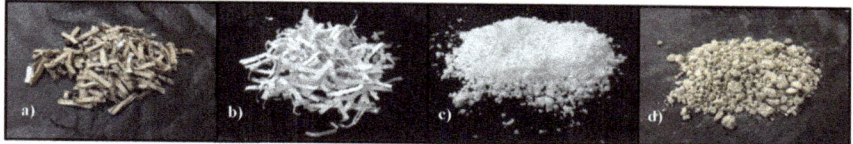

Fig. 1 - *Aggregate appearance before mixing – a) cardboard, b) paper c) EPS and d) clay.*

Using PET plastic as the variant in the mix design, the parts of paper, cardboard and earth were maintained consistently at a ratio of 1, 1, and 3, respectively. PET from recycled water and soft drink bottles were broken down with a paper shredder. Three separate mix designs were made with 1, 1.5 and 2 parts of plastic each. Every mix design used a cement binder at 10% of specimen weight. One universal measurement container was used to measure material amounts and the weight of a full container was taken for each material in order to maintain consistent standards for a full container. Five 100 mm x 200 mm cylinders were prepared for each mix design. Figures 2a and 2b show the materials being mixed and the cylinders in their forms respectively.

Fig. 2 - a) plastic earth mix, b) cylinders

TEST Methods
Mix Designs
For the preparation of the earth cylinders, paper, cardboard and earth aggregates mixed into each mixture remained at a ratio of 1:1:3 respectively, on a volume basis, for all three mixes. The total mass of each part was also recorded, using a Ohaus Adventurer-Pro scale with a 0.1 g readability, to remain as consistent as possible. The volume of EPS increased by 1-part volume for each mix, therefore resulting in three mixes including 1 part EPS, 2 parts EPS, and 3 parts EPS. A control mix containing 0 parts EPS was also created. The amount of added cement was based on 10% of the total mass of all aggregates.

Following the method discussed by Kazemi (1987), aggregates were mixed thoroughly until the cementitious material was added. Mixing of materials, confined inside of a metal container, was completed by hand using a hoe. The dry mixture was mixed for approximately one minute before adding water.

The water requirement for the mix depended on the volume of EPS added, and thus, was governed by the stability and consistency of the mixture. Potable water was added evenly to the mix on a mass basis. Mixing of the wet mixture was performed for two to four minutes to ensure proper distribution of the aggregates with the water. If required, additional water was added to the mix, and remixing of the product was completed until the right consistency was achieved, determined based on the appearance and feel of the material. This form of mixing is quite subjective but is a practice followed regularly during the mixing process (Kan and Demirboğa. 2009; Kazemi 1987; Ramamurthy et al. 2009). Table I summarizes the mix proportions used for the test specimens. In total, enough mix was made for each of the specified mixes to create three cylinders for compression testing and two cylinders that will be exposed to heat prior to compression testing.

Table I - Mix proportions of aggregate and cement –eps specimens

Mix #	Ratio[a] (EPS:P:CB:C)	EPS (kg)	Paper (kg)	Cardboard (kg)	Clay (kg)	Cement (kg)[b]	Water (kg)	% Recycled Material[c]
Control	0:1:1:3	0	0.020	0.052	2.472	0.254	0.418	40
1	1:1:1:3	0.006	0.020	0.052	2.472	0.255	0.420	50
2	2:1:1:3	0.012	0.020	0.052	2.472	0.256	0.430	57
3	3:1:1:3	0.018	0.020	0.052	2.472	0.256	0.463	63

[a]EPS = Expanded Polystyrene, P = Paper, CB = Cardboard, C = Clay
[b]Cement calculated based on 10% of the total mass
[c]Based on total volume of aggregates

Preparation of the Cylinders

Mixing was done by hand with a hoe in a large, steel trough. All parts were mixed together and then water was added until the desired consistency was reached. If a ball of the mixture could be dropped from shoulder height and only minimally break apart the consistency was deemed acceptable. This consistency was found to be similar to a wet rammed earth. Once the mixture was ready, handfuls were tamped into the forms and levelled with the top of the specimen mould.

The specimens were rammed into 100 x 200 mm cylindrical forms. The 100 mm diameter PVC tubes in half longitudinally. To fasten the two halves together, a 100 mm end cap was fastened to the bottom of the two pieces and reinforced with tape at the top. After initial curing the forms could be easily split by removing the tape.

Curing Period

To provide air-curing specimens, they were removed from the forms after 9 days, following packing and stored on shelving that provided air movement until day 20.

Two specimens from each mixture were then placed in an oven (Jelo Tech Lab Companion, Model OF-11E) and heated to determine whether the heat would modify the properties of the EPS and PET. Some studies suggest that by exposing EPS to heat treatment, service life can be maximized by improving strength properties and density (Kan and Demirboğa. 2009). Currently, there is no literature describing how the properties of EPS are affected by heat once already mixed with other aggregates, cement and water. Instead, discussion on how EPS is impacted when exposed to heat treatment on its own is common. Based on the data collected by heating EPS alone, placing the EPS in an oven at a temperature of 130°C for 15 minutes was found to be ideal. By heating the EPS for this specified temperature and duration, the EPS was expected to transform from a foamy state to a plastic state (Kan and Demirboğa. 2009).

For this reason, once cured, the goal was the keep the cylinders in the oven at 130°C for 15 minutes. Due to the fact that the cylinder's initial temperatures were at room temperature, once placed in the oven, it took approximately 30 minutes for the oven to equilibrate back at 130°C. To ensure the EPS at the center of the cylinders reached a temperature of 130°C, 15 minutes was counted from the point the oven reached 130°C with the cylinders inside. Therefore, in total, the

Non-Conventional Materials and Technologies – NOCMAT for XXI Century   Materials Research Forum LLC
Materials Research Proceedings 7 (2018) 775-784   doi: http://dx.doi.org/10.21741/9781945291838-76

cylinders were in the oven for a total of 44 minutes and 27 seconds. Cylinders were left to cool at room temperature for 3 days prior to testing.

To observe how this heat would affect an unmodified piece of EPS, a sample size of 10 x 10 x 2.3 cm was cut, with a density of 20.87 kg/m³. After removing it from the heat, the EPS was hard, decreased in volume by approximately 90% and had a density of 208.3 kg/m³.

After curing the PET specimens selected for heat treatment were placed in an oven for five minutes at 260°C corresponding to the melting point of PET. The purpose for this was to attempt to partially melt the PET to see if there would be any impact on compressive strength.

Testing Procedure

After 20 days, the resulting molded cylinders were tested under axial compression using a loading frame at the Alternative Village at the University of Manitoba. A load rate of 1.4 mm/min was used. Linear potentiometers were used to monitor the vertical and lateral movement of the specimen. A 22 kN load cell was used to measure the applied axial compression force. A computer controlled data acquisition system was used to record the load-deformation measurements.

Given that the resulting densities of the cylinders were in excess of 800 kg/m3, the Standard Test Method for Compressive Strength of Cylindrical Concrete Specimens (ASTM C39/C39M-16b) was followed for each mixture, replicated three times. This method consisted of applying a compressive axial load until failure occurred. For the specimens placed in the oven, the ASTM C39/C39M-16b standard was also followed to determine the compressive strength. The compressive axial load was applied to each of the two heat-treated cylinders for each mixture.

Results and Discussion

The aim of this study was to evaluate the effects of EPS and PET with paper and cardboard as aggregates on physical properties including density, moisture content, lateral movement, and compressive strength in both the unmodified and modified heat-treated states. This study also assessed whether the recyclable materials can be used in concrete to create a material that is suitable to replace conventional building materials.

Density

Table 2 - *Average densities for each mixture after packing, curing, and heat-treatment.*

Mix #	After packing (kg/m³)	After curing (kg/m³)	After heat-treatment (kg/m³)
Control	2226.3	1988.5	1964.0
EPS SPECIMENS			
1	1950.6	1732.2	1706.1
2	1749.4	1563.5	1534.9
3	1570.5	1397.4	1373.8
PET SPECIMENS			
1	2206.0	2088.6	2042.7
2	2123.0	2010.0	2011.0
3	2127.1	2013.9	1932.3

The specimen bulk density was determined during each stage of the cylinder preparation process, including immediately following ramming, after 20 days of curing, and following exposure of heat. The densities of the EPS cylinders after ramming, to the cured cylinders decreased by 10.7, 11.2,

Non-Conventional Materials and Technologies – NOCMAT for XXI Century   Materials Research Forum LLC
Materials Research Proceedings 7 (2018) 775-784   doi: http://dx.doi.org/10.21741/9781945291838-76

10.6, and 11.3 % for the control, mix 1, mix 2, and mix 3, respectively. Similarly, the density of the cured cylinders to the heat-treated cylinders decreased by 1.23, 1.51, 1.83, and 1.69 % for the control, mix 1, mix 2, and mix 3, respectively. Table 2 provides a summary of the densities for the specimen types.

Compressive Strength
The axial compressive strength was determined for the maximum load attained from each test.

Non-Heat Treated EPS Specimens
The compressive strength values for the cylinders not subjected to heat were determined by taking the average of the three cylinders not placed in the oven. After conducting the compressive strength tests, the results showed that a higher EPS content, therefore a lower density, resulted in a lower compressive strength. Compared to the control, on average there was a 29.7, 52.4, and 69.8 % decrease in compressive strength for mix 1, mix 2, and mix 3, respectively. Thus, increasing the volume of EPS in the mix directly affects the compressive strength due to EPS having little to no strength alone. EPS also displays high compressibility behavior, resulting in the formation of micro cracks at the points where the aggregate meets with the binding material (Sayadi 2016).

Heat Treatment Specimens
Similar to mixtures that obtained no heat treatment, an increase in EPS volume resulted in a decrease in density and a decrease in compressive strength for the specimens exposed to heat. Compared to the control, on average there was an 18.7, 42.1. and 61.4 % decrease in compressive strength for mix 1, mix 2, and mix 3, respectively. Compared to the non-heat treated specimens, there was a 2.81, 8.14, and 16.6 % increase in compressive strength for mix 1, mix 2, and mix 3, respectively, and an 11.1 % decrease for the control. Unlike the specimens not subjected to heat treatment, the samples containing EPS that were exposed to heat prior to testing, demonstrated higher compressive strengths. The heat treatment worked to toughen up the EPS within the earth by transforming the foam state into a hardened plastic state. The control specimens illustrated a different behavior and it is likely that the heat dried the specimens to the point where the strength was compromised as a result.

A summary of the average compressive strength of the various mixtures, containing the recycled aggregates and EPS, is presented in Table 3 while the PET specimen values are summarized in Table 4. Figure 3 provides a graphical comparison of the compressive strengths.

Table 3 - Compressive Strength of Control Mix and EPS Specimens

Specimen Type [1]	Average Compressive Strength (Mpa)	
	Non Heated	Heat Treatment
Control	1.94	1.73
Mix 1	1.36	1.40
Mix 2	0.93	1.00
Mix 3	0.59	0.68

Note: 1. Control mix – no EPS content, Mix 1 – 1 part EPS, Mix 2 – 2 parts EPS, Mix 3 – 3 parts EPS

Non-Conventional Materials and Technologies – NOCMAT for XXI Century   Materials Research Forum LLC
Materials Research Proceedings 7 (2018) 775-784   doi: http://dx.doi.org/10.21741/9781945291838-76

TABLE 4 - COMPRESSIVE STRENGTH OF CONTROL MIX AND PET SPECIMENS

Specimen Type [1]	Average Compressive Strength (Mpa)	
	Non Heated	Heat Treatment
Control	1.94	1.73
Mix 1	1.60	1.19
Mix 2	1.78	1.82
Mix 3	1.69	1.01

Note: 1. Control mix – no EPS/PET content, Mix 1 – 1 part PET, MIX 2 – 1.5 parts PET, Mix 3 – 2 parts PET

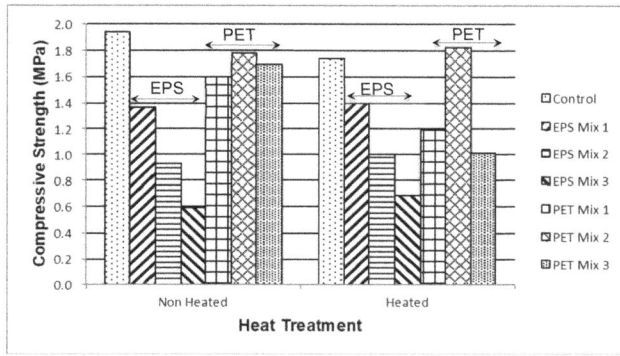

Figure 3 - Compressive Strength Comparison

Poisson's Ratio

It was postulated prior to testing that the incorporation of EPS might have an impact on the transverse deformation thus determining Poisson's ratio was of interest. Potentiometers were used to measure transverse expansion of the test cylinder in two orthogonal directions as well as the vertical direction. The ratio between the lateral and axial strain was used to obtain Poisson's ratio.

As anticipated, bulging of the specimen during compression was clearly noticeable during testing. Although this is the case, on average, the specimens not subjected to heat for mixture 3 resulted in the highest Poisson's ratio for the EPS specimens. This was expected as the high EPS content allowed the load to compress the foam within the specimen. Also, specimens subjected to heat treatment resulted in lower Poisson's values for all mixes containing EPS, including, mix 1, mix 2, and mix 3. Mix 3 specimens did not exhibit shearing but rather separated horizontally when the load was relaxed. The ability to deform without shearing has the potential to be used as an energy-dissipating element in a structure. Lastly, based on average values for this limited sample, no apparent difference in Poisson's ratio for the control specimens occurred between the non-heat treated specimens and heat-treated specimens. The PET specimens did not exhibit the same bulging behaviour without cracking. The ratios for the PET specimens were a result of specimen shearing and breaking apart as opposed to bulging out and were thus not considered to be valid at maximum load. Table 5 provides a summary of results for the EPS specimens.

Table 5 - Average Poisson's Ratio for Control and EPS Specimens at Maximum Load

Specimen Type	Average Poisson's Ratio	
	Non Heated	Heat Treated
Control	0.1420	0.1428
EPS		
Mix 1	0.0986	0.0846
Mix 2	0.0911	0.0829
Mix 3	0.2619	0.0979

Conclusion

By incorporating waste EPS, recycled cardboard and paper into earth test cylinders as aggregate, a new material that demonstrates properties suitable for construction has been developed. The properties, including the density, Poisson's ratio, stiffness, and compressive strength of the specimens, containing a minimum of 50% recycled material based on volume, were evaluated.

Densities for EPS specimens, following packing, for the control, mix 1, mix 2, and mix 3 were 2226.3, 1950.6, 1749.4, and 1570.5 kg/m^3, respectively. On average, axial deformation and Poisson's ratio increased with increasing amounts of EPS. Based on load-deformation plots the stiffness of the heat-treated specimens was greater than the non-heat treated specimens. It is assumed that the transformation of the EPS from a foam state to a plastic state created this difference. Also, on average, the stiffness decreased as the EPS content increased. The compressive strengths recorded for the specimens were 1.94, 1.36, 0.93, and 0.59 MPa for the control, mix 1, mix 2, and mix 3, respectively, with no heat-treatment and 1.73, 1.40, 1.00, and 0.68 MPa for the control, mix 1, mix 2, and mix 3, respectively, for the specimens subjected to heat. The results of this study indicated that an increase in the EPS content, resulted in a decreased in the density, stiffness, and compressive strength, whereas axial deformation, and Poisson's ratio increased. Also, subjecting the cylinders to heat prior to testing resulted in higher stiffness values and compressive strengths, but decreased the average density, axial deformation and Poisson's ratio for the mixtures.

Although this is the case, the combination of EPS with recycled aggregates has the potential to be used as a suitable building material. Qin (2014) states that the compressive strength of adobe in rural areas of China resulted in a compressive strength ranging from 0.8-2.0 MPa. Therefore, based on the compressive strengths noted, cylinders containing 1 part and 2 parts of EPS resulted in compressive strengths within this range. Qin (2014) also discussed that the strength of adobe, with the addition of 10% cement, resulted in compressive strengths within the range of 5.5-6.5 MPa, much higher than the compressive strengths observed in this study.

In conclusion, a method of using cardboard, paper, and EPS in lower quantities, as a sustainable material in the construction industry, is still presented. Although compressive strengths were lower than those noted in adobe with the addition of cement, earth blocks containing EPS still resulted in compressive strengths similar to abode without cement. EPS earth blocks also have the ability to compress a substantial amount before failing, a phenomenon not present in ordinary concrete. Based solely on the results obtained in this study, it is recommended that mixture 1 with 1 part EPS, without heat-treatment, be investigated further. This is suggested as mixture 1 with no heat-treatment resulted in a compressive strength of 1.36 MPa, which is similar to adobe. Mixture 1 also demonstrated an axial deformation of 5.46 mm, a Poisson's ratio of 0.099 (similar to concrete at 0.1) and a stiffness of 2.69 GPa (lower than concrete at 30 GPa), indicating the mixtures ability to compress under a load. Although the heat improved the compressive strength of mixture 1 by 2.81 %, the average stiffness increased by 25.2 %.

Non-Conventional Materials and Technologies – NOCMAT for XXI Century   Materials Research Forum LLC
Materials Research Proceedings 7 (2018) 775-784   doi: http://dx.doi.org/10.21741/9781945291838-76

If further testing is conducted to investigate additional properties of the specified mixes, several modifications to the experimental methods discussed within this paper are recommended. First , it is recommended that each mixture for each individual cylinder is mixed separately. This is recommended because variations in cylinders for the same mix were noted, likely due to the segregation of the EPS that occurred in the mix as a whole. Although more time consuming, by mixing each cylinder individually, cylinders would be more consistent. Also, it is recommended that the sample size is increased to allow for outliers to be eliminated. Another limitation of the research, related to the set oven temperature used to modify the EPS within the cylinders, was identified. Further research on various temperatures, investigating the internal properties of each cylinder after heat treatment, is recommended. Lastly, the behavior of each mixture in regards to the vertical deformation during compression and the rebound after removing the load is recommended. The significant deformation of the EPS Mix 3 is of particular interest to investigate the potential for use in energy-dissipation in a structural system. While the compressive strength is low, if used in a framing system then it is postulated that this block could tolerate racking loads while still remaining intact.

Unrelated to the methods discussed in this paper, it is recommended that additional research into the heat transfer characteristics of the cylinders containing EPS and recycled aggregates is investigated before finalizing a mixture that is optimal for building.

Acknowledgements:
Authors would like to thank the Alternative Village at the University of Manitoba for supporting this research.

References

[1] ASTM International. 2016. Standard test method for compressive strength of cylindrical concrete specimens – C39/C39M-16b. West Conshohocken, PA.

[2] Babu, D.S., K.G. Babu and W. Tiong-Huan. 2006. Effect of polystyrene aggregate size on strength and migration characteristics of lightweight concrete. *Cement & Concrete Composites* 28:520–527. https://doi.org/10.1016/j.cemconcomp.2006.02.018

[3] Balaban, O. and J.A.Puppim de Oliveira. 2016. Sustainable cities for healthier cities: assessing the co-benefits of green buildings in Japan. *Journal of Cleaner Production.*

[4] Collet, F. and C. Lanos. 2011. Mechanical properties of hempcrete. Équipe Matériaux Thermo-*Rhéologie*. Rennes, France.

[5] Demirbaş, A. 1999. Physical properties of briquettes from waste paper and wheat straw mixtures. *Energy Conversion & Management.* 40: 437-445. https://doi.org/10.1016/S0196-8904(98)00111-3

[6] EPSA Ltd. 2014. About EPS. http://epsa.org.au/about-eps/. 2016.

[7] Griffith, K. 2007. Physical properties of an Anola soil based cob re-inforced with chicken feathers. Unpublished senior thesis Biosystems Engineering, University of Manitoba, Advisor K. Dick, Winnipeg, Manitoba, Canada. University Department, University, City, Province.

[8] Kan, A., R. Demirboğa. 2009. A new technique of processing for waste-expanded polystyrene foams as aggregates. *Journal of Materials Processing Technology* 209:2994-3000. https://doi.org/10.1016/j.jmatprotec.2008.07.017

[9] Kariyawasam, K.K.G.K.D., and C. Jayasinghe. 2016. Cement stabilized rammed earth as a suitable construction material. *Construction and Building Materials* 105:519–527. https://doi.org/10.1016/j.conbuildmat.2015.12.189

Non-Conventional Materials and Technologies – NOCMAT for XXI Century   Materials Research Forum LLC
Materials Research Proceedings 7 (2018) 775-784   doi: http://dx.doi.org/10.21741/9781945291838-76

[10] Kazemi, A. 1987. Strength development in concrete blocks containing flyash. Unpublished M.Sc. thesis, Department of Civil Engineering, University of Manitoba, Winnipeg, MB.

[11] Krundaeva, A., G. De Bruyne, F. Gagliardi and W. Van Parpegem. 2016. Dynamic compressive   strength and crushing properties of expanded polystyrene foam for different strain rates and   different temperatures. *Polymer Testing* 55: 61-68. https://doi.org/10.1016/j.polymertesting.2016.08.005

[12] National Ready Mix Concrete Association. 2003. Testing compressive strength of concrete. *Concrete in Practice.* CIP-35.

[13] Qin, L., W. Chen, and X. Li. 2014. Experimental research on compressive strength of adobe with cement. *Applied Mechanics and Materials* 507:217-221. https://doi.org/10.4028/www.scientific.net/AMM.507.217

[14] Ramamurthy, K., E.K. Kunhanandan Nambiar and G. Indu Siva Ranjani. 2009. A classification of studies on properties of foam concrete. Cement and Concrete Composites 31: 388-396. https://doi.org/10.1016/j.cemconcomp.2009.04.006

[15] Roylance, D. 2008. *Mechanical Properties of Materials.* Cambridge, Massachussetts: MIT.

[16] Saikia, N. and J. de Brito. 2012. Use of plastic waste as aggregate in cement mortar and concrete preparation: A review. *Construction and Building Materials* 34: 385-401. https://doi.org/10.1016/j.conbuildmat.2012.02.066

[17] Sayadi, A.A., J.V. Tapia, T.R. Neitzert and G.C. Clifton. 2016. Effects of expanded polystyrene (EPS) particles on fire resistance, thermal conductivity and compressive strength of foamed concrete. *Construction and Building Materials* 112:716-724. https://doi.org/10.1016/j.conbuildmat.2016.02.218

[18] Singh, K.D. and Tripura, D.D. 2014. Behaviour of cement-stabilized rammed earth circular column under axial loading. *Materials and Structures.* 49: 371-382.

Non-Conventional Materials and Technologies – NOCMAT for XXI Century   Materials Research Forum LLC
Materials Research Proceedings 7 (2018) 785-790   doi: http://dx.doi.org/10.21741/9781945291838-77

Evaluation of EPS Beads Inclusion on Strength and Stiffness of Soil

Mariana Vela Silveira[a, *], Alena Vitková Calheiros[a], Michéle Dal Toé Casagrande[a]

[a] PUC-Rio, Brasil

vela.silveira@gmail.com, alenacalheiros@hotmail.com, michele_casagrande@puc-rio.br

Abstract. EPS is the international acronym for Expanded Polystyrene, which is a thermoplastic derived from petroleum. Each EPS bead is composed of 98% of air and 2% of raw material (in mass) therefore, considered a villain in matters of waste because it occupies a lot of space in landfill sites. An alternative for reducing the disposal of this product is using it as a soil reinforcement material in earthworks. This experimental study reports the influence of the addition of EPS beads on the parameters of the colluvial clayey soil assessed through the compaction test and the triaxial test. A series of triaxial compression isotropic drained test was performed to seek to establish patterns of behavior that might explain the influence of the addition of EPS beads, relating it to the shear strength and deformation parameters of the soil. Effects of adding proportionate quantities of EPS beads (i.e., 0.25, 0.5, 0.75 and 1.0% by dry weight of the soil) were investigated and evaluated. Through the obtained results, it was observed that the addition of EPS beads to the clayey soil did not entail a worsening in the soil's behavior in the four performed mixtures, since it was observed an increase in the cohesion or in the angle of friction of the mixtures compared to the values obtained for the pure soil. The results show that the use of EPS beads in geotechnical works of elastic loads could be an alternative of an environmentally correct destination for this material.

Keywords: Expanded Polystyrene Beads, Soil Reinforcement, Triaxial Test, Mechanical Behavior

Introduction

EPS is the international acronym for Expanded Polystyrene, which is a thermoplastic derived from petroleum. It is composed of a styrene hydrocarbon polymer (polystyrene) that undergoes expansion through the use of pentane gas (another hydrocarbon). Therefore, the EPS is chemically composed of only two elements: carbon and hydrogen. An EPS bead is composed of 98% of air and 2% of raw material (in mass).

The EPS applications comprise several different areas, such as: packages for electronic equipment, household appliances such as thermal boxes, thermal and acoustic insulation, trays for food packaging etc.

Civil construction is a sector responsible for a large portion of EPS consumption. The wide usage in civil construction occurs due to the material's insulating characteristics, its lightweight aspect, and its resistance.

The disposal of EPS is an issue faced by many cities. EPS alone does not pollute or contaminate the soil, but, since its decomposition takes hundreds of years, it ends up taking too much space and reducing the usable area of landfill sites. The lightweight aspect and the low density, which are the main characteristics of the EPS, are responsible for hindering its recycling.

The reutilization of this residue can serve as contribution for minimizing environmental liabilities and for adding value to the material, thus eliminating current problems regarding the disposal of residues in landfill sites. An alternative would consist in the use of EPS beads as a reinforcement material in earthworks.

Soil improvement or reinforcement comprises the use of physical and/or chemical procedures to improve the soil's mechanical properties. The technique to reinforce soil has been widely used, and a

Non-Conventional Materials and Technologies – NOCMAT for XXI Century   Materials Research Forum LLC
Materials Research Proceedings 7 (2018) 785-790   doi: http://dx.doi.org/10.21741/9781945291838-77

few of the materials used to this end are: fibres (PET, glass, polypropylene), crumb rubber, geosynthetics, etc.

Reference [1]-[3] performed experimental tests with EPS beads to assess the aspects of reinforced soils. In the present study, a series of mechanical tests sought to determine the optimum content for the addition of EPS beads as a soil reinforcement.

The aim of these tests was to establish a maximum improvement with the highest volume of residue, since one of the goals of using this material as reinforcement is to provide an environmentally correct destination for the largest possible amount of residue.

Experimental program

The experimental program consisted in the execution of compaction test and triaxial compression tests to determine the strength parameters that were the basis for identifying and evaluating the effect of dispersed inclusions of EPS beads on the mechanical properties of a clayey soil.

Materials and Methods

Clayey Soil

The colluvial clayey soil used in this study showed a specific gravity (Gs) of 2.72. The obtained consistency limits were as follows: Liquid Limit of 53%, Plastic Limit of 39%, and Plasticity Index of 14%. The Unified Soil Classification System (USCS) categorizes this soil as CH, which corresponds to a sandy clay of medium plasticity.

EPS Beads

The EPS beads used as soil reinforcement element were commercially acquired and had an average diameter of approximately 1 mm. In practice, these beads can originate from the milling of discarded polystyrene sheets, which turns them into smaller bits.

Sample Preparation

The utilized mixtures with the clayey soil used contents of 0%, 0.25%, 0.50%, 0.75%, and 1.0% of EPS beads in relation to the soil's dry weight. The compaction test used Standard Proctor energy, with reutilization of material, to determine the optimum moisture content and the maximum dry specific weight for each mixture.

To determine the resistance parameters, isotropically consolidated drained triaxial compression tests were performed. The specimens of the clayey soil and of the soil-EPS mixtures had dimensions of 7.82cm of height and 3.80cm of diameter, and were manufactured using the values of optimum moisture and of maximum dry specific weight obtained from the compaction test for each composite.

The Table 1 summarizes the variables investigated and the employed abbreviations.

Table 1 - VARIABLES INVESTIGATED AND THE EMPLOYED ABBREVIATIONS.

Soil	EPS content (%)	Abbreviations
	0.0	C0
	0.25	C-EPS 0.25
Clayey	0.50	C-EPS 0.50
	0.75	C-EPS 0.75
	1.0	C-EPS 1.0

Non-Conventional Materials and Technologies – NOCMAT for XXI Century   Materials Research Forum LLC
Materials Research Proceedings 7 (2018) 785-790   doi: http://dx.doi.org/10.21741/9781945291838-77

Triaxial Tests

The triaxial tests performed in this research pertain to the isotropically consolidated drained (CID) type and were carried out using the clayey soil and clayey-EPS mixtures aiming to determine their resistance parameters.

The tests were performed using a press of the Wykeham Farrance brand, with controlled displacement speed and under effective confining stress of 50, 150, and 300 kPa. These values are coherent with the realistic assumptions made in some engineering applications, such as shallow foundations placed on the improved layers of soil.

Saturation was monitored in each test, ensuring B values of at least 0.95. The criterion presented in [4] was employed to define the shearing speed, which means that the speed of 0.030 mm/min was used for all samples

Results and analyses

The Proctor compaction curves, as well as the values of optimum moisture content (w) and the maximum specific dried mass (ρ_s) for the clayey soil and for the clay-EPS mixtures, are shown in Figure 1.

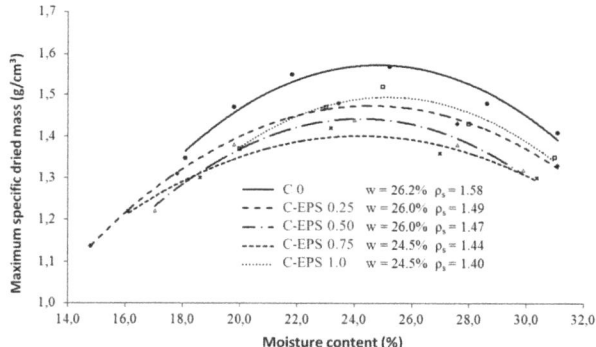

FIGURE 1 – *STANDARD PROCTOR COMPACTION CURVES AND PARAMETERS FOR THE CLAYEY SOIL AND FOR THE CLAYEY-EPS MIXTURES.*

It is possible to observe that the addition of EPS beads decreased the materials maximum specific dried mass, and this value declined as the EPS content increased. The optimum moisture showed the same behavior. This behavior can be justified by the EPS beads low apparent density and low moisture absorption.

This behavior was also observed by [3].

The curves of deviatoric stress (σ_d) and volumetric strain (ε_v) versus axial strain (ε_a) obtained from both reinforced and nonreinforced clayey soil, under triaxial compression, are shown in Figure 2.

Analysing the curves deviatoric stress vs. axial deformation, it is possible to observe that the leading effect of adding EPS beads to the clayey soil is the change in this material's stress vs. deformation behaviour pattern at higher initial effective stress.

This change is indicated by the constant resistance increase along with the rise of the axial deformation, up to the assessed axial deformation, which characterizes an elastic-plastic hardening behaviour. Regardless of the beads content, this change is greater as the greater initial confining stress.

Non-Conventional Materials and Technologies – NOCMAT for XXI Century   Materials Research Forum LLC
Materials Research Proceedings 7 (2018) 785-790   doi: http://dx.doi.org/10.21741/9781945291838-77

At lower effective confining stresses, this change in the stress vs. deformation behaviour pattern does not occur and EPS addition did not affect the clay's shearing resistance. This indicates that the addition of EPS beads to a clayey soil shows a better resistance performance when the initial effective confining stresses are low.

The addition of EPS beads to the clayey soil, regardless of the amount, affected the soil's stiffness. The samples with EPS showed less stiffness than the non-reinforced samples specially at higher initial confining stress.

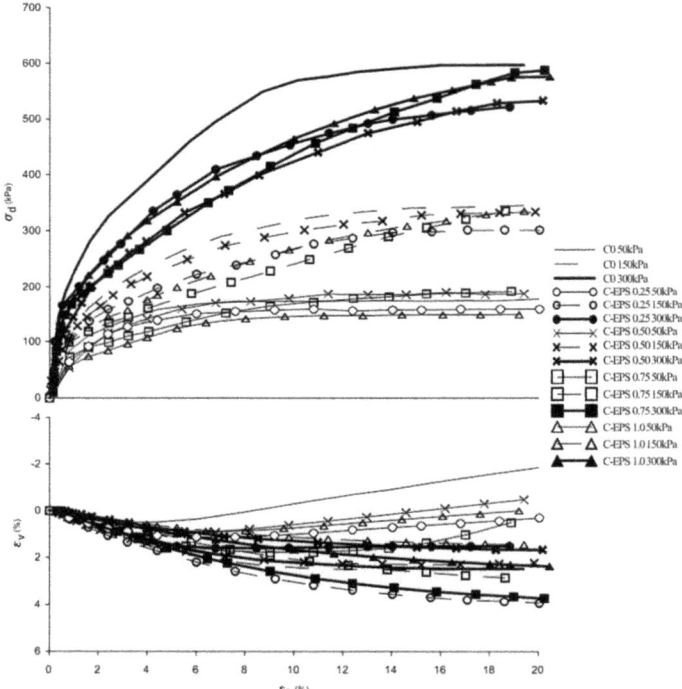

FIGURE 2 – STRESS-STRAIN-VOLUMETRIC RESPONSE OF THE CLAYEY SOIL AND THE CLAY-EPS MIXTURES IN CONVENTIONAL TRIAXIAL COMPRESSION TESTS.

In the curves associated with the volumetric deformation vs. the axial deformation, it is possible to notice that the addition of EPS beads to the Colluvial clayey soil, regardless of the amount, generally resulted in a decrease in the expansion and contraction behaviour.

To verify the effect of the addition of EPS beads on resistance parameters of the colluvial clayey soil, Figure 3 shows the resistance envelopes in p'-q space, as well as the values regarding the cohesion intercept and the internal friction angle obtained for the colluvial clayey soil and for this soil reinforced with EPS beads with an axial deformation of 18%. This was because the curves σ_d-ε_a were generally experiencing slight strain hardening until the maximum strain that the apparatus could

Non-Conventional Materials and Technologies – NOCMAT for XXI Century   Materials Research Forum LLC
Materials Research Proceedings 7 (2018) 785-790   doi: http://dx.doi.org/10.21741/9781945291838-77

reach, as can be seen in the stress-strain data in Figure 2, so that a "strength" had to be defined at a particular strain.

The addition of 0.2%, 0.5%, 0.75%, and 1.0% of EPS beads to the colluvial clayey soil affected this soil's resistance parameters, and the influence depended on the added content. While, for certain contents, the cohesion intercept value increased as the friction angle decreased, the opposite occurred for others.

Relating Fig. 1, 2 and 3, it can be observed that the changes verified in the stress vs. deformation behaviour pattern and in the resistance parameters in performing triaxial tests are in agreement with the decreased of the material's maximum specific dried mass in the compaction test caused by the addition of EPS beads to the colluvial clayey soil. However, the addition of EPS beads to the clayey soil did not entail a worsening in the soil's behaviour regarding the four utilized mixtures, seeing as one of the resistance parameters always increased with the EPS addition.

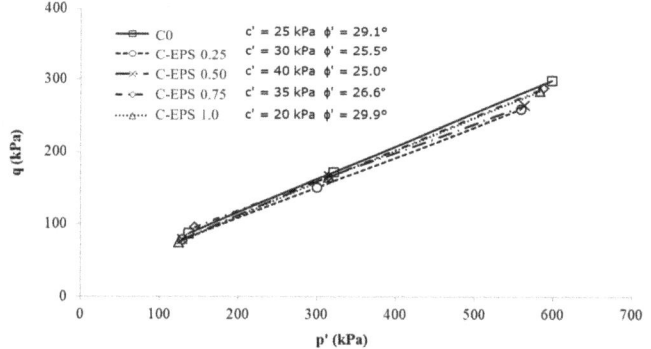

FIGURE 3 – RESISTANCE PARAMETERS AND STRESS ENVELOPES FOR THE CLAYEY SOIL AND THE CLAY-EPS MIXTURES WITH AN AXIAL DEFORMATION OF 18%

Final considerations
From the compaction tests and the conventional triaxial compression tests performed on clayey soil and clayey-EPS beads mixtures with content of 0, 0.25, 0.50, 0.75 and 1.0% and initial effective stress of 50, 150 and 300kPa were established some conclusions reported below:

The mechanical behavior of the composites regarding the resistance parameters is influenced by the EPS beads' content, and the level of confining stress.

The addiction of EPS beads decreased the material's maximum specific dried mass and optimum moisture, and both values declined as the EPS content increased. This behaviour can be justified by the EPS beads low apparent density and low moisture absorption.

The EPS addiction affected the material's stress vs. strain behaviour pattern at higher initial effective stress, characterizing an elastic-plastic hardening behaviour.

Assessing the volumetric strain and vertical displacement behaviour, the addiction of EPS beads, regardless of the amount, generally resulted in a decrease in the expansion and contraction behaviour.

The addition of EPS beads to the clayey matrix did not entail a worsening in the material's behaviour in all performed mixtures, seeing as one of the resistance parameters always increased with the EPS addition.

Non-Conventional Materials and Technologies – NOCMAT for XXI Century   Materials Research Forum LLC
Materials Research Proceedings 7 (2018) 785-790   doi: http://dx.doi.org/10.21741/9781945291838-77

The results contribute to a better interpretation of the mechanical behaviour of soil reinforced with EPS beads and show the potential of this material being used as reinforcement of soils in earthwork subjected to static loads, offering an alternative environmentally correct destination for this material.

Acknowledgements

The writers wish to express their gratitude to CNPq-National Council of Scientific and Technological Research - Brazil for financial support to the research group and the LGMA laboratory at the Pontificia Universidade Católica do Rio de Janeiro for the physical and technical support to carry out the research.

References

[1] ABRAPEX, O eps na construo civil: Características do poliestireno expandido para utilizao em edificações., 2000.

[2] Abdelrahman G, Lightweight mixture using clay, eps-beads and cement. In: Seventeen International Conference in Soil Mechanics and Geotechnical Engineering (17 ICSMGE 2009), 2009.

[3] Abdelrahman G, Mohamed H, Ahmed H, New replacement formations on expansive soils using recycled eps beads, Proceedings of 18th ICSMGE (2013).

[4] Nataatmadja A, Illuri H, Sustainable backfill materials made of clay and recycled eps, in: Proc. of the 3rd CIB Intl. Conf. on Smart and Sustainable Build Environments (SASBE 2009), 2009.

[5] Head K. Manual of soil laboratory testing: Effective stress tests., 1986.

Non-Conventional Materials and Technologies – NOCMAT for XXI Century   Materials Research Forum LLC
Materials Research Proceedings 7 (2018) 791-798   doi: http://dx.doi.org/10.21741/9781945291838-78

Evaluation of Sisal and Curauá Fibers Inclusions on Strength and Stiffness Response of Soil

Mariana Vela Silveira [a, *], Michéle Dal Toé Casagrande [a]

[a] PUC-Rio, Brasil

vela.silveira@gmail.com, michele_casagrande@puc-rio.br

Abstract. This experimental study reports the behavior of a reinforced and unreinforced granular soil with the addition of short sisal and curauá fibers. The fibers used as soil reinforcing element are extracted from the leaves of the plants curauá (*Ananás erectifolius*) and sisal (*Agave sisalana*). These vegetal fibers were chosen because they have good mechanical properties and the need for new renewable materials. The sisal and curauá fibers were mixed with granular soil in a randomly distributed form to evaluate its influence on the mechanical properties of the soil. A series of triaxial compression isotropic drained test was performed to seek to establish patterns of behavior that might explain the influence of the addition of vegetal fibers, relating it to the shear strength and deformation parameters of the soil. The tests were performed on samples subjected to a relative density of 50%, moisture content of 10%, fibers content in proportions of 0 and 0.5% of the dry weight of the soil and the fibers lengths of 25mm. Through the obtained results, it was observed that the addition of sisal and curauá fibers randomly distributed leads to significant improvements in the mechanical properties of the soil, since it was observed an increase in the cohesion and angle of friction of the mixtures compared to the values obtained for the pure soil. The results show the potential of sisal and curauá fibers used as reinforcement of soils in earthwork layers subjected to static loads, such as landfills and embankments on soft soils.

Keywords: Vegetal Fibers, Soil Reinforcement, Triaxial Tests, Mechanical Behavior

Introduction

The technique of improving the mechanical properties of the soil with addition of fibers is long known and used by mankind. There are records of vegetal materials being used to give greater resistance to clay bricks dating back to ages before Christ [1].

The technique of soil reinforced with fibers is inserted in the technology of composite materials. According to [2], a composite material is a combination of two or more materials that together have properties that the components do not have by themselves.

For [3], [4] and [5] is a consensus that the greatest potential of fiber composite materials is in the post-peak strength where the fibers contribute more effectively in the strength of the material, thereby increasing its energy absorption capacity.

Reinforcements in the form of steel strips, wire mesh and various types of synthetic materials have been widely used in geotechnical works. Such applications range from the conventional structures to stabilizing embankments on soft soil, passing through slope reinforcements, increase the bearing capacity of foundations and reinforcement of pavements [6].

The development of composite materials reinforced by vegetal fibers is largely motivated by the greater environmental awareness due to waste disposal problems and petrochemical resources exhaustion [7].

Vegetable fibers, compared to synthetic fibers are inexpensive, easy to obtain, widely available, easy to handle, have good mechanical properties, do not generate excessive amounts of residues, employ relatively simple technology and require less energy in the production process, besides being renewable sources [8], [7] and [9].

Non-Conventional Materials and Technologies – NOCMAT for XXI Century   Materials Research Forum LLC
Materials Research Proceedings 7 (2018) 791-798   doi: http://dx.doi.org/10.21741/9781945291838-78

The improvement or alteration of mechanical properties of soil reinforced with fibers depends on characteristics of the fiber (tensile strength, modulus of elasticity, length, content and surface roughness), of the soil (degree of cementation, size, shape and particle size, ratio voids, etc.) of the confinement tension and load mode [5].

This work seeks to contribute to a better interpretation of the mechanical behavior of soil reinforced with natural vegetal fibers, aiming to enhance the use of these composite in earthworks.

Experimental program

The experimental program consisted in the execution of triaxial compression tests to determine the strength parameters that were the basis for identifying and evaluating the effect of dispersed inclusions of curauá and sisal fibers on the mechanical properties of a granular soil.

Materials and Methods

Sandy Soil

A material with well-defined mechanical behavior was chosen in order to facilitate the interpretation of the effect caused by the addition of fibers. It was used a sand with uniform particle size from a deposit located in the city of Itaboraí in the state of Rio de Janeiro.

The particle size curve (Figure 1), and the physical indices of the material (Table 1) were determined at the Laboratory of Geotechnics and Environment (LGMA) from the Pontifícia Universidade Católica do Rio de Janeiro (PUC-Rio).

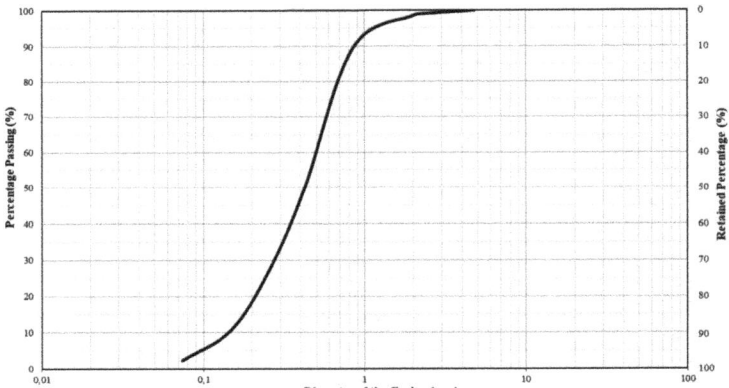

FIGURE 1 - PARTICLE SIZE CURVE OF THE SANDY SOIL.

Non-Conventional Materials and Technologies – NOCMAT for XXI Century   Materials Research Forum LLC
Materials Research Proceedings 7 (2018) 791-798   doi: http://dx.doi.org/10.21741/9781945291838-78

TABLE 1 - PHYSICS INDEX OF THE SANDY SOIL.

Physics Index	Values
Specific Weight (Gs)	2,65
Effective Diameter D_{10}	0,22 mm
Average Diameter D_{50}	0,58 mm
Uniformity Coefficient (Cu)	3,27
Coefficient of Curvature (Cc)	0,87
Maximum voids index (e_{max})	0,96
Minimum voids index (e_{min})	0,71

Vegetal Fibers

The fibers used as reinforcement element are curauá (*Ananas erectifolius*) and sisal (*Agave sisalana*). The curauá fibers were purchased from Triangel Pematec of Brazil Company, located in Santarém/Pará and sisal fibers were purchased from Sustainable Development Association and Solidarity of the Sisal Region (APAEB), located in Valente/ Bahia.

These vegetal fibers were chosen because they have good mechanical properties and the need for new raw materials from renewable sources, which can successfully replace synthetic fibers. The use of vegetal fibers, beyond ensure the production of an environmentally friendly product, can guarantee a new alternative income for farmers in producing regions when produced on an industrial scale.

The Table 2 shows values available in the literature for the main characteristics of the fibers used, and ratios concerning relative sizes of fibers and the sand grains are given in Table 3.

TABLE 2 - PROPERTIES OF THE CURAUÁ AND SISAL FIBERS.

Fiber	Diameter (D_f) (mm)	Tensile strength (MPa)	Young's Modulus (GPa)	Strain at rupture (%)	Reference
	0.071	543	63.7	1.0	[10]
Curauá	0.092 - 0.127	492.62	11.54	3.02	[11]
	0.09	605	23	2.5	[12]
	-	80 - 855	9.0 - 38	1.9 - 14	[13]
Sisal	0.08 - 0.30	227.8 - 1002.3	10.9 - 26.7	2.8 - 4.2	[14]
	0.17	484	19.5	3.3	[10]
	0.15	340	12	3.3	[12]

TABLE 3 - RATIOS RELATED WITH FIBER GEOMETRY AND SAND GRANULOMETRY

Fiber	Aspect Ratio (L_f / D_f)	D_f / D_{50}
Curauá	352,11	0,122
Sisal	146,11	0,295

Non-Conventional Materials and Technologies – NOCMAT for XXI Century   Materials Research Forum LLC
Materials Research Proceedings 7 (2018) 791-798   doi: http://dx.doi.org/10.21741/9781945291838-78

Sample Preparation

Composites of sandy soil reinforced with sisal and curauá fibers dispersed in the matrix were tested. The variables investigated were type of fiber, fiber content (0 and 0.5%), fiber length (25mm) and initial effective stress (50, 100 and 150 kPa).

The parameters for molding the soil and soil-fiber samples were the same used by [15], [16], [17], [18] and [12] who also worked with sandy soil, being moisture content of 10% and a relative density of 50%.

The preparation of the test specimens of pure sandy soil and soil-fiber mixture for conducting triaxial tests were made by compaction in a cylindrical tripartite mold directly positioned in the press where the test was performed. Samples were manually compacted in three layers, the added weight of the mixture and the height of the layers were controlled to obtain the desired relative density.

Triaxial Tests

Isotropic consolidated drained (CID) triaxial test were performed to obtain the strength parameters in sandy soil and soil-fiber composites samples. The available press at PUC -Rio is of controlled displacement velocity.

Samples dimensions of 4.0 cm diameter and 8.6 cm height were saturated using the techniques of backpressure and percolation of water. To ensure that the samples were saturated, it was used the Skempton's parameter B. The criteria presented in [19] was used to define the shear velocity, with the rate of 0,030 mm / min used for all samples.

In the tests, the rupture was defined from the maximum value of the deviator stress (σ_d), in the case that the stress-strain curve showed a peak. When the soil behaved in a strain-hardening manner, the criterion of rupture used was the one proposed in [20], who is based on the slope of the stress-strain curve. In these cases, the rupture was assumed when the stress-strain curve started to have a constant slope.

Results and analyses

Figure 2 shows the deviator stress (σ_d) and volume strain (ε_v) versus axial strain (ε_a) curves, corresponding to the CID triaxial test type with initial effective stress of 50, 100 and 150kPa related to the sandy soil and the sandy soil reinforced with curauá and sisal fibers randomly distributed.

Based on the results shown, for the materials tested, it is observed that the resistance increases with the increase of the confining pressure. It is possible to see that the shear strength increase due to the addition of the fibers in relation to the unreinforced material.

Also is possible to notice that the fibers begin to contribute in the resistance increase of the material at the start of the test, when the axial deformation is lower than 2.5%. After this deformation becomes apparent the difference in behavior between the deviator tension vs axial strain curves of the tested materials, and the contribution of the fibers addition remains visible until the measured axial strain limit of 20%.

Strain-hardening behavior (a constant increase in resistance with increasing axial deformation) was observed for all confining stresses for the soil reinforced with sisal fiber. For the soil reinforced with curauá this behavior was just observed at initial confining stress of 50 kPa, at higher confining stresses the curves show samples with ductile rupture, without the occurrence of a clear peak of resistance, behavior equal to that of sand without reinforcement.

Comparing the results of the two fibers at low initial confining stress (50 and 100kPa), the samples reinforced with curauá fiber presented higher shear strength. However in the stress of 150 kPa the difference between both relies on the fact that sisal presents the strain-hardening behavior after 10% of axial deformation.

The two fibers used as reinforcement in this study have as their origin the leaf of the sisal and curauá plants. Both were added to the matrix in length of 25mm and one of the main characteristics

Non-Conventional Materials and Technologies – NOCMAT for XXI Century   Materials Research Forum LLC
Materials Research Proceedings 7 (2018) 791-798   doi: http://dx.doi.org/10.21741/9781945291838-78

of the natural fibers is the variability of the mechanical properties, what is possible to observe in the Table 2 but even so, both fibers have the same magnitude of these properties, so this shouldn't be used to justify the observed difference between the two composites.

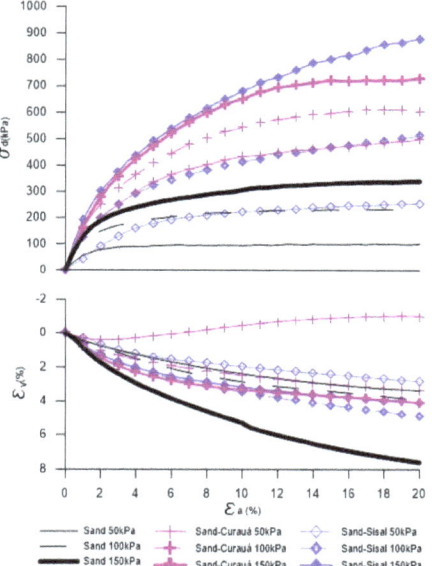

FIGURE 2 – *STRESS-STRAIN-VOLUMETRIC RESPONSE OF SANDY SOIL AND SANDY SOIL REINFORCED WITH CURAUÁ AND SISAL FIBERS.*

It has been reported that the fine fibers, that is, those of high aspect ratio have a greater contribution in increasing the shear strength than when compared to fibers with lower aspect ratio. Among the two fibers used, the curauá fiber is the thinnest one and, because they have the same length, it presents a higher aspect ratio, as can be seen in Table 3, justifying its greater resistance to shearing in the initial confining stress of 50 and 100 kPa [21, 22].

Also, the progressive decrease of the fiber contribution in the composite strength with the increase of the initial confining stress, that is, the fibers have a much higher performance in the composite strength when requested at low initial confining stress. The lower influence of the fibers in higher confining stress could justify the same strength magnitude between the two composites in the tension of 150kPa [23].

For very thin and long fibers have problems related to tangling and reduced effectiveness when fibers used are longer than about 24 mm. Such an assertion would justify the absence of strain-hardening behavior in the stress-strain curves of the curauá composites. A fiber content above the optimum one could also justify the absence of this behavior in the curauá composites [24].

In the volumetric variation curves, with the exception of the sample sand-curauá fiber under initial effective stress of 50 kPa, there is a contraction tendency for all samples in all employed confining stresses. In general, the addition of fibers decreased the contraction behavior, indicating that the fibers "sew" the sandy matrix making their deformation difficult.

The shear strength envelopes obtained from triaxial tests with sisal and curauá fibers-reinforced samples and non-reinforced samples as well as the values of cohesive intercept and the internal friction angle obtained are shown in Figure 3, where the deviator stress (q) is plotted against the corresponding mean effective stress (p').

The addition of curauá and sisal fibers in a sandy soil caused an increase in the cohesive intercept (c') and/or the friction angle (φ') of the material. During the mixing process of the soil with fiber randomly distributed, it was clearly noted that the fibers afforded a tangle involving and somehow tied the soil grains, promoting an anchoring effect. This observation has been translated in the increase of the cohesive intercept and/or friction angle and proven by the values obtained.

It has been justified that an improvement in the load-carrying capacity of the sand by adding randomly distributed fibers is a result of the distributions of loads (reduction in the strains induced by applied loads) throughout the soil mass through the interlocking mesh of fibers created and the fiber/sand particle surface friction [24].

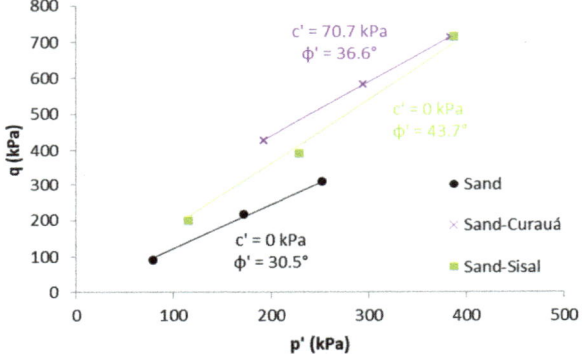

FIGURE 3 – SHEAR STRENGTH ENVELOPES OF NON-REINFORCED SAND AND FIBER-REINFORCED SAND (SISAL AND CURAUÁ FIBER).

The values of c' and φ' found are consistent with the stress-strain curves shown in Figure 2. A high value of c' for the sand-curauá composite occurs because of the greater aspect ratio of this fiber, which contributed significantly to the increase in resistance at low initial confining stress. The value of φ' found for this composite was lower than that found for the sand-sisal composite due to greater loss of the influence of the curauá fiber on the shear strength at the higher initial confining stress evaluated in this study. Unlike the curauá fiber, the addition of sisal fibers did not contribute to the increase of c', probably due to the smaller aspect ratio of this fiber, impacting only in the increase of φ'.

Final considerations

From the conventional triaxial compression tests performed on sandy soil, soil-curauá fiber and soil-sisal fiber mixtures with 25mm in length, with contents of 0 and 0.5% and initial effective stress of 50, 100 and 150kPa were established some conclusions reported below:

The inclusion of vegetal fibers, regardless of the type (sisal and curauá) caused an increase in cohesive intercept and/or friction angle of the material and consequently the material strength.

The stress vs strain behaviour of the soil reinforced with curauá fibers is influenced by the initial effective stress. The fibers showed a higher performance in the composite strength when requested at low initial effective stress.

Non-Conventional Materials and Technologies – NOCMAT for XXI Century   Materials Research Forum LLC
Materials Research Proceedings 7 (2018) 791-798   doi: http://dx.doi.org/10.21741/9781945291838-78

The fiber aspect ratio had a major role in the mechanical behaviour difference between the two composites. For a same length, the fiber with smaller diameter (curauá fiber) had greater contribution in the increase of the shear strength in the initial confinement stress of 50 and 100 kPa.

For the soil reinforced with sisal fibers a constant increase of the resistance with the increase of the axial deformation is observed in all the initial confinement stress evaluated, characterizing a strain-hardening behaviour.

Strain-hardening behaviour was not observed in the sand- curauá fiber composite. Its smaller diameter and length of 25mm may have caused problems related to entanglement and because of this it has lost effectiveness.

The results contribute to a better interpretation of the mechanical behavior of soil reinforced with vegetal fibers (sisal and curauá) and show the potential of this material being used as reinforcement of soils in earthwork subjected to static loads, offering an environmentally correct alternative.

Acknowledgements

The authors acknowledge the National Counsel of Technological and Scientific Development (CNPq) for supporting the doctoral research of the first author and the LGMA laboratory at the Pontifícia Universidade Católica do Rio de Janeiro for the physical and technical support to carry out the research.

References

[1] Palmeira EM. Geossintéticos: tipos e evolução nos últimos anos. Seminário sobre Aplicações de Geossintéticos em Geotecnia. Geossintéticos 92. Brasília. 1992. pp.1-20.

[2] Budinski KG. Engineering materials, properties and selection. 5ed. New Jersey: Prentice Hall International. 1996. 653p.

[3] Taylor GD. Materials in construction. 2ed. London: Longman Scientific & Technical. 1994. 284p.

[4] Illston JM. Constrution materials; their nature and behavior. 2ed. London: E & FN Spon. 1994. 518p

[5] Hannant L. Polymers and polymers composites. In: J.M. ILLSTON. Construction materials: their nature and behavior. 2ed., London: J.M. Illston/E & FN Spon. 1994. pp. 359-403

[6] Palacios MAP. Comportamento de uma Areia Reforçada com Fibras de Polipropileno Submetida a Ensaios Triaxiais de Extensão. MSc Dissertation – Pontifícia Universidade Católica do Rio de Janeiro, Rio de Janeiro. 2012.

[7] Dittenber DB, GangaRao HVS. Critical Review of Recent Publications on Use of Natural Composites in Infrastructure. Composites Part A: Applied Science and Manufacturing, 43(8). 2012. 1419-1429. https://doi.org/10.1016/j.compositesa.2011.11.019

[8] Ghavami K, Toledo Filho RD, Barbosa NP. Behaviour of composite soil reinforced with natural fibers. 1999. Cement and Concrete Composites 21.

[9] Martins APS. Desenvolvimento, caracterização mecânica e durabilidade de compósitos solo-cimento autoadensáveis reforçados com fibras de sisal. 2014. PhD Thesis – UFRJ/COPPE. Rio de Janeiro

[10] Fidelis MEA. Desenvolvimento e caracterização mecânica de compósitos cimentícios têxteis reforçados com fibras de juta. 2014. PhD Thesis - Universidade Federal do Rio de Janeiro. Programa de Pós-graduação em Engenharia Civil, COPPE. Rio de Janeiro

[11] Picanço MS. Compósitos cimentícios reforçados com fibras de curauá. 2005. MSc Dissertation – Pontifícia Universidade Católica do Rio de Janeiro, Departamento de Engenharia Civil. Rio de Janeiro.

[12] Santiago GA. Estudo do Comportamento Mecânico de Compósitos Solo-Fibras Vegetais Impermeabilizadas com Solução de Poliestireno Expandido (EPS) e Cimento Asfáltico de Petróleo (CAP). 2011. PhD Thesis – Universidade Federal de Ouro Preto. Escola de Minas.

[13] Müssig, J. Industrial Applications of Natural Fibres. 2010. Chichester, UK: John Wiley & Sons, Ltd. https://doi.org/10.1002/9780470660324

[14] Tolêdo Filho RD. Materiais compósitos reforçados com fibras naturais: caracterização experimental. 1997. PhD Thesis - Pontifícia Universidade Católica do Rio de Janeiro. Departamento de Engenharia Civil. Rio de Janeiro.

[15] Casagrande MDT. Comportamento de solos reforçados com fibras submetidos a grandes deformações. 2005. PhD Thesis - Universidade Federal do Rio Grande do Sul. Escola de Engenharia. Programa de Pós-Graduação em Engenharia Civil. Rio Grande do Sul.

[16] Girardello V. Ensaios de placa em areias não saturadas reforçada com fibras. 2010. MSc Dissertation – Universidade Federal do Rio Grande do Sul. Escola de Engenharia. Programa de Pós-Graduação em Engenharia Civil. Rio Grande do Sul.

[17] Sotomayor JMG. Avaliação do comportamento carga-recalque de uma areia reforçada com fibra de coco submetida a ensaios de placa em verdadeira grandeza. 2014. MSc Dissertation – Pontifícia Universidade Católica do Rio de Janeiro, Rio de Janeiro.

[18] Louzada, NSL. Experimental Study of Soils Reinforced with Crushed Polyethylene Terephthalate (PET) Residue. 2015. MSc Dissertation – Pontifícia Universidade Católica do Rio de Janeiro, Rio de Janeiro.

[19] Head KH. Manual of Soil Laboratory Testing: Effective Stress Test. 1986. Wiley, 2nd ed., v.3, West Sussex, Inglaterra, p.227.

[20] De Campos TMP, Carrillo CW. Direct Shear Testing on an Unsaturated Soil from Rio de Janeiro. 1995. Unsaturated Soils, Alonso & Delage (eds), pp. 31-38.

[21] Michalowski RL, Cermák J. Triaxial compression of sand reinforced with fibres. 2003. Journal of Geotechnical and Geoenvironmental Engineering, 129(2), 125-136. https://doi.org/10.1061/(ASCE)1090-0241(2003)129:2(125)

[22] Ibraim E, Diambra A, Muir Wood D, Russell AR. Static liquefation of fibre reinforced sand under monotonic loading. Geotextiles and Geomembranes 28. 2010. 374-385. https://doi.org/10.1016/j.geotexmem.2009.12.001

[23] Silva dos Santos AP, Consoli NC, Baudet BA. The mechanics of fibre-reinforced sand. 2010. Géotechnique 60 (10), 791-799. https://doi.org/10.1680/geot.8.P.159

[24] Consoli NC, Festugato L, Heineck KS. Strain-hardening behaviour of fibre-reinforced sand in view of filament geometry. 2009. Geosynthetics Int. 16, No. 2, 109–115. https://doi.org/10.1680/gein.2009.16.2.109

Non-Conventional Materials and Technologies – NOCMAT for XXI Century   Materials Research Forum LLC
Materials Research Proceedings 7 (2018) 799-803   doi: http://dx.doi.org/10.21741/9781945291838-79

Experimental and Numerical Study of the Mechanical Behavior of a Commercial Polypropylene Woven Fabric

E. Agaliotis [a, b], C. Bernal [a, b], P.J. Herrera-Franco [c], E.A. Flores-Johnson [d, *]

[a] Universidad de Buenos Aires, Facultad de Ingeniería, Buenos Aires, Argentina;
eagaliotis@fi.uba.ar, cbernal@fi.uba.ar

[b] CONICET-Universidad de Buenos Aires, Instituto de Tecnología en Polímeros y
Nanotecnología (ITPN), Av. Las Heras 2214, C1127AAR, Buenos Aires, Argentina

[c] Unidad de Materiales, Centro de Investigación Científica de Yucatán, Calle 43, No. 130 Col.
Chuburná de Hidalgo, Mérida, Yucatán 97205, México; pherrera@cicy.mx

[d] CONACYT – Unidad de Materiales, Centro de Investigación Científica de Yucatán, Calle 43,
No. 130 Col. Chuburná de Hidalgo, Mérida, Yucatán 97205, México; emmanuel.flores@cicy.mx

Abstract. Heterogeneous conventional composites have limited recyclability and their end-of-life processes require high energy thus, they represent a recycling challenge. A promising approach to composites recycling is to employ thermoplastic self-reinforced composites (SRCs) for structural building applications. SRCs have specific economic and ecological advantages and can be recycled. Similar to traditional composites, the stress transfer in self-reinforced composites occurs via an interface/interphase, but in this case, molecular entanglements, favorable amorphous/crystalline superstructures, and H-bonding may serve for improved adhesion and hence for stress transfer between matrix and reinforcement via the interphase. In addition, their low density and preferred recycling make them an appealing alternative over traditional composites in many structural applications. As in conventional composites, one of the major determining factors of SRCs final performance is reinforcement behavior. Hence, the evaluation of reinforcement mechanical properties seems to be essential.

In this work, the tensile properties of a low-cost commercial polypropylene (PP) woven fabric composed of highly stretched PP tapes, were investigated. A 3D finite-element model for uniaxial tensile tests of the PP woven fabric was built in Abaqus/Explicit by modeling individual tapes and taking into account friction among them. An orthotropic elastic model was used. In order to obtain an optimal model design for the woven fabric, all the geometrical parameters were experimentally obtained and the tensile properties of individual tapes in both warp and weft directions were also determined. It was observed that the fabric mechanical behavior is mainly affected by the internal geometry of the woven fabric. The simulation results were in good agreement with the fabric mode of failure observed in the experiments.

Keywords: Polypropylene, Woven Fabric, Finite-Element Modeling, Recycling

Introduction

The use of self-reinforcement composites (SRCs) have been growing rapidly during the last decades leading to an important technological development in structural applications. They offer the opportunity to achieve a unique combination that may compete with traditional composites in various application fields based on their performance/cost balance. Owing to their lightweight, balanced properties and excellent impact resistance, they are used as components in aerospace, automotive, energy and marine industries over traditional composites. In addition, self-reinforced composites compared to conventional composites, can be considered environmentally friendly materials due to their easy recyclability [1-4].

Non-Conventional Materials and Technologies – NOCMAT for XXI Century   Materials Research Forum LLC
Materials Research Proceedings 7 (2018) 799-803   doi: http://dx.doi.org/10.21741/9781945291838-79

Comparative studies have been developed to determine the effect of fabric weaves on the mechanical properties of SRCs. For example, Rouf *et al.* [5] have shown how the difference in yarn undulation affects the mechanical properties by changing woven structures. The effect of woven structure was also studied to determine its influence in storage modulus [6], and in the dynamic mechanical analysis of natural hybrid polymer composites [7]. Hence, it is essential to assess woven geometries as reinforcement before developing SRCs. Numerical models offer a practical approach to characterize the behavior of different types of reinforcements; however, the development of a model for woven fabrics is a complex task.

The aim of this work is to study the failure of a commercial polypropylene woven fabric which includes the definition of geometry, morphology and weave pattern for the development of a numerical model. Uniaxial tensile behavior of the plain woven was obtained from the work of Lucchetta *et al.* [8], whereas the response of individual tapes was experimentally determined in this work. These results were used as input parameters in a 3D finite element model of the plain woven fabric tensile test. The model was performed in Abaqus/Explicit by considering individual tapes and taking into account friction among them. Numerical and experimental results were compared and discussed.

Materials and methods
Materials and mechanical testing
The material under investigation was a commercial plain woven fabric composed of highly stretched PP tapes (areal weight of 190 g/m^2) supplied by Politejidos S.R.L (Argentina). The mechanical properties of the plain woven fabric have been reported by Lucchetta *et al.* [8] and they are listed in Table 1. In this work, the tensile properties of individual tapes were determined in accordance to ASTM D2256. Uniaxial tensile tests were performed in a universal testing machine Instron Model 5982 at a crosshead speed of 5 mm/min.

Morphology
The woven fabric used is sketched in Figure 1. It is a balanced polypropylene woven fabric since warp and weft tapes are geometrically similar. In order to accurately calculate the dimensions of each tape, measurements were performed by means of optical microscopy (OM) and scanning electron microscopy (SEM). The analysis of the results was made with the help of the processing image software *ImageJ*. Characteristic dimensions were tape width *l*, tape thickness *e*, air-gap between tapes *d*, distance between tapes *L* (Figure 2) [9]. The tape wave was experimentally obtained by recording the curvature coordinate at different equidistant points.

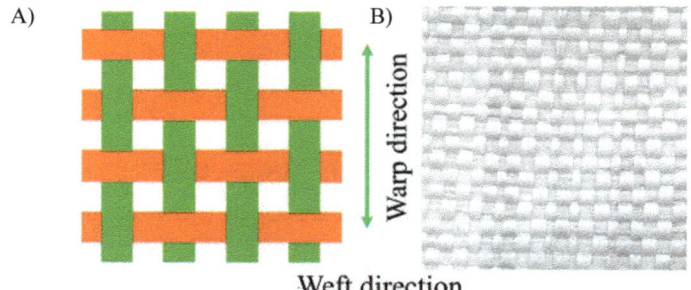

A)   B)

Warp direction

Weft direction

FIGURE 1 *– A) SCHEMATIC REPRESENTATION OF THE WOVEN FABRIC. B) MACROPHOTOGRAPH OF THE WOVEN FABRIC.*

Non-Conventional Materials and Technologies – NOCMAT for XXI Century   Materials Research Forum LLC
Materials Research Proceedings 7 (2018) 799-803   doi: http://dx.doi.org/10.21741/9781945291838-79

A)
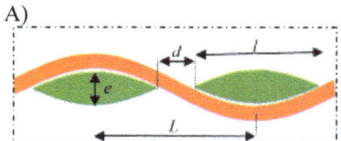

B)
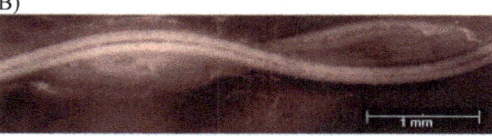

FIGURE 2 – *A) SCHEMATIC REPRESENTATION OF A SINGLE WAVE WITH CHARACTERISTIC DIMENSION: L= TAPE WIDTH, E=TAPE THICKNESS, D=AIR-GAP BETWEEN TAPES, L=DISTANCE BETWEEN TAPES. B) OPTICAL MICROGRAPH OF A SINGLE WAVE.*

Numerical simulations: Finite element -model

The mechanical behavior of the plain woven fabric was modelled using a 3D finite element model built with the commercial package Abaqus/Explicit [10]. Taking into account the symmetry of the problem, a representative half-model of the woven fabric was developed to reduce computational expenses. Tensile properties used in the model were determined on samples of 25 mm in width and 180 mm in length cut out from the plain woven fabric in both warp and weft directions [8]. Hence, the dimensions of the model were 25 mm (14 tapes) and 90 mm in length. The total number of elements are 286720 and the type of elements used was 8-node brick elements C3D8 [10]. Friction among tapes was considered by using the penalty contact algorithm and defining a constant coefficient of friction μ=0.2 [10]. The elastic behavior of the tapes was described by assuming an orthotropic elastic material. In addition, the orientation of the material was defined with E_{11} along the tape direction using a local coordinate system. Small values of shear moduli, Poisson´s ratio and transverse elastic moduli were used [11-12]. It was assumed that the material fails when the equivalent stress exceeds the yield stress at a strain of approximately 0.024.

Results and discussion
Experimental results

Eight individual tapes of 250 mm in length were obtained from the warp and weft directions and subsequently tested under tension. The results are shown in Table 1. It can be observed that there is a difference of about 25% in both tensile strength and failure strain values between warp and weft directions. Nevertheless, the difference in the longitudinal Young´s modulus is less than 4%. The small difference observed (1%) between tensile parameters for the tapes and the fabric could be attributed to decrimping of the tapes in the fabric during tensile tests [13].

TABLE 1 – MATERIAL PROPERTIES

Property	PP Tapes	Plain woven fabric [8]
Density ρ (kg/m^3)	900	900
Longitudinal Young´s modulus E_{11} (MPa)	1564 (warp), 1569 (weft)	1629 (warp), 1338 (weft)
Tensile strength σ_t (MPa)	149.6 (warp), 118.5 (weft)	153.4 (warp), 117.4 (weft)
Failure strain ε_f	0.15 (warp), 0.11 (weft)	0.138 (warp), 0.138 (weft)

Numerical results

The results of geometrical characterization were used to build the plain woven fabric model. Material properties are listed in Table 1. Figure 3 and Figure 4 show simulation and experimental results, respectively. A good agreement between both results was obtained for the woven fabric in the

Non-Conventional Materials and Technologies – NOCMAT for XXI Century   Materials Research Forum LLC
Materials Research Proceedings **7** (2018) 799-803   doi: http://dx.doi.org/10.21741/9781945291838-79

deformed state. As it can be clearly observed in Figure 3, simulation results closely represented the actual fabric mode of failure characterized by extensive deformation in the middle of the sample and separation of the tapes from the fabric at the edges and defibrillation (Figure 4).

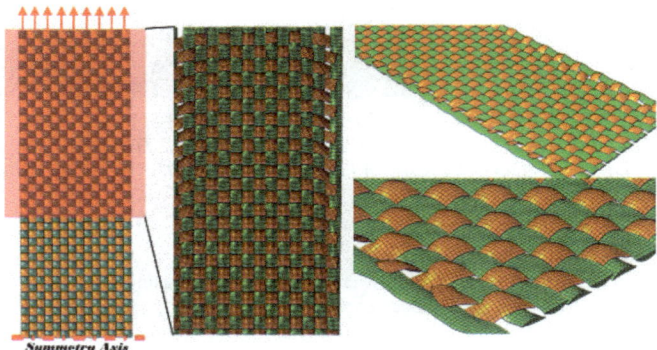

FIGURE 3 – *SIMULATION OF UNIAXIAL TENSILE TEST ON THE PLAIN WOVEN FABRIC.*

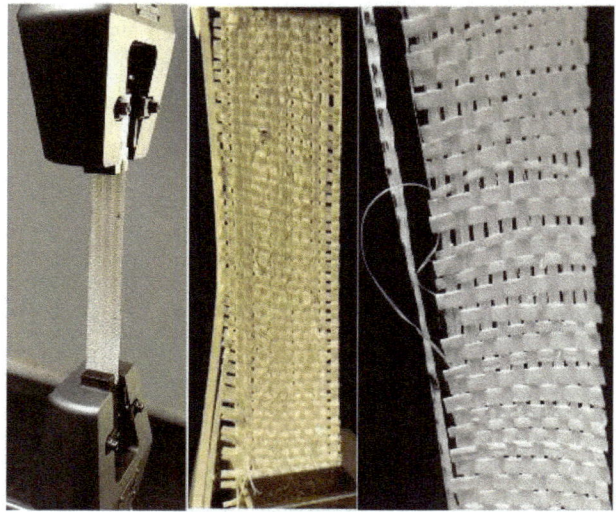

FIGURE 4 – *UNIAXIAL EXPERIMENTAL TENSILE TEST ON THE PLAIN WOVEN FABRIC*

Summary and conclusions
The uniaxial tensile response of a commercial plain woven fabric was investigated both experimentally and numerically. The results of the tensile tests on the fabric and also on individual tapes indicated that significant differences (25%) exist between strength and ductility values for the warp and weft directions. Stiffness values, on the other hand, were roughly similar in both directions.

Non-Conventional Materials and Technologies – NOCMAT for XXI Century   Materials Research Forum LLC
Materials Research Proceedings 7 (2018) 799-803   doi: http://dx.doi.org/10.21741/9781945291838-79

Morphology and mechanical properties experimentally obtained were used to build a 3D finite-element model. This model takes into account the effect of the friction between tapes within the plain woven fabric. The numerical model accurately reproduced visual observations of the fabric mode of failure, which is an extensive deformation in the middle of the sample and separation of the tapes from the fabric at the edges and defibrillation. Further experimental work and simulations are needed to validate the numerical model.

References

[1] Matabola KP, De Vries AR, Moolman FS, Luyt AS. Single polymer composites: A review. J Mater Sci 2009;44:6213-6222. https://doi.org/10.1007/s10853-009-3792-1

[2] Karger-Kocsis J, Bárány T. Single-polymer composites (SPCs): Status and future trends. Compos Sci Technol 2014;92:77-94. https://doi.org/10.1016/j.compscitech.2013.12.006

[3] Kmetty Á, Bárány T, Karger-Kocsis, J. Self-reinforced polymeric materials: a review. Prog Polym Sci 2010;35(10):1288-1310. https://doi.org/10.1016/j.progpolymsci.2010.07.002

[4] Chawla KK. Composite Materials: Science and Engineering. New York: Springer, 2012. https://doi.org/10.1007/978-0-387-74365-3

[5] Rouf K, Denton NL, French RM. Effect of fabric weaves on the dynamic response of two-dimensional woven fabric composites. J Mater Sci 2017;52: 10581-10591. https://doi.org/10.1007/s10853-017-1183-6

[6] Houshyar S, Shanks RA, Hodzic A. Influence of different woven geometry in poly(propylene) woven composites. Macromol Mater Eng 2005;290:45-52. https://doi.org/10.1002/mame.200400158

[7] Rajesh M, Pitchaimani J. Dynamic mechanical analysis and free vibration behavior of intra-ply woven natural fiber hybrid polymer composite. J Reinf Plast Compos 2016;35(3):228-242. https://doi.org/10.1177/0731684415611973

[8] Lucchetta MC, Morales Arias JP, Mollo M, Bernal CR. Self-reinforced composites based on commercial PP woven fabrics and a random PP copolymer modified with quartz. Polym Adv Technol 2016;27(8):1072-1081. https://doi.org/10.1002/pat.3772

[9] Loix F, Badel P, Orgéas L, Geindreau C, Boisse P. Woven fabric permeability: From textile deformation to fluid flow mesoscale simulations. Compos Sci Technol 2008;68(7):1624-1630. https://doi.org/10.1016/j.compscitech.2008.02.027

[10] Abaqus Analysis User's Guide, Dassault Systèmes, Providence, RI, USA (2014).

[11] Duan Y, Keefe M, Bogetti TA, Cheeseman BA. Modeling friction effects on the ballistic impact behavior of a single-ply high-strength fabric. Int J Impact Eng 2005;31:996-1012. https://doi.org/10.1016/j.ijimpeng.2004.06.008

[12] Flores-Johnson, EA, Carrillo, JG, Gamboa, RA, Shen, L. Experimental and numerical study of plain-woven aramid fabric. Adv Mater Res 2014;856:74-78. https://doi.org/10.4028/www.scientific.net/AMR.856.74

[13] Cavallaro PV, Sadegh AM, Quigley CJ. Decrimping Behavior of Uncoated Plain-woven Fabrics Subjected to Combined Biaxial Tension and Shear Stresses. Text Res J 2007;77(6):403-416. https://doi.org/10.1177/0040517507080258

Non-Conventional Materials and Technologies – NOCMAT for XXI Century   Materials Research Forum LLC
Materials Research Proceedings 7 (2018) 804-809   doi: http://dx.doi.org/10.21741/9781945291838-80

Mechanical Characterization of Plaster Reinforced with Recycled Cellulose Fiber from Multi-Layer Packaging Waste for Construction Applications

R.A. Gamboa [a], C.M. Moo-Chalé [a], E.A. Flores-Johnson [b], J.G. Carrillo [c, *]

[a] Instituto Tecnológico Superior de Motul. Motul de Carrillo Puerto, Yucatán;
ricardo.gamboa@itsmotul.edu.mx, carlos.moo@itsmotul.edu.mx

[b] CONACYT – Unidad de Materiales, Centro de Investigación Científica de Yucatán, Calle 43, No. 130 Col. Chuburná de Hidalgo, Mérida, Yucatán 97205, México; emmanuel.flores@cicy.mx

[c] Unidad de Materiales, Centro de Investigación Científica de Yucatán, Calle 43, No. 130 Col. Chuburná de Hidalgo, Mérida, Yucatán 97205, México; jgcb@cicy.mx

Abstract. To address the current demand for materials with good mechanical performance for construction applications that are environmentally friendly, the improvement of the mechanical properties of gypsum plaster using recycled cellulose obtained from multi-layer waste containers is investigated. Gypsum is an important material in construction that is used as a retardant agent for setting concrete, plaster and bonding material. Gypsum is also used for prefabricated panels, which are good for flame resistance. Multilayer carton (i.e. Tetra Brik®) is an important part of municipal solid waste; however, it is not recycled in large volumes due to the complexity of its recovering. In recent years, new methods have been developed to recycle multilayer carton, which includes the hydropulping process. This method allows recovering the cellulose, which can be used as reinforcement for plaster made of gypsum. In this work, a composite material based on gypsum reinforced with recycled cellulose, which is obtained from multilayer carton containers using the hydropulping process, is investigated. Gypsum/cellulose plaster samples were fabricated at fiber concentrations of 0%, 1%, 2% and 3%, in weight fraction. Flexural tests performed on the plaster showed an improvement of the mechanical properties with low proportion of cellulose (<4%) when compared to plain gypsum plaster. These results show that recycled cellulose/gypsum plaster is an interesting option for construction applications with improved flexural strength.

Keywords: Tetra Brik®, Cellulose, Gypsum, Flexural Test, Composite

Introduction

Gypsum is characterized as one of the most versatile materials in construction due to its good drying time, easy application and preparation and the quality of its finish. This makes this material a very good candidate for various applications in the areas of architecture and construction. One of the most notorious uses of gypsum is for prefabricated gypsum panels to be used for facades or false walls in buildings [1]. Tetra Brik® packaging is widely used worldwide because of its properties for preserving food. Its extensive use has led to a problem of recycling in developing countries, as it is not easily recyclable due to the layering of its three main components, low density polyethylene, Kraft paper and aluminum; however, recent methods have been developed that can be used to recycle this type of packaging such as the hydropulping process [2-5].

The use of gypsum with natural fibers has been previously investigated; however, most of the studies are related to gypsum sandwich panels, where the cellulose is in the form of outer coating sheets to protect the gypsum, thus, studies that present a composite material reinforced with cellulose fibers are scarce. Carvalho *et. al.* studied a gypsum composite material reinforced with recycled cellulose pulp obtained from cement bags, using different fiber concentrations from 5% to 15%. They demonstrated that the presence of cellulose improves the kinetics of gypsum curing, which

Non-Conventional Materials and Technologies – NOCMAT for XXI Century   Materials Research Forum LLC
Materials Research Proceedings **7** (2018) 804-809   doi: http://dx.doi.org/10.21741/9781945291838-80

positively influences its mechanical properties; they showed an increase of the flexural strength of 160%, using 12.5% of fiber fraction [6]. Nazerian and Kamyab studied the effect of adding bagasse (*Saccharum officinarum*) fibers and wheat straw (*Triticum aestivum*) to a plaster mixture and evaluated physical and mechanical properties. They found that by adding 6-12% of fiber, the mechanical properties improved [1]. The present study contributes to the knowledge of gypsum plaster reinforced with recycled fiber by analyzing the flexural mechanical behavior of gypsum panels with recycled cellulose aggregates recovered from multilayer cartons. The percentage of cellulose content with respect to gypsum content is varied and experimental results from flexural tests are reported.

Experimental part
Materials
The multilayer cartons are composed of 5% aluminum, 20% LDPE and 75% Kraft paper [2-3, 7]. For the separation of these three constituents the technique of hydropulping is used, which consists of submerging in water the multilayer containers, which were previously sectioned in small squares of 8 cm x 8 cm, to hydrate for 24 hours. Subsequently, the squares were introduced to a hydropulping system, GUSTAV SPANGENBERG model LSD20S, which effectively agitates the aqueous medium with the material to separate the constituents of the Tetra Brik®. The Kraft paper separation is completed by hand, and then dried out in a forced convection oven at 65 °C for 24 hours. After these procedures, we obtained dry cellulose ready to be used in the material formulation with gypsum [5, 8]. Figure 1a shows the cellulose obtained from the Tetra Brik® cartons, and Figure 1b shows the mixing process of the cellulose and gypsum.

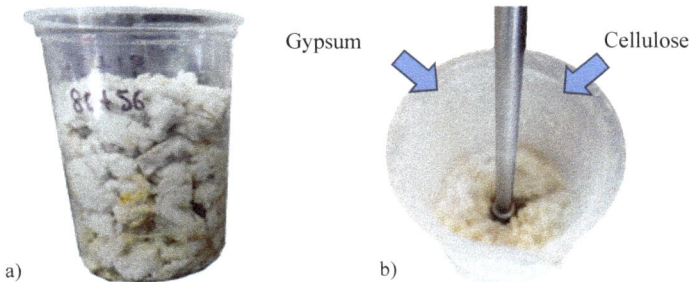

FIGURE 1 – MIXING OF THE MATERIAL: A) MULTILAYER CARTON CELLULOSE, B) MIXING OF GYPSUM AND CELLULOSE.

Plaster samples
The gypsum/cellulose mixtures were made using 1%, 2% and 3% of fiber content (weight %), including as reference material gypsum without fiber (0%). To make plaster mixtures with cellulose, it is necessary to weigh both parts separately; then the cellulose is soaked in water 24 hours for its proper dispersion before adding the gypsum. The water used with the fiber is only the necessary amount for the subsequent mixing with the gypsum according to the global formulation. A quantity of gypsum per sample of about 650 g is used. An equal part of water is also used (650 g).

Once the mixture is ready, the material is poured into a mold with acrylic walls, which has the dimensions recommended by ASTM C348. To prevent samples from being damaged at the time of demolding, a wax release agent is used in the mold, which does not contaminate the samples or affect their texture. Figure 2 shows a schematic of the mold where the mixture is poured for consolidation.

Non-Conventional Materials and Technologies – NOCMAT for XXI Century   Materials Research Forum LLC
Materials Research Proceedings 7 (2018) 804-809   doi: http://dx.doi.org/10.21741/9781945291838-80

FIGURE 2 – SCHEMATIC OF PLASTER WITH CELLULOSE SAMPLE FABRICATION.

Flexural test

The flexural test was performed according to ASTM C348 Standard Test Method for Flexural Strength of Hydraulic Cement Mortars, in which, the methodology for preparation of mixture, the dimensions of the samples and testing procedures are specified [9]. The samples were kept under laboratory conditions until the testing. The samples were tested on a Shimadzu model AG-1 universal testing machine with a 20 kN load cell (Figure 3a), using a three-point bending rig (Figure 3b).

a) 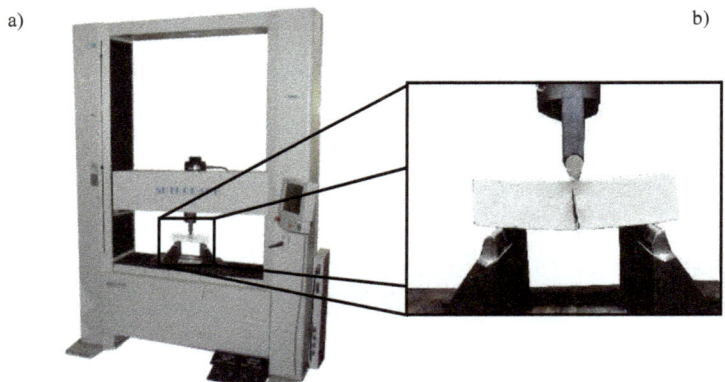 b)

FIGURE 3 – A) FLEXURAL TEST USING A UNIVERSAL TESTING MACHINE, B) CLOSE-UP IMAGE OF FLEXURAL TEST RIG.

The data obtained from the machine are load and displacement. By using the dimensions established in the ASTM C348 and the equation $S_f = 0.0028P$, where S_f is the flexural strength in MPa and P is the maximum load reached [9], the flexural strength can be obtained.

Non-Conventional Materials and Technologies – NOCMAT for XXI Century   Materials Research Forum LLC
Materials Research Proceedings 7 (2018) 804-809   doi: http://dx.doi.org/10.21741/9781945291838-80

Results
Flexural test

The results of the flexural tests are shown in Table 1. It can be seen an increase of the flexural strength of the samples with the increase of fiber content. This increase in the strength is attributed to the fiber interaction with the gypsum, which contributes to the redistribution of the load by means of multidirectional bridges, since the fibers are randomly distributed within the material. However, as shown in Table 1, a reduction of the maximum displacement with the increase of fiber content is observed, which is attributed to the effect of the fibers on the ductility of the material by restricting the deformation of the material. It is noted that the fibers kept the two blocks formed by the middle fracture (Fig. 4) together at the time of loading, which could contribute to the reduction of catastrophic failure, a common behavior of fragile materials. In Figure 4 it is possible to observe an image of a sample where the small cellulose fibers keep the two blocks together after fracture. This suggests that an important contribution of fibers to the global mechanical performance is the redistribution of load via bridging the two blocks.

TABLE 1 –MECHANICAL RESPONSE IN FLEXURAL TEST SAMPLES FOR DIFFERENT CELLULOSE CONCENTRATIONS

Fiber content (wt %)	Flexural strength (MPa)	Maximum displacement (mm)
0	1.13 ± 0.11	1.06 ± 0.60
1	1.15 ± 0.12	0.80 ± 0.23
2	1.23 ± 0.08	0.41 ± 0.06
3	1.34 ± 0.15	0.59 ± 0.20

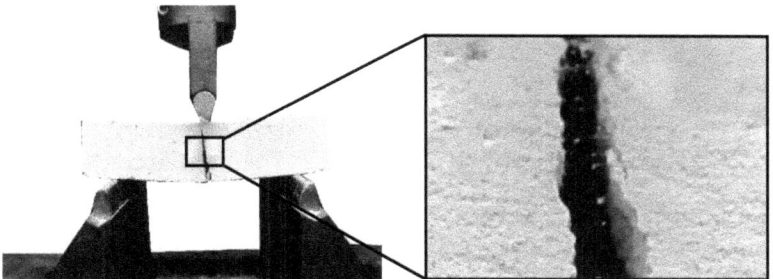

FIGURE 4 –FLEXURAL TEST OF A PLASTER SAMPLE WITH CELLULOSE

Figure 5 shows flexural strength vs fiber concentration for cellulose percentages of 0%, 1%, 2% and 3% w/w. Adding as little as 1% of fiber to the material, the flexural strength of the plaster increased. At 3% of cellulose content, an increase of ~18% in the strength is observed when compared to the reference material (plain gypsum).

Non-Conventional Materials and Technologies – NOCMAT for XXI Century   Materials Research Forum LLC
Materials Research Proceedings **7** (2018) 804-809   doi: http://dx.doi.org/10.21741/9781945291838-80

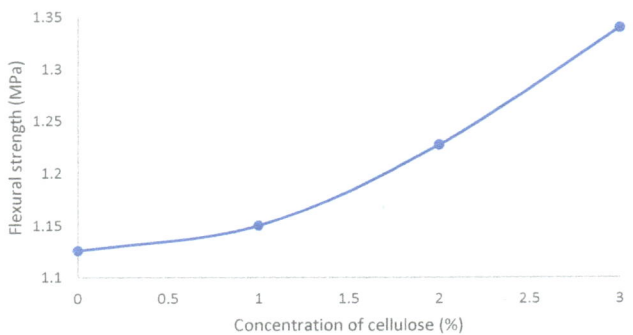

FIGURE 5 – FLEXURAL STRENGTH OF PLASTER SAMPLES.

Conclusions

The tests carried out on reinforced plaster with recycled cellulose from multilayer cartons show that an improvement in the flexural strength is obtained by adding fiber content of up to 3%. This is a promising result that warrants further investigation. The recycled cellulose fiber not only increased the flexural strength but also provided some integrity to the material after fracture. It is noted that this reinforcing material comes from a waste material, which provides a new way for utilizing the recycling products of multilayer containers. More detailed studies should be performed on this material to fully understand its possible applications for the construction industry.

References

1.   Nazerian M. and Kamyab M. Gypsum-bonded particleboard manufactured from agricultural based material. Forest Science and Practice, 2013. 15 p. 325-331. https://doi.org/10.1007/s11632-013-0420-6

2.   Cruz I.M. Caracterización mecánica de un material compuesto tipo sándwich reciclado del Tetra Brik, Engineering Thesis, 2012 Instituto Tecnológico Superior de Motul: Motul, Yucatán.

3.   Pool DA, Carrillo J.G. and Gamboa R.A. Absorción de agua en aglomerados de Tetra Brik con matriz de polietileno de alta densidad a diferente fracción volumen, Congreso Nacional de la Sociedad Polimérica de México. 2012: Mérida, Yucatán.

4.   Figen, A.K., Terzi, E. Yilgör, N. Kartal. S, Pişkin, S. Thermal degradation characteristic of Tetra Pak panel boards under inert atmosphere. Korean Journal of Chemical Engineering, 2013. 30: p. 878-890. https://doi.org/10.1007/s11814-012-0185-y

5.   Moo J.A. Elaboración y caracterización mecánica de un material compuesto tipo sándwich con núcleo particulado tetra pak y caras de aluminio, Engineering Thesis. 2013, Instituto Tecnológico Superior de Motul: Motul, Yucatán.

6.   Magaly AC. Carlito CJ. Holmer SJ. Rejane T. Michelle TC. Microstructure and mechanical properties of gypsum composites reinforced with recycled cellulose pulp. Materials Research, 2008. 11: p. 7.

Non-Conventional Materials and Technologies – NOCMAT for XXI Century   Materials Research Forum LLC
Materials Research Proceedings 7 (2018) 804-809   doi: http://dx.doi.org/10.21741/9781945291838-80

7.   Carrillo, JG. Ventura, DA. Gamboa, RA. Cruz-Estrada, RH. Improvement on Mechanical
Properties of a Particle Board Made of Recycled Material Based on Tetra Brik®. MRS Online
Proceedings Library, 2014. 1611: p. null-null.

8.   Euan M.Z. Optimización de un proceso de hidropulpeo para elaboración de contenedores a
base de celulosa, Engineering Thesis. 2014, Instituto Tecnológico Superior Progreso: Progreso,
Yucatán.

9.   ASTM Standard C 348. Standard Test Method for Flexural Strength of Hydraulic-Cement
Mortars. IB. 2014, American Society for Testing and Materials: West Conshohocken, PA.

Keyword Index

Lightning Source UK Ltd.
Milton Keynes UK
UKHW020632080321
379974UK00005B/82